No. 3138
$60.00

Encyclopedia of
ELECTRONIC CIRCUITS
Volume 2

Rudolf F. Graf

FIRST EDITION
THIRD PRINTING

Library of Congress Cataloging in Publication Data

(Revised for v. 2)

Graf, Rudolf F.
Encyclopedia of electronic circuits—Volume 2.

Bibliography: v. 1, p.
Includes indexes.
1. Electronic circuits. I. Title.
TK7867.G66 1985 621.3815′3 84-26772
ISBN 0-8306-0938-5 (v. 1)
ISBN 0-8306-1938-0 (pbk. : v. 1)
ISBN 0-8306-9138-3 (v. 2)
ISBN 0-8306-3138-0 (pbk. : v. 2)

Questions regarding the content of this book should be addressed to:

Reader Inquiry Branch
TAB BOOKS Inc.
Blue Ridge Summit, PA 17294-0214

CONTENTS

Introduction

Encyclopedia of Electronic Circuits—Volume 2, a companion to Volume 1 published in 1985, contains well over 1400 not-previously covered circuits organized into 108 chapters. For each reference, circuits are listed at the beginning of each chapter. The extensive index further enhances the usefulness of this new work. The browser, as well as the serious researcher looking for a very specific circuit, will be richly rewarded by the context of this volume. A brief explanatory text accompanies almost every entry. The original source for each item is also given so that the reader requiring additional data will know where to find it.

I am most grateful to William Sheets for his many and varied contributions to this book, and to Mrs. Stella Dillon for her fine work at the word processor. These friends and associates of long standing have my sincere thanks for contributing to the successful completion of this book.

To Danny and David,
the newest stars on the horizon
From Popsi

1

Alarm and Security Circuits

The sources of the following circuits are contained in the Sources section beginning on page 694. The figure number contained in the box of each circuit correlates to the source entry in the Sources section.

Auto Burglar Alarm
Multiple Alarm Circuit
Differential-Voltage or Current Alarm
Trouble Tone Alert
Photoelectric Alarm System
Alarm Circuit

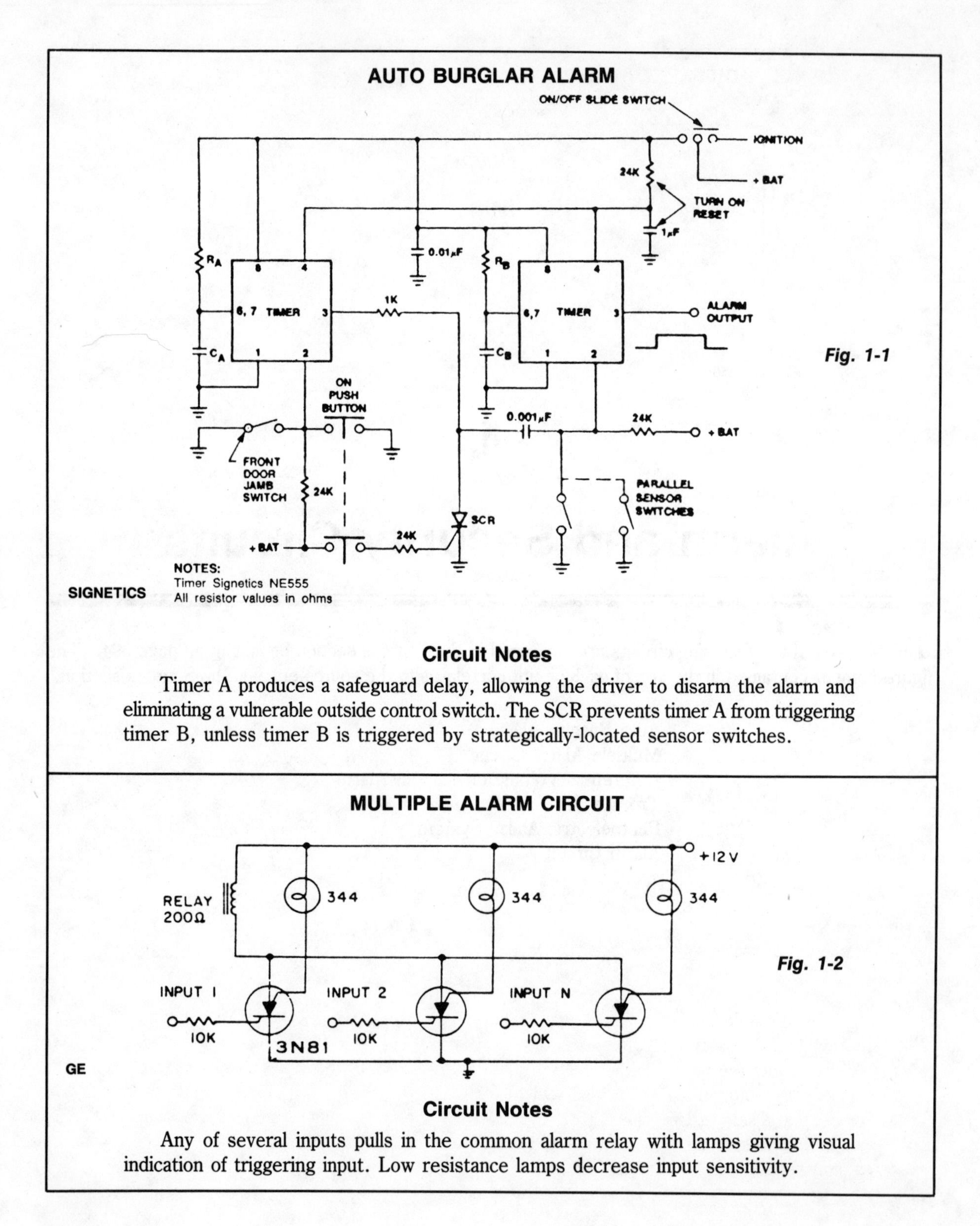

Circuit Notes

Timer A produces a safeguard delay, allowing the driver to disarm the alarm and eliminating a vulnerable outside control switch. The SCR prevents timer A from triggering timer B, unless timer B is triggered by strategically-located sensor switches.

Circuit Notes

Any of several inputs pulls in the common alarm relay with lamps giving visual indication of triggering input. Low resistance lamps decrease input sensitivity.

DIFFERENTIAL VOLTAGE OR CURRENT ALARM

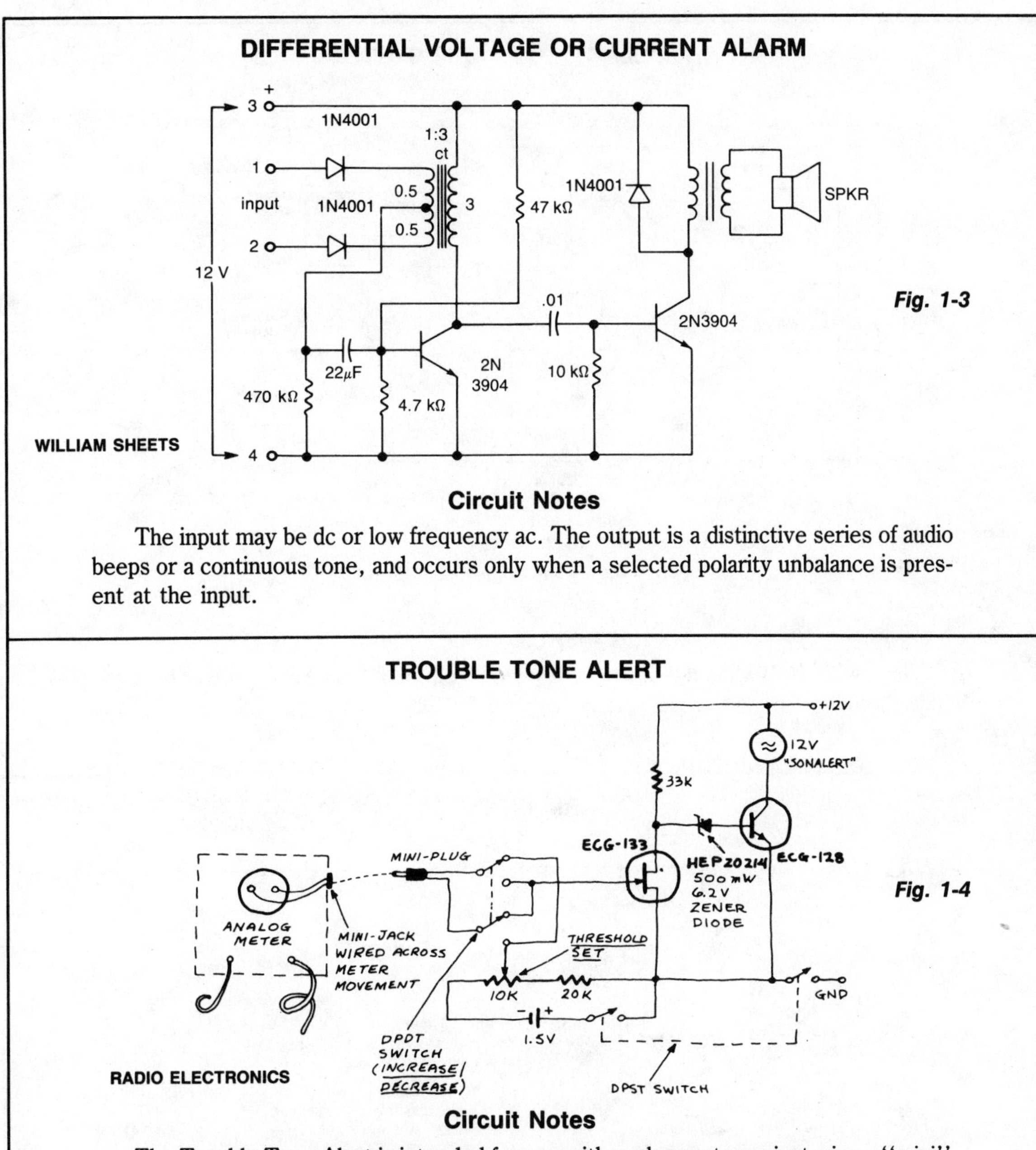

WILLIAM SHEETS

Fig. 1-3

Circuit Notes

The input may be dc or low frequency ac. The output is a distinctive series of audio beeps or a continuous tone, and occurs only when a selected polarity unbalance is present at the input.

TROUBLE TONE ALERT

RADIO ELECTRONICS

Fig. 1-4

Circuit Notes

The Trouble Tone Alert is intended for use with analog meters—just wire a "mini" earphone jack directly across the meter movement, plug it in, and you're all set. This device reacts the to the meter-movement driving voltage. It will respond to a change in ac or dc voltage, current, or in resistance. The circuit will respond to an increase or decrease selected by the DPDT switch and is adjusted with the threshold control until the tone from the Sonalert just disappears (with the meter in the circuit being tested, of course).

PHOTOELECTRIC ALARM SYSTEM

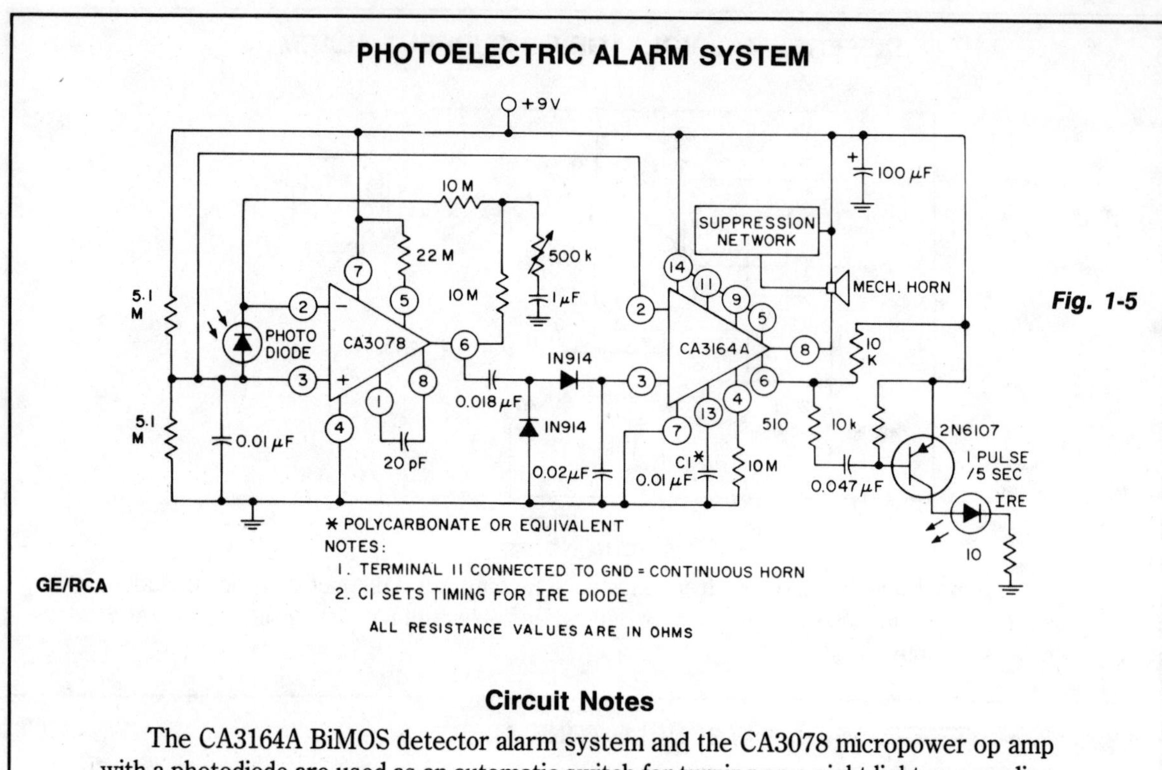

Fig. 1-5

GE/RCA

Circuit Notes

The CA3164A BiMOS detector alarm system and the CA3078 micropower op amp with a photodiode are used as an automatic switch for turning on a night light or sounding a mechanical horn.

ALARM CIRCUIT

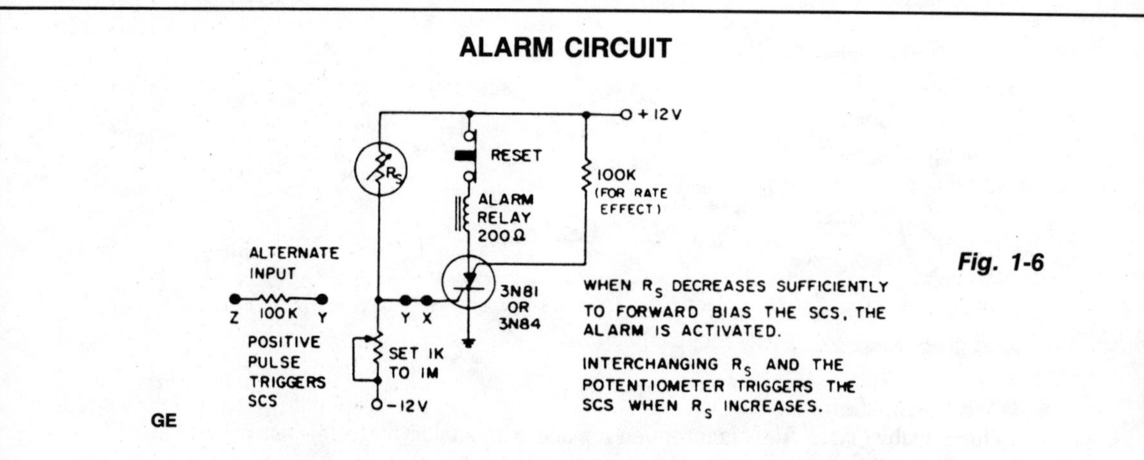

Fig. 1-6

GE

Circuit Notes

Temperature, light, or radiation sensitive resistors up to 1 megohm readily trigger the alarm when they drop below the value of the preset potentiometer. Alternately, 0.75 V at the input to the 100 kΩ triggers the alarm. Connecting SCS between ground and −12 V permits triggering on negative input to G_A.

2

Amplifiers

The sources of the following circuits are contained in the Sources section beginning on page 694. The figure number contained in the box of each circuit correlates to the source entry in the Sources section.

STABLE UNITY GAIN BUFFER WITH GOOD SPEED AND HIGH INPUT IMPEDANCE

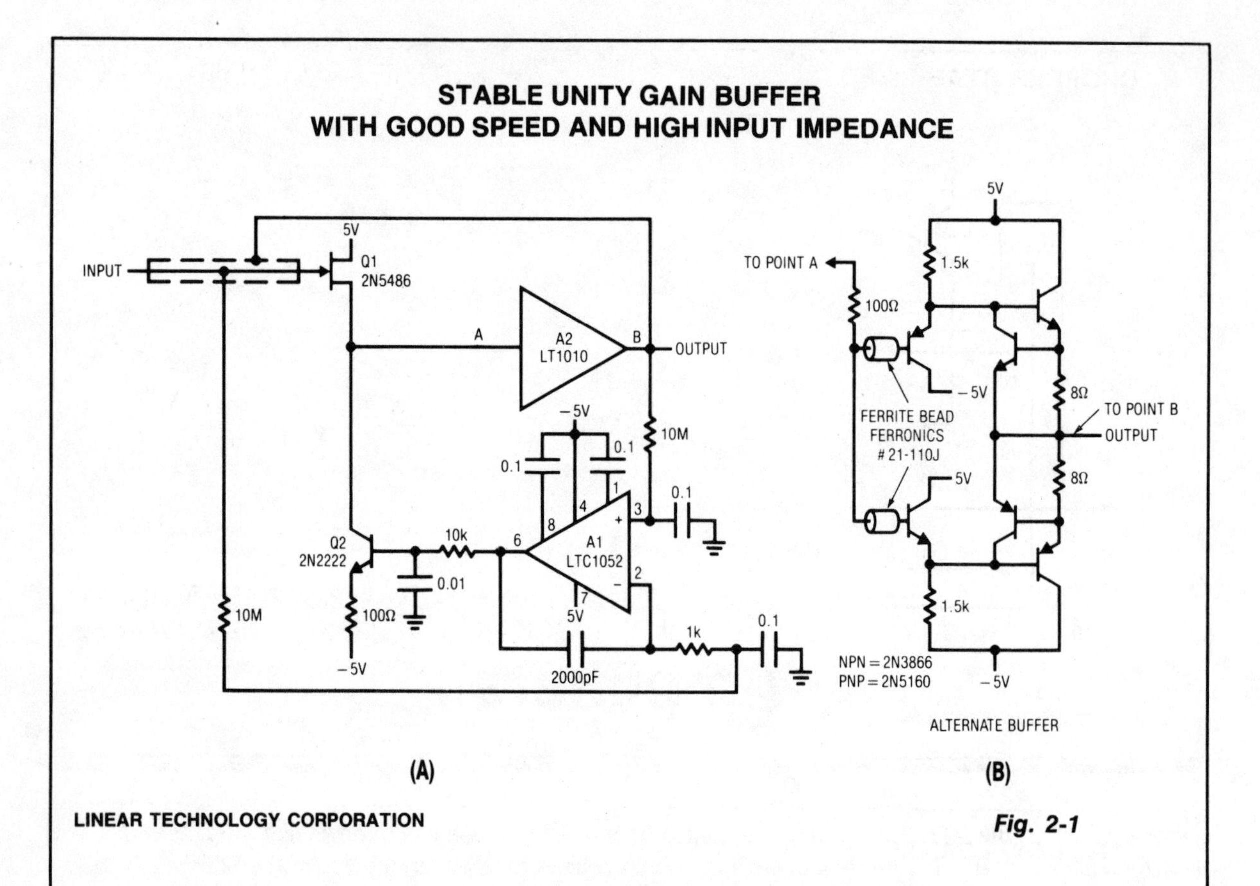

LINEAR TECHNOLOGY CORPORATION

Fig. 2-1

Circuit Notes

Q1 and Q2 constitute a simple, high speed FET input buffer. Q1 functions as a source follower, with the Q2 current source load setting the drain-source channel current. Normally, this open loop configuration would be quite drifty because there is no dc feedback. The LTC1052 contributes this function to stabilize the circuit by comparing the filtered circuit output to a similarly filtered version of the input signal. The amplified difference between these signals is used to set Q2's bias and hence Q1's channel current. This forces Q1's V_{GS} to whatever voltage is required to match the circuit's input and output potentials. The 2000 pF capacitor at A1 provides stable loop compensation. The RC network in A1's output prevents it from seeing high speed edges coupled through Q2's collector-base junction. A2's output is also fed back to the shield around Q1's gate lead, bootstrapping the circuit's effective input capacitance down to less than 1 pF. For very fast requirements, the alternate discrete component buffer shown will be useful. Although its output is current limited at 75 mA, the GHz range transistors employed provide exceptionally wide bandwidth, fast slewing and very little delay.

CHOPPER STABILIZED AMPLIFIER

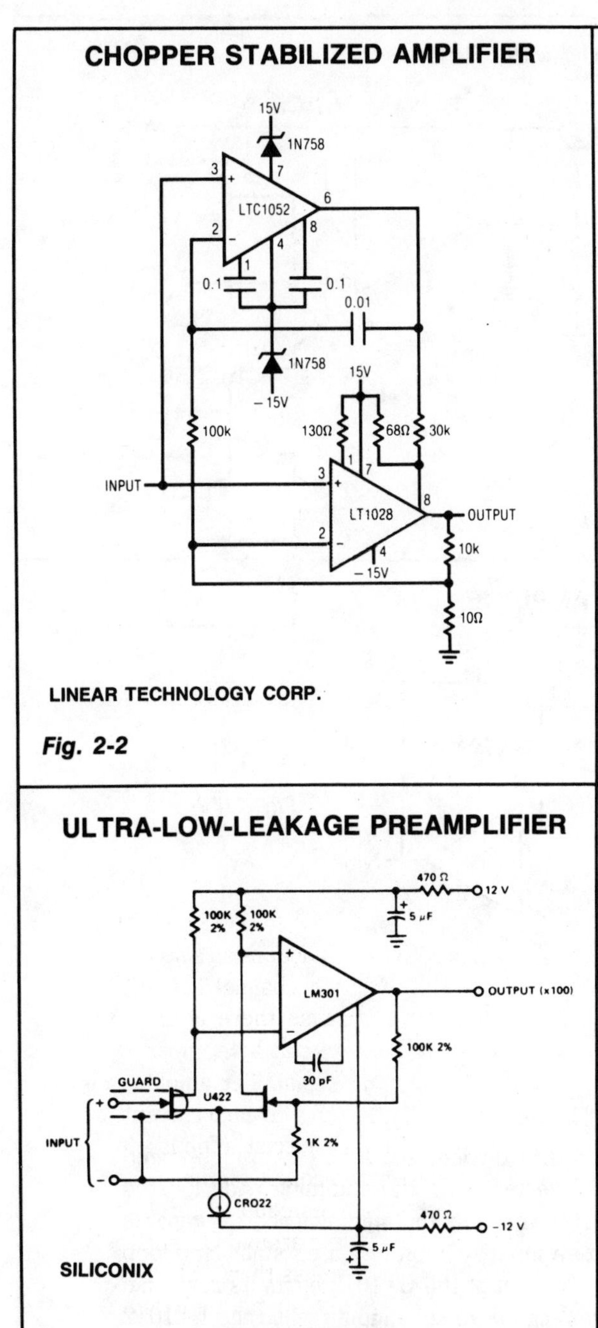

LINEAR TECHNOLOGY CORP.

Fig. 2-2

ULTRA-LOW-LEAKAGE PREAMPLIFIER

SILICONIX

Circuit Notes

The circuit has an input leakage of only 2 pA typical at 75°C and would be usable with 1 M ohm input resistance.

Fig. 2-3

FET INPUT AMPLIFIER

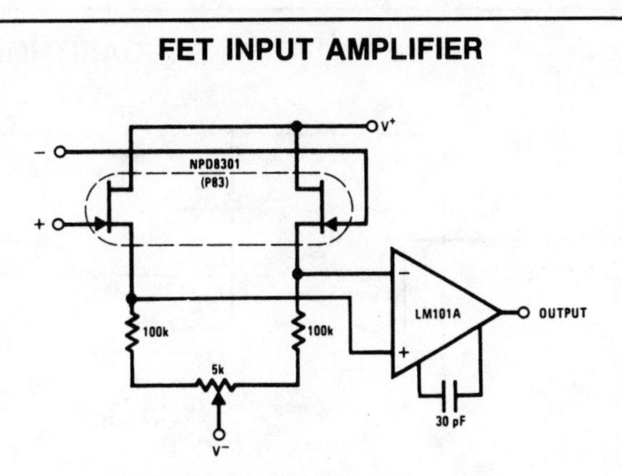

NATIONAL SEMICONDUCTOR CORP.

Circuit Notes

The NPD8301 monolithic-dual provides an ideal low offset, low drift buffer function for the LM101A op amp. The excellent matching characteristics of the NPD8301 track well over its bias current range, thus improving common-mode rejection.

Fig. 2-4

ULTRA-HIGH Z_{in} AC UNITY GAIN AMPLIFIER

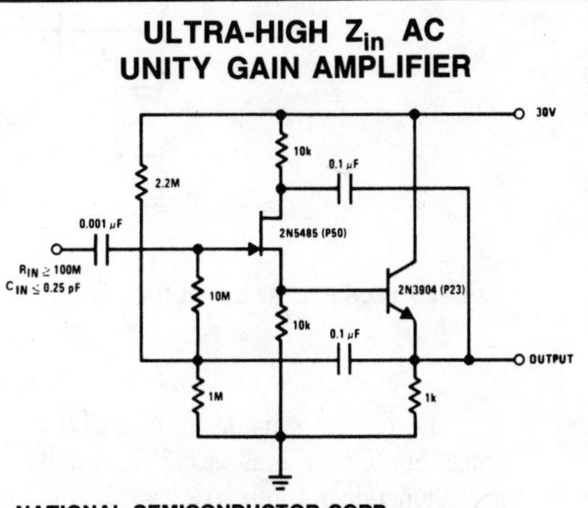

NATIONAL SEMICONDUCTOR CORP.

Circuit Notes

Nothing is left to chance in reducing input capacitance. The 2N5485, which has low capacitance in the first place, is operated as a source follower with bootstrapped gate bias resistor and drain.

Fig. 2-5

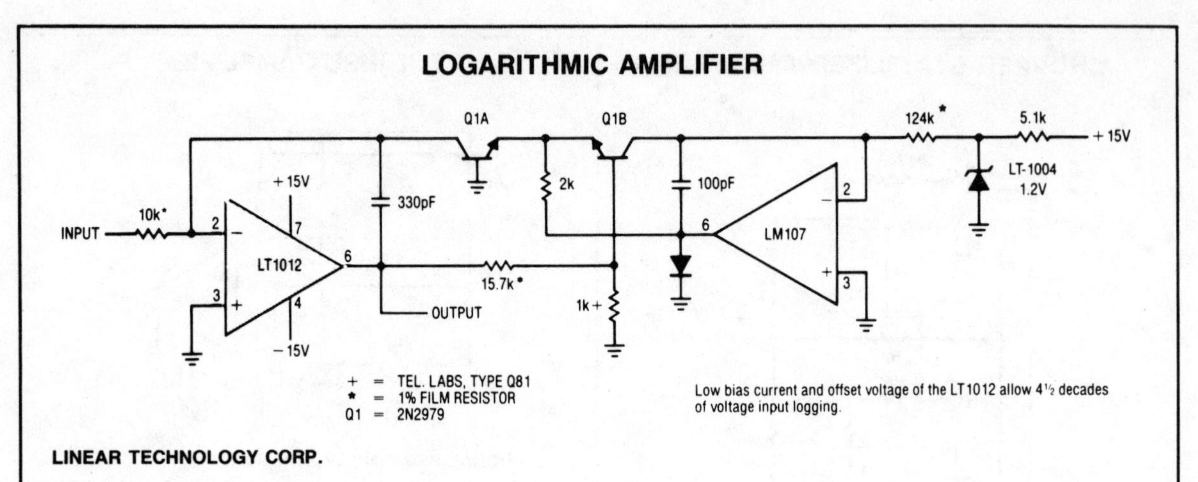

LINEAR TECHNOLOGY CORP.

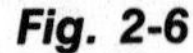

Fig. 2-6

COMPOSITE AMPLIFIER

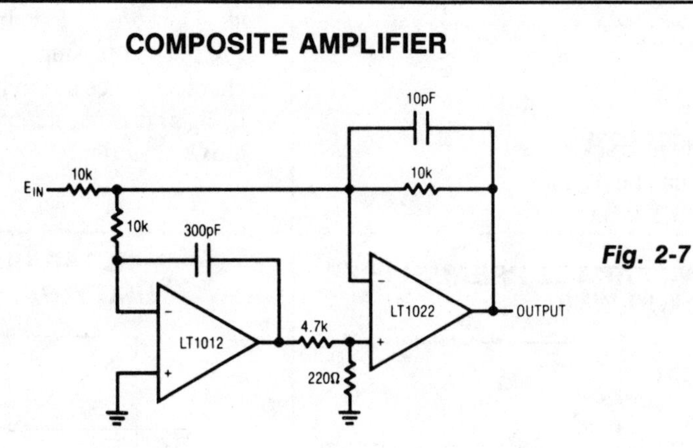

Fig. 2-7

LINEAR TECHNOLOGY CORPORATION

Circuit Notes

The circuit is made up of an LT1012 low drift device, and an LT1022 high speed amplifier. The overall circuit is a unity gain inverter, with the summing node located at the junction of three 10 k resistors. The LT1012 monitors this summing node, compares it to ground, and drives the LT1022's positive input, completing a dc stabilizing loop around the LT1022. The 10 k - 300 pF time constant at the LT1012 limits its response to low frequency signals. The LT1022 handles high frequency inputs while the LT1012 stabilizes the dc operating point. The 4.7 k - 220 ohm divider at the LT1022 prevents excessive input overdrive during start-up. This circuit combines the LT1012's 35 μV offset and 1.5 V/°C drift with the LT1022's 23 V/μs slew rate and 300 kHz full power bandwidth. Bias current, dominated by the LT1012, is about 100 pA.

STEREO AMPLIFIER WITH GAIN CONTROL

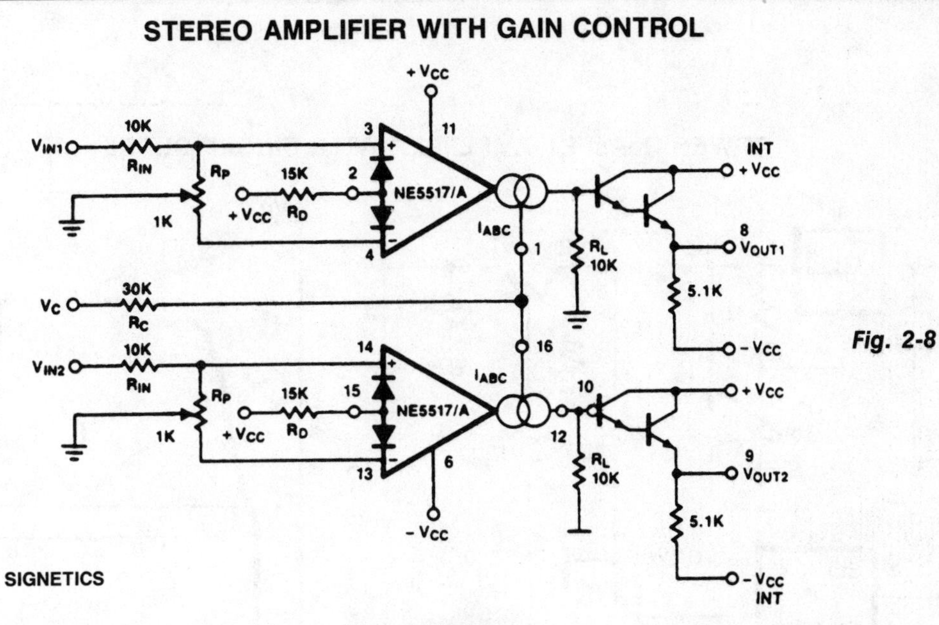

Fig. 2-8

SIGNETICS

Circuit Notes

Excellent tracking of typical 0.3 dB is easy to achieve. With the potentiometer, R_p, the offset can be adjusted. For ac-coupled amplifiers, the potentiometer may be replaced with two 5.1 k ohm resistors.

PRECISION-WEIGHTED RESISTOR PROGRAMMABLE-GAIN AMPLIFIER

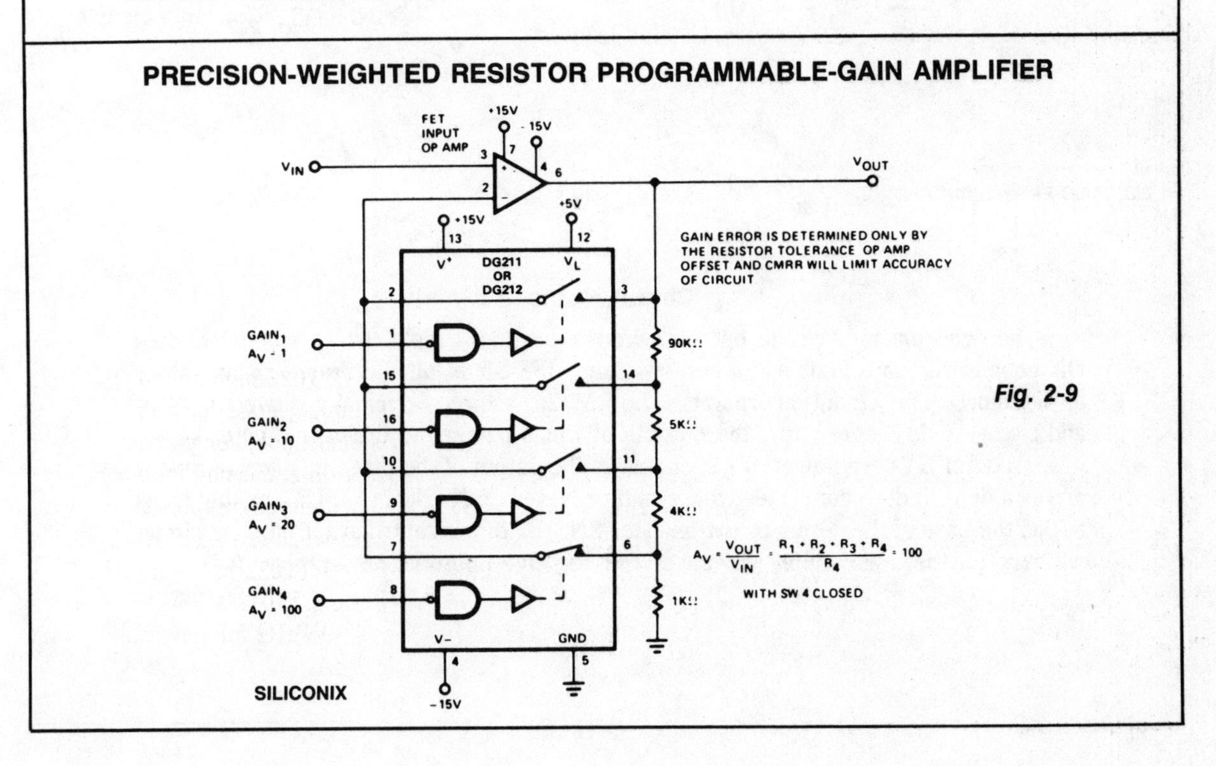

Fig. 2-9

SILICONIX

POWER GaAsFET AMPLIFIER WITH SINGLE SUPPLY

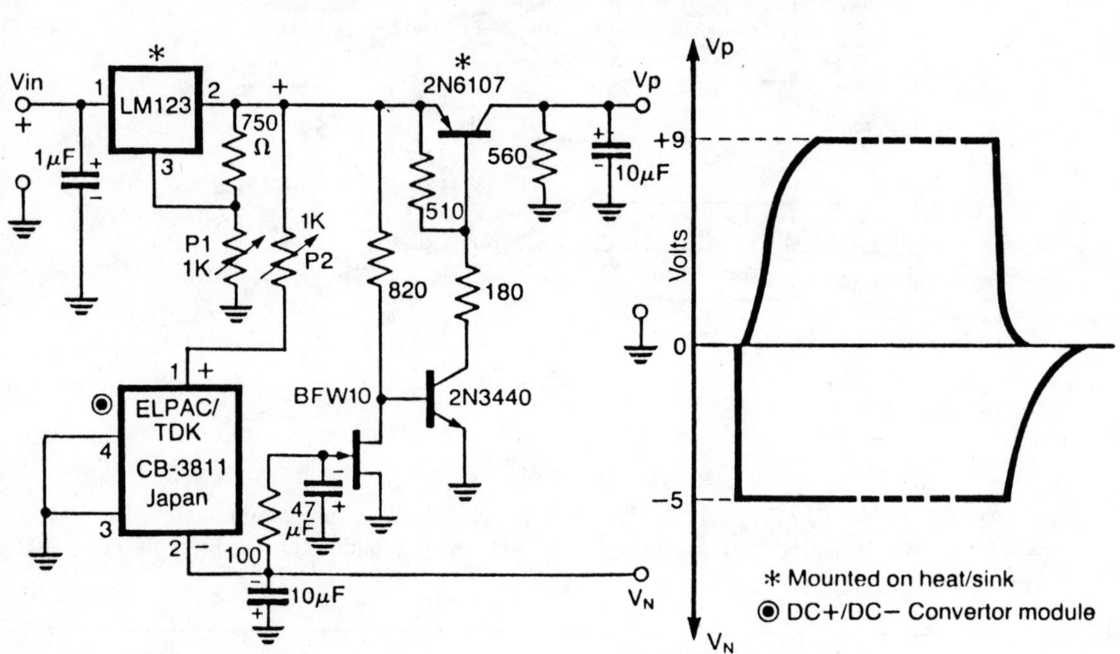

ELECTRIC ENGINEERING

Fig. 2-10

Circuit Notes

The dual regulator circuit operates from a positive supply, which when switched ON powers the gate first, and when switched OFF shuts off the drain first as shown in the figure. This circuit incorporates the LM123, a three terminal positive regulator and a dc+ to dc− converter, the outputs of which power the drains and gates of the power GaAsFETs in a power amplifier relay. The output of the three terminal regulator drives a dc+ to dc− converter whose output biases an N-channel JFET suitably so as to pull the base of the series pass transistor 2N6107 to a level to turn it on. The circuit will turn off the drain supply whenever the negative potential on the Gate fails.

LINEAR AMPLIFIERS FROM CMOS INVERTERS

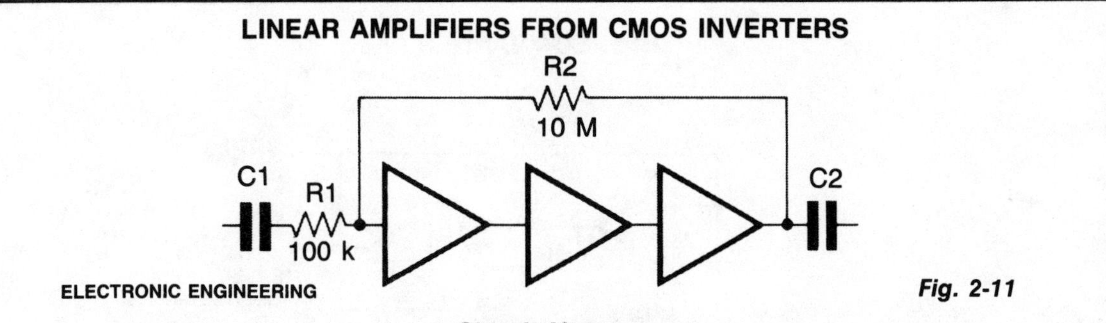

ELECTRONIC ENGINEERING

Fig. 2-11

Circuit Notes

CMOS inverters can be used as linear amplifiers if negative feedback is applied. Best linearity is obtained with feedback applied around three inverters which gives almost perfect linearity up to an output swing of 5 V p-p with a 10 V supply rail. The gain is set by the ratio of R1 and R2 and the values shown are typical for a gain of 100. The high frequency response with the values shown is almost flat to 20 kHz. The frequency response is determined by C1 and C2. This circuit is not suitable for low level signals because the signal-to-noise ratio is only approx. 50 dB with 5 V p-p output with the values shown.

CURRENT-COLLECTOR HEAD-AMPLIFIER

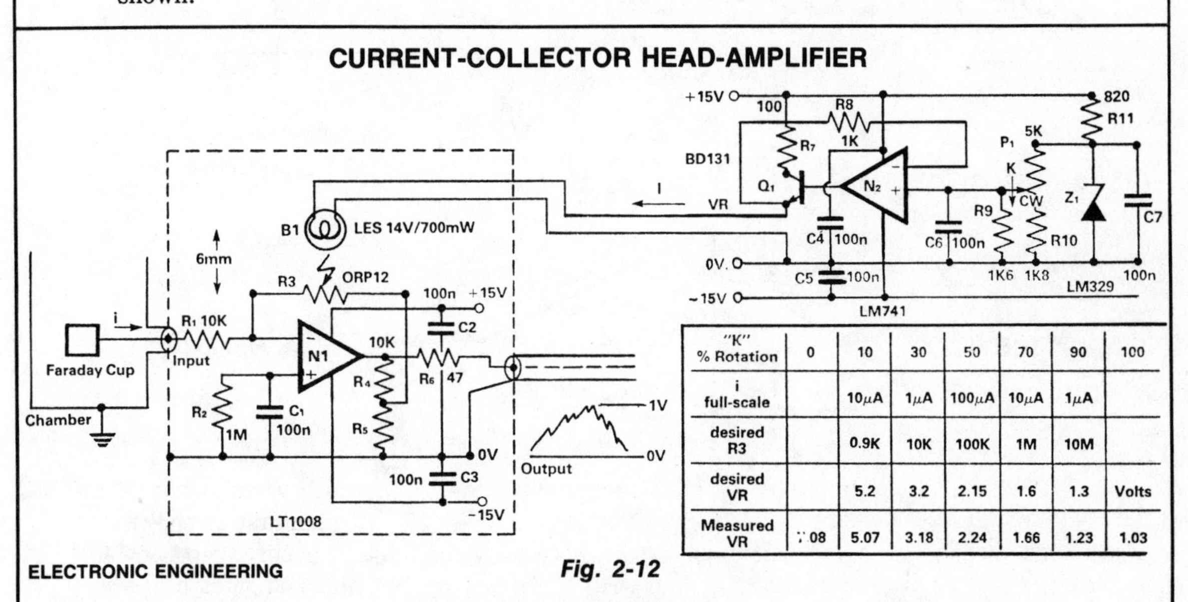

"K" % Rotation	0	10	30	50	70	90	100
i full-scale		10μA	1μA	100μA	10μA	1μA	
desired R3		0.9K	10K	100K	1M	10M	
desired VR		5.2	3.2	2.15	1.6	1.3	Volts
Measured VR	[illegible].08	5.07	3.18	2.24	1.66	1.23	1.03

ELECTRONIC ENGINEERING

Fig. 2-12

Circuit Notes

To amplify small current signals such as from an electron-collector inside a vacuum chamber, it is convenient for reasons of noise and bandwidth to have a ''head-amplifier'' attached to the chamber. The op-amp N_1 is a precision bipolar device with extremely low bias current and offset voltage (1) as well as low noise, which allows the 100:1 feedback attenuator to be employed. The resistance of R_3 can be varied from above 10 M to below 1 k, and so the nominal 0 to 1 V-peak output signal corresponds to input current ranges of 1 nA to 10 μA.

HI-FI COMPANDER

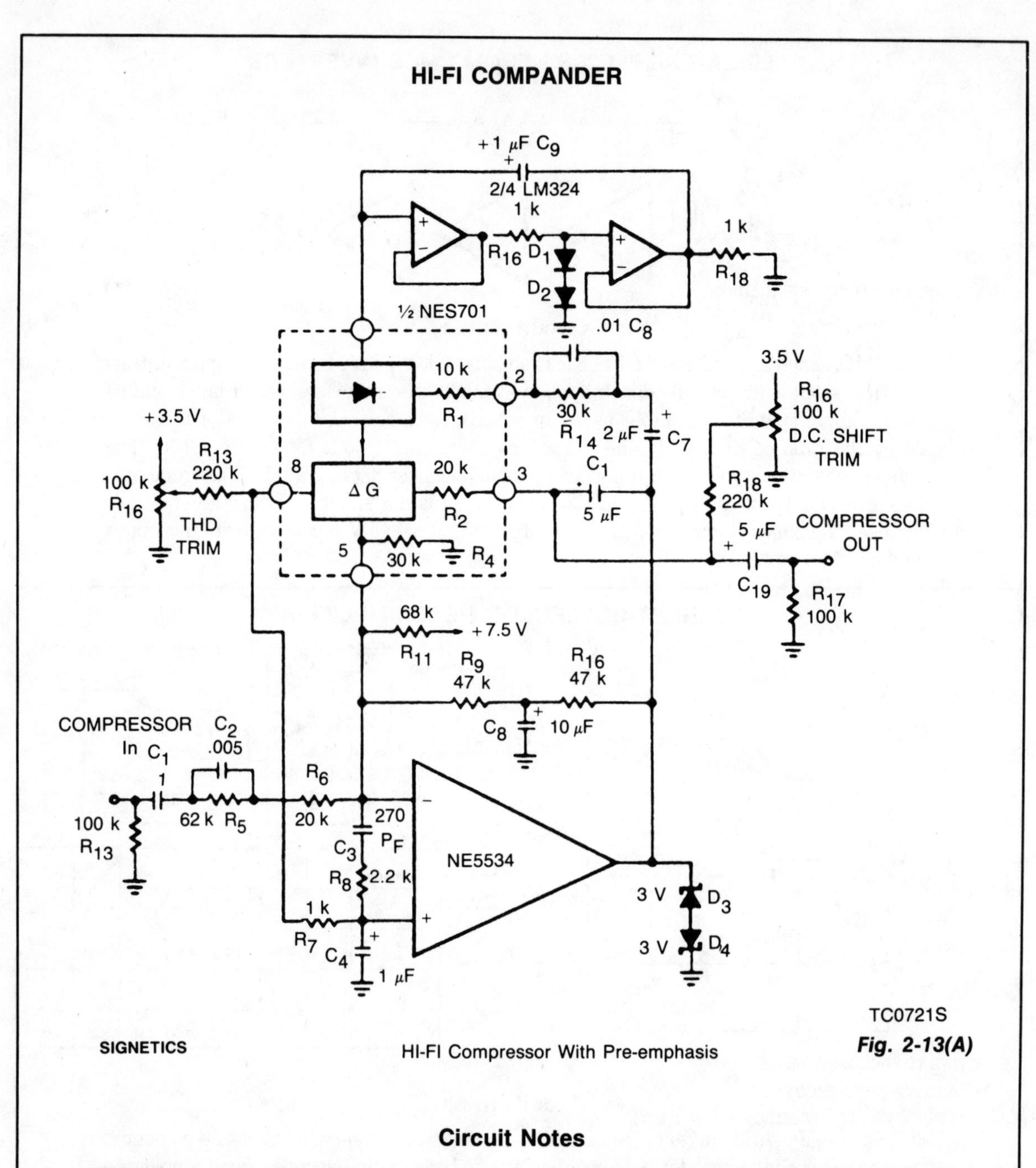

HI-FI Compressor With Pre-emphasis *Fig. 2-13(A)*

Circuit Notes

This circuit for a high fidelity compressor uses an external op amp, and has a high gain and wide bandwidth. An input compensation network is required for stability. The rectifier capacitor (C_9) is not grounded, but is tied to the output of an op amp circuit. When a compressor is operating at high gain, (small input signal), and is suddenly hit with a signal, it will overload until

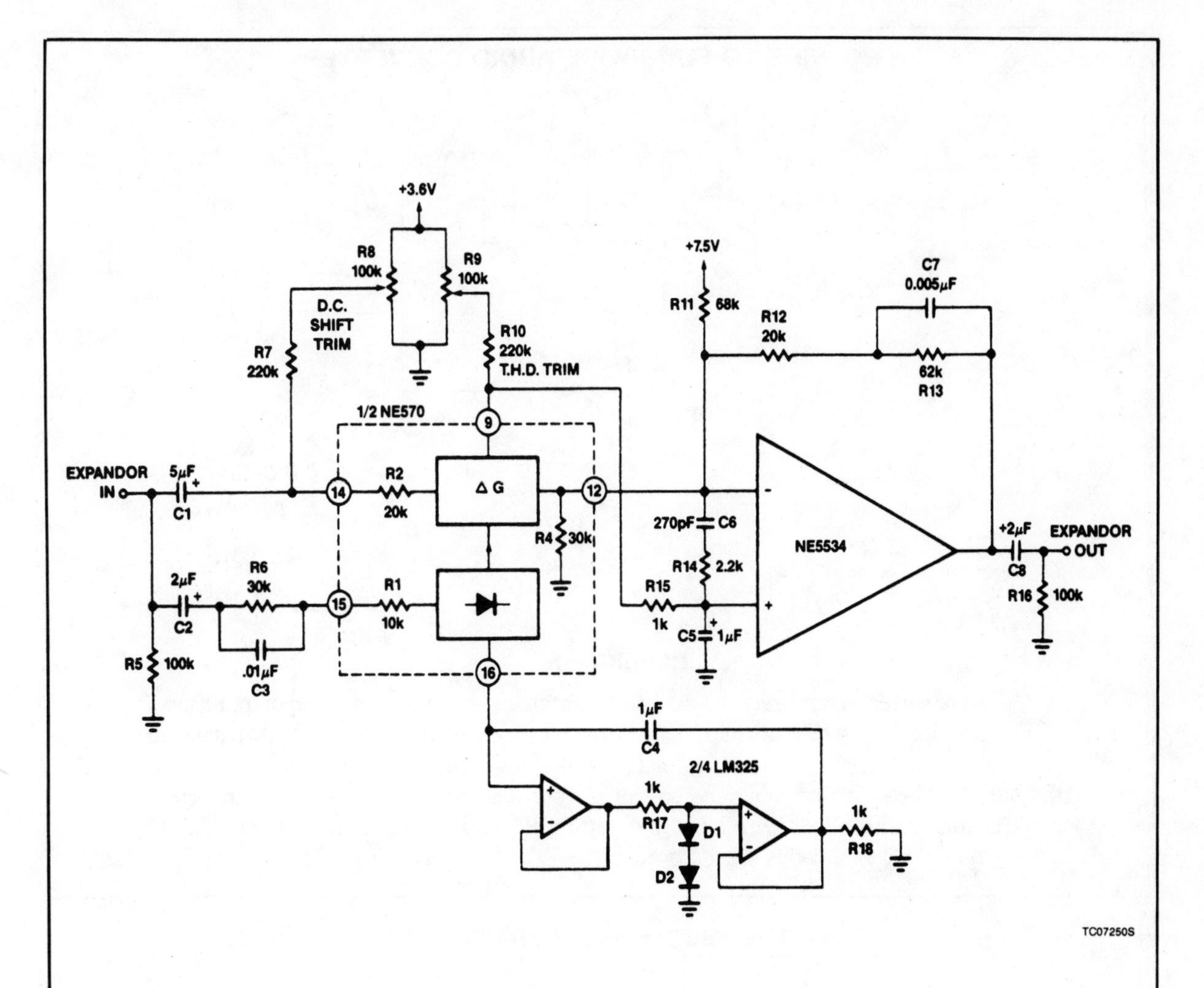

Hi-Fi Expandor With De-emphasis *Fig. 2-13(B)*

it can reduce its gain. The time it takes for the compressor to recover from overload is determined by the rectifier capacitor C_9. The expandor to complement the compressor is shown in Fig. 2-13B. Here an external op amp is used for high slew rate. Both the compressor and expandor have unity gain levels of 0 dB. Trim networks are shown for distortion (THD) and dc shift. The distortion trim should be done first, with an input of 0 dB at 10 kHz. The dc shift should be adjusted for minimum envelope bounce with tone bursts.

TWO-WIRE TO FOUR-WIRE AUDIO CONVERTER

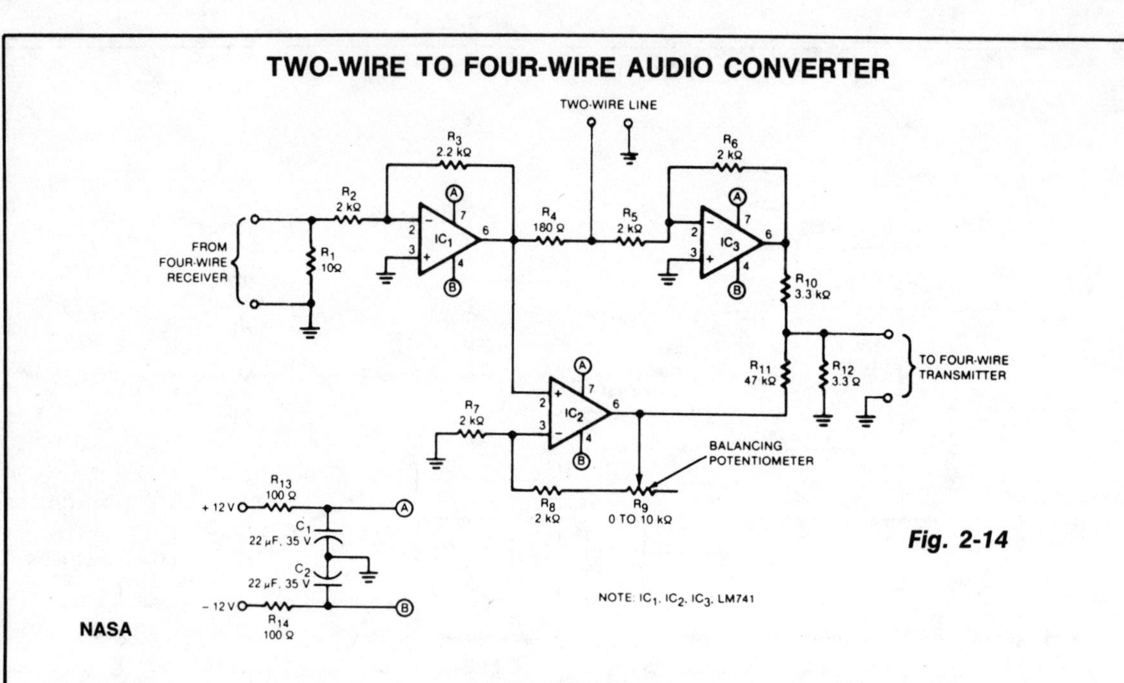

Fig. 2-14

Circuit Notes

This converter circuit maintains 40 dB of isolation between the input and output halves of a four-wire line while permitting a two-wire line to be connected. A balancing potentiometer, Rg, adjusts the gain of IC2 to null the feed-through from the input to the output. The adjustment is done on the workbench just after assembly by inserting a 1 kHz tone into the four-wire input and setting R_g for minimum output signal. An 82 ohm dummy-load resistor is placed across the two wire terminals.

THERMOCOUPLE AMPLIFIER

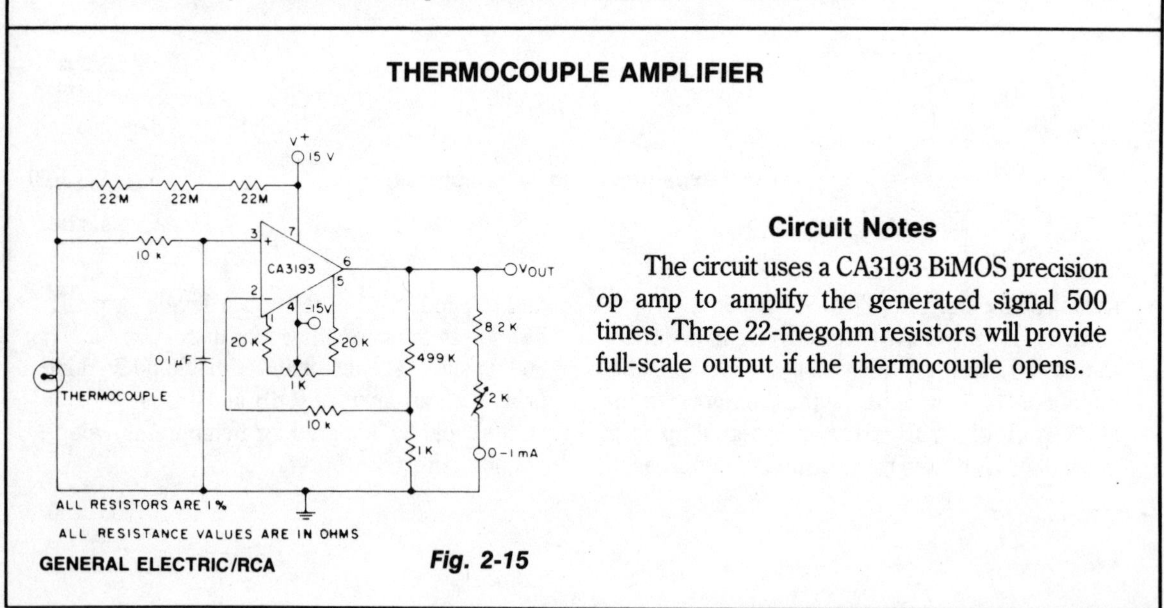

Fig. 2-15

Circuit Notes

The circuit uses a CA3193 BiMOS precision op amp to amplify the generated signal 500 times. Three 22-megohm resistors will provide full-scale output if the thermocouple opens.

LOW-DISTORTION AUDIO LIMITER

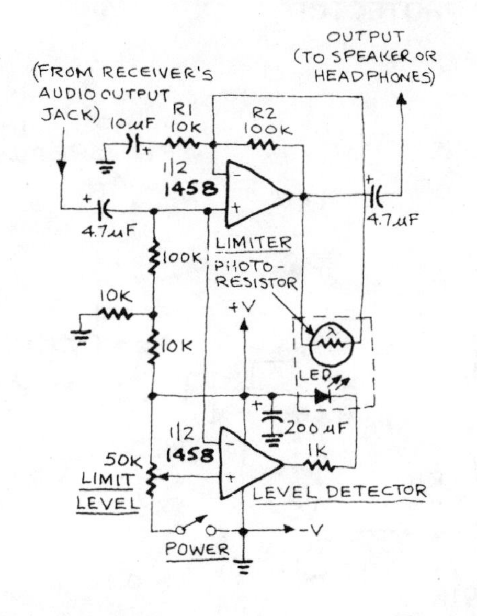

RADIO-ELECTRONICS

Fig. 2-16

Circuit Notes

The level at which the audio limiter comes into action can be set with the LIMIT LEVEL trimmer potentiometer. When that level is exceeded, the output from the LIMITER-DETECTOR half of the op-amp (used as a comparator) turns the LED which causes the resistance of the photoresistor to decrease rapidly. That in turn causes the gain of the LIMITER half of the op-amp to decrease. When the signal drops below the desired limiting level, the LED turns off, the resistance of the photoresistor increases, and the gain of the LIMITER op-amp returns to its normal level—that set by the combination of resistors R1 and R2. A dual-polarity power supply (±12 volts is desirable) is needed for the op-amp.

SPEECH COMPRESSOR

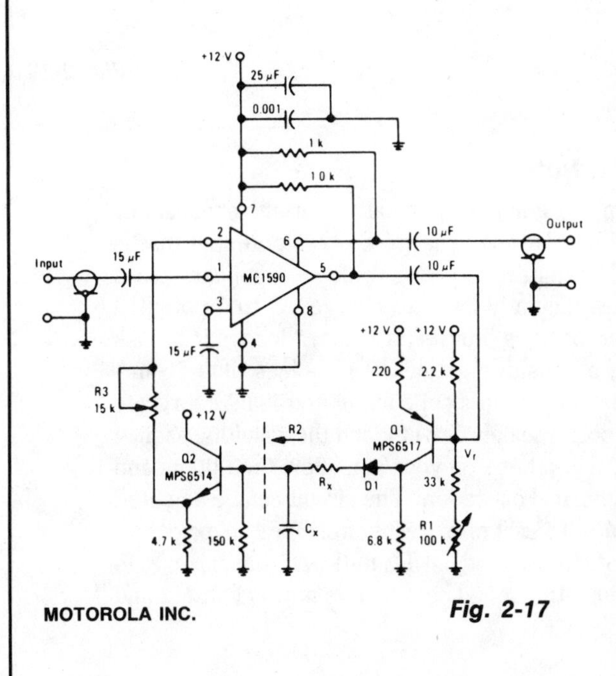

MOTOROLA INC.

Fig. 2-17

Circuit Notes

The amplifier drives the base of a pnp MPS6517 operating common-emitter with a voltage gain of approximately 20. The control R1 varies the quiescent Q point of this transistor so that varying amounts of signal exceed the level V_r. Diode D1 rectifies the positive peaks of Q1's output only when these peaks are greater than $V_r \simeq 7.0$ volts. The resulting output is filtered C_x, R_x. R_x controls the charging time constant or attack time. C_x is involved in both charge and discharge. R2 (150 K, input resistance of the emitter-follower Q2) controls the decay time. Making the decay long and attack short is accomplished by making R_x small and R2 large. (A Darlington emitter-follower may be needed if extremely slow decay times are required.) The emitter-follower Q2 drives the AGC Pin 2 of the MC1590 and reduces the gain. R3 controls the slope of signal compression.

SPEAKER OVERLOAD PROTECTOR

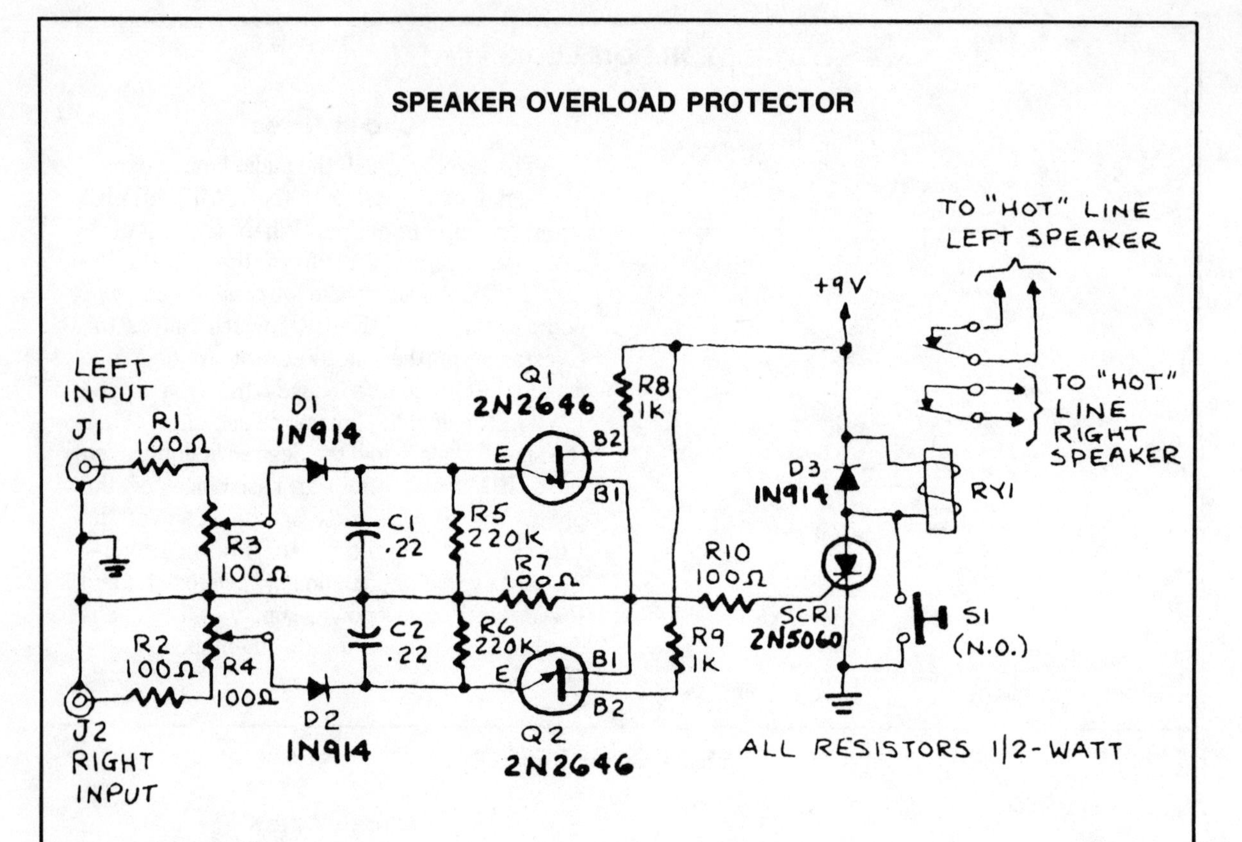

RADIO ELECTRONICS *Fig. 2-18*

Circuit Notes

The input to the circuit is taken from the amplifier's speaker-output terminals or jacks. If the right-channel signal is sufficiently large to charge C1 to a potential that is greater than the breakdown voltage of Q1's emitter, a voltage pulse will appear across R7. Similarly, if the left-channel signal is sufficiently large to charge C2 to a potential that is greater than the breakdown voltage of Q2's emitter, a pulse will appear across R7. The pulse across R7 triggers SCR1, a sensitive gate SCR (I_{GT} less than 15 mA where I_{GT} is the gate trigger-current), that latches in a conducting state and energizes RY1. The action of the relay will interrupt both speaker circuits, and the resulting silence should alert you to the problem. Cut back the volume on your amplifier, then press and release S1 to reset the circuit and restore normal operation. The circuit can be adjusted to trip at any level from 15 to 150 watts RMS. To calibrate, deliberately feed an excessive signal to the right input of the speaker protector and adjust R3 until RY1 energizes. Do the same with the left channel, this time adjusting R4. The circuit is now calibrated and ready for use.

AUDIO AUTOMATIC GAIN CONTROL

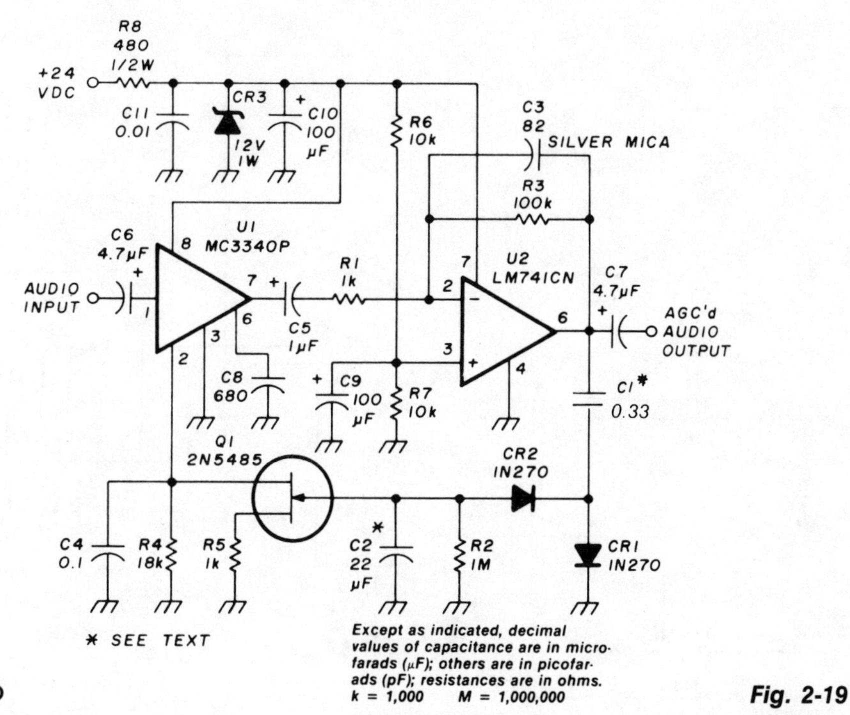

HAM RADIO

Fig. 2-19

Circuit Notes

An audio signal applied to U1 is passed through to the 741 operational amplifier, U2. After being amplified, the output signal of U2 is sampled and applied to a negative voltage doubler/rectifier circuit composed of diodes CR1 and CR2 along with capacitor C1. The resulting negative voltage is used as a control voltage that is applied to the gate of the 2N5485 JFET Q1. Capacitor C2 and resistor R2 form a smoothing filter for the rectified audio control voltage.

The JFET is connected from pin 2 of the MC3340P to ground through a 1 kilohm resistor. As the voltage applied to the gate of the JFET becomes more negative in magnitude, the channel resistance of the JFET increases causing the JFET to operate as a voltage controlled resistor. The MC3340P audio attenuator is the heart of the AGC. It is capable of 13 dB gain or nearly −80 dB of attenuation depending on the external resistance placed between pin 2 and ground. An increase of resistance decreases the gain achieved through the MC3340P. The circuit gain is not entirely a linear function of the external resistance but approximates such behavior over a good portion of the gain/attenuation range. An input signal applied to the AGC input will cause the gate voltage of the JFET to become proportionally negative. As a result the JFET increases the resistance from pin 2 to ground of the MC3340P causing a reduction in gain. In this way the AGC output is held at a nearly constant level.

VOLTAGE-CONTROLLED ATTENUATOR

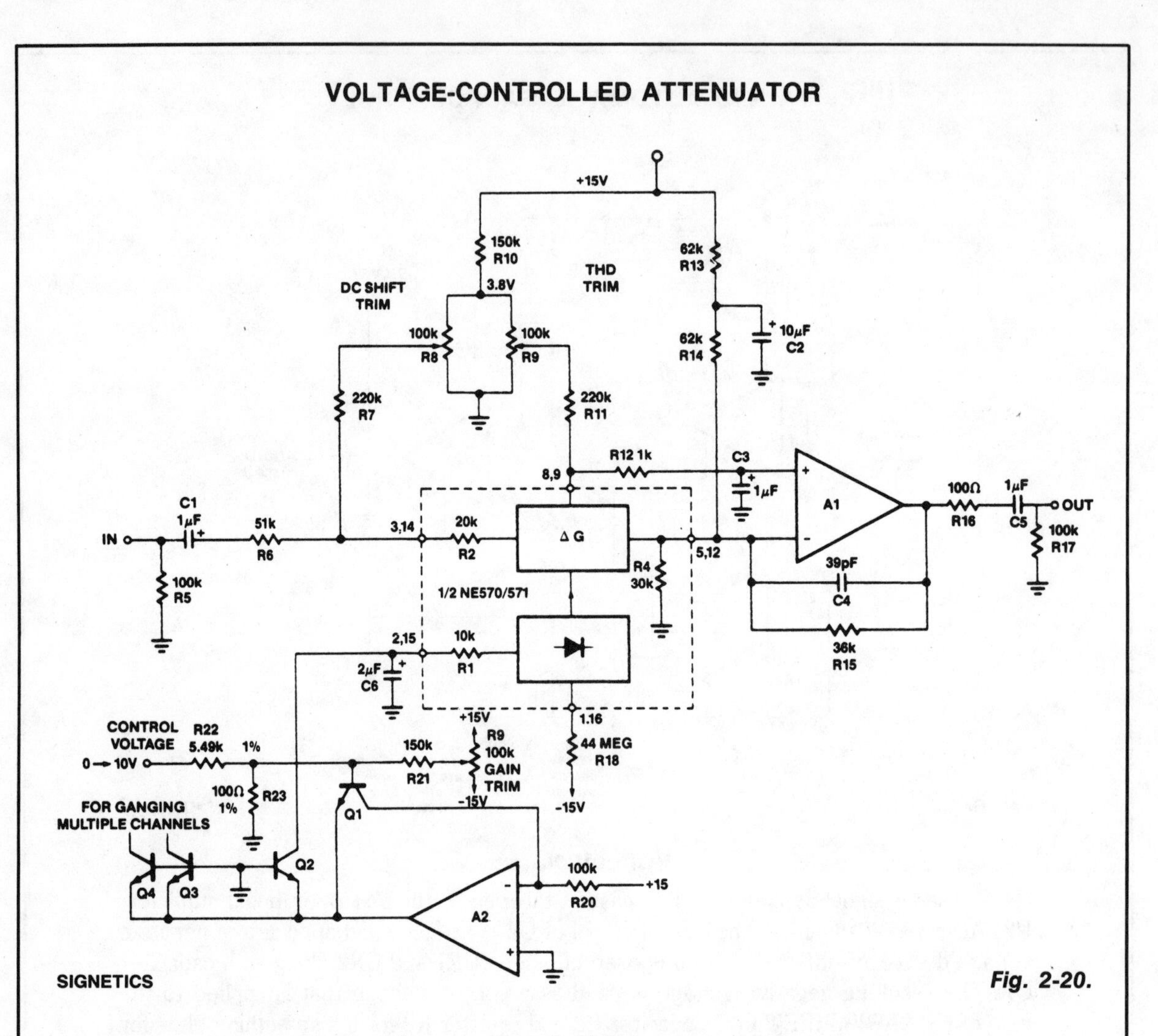

Fig. 2-20.

Circuit Notes

Op amp A_2 and transistors Q_1 and Q_2 form the exponential converter generating an exponential gain control current, which is fed into the rectifier. A reference current of 150 μA, (15 V and R_{20} = 100 k), is attenuated a factor of two (6 dB) for every volt increase in the control voltage. Capacitor C_6 slows down gain changes to a 20 ms time constant ($C_6 \times R_1$) so that an abrupt change in the control voltage will produce a smooth sounding gain change. R_{18} ensures that for large control voltages the circuit will go to full attentuation. The rectifier bias current would normally limit the gain reduction to about 70 dB. R_{16} draws excess current out of the rectifier. After approximately 50 dB of attentuation at a −6 dB/V slope, the slope steepens and attenuation becomes much more rapid until the circuit totally shuts off at about 9 V of control voltage. A_1 should be a low-noise high slew rate op amp. R_{13} and R_{14} establish approximately a 0 V bias at A_1's output.

HIGH-INPUT-IMPEDANCE DIFFERENTIAL AMPLIFIER

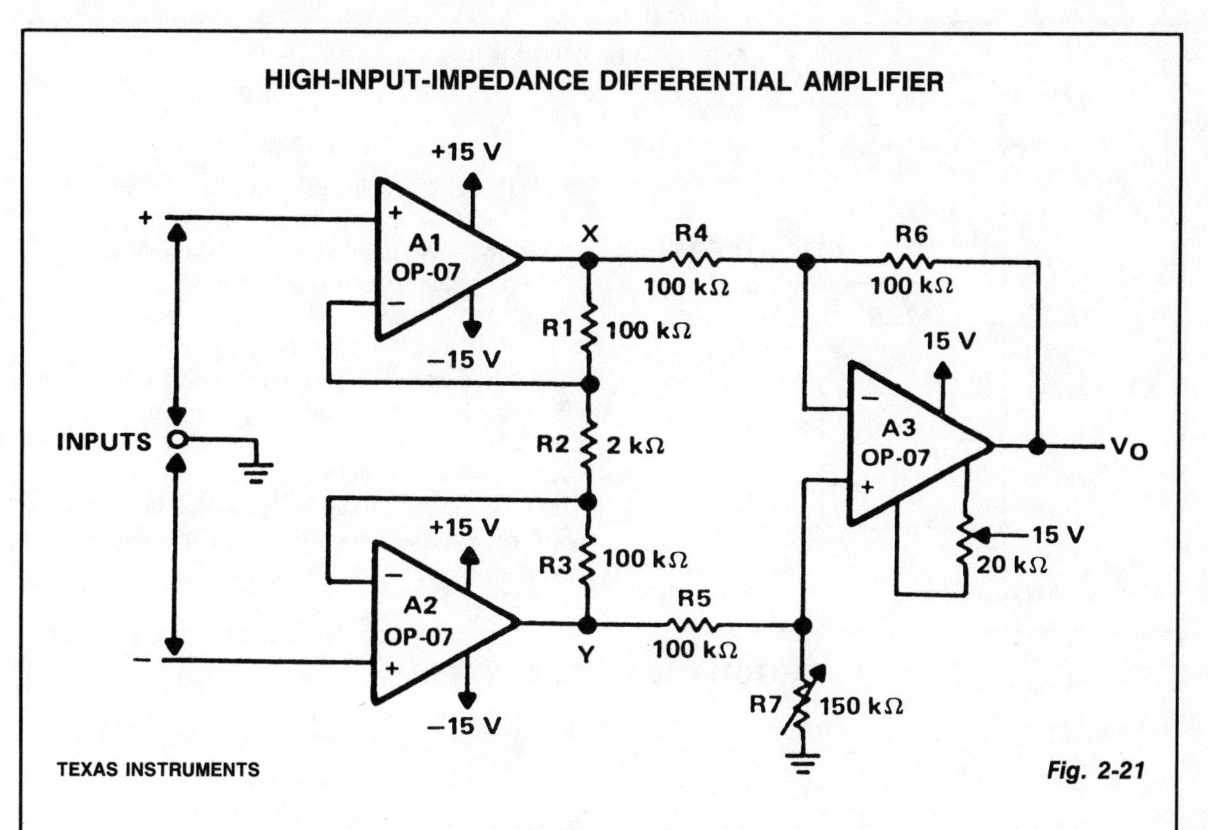

TEXAS INSTRUMENTS

Fig. 2-21

Circuit Notes

Operational amplifiers A1 and A2 are connected in a non-inverting configuration with their outputs driving amplifier A3. Operational amplifier A3 could be called a subtractor circuit which converts the differential signal floating between points X and Y into a single-ended output voltage. Although not mandatory, amplifier A3 is usually operated at unity gain and R4, R5, R6, and R7 are all equal.

The common-mode-rejection of amplifier A3 is a function of how closely the ratio R4:R5 matches the ratio R6:R7. For example, when using resistors with 0.1% tolerance, common-mode rejection is greater than 60 dB. Additional improvement can be attained by using a potentiometer (slightly higher in value than R6) for R7. The potentiometer can be adjusted for the best common-mode rejection. Input amplifiers A1 and A2 will have some differential gain but the common-mode input voltages will experience only unity gain. These voltages will not appear as differential signals at the input of amplifier A3 because, when they appear at equal levels on both ends of resistor R2, they are effectively canceled.

This type of low-level differential amplifier finds widespread use in signal processing. It is also useful for dc and low-frequency signals commonly received from a transducer or thermocouple output, which are amplified and transmitted in a single-ended mode. The amplifier is powered by ±15 V supplies. It is only necessary to null the input offset voltage of the output amplifier A3.

AUDIO Q-MULTIPLIER

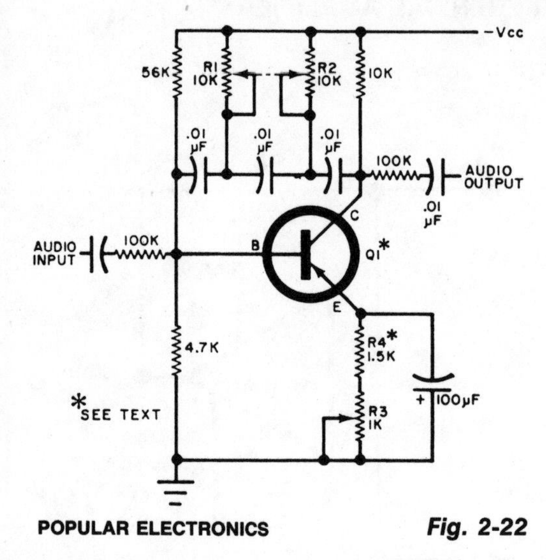

POPULAR ELECTRONICS

Fig. 2-22

Circuit Notes

This circuit is for selective tuning between two closely spaced audio tones. The selective frequency is dependent on the value of capacitors and resistors in the feedback circuit between the collector and base of Q1. With the values shown, the frequency can be "tuned" a hundred cycles or so around 650 Hz. R1 and R2 must be ganged. Emitter potentiometer R3 determines the sharpness of response curve. Any transistor having a beta greater than 50 can be used. Select a value for R4 so that the circuit will not oscillate when R3 is set for minimum bandwidth (sharpest tuning).

AUTOMATIC LEVEL CONTROL

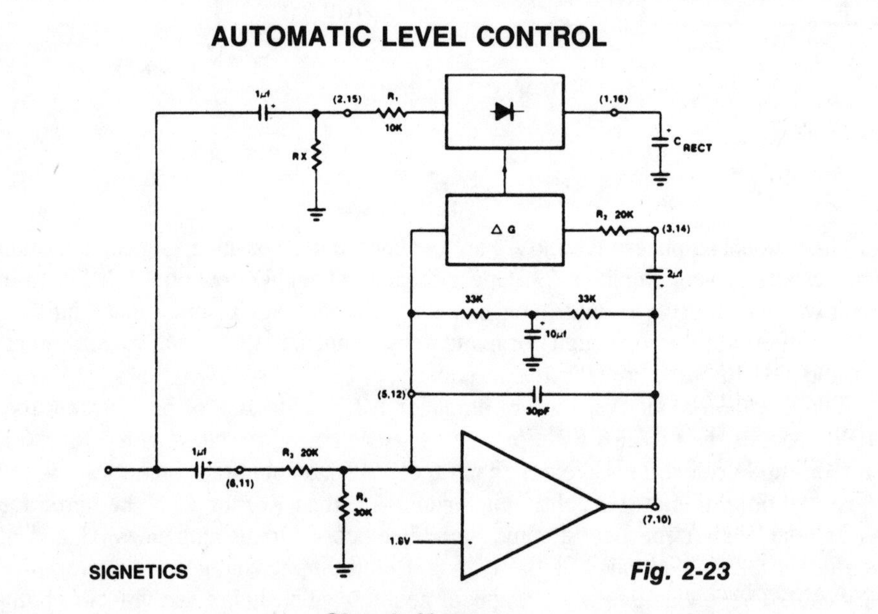

SIGNETICS

Fig. 2-23

Circuit Notes

The NE570 can be used to make a very high performance ALC compressor, except that the rectifier input is tied to the input. This makes gain inversely proportional to input level so that a 20 dB drop in input level will produce a 20 dB increase in gain. The output will remain fixed at a constant level. As shown, the circuit will maintain an output level of ± 1 dB for an input range of $+14$ to -43 dB at 1 kHz. Additional external components will allow the output level to be adjusted.

PULSE-WIDTH PROPORTIONAL-CONTROLLER CIRCUIT

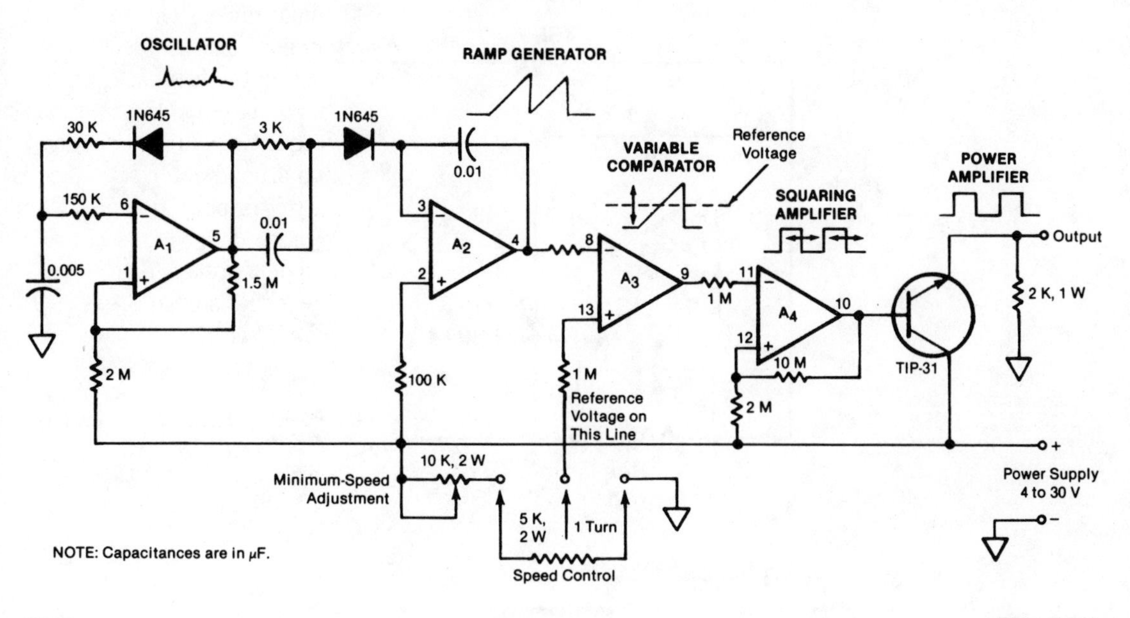

NASA *Fig. 2-24*

Circuit Notes

The quad operational amplifier circuit yields full 0 to 100 percent pulse-width control. The controller uses an LM3900 that requires only a single supply voltage of 4 to 30 V. The pulse-repetition rate is set by a 1 kHz oscillator that incorporates amplifier A_1. The oscillator feeds ramp generator A_2, which generates a linear ramp voltage for each oscillator pulse. The ramp signal feeds the inverting input of comparator A_3; the speed-control voltage feeds the noninverting input. Thus, the output of the comparator is a 1 kHz pulse train, the pulse width of which changes linearly with the control voltage. The control voltage can be provided by an adjustable potentiometer or by an external source of feedback information such as a motor-speed sensing circuit. Depending on the control-voltage setting, the pulse duration can be set at any value from zero (for zero average dc voltage applied to the motor) to the full pulse-repetition period (for applied motor voltage equal to dc power-supply voltage). An amplifier stage (A_4) with a gain of 10 acts as a pulse-squaring circuit. A TIP-31 medium-power transistor is driven by A_4 and serves as a separate power-amplifier stage.

OP AMP CLAMPING

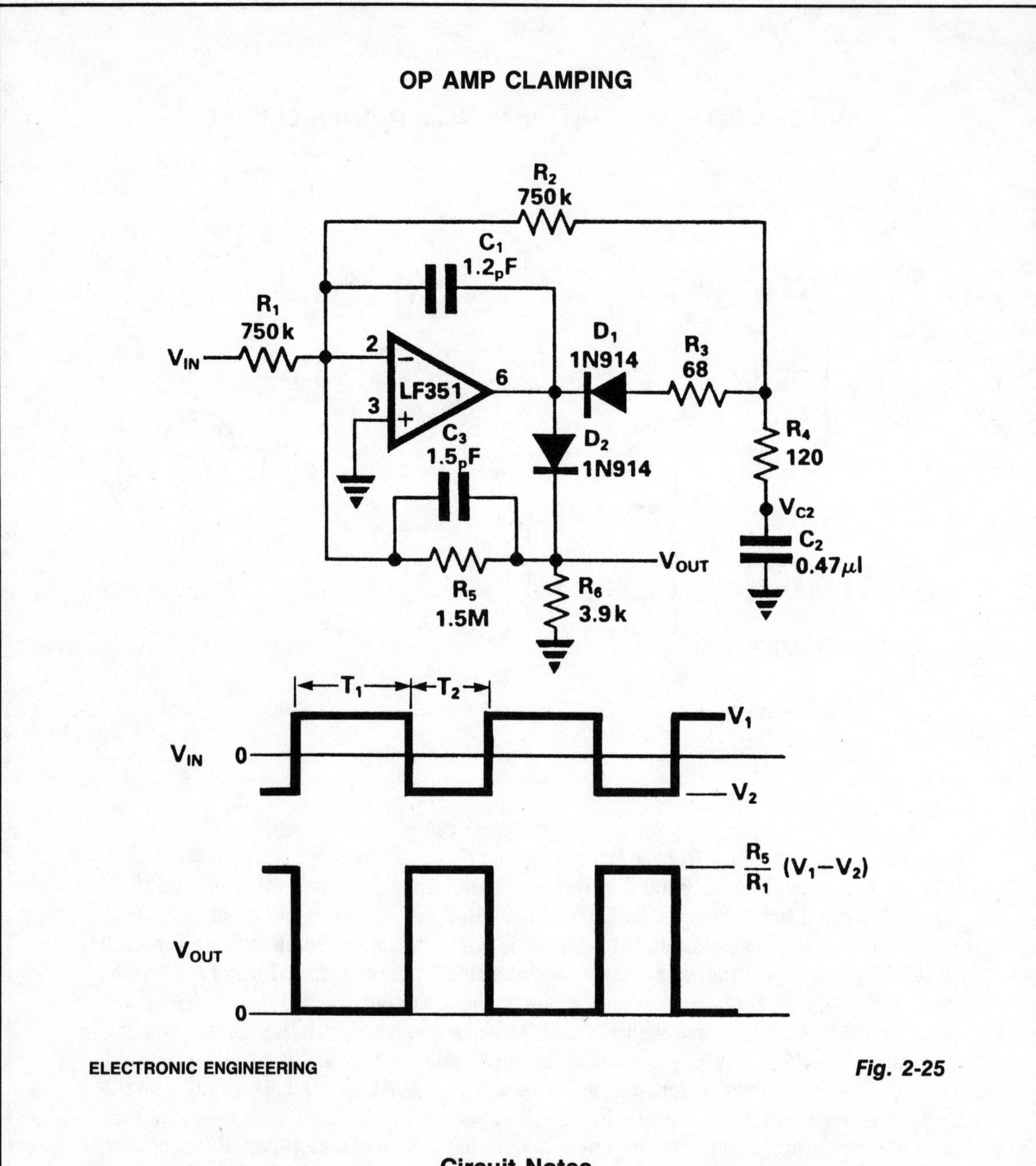

ELECTRONIC ENGINEERING

Fig. 2-25

Circuit Notes

The circuit clamps the most positive value of the input pulse signal to the zero base level. Additionally, the circuit inverts and amplifies the input signal by the factor of R_5/R_1. The waveforms are shown in the bottom of Fig. 2-24.

3

Analog-to-Digital Converters

The sources of the following circuits are contained in the Sources section beginning on page 694. The figure number contained in the box of each circuit correlates to the source entry in the Sources section.

SUCCESSIVE APPROXIMATION A/D CONVERTERS

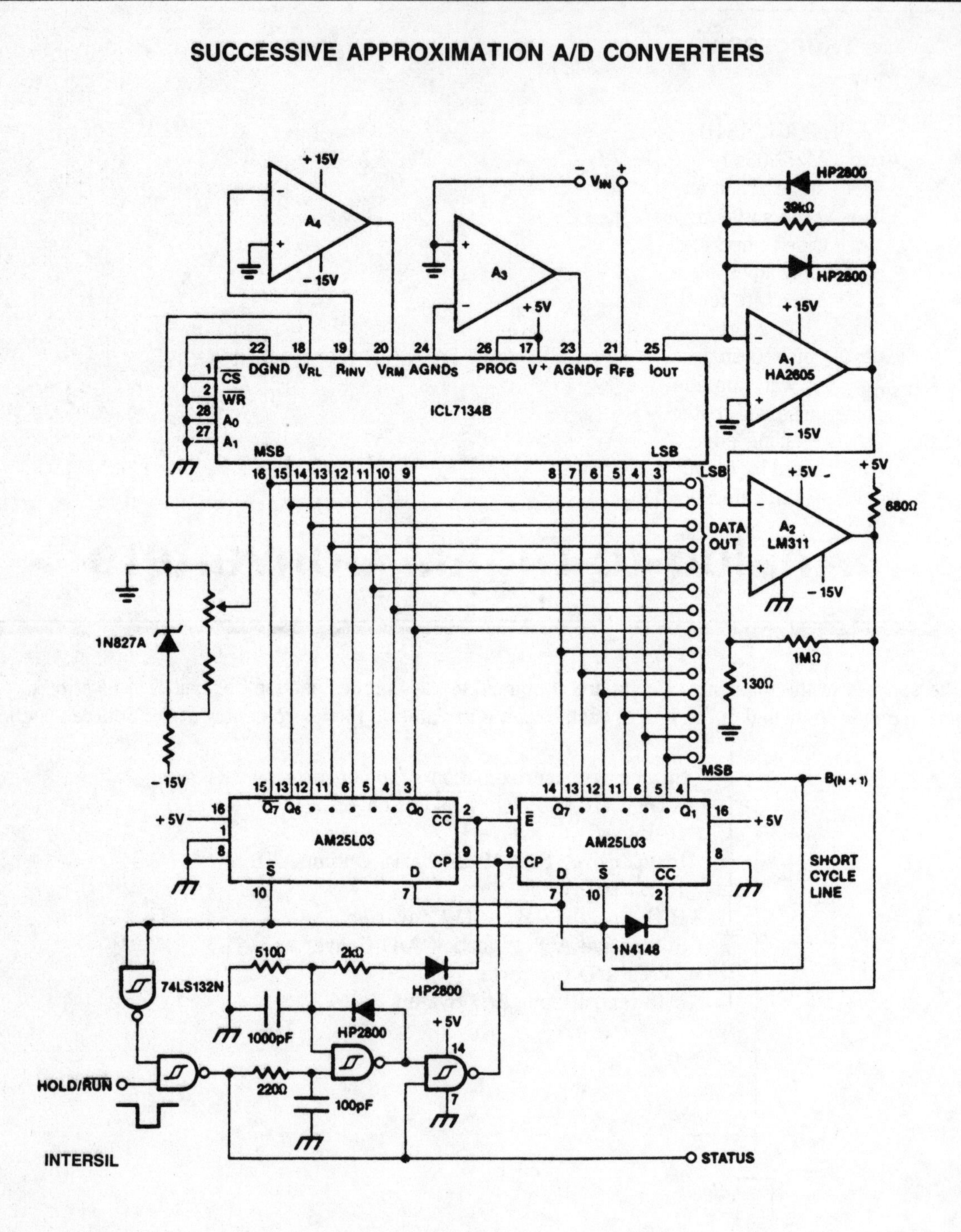

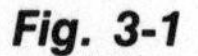

Fig. 3-1

SUCCESSIVE APPROXIMATION A/D CONVERTERS, Continued.

Circuit Notes

The ICL7134B-based circuit is for a bipolar-input high-speed A/D converter, using two AM25L03s to form a 14-bit successive approximation register. The comparator is a two-stage circuit with an HA2605 front-end amplifier, used to reduce settling time problems at the summing node (see A020). Careful offset-nulling of this amplifier is needed, and if wide temperature range operation is desired, an auto-null circuit using an ICL7650 is probably advisable (see A053). The clock, using two Schmitt trigger TTL gates, runs at a slower rate for the first 8 bits, where settling-time is most critical than for the last 6 bits. The short-cycle line is shown tied to the 15th bit; if fewer bits are required, it can be moved up accordingly. The circuit will free-run if the HOLD/RUN input is held low, but will stop after completing a conversion if the pin is high at that time. A low-going pulse will restart it. The STATUS output indicates when the device is operating, and the falling edge indicates the availability of new data. A unipolar version can be constructed by typing the MSB (D13) on an ICL7134U to pin 14 on the first AM25L03, deleting the reference inversion amplifier A4, and tying V_{RFM} to V_{RFL}.

4 DIGIT (10,000 COUNT) A/D CONVERTER

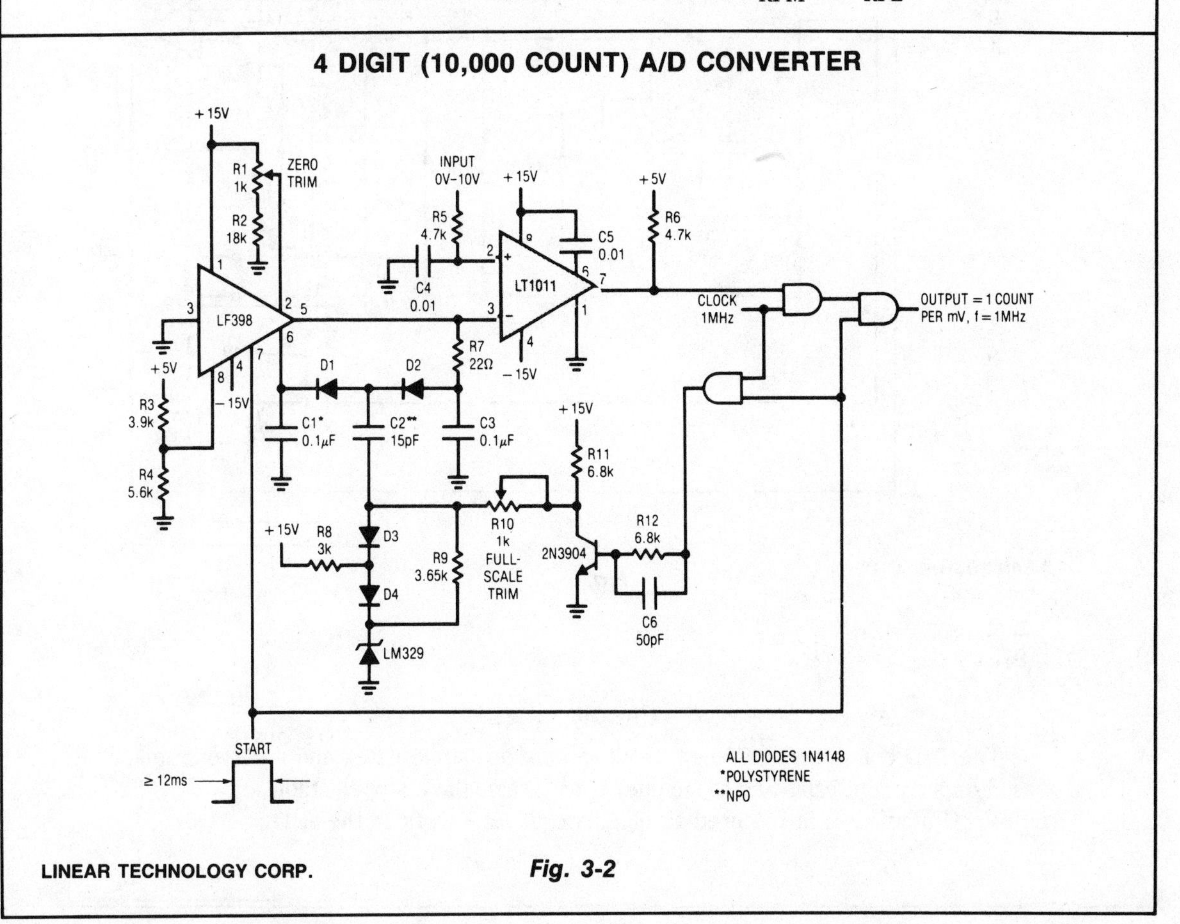

LINEAR TECHNOLOGY CORP.

Fig. 3-2

16-BIT A/D CONVERTER

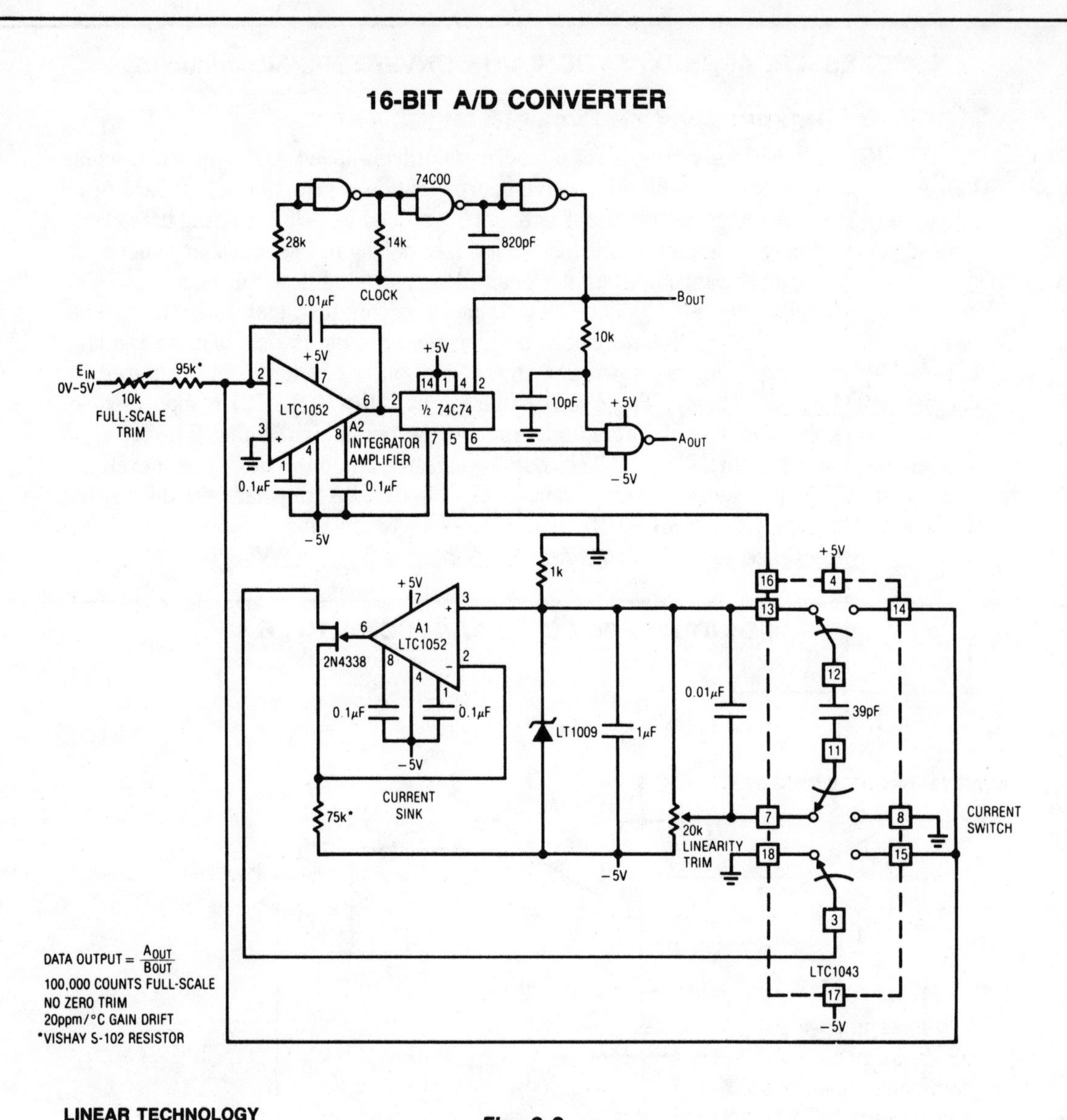

LINEAR TECHNOLOGY

Fig. 3-3

Circuit Notes

The A/D converter, made up of A2, a flip-flop, some gates and a current sink, is based on a current balancing technique. Once again, the chopper-stabilized LTC1052's 50 nV/°C input drift is required to eliminate offset errors in the A/D.

INEXPENSIVE, FAST 10-BIT SERIAL OUTPUT A/D

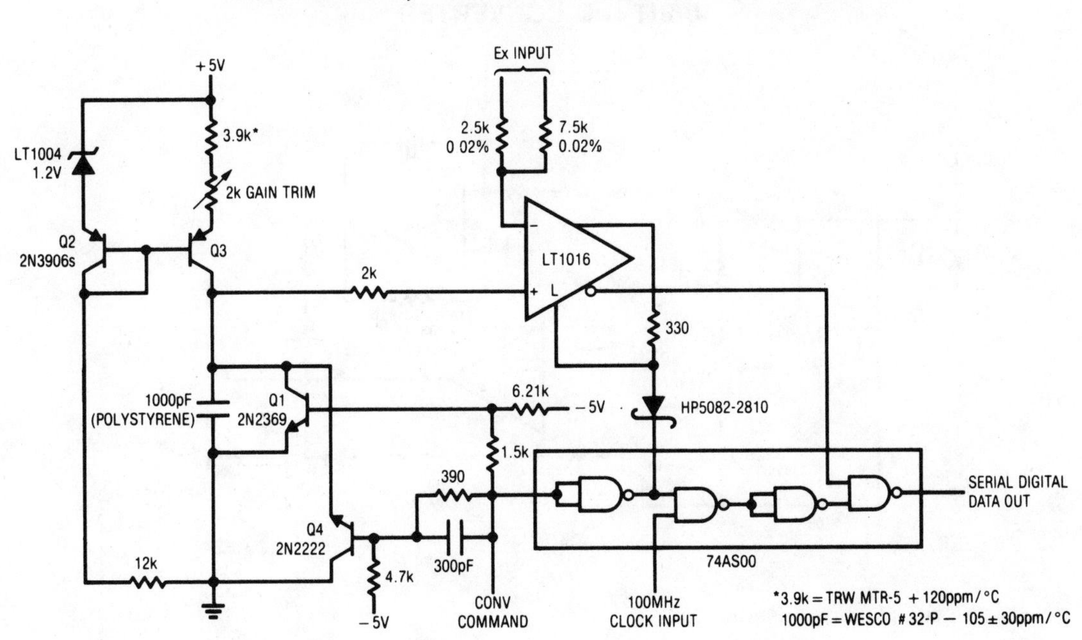

LINEAR TECHNOLOGY CORPORATION

Fig. 3-4

Circuit Notes

Everytime a pulse is applied to the convert command input, Q1 resets the 1000 pF capacitor to 0 V. This resetting action takes 200 ns of the falling edge of the convert command pulse, the capacitor begins to charge linearly. In precisely 10 microseconds, it charges to 2.5 V. The 10 microseconds ramp is applied to the LT1016's positive input. The LT1016 compares the ramp to Ex, the unknown, at its negative input. For a 0 V - 2.5 V range, Ex is applied to the 2.5 k ohm resistor. From a 0 V - 10 V range, the 2.5 k ohm resistor is grounded and Ex is applied to the 7.5 k ohm resistor. Output of the LT1016 is a pulse whose width is directly dependent on the value of Ex. This pulse width is used to gate a 100 MHz clock. The 100 MHz clock pulse bursts that appear at the output are proportional to Ex. For a 0 V - 10 V input, 1024 pulses appear at full-scale, 512 at 5.00 V, etc.

10-BIT A/D CONVERTER

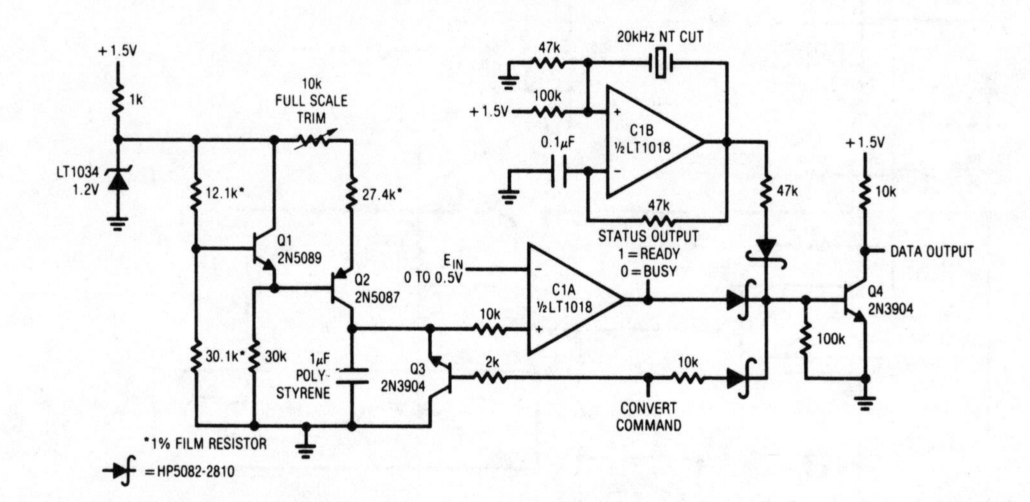

LINEAR TECHNOLOGY CORPORATION

Fig. 3-5

Circuit Notes

The converter has a 60 ms conversion time, consumes 460 μA from its 1.5 V supply and maintains 10 bit accuracy over a 15°C to 35°C temperature range. A pulse applied to the convert command line causes Q3, operating in inverted mode, to discharge through the 10 kΩ diode path, forcing its collector low. Q3's inverted mode switching results in a capacitor discharge within 1 mV of ground. During the time the ramps' value is below the input voltage, CIA's output is low. This allows pulses from C1B, a quartz stabilized oscillator, to modulate Q4. Output data appears at Q4's collector. When the ramp crosses the input voltages value C1A's output goes high, biasing Q4 and output data ceases. The number of pulses at the output is directly proportional to the input voltage. To calibrate apply 0.5 V to the input and trim the 10 kΩ potentiometer for exactly 1000 pulses out each time the convert command line is pulsed.

HIGH SPEED 12-BIT A/D CONVERTER

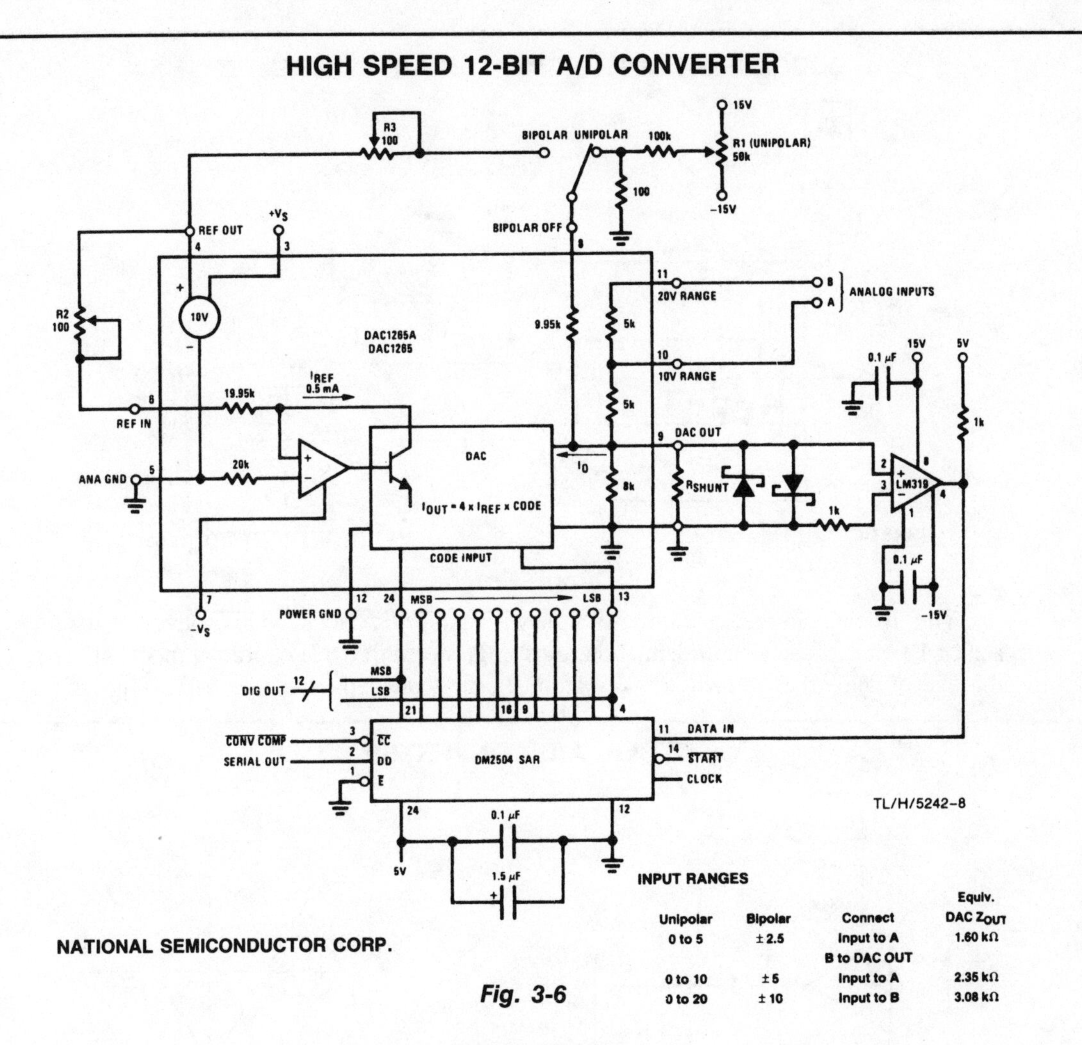

INPUT RANGES

Unipolar	Bipolar	Connect	Equiv. DAC Z_{OUT}
0 to 5	±2.5	Input to A B to DAC OUT	1.60 kΩ
0 to 10	±5	Input to A	2.35 kΩ
0 to 20	±10	Input to B	3.08 kΩ

NATIONAL SEMICONDUCTOR CORP.

Fig. 3-6

Circuit Notes

This system completes a full 12-bit conversion in 10 μs unipolar or bipolar. This converter will be accurate to ± ½ LSB of 12 bits and have a typical gain TC of 10 ppm/°C. In the unipolar mode, the system range is 0 V to 9.9976 V, with each bit having a value of 2.44 mV. For the true conversion accuracy, an A/D converter should be trimmed so that given bit code output results from input levels from ½ LSB below to ½ LSB above the exact voltage which that code represents. Therefore, the converter zero point should be trimmed with an input voltage of 1.22 mV; trim R1 until the LSB just begins to appear in the output code (all other bits "0"). For full-scale, use an input voltage of 9.9963 V (10 V-1 LSB-½ LSB); then trim R2 until the LSB just begins to appear (all other bits "1"). The bipolar signal range is −5.0 V to 4.9976 V. Bipolar offset trimming is done by applying a −4.9988 V input signal and trimming R3 for the LSB transition (all other bits "0"). Full-scale is set by applying 4.9963 V and trimming R2 for the LSB transition (all other bits "1").

SUCCESSIVE APPROXIMATION A/D CONVERTER

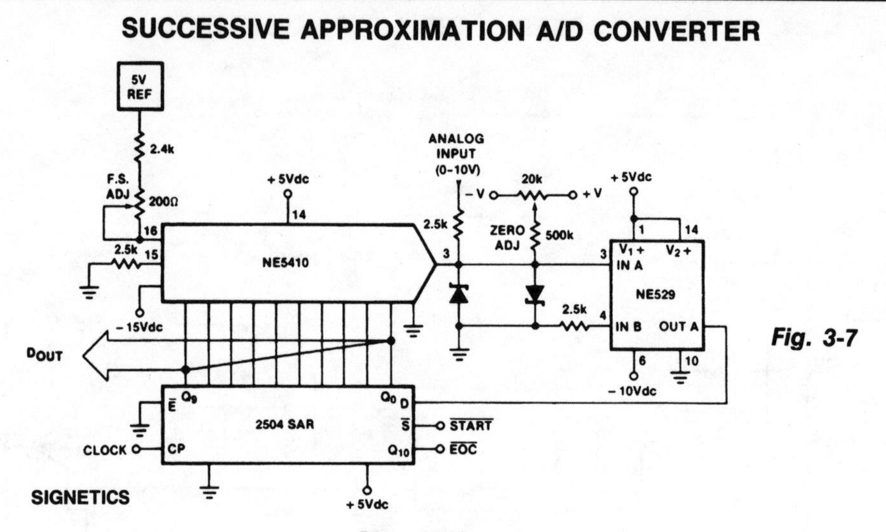

SIGNETICS

Fig. 3-7

Circuit Notes

The 10-bit conversion time is 3.3 μs with a 3 MHz clock. This converter uses a 2504 12-bit successive approximation register in the short cycle operating mode where the end of conversion signal is taken from the first unused bit of the SAR (Q_{10}).

CYCLIC A/D CONVERTER

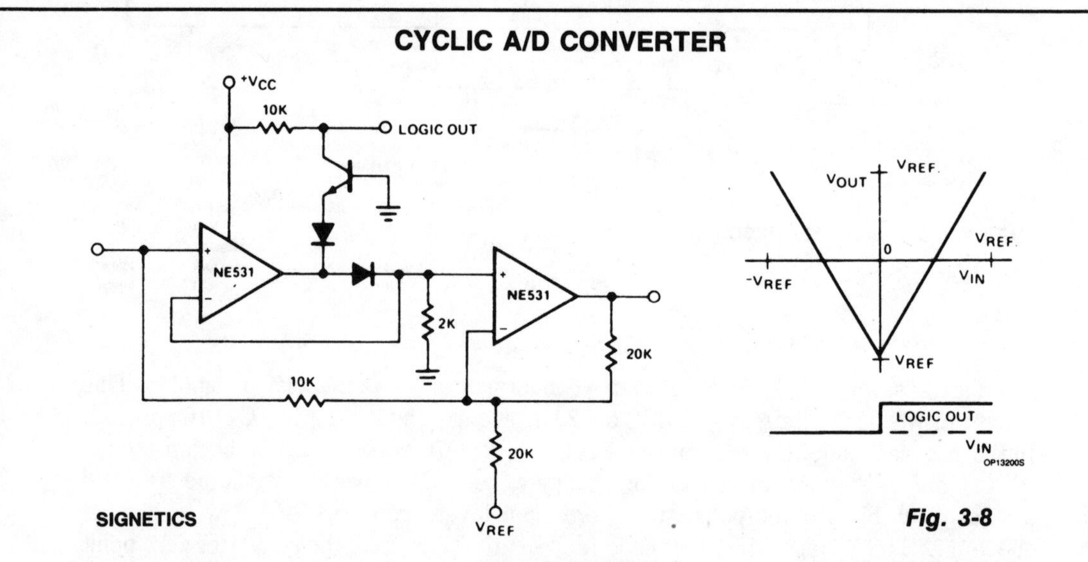

SIGNETICS

Fig. 3-8

Circuit Notes

The cyclic converter consists of a chain of identical stages, each of which senses the polarity of the input. The stage then subtracts V_{REF} from the input and doubles the remainder if the polarity was correct. The signal is full-wave rectified and the remainder of $V_{IN} - V_{REF}$ is doubled. A chain of these stages gives the gray code equivalent of the input voltage in digitized form related to the magnitude of V_{REF}. Possessing high potential accuracy, the circuit using NE531 devices settles in 5 μs.

DIFFERENTIAL INPUT A/D SYSTEM

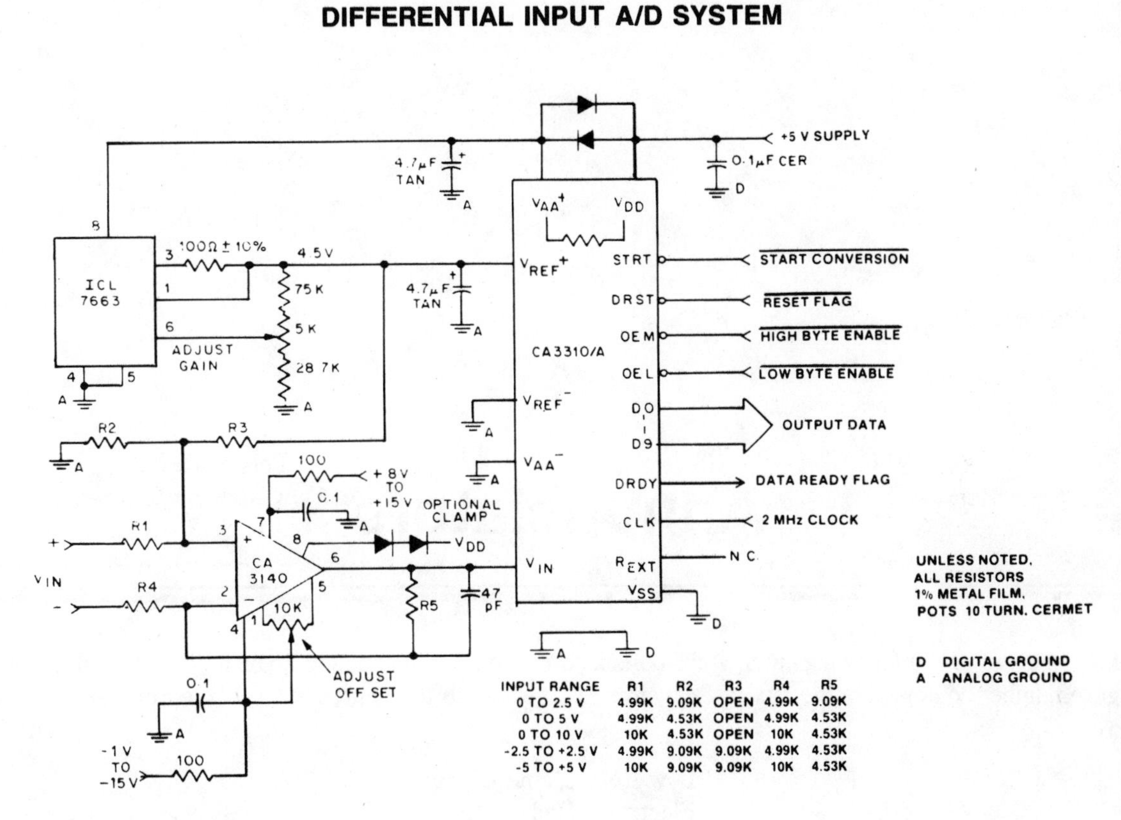

INPUT RANGE	R1	R2	R3	R4	R5
0 TO 2.5 V	4.99K	9.09K	OPEN	4.99K	9.09K
0 TO 5 V	4.99K	4.53K	OPEN	4.99K	4.53K
0 TO 10 V	10K	4.53K	OPEN	10K	4.53K
-2.5 TO +2.5 V	4.99K	9.09K	9.09K	4.99K	4.53K
-5 TO +5 V	10K	9.09K	9.09K	10K	4.53K

GENERAL ELECTRIC/RCA

Fig. 3-9

Circuit Notes

Using a CA3140 BiMOS op amp provides good slewing capability for high bandwidth input signals, and can quickly settle energy that the CA3310 outputs at its V_{IN} terminal. The CA3140 can also drive close to the negative supply rail. If system supply sequencing or an unknown input voltage is likely to cause the op amp to drive above the V_{DD} supply, a diode clamp can be added from pin 8 of the op amp to the V_{DD} supply.

4

Annunciators

The sources of the following circuits are contained in the Sources section beginning on page 694. The figure number contained in the box of each circuit correlates to the source entry in the Sources section.

LOW-COST CHIME CIRCUIT

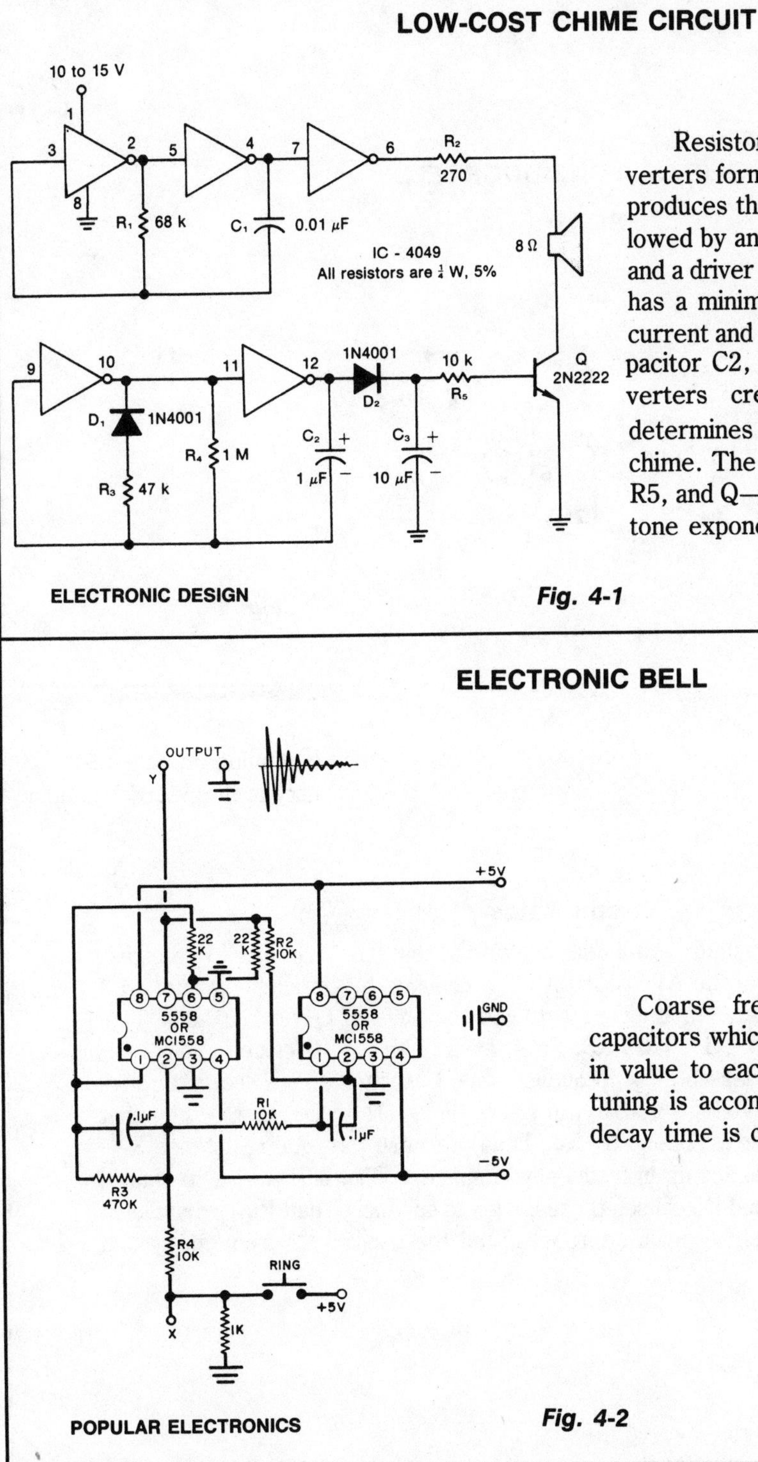

Circuit Notes

Resistor R1, capacitor C1, and two inverters form a square wave generator, which produces the basic tone. The generator is followed by an inverter that acts as both a buffer and a driver for the speaker. Resistor R2, which has a minimum value of 100 ohms, limits the current and controls the volume. Diode D1, capacitor C2, resistors R3 and R4, and two inverters create the pulse generator that determines the turn-on and decay times of the chime. The decay circuit—formed by D2, C3, R5, and Q—reduces the amplitude of the chime tone exponentially as a function of time.

ELECTRONIC DESIGN

Fig. 4-1

ELECTRONIC BELL

Circuit Notes

Coarse frequency is controlled by the capacitors which must be kept nearly identical in value to each other for best results. Fine tuning is accomplished with R1 and R2. The decay time is controlled by R3.

POPULAR ELECTRONICS

Fig. 4-2

SLIDING-TONE DOORBELL

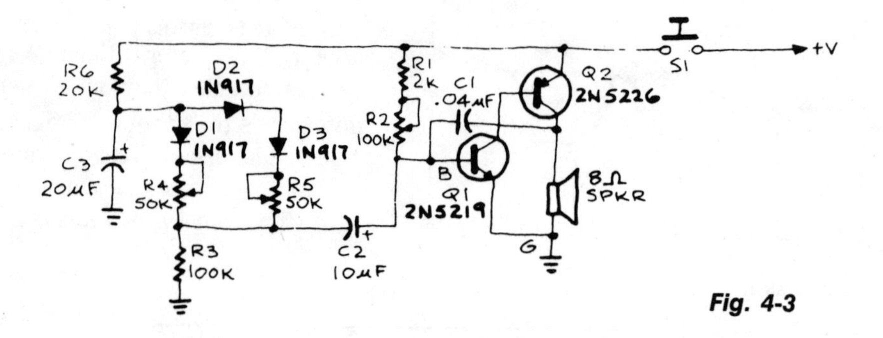

Fig. 4-3

RADIO ELECTRONICS

Circuit Notes

When the doorbell is pushed, you'll hear a low tone that will ''slide up'' to a higher frequency. The frequency of the AF oscillator is determined by coupling capacitor, C1 and the value of the resistance connected between the base of Q1 and ground. That resistance, R_{BG} is equal to (R1 + R2) R3. First, assume that S1 is closed and R2 has been adjusted to produce a pleasant, low-frequency tone. Capacitor C3 will charge through R6 until it reaches such a voltage that it will cause diode D1 to conduct. When that happens, the value of R_{BG} is paralleled by R4. Thus, because the total resistance R_{BG} decrease, the output tone slides up in frequency. Capacitor C3 will continue to charge until the voltage across D2 and D3 causes those diodes to conduct. Then R_{BG} is paralleled also by R5, the total resistance again decreases, and the oscillator's frequency again increases.

5

Audio Mixers, Crossovers and Distribution Circuits

The sources of the following circuits are contained in the Sources section beginning on page 694. The figure number contained in the box of each circuit correlates to the source entry in the Sources section.

Electronic Crossover Circuit
Sound Mixer Amplifier
Microphone Mixer
Low Distortion Input Selector for Audio Use
Audio Distribution Amplifier
Four Channel Four Track Mixer

ELECTRONIC CROSSOVER CIRCUIT

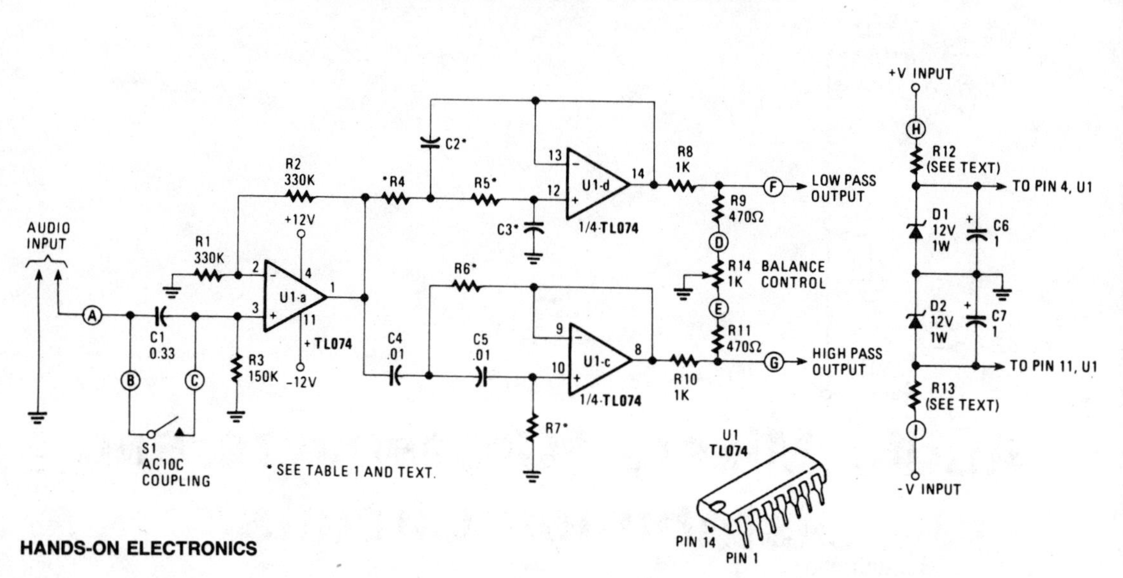

Fig. 5-1

Circuit Notes

An audio source, such as a mixer, preamplifier, equalizer, or recorder, is fed to the Electronic Crossover Circuit's input. That signal is either ac- or dc-coupled, depending on the setting of switch S1, to the non-inverting input of buffer-amplifier U1a, one section of a quad, BIFET, low-noise TL074 op amp made by Texas Instruments. That stage has a gain of 2, and its output is distributed to both a lowpass filter made by R4, R5, C2, C3, and op-amp U1d, and a highpass filter made by R6, R7, C4, C5, and op amp U1c. Those are 12 dB/octave Butterworth-type filters. The Butterworth filter response was chosen because it gives the best compromise between damping and phase shift. Values of capacitors and resistors will vary with the selected crossover at which your unit will operate. The filter's outputs are fed to a balancing network made by R8, R9, R10, R11 and balance potentiometer R14. When the potentiometer is at its mid-position, there is unity gain for the passbands of both the high and low filters. Dc power for the Electronic Crossover Circuit is regulated by R12, R13, D1, and D2, and decoupled by C6 and C7.

SOUND MIXER/AMPLIFIER

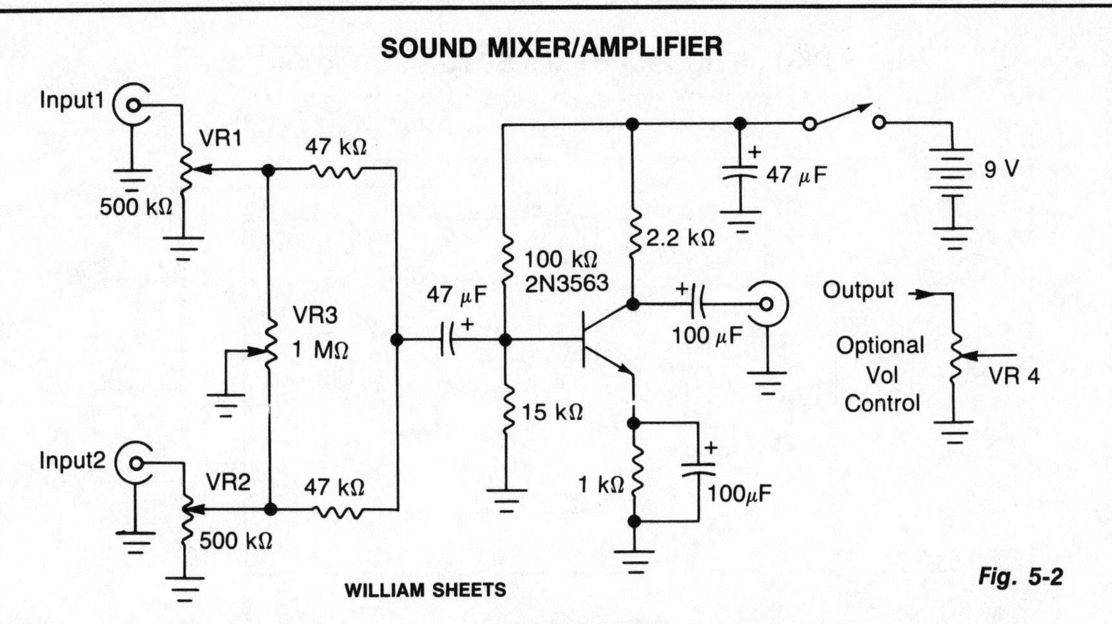

Fig. 5-2

Circuit Notes

Both input signals can be independently controlled by VR1 and VR2. The balance control VR3 is used to fade out one signal while simultaneously fading in the other. The transistor provides gain and the combined output signal level is controlled by VR4 (optional).

MICROPHONE MIXER

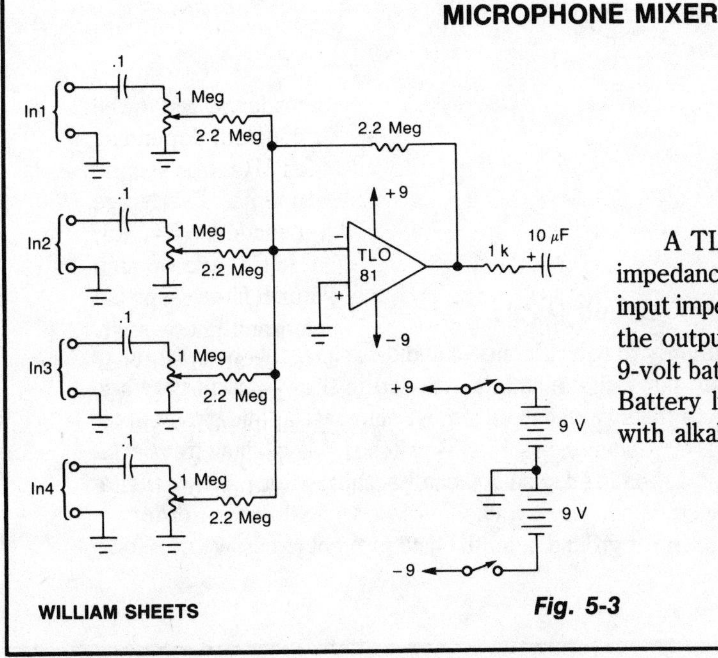

Circuit Notes

A TL081 op amp is used as a high-to-low impedance converter and signal mixer. The input impedance is approximately 1 megohm and the output impedance is about 1 kilohm. Two 9-volt batteries are used as the power source. Battery life should be several hundred hours with alkaline batteries.

WILLIAM SHEETS

Fig. 5-3

LOW DISTORTION INPUT SELECTOR FOR AUDIO USE

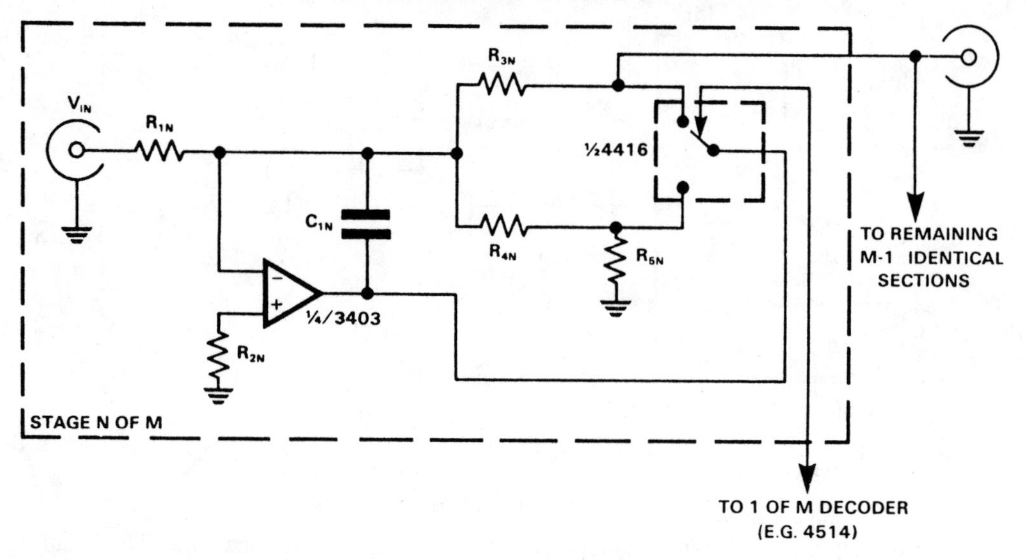

EQUIVALENT CIRCUIT OF EACH STAGE:

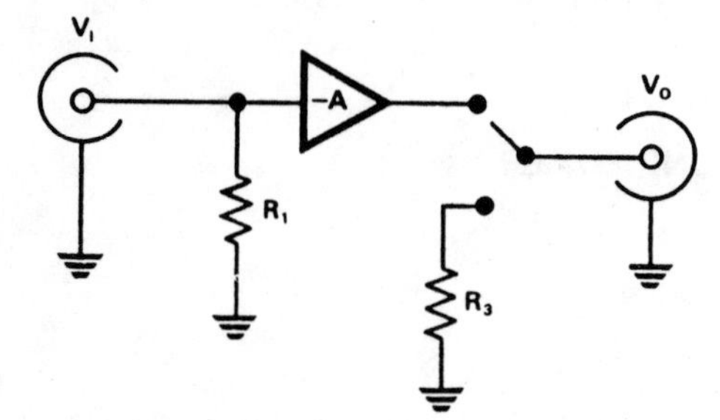

$$R_{3N} = R_{4N} = AR_{1N}$$

$$R_{2N} = (R_{1N} + R_S) // R_{3N} // (R_{4N} + R_{5N})$$

$$\frac{1}{2\pi f_{MAX}} \gg R_{3N} C_{1N} \gg \tau_s$$

$$R_{5N} = \frac{1}{R_L^{-1} + \sum_{i=1}^{M} (R_{3i})^{-1}}$$

ELECTRONIC ENGINEERING

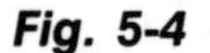

Fig. 5-4

Circuit Notes

CMOS switches are used directly to select inputs in audio circuits, this can introduce unacceptable levels of distortion, but if the switch is included in the feedback network of an op amp, the distortion due to the switch can be almost eliminated. The circuit uses a 4416 CMOS switch, arranged as two independent SPDT switches. If switching transients are unimportant, R5 and C1 can be omitted, and R4 can be shorted out. However, a feedback path must be maintained, even when a channel is switched out, in order to keep the inverting input of the op amp at ground potential, and prevent excessive crosstalk between channels.

AUDIO DISTRIBUTION AMPLIFIER

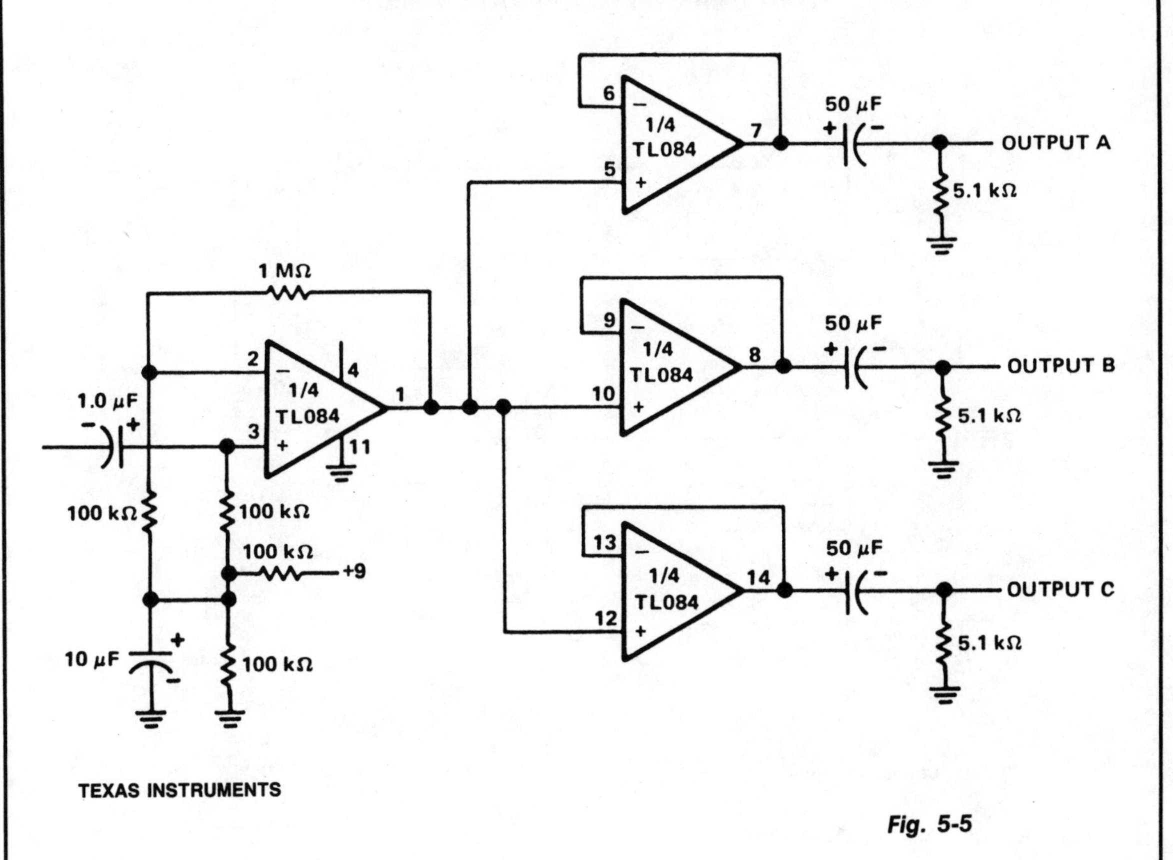

Fig. 5-5

Circuit Notes

The three channel output distribution amplifier uses a single TL084. The first stage is capacitively coupled with a 1.0 μF electrolytic capacitor. The inputs are at ½ V_{CC} rail or 4.5 V. This makes it possible to use a single 9 V supply. A voltage gain of 10 (1 M ohm/100 k ohm) is obtained in the first stage, and the other three stages are connected as unity-gain voltage followers. Each output stage independently drives an amplifier through the 50 μF output capacitor to the 5.1 k ohm load resistor. The response is flat from 10 Hz to 30 kHz.

FOUR CHANNEL FOUR TRACK MIXER

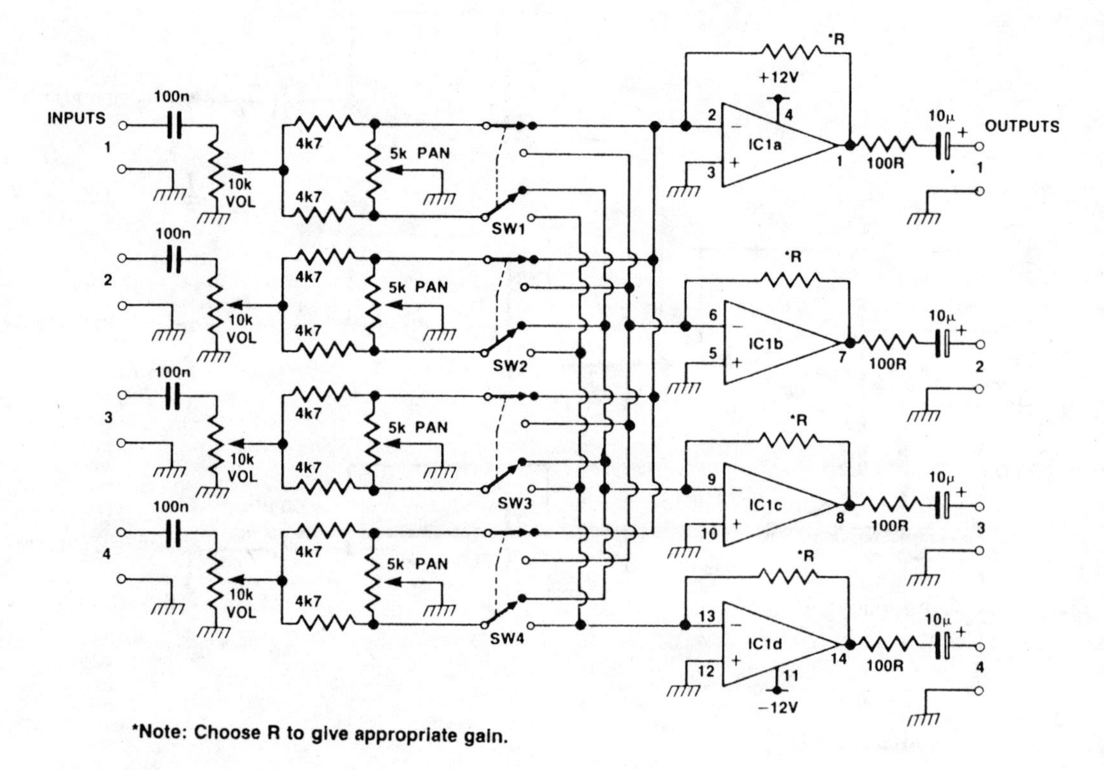

ELECTRONICS TODAY INTERNATIONAL

Fig. 5-6

Circuit Notes

This circuit can be used as a stereo mixer as well as a four track. The quad op-amp IC gives a bit of gain for each track. The pan control allows panning between tracks one and two with the switch in the up position, and with the switch in the down position, it makes possible panning between tracks three and four. Extra channels can be added. A suitable op amp for IC1 is TL074 or similar.

6

Audio Signal Amplifiers

The sources of the following circuits are contained in the Sources section beginning on page 694. The figure number contained in the box of each circuit correlates to the source entry in the Sources section.

Auto Fade
Transistor Headphone Amplifier
Stereo Preamplifier
Audio Compressor
Micropower High-Input-Impedance 20-dB Amplifier
Stereo Preamplifier
Microphone Preamplifier
Volume, Balance, Loudness & Power Amps
Balance and Loudness Amplifier

AUTO FADE

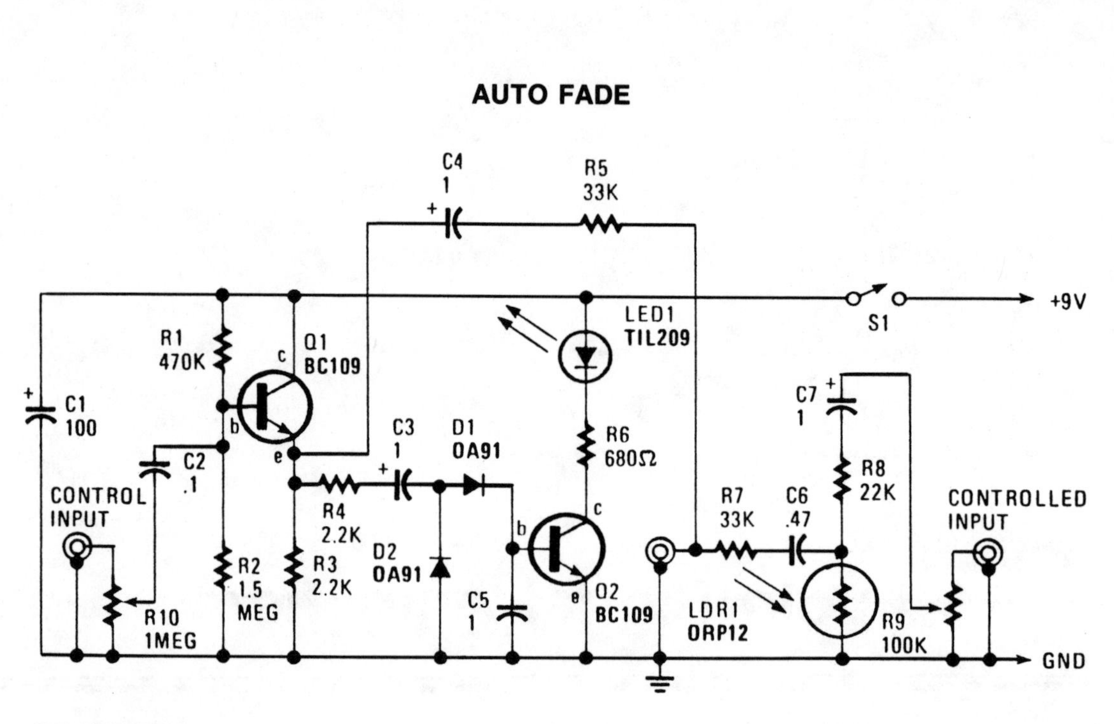

TAB BOOKS INC.

Fig. 6-1

Circuit Notes

The automatic fader drops the level of the background music when the narration comes up. The control input goes through R10, a preset audio level control, to the input of an emitter-follower buffer stage (Q1). The buffer offers a high input impedance and makes sure that the source impedance is low enough to drive the rectifier and smoothing circuit, which consist of D1, D2, and C5. The smoothed output drives a simple LED circuit. R8 and LDR1 form an input attenuator across which the output is fed via C6 and C7 to the output jack. The output at the emitter of Q1 couples to this socket through C4 and R5. R5 and R7 are a passive mixer. With 200 mV or less at the input, there isn't sufficient voltage across C5 to make Q2 turn on. Over 200 mV, Q2 does turn on to a limit, and the LED gets power. That makes the LDR's resistance fall, and signal loss through the attenuator increases. Increase the input to 350 mV rms, and you get a signal reduction of better than 20 dB.

TRANSISTOR HEADPHONE AMPLIFIER

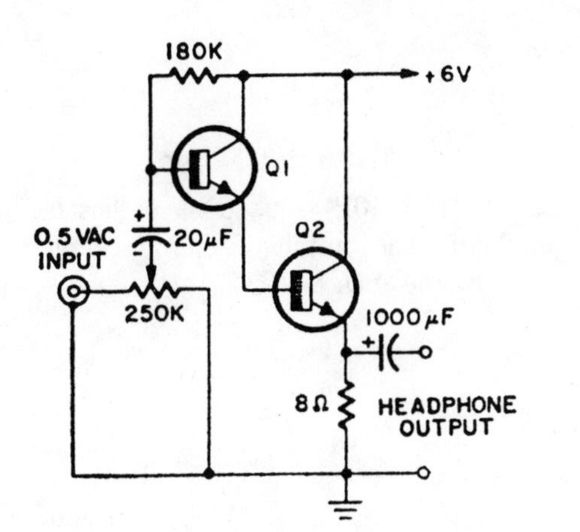

RADIO ELECTRONICS

Fig. 6-2

STEREO PREAMPLIFIER

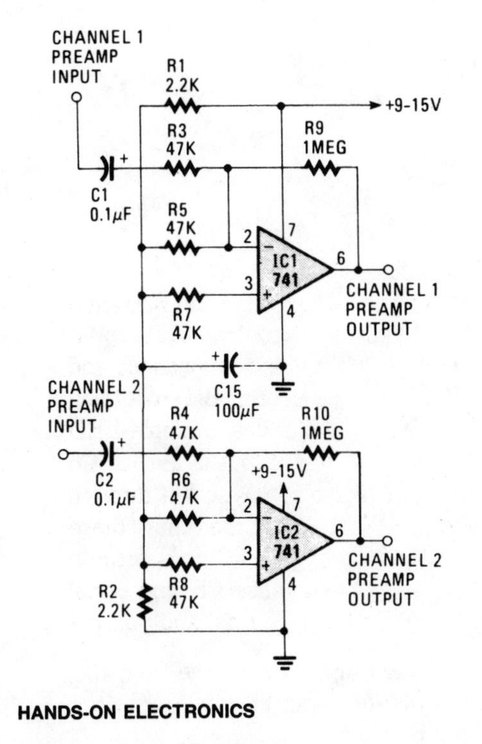

Circuit Notes

The circuit provides better than 20-dB gain in each channel. A better op-amp type will give a better noise figure and bandpass. In this circuit the roll-off is acute at 20,000 Hertz.

HANDS-ON ELECTRONICS

Fig. 6-2

AUDIO COMPRESSOR

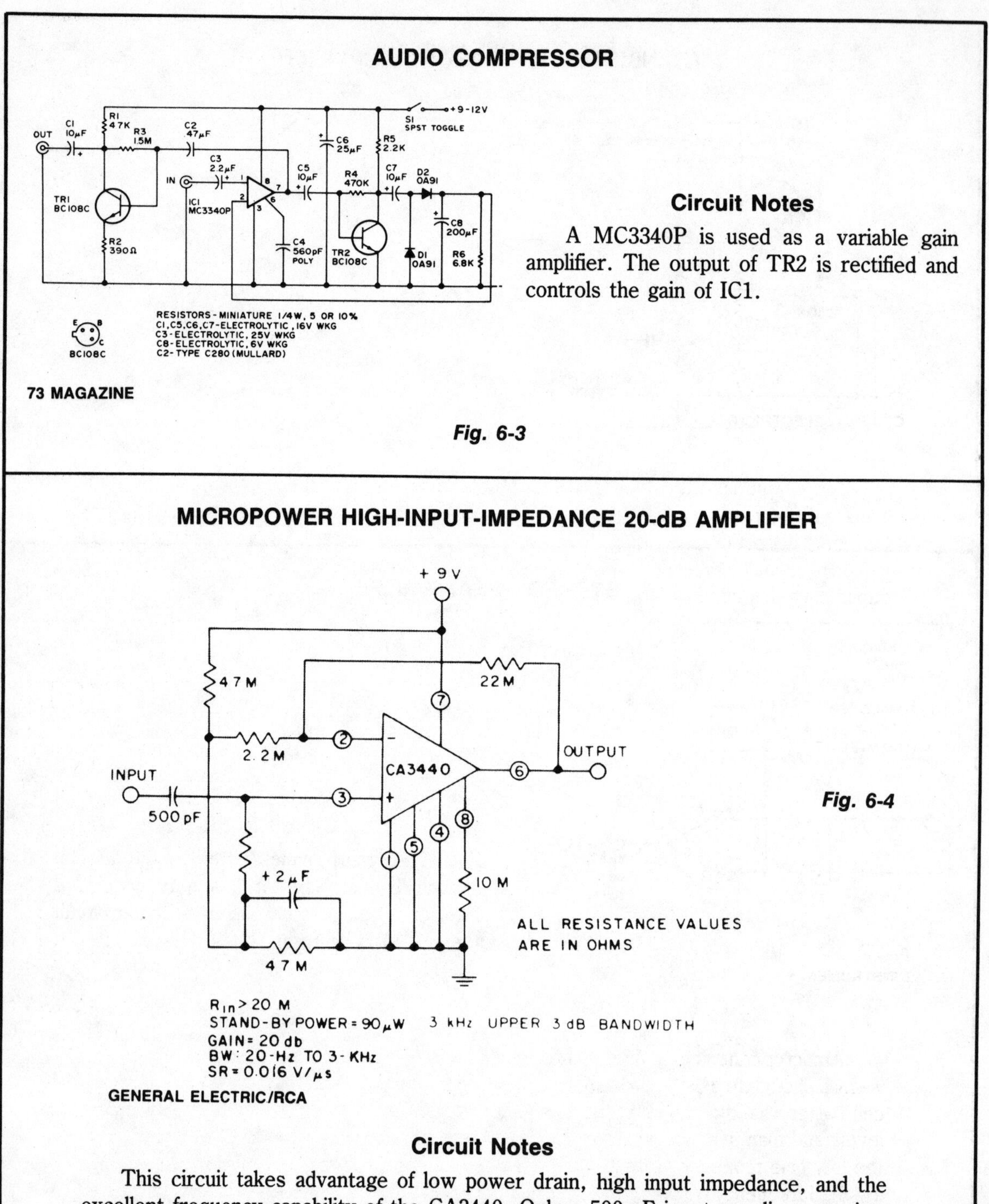

73 MAGAZINE

Circuit Notes

A MC3340P is used as a variable gain amplifier. The output of TR2 is rectified and controls the gain of IC1.

Fig. 6-3

MICROPOWER HIGH-INPUT-IMPEDANCE 20-dB AMPLIFIER

Fig. 6-4

GENERAL ELECTRIC/RCA

Circuit Notes

This circuit takes advantage of low power drain, high input impedance, and the excellent frequency capability of the CA3440. Only a 500-pF input coupling capacitor is needed to achieve a 20 Hz, −3 dB low-frequency response.

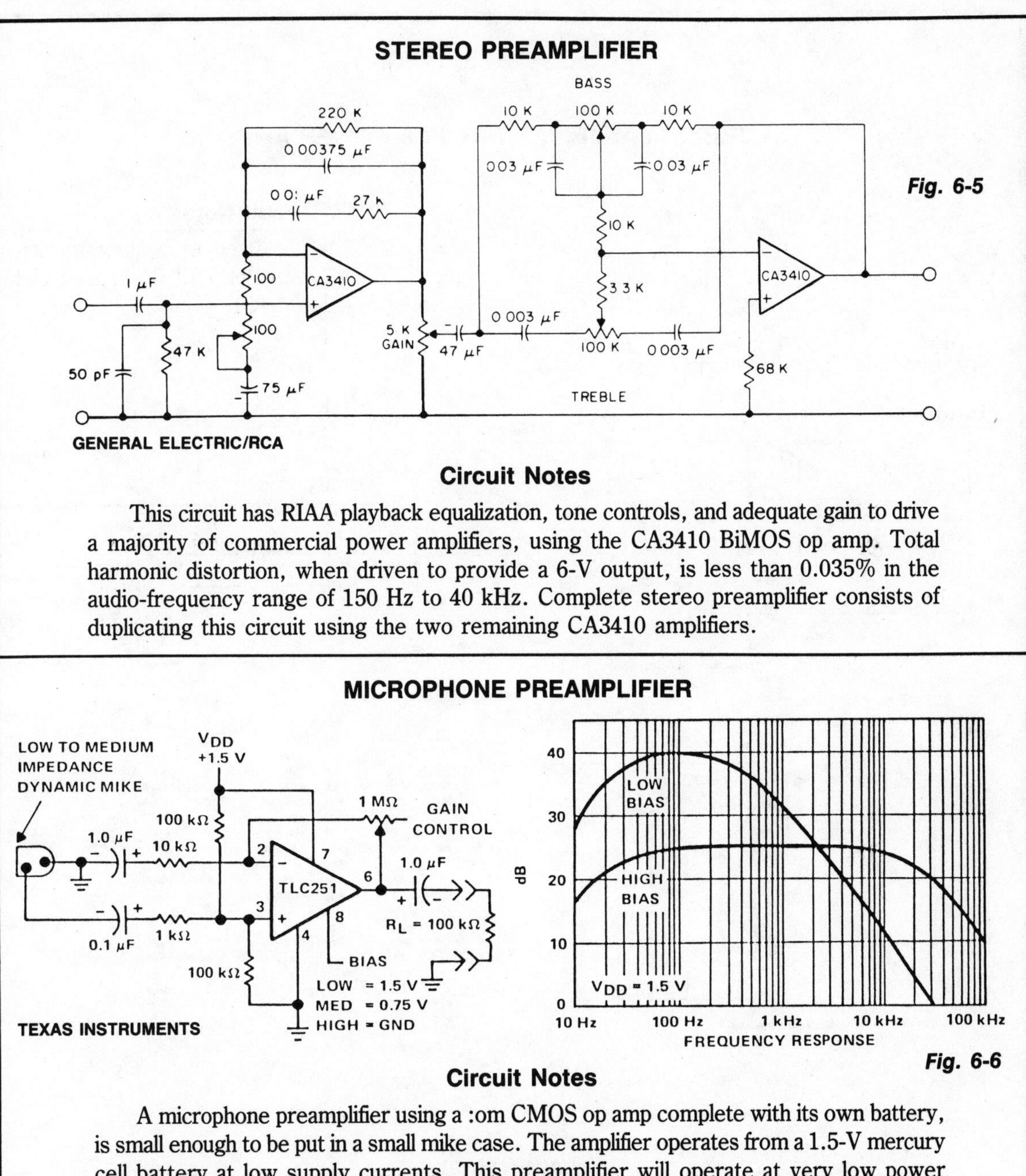

STEREO PREAMPLIFIER

Fig. 6-5

GENERAL ELECTRIC/RCA

Circuit Notes

This circuit has RIAA playback equalization, tone controls, and adequate gain to drive a majority of commercial power amplifiers, using the CA3410 BiMOS op amp. Total harmonic distortion, when driven to provide a 6-V output, is less than 0.035% in the audio-frequency range of 150 Hz to 40 kHz. Complete stereo preamplifier consists of duplicating this circuit using the two remaining CA3410 amplifiers.

MICROPHONE PREAMPLIFIER

TEXAS INSTRUMENTS

Fig. 6-6

Circuit Notes

A microphone preamplifier using a :om CMOS op amp complete with its own battery, is small enough to be put in a small mike case. The amplifier operates from a 1.5-V mercury cell battery at low supply currents. This preamplifier will operate at very low power levels and maintain a reasonable frequency response as well. The TLC251 operated in the low bias mode (operating at 1.5 V) draws a supply current of only 10 μA and has a −3 dB frequency response of 27 Hz to 4.8 kHz. With pin 8 grounded, which is designated as the high bias condition, the upper limit increases to 25 kHz. Supply current is only 30 μA under those conditions.

VOLUME, BALANCE, LOUDNESS & POWER AMPS

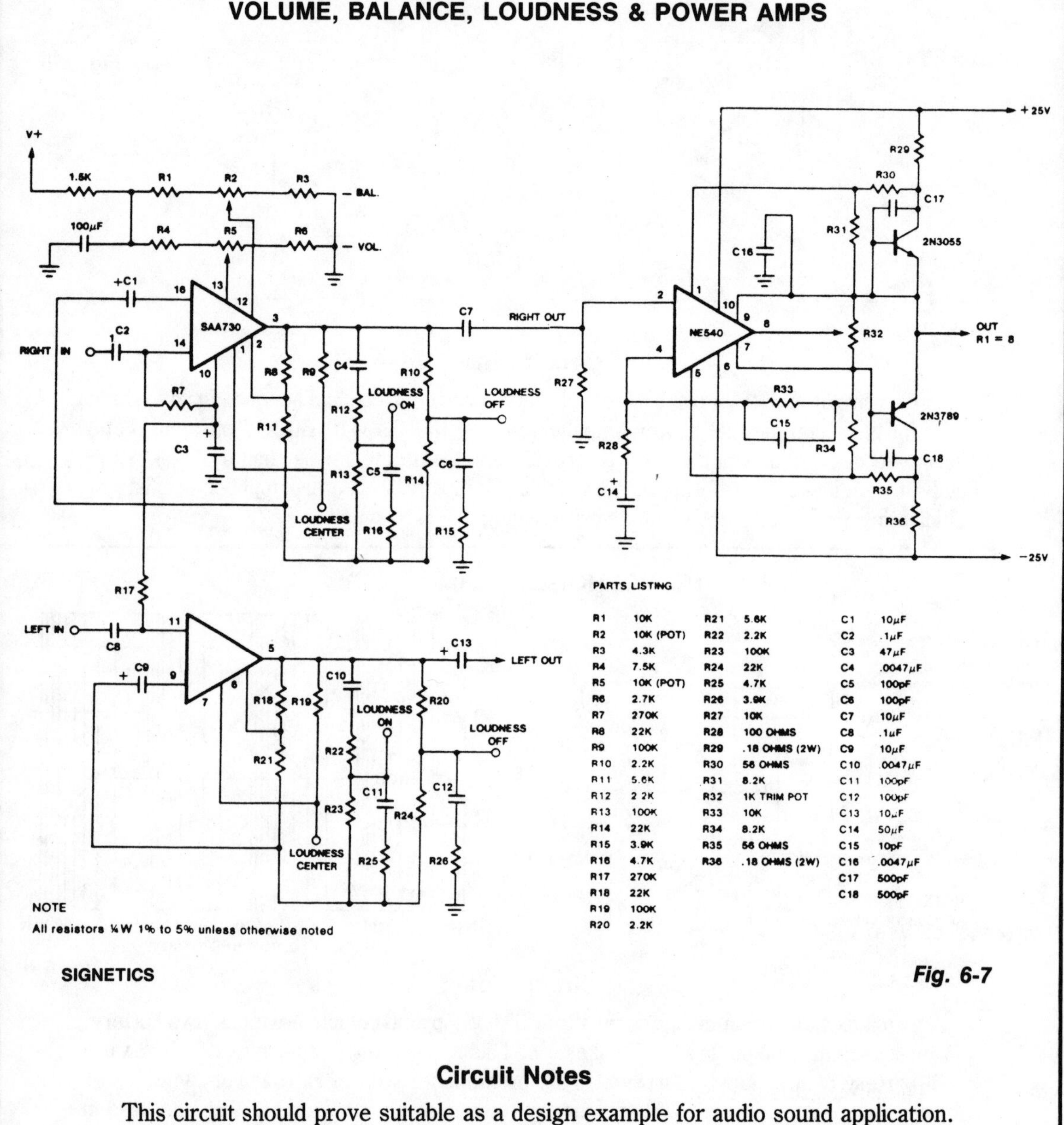

PARTS LISTING

R1	10K	R21	5.6K	C1	10μF
R2	10K (POT)	R22	2.2K	C2	.1μF
R3	4.3K	R23	100K	C3	47μF
R4	7.5K	R24	22K	C4	.0047μF
R5	10K (POT)	R25	4.7K	C5	100pF
R6	2.7K	R26	3.9K	C6	100pF
R7	270K	R27	10K	C7	10μF
R8	22K	R28	100 OHMS	C8	.1uF
R9	100K	R29	.18 OHMS (2W)	C9	10μF
R10	2.2K	R30	56 OHMS	C10	.0047μF
R11	5.6K	R31	8.2K	C11	100pF
R12	2.2K	R32	1K TRIM POT	C12	100pF
R13	100K	R33	10K	C13	10μF
R14	22K	R34	8.2K	C14	50μF
R15	3.9K	R35	56 OHMS	C15	10pF
R16	4.7K	R36	.18 OHMS (2W)	C16	.0047μF
R17	270K			C17	500pF
R18	22K			C18	500pF
R19	100K				
R20	2.2K				

Fig. 6-7

Circuit Notes

This circuit should prove suitable as a design example for audio sound application.

BALANCE AND LOUDNESS AMPLIFIER

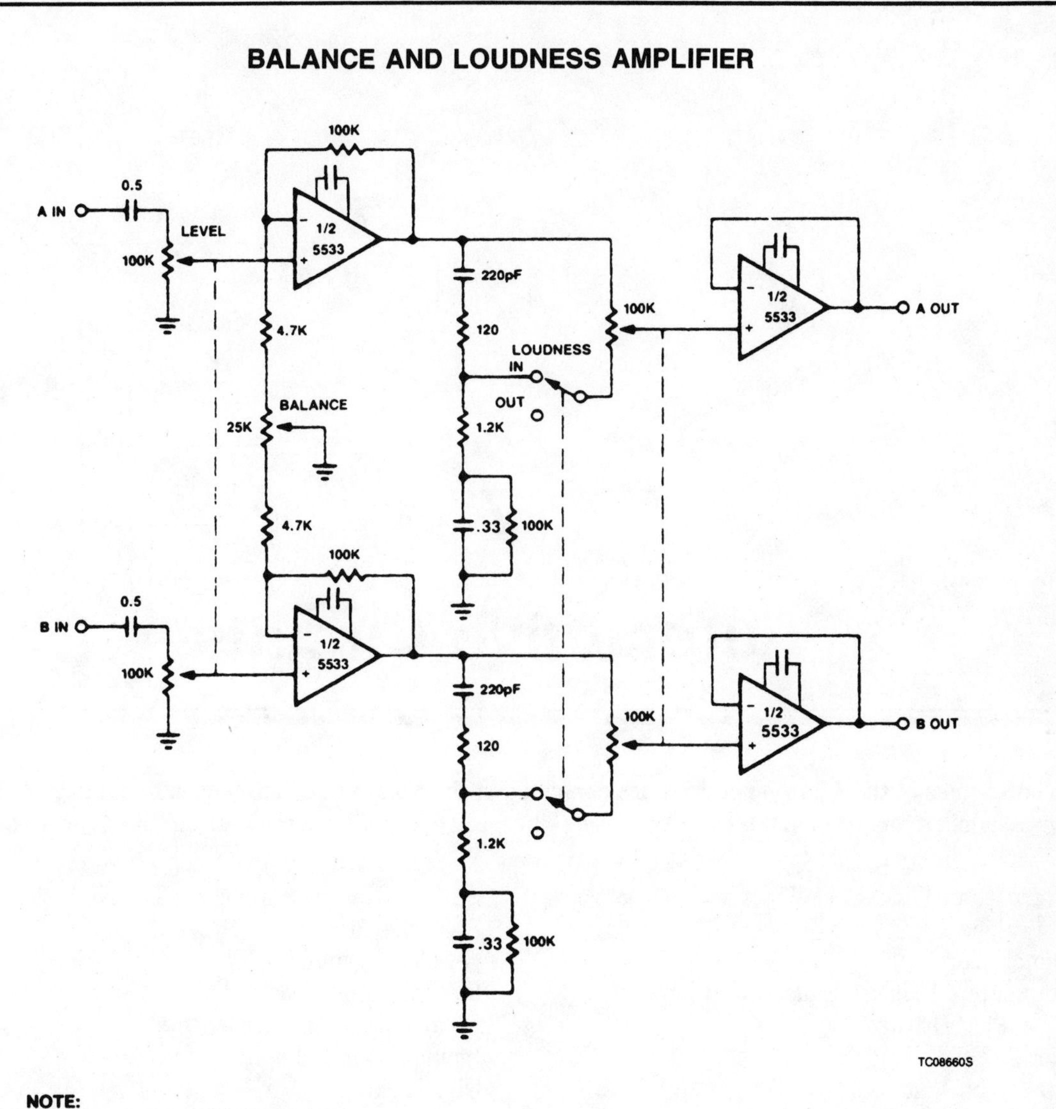

NOTE:
All resistor values are in ohms.

SIGNETICS

Fig. 6-7

Circuit Notes

The circuit shows a combination of balance and loudness controls. Due to the nonlinearity of the human hearing system, the low frequencies must be boosted at low listening levels. Balance, level, and loudness controls provide all the listening controls to produce the desired music response.

7

Automotive Circuits

The sources of the following circuits are contained in the Sources section beginning on page 694. The figure number contained in the box of each circuit correlates to the source entry in the Sources section.

INTERMITTENT WINDSHIELD WIPER WITH DYNAMIC BRAKING

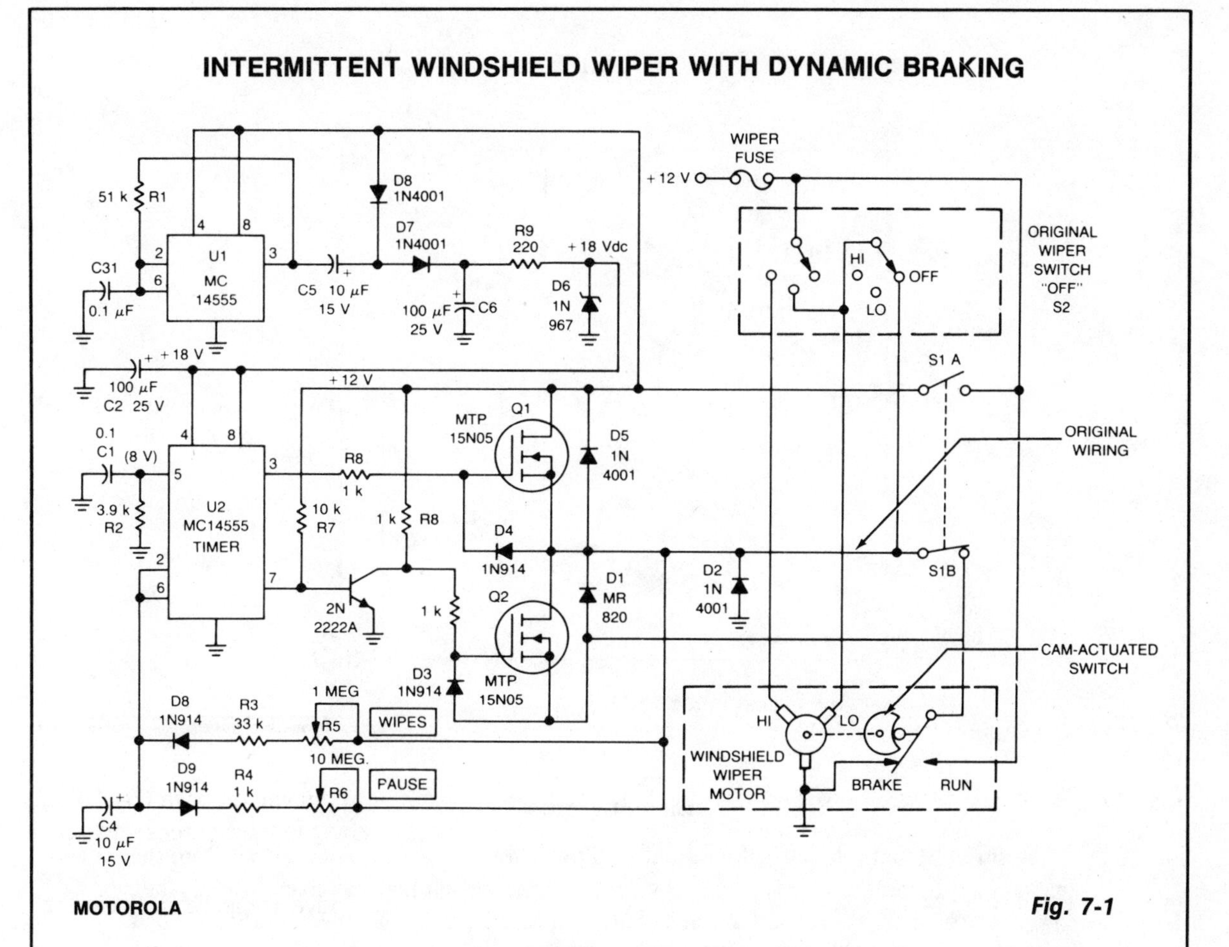

MOTOROLA

Fig. 7-1

Circuit Notes

The circuit provides a delayed windshield wiping, and dynamic braking of wiper blades when they reach the rest position. This prevents the blades from overshooting, which might cause them to stop at a point where they interfere with the drivers' vision.

With the original wiper switch off, switch S1A turns on the delay circuit and S1B disconnects the original automotive wiring. When S1 is turned off, the original wiring controls the system and the delay circuit is bypassed.

Turning S1 on applies the +12-V battery to U1 which is a voltage doubler that produces +18 V. This higher voltage supply is necessary to ensure reliable turn on of Q1 by multivibrator U2. This arrangement provides about +18 V to the gate of Q1, whose source is +12 V minus the V_{DS} drop of Q1.

Q1 remains on for a time determined by the WIPES potentiometer. The interval between wipes is controlled by the PAUSE control. When C1 drops below +4 V, U2 fires, turning Q1 on and restarting the cycle.

IMMOBILIZER

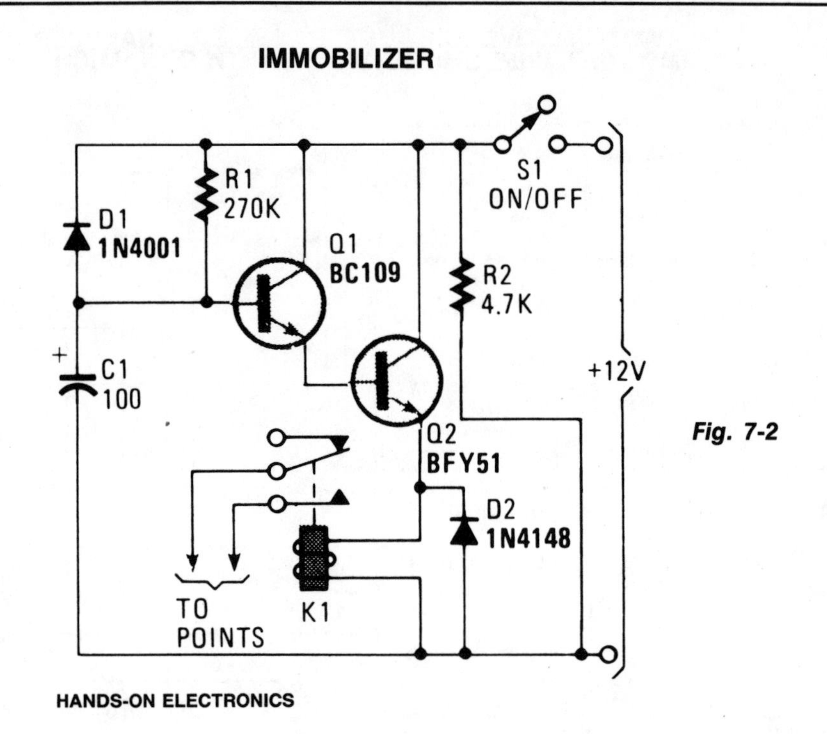

Fig. 7-2

Circuit Notes

A flip of S1 puts the circuit into action. Power for the circuit is picked up from the ignition switch, and the circuit receives no power until the ignition switch is closed. When power is turned on, capacitor C1 is not charged and the emitter-follower Darlington pair (formed by Q1 and Q2) are cutoff, thus no power is applied to the relay (K1), which serves as Q1's emitter load. The relay's normally-open contacts are connected across the vehicle's points. (At this time, the relay contacts are open and have no effect on the ignition system). C1 charges by way of R1, causing the voltage at the base of Q1 to rise steadily. That creates a similar rise in the voltage at the emitter of Q2. A Darlington pair is used to provide a high input-impedance, buffer stage so that the voltage across C2 is free to rise almost to the full supply potential. Loading effects do not limit the charge potential to just a few volts. Eventually, the voltage applied to the relay becomes sufficient to activate it. The contacts close and short out the points. The ignition system now doesn't act properly and the vehicle is disabled. If the ignition is switched off, power is removed from the circuit and diode D1, which was previously reverse-biased, is now forward biased by the charge on C1. D1 allows C1 to rapidly discharge through R2 (and any other dc paths across the supply lines). The circuit is ready to operate when the ignition is again turned on. The engine will operate, but not for very long. The values of R1 and C1 provides a delay of about 25 to 30 seconds. Increase R1's value to provide a longer delay.

AUTOMOTIVE EXHAUST EMISSIONS ANALYZER

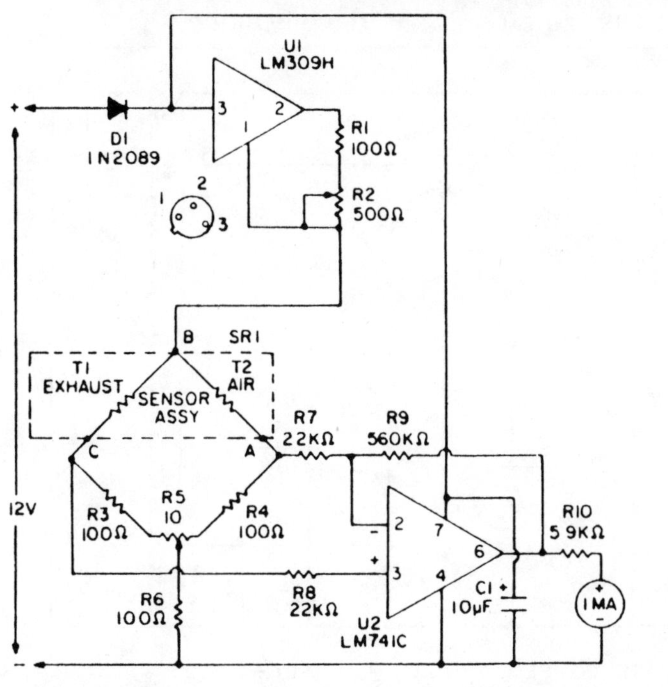

Fig. 7-3

C1—10-uF, 25 volt electrolytic capacitor
D1—Silicon diode, general purpose, 1N2069 or equivalent
M1—0.1 ma ammeter
R1, R3, R4, R6—100-ohm 10% ¼ watt resistor (All resistors are 10%, ¼ watt)
R2—500-ohm potentiometer, PC mount
R5—10-ohm potentiometer, front panel mount
R7, R8—22,000-ohm resistor
R9—570,000-ohm resistor
R10—5,900-ohm resistor
SR1—thermister sensor assembly, part number 100-1648 from Heathkit model CI-1080, Heath Company Benton Harbor, MI, 49022
U1—LM309H 5-volt regulator integrated circuit
U2—LM741TC op amp intergrated circuit
Misc.—PC Board, screws, cabinet, solder, hardware, 10-foot plastic tube, etc.

TAB BOOKS, INC.

Circuit Notes

A bridge circuit contains two 100-ohm resistors (R3 and R4), and two thermistors (T1 and T2). At room temperature the resistance of T1 and T2 is about 2000 ohms. When they are each heated to 150°C by a 10 mA current, the resistance value decreases to 100 ohms. Thus, the four elements comprise a bridge circuit. A characteristic of CO is that it conducts heat away from a thermistor at a different rate than air. One thermistor, T1, is exposed to the automobile exhaust while the other, T2, is isolated in a pure air environment. The difference in thermal conduction unbalances the bridge. A voltage difference is caused between points A and C. A differential amplifier, U1, amplifies this difference and drives the meter with sufficient current to read out the percentage of CO and the air-fuel ratio. A front panel balance control, R5, balances the bridge and calibrates the instrument. Calibration is performed when both thermistors are exposed to the outside air.

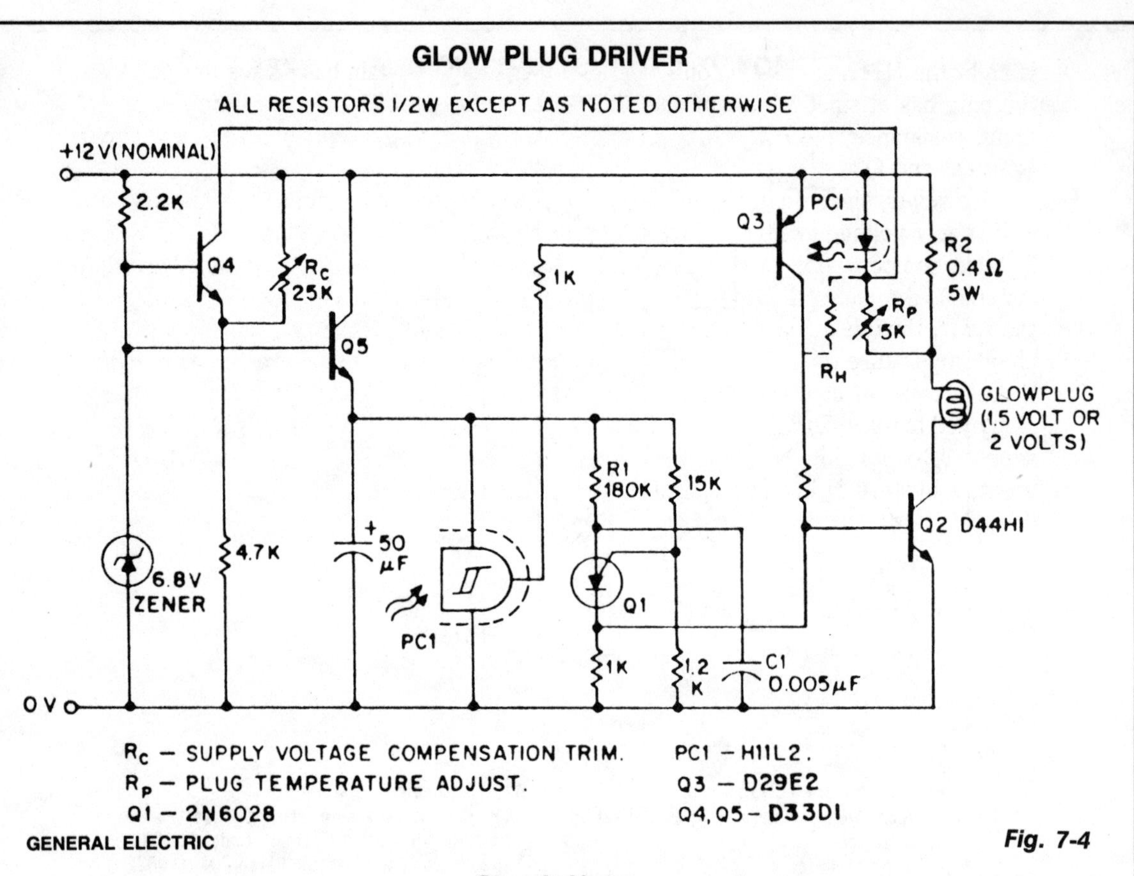

GENERAL ELECTRIC

Fig. 7-4

Circuit Notes

Model airplanes, boats, and cars use glow plug ignitions for their miniature (0.8cc to 15cc) internal combustion engines. Such engines dispense with the heavy on-board batteries, H.T. coil, and "condenser" required for conventional spark ignition, while simultaneously developing much higher RPM (hence power) than the compression ignition (diesel) motors. The heart of a glow plug is a platinum alloy coil heated to incandescence for engine starting by an external battery, either 1.5 volts or 2 volts. Supplementing this battery, a second 12-volt power supply is frequently required for the engine starter, together with a third 6 volt type for the electrical fuel pump.

Rather than being burdened by all these multiple energy sources, the model builder would prefer to carry (and buy) a single 12-volt battery, deriving the lower voltages from this by use of suitable electronic step-down transformers (choppers). The glow driver illustrated does this and offers the additional benefit of (through negative feedback) maintaining constant plug temperature independent of engine flooding, or battery voltage while the starter is cranking.

In this circuit, the PUT relaxation oscillator Q1 turns on the output chopper transistor Q2 at a fixed repetition rate determined by R1 and C1. Current then flows through the glow plug and the parallel combination of the current sense resistor R2 and the LED associated with the H11L Schmitt trigger. With the plug cold (low resistance), current is high, the H11L is biased "on", and Q3 conducts to sustain base drive to Q2. Once the plug has attained optimum operating temperature, which can be monitored by its ohmic resistance, the H11L is programmed (via R_p) to switch off, removing base drive from Q3 and Q2.

However, since the H11L senses glow plug current, not resistance, this is only valid if supply voltage is constant, which is not always the case. Transistor Q4 provides suitable compensation in this case; if battery voltage falls (during cold cranking, for instance), the collector current of Q4 rises, causing additional current to flow through the LED, thus delaying the switch-off point for a given plug current. The circuit holds plug temperature relatively constant, with the plug either completely dry or thoroughly "wet", over an input voltage range of 8 to 16 volts. A similar configuration can be employed to maintain constant temperature for a full size truck diesel glow plug (28-volts supply, 12-volts glow plug); in this case, since plug temperature excursions are not so great, a hysteresis expansion resistor R_H may be required.

Fig. 7-4 Continued

GARAGE STOP LIGHT

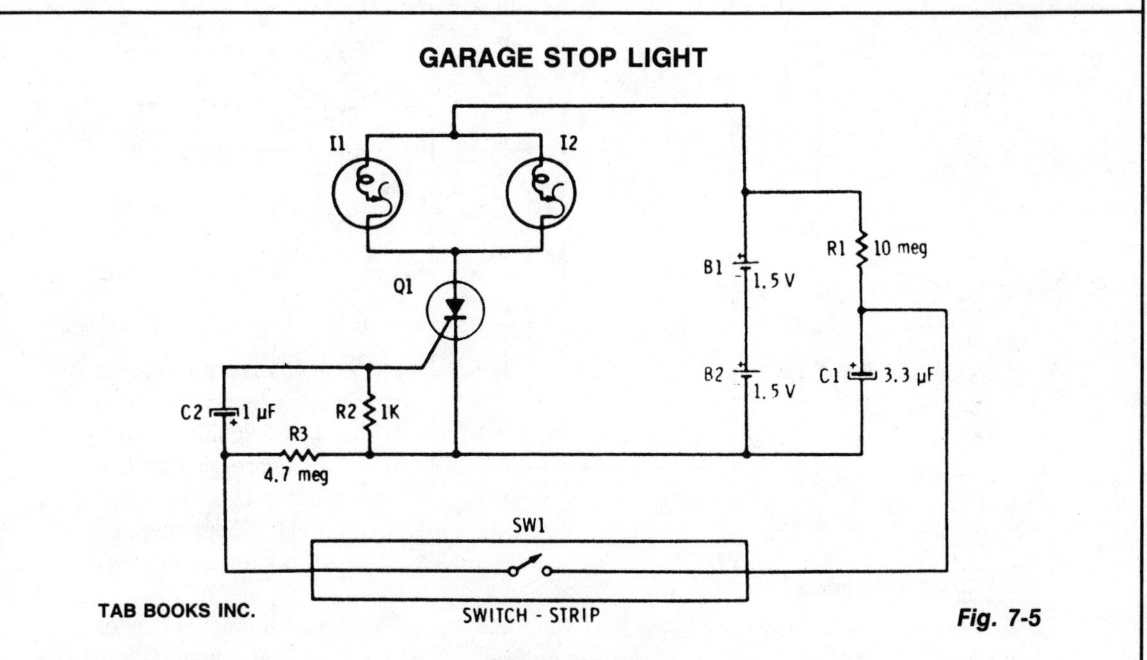

TAB BOOKS INC.

Fig. 7-5

Circuit Notes

Capacitor C1 is permanently connected across the 3-volt supply through 10 megohm resistor R1. The capacitor charges (relatively slowly) to 3 volts. The instant switch SW1 is closed, it connects the charged capacitor (C1) in series with C2 and R2. Capacitor C2 starts to charge, placing a positive-going voltage on the gate of the SCR and causing it to turn on. The two parallel-connected "self-flashing" bulbs I1 and I2 turn on. They flash and turn off the SCR and the circuit is off until car is driven off the switch and C1 can recharge.

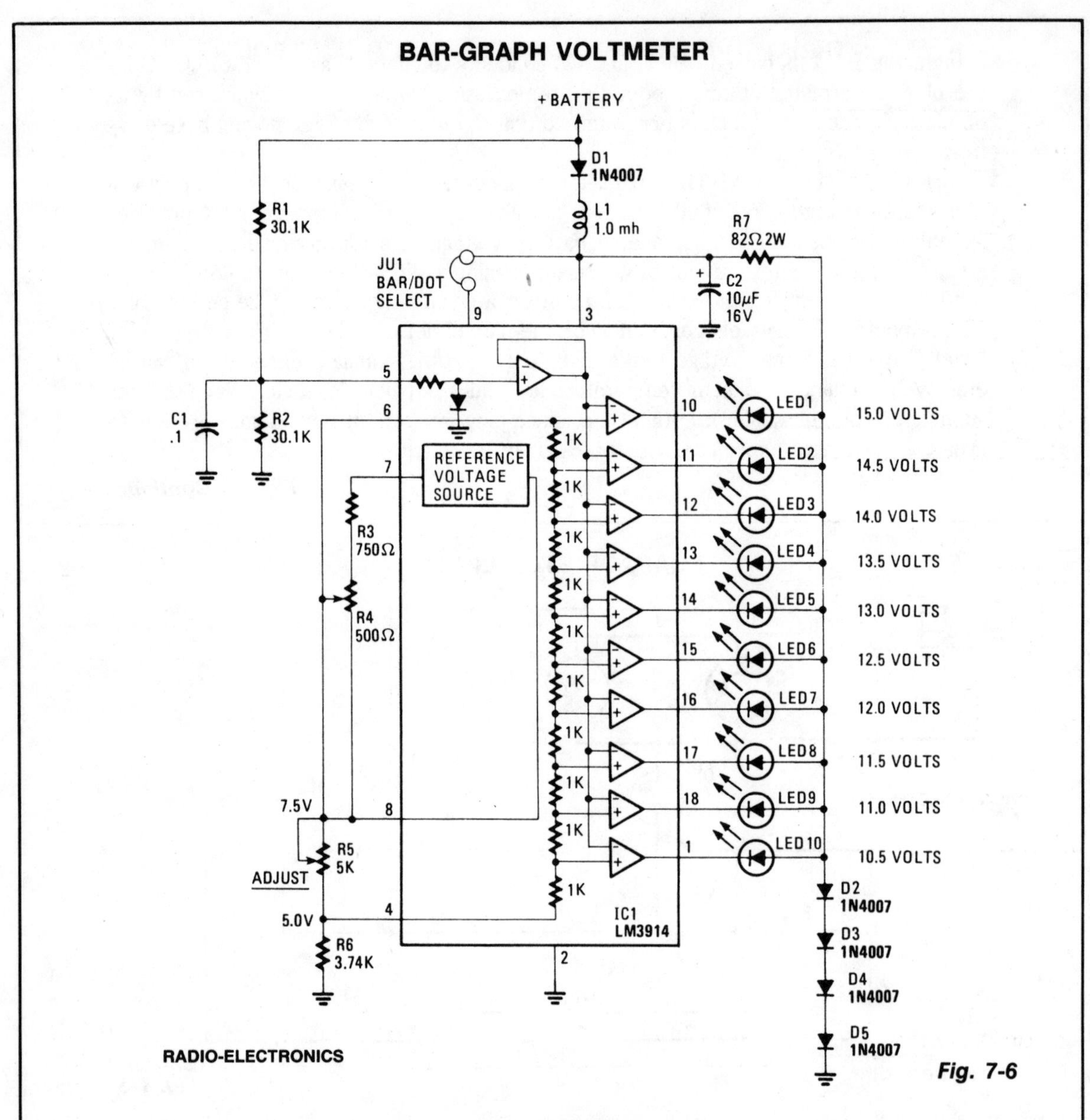

Fig. 7-6

Circuit Notes

This display uses ten LED's to display a voltage range from 10.5 to 15 volts. Each LED represents a 0.5-volt step in voltage. The heart of the circuit is the LM-3914 dot/bar display driver. Trimmer potentiometer R5 is adjusted so that 7.5 volts is applied to the top side of the divider. Resistor R7 and diodes D2 through D5 clamp the voltage applied to the LED's to about 3 volts. A lowpass filter made up of L1 and C2 guards against voltage spikes. Diode D1 is used to protect against reverse voltage in case the voltmeter is hooked up backward.

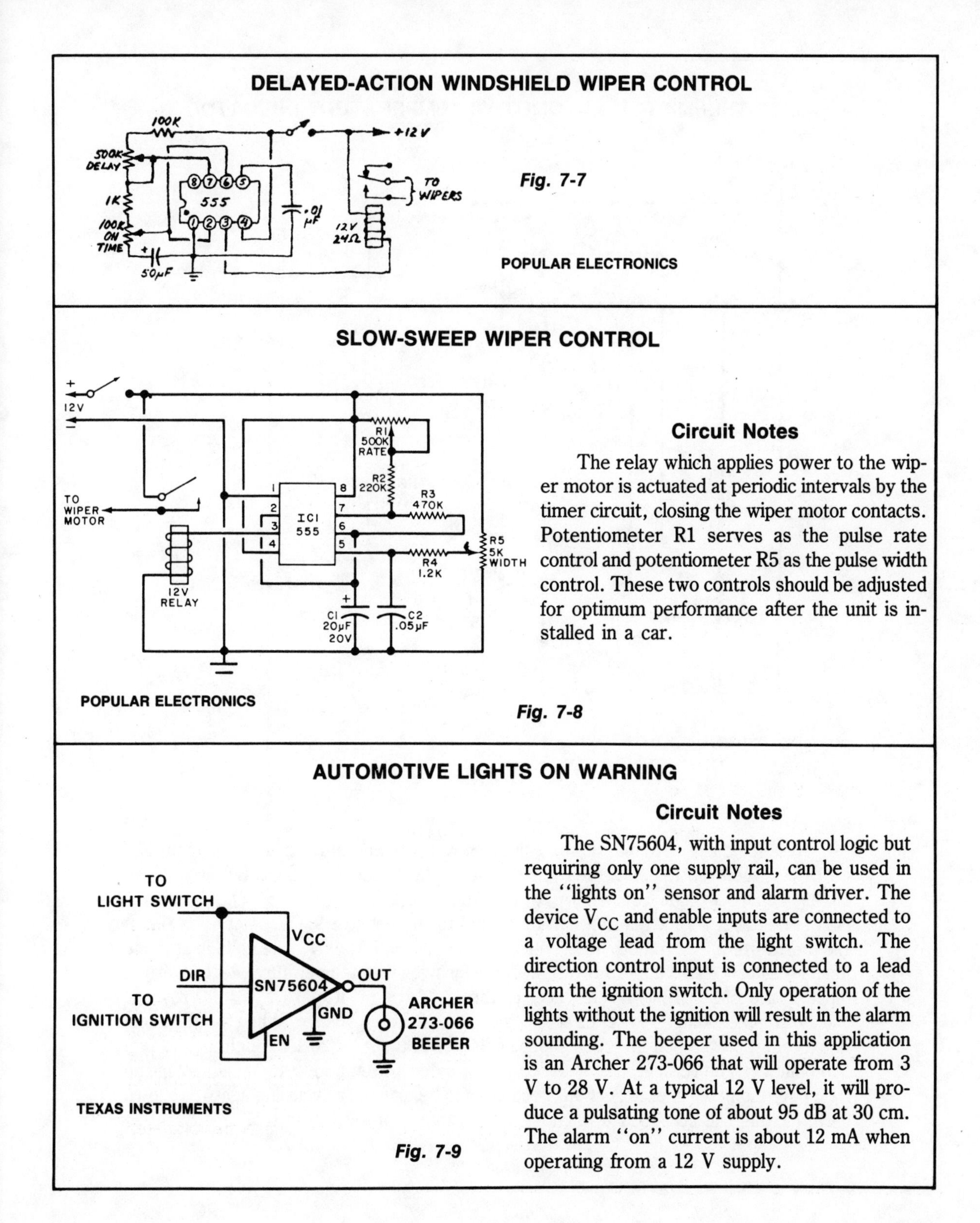

DELAYED-ACTION WINDSHIELD WIPER CONTROL

Fig. 7-7

POPULAR ELECTRONICS

SLOW-SWEEP WIPER CONTROL

POPULAR ELECTRONICS

Circuit Notes

The relay which applies power to the wiper motor is actuated at periodic intervals by the timer circuit, closing the wiper motor contacts. Potentiometer R1 serves as the pulse rate control and potentiometer R5 as the pulse width control. These two controls should be adjusted for optimum performance after the unit is installed in a car.

Fig. 7-8

AUTOMOTIVE LIGHTS ON WARNING

TEXAS INSTRUMENTS

Circuit Notes

The SN75604, with input control logic but requiring only one supply rail, can be used in the "lights on" sensor and alarm driver. The device V_{CC} and enable inputs are connected to a voltage lead from the light switch. The direction control input is connected to a lead from the ignition switch. Only operation of the lights without the ignition will result in the alarm sounding. The beeper used in this application is an Archer 273-066 that will operate from 3 V to 28 V. At a typical 12 V level, it will produce a pulsating tone of about 95 dB at 30 cm. The alarm "on" current is about 12 mA when operating from a 12 V supply.

Fig. 7-9

PTC THERMISTOR AUTOMOTIVE TEMPERATURE INDICATOR

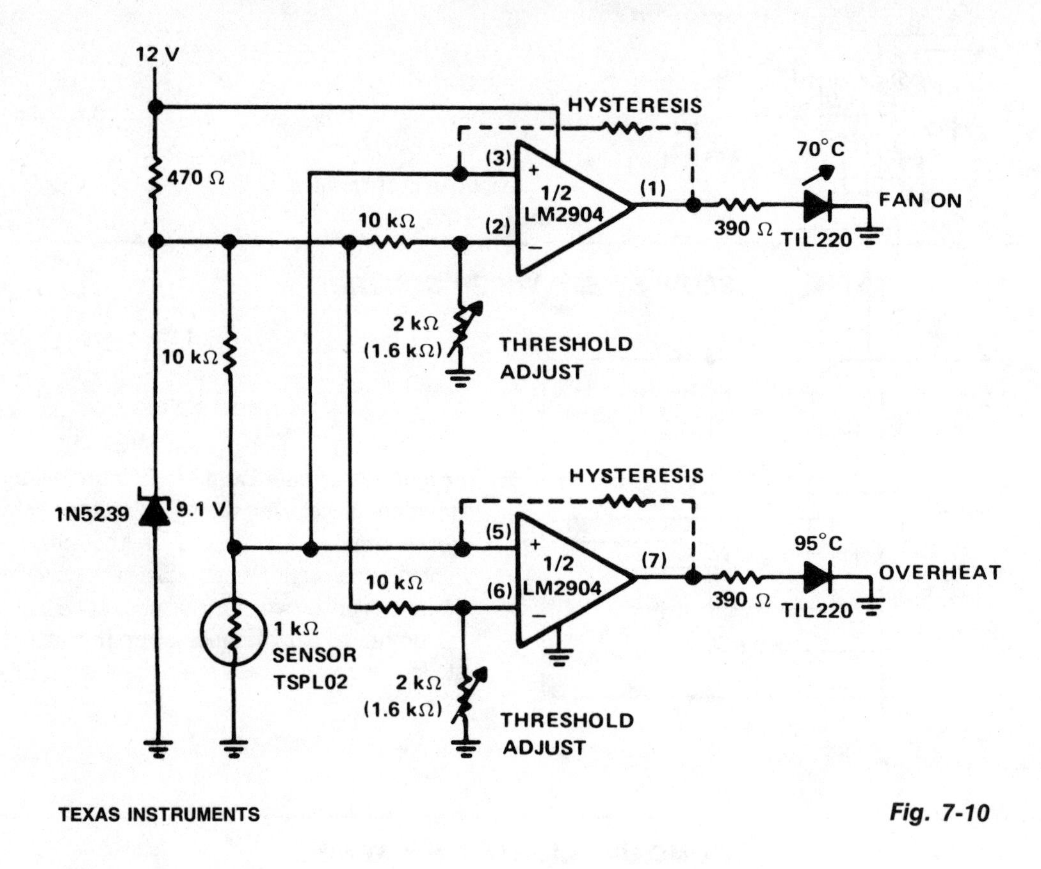

TEXAS INSTRUMENTS

Fig. 7-10

Circuit Notes

The circuit is used to indicate two different water temperature trip points by turning on LEDs when the temperatures are reached. The circuit is constructed around the LM2904 dual operational amplifier powered from the 12 V auto system. The thermistor is in series with a 10 kΩ resistor from ground to the positive 9.1 V point. The top of the thermistor is tied to both non-inverting inputs of the LM2904. The voltage at these inputs will change as the thermistor resistance changes with temperature. Each inverting input on the LM2904 has a reference, or threshold trip point, set by a 10 kΩ resistor and a 2 kΩ potentiometer in series across the 9.1 V regulated voltage. When this threshold is exceeded on the non-inverting input of LM2904, the TIL220 LED lights. The two trip points can be recalibrated or set to trip at different temperatures by adjusting the 2 kΩ potentiometer in each section. In addition to being used as warning lights as shown here, circuits can be added to turn on the fan motor or activate a relay.

ROAD ICE ALARM

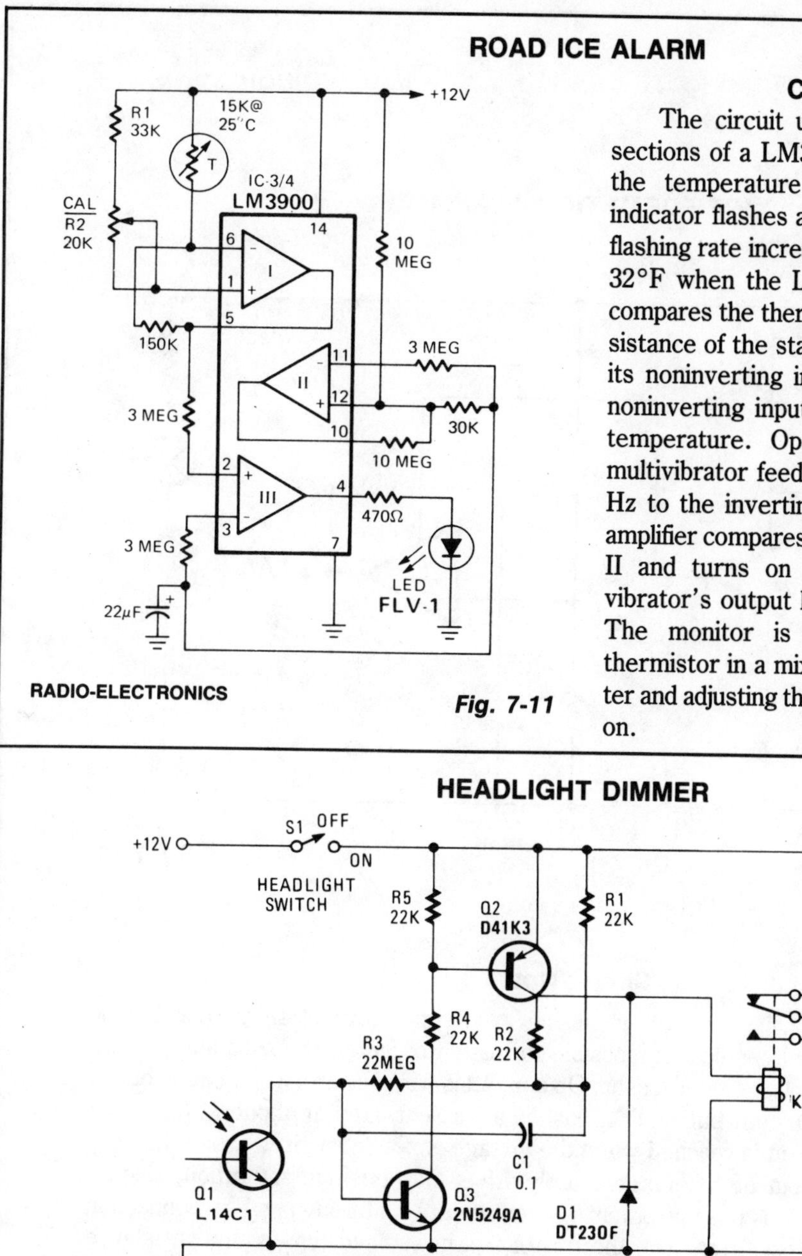

Fig. 7-11

Circuit Notes

The circuit uses a thermistor and three sections of a LM3900 quad op amp IC. When the temperature drops to 36°F the LED indicator flashes about once each second. The flashing rate increases as temperature drops to 32°F when the LED remains on. Amplifier I compares the thermistor's resistance to the resistance of the standard network connected to its noninverting input. Its output—fed to the noninverting input of op amp III—varies with temperature. Op amp II is a free-running multivibrator feeding a pulse signal of about 1 Hz to the inverting input of op amp III. This amplifier compares the outputs of op amps I and II and turns on the LED when the multivibrator's output level drops below op amp I. The monitor is calibrated by placing the thermistor in a mixture of crushed ice and water and adjusting the 20 kΩ pot so the LED stays on.

HEADLIGHT DIMMER

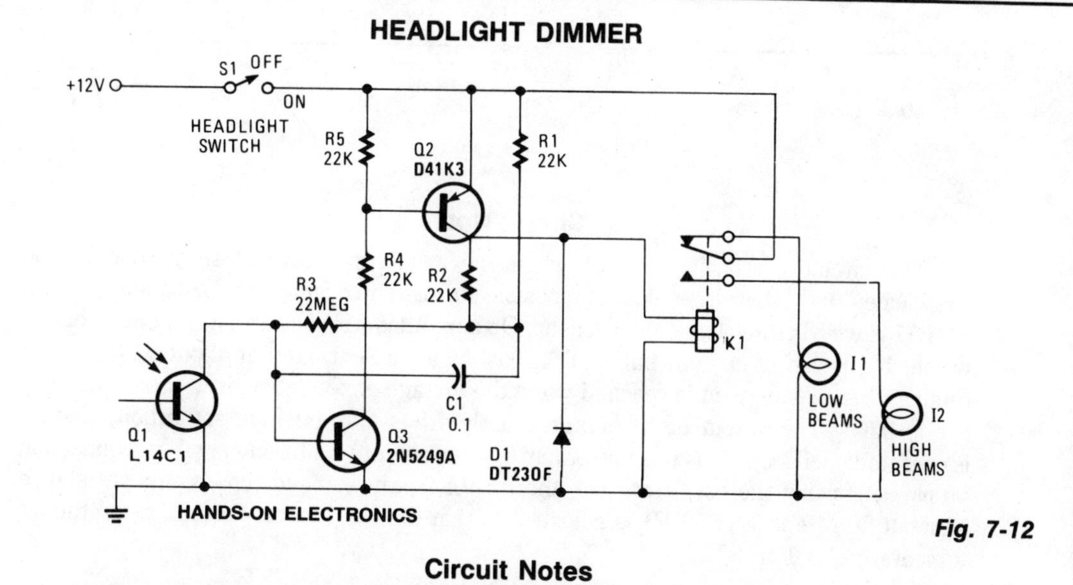

Fig. 7-12

Circuit Notes

When the lights of an on-coming car are sensed by photo-transistor Q1, things get going. Sensitivity is set by the 22-megohm resistor, R5, to about half a foot-candle. The relay used has a 12-volt, 0.3A coil. The L14C1 is complete with a lens that has a diameter of one inch for a 10° viewing angle.

ICE FORMATION ALARM

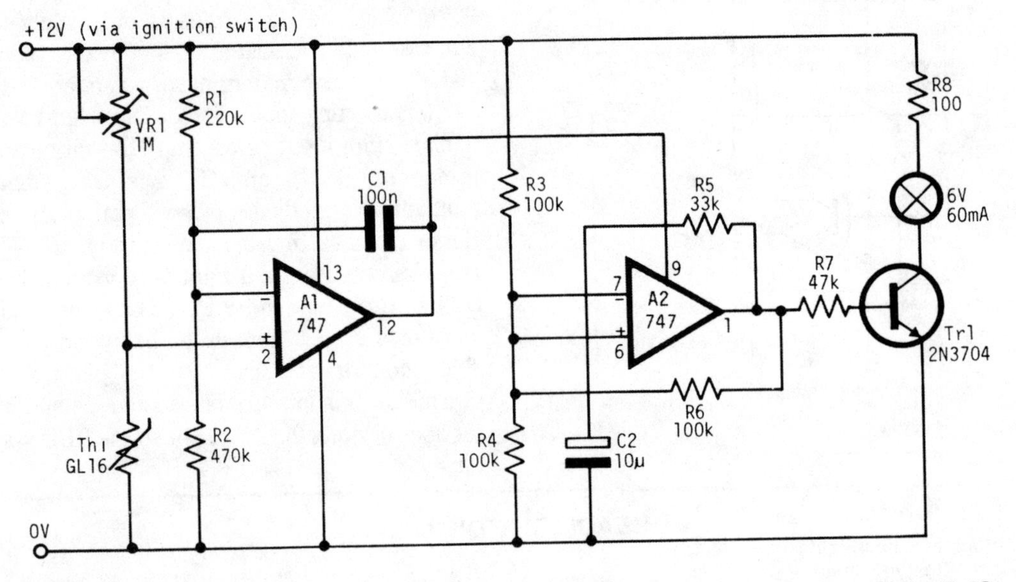

ELECTRONIC ENGINEERING

Fig. 7-13

Circuit Notes

The circuit warns car drivers when the air temperature close to the ground approaches 0°C, thereby indicating possible formation of ice on the road surface. Op amp A1 is wired as a voltage level sensor. Op amp A2 is wired as an astable multivibrator which, by means of current buffer Tr1, flashes a filament lamp at about 1 Hz. As air temperature falls, a point is reached when the voltage at pin 2 just rises above the voltage at pin 1. The output of A1 is immediately driven into positive saturation, since it is operated open loop. This positive output voltage powers A2 through its V + connection on pin 9, starting the oscillator. The thermistor is a glass bead type with a resistance of about 20 MΩ at 20°C. VR1 is adjusted so that the lamp starts flashing when the air temperature is 1 to 2°C.

DELAY CIRCUITS FOR HEADLIGHTS

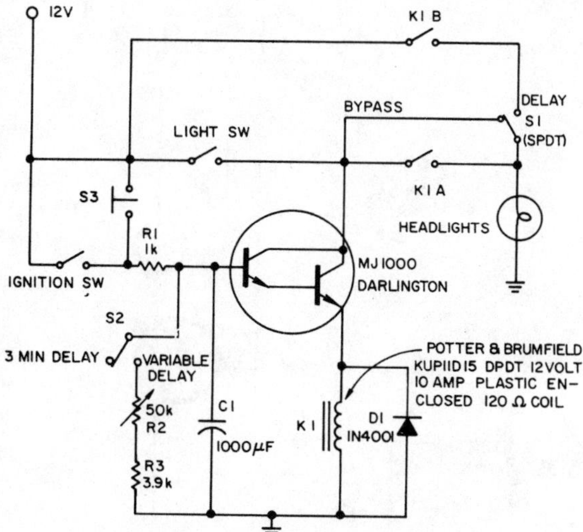

1. **Automobile headlights may be kept on** up to 3 minutes after you leave the car with this Darlington time-delay circuit.

2. **A FET version of the delay circuit** allows the use of a smaller timing capacitor, C_1, for a given delay, and almost instantaneous reset with S_3; the Darlington circuit needs almost 2 s.

ELECTRONIC DESIGN

Fig. 7-14

Circuit Notes

This circuit keeps an automobile's headlights on temporarily. It also will turn the lights off, even if you forget to flip the light switch. The circuit's shut-off delay is actuated only after both the ignition and light switches have been on, and only if the ignition switch is turned off first. If the light switch is turned off first, no delay results. Parking and brake-light operation is not affected. The maximum time out can be up to 3 minutes in part 1 and hours with the circuit in part 2, depending on the relay selected and the value of R2. A switch S2 can be used to permit selection of either a short or long delay. Momentary switch S3 can restart circuit timing before the time-out is completed. A bypass switch, S1 removes the delay action.

IGNITION TIMING LIGHT

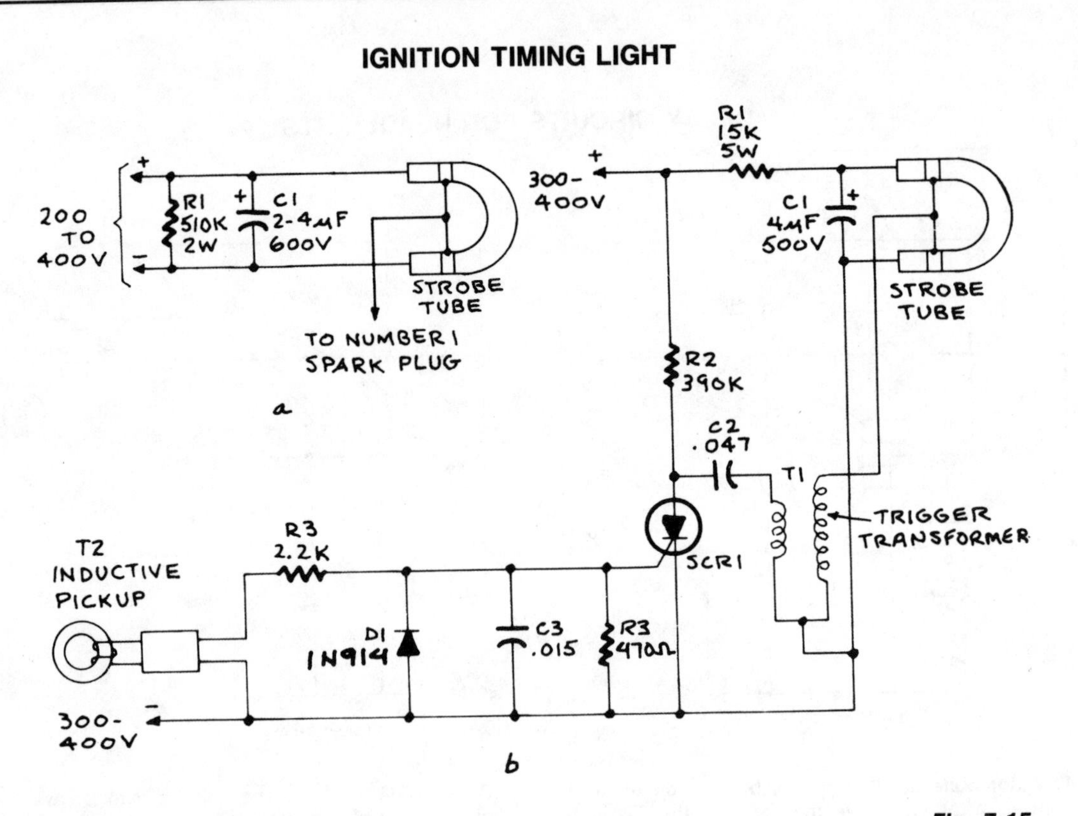

RADIO-ELECTRONICS

Fig. 7-15

Circuit Notes

Figure *A* shows the circuit of a direct-trigger timing light. The trigger voltage is taken from the car's ignition circuit by a direct connection to a spark plug. A circuit using an inductive pickup is shown in Fig. *B*. A trigger transformer is used to develop the high-voltage pulse for triggering. The triggering circuit consists of T1, C1, SCR1, inductive pickup coil T2, and the waveshaping components in the SCR's gate circuit.

When the spark plug fires, it induces a pulse in pickup coil T2 that triggers the SCR gate. The SCR fires and discharges C2 through the primary of T1. The secondary of T1 feeds a high-voltage pulse to the trigger electrode of the flash tube. That pulse causes the gas—usually neon or xenon—to ionize. The ionized gas provides a low-resistance path for C1 to discharge, thereby creating a brilliant flash of light.

Resistor R1 limits current from the supply as the tube fires. When C1 is fully discharged the strobe tube cuts off and returns to its "high-resistance" state. The current through R2 is not enough to sustain conduction through SCR1, so it cuts off and remains off until it is re-triggered by a gate pulse.

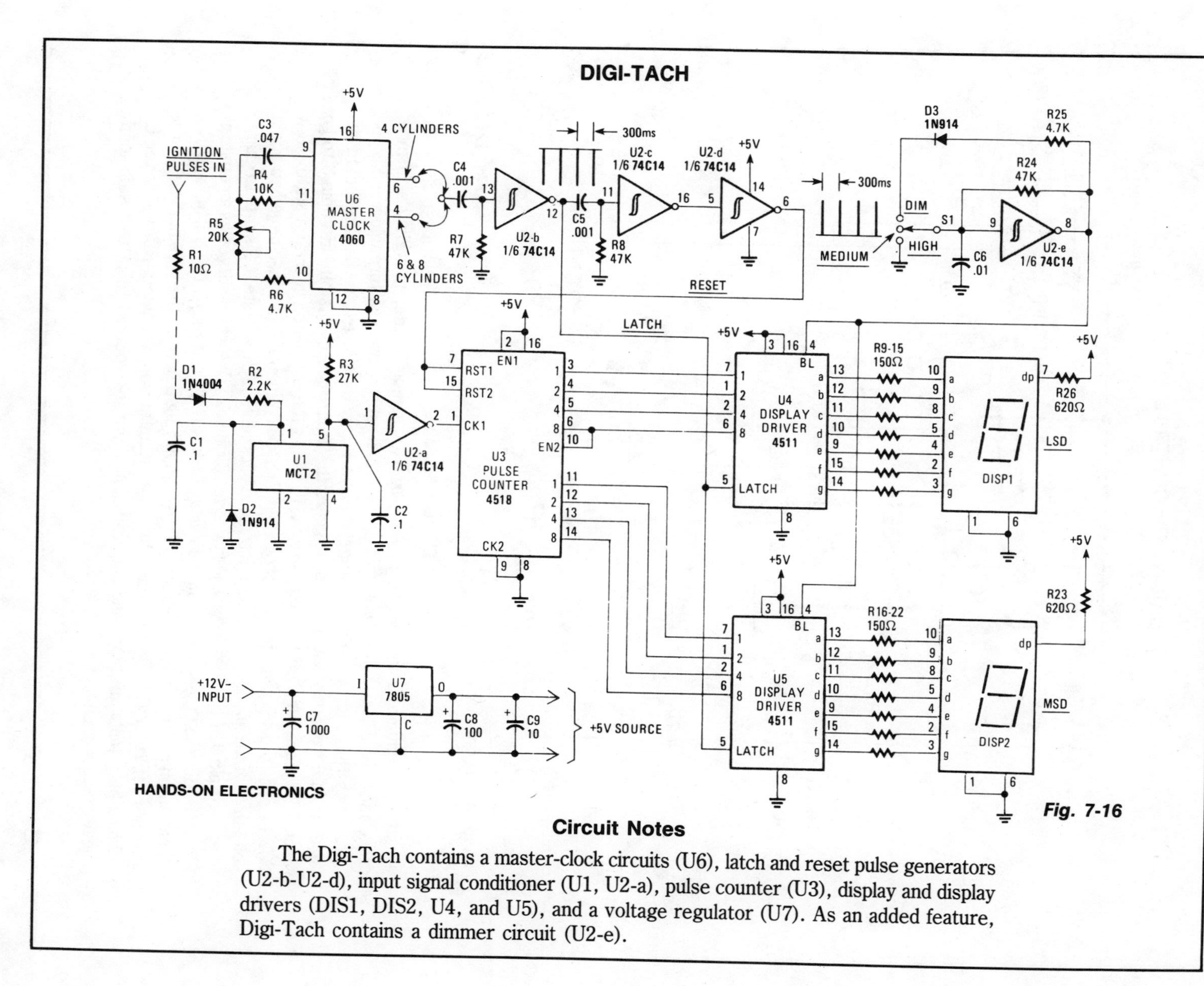

DIGI-TACH

Fig. 7-16

Circuit Notes

The Digi-Tach contains a master-clock circuits (U6), latch and reset pulse generators (U2-b-U2-d), input signal conditioner (U1, U2-a), pulse counter (U3), display and display drivers (DIS1, DIS2, U4, and U5), and a voltage regulator (U7). As an added feature, Digi-Tach contains a dimmer circuit (U2-e).

CAR-WIPER CONTROL

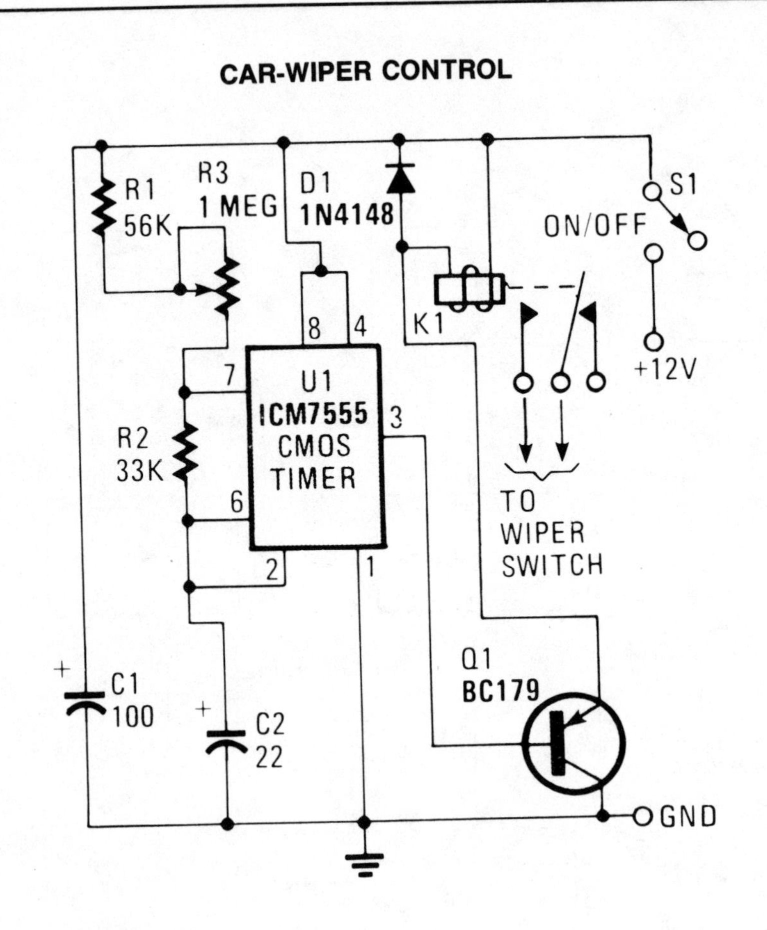

HANDS-ON ELECTRONICS

Fig. 7-17

Circuit Notes

U1 is configured to operate in the standard astable mode, providing a form of relaxation oscillator. When power is applied, C2 initially charges through R1, R2 and R3 to two-thirds of the supply voltage. At that point, U1 senses that its threshold voltage at pin 6 has been reached, and triggers the timer, causing its output at pin 3 to go high. That high, applied to the base of Q1, keeps the transistor in the off state. Now C2 begins to discharge through R2 to pin 7 of U1. When C2 has discharged to about one-third of the supply voltage, U1 is toggled back to its original state. C2 starts to charge again, as pin 3 of U1 goes low. The low at pin 3 causes Q1—which serves as an emitter-follower buffer stage—to turn on, allowing current to flow through the coil of relay K1. That, in turn, causes K1's contacts to close, applying power to the wipers. The charge time of capacitor C2 is determined by the setting of potentiometer R3. Capacitor C2 should be a tantalum type, and actually, almost any 12-volt coil relay with sufficiently heavy contacts should serve well.

AUTOMATIC HEADLIGHT DIMMER

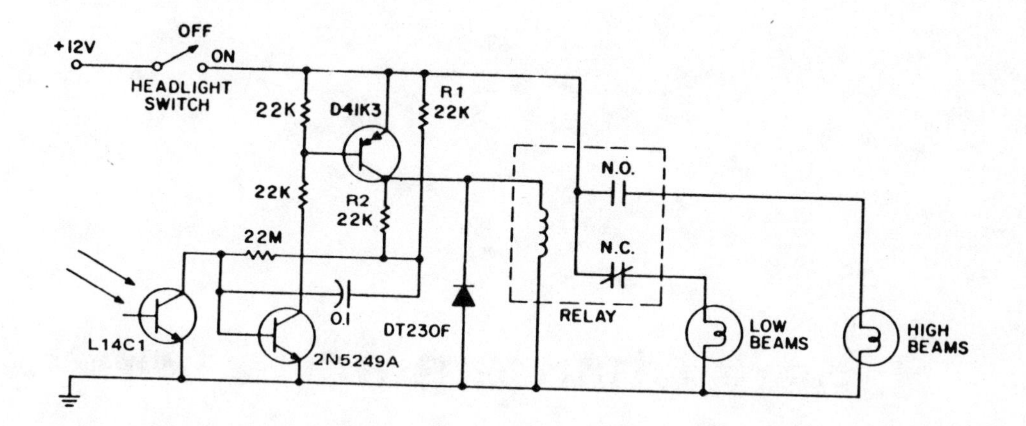

RELAY: 12V, 0.3A COIL: 20A, FORM C, CONTACTS OR SOLID-STATE SWITCHING OF 16A STEADY-STATE 150A COLD FILAMENT SURGE, RATING.

LENS: MINIMUM 1" DIAMETER, POSITIONED FOR ABOUT 10° VIEW ANGLE.

GENERAL ELECTRIC

Fig. 7-18

Circuit Notes

This circuit switches car headlights to the low beam state when it senses the lights of an on-coming car. The received light is very low level and highly directional, indicating the use of a lens with the detector. A relatively large amount of hysteresis is built into the circuit to prevent "flashing lights." Sensitivity is set by the 22 megohm resistor to about 0.5 ft. candle at the transistor (0.01 at the lens), while hysteresis is determined by the R1, R2 resistor voltage divider, parallel to the D41K3 collector emitter, which drives the 22 megohm resistor; maximum switching rate is limited by the 0.1 μF capacitor to 15/minute.

8

Battery Chargers and Zappers

The sources of the following circuits are contained in the Sources section beginning on page 694. The figure number contained in the box of each circuit correlates to the source entry in the Sources section.

Rapid Battery Charger for ICOM IC-2A
Gel Cell Charger
Ni-Cad Battery Zapper
Lithium Battery Charger
Thermally Controlled Ni-Cad Charger
Ni-Cad Battery Zapper II
Battery Charger
Wind Powered Battery Charger
Battery Charger Operates On Single Solar Cell
Versatile Battery Charger
14-Volt, 4-Amp Battery Charger/Power Supply

RAPID BATTERY CHARGER FOR ICOM IC-2A

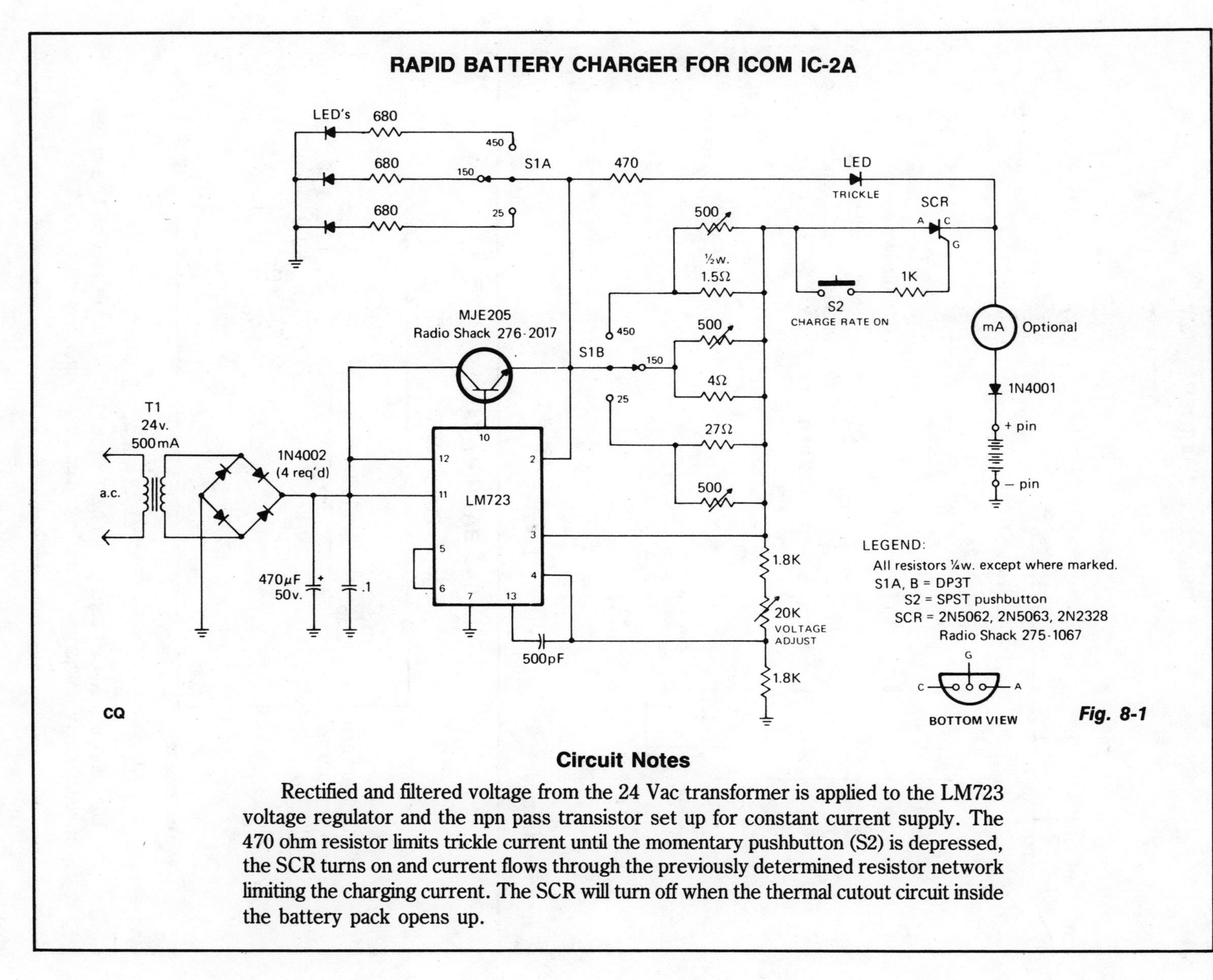

Fig. 8-1

Circuit Notes

Rectified and filtered voltage from the 24 Vac transformer is applied to the LM723 voltage regulator and the npn pass transistor set up for constant current supply. The 470 ohm resistor limits trickle current until the momentary pushbutton (S2) is depressed, the SCR turns on and current flows through the previously determined resistor network limiting the charging current. The SCR will turn off when the thermal cutout circuit inside the battery pack opens up.

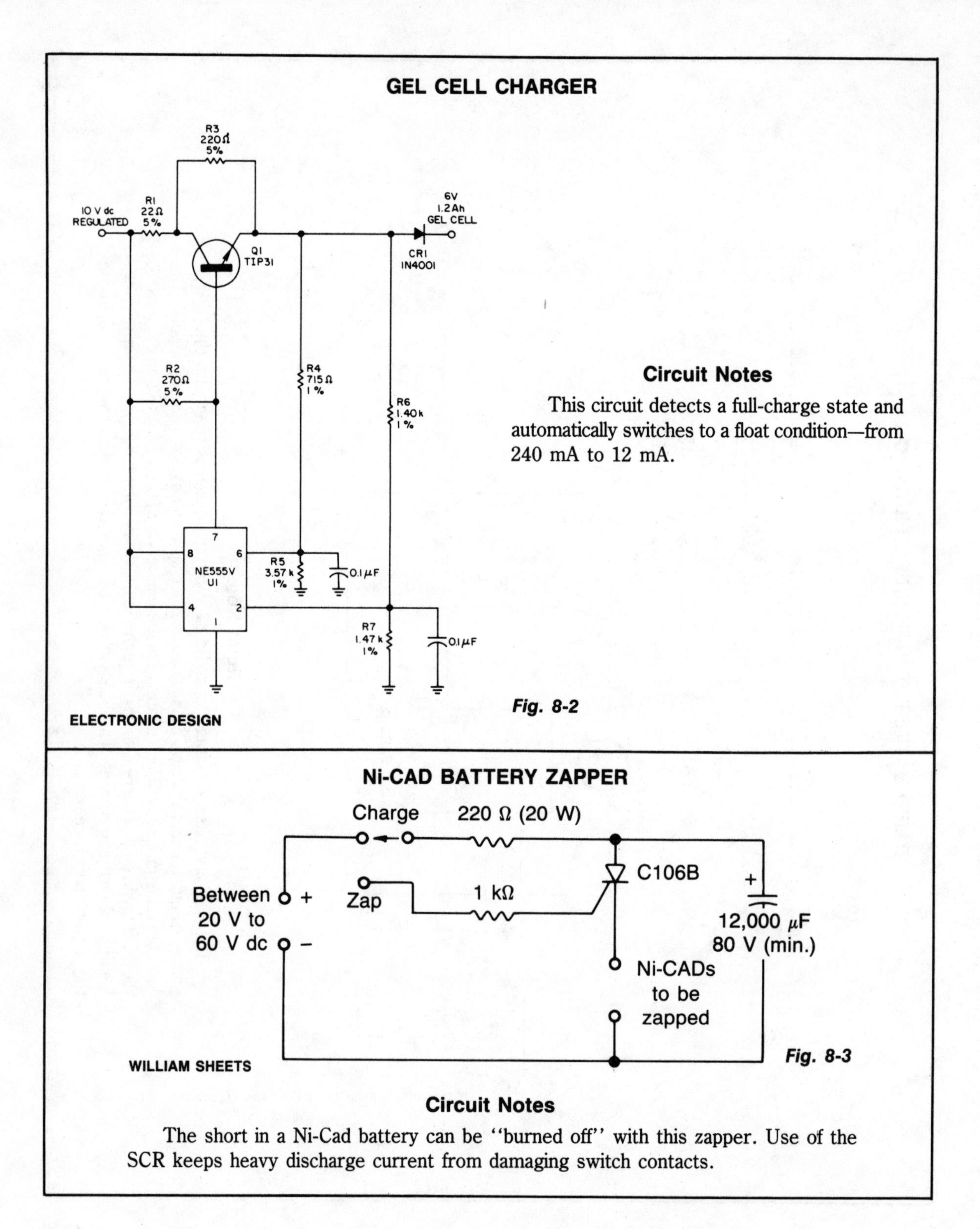

GEL CELL CHARGER

ELECTRONIC DESIGN

Circuit Notes

This circuit detects a full-charge state and automatically switches to a float condition—from 240 mA to 12 mA.

Fig. 8-2

Ni-CAD BATTERY ZAPPER

WILLIAM SHEETS

Fig. 8-3

Circuit Notes

The short in a Ni-Cad battery can be "burned off" with this zapper. Use of the SCR keeps heavy discharge current from damaging switch contacts.

LITHIUM BATTERY CHARGER

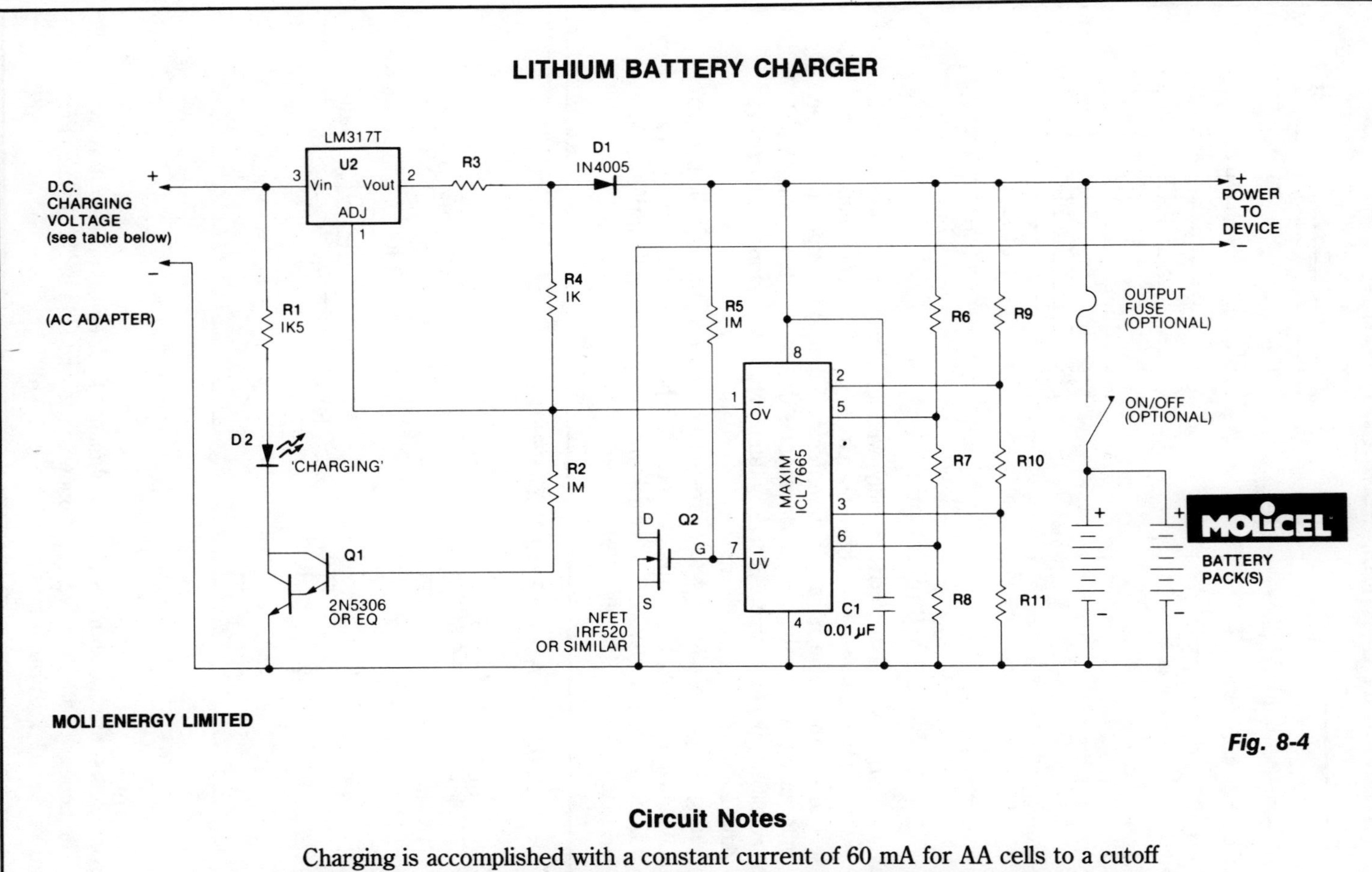

Fig. 8-4

Circuit Notes

Charging is accomplished with a constant current of 60 mA for AA cells to a cutoff voltage of 2.4 V per cell at which point the charge must be terminated. The charging system shown is designed for multi-cell battery packs of 2 to 6 series-connected cells or series/parallel arrangements. It is essential that all cells assembled in the pack be at an identical state-of-charge (voltage) prior to charging. The maximum upper cut-off voltage is 15.6 volts (6 × 2.6 V).

THERMALLY CONTROLLED Ni-CAD CHARGER

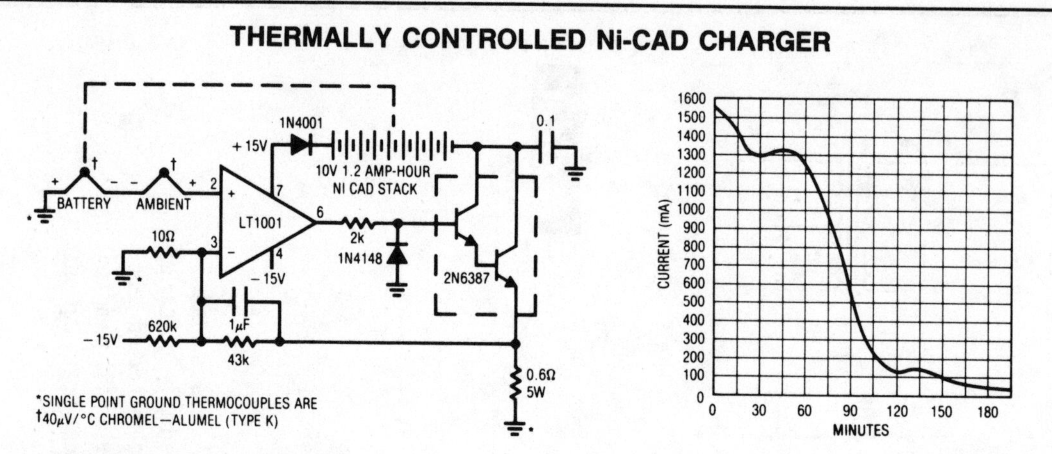

LINEAR TECHNOLOGY CORPORATION

Fig. 8-5

Circuit Notes

One way to charge Ni-Cad batteries rapidly without abuse is to measure cell temperature and taper the charge accordingly. The circuit uses a thermocouple for this function. A second thermocouple nulls out the effects of ambient temperature. The temperature difference between the two thermocouples determines the voltage which appears at the amplifier's positive input. As battery temperature rises, this small negative voltage (1°C difference between the thermocouples equals 40 μV) becomes larger. The amplifier, operating at a gain of 4300, gradually reduces the current through the battery to maintain its inputs at balance. The battery charges at a high rate until heating occurs and the circuit then tapers the charge. The values given in the circuit limit the battery surface temperature rise over ambient to about 5°C.

Ni-CAD BATTERY ZAPPER II

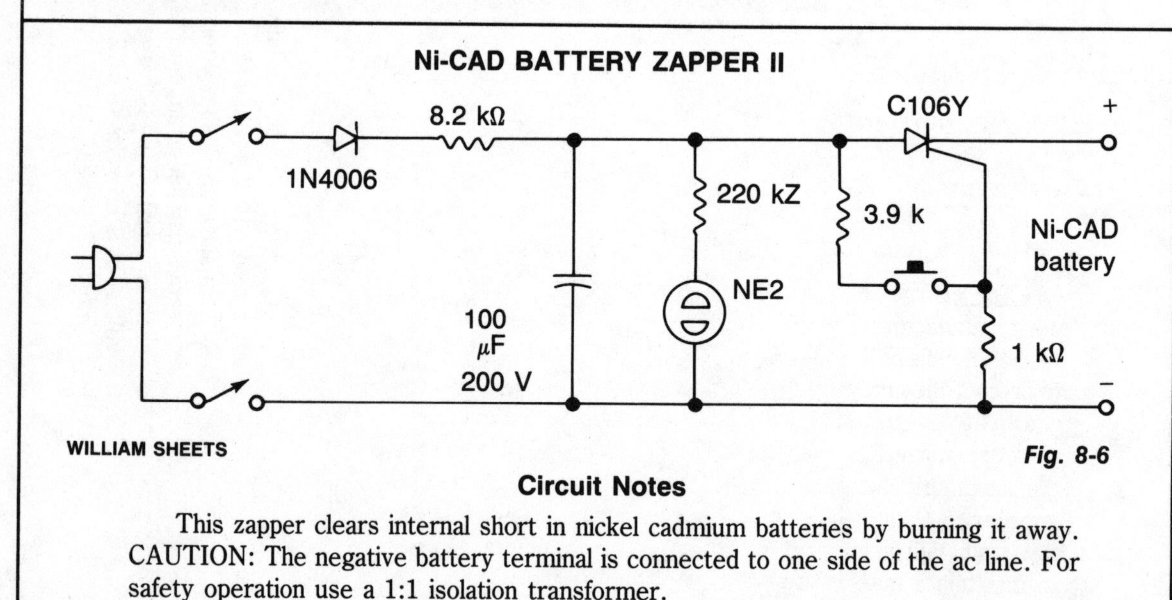

WILLIAM SHEETS

Fig. 8-6

Circuit Notes

This zapper clears internal short in nickel cadmium batteries by burning it away. CAUTION: The negative battery terminal is connected to one side of the ac line. For safety operation use a 1:1 isolation transformer.

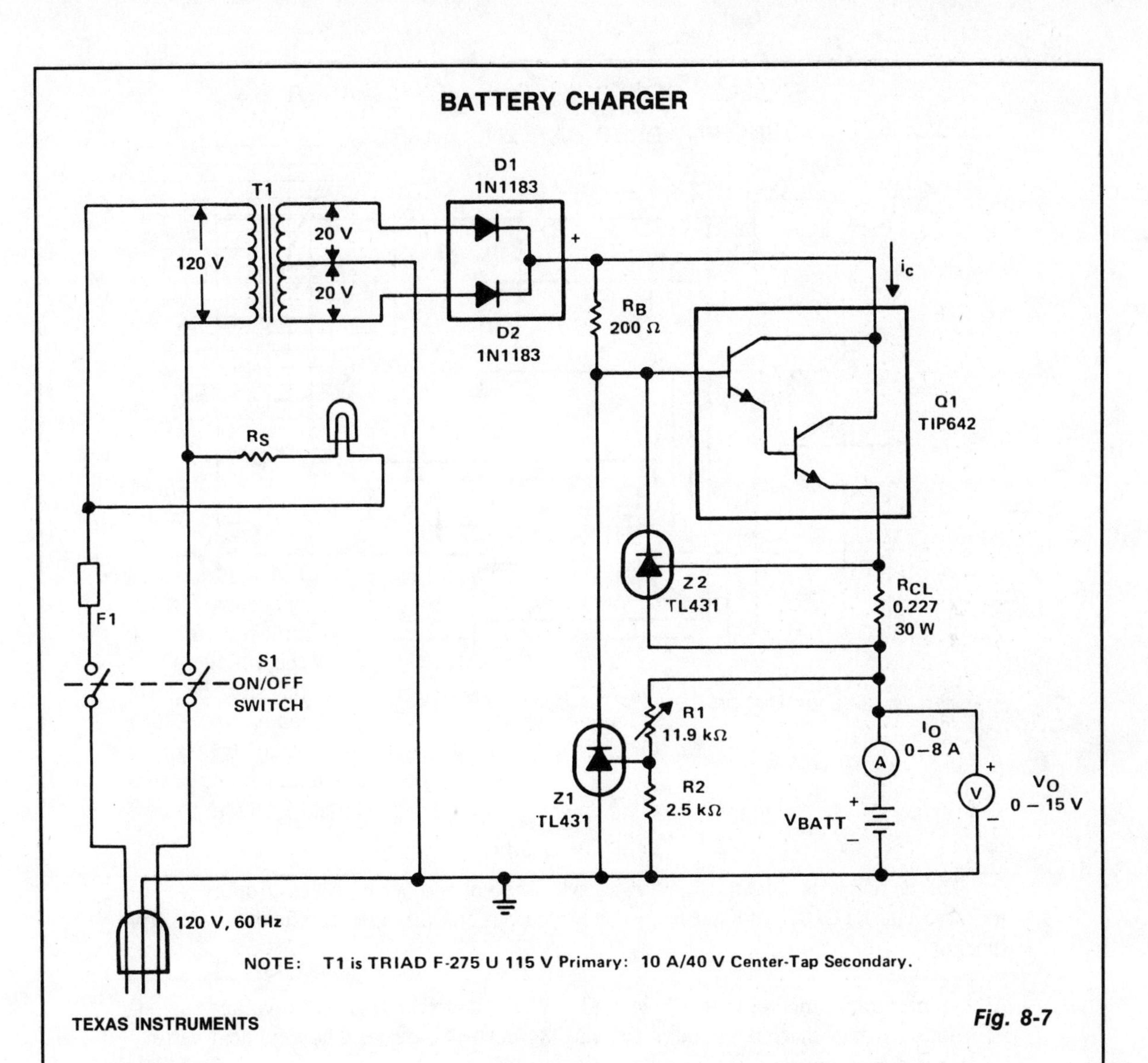

Fig. 8-7

Circuit Notes

The charger is based on a charging voltage of 2.4 V per cell, in accordance with most manufacturers' recommendations. The circuit pulses the battery under charge with 14.4 V (6 cells × 2.4 V per cell) at a rate of 120 Hz. The design provides current limiting to protect the charger's internal components while limiting the charging rate to prevent damaging severely discharged lead-acid batteries. The maximum recommended charging current is normally about one-fourth the ampere-hour rating of the battery. For example, the maximum charging current for an average 44 ampere-hour battery is 11 A. If the impedance of the load requires a charging current greater than the 11 A current limit, the circuit will go into current limiting. The amplitude of the charging pulses is controlled to maintain a maximum peak charging current of 11 A (8 A average).

WIND POWERED BATTERY CHARGER

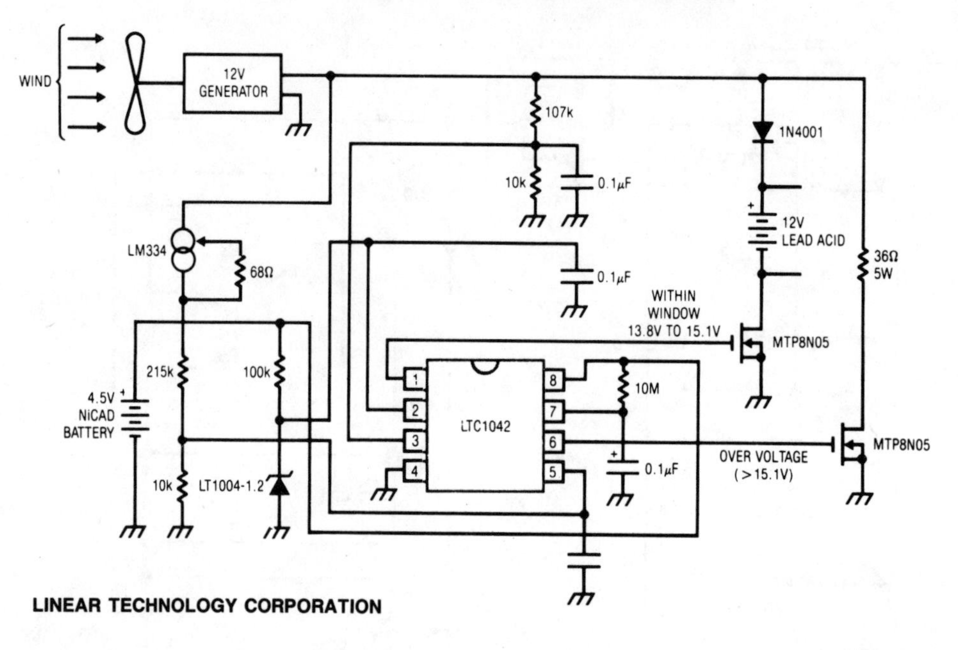

Fig. 8-8

Circuit Notes

The dc motor is used as a generator with the voltage output being proportional to its rpm. The LTC1042 monitors the voltage output and provides the following control functions.

1. If generator voltage output is below 13.8 V, the control circuit is active and the Ni-Cad battery is charging through the LM334 current source. The lead acid battery is not being charged.
2. If the generator voltage output is between 13.8 V and 15.1 V, the 12 V lead acid battery is being charged at about 1 amp/hour rate (limited by the power FET).
3. If generator voltage exceeds 15.1 V (a condition caused by excessive wind speed or 12 V battery being fully charged) then a fixed load is connected limiting the generator rpm to prevent damage.

This charger can be used as a remote source of power where wind energy is plentiful such as on sailboats or remote radio repeater sites. Unlike solar powered panels, this system will function in bad weather and at night.

BATTERY CHARGER OPERATES ON SINGLE SOLAR CELL

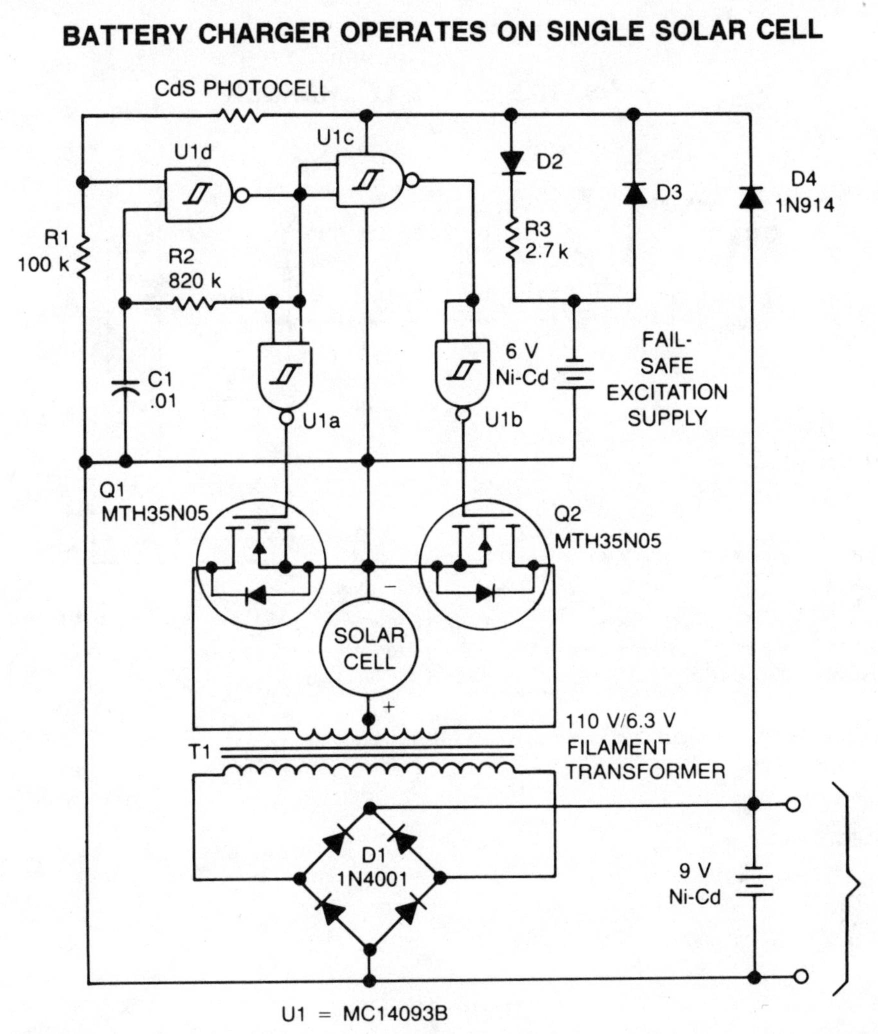

MOTOROLA

Fig. 8-9

Circuit Notes

The circuit charges a 9-V battery at about 30 mA per input ampere at 0.4 V. U1, a quad Schmitt trigger, operate as an astable multivibrator to drive push-pull TMOS devices Q1 and Q2. Power for U1 is derived from the 9-V battery via D4; power for Q1 and Q2 is supplied by the solar cell. The multivibrator frequency, determined by R2-C1, is set to 180 Hz for maximum efficiency from a 6.3-V filament transformer, T1. The secondary of the transformer is applied to a full wave bridge rectifier, D1, which is connected to the batteries being charged. The small Ni-Cad battery is a fail-safe excitation supply to allow the system to recover if the 9-V battery becomes fully discharged.

A CdS photocell shuts off the oscillator in darkness to preserve the fail-safe battery during shipping and storage, or prolonged darkness.

VERSATILE BATTERY CHARGER

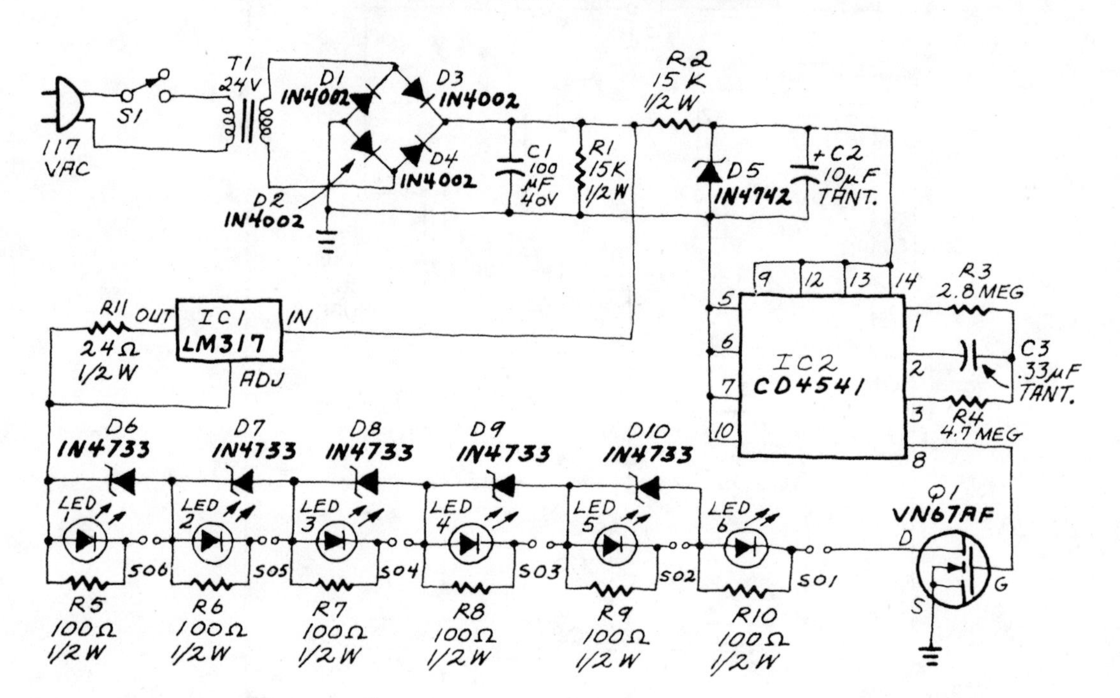

RADIO ELECTRONICS

Fig. 8-10

Circuit Notes

An LM317 voltage regulator is configured as a constant-current source. It is used to supply the 50 mA charging current to S01-S06, an array of AA-cell battery holders. Each of the battery holders is wired in series with an LED and its associated shunt resistor. When the battery holder contains a battery, the LED glows during charging. Each battery holder/LED combination is paralleled by a 5.1-volt Zener diode. If the battery holder is empty, the Zener conducts the current around the holder.

A timing circuit prevents overcharging. When power is applied to the circuit, timing is initiated by IC2, a CD4541 oscillator/programmable timer. The output of IC2 is fed to Q1. When that output is high, the transistor is on, and the charging circuit is completed. When the output is low, the transistor is off, and the path to ground is interrupted.

14-VOLT, 4-AMP BATTERY CHARGER/POWER SUPPLY

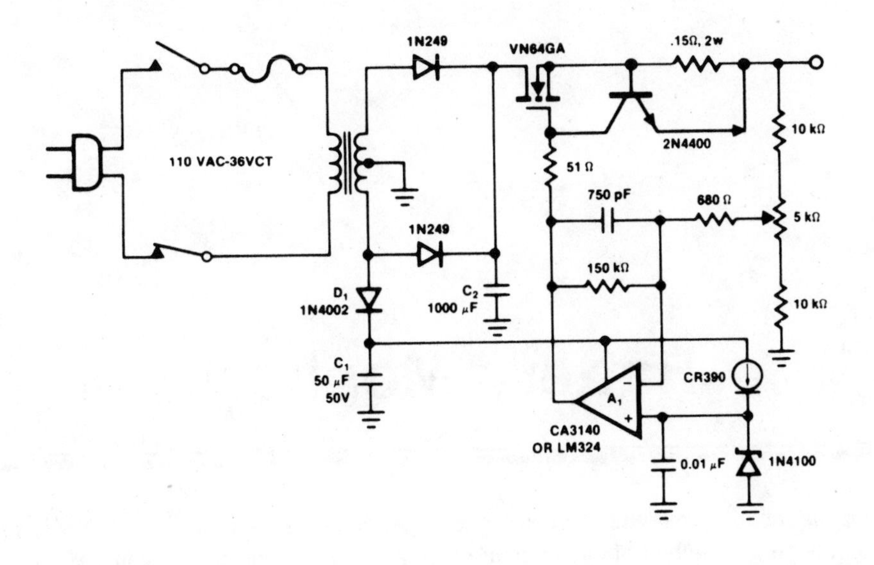

SILICONIX, INC.

Fig. 8-11

Circuit Notes

Operation amplifier A1 directly drives the VN64GA with the error signal to control the output voltage. Peak rectifier D1, C1 supplies error amplifier A1 and the reference zener. This extra drive voltage must exceed its source voltage by several volts for the VN64GA to pass full load current. The output voltage is pulsating dc which is quite satisfactory for battery charging. To convert the system to a regulated dc supply, capacitor C2 is increased and another electrolytic capacitor is added across the load. The response time is very fast, being determined by the op-amp. The 2N4400 current limiter circuit prevents the output current from exceeding 4.5 A. However, maintaining a shorted condition for more than a second will cause the VN64GA to exceed its temperature ratings. A generous heat sink, on the order of 1°C/W, must be used.

9

Battery Monitors

The sources of the following circuits are contained in the Sources section beginning on page 694. The figure number contained in the box of each circuit correlates to the source entry in the Sources section.

DYNAMIC, CONSTANT CURRENT LOAD FOR FUEL CELL/BATTERY TESTING

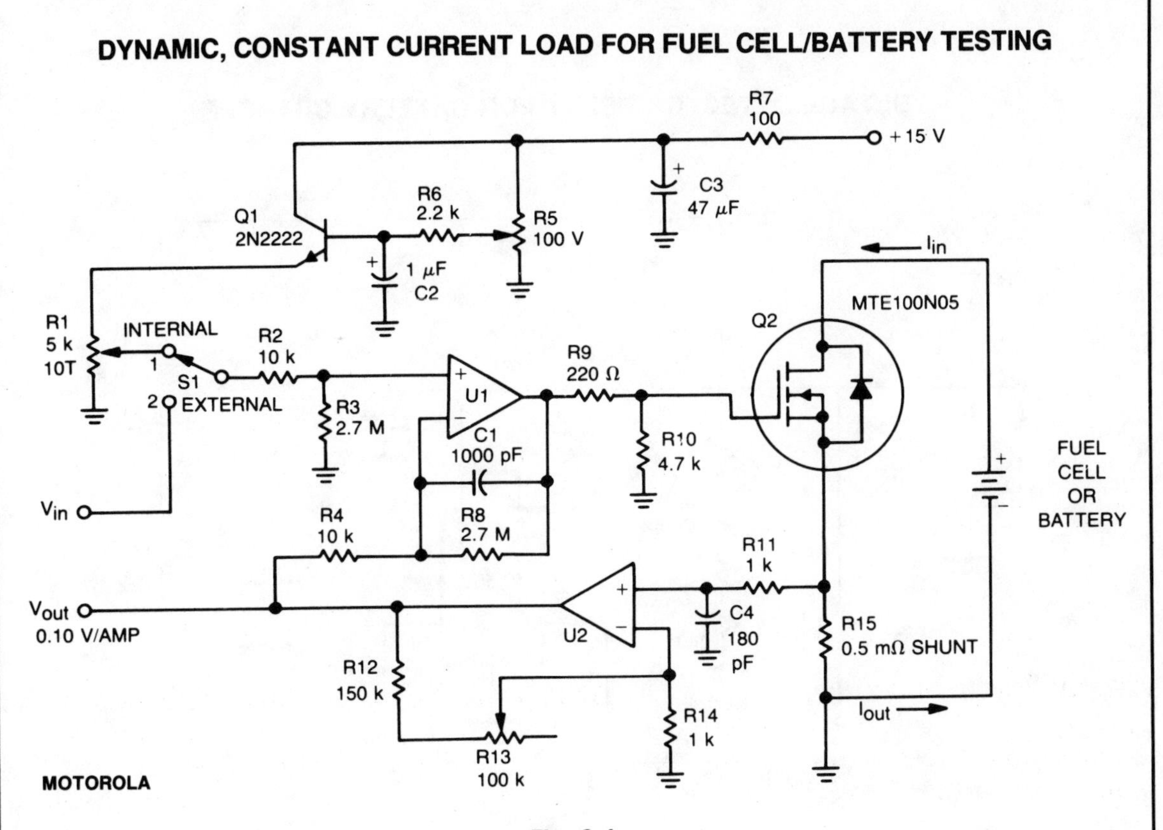

Fig. 9-1

Circuit Notes

This circuit was designed for testing fuel cells, but it could also be used for testing batteries under a constant current load. It provides a dynamic, constant current load, eliminating the need to manually adjust the load to maintain a constant load.

For fuel cell application, the load must be able to absorb 20-40 A, and since a single cell develops only 0.5 to 1.0 V, bipolar power devices (such as a Darlington) are impractical. Therefore, this dynamic load was designed with a TMOS Power FET (Q2).

With switch S1 in position 1, emitter follower Q1 and R1 establish the current level for the load. In position 2, an external voltage can be applied to control the current level.

Operational amplifier U1 drives TMOS device Q1, which sets the load current seen by the fuel cell or battery. The voltage drop across R15, which is related to the load current, is then applied to U2, whose output is fed back to U1. Thus, if the voltage across R15 would tend to change, feedback to the minus input of U1 causes that voltage (and the load current) to remain constant. Adjustment of R13 controls the volts/amp of feedback. The V_{OUT} point is used to monitor the system.

VOLTAGE DETECTOR RELAY FOR BATTERY CHARGER

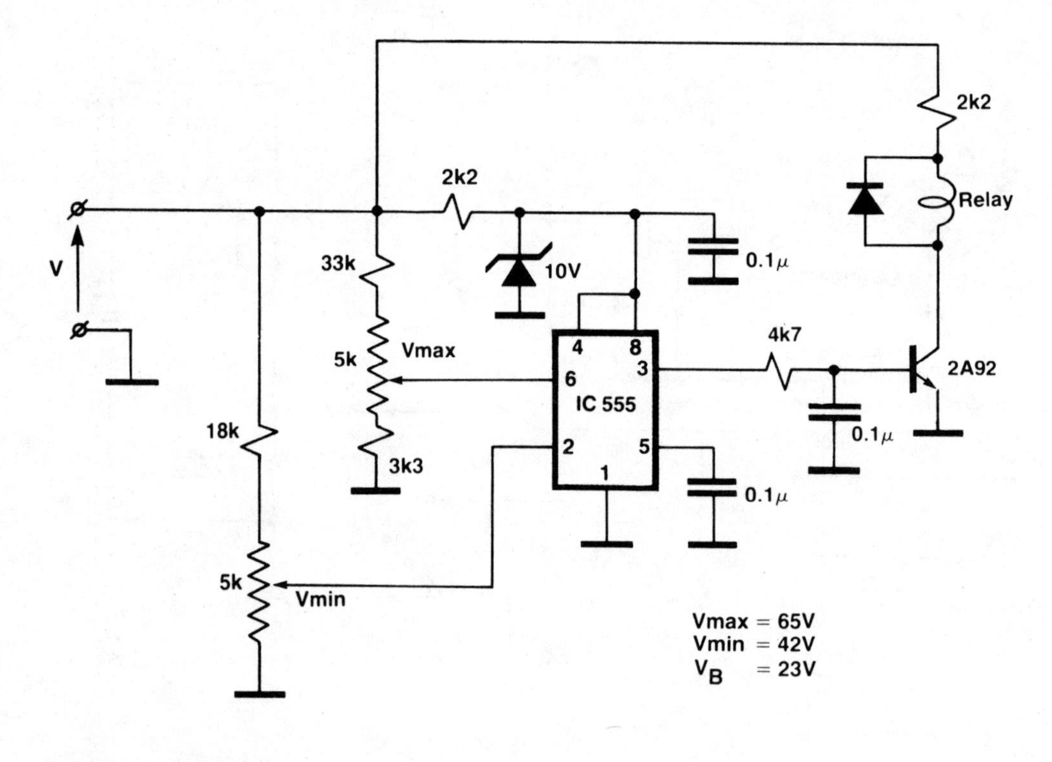

ELECTRONIC ENGINEERING

Fig. 9-2

Circuit Notes

While the battery is being charged, its voltage is measured at V. If the measured voltage is lower than the minimum the relay will be energized, that will connect the charger circuit. When the battery voltage runs over the maximum set point, the relay is deenergized and it will be held that way until the voltage decreases below the minimum when it will be connected again. The voltage is lower than a threshold V_B (low breaking voltage) the relay will be assumed that such a low voltage is due to one or several damaged battery components. Of course V_B is much lower than the minimum set point.

BATTERY STATUS INDICATOR

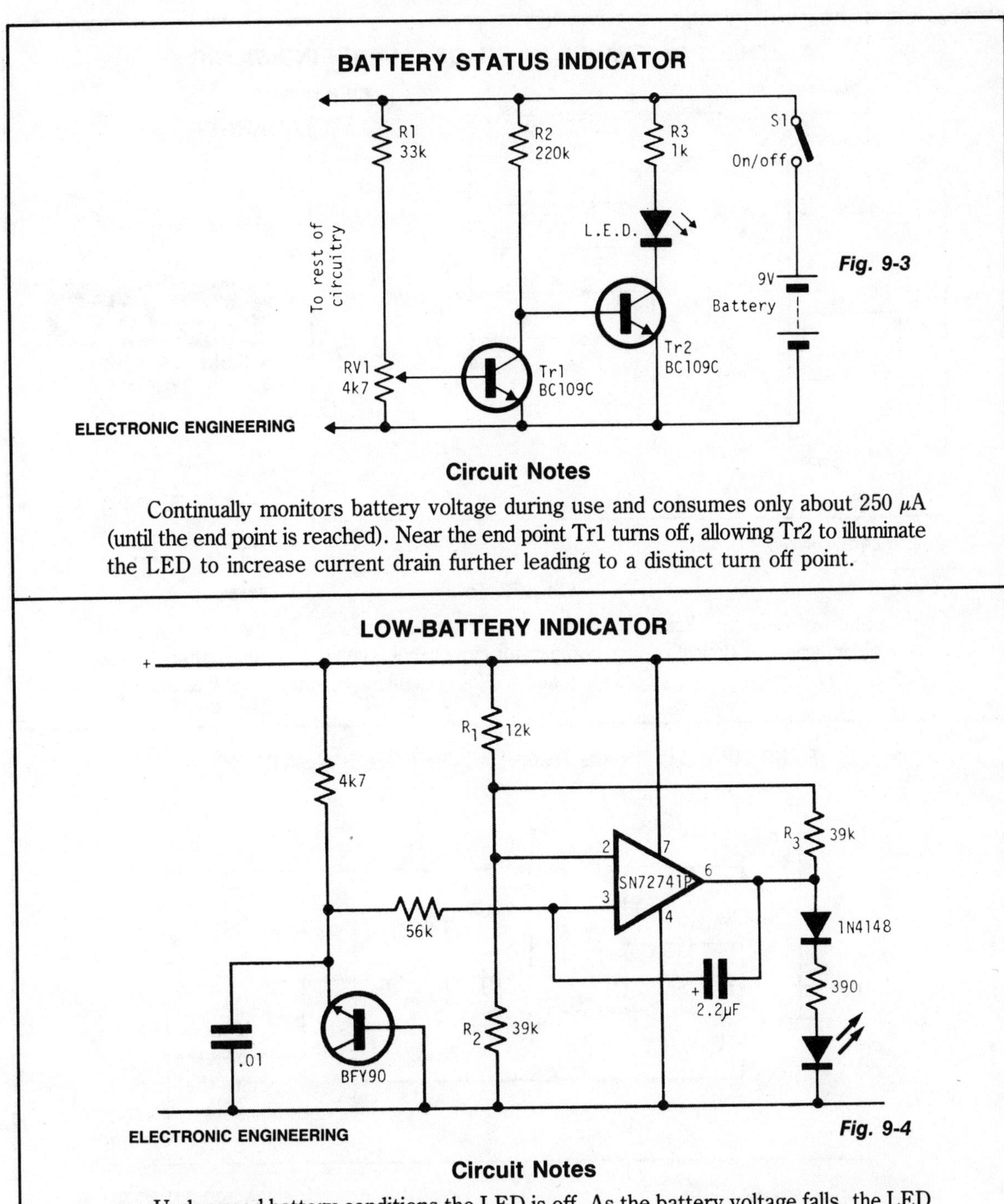

Circuit Notes

Continually monitors battery voltage during use and consumes only about 250 μA (until the end point is reached). Near the end point Tr1 turns off, allowing Tr2 to illuminate the LED to increase current drain further leading to a distinct turn off point.

LOW-BATTERY INDICATOR

Circuit Notes

Under good battery conditions the LED is off. As the battery voltage falls, the LED begins to flash until, in the low battery condition, the LED lights continuously. Designed for a 9-volt battery, with the values shown the LED flashes from 7.5 to 6.5 volts.

A LITHIUM BATTERY'S STATE-OF-CHARGE INDICATOR

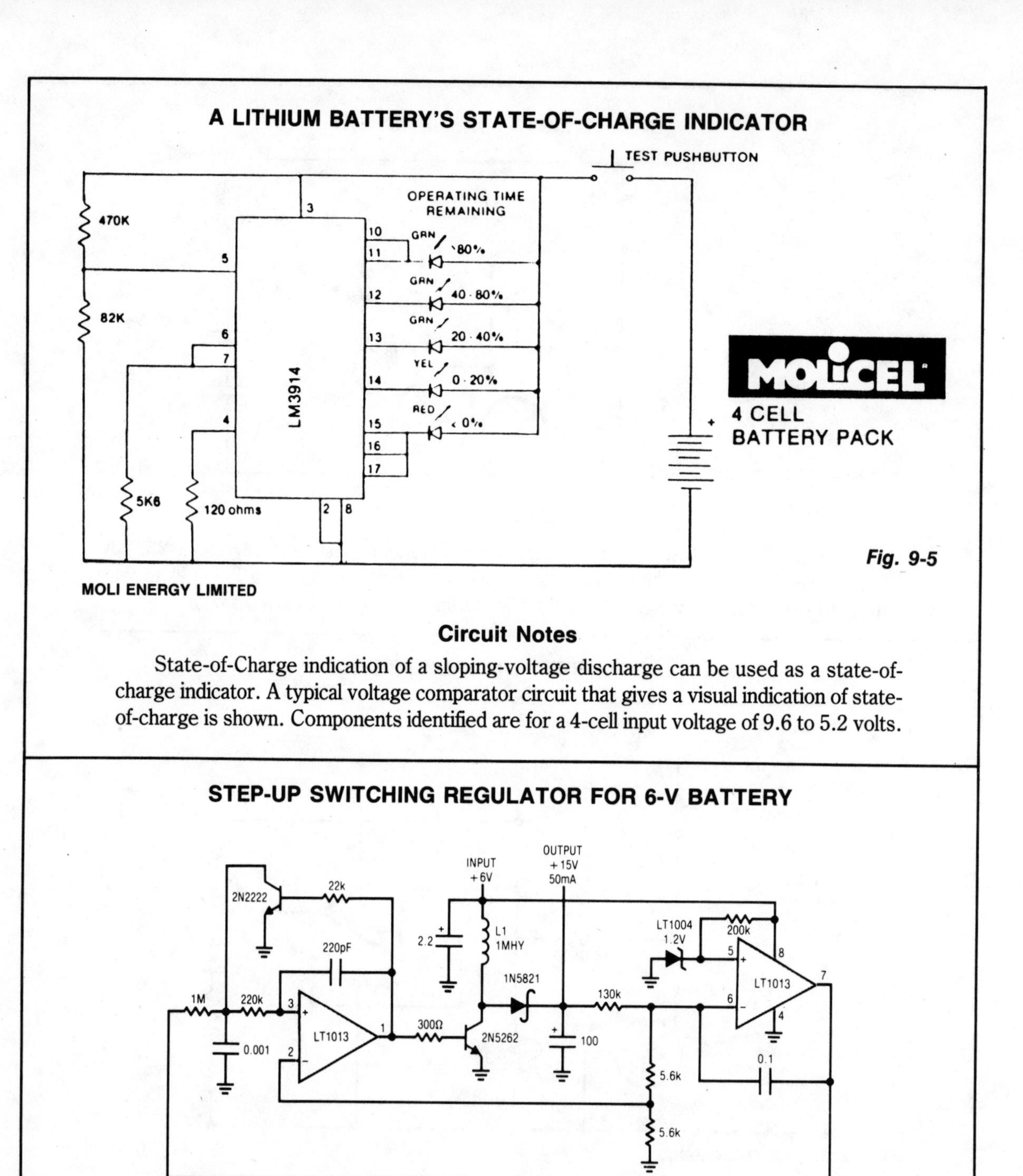

MOLI ENERGY LIMITED

Fig. 9-5

Circuit Notes

State-of-Charge indication of a sloping-voltage discharge can be used as a state-of-charge indicator. A typical voltage comparator circuit that gives a visual indication of state-of-charge is shown. Components identified are for a 4-cell input voltage of 9.6 to 5.2 volts.

STEP-UP SWITCHING REGULATOR FOR 6-V BATTERY

L1 = AIE—VERNITRON 24-104
78% EFFICIENCY

LINEAR TECHNOLOGY CORP.

Fig. 9-6

BATTERY VOLTAGE MONITOR

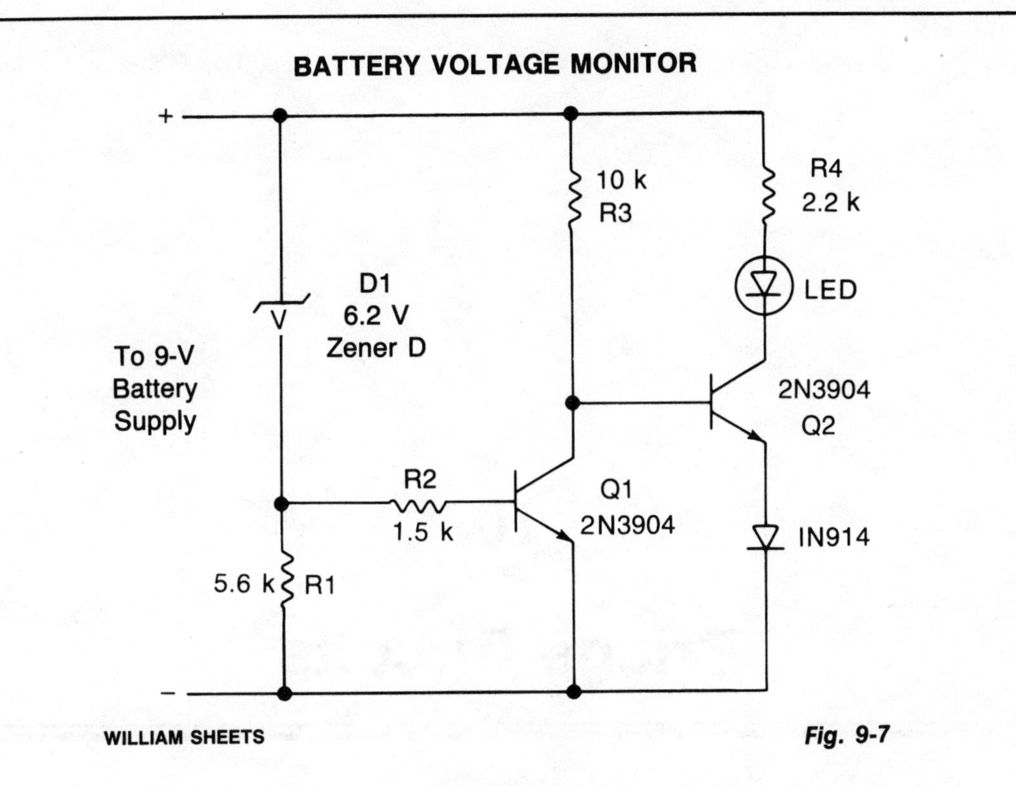

WILLIAM SHEETS

Fig. 9-7

Circuit Notes

This circuit gives an early warning of the discharge of batteries. Zener diode D1 is chosen for the voltage below which an indication is required (9 V). Should the supply drop to below 7 V, D1 will cease conducting causing Q1 to shut off. Its collector voltage will now increase causing Q2 to start conducting via LED1 and its limiting resistor R4.

BATTERY MONITOR

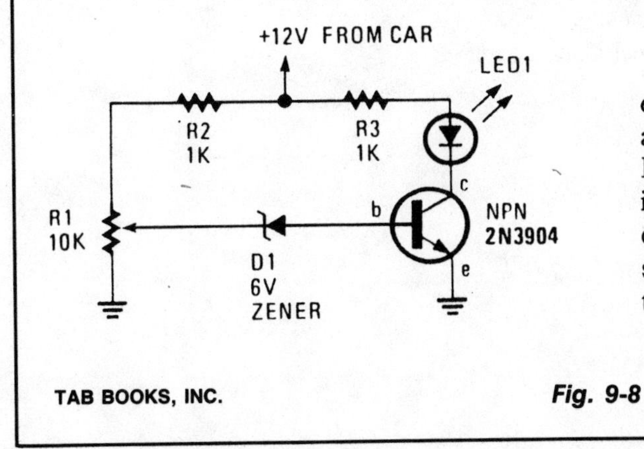

Circuit Notes

The circuit is quick and easy to put together and install, and tells you when battery voltage falls below the set limit as established by R1 (a 10,000-ohm potentiometer). It can indicate, via LED1, that the battery may be defective or in need of change if operating the starter causes the battery voltage to drop below the present limit.

TAB BOOKS, INC.

Fig. 9-8

10

Bridge Circuits

The sources of the following circuits are contained in the Sources section beginning on page 694. The figure number contained in the box of each circuit correlates to the source entry in the Sources section.

AC BRIDGE

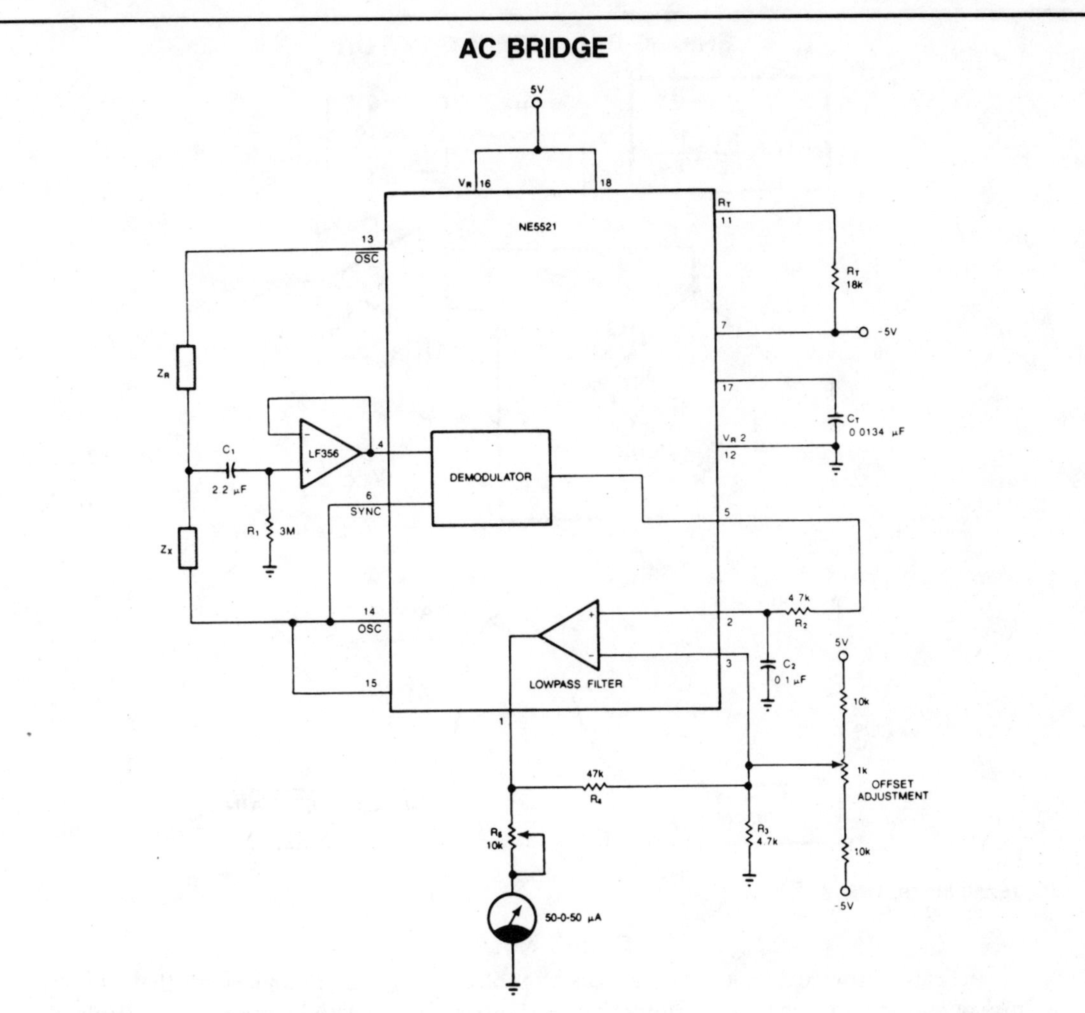

SIGNETICS

Fig. 10-1

Circuit Notes

The circuit provides a simple and cost-effective solution to matching resistors and capacitors. Impedances Z_R and Z_X form a half-bridge, while OSC and $\overline{\text{OSC}}$ excite the bridge differentially. The external op amp is a FET input amplifier (LF356) with very low input bias current on the order of 30 pA (typical). C1 allows ac coupling by blocking the dc common mode voltage from the bridge, while R1 biases the output of LF356 to 0 V at dc. Use of FET input op amp insures that dc offset due to bias current through R1 is negligible. Ac output of the demodulator is filtered via the uncommitted amp to provide dc voltage for the meter. The 10 k potentiometer, R5, limits the current into the meter to a safe level. Calibration begins by placing equal impedances at Z_R and Z_X, and the system offset is nulled by the offset adjust circuit so that Pin 1 is at 0 V. Next, known values are placed at Z_X and the meter deviations are calibrated. The bridge is now ready to measure an unknown impedance at Z_X with $\pm 0.05\%$ accuracy or better.

BRIDGE-BALANCE INDICATOR

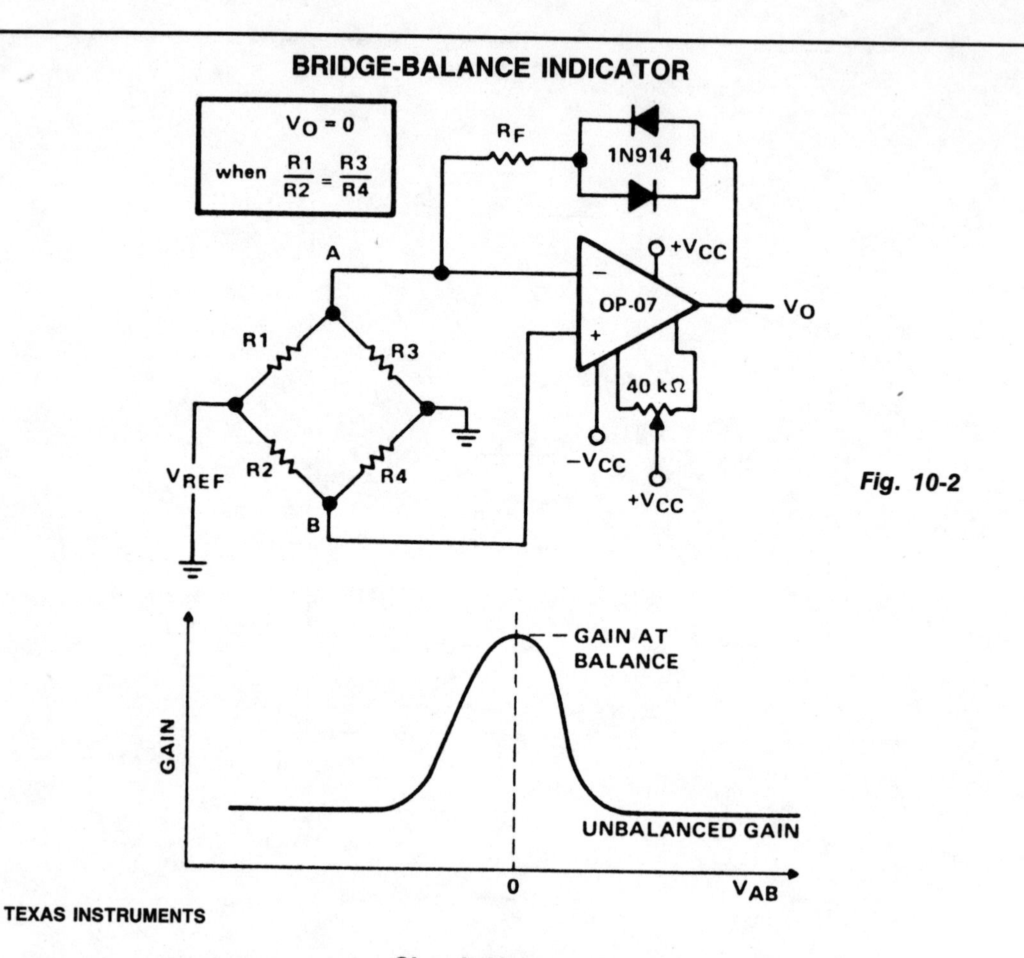

Fig. 10-2

TEXAS INSTRUMENTS

Circuit Notes

Indicator provides an accurate comparison of two voltages by indicating their degree of balance (or imbalance). Detecting small variations near the null point is difficult with the basic Wheatstone bridge alone. Amplification of voltage differences near the null point will improve circuit accuracy and ease of use.

The 1N914 diodes in the feedback loop result in high sensitivity near the point of balance ($R1/R2 = R3/R4$). When the bridge is unbalanced the amplifier's closed-loop gain is approximately R_F/r, where r is the parallel equivalent of R1 and R3. The resulting gain equation is $G = R_F(1/R1 + 1/R3)$. During an unbalanced condition the voltage at point A is different from that at point B. This difference voltage (V_{AB}), amplified by the gain factor G, appears as an output voltage, As the bridge approaches a balanced condition ($R1/R2 = R3/R4$), V_{AB} approaches zero. As V_{AB} approaches zero the 1N914 diodes in the feedback loop lose their forward bias and their resistance increases, causing the total feedback resistance to increase. This increases circuit gain and accuracy in detecting a balanced condition. The figure shows the effect of approaching balance on circuit gain. The visual indicator used at the output of the OP-07 could be a sensitive voltmeter or oscilloscope.

BRIDGE CIRCUIT

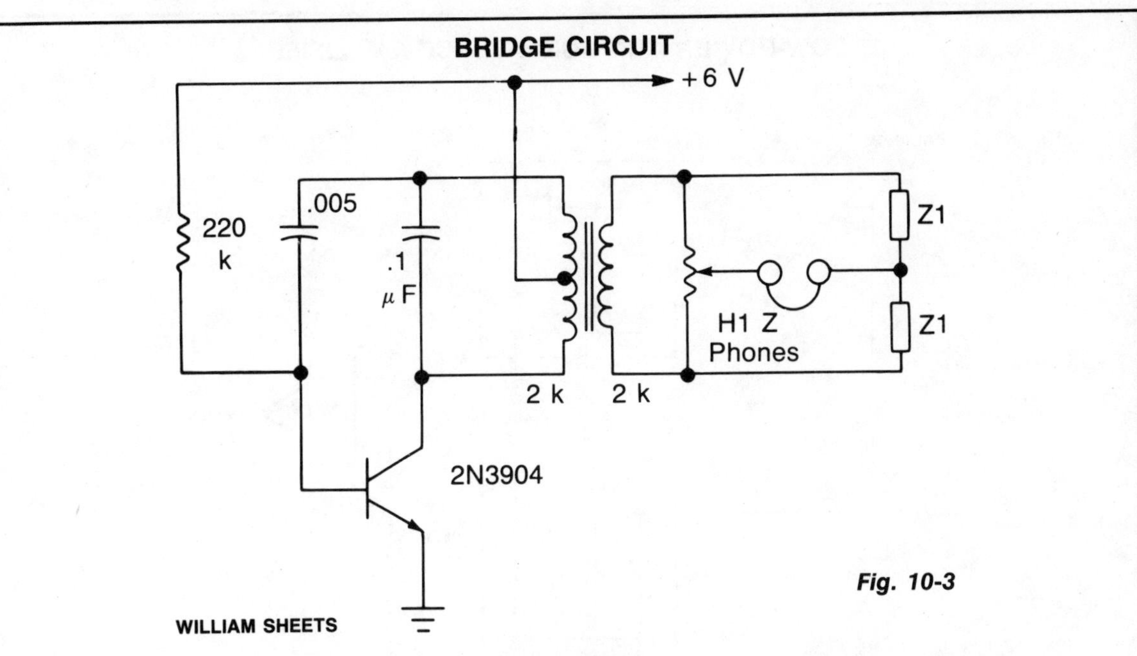

Fig. 10-3

Circuit Notes

The transistor is connected as an audio oscillator, using an audio transformer in the collector. The secondary goes to a linear pot. The ratio between the two parts of the pot from the slider is proportional to the values of Z1 and Z2 when no signal is heard in the phones.

TYPICAL TWO OP AMP BRIDGE-TYPE DIFFERENTIAL AMPLIFIER

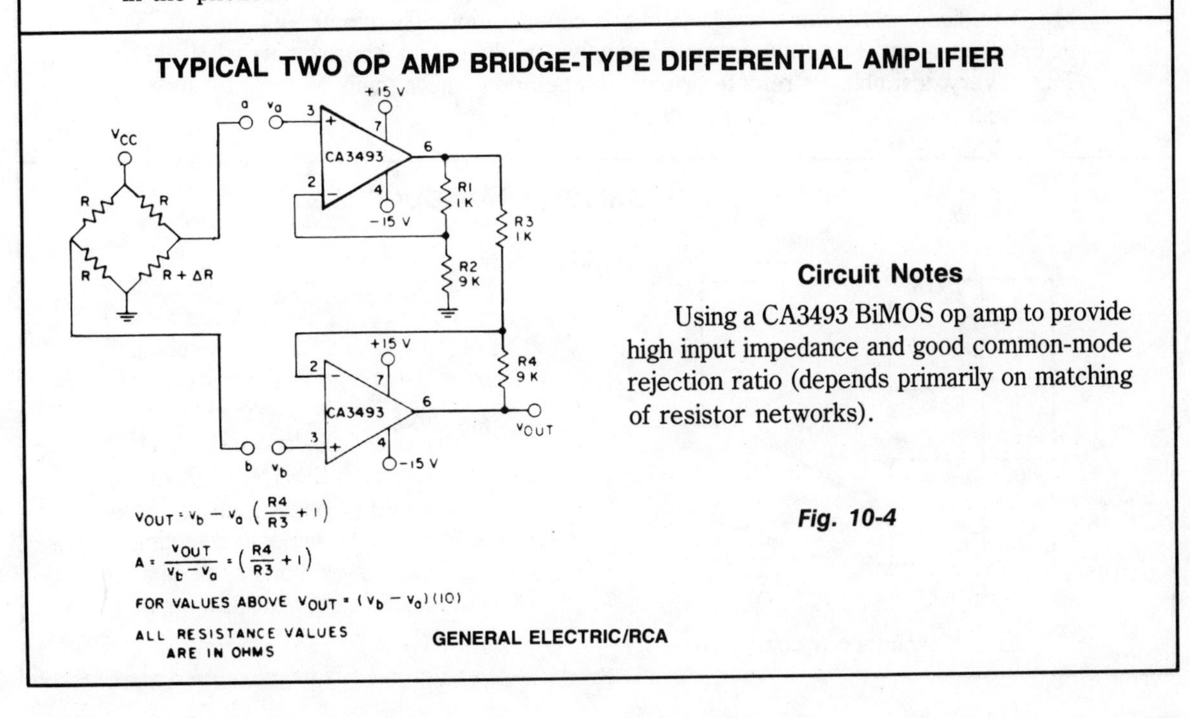

Circuit Notes

Using a CA3493 BiMOS op amp to provide high input impedance and good common-mode rejection ratio (depends primarily on matching of resistor networks).

Fig. 10-4

LOW-POWER COMMON SOURCE AMPLIFIER

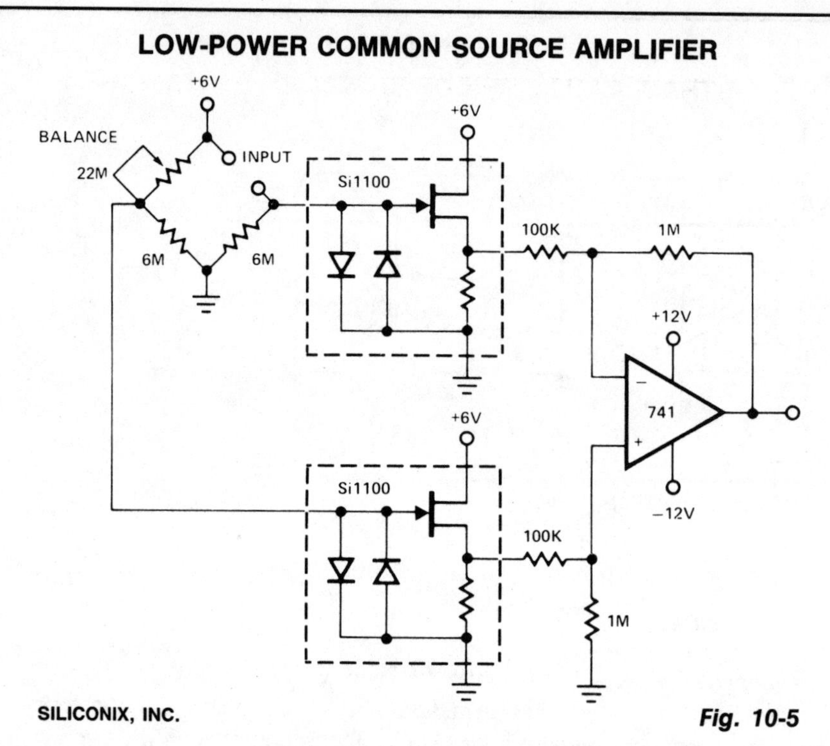

SILICONIX, INC.

Fig. 10-5

Circuit Notes

A circuit that will operate in the 10- to 20- microamp range at a 12-volt supply voltage. The diode protection is available in this configuration. The circuit voltage gain will be between 10 and 20, with extremely low power consumption (approximately 250 μW). This is very desirable for remote or battery operation where minimum maintenance is important.

AMPLIFIER FOR BRIDGE TRANSDUCERS

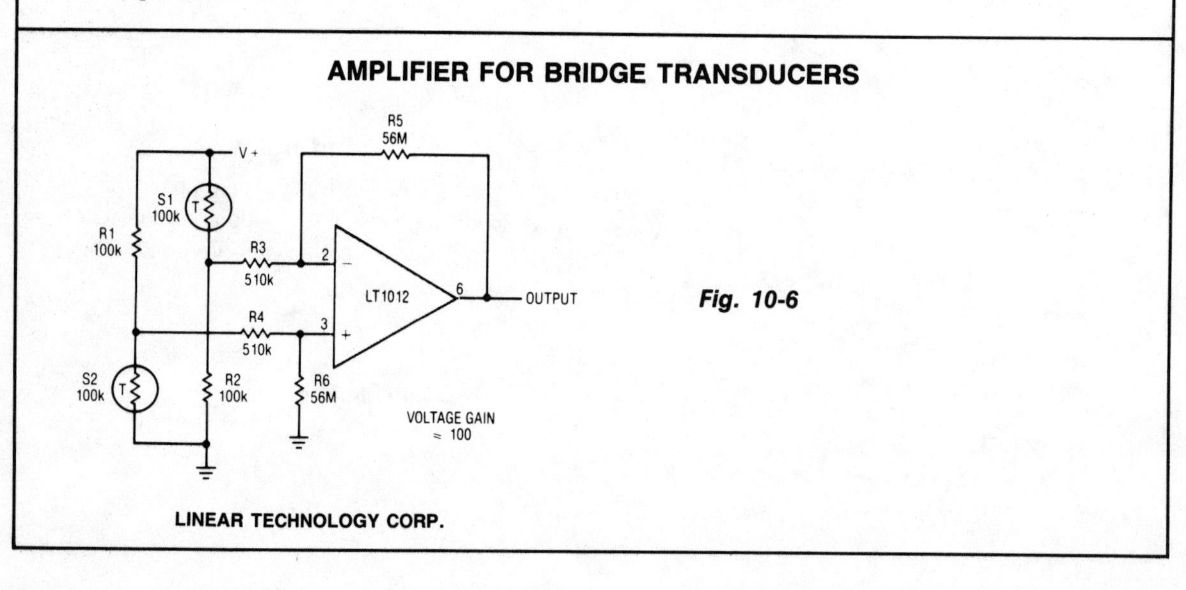

Fig. 10-6

LINEAR TECHNOLOGY CORP.

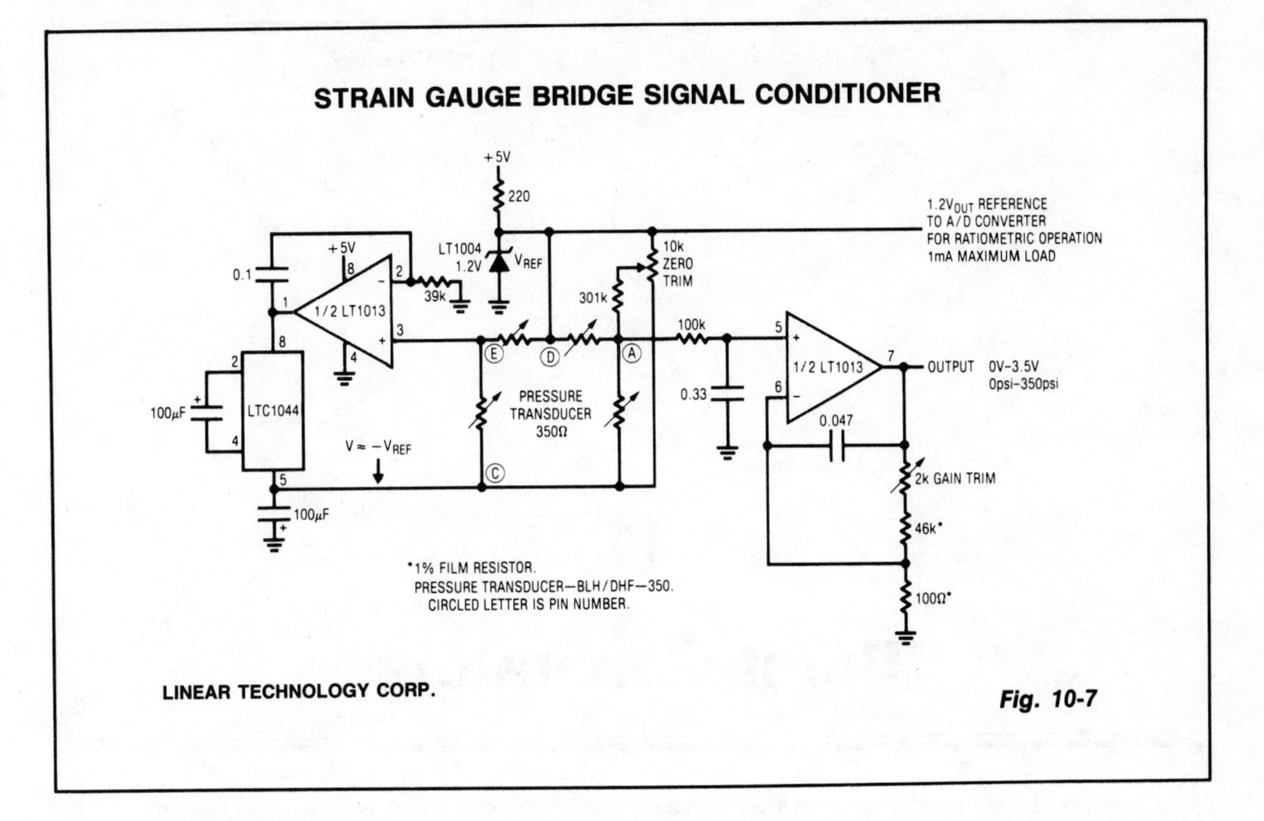
STRAIN GAUGE BRIDGE SIGNAL CONDITIONER
+5V
220
1.2VOUT REFERENCE
TO A/D CONVERTER
FOR RATIOMETRIC OPERATION
1mA MAXIMUM LOAD
LT1004
1.2V
VREF
10k
ZERO
TRIM
0.1
+5V
1/2 LT1013
39k
301k
100k
LTC1044
100µF
PRESSURE
TRANSDUCER
350Ω
0.33
1/2 LT1013
OUTPUT 0V–3.5V
0psi–350psi
0.047
V ≈ −VREF
2k GAIN TRIM
46k*
100µF
*1% FILM RESISTOR.
PRESSURE TRANSDUCER—BLH/DHF—350.
CIRCLED LETTER IS PIN NUMBER.
100Ω*
LINEAR TECHNOLOGY CORP.

Fig. 10-7

11

Burst Generators

The sources of the following circuits are contained in the Sources section beginning on page 694. The figure number contained in the box of each circuit correlates to the source entry in the Sources section.

SINGLE-TONE BURST GENERATOR

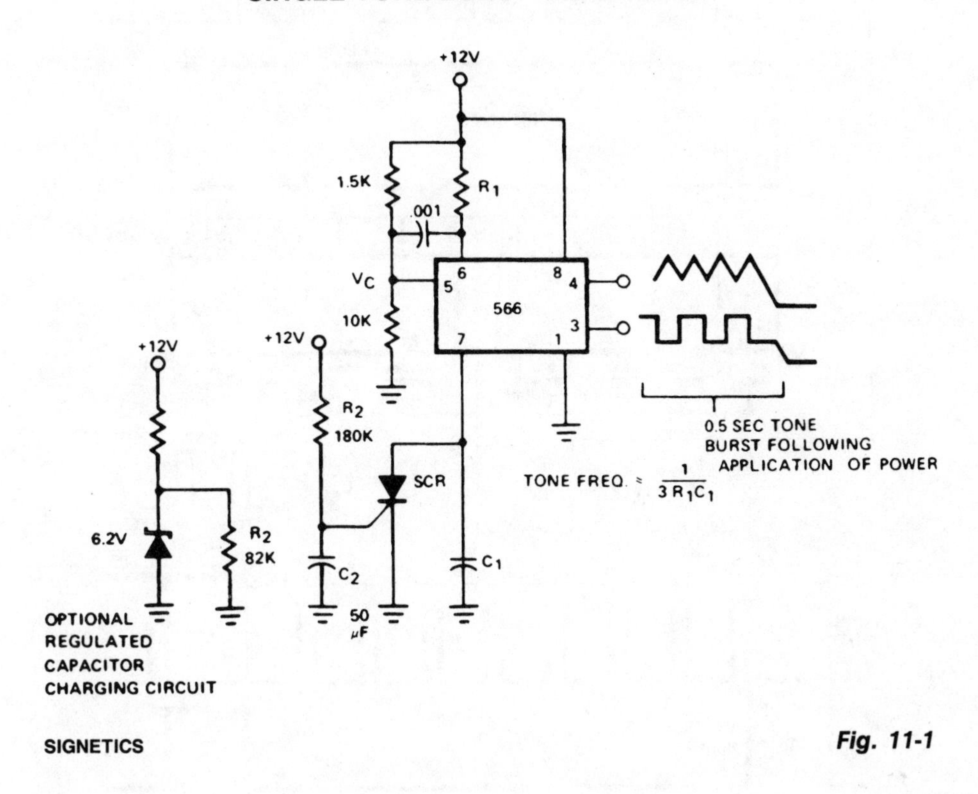

SIGNETICS

Fig. 11-1

Circuit Notes

The tone burst generator supplies a tone for one-half second after the power supply is activated; its intended use is a communications network alert signal. Cessation of the tone is accomplished at the SCR, which shunts the timing capacitor C1 charge current when activated. The SCR is gated on when C2 charges up to the gate voltage which occurs in 0.5 seconds. Since only 70 μA are available for triggering, the SCR must be sensitive enough to trigger at this level. The triggering current can be increased, of course, by reducing R2 (and increasing C2 to keep the same time constant). If the tone duration must be constant under widely varying supply voltage conditions, the optional Zener diode regulator circuit can be added, along with the new value for R_2 $R_2' = 82\ k\Omega$. If the SCR is replaced by an npn transistor, the tone can be switched on and off at will at the transistor base terminal.

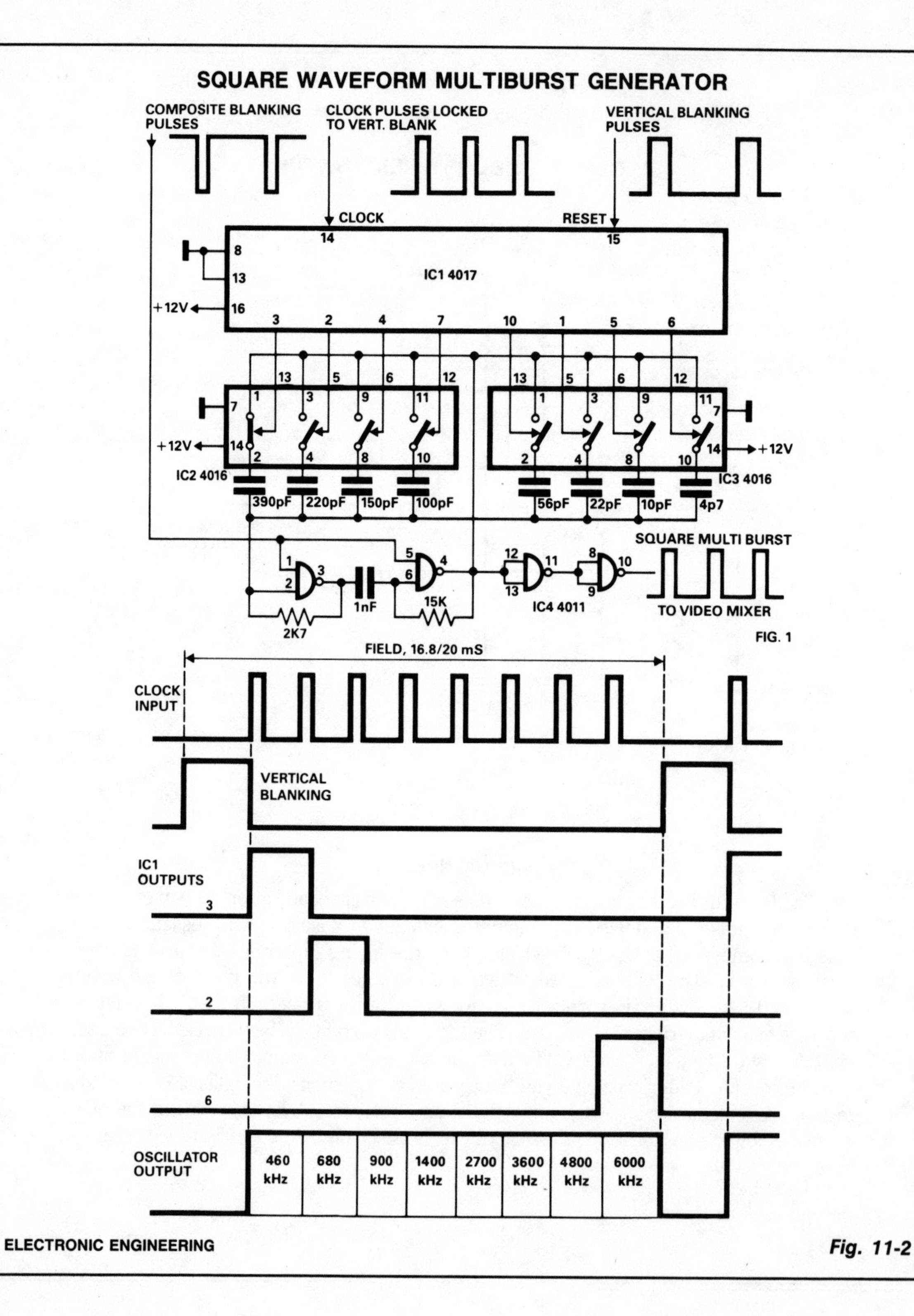
SQUARE WAVEFORM MULTIBURST GENERATOR
COMPOSITE BLANKING PULSES
CLOCK PULSES LOCKED TO VERT. BLANK
VERTICAL BLANKING PULSES
CLOCK
RESET
IC1 4017
+12V
IC2 4016
IC3 4016
390pF
220pF
150pF
100pF
56pF
22pF
10pF
4p7
1nF
15K
2K7
IC4 4011
SQUARE MULTI BURST
TO VIDEO MIXER
FIG. 1
FIELD, 16.8/20 mS
CLOCK INPUT
VERTICAL BLANKING
IC1 OUTPUTS
OSCILLATOR OUTPUT
460 kHz
680 kHz
900 kHz
1400 kHz
2700 kHz
3600 kHz
4800 kHz
6000 kHz
ELECTRONIC ENGINEERING

Fig. 11-2

SQUARE WAVEFORM MULTIBURST GENERATOR, Continued.

Circuit Notes

The generator described here is intended for multiburst signal square waveform generation and can be used as a device for characterizing the response of TV monitor amplifiers as shown. The circuit is an RC oscillator with NAND gates (IC4-4011), with its capacitor C changed periodically by means of bilateral switches (IC2, IC3-4016). The control inputs of bilateral switches are driven by the outputs of a counter/decoder (IC1-4017) the operation of which is determined by generated clock pulses, so that they occur eight times at half-picture (field). These pulses are locked to vertical blank pulses.

Horizontal synchronization is achieved by means of composite blanking pulses (negative polarization) applied to pins 1 and 5 of IC4. The oscillator frequency changes in the following discrete steps: 460 kHz, 680 kHz, 900 kHz, 1400 kHz, 2700 kHz, 3600 kHz, for the time of one frame. The video signal is fed on a mixer where it is superimposed with a composite sync signal.

SINGLE-TIMER IC PROVIDES SQUARE-WAVE TONE BURSTS

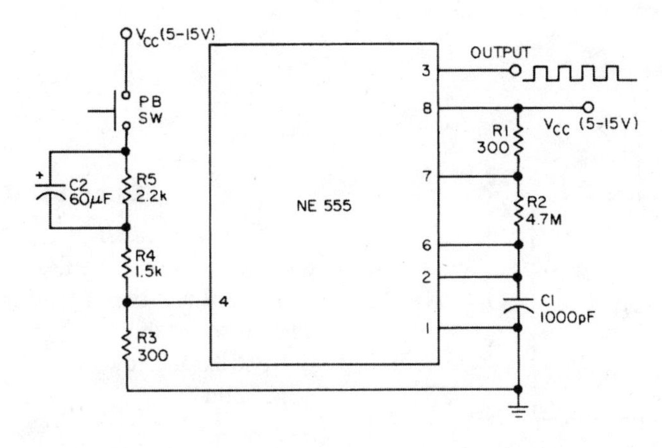

ELECTRONIC DESIGN

Fig. 11-3

Circuit Notes

The tone-burst generator gives a 50-ms burst of 1.5 kHz square waves with each operation of the pushbutton and can source or sink 200 mA.

STROBE-TONE BURST GENERATOR

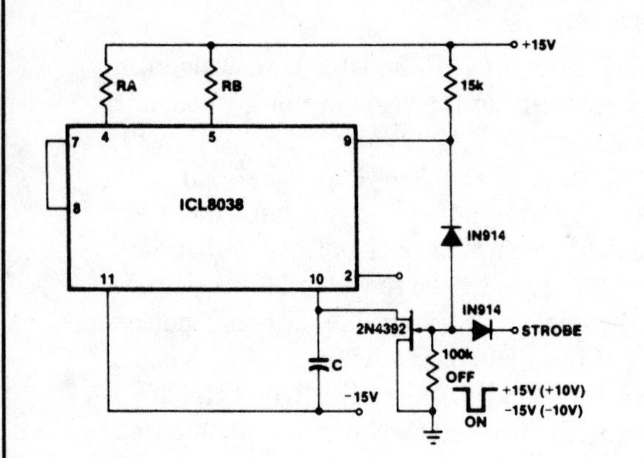

Fig. 11-4

INTERSIL

TONE BURST GENERATOR

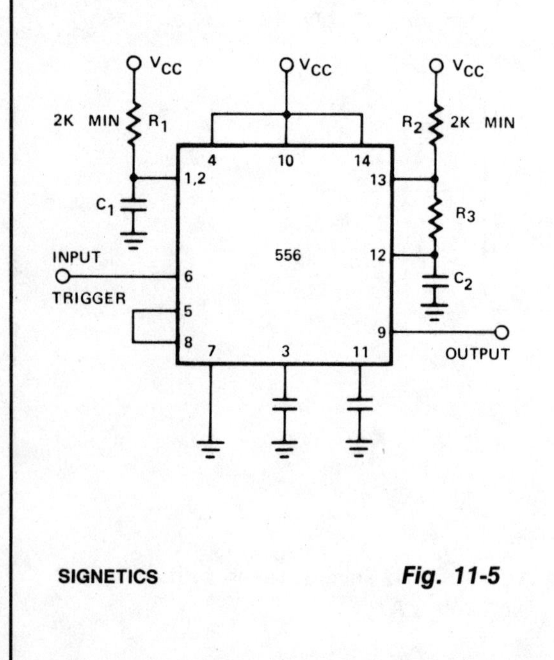

SIGNETICS

Fig. 11-5

Circuit Notes

The dual timer makes an excellent tone burst generator. The first half is connected as a one shot and the second half as an oscillator. The pulse established by the one shot turns on the oscillator allowing a burst of pulses to be generated.

12

Capacitance Meters

The sources of the following circuits are contained in the Sources section beginning on page 694. The figure number contained in the box of each circuit correlates to the source entry in the Sources section.

CAPACITANCE-TO-VOLTAGE METER

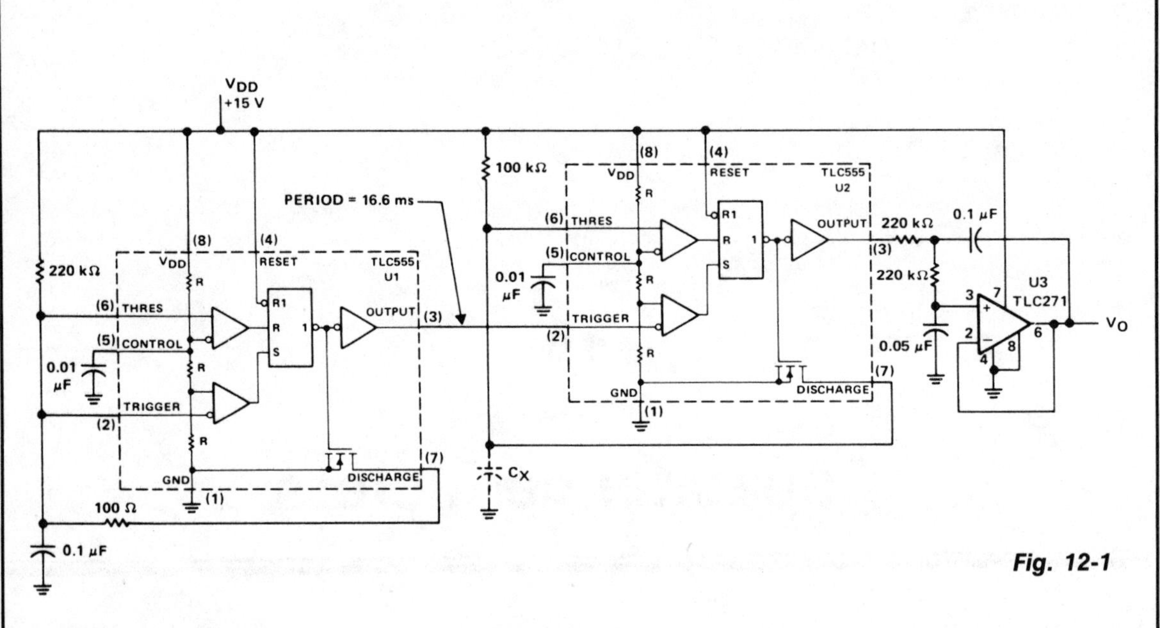

Fig. 12-1

TEXAS INSTRUMENTS

Circuit Notes

Timer U1 operates as a free-running oscillator at 60 Hz, providing trigger pulses to timer U2 which operates in the monostable mode. Resistor R1 is fixed and capacitor Cx is the capacitor being measured. While the output of U2 is 60 Hz, the duty cycle depends on the value of Cx. U3 is a combination low-pass filter and unity-gain follower whose dc voltage output is the time-averaged amplitude of the output pulses of U2, as shown in the timing diagram.

The diagram shows when the value of Cx is small the duty cycle is relatively low. The output pulses are narrow and produce a lower average dc voltage level at the output of U3. As the capacitance value of Cx increases, the duty cycle increases making the output pulses at U2 wider and the average dc level output at U3 increases. The graph illustrates capacitance values of 0.01 μF to 0.1 μF plotted against the output voltage of U3. Notice the excellent linearity and direct one-to-one scale calibration of the meter. If this does not occur the 100 k ohm resistor, R1, can be replaced with a potentiometer which can be adjusted to the proper value for the meter being used.

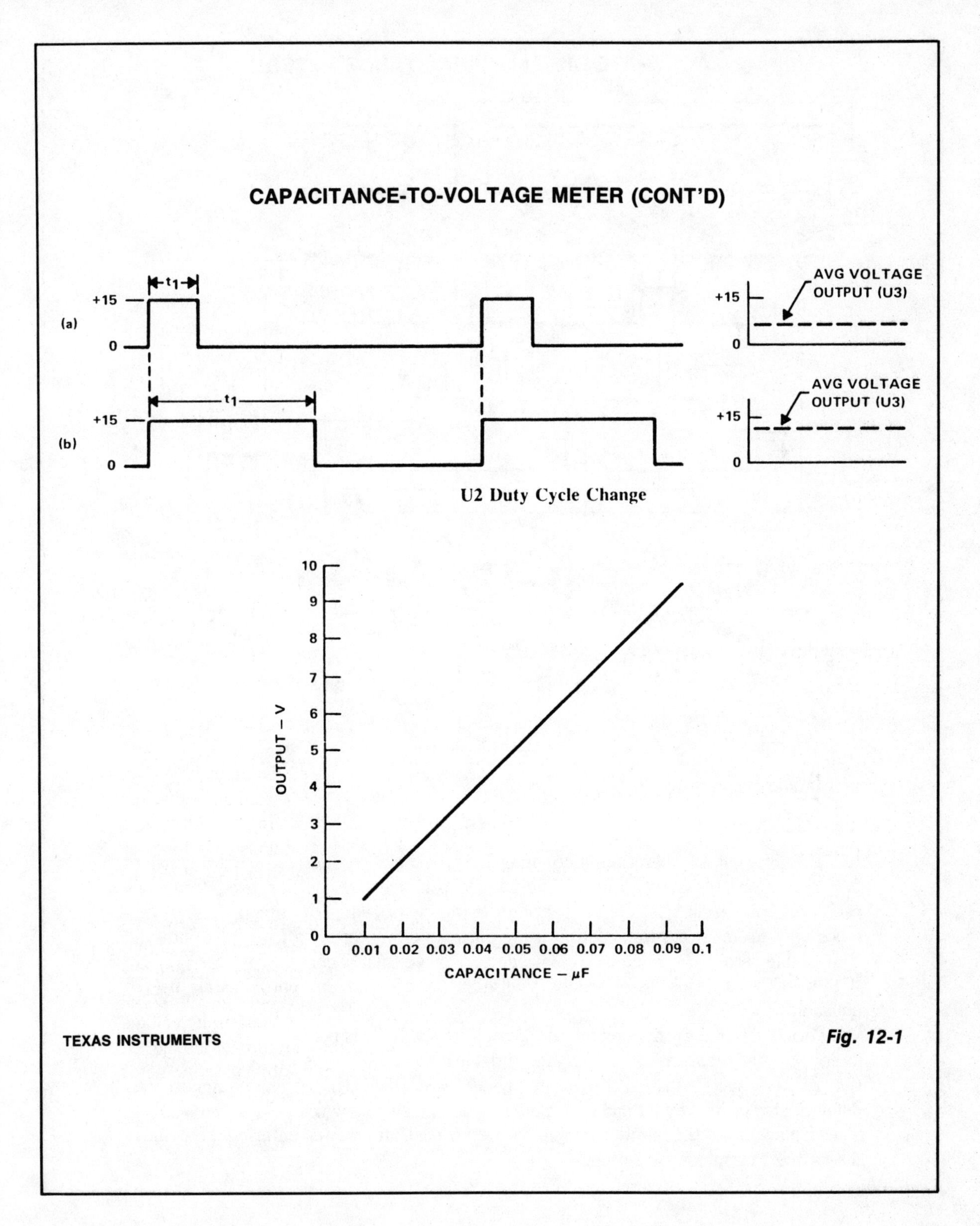
CAPACITANCE-TO-VOLTAGE METER (CONT'D)
t1
+15
(a)
0
AVG VOLTAGE
OUTPUT (U3)
+15
0
t1
+15
(b)
0
AVG VOLTAGE
OUTPUT (U3)
+15
0
U2 Duty Cycle Change
10
9
8
7
6
5
4
3
2
1
0
OUTPUT — V
0
0.01 0.02 0.03 0.04 0.05 0.06 0.07 0.08 0.09 0.1
CAPACITANCE — μF
TEXAS INSTRUMENTS

Fig. 12-1

ACCURATE DIGITAL CAPACITANCE METER

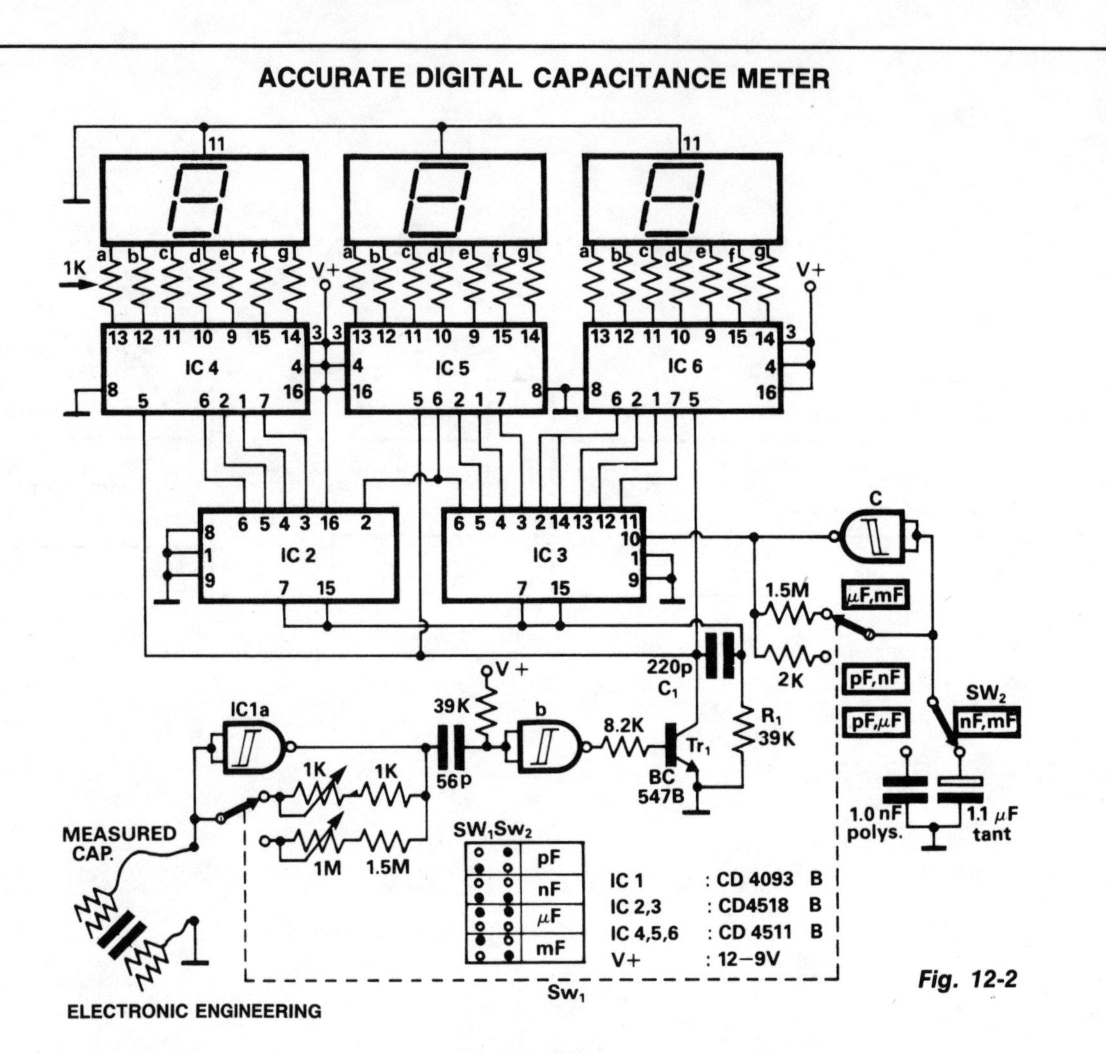

Fig. 12-2

Circuit Notes

The principle of operation is counting the pulse number derived from a constant frequency oscillator during a fixed time interval produced by another lower frequency oscillator. This oscillator uses the capacitor being measured as the timing. The capacitance measurement is proportional during pulse counting during a fixed time interval. The astable oscillator formed by IC1c produces a pulse train of constant frequency. Gate IC1a also forms an oscillator whose oscillation period is given approximately by the equation: $T = 0.7\ RC$.

Period T is linearly dependent on the capacitance C. This period is used as the time interval for one measurement. The differentiator network following the oscillator creates the negative spikes shaped in narrow pulses by IC1b NAND Schmitt Trigger. The differentiator formed by R1 and C1 produces a negative spike which resets the counters. The display shows the number of high frequency oscillator pulses entering the counter during the measurement period.

13

Circuit Protection Circuits

The sources of the following circuits are contained in the Sources section beginning on page 694. The figure number contained in the box of each circuit correlates to the source entry in the Sources section.

Overvoltage Protector
High Speed Electronic Circuit Breaker
12 ns Circuit Breaker
Low Voltage Power Disconnector
Automatic Power-Down Protection Circuit
Line Dropout Detector
Electronic Crowbar

OVERVOLTAGE PROTECTOR

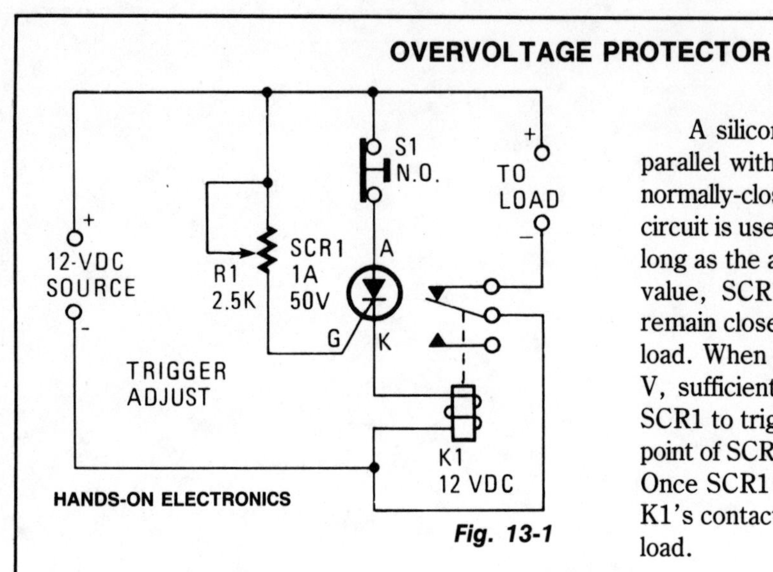

Fig. 13-1

Circuit Notes

A silicon-controlled rectifier is installed in parallel with the 12-V line and connected to a normally-closed 12-V relay, K1. The SCR's gate circuit is used to sample the applied voltage. As long as the applied voltage stays below a given value, SCR1 remains off and K1's contacts remain closed, thereby supplying power to the load. When the source voltage rises above 12 V, sufficient current is applied to the gate of SCR1 to trigger it into conduction. The trigger point of SCR1 is dependent on the setting of R1. Once SCR1 is triggered (activating the relay), K1's contacts open, halting current flow to the load.

HIGH SPEED ELECTRONIC CIRCUIT BREAKER

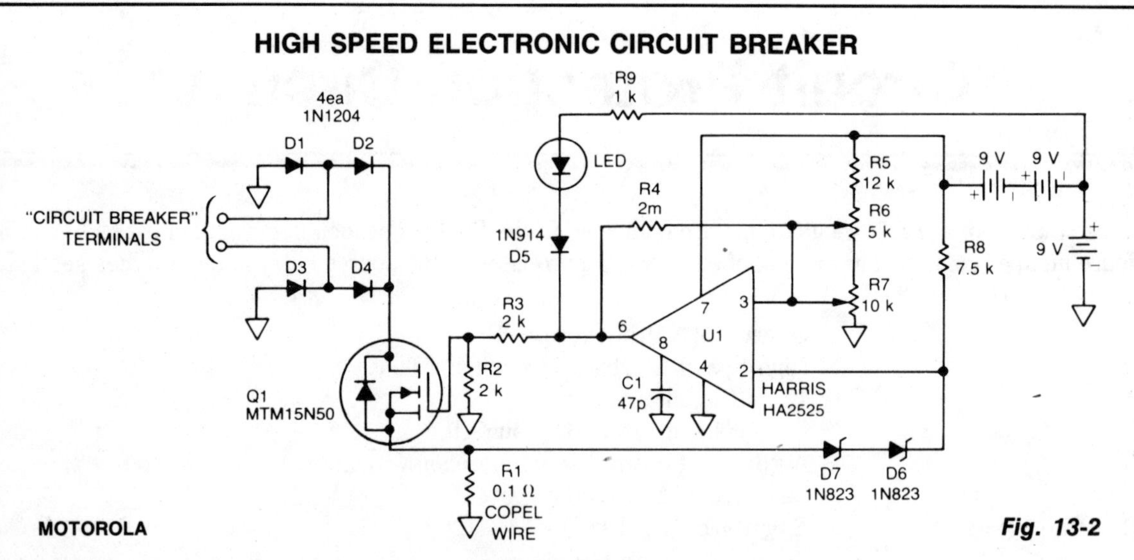

Fig. 13-2

Circuit Notes

This 115 Vac, electronic circuit breaker uses the low drive power, low on resistance and fast turn off of the TMOS MTM15N50. The trip point is adjustable, LED fault indication is provided and battery power provides complete circuit isolation.

The two "circuit breaker" terminals are across one leg of a full wave diode bridge consisting of D1–D4. Normally, Q1 is turned ON so that the circuit breaker looks like a very low resistance. One input to comparator U1 is a fraction of the internal battery voltage and the other input is the drop across zeners D6 and D7 and the voltage drop across R1. If excessive current is drawn, the voltage drop across R1 increases beyond the comparator threshold (determined by the setting of R6), U1 output goes low, Q1 turns OFF, and the circuit breaker "opens." When this occurs, the LED fault indicator is illuminated.

12 ns CIRCUIT BREAKER

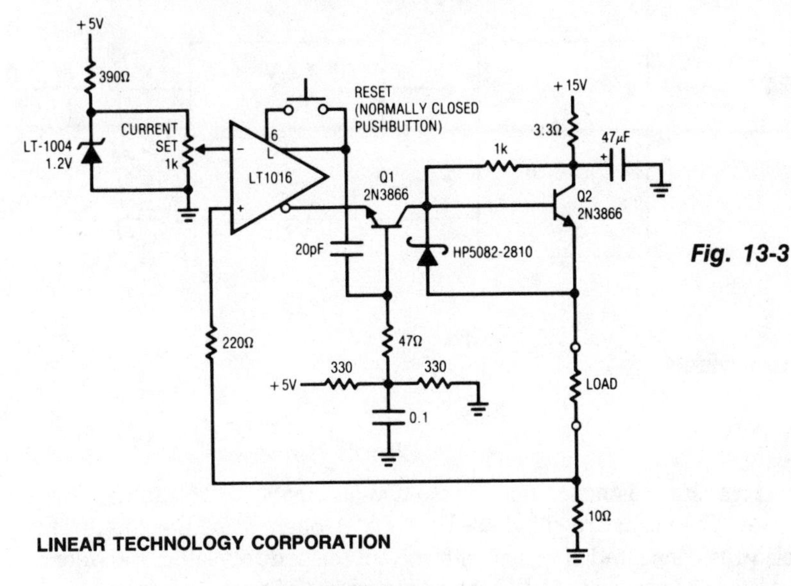

Fig. 13-3

LINEAR TECHNOLOGY CORPORATION

Circuit Notes

This circuit will turn off current in a load 12 ns after it exceeds a preset value. Under normal conditions the voltage across the 10 ohm shunt is smaller than the potential at the LT1016's negative input. This keeps Q1 off and Q2 receives bias, driving the load. When an overload occurs the current through the 10 ohm sense resistor begins to increase. When this current exceeds the preset value, the LT1016's outputs reverse. This provides ideal turn-on drive for Q1 and it cuts off Q2 in 5 ns. The delay from the onset of excessive load current to complete shutdown is just 13 ns. Once the circuit has triggered, the LT1016 is held in its latched state by feedback from the non-inverting output. When the load fault has been cleared the pushbutton can be used to reset the circuit.

LOW VOLTAGE POWER DISCONNECTOR

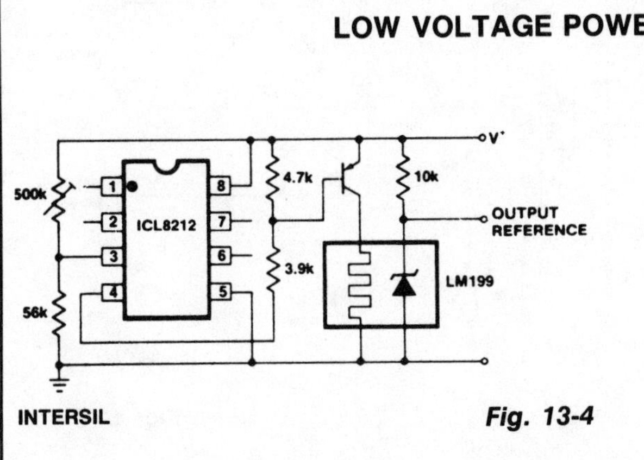

INTERSIL

Fig. 13-4

Circuit Notes

There are some classes of circuits that require the power supply to be disconnected if the power supply voltage falls below a certain value. As an example, the National LM199 precision reference has an on chip heater which malfunctions with supply voltages below 9 volts causing an excessive device temperature. The ICL8212 can be used to detect a power supply voltage of 9 volts and turn the power supply off to the LM199 heater section below that voltage.

AUTOMATIC POWER-DOWN PROTECTION CIRCUIT

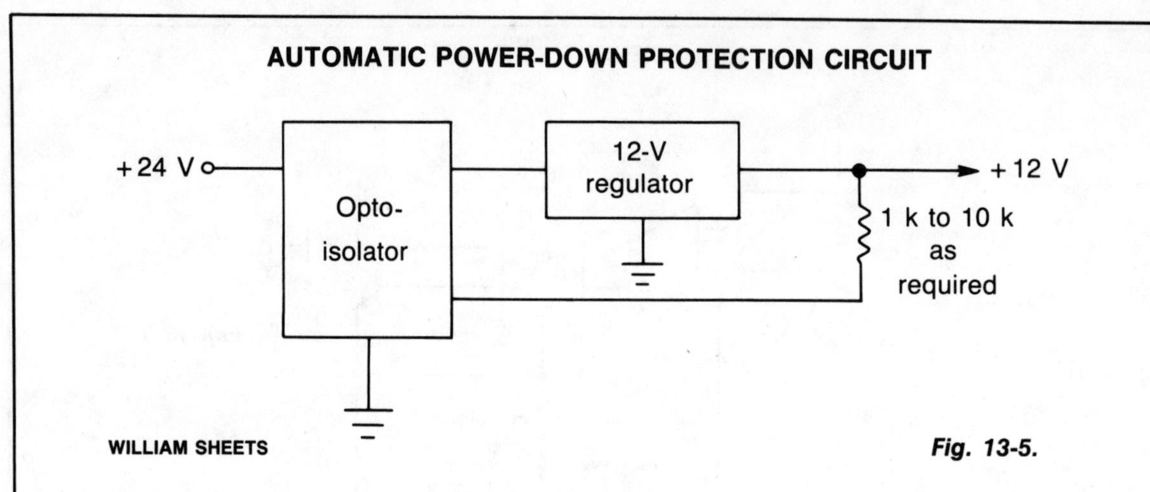

WILLIAM SHEETS

Fig. 13-5.

Circuit Notes

This circuit is faster than a fuse and automatically resets itself when a short is removed. The normal regulated dc input line is opened and the phototransistor of the opto isolator is connected in series with the source and regulator. Between the output of the regulator and ground is a LED and an associated current-limiting resistor, placed physically close to the surface of the photosensitive device. As long as the regulator is delivering its rated output, the LED glows and causes the photo device to have a low resistance. Full current is thus allowed to flow. If a short circuit occurs on the output side of the regulator, the LED goes dark, the resistance of the photo device increases, and the regulator shuts off. When the short is removed, the LED glows, and the regulator resumes operation.

LINE DROPOUT DETECTOR

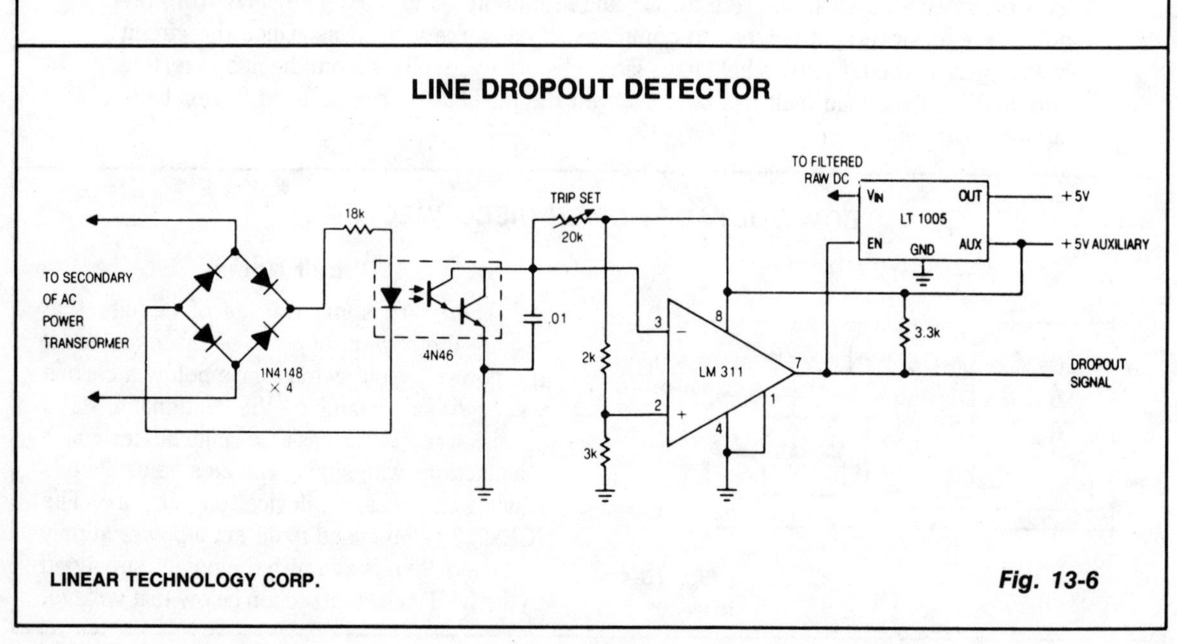

LINEAR TECHNOLOGY CORP.

Fig. 13-6

ELECTRONIC CROWBAR

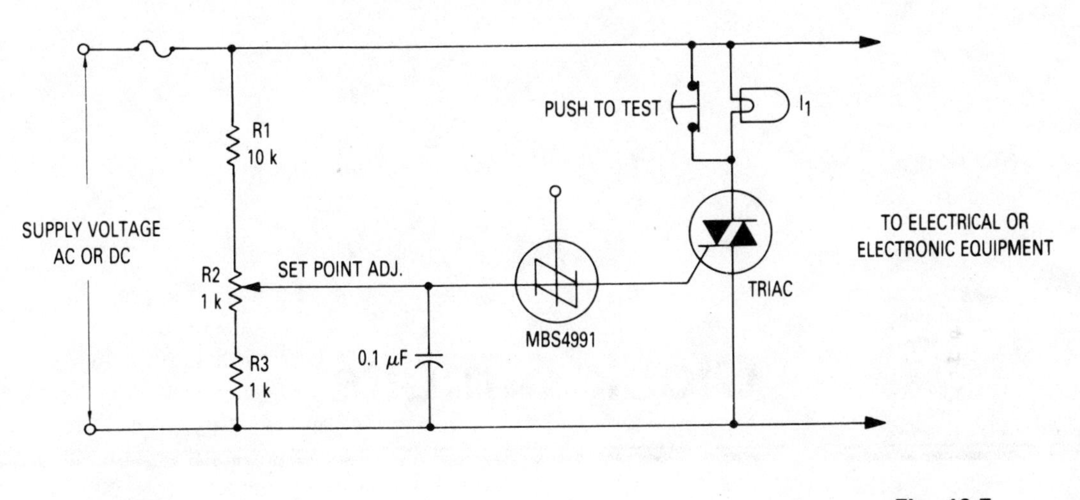

MOTOROLA

Fig. 13-7

Circuit Notes

Where it is desirable to shut down equipment rather than allow it to operate on excessive supply voltage, an electronic "crowbar" circuit can be employed to quickly place a short-circuit across the power lines, thereby dropping the voltage across the protected device to near zero and blowing a fuse. Since the TRIAC and SBS are both bilateral devices, the circuit is equally useful on ac or dc supply lines. With the values shown for R1, R2 and R3, the crowbar operating point can be adjusted over the range of 60 to 120 volts dc or 42 to 84 volts ac. The resistor values can be changed to cover a different range of supply voltages. The voltage rating of the TRIAC must be greater than the highest operating point as set by R2. I_1 is a low power incandescent lamp with a voltage rating equal to the supply voltage. It may be used to check the set point and operation of the unit by opening the test switch and adjusting the input or set point to fire the SBS.

14

Clock Circuits

The sources of the following circuits are contained in the Sources section beginning on page 694. The figure number contained in the box of each circuit correlates to the source entry in the Sources section.

THREE PHASE CLOCK FROM A REFERENCE CLOCK

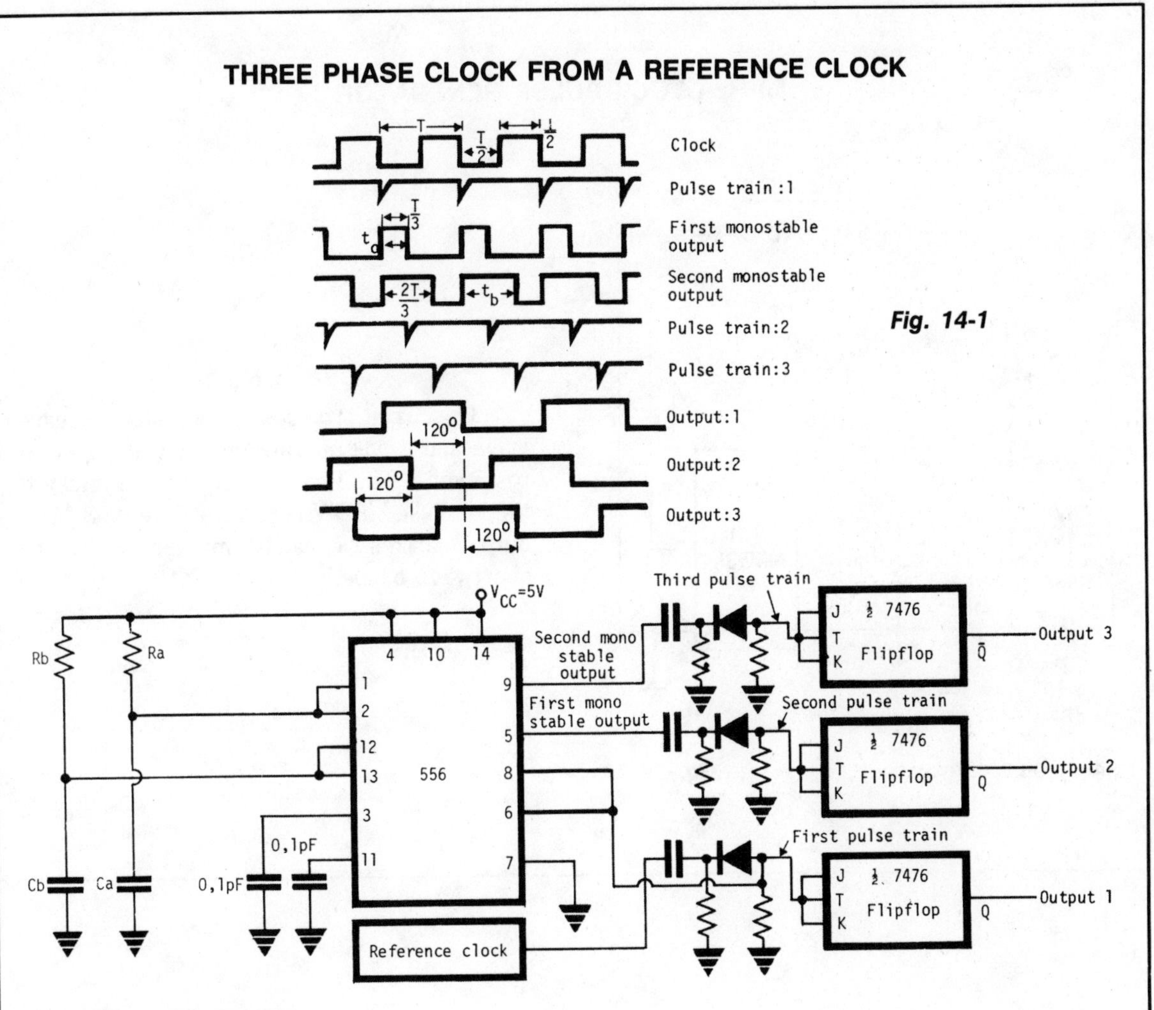

Fig. 14-1

ELECTRONIC ENGINEERING

Circuit Notes

The circuit provides three square wave outputs with 120° of phase difference between each other. Reference clock frequency is twice that of the required frequency. This can be obtained from a crystal oscillator with a chain of dividers or by using LM 555 in 50% duty cycle astable mode. If 1/T is the frequency of the reference clock, the dual timer 556 is connected to give two mono-stable output pulses of duration T/3 and 2T/3. The first timer R and C value are adjusted so that $t_a = 1.1RaCa = T/3$ and the second timer R and C values so that $t_b = 1.1RbCb = 2T/3$. For triggering the two monostables a negative pulse train (1st) is derived from the reference clock with a differentiator and a clipper combination as shown. The three pulse trains trigger three JK flip flops giving three phase square wave outputs.

60 Hz CLOCK PULSE GENERATOR

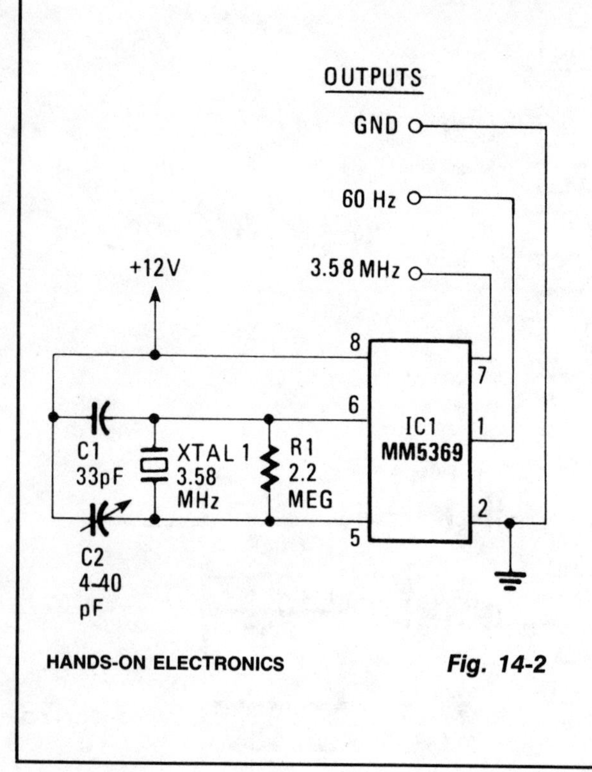

HANDS-ON ELECTRONICS

Fig. 14-2

Circuit Notes

The circuit provides a clean, stable square wave and it will operate on anywhere from 6 to 15 volts. The IC and color-burst crystal are the kind used in TV receivers. The 3.58 MHz output makes a handy marker signal for shortwave bands.

15

Comparators

The sources of the following circuits are contained in the Sources section beginning on page 694. The figure number contained in the box of each circuit correlates to the source entry in the Sources section.

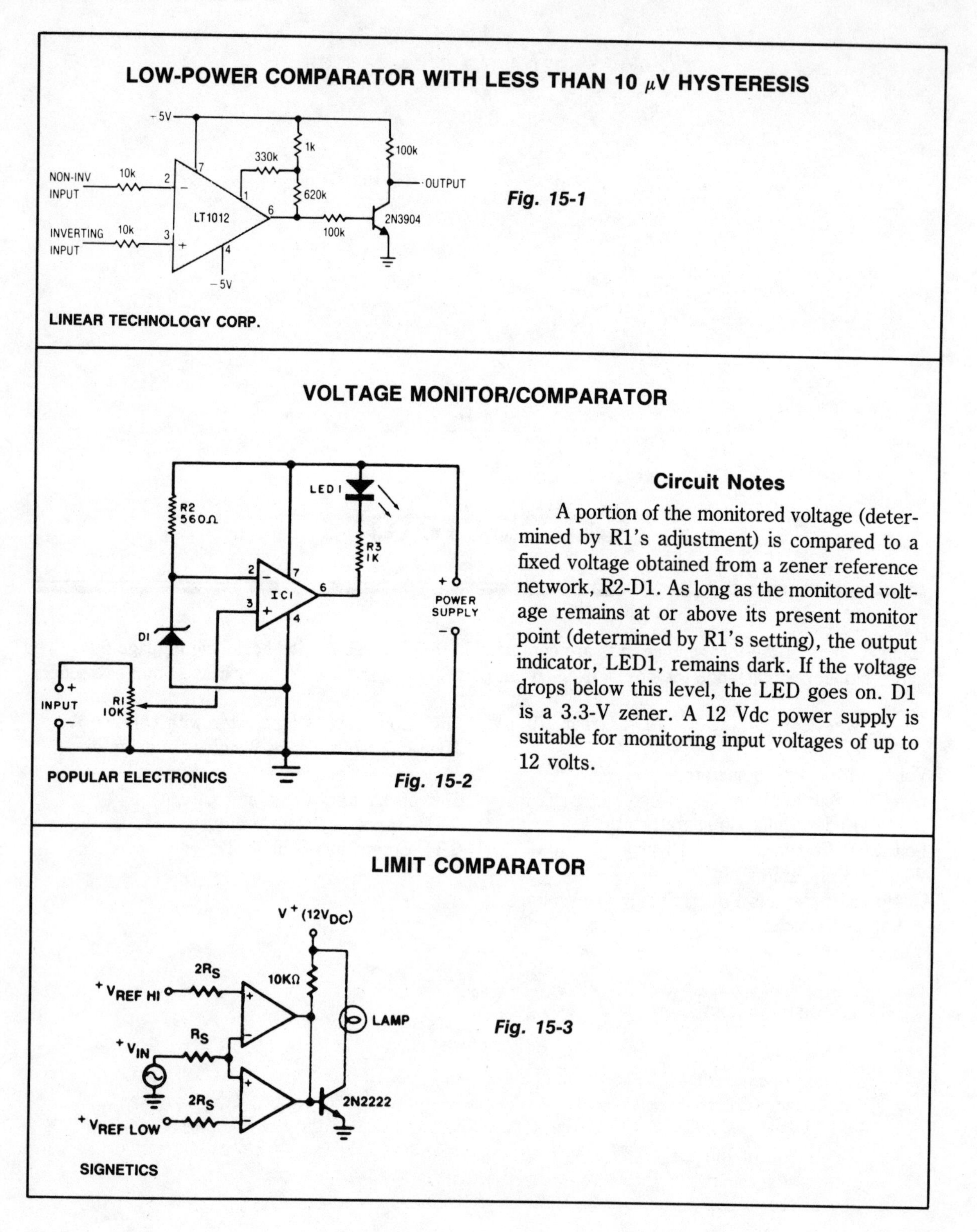

LOW-POWER COMPARATOR WITH LESS THAN 10 μV HYSTERESIS

LINEAR TECHNOLOGY CORP.

Fig. 15-1

VOLTAGE MONITOR/COMPARATOR

POPULAR ELECTRONICS

Fig. 15-2

Circuit Notes

A portion of the monitored voltage (determined by R1's adjustment) is compared to a fixed voltage obtained from a zener reference network, R2-D1. As long as the monitored voltage remains at or above its present monitor point (determined by R1's setting), the output indicator, LED1, remains dark. If the voltage drops below this level, the LED goes on. D1 is a 3.3-V zener. A 12 Vdc power supply is suitable for monitoring input voltages of up to 12 volts.

LIMIT COMPARATOR

SIGNETICS

Fig. 15-3

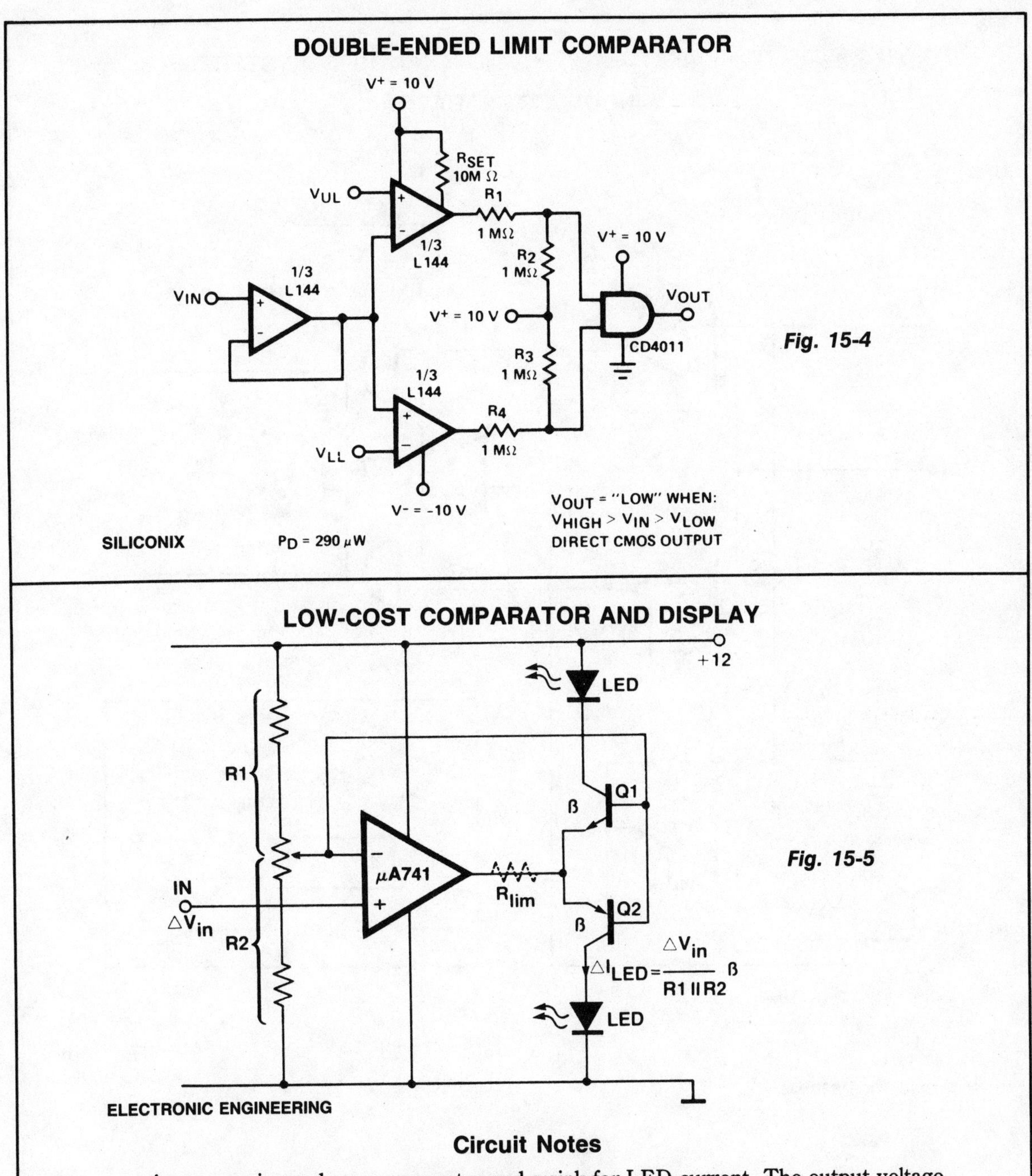

Fig. 15-4

Fig. 15-5

Circuit Notes

An op amp is used as a comparator and a sink for LED current. The output voltage of the amplifier changes about 1.4 V depending on the direction of the current. Only one transistor is on at any time. Maximum LED current is limited to 25 mA by overcurrent protection of the uA741. If LEDs are not capable of carrying such a current or an alternative op amp is used and an additional resistor R_{lim} is necessary.

WINDOW COMPARATOR

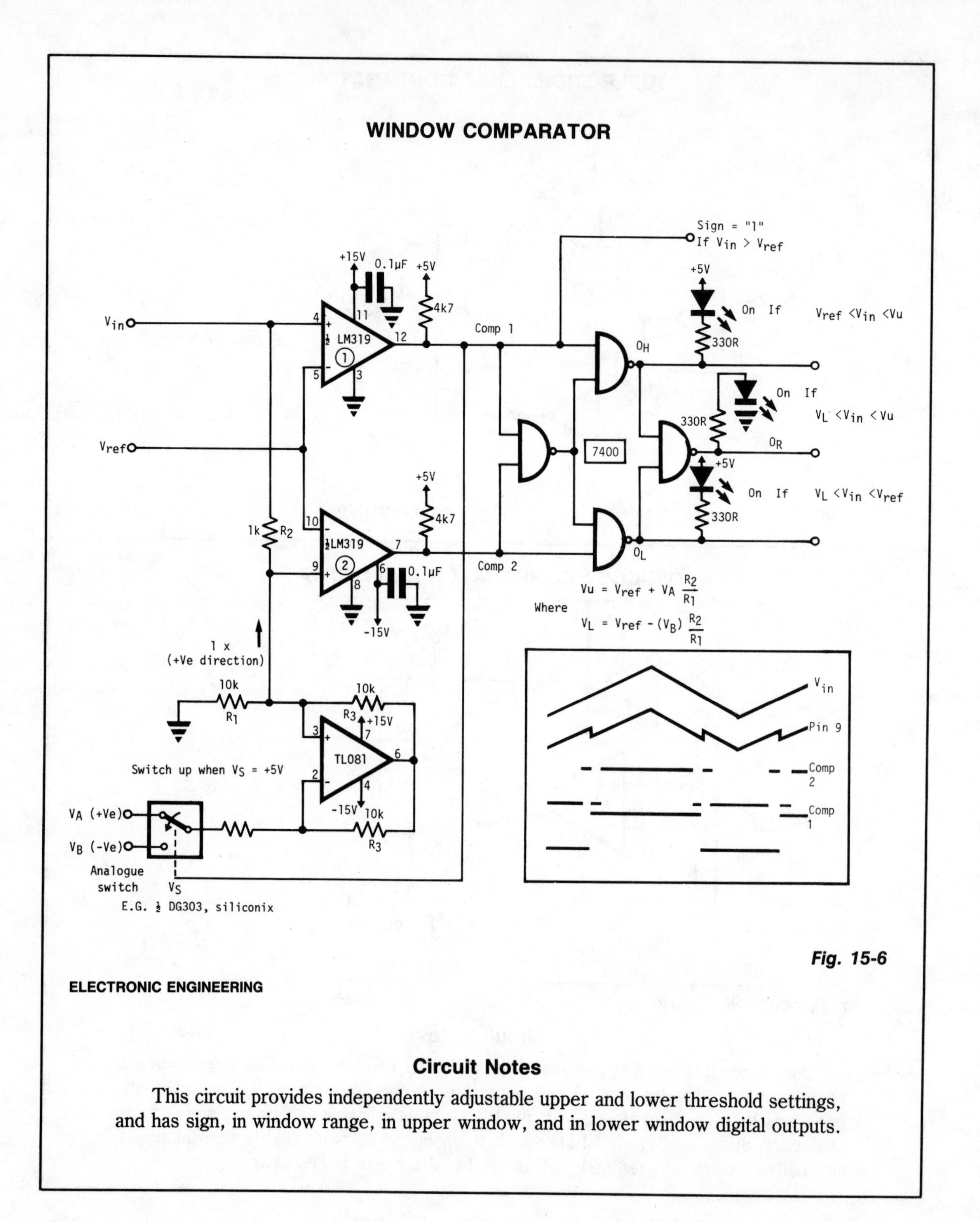

Fig. 15-6

ELECTRONIC ENGINEERING

Circuit Notes

This circuit provides independently adjustable upper and lower threshold settings, and has sign, in window range, in upper window, and in lower window digital outputs.

COMPARATOR DETECTS POWER SUPPLY OVERVOLTAGES, CATCHES GLITCHES

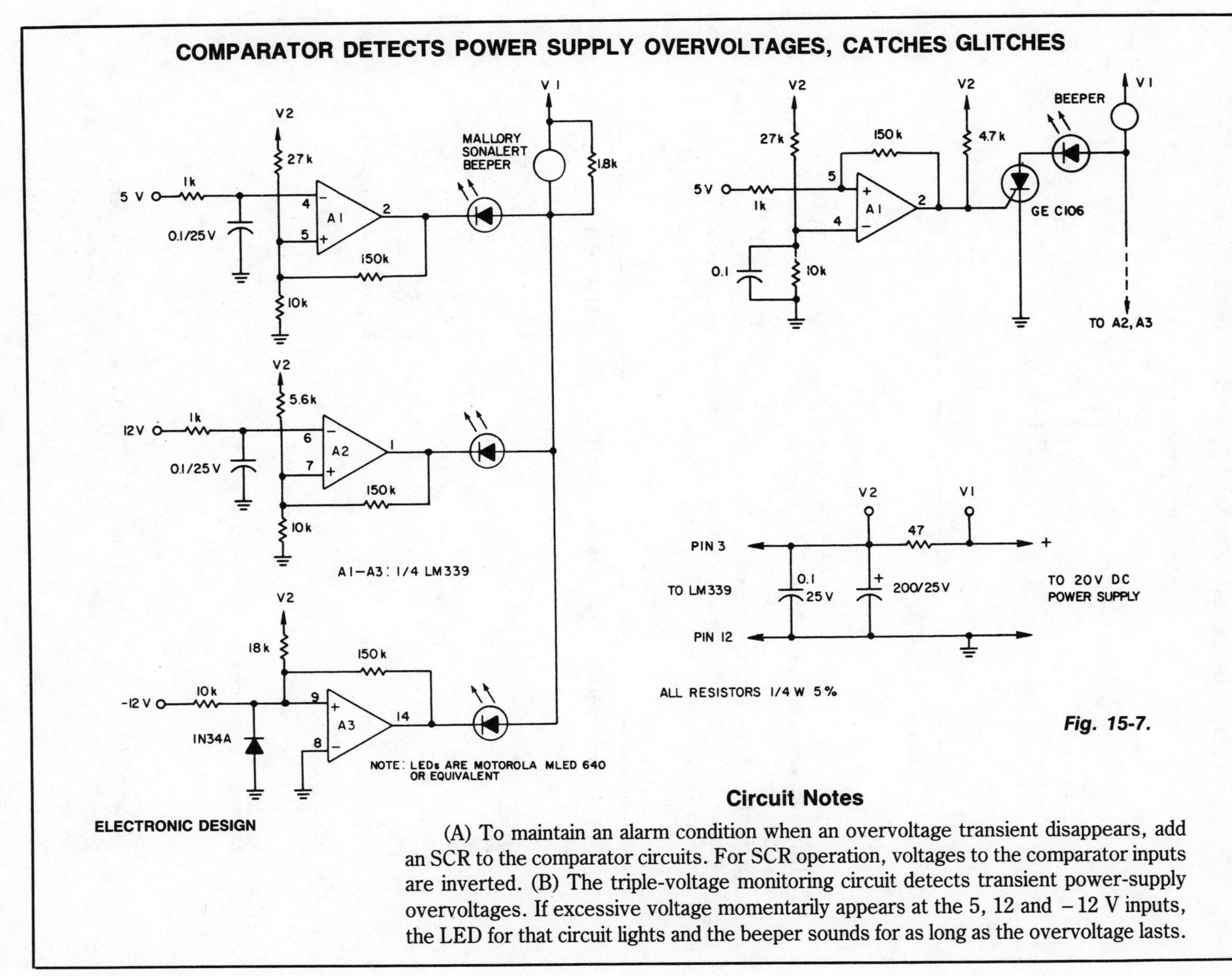

Fig. 15-7.

ELECTRONIC DESIGN

Circuit Notes

(A) To maintain an alarm condition when an overvoltage transient disappears, add an SCR to the comparator circuits. For SCR operation, voltages to the comparator inputs are inverted. (B) The triple-voltage monitoring circuit detects transient power-supply overvoltages. If excessive voltage momentarily appears at the 5, 12 and −12 V inputs, the LED for that circuit lights and the beeper sounds for as long as the overvoltage lasts.

HIGH-LOW LEVEL COMPARATOR WITH ONE OP AMP

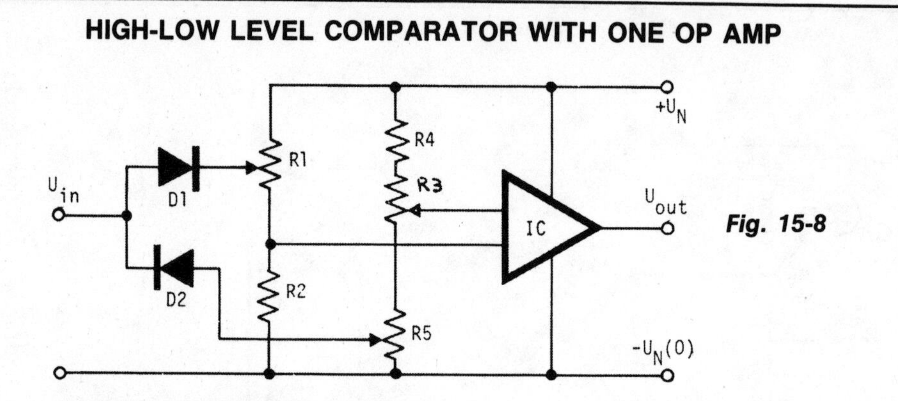

Fig. 15-8

ELECTRONIC ENGINEERING

Circuit Notes

The voltage to be compared is fed through diode D1 and D2 to the voltage dividers R1 and R5 where the low and high limits are present. When the voltage level of an input signal exceeds the high threshold limit set with potentiometer R1, the diode D1 becomes forward biased and the increased voltage on the inputs of the op amp drives it into positive saturation. Similarly, a decrease of the input voltage at the op amp inputs turns the op amp to positive saturation. Potentiometer R3 is used for zeroing the op amp in the off state.

HIGH-INPUT-IMPEDANCE WINDOW COMPARATOR

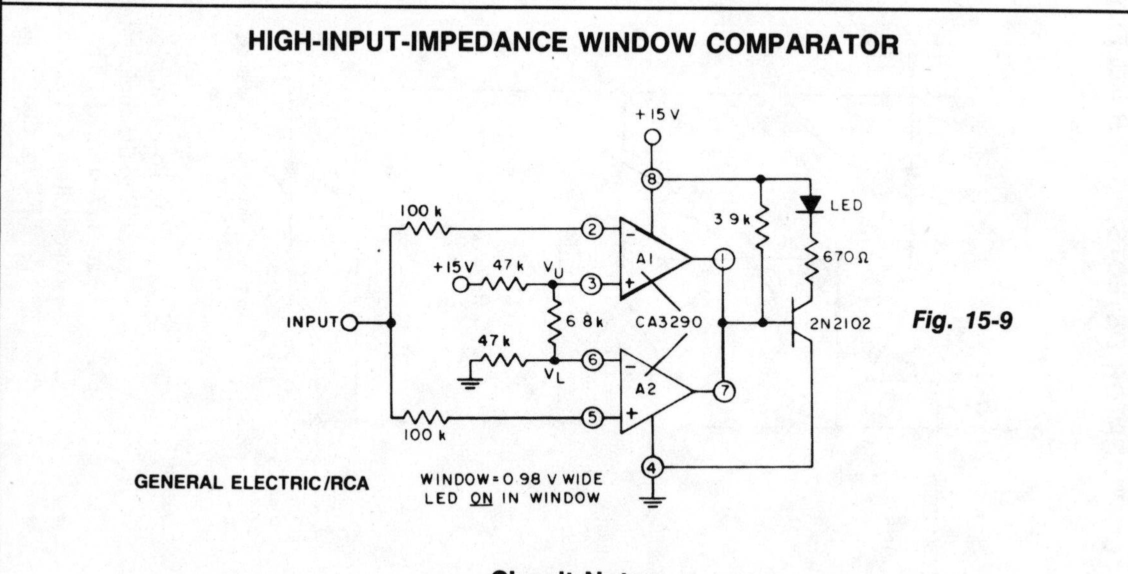

Fig. 15-9

GENERAL ELECTRIC/RCA

Circuit Notes

The circuit uses both halves of the CA3290 BiMOS dual voltage comparator. The LED will be turned "ON" whenever the input signal is above the lower limit (V_L) but below the upper limit (V_U).

FREQUENCY COMPARATOR

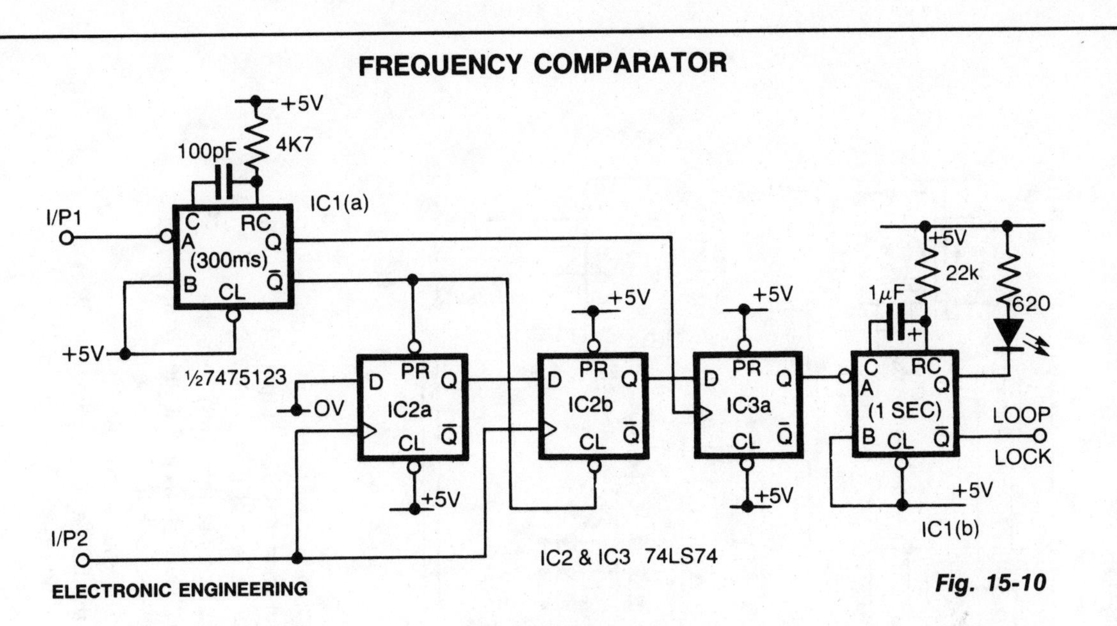

Fig. 15-10

Circuit Notes

Input 1 is used as a gating period, during which a single rising edge on input 2 will cause a logic 1 output-any other number, indicating non-identical frequencies causes a logic 0 output.

IC1a converts input 1 to a narrow pulse which initializes IC2 which forms a two-stage shift register clocked by input 2. On the first edge of input 2 a logic 1 appears on the output of IC2b and for all subsequent inputs a logic 0 is present. At the end of the gating period this output is latched by IC3 forming the lock output. As this is only valid for one input period a monostable is added to the output to enable, for example, visual monitoring of the output. Either output from IC3 can be used depending on which state is most important. As connected the failure state is indicated.

DEMONSTRATION COMPARATOR CIRCUIT

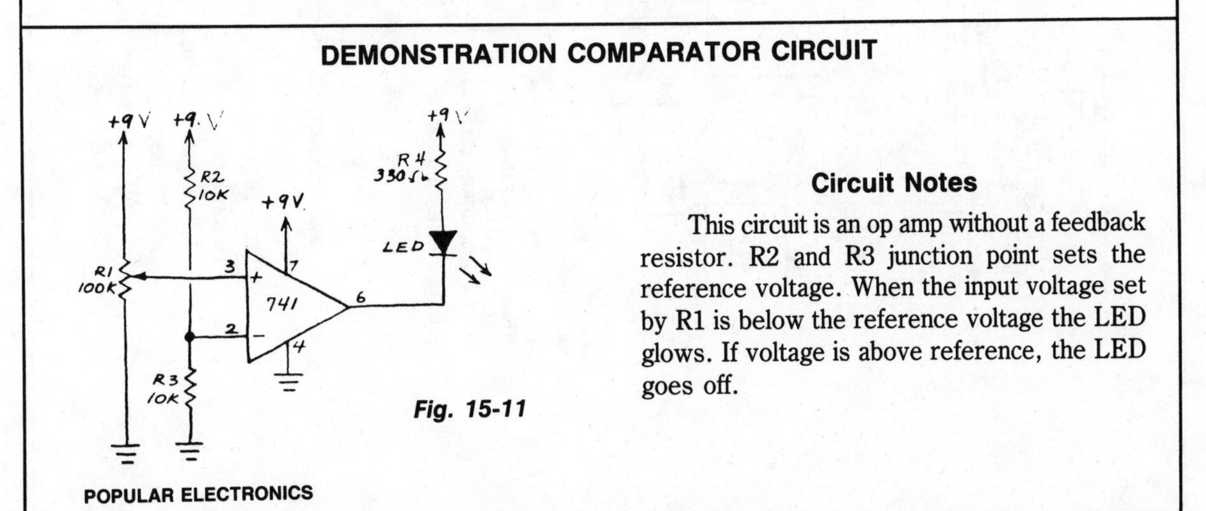

Fig. 15-11

Circuit Notes

This circuit is an op amp without a feedback resistor. R2 and R3 junction point sets the reference voltage. When the input voltage set by R1 is below the reference voltage the LED glows. If voltage is above reference, the LED goes off.

LED FREQUENCY COMPARATOR

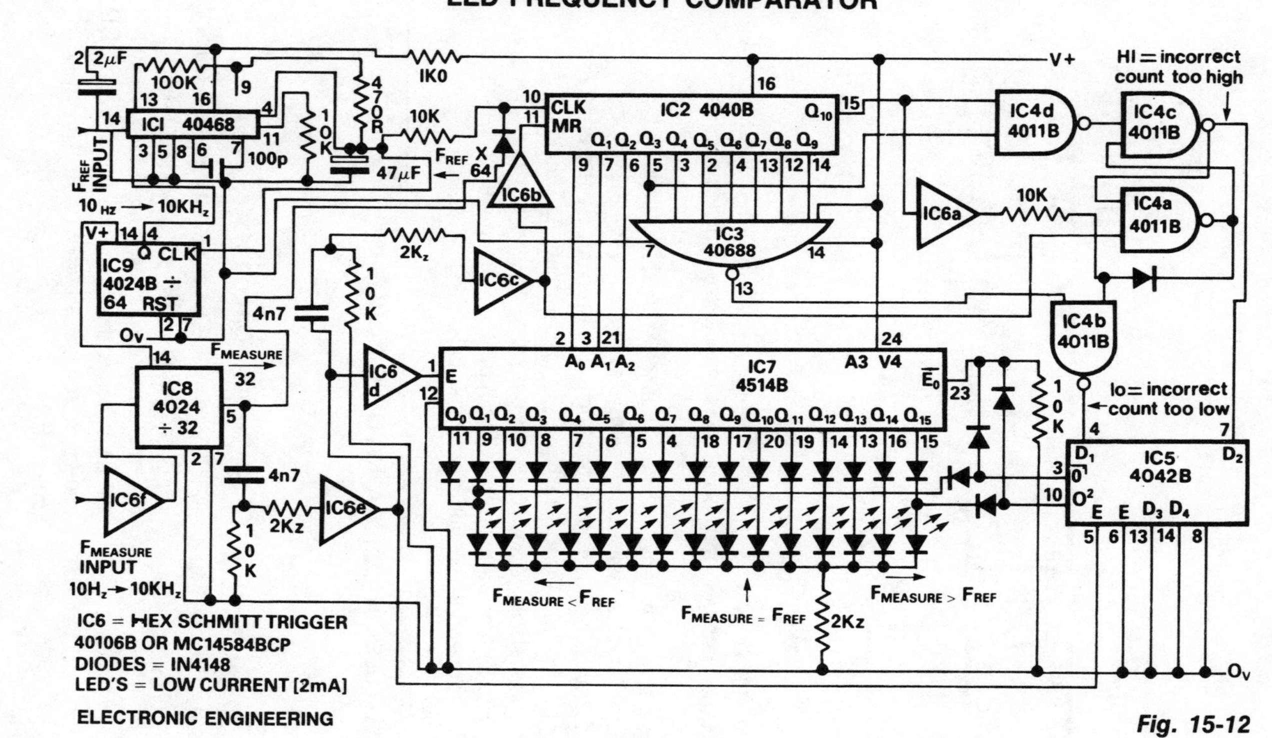

Fig. 15-12

Circuit Notes

The circuit provides unambiguous LED + or − bar readout with steps of 0.1%. The reference frequency is multiplied by the PLLIC1 and divider IC9 to output 64 × F (ref) and this is then gated by dividing F (measure) by 32 in IC8 thus is F (ref) = (measure) then IC2 counts 1024 pulses. Should the count be more than 1031 than the latch IC4c/IC4a is set to indicate count too high (F (measure) F (ref)) and if the count is less than 1017 then IC3/IC4b indicate count too low (F (measure) F (ref). These signals are latched by IC5 at the end of each period by the latch signal from IC6e.

When the two frequencies are within + or − 0.6% the LSB's of the counter IC2 are decoded and latched by IC7 and displayed on LED's IC6c resets the counter after latching the data.

TTL-COMPATIBLE SCHMITT TRIGGER

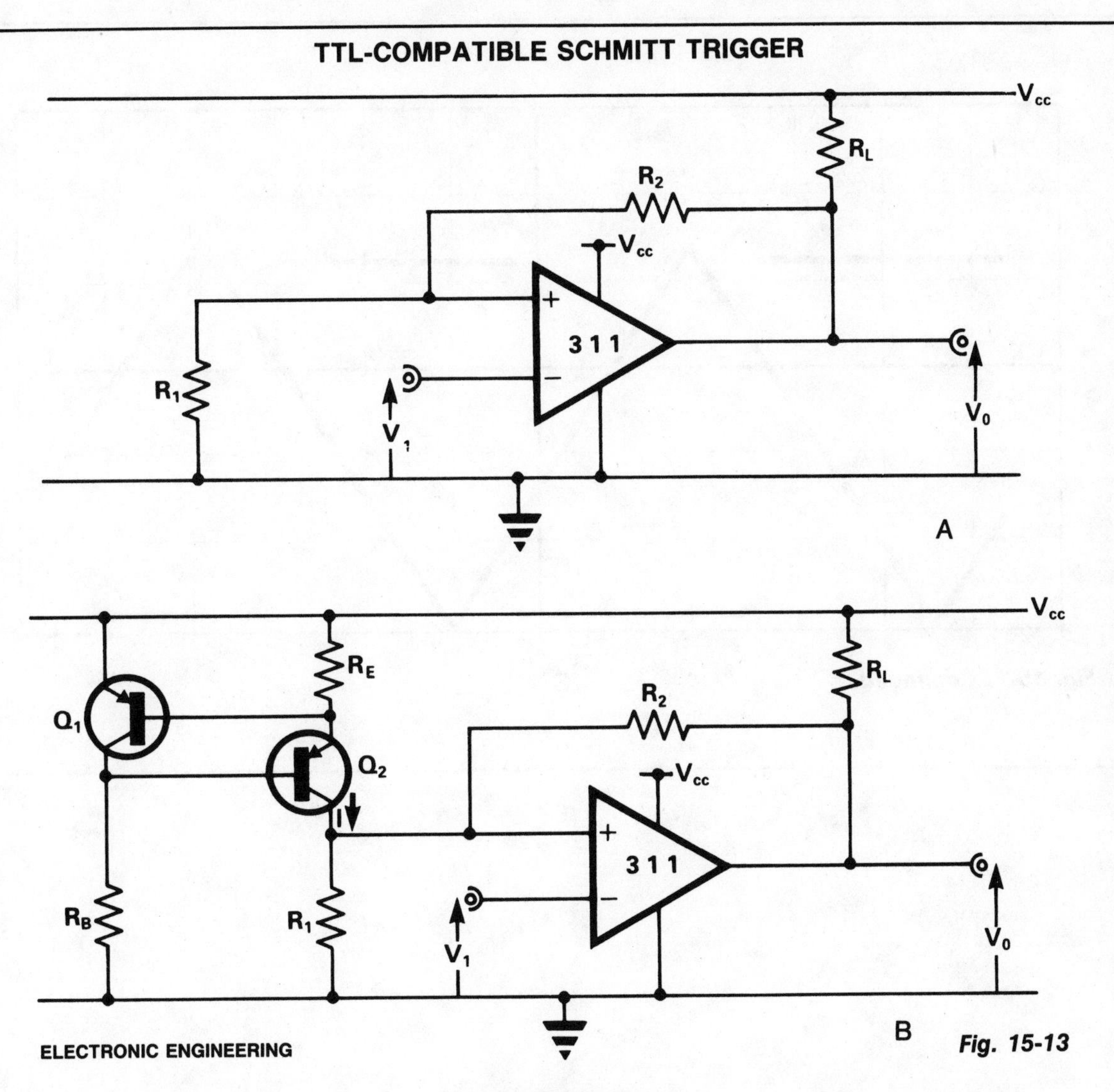

ELECTRONIC ENGINEERING

Fig. 15-13

Circuit Notes

The comparator has an output pull-up resistor R_L and is connected up to operate as a Schmitt trigger using the single rail supply V_{CC}. The feedback resistors R1 and R2 give upper and lower threshold levels V_{T+}, VVV_{TW}, respectively. V_{T+} is easily set by suitable resistor selection but there is little independent choice of V_{T-} because V_{T-} cannot exceed $V_{CE(SAT)}$. In Fig. 15-13B current-source, comprising the transistors R_E, R_B produces a current $I \sim: (V_{EB}/R_E)$, V_{EB} ($\sim$ 0.65 V) being the emitter-base voltage of Q1 and Q2. Fig. 15-13C shows the results of a practical test using the circuit of Fig. 15-13B, and the following operating and component data:

V_{CC} = 5 V; R_L = 1 K ohm; R_1 = R_2 = 10 K ohm;
R_B = 3.6 K ohm; R_E = 1 K ohm + 10 K ohm pot;
Q_1 = ZTX500; Q_2 = ZTX500.

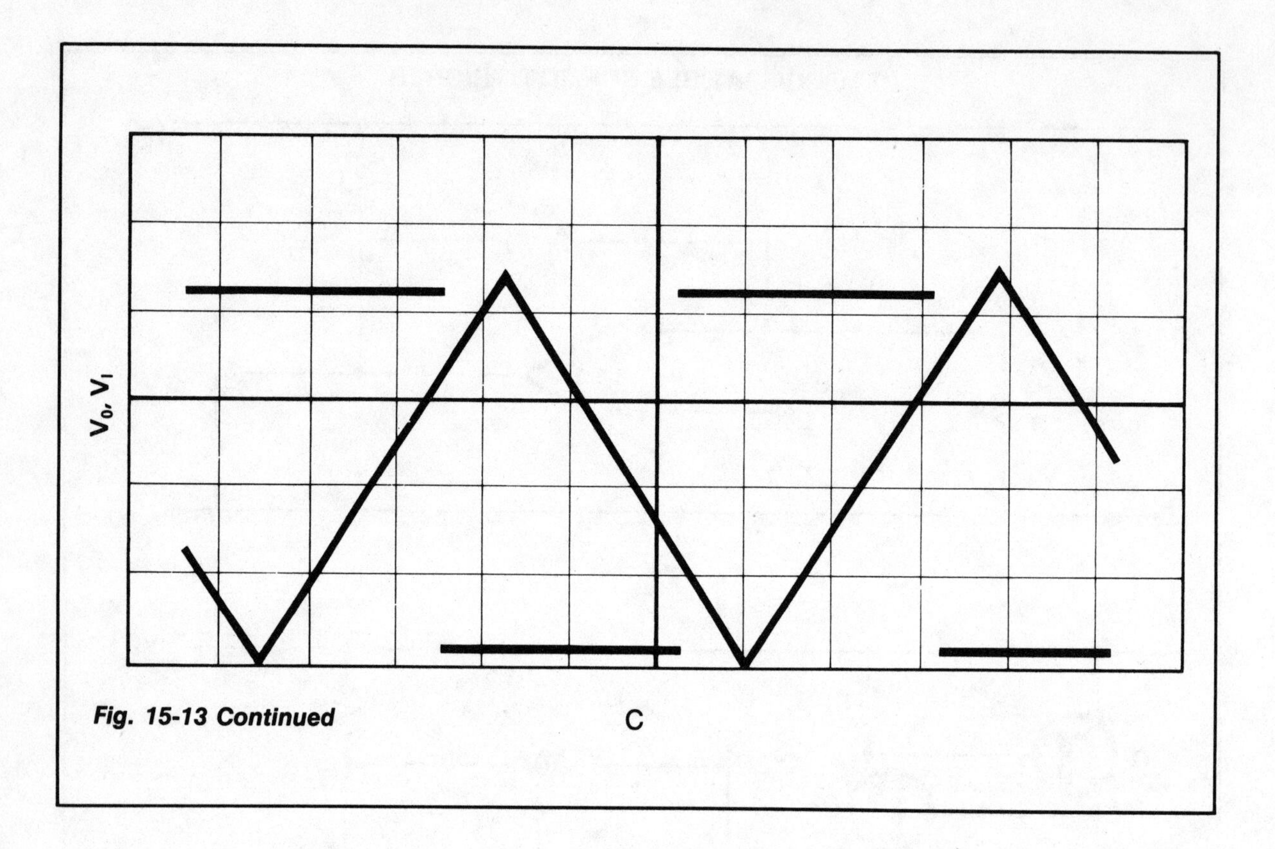

Fig. 15-13 Continued

16

Computer Circuits

The sources of the following circuits are contained in the Sources section beginning on page 694. The figure number contained in the box of each circuit correlates to the source entry in the Sources section.

8-BIT µP BUS INTERFACE

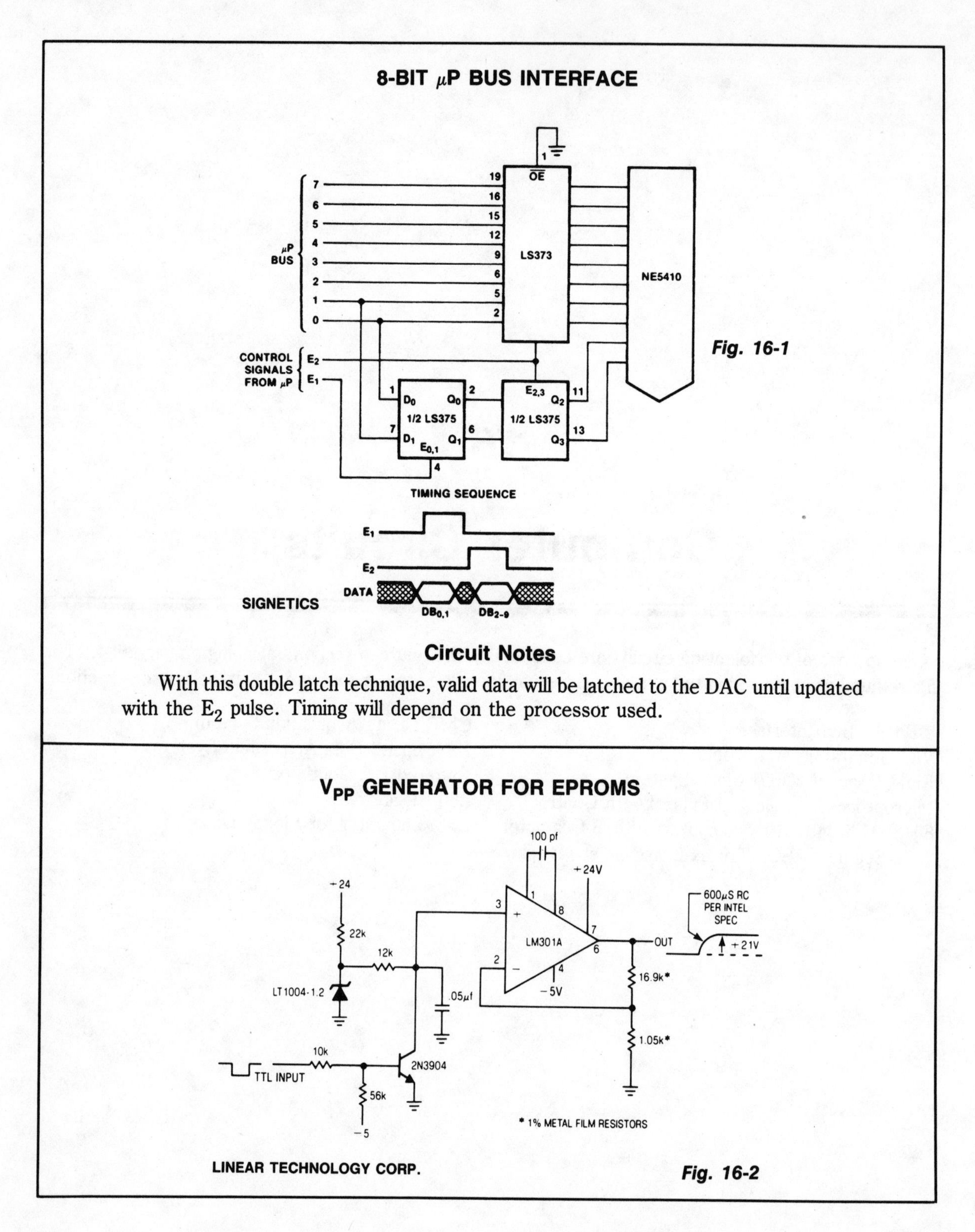

SIGNETICS

Fig. 16-1

Circuit Notes

With this double latch technique, valid data will be latched to the DAC until updated with the E_2 pulse. Timing will depend on the processor used.

V_{PP} GENERATOR FOR EPROMS

LINEAR TECHNOLOGY CORP.

Fig. 16-2

EIGHT-CHANNEL MUX/DEMUX SYSTEM

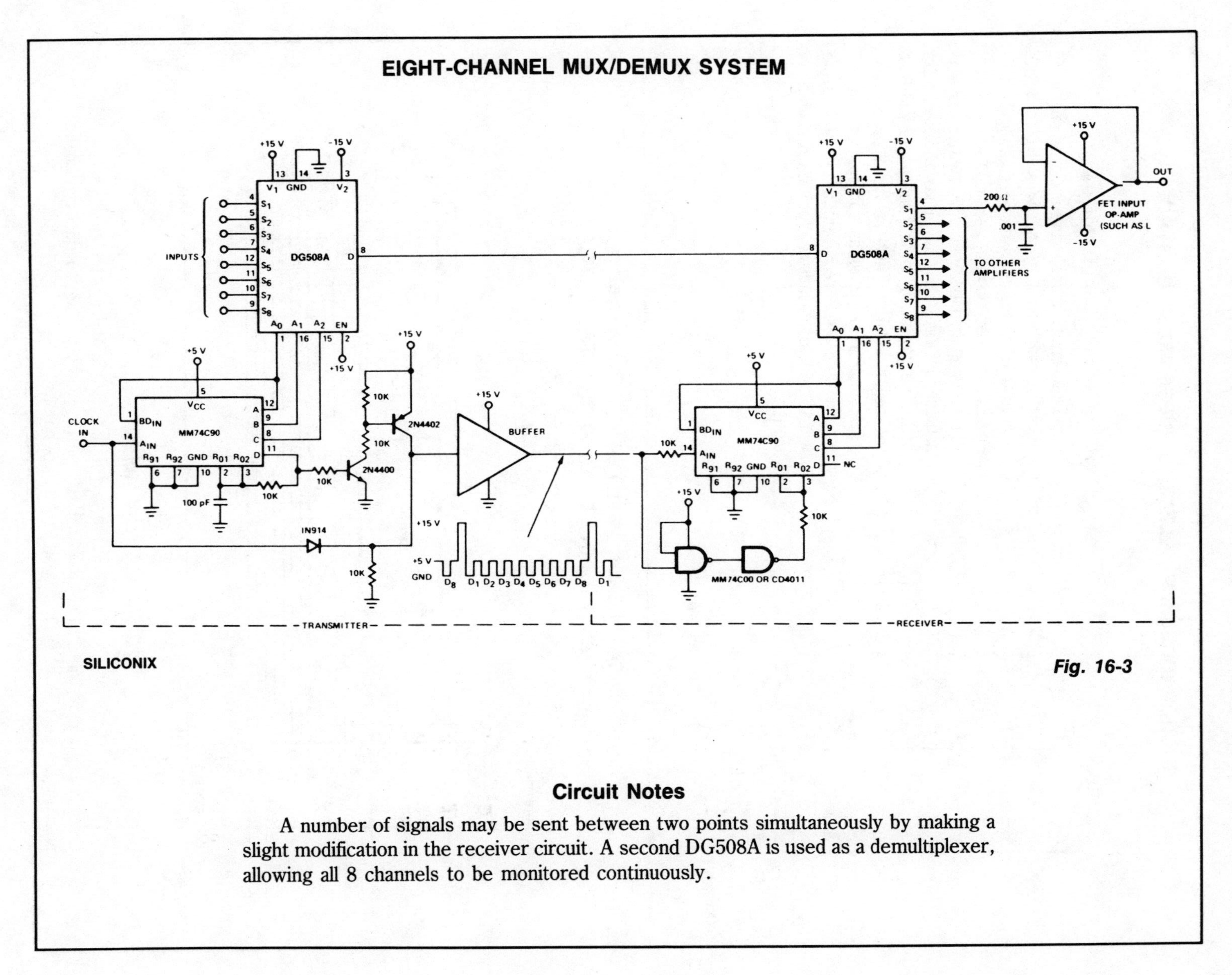

SILICONIX

Fig. 16-3

Circuit Notes

A number of signals may be sent between two points simultaneously by making a slight modification in the receiver circuit. A second DG508A is used as a demultiplexer, allowing all 8 channels to be monitored continuously.

MICROPROCESSOR SELECTED PULSE WIDTH CONTROL

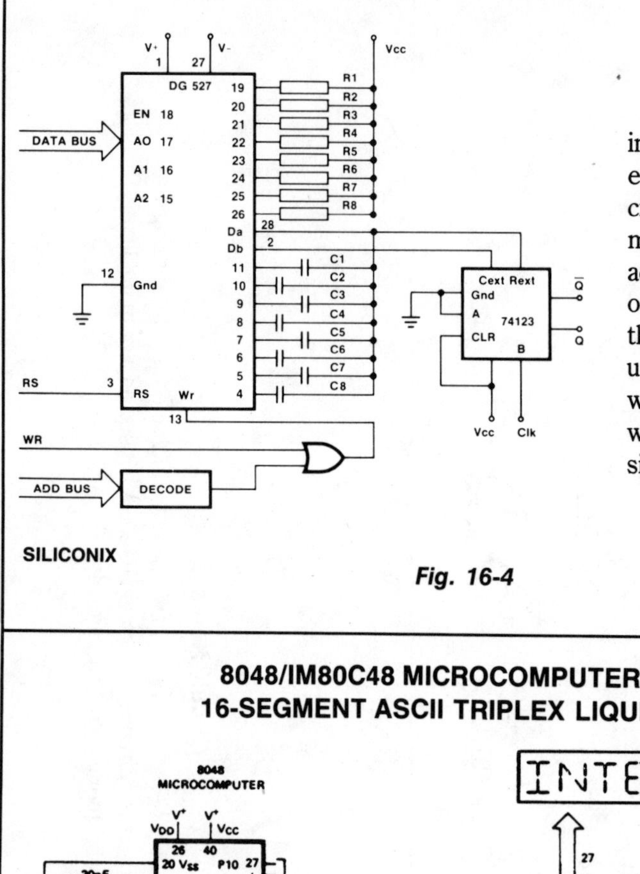

Fig. 16-4

Circuit Notes

Differential multiplexers are generally used in process control applications to eliminate errors due to common mode signals. In this circuit however, advantage is taken of the dual multiplexing capability of the switch. This is achieved by using the multiplexer to select pairs of RC networks to control the pulse width of the multivibrator. This can be a particularly useful feature in process control applications where there is a requirement for a variable width sample "window" for different control signals.

8048/IM80C48 MICROCOMPUTER WITH 8-CHARACTER 16-SEGMENT ASCII TRIPLEX LIQUID CRYSTAL DISPLAY

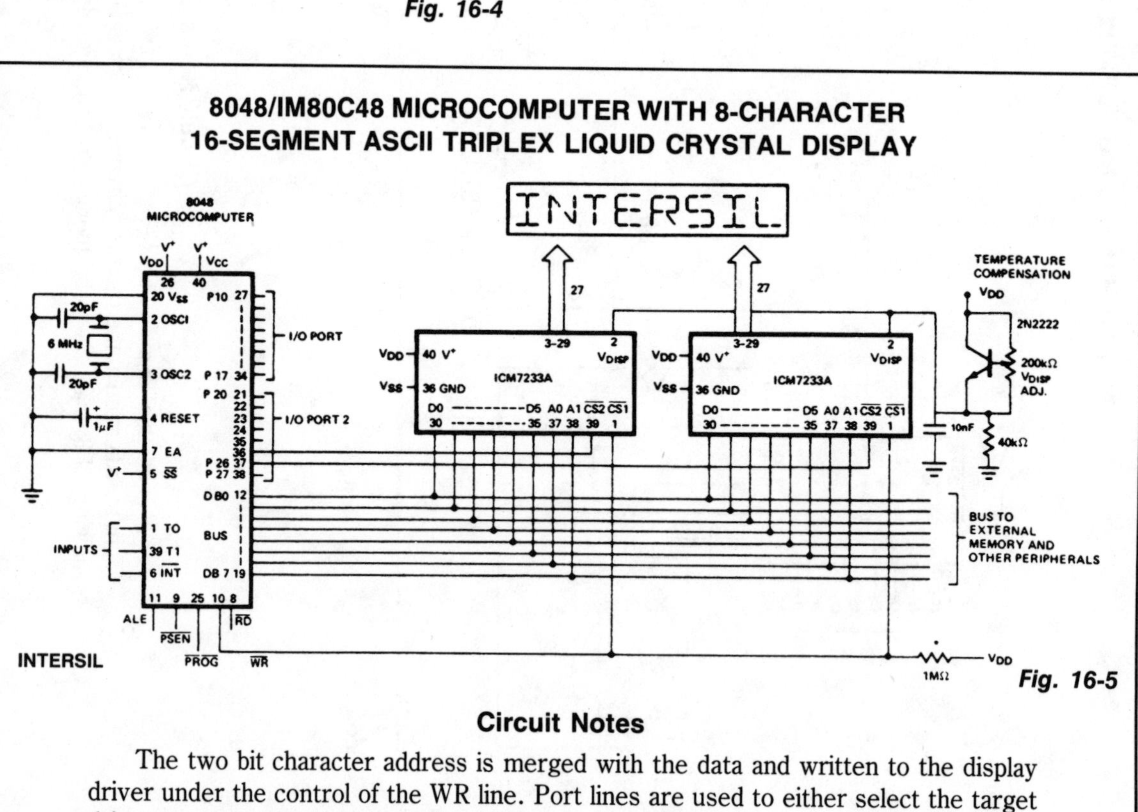

INTERSIL

Fig. 16-5

Circuit Notes

The two bit character address is merged with the data and written to the display driver under the control of the WR line. Port lines are used to either select the target driver, or deselect all of them for other bus operations.

CMOS DATA ACQUISITION SYSTEM

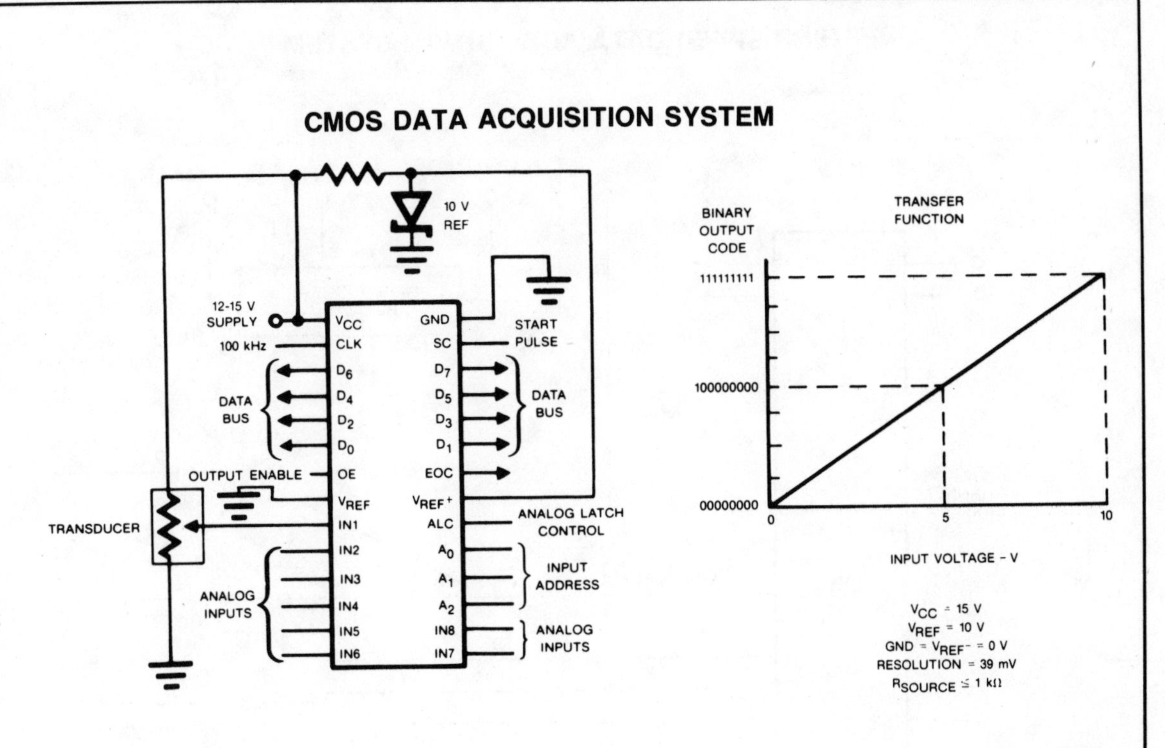

Fig. 16-6

SILICONIX

Circuit Notes

Charge redistribution to achieve A/D conversion. In typical applications, as a ratiometric conversion system for a microprocessor, V_{REF-} will be connected to ground and V_{REF+} will be connected to V_{CC}. The output will then be a simple proportional ratio between analog input voltage and V_{CC}. The general relationship is:

$$\frac{D_{OUT}}{2^8} = \frac{V_{IN}}{V_{REF+} - V_{REF-}}$$

Where D_{OUT} = Digital Output
V_{IN} = Analog Input
V_{REF} = Positive Reference Potential
V_{REF} = Negative Reference Potential

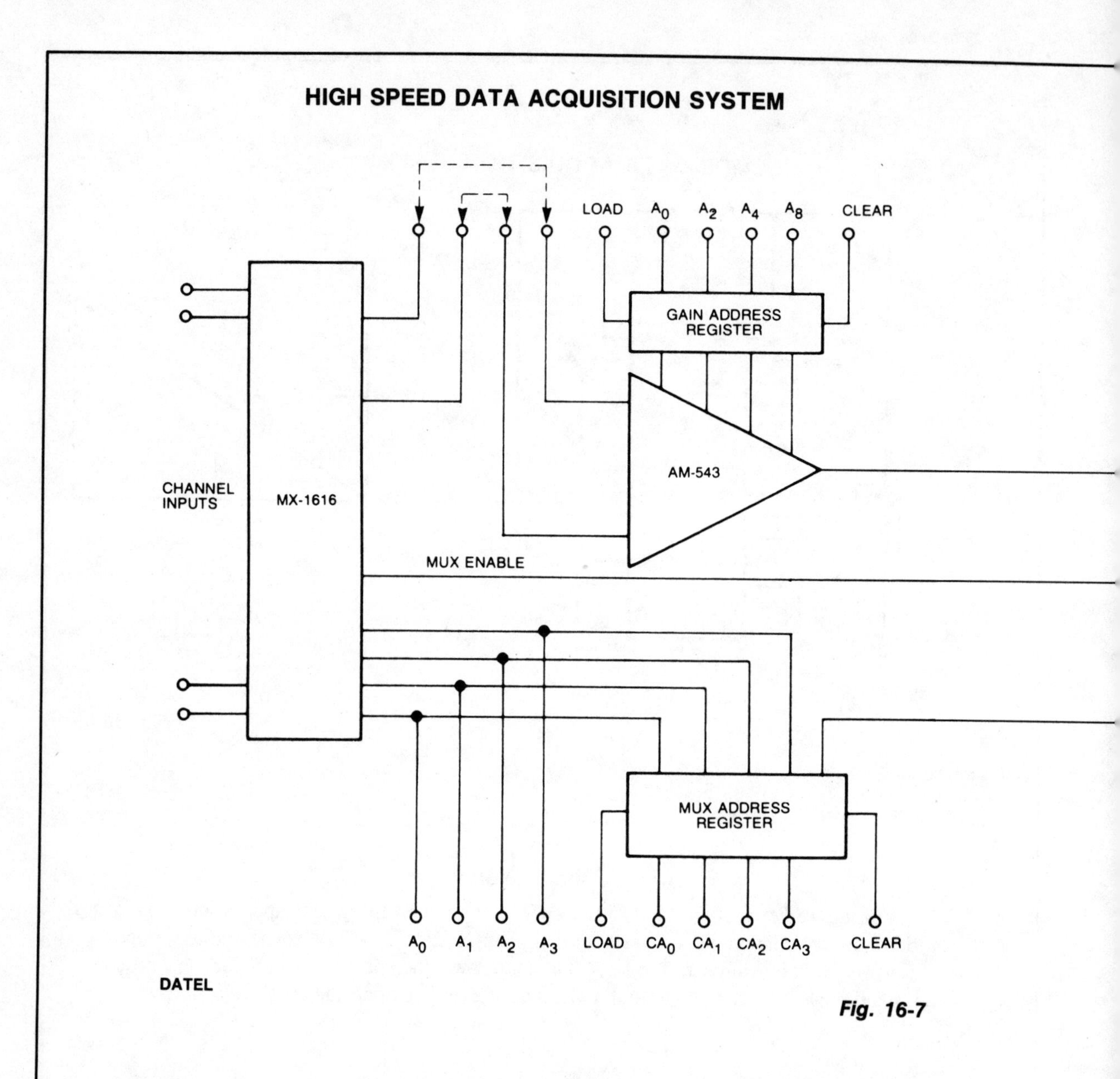

Fig. 16-7

Circuit Notes

This diagram shows a high-speed data acquisition system with 8 differential inputs and 12-bit resolution using the AM-543. If the control logic is timed so that the Sample-Hold-ADC section is converting one analog value while the mux-amplifier section is allowed to settle to the next input value, throughout rates greater than 156 KHz can be achieved. The AM-543 is used with Datel's ADV-817, a 12-bit hybrid A/D with a 2 μsec conversion rate, the SHM-6, a 0.01%, 1 μsec hybrid Sample-Hold, and the MX-1616,

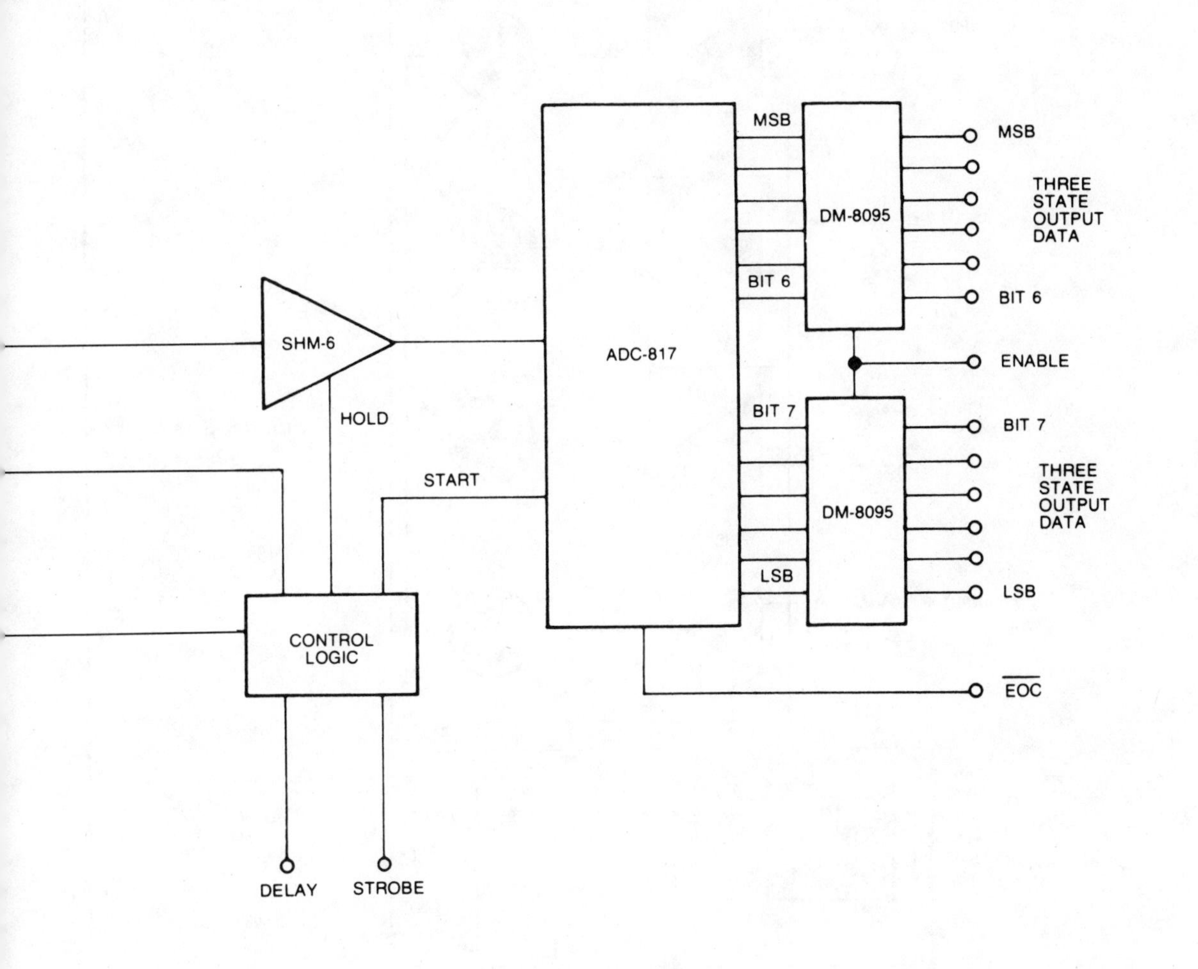

a low cost, high-speed monolithic analog multiplexer. The system works as follows:

The μP selects a channel and initiates a conversion at G=1 and then looks at the MSB of the conversion result. If the MSB = 1, the μP will store the value. If the MSB = 0, the μP will select G = 2. The μP will repeat the cycle of gain incrementing, comparison, and analog-to-digital conversion until the MSB = 1. The μP will then test for an output of all 1's, as this is the full-scale output of the A/D. If the output is all 1's, the μP will decrement the gain by 1 step and perform the final conversion.

BUFFERED BREAKOUT BOX

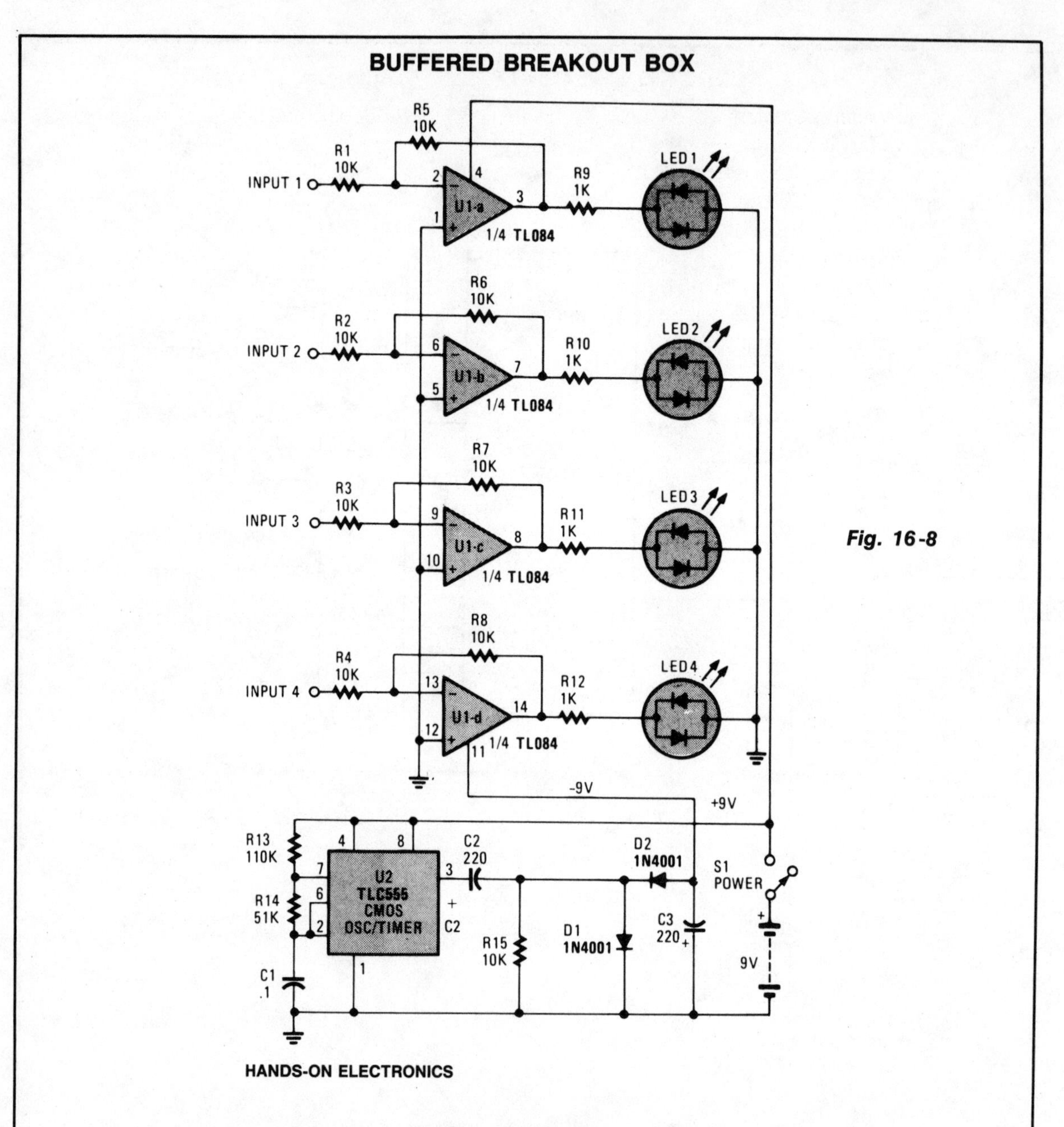

Fig. 16-8

Circuit Notes

The monitoring circuit consists of four tri-color LEDs driven by an equal number of op amps configured as gain-of-one inverting amplifiers. Each LED is wired in the circuit so that it glows red when the input to the op amp is high, and green when the input is low. The LED remains off when the input is disconnected from a circuit, when it's at ground potential, and when it's connected to a 3-state output that's in the high-impedance state. Each input has an impedance of 10,000 ohms preventing the circuit

BUFFERED BREAKOUT BOX, Continued.

from loading communication lines. The op amp requires both positive and negative supply voltages to properly drive the LEDs. Both voltages are supplied by a single, nine-volt battery. The battery supplies the positive source directly. The negative source is supplied via a CMOS 555 oscillator/timer that's configured as an astable oscillator, which is used to drive a standard diode/capacitor voltage doubler. When the 555 is connected to the monitoring circuit, the output voltage is not 18 volts (2 × 9), but a little under nine volts, due to loading. The circuit draws about 16 mA with all LEDs off; with all four on, it draws between 20 and 30 mA, depending on how many LEDs are high, and how many are low. The use of CMOS op amps reduces quiescent current drain considerably.

Z80 CLOCK

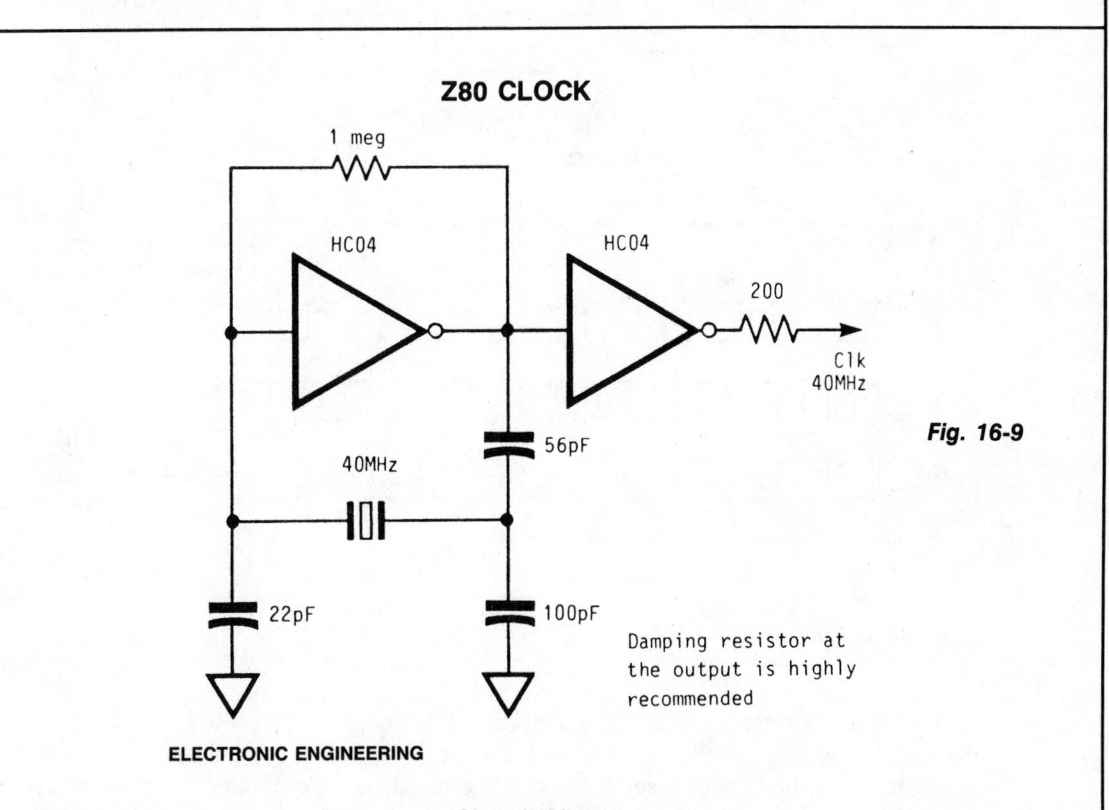

Fig. 16-9

Circuit Notes

The circuit will operate reliably from below 1 MHz to above 400 MHz. With V_{CC} = 5 V the output of the second inverter essentially attains a full swing from 0 V to 5 V. Such large logic output levels and broad frequency range capabilities make this oscillator quite suitable for driving MOS components such as CPU, controller chip, peripheral devices, as well as other TTL products. A damping resistor in series between the clock output of the oscillator and the input of the device being driven will remove the undesirable undershoot and ringing caused by the high speed CMOS part.

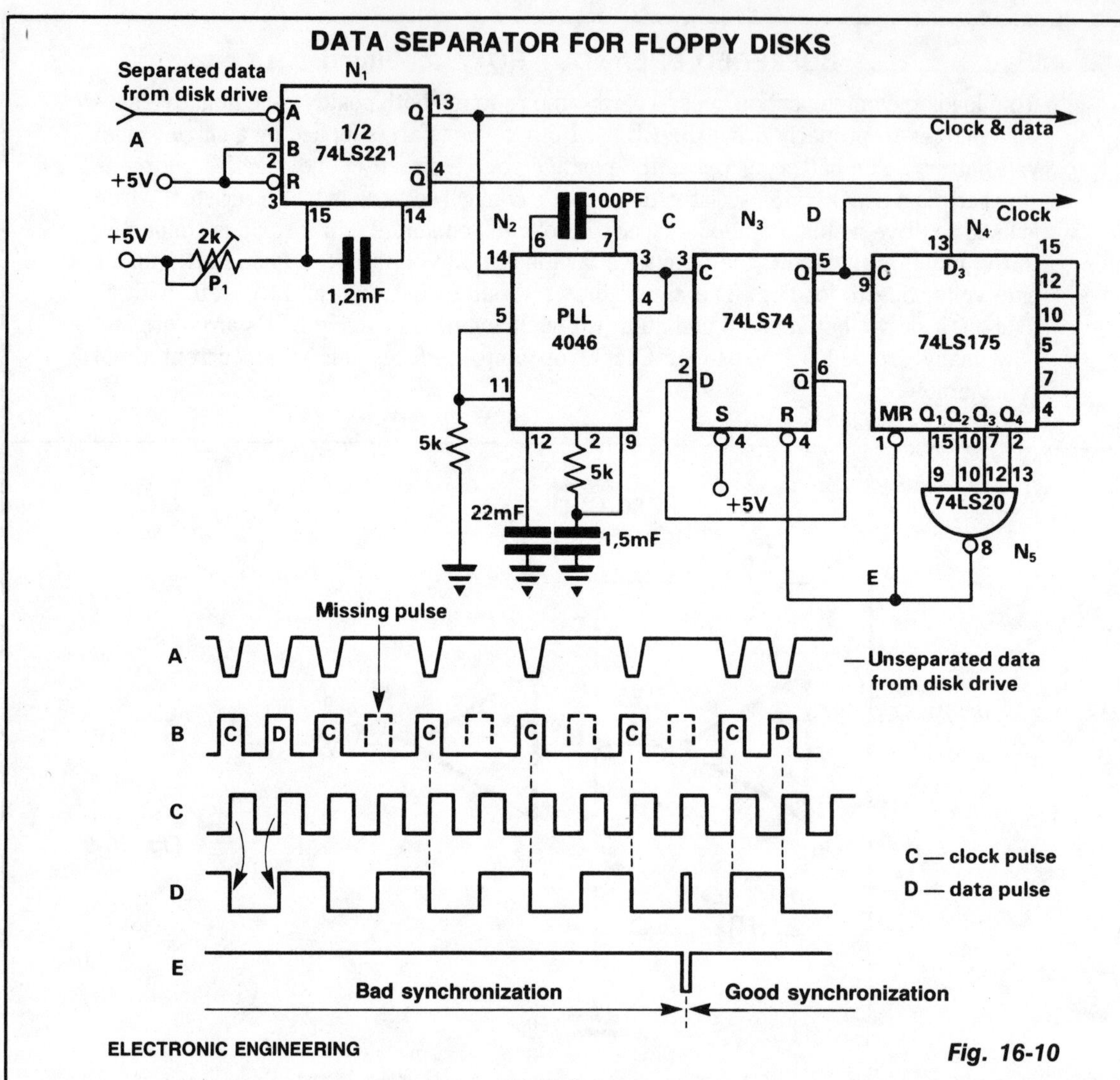

ELECTRONIC ENGINEERING

Fig. 16-10

Circuit Notes

The data separator is intended for use with 8″ flexible diskettes with IBM 3870 soft sectored format. The circuit delivers data and clock (B) and clock pulses (D). These two signals must be in such a sequence that the negative edge of the clock pulse is at the middle of a data cell.

Unseparated data (A) from the floppy unit is shaped with one shot N1. Trimmer P1 should be adjusted so that pulses (B) are 1 μs wide. This signal synchronizes PLL N2 with a free running frequency adjusted to 500 kHz. The output of the PLL is 90° out of phase with its input. D-type flip-flop N3 is connected as a divider by two and changes state at each positive edge of (C). N4, connected as a shift register, looks for four consecutive missing pulses. When this happens, the circuit is resynchronized with (E) so that the negative edge of (D) is in the middle of a data cell.

17

Converters

The sources of the following circuits are contained in the Sources section beginning on page 694. The figure number contained in the box of each circuit correlates to the source entry in the Sources section.

VOLTAGE-TO-PULSE DURATION CONVERTER

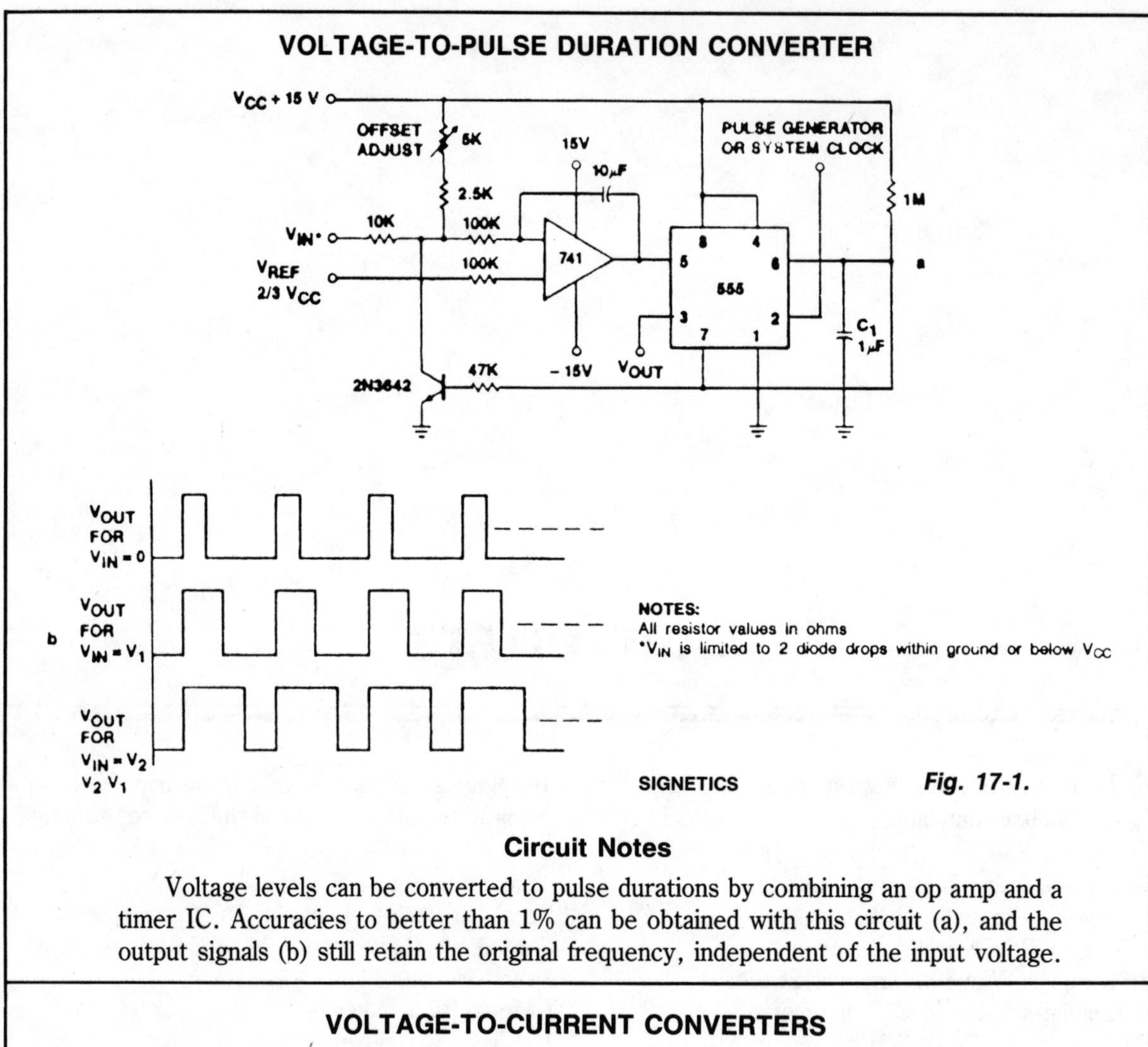

Fig. 17-1.

Circuit Notes

Voltage levels can be converted to pulse durations by combining an op amp and a timer IC. Accuracies to better than 1% can be obtained with this circuit (a), and the output signals (b) still retain the original frequency, independent of the input voltage.

VOLTAGE-TO-CURRENT CONVERTERS

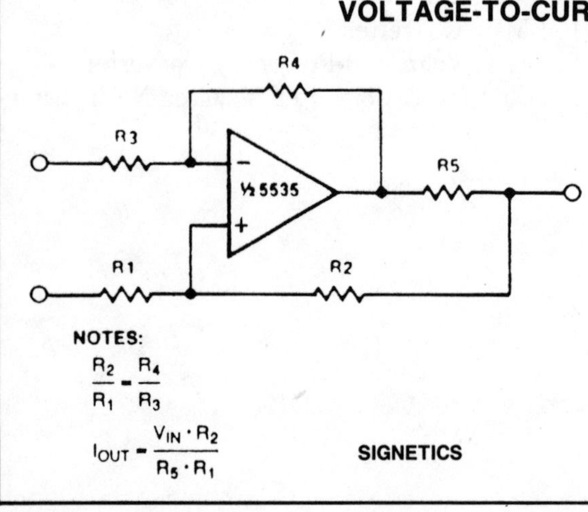

NOTES:

$$\frac{R_2}{R_1} = \frac{R_4}{R_3}$$

$$I_{OUT} = \frac{V_{IN} \cdot R_2}{R_5 \cdot R_1}$$

SIGNETICS

Circuit Notes

A simple voltage-to-current converter is shown in the figure. The current out is I_{OUT} or V_{IN}/R. For negative currents, a pnp can be used and, for better accuracy, a Darlington pair can be substituted for the transistor. With careful design, this circuit can be used to control currents of many amps. Unity gain compensation is necessary.

Fig. 17-2

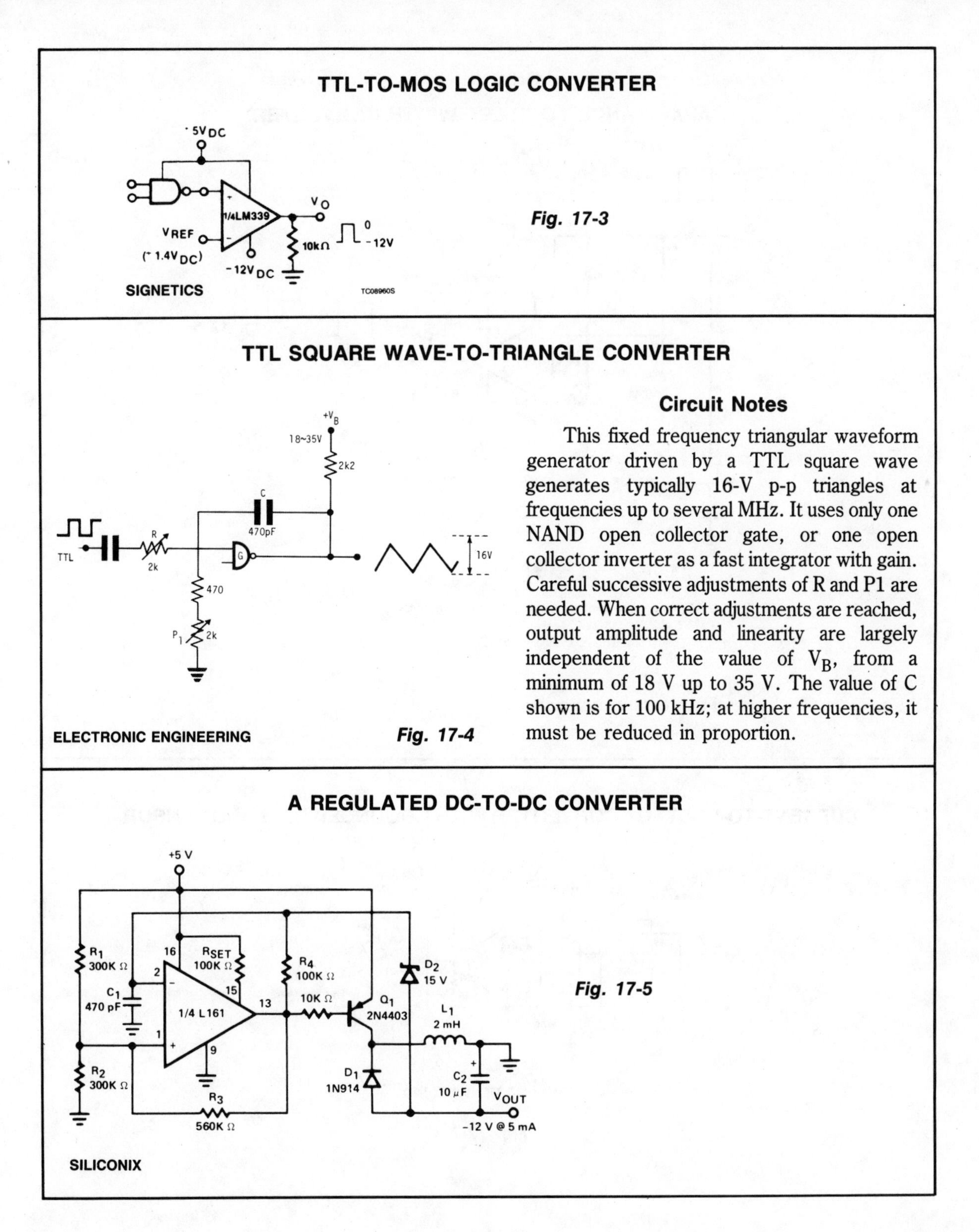

TTL-TO-MOS LOGIC CONVERTER

Fig. 17-3

TTL SQUARE WAVE-TO-TRIANGLE CONVERTER

Circuit Notes

This fixed frequency triangular waveform generator driven by a TTL square wave generates typically 16-V p-p triangles at frequencies up to several MHz. It uses only one NAND open collector gate, or one open collector inverter as a fast integrator with gain. Careful successive adjustments of R and P1 are needed. When correct adjustments are reached, output amplitude and linearity are largely independent of the value of V_B, from a minimum of 18 V up to 35 V. The value of C shown is for 100 kHz; at higher frequencies, it must be reduced in proportion.

Fig. 17-4

A REGULATED DC-TO-DC CONVERTER

Fig. 17-5

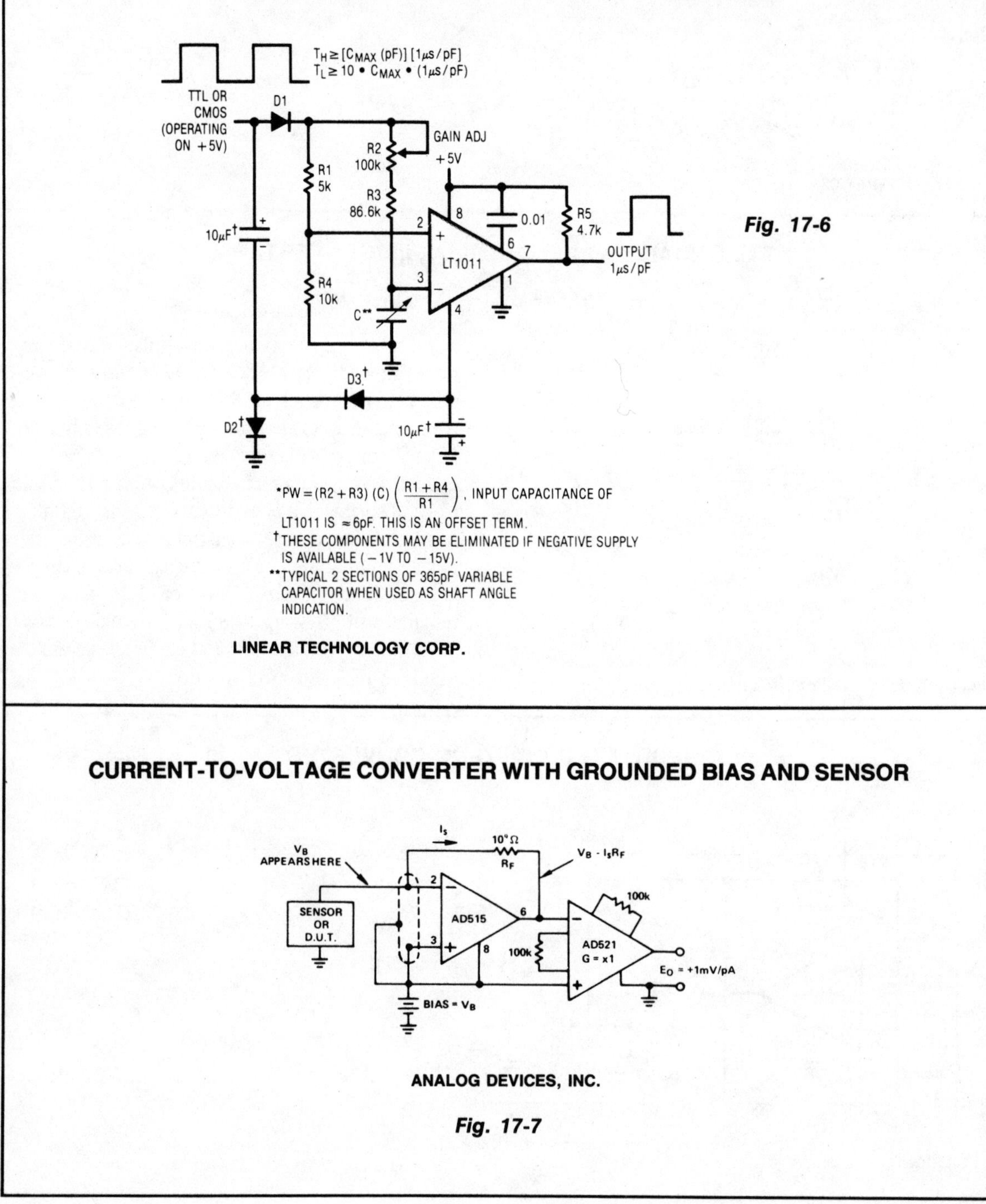

CAPACITANCE TO PULSE WIDTH CONVERTER

*PW = (R2 + R3) (C) $\left(\frac{R1 + R4}{R1}\right)$, INPUT CAPACITANCE OF LT1011 IS ≈ 6pF. THIS IS AN OFFSET TERM.

†THESE COMPONENTS MAY BE ELIMINATED IF NEGATIVE SUPPLY IS AVAILABLE (−1V TO −15V).

**TYPICAL 2 SECTIONS OF 365pF VARIABLE CAPACITOR WHEN USED AS SHAFT ANGLE INDICATION.

LINEAR TECHNOLOGY CORP.

Fig. 17-6

CURRENT-TO-VOLTAGE CONVERTER WITH GROUNDED BIAS AND SENSOR

ANALOG DEVICES, INC.

Fig. 17-7

TRIANGLE-TO-SINE CONVERTERS

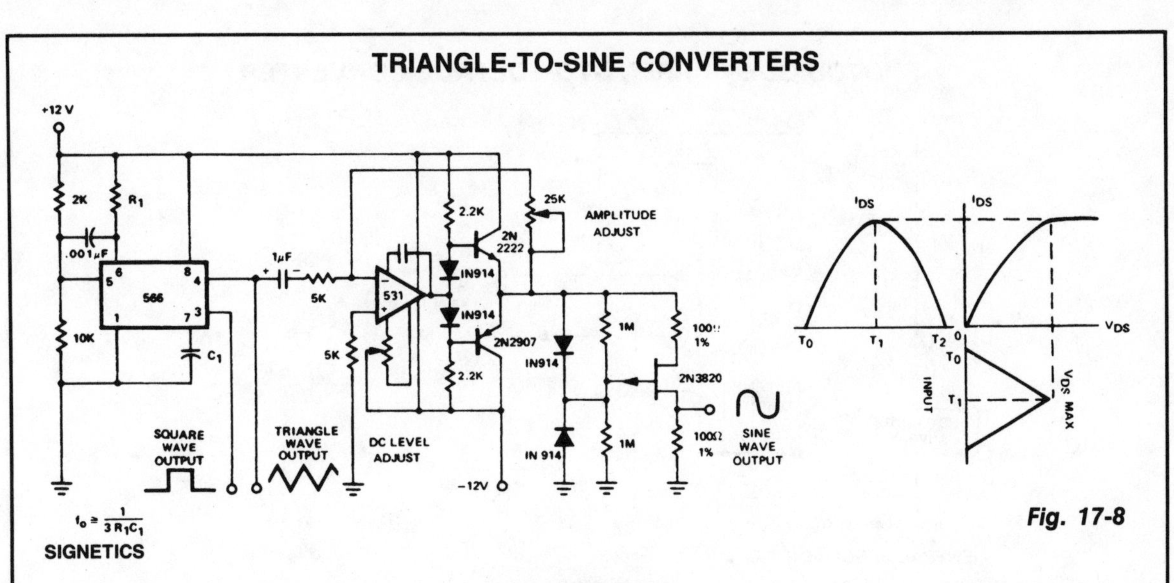

Fig. 17-8

Circuit Notes

Conversion of triangle wave shapes to sinusoids is usually accomplished by diode-resistor shaping networks, which accurately reconstruct the sine wave segment by segment. Two simpler and less costly methods may be used to shape the triangle waveform of the 566 into a sinusoid with less than 2% distortion. The non-linear $I_{DS}V_{DS}$ transfer characteristic of a P-channel junction FET is used to shape the triangle waveform. The amplitude of the triangle waveform is critical and must be carefully adjusted to achieve a low distortion sinusoidal output. Naturally, where additional waveform accuracy is needed, the diode-resistor shaping scheme can be applied to the 566 with excellent results since it has very good output amplitude stability when operated from a regulated supply.

PRECISION PEAK-TO-PEAK AC-DC CONVERTER

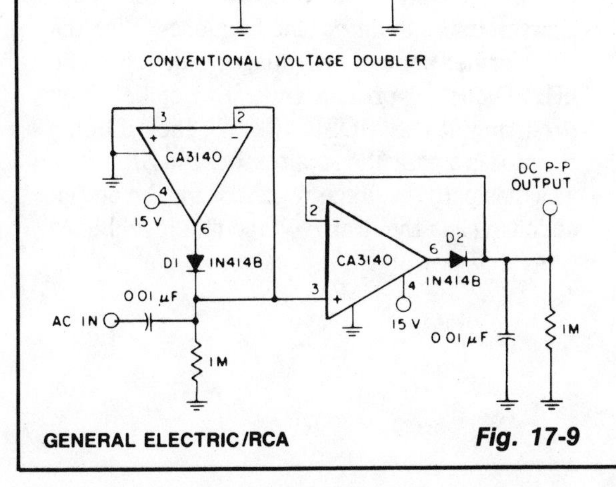

GENERAL ELECTRIC/RCA

Fig. 17-9

Circuit Notes

Using a CA3140 BiMOS op amp and a single positive supply converts a conventional voltage doubler with two precision diodes into a precision peak-to-peak ac-to-dc voltage converter having wide dynamic range and wide bandwidth.

ALL RESISTANCE VALUES ARE IN OHMS

PHOTODIODE CURRENT-TO-VOLTAGE CONVERTER

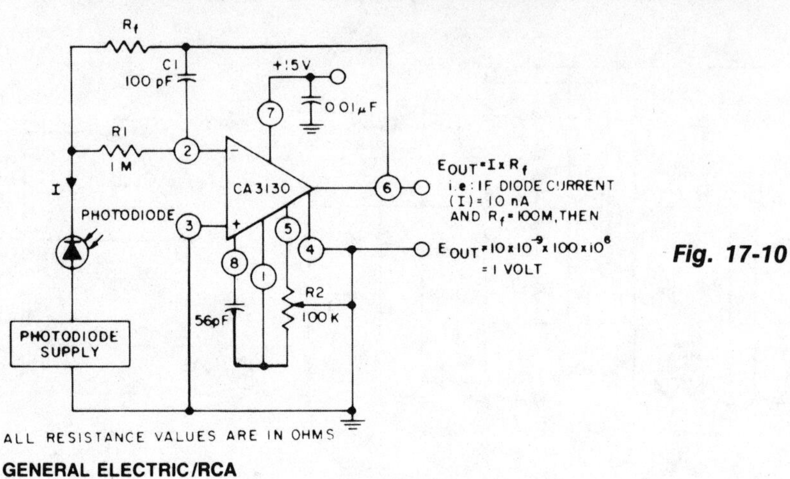

GENERAL ELECTRIC/RCA

Fig. 17-10

Circuit Notes

The circuit uses three CA3130 BiMOS op amps in an application sensitive to subpicoampere input currents. The circuit provides a ground-referenced output voltage proportional to input current flowing through the photodiode.

SELF OSCILLATING FLYBACK CONVERTER

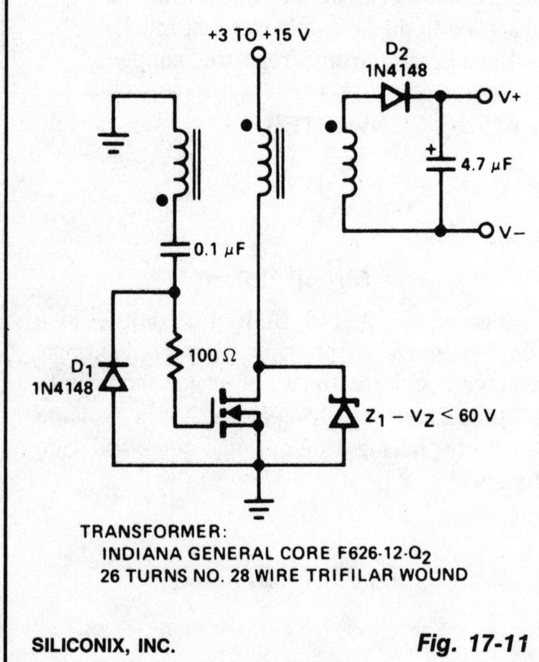

Circuit Notes

Low-power converter uses the core characteristics to determine frequency. With the transformer shown, operating frequency is 250 kHz. Diode D1 prevents negative spikes from occurring at the MOSFET gate, the 100 ohm resistor is a parasitic suppressor, and Z1 serves as a dissipative voltage regulator for the output and also clips the drain voltage to a level below the rated power FET breakdown voltage.

SILICONIX, INC.

Fig. 17-11

RMS-TO-DC CONVERTER

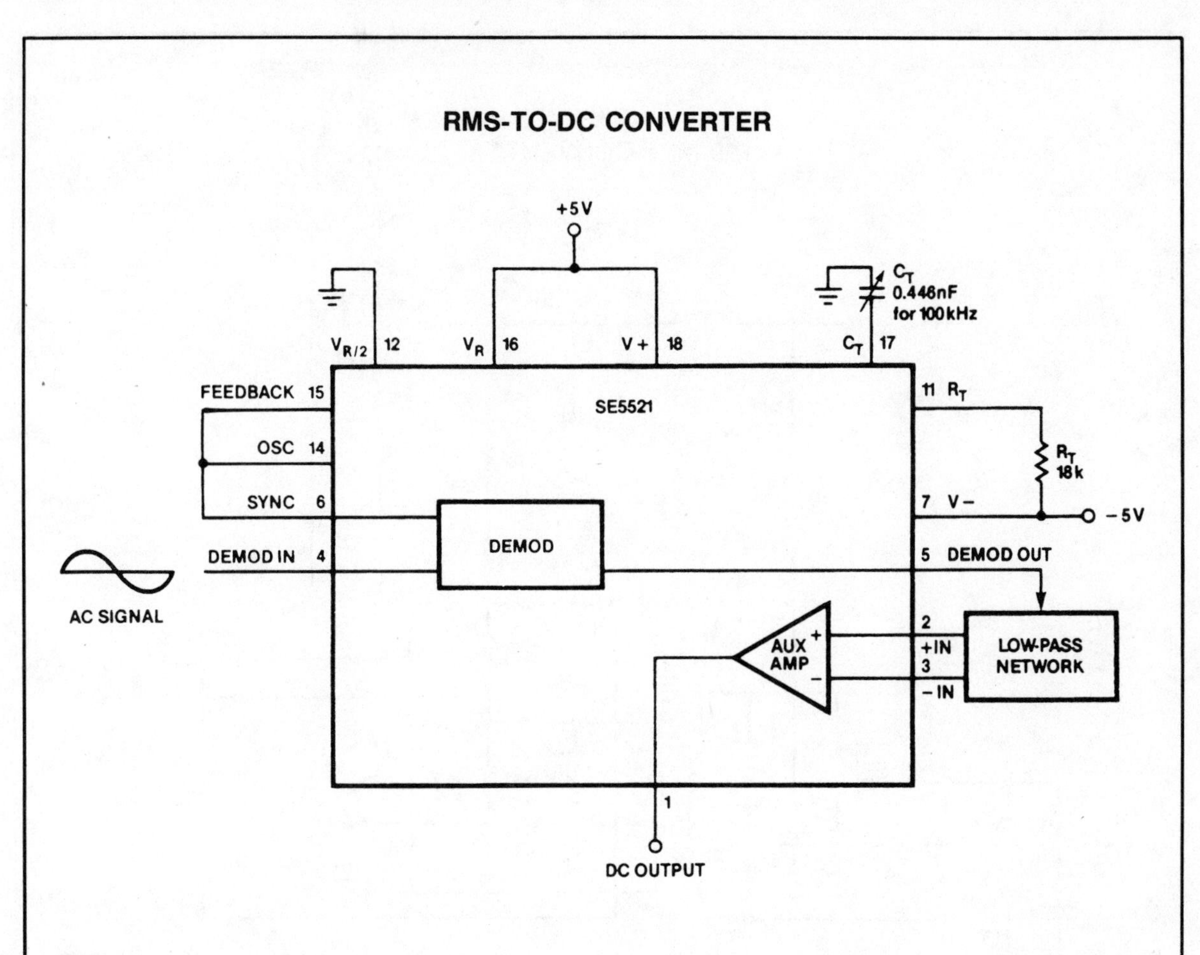

NOTE:
1. The DC output at Pin 1 varies linearly with the RMS input at Pin 4.
2. C_T is tweaked until the sync signal is in phase with the AC signal.

SIGNETICS *Fig. 17-12*

Circuit Notes

An ac voltmeter may be easily constructed. Simplicity of the circuit and low component count make it particularly attractive. The demodulator output is a full-wave rectified signal from the ac input at Pin 4. The dc component on the rectified signal at Pin 5 varies linearity with the rms input at Pin 4 and thus provides an accurate rms-to-dc conversion at the output of the filter (Pin 1). C_T is a variable capacitor that is tweaked until the oscillator signal to the sync input of the demodulator is in phase with the ac signal at Pin 4.

100 MHz CONVERTER

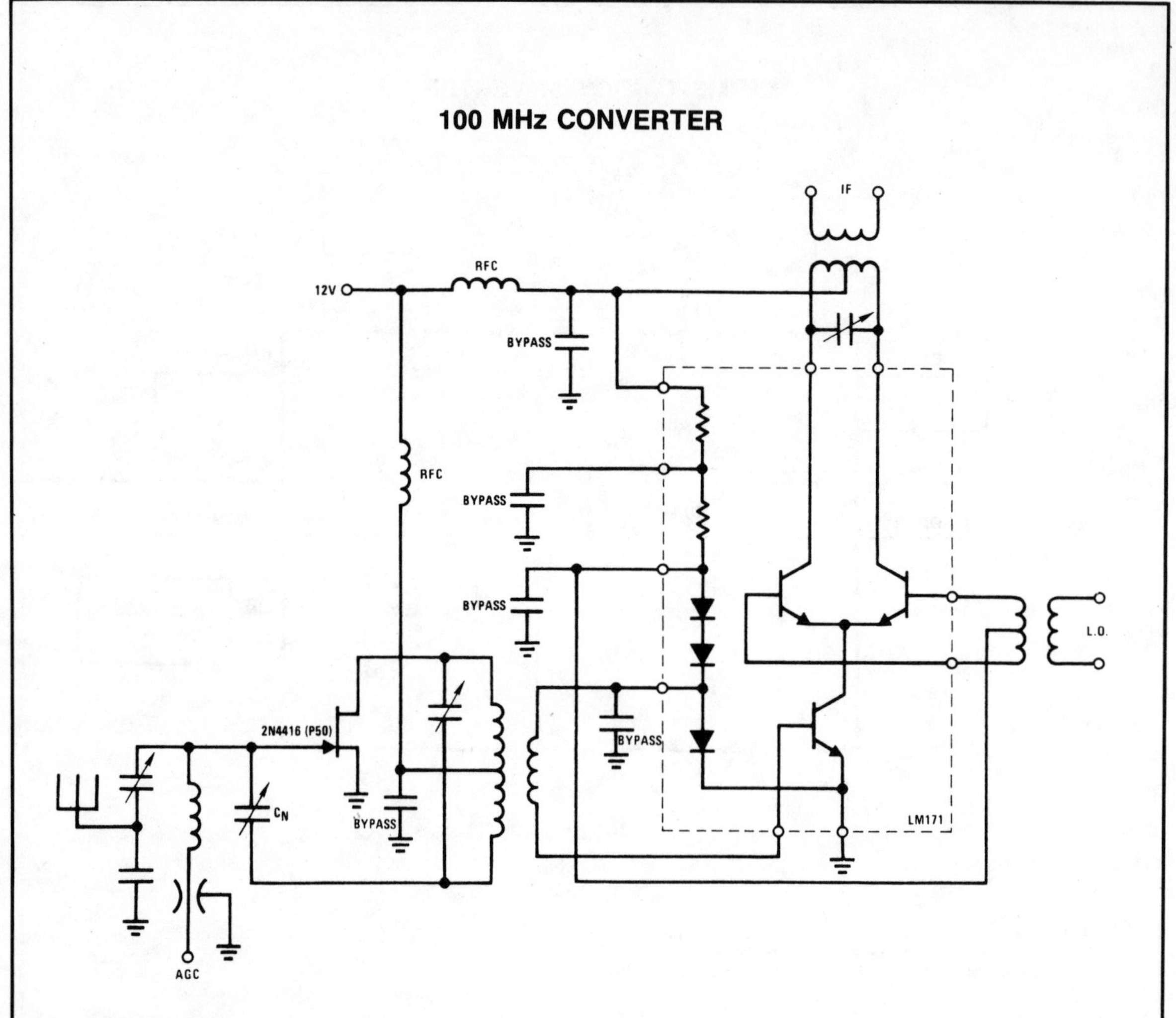

NATIONAL SEMICONDUCTOR CORP.

Fig. 17-13

Circuit Notes

The 2N4416 JFET will provide noise figures of less than 3 dB and power gain of greater than 20 dB. The JFET's outstanding low crossmodulation and low intermodulation distortion provides an ideal characteristic for an input stage. The output feeds into an LM171 used as a balanced mixer. This configuration greatly reduces local oscillator radiation both into the antenna and into the *if* strip and also reduces *rf* signal feedthrough.

PRECISION VOLTAGE-TO-FREQUENCY CONVERTER

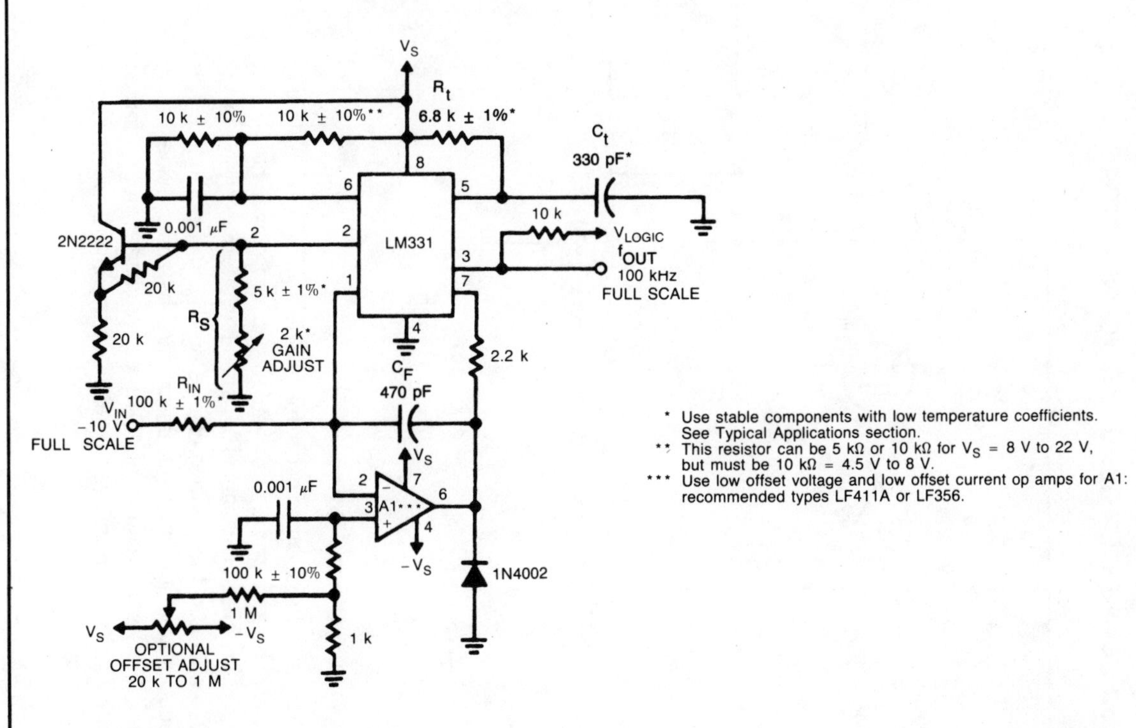

NATIONAL SEMICONDUCTOR CORP. ***Fig. 17-14***

Circuit Notes

In this circuit, integration is performed by using a conventional operational amplifier and feedback capacitor, C_F. When the integrator's output crosses the nominal threshold level at pin 6 of the LM131, the timing cycle is initiated. The average current fed into the op amp's summing point (pin 2) is i × (1.1 R_tC_t) × f which is perfectly balanced with $-V_{IN}/R_{IN}$. In this circuit, the voltage offset of the LM131 input comparator does not affect the offset or accuracy of the V-to-F converter as it does in the stand-alone V-to-F converter, nor does the LM131 bias current or offset current. Instead, the offset voltage and offset current of the operational amplifier are the only limits on how small the signal can be accurately converted.

BIPOLAR DC-DC CONVERTER REQUIRES NO INDUCTOR

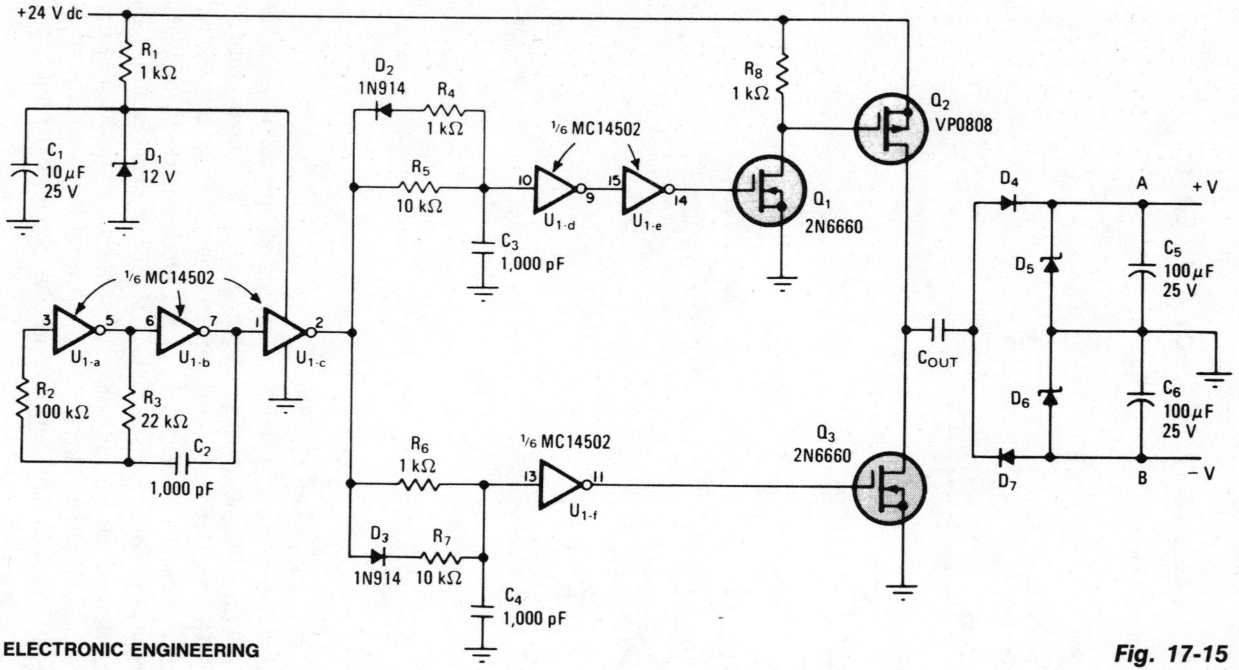

ELECTRONIC ENGINEERING

Fig. 17-15

Circuit Notes

Inverters U1a and U1b form a 20-kilohertz oscillator whose square wave output—further shaped by D2, R4, and R5 and by D3, R6, and R7—drives power field-effect transistors Q2 and Q3. The p-channel and n-channel FETs conduct alternately, in a push-pull configuration. When Q2 conducts, the positive charge on C_{out} forces diode D4 to conduct as well, which produces a positive voltage, determined by zener diode D5, at terminal A. Similarly, when Q3, in its turn conducts, the negative charge on C_{out} forces D7 to do so as well. A negative voltage, therefore, develops at terminal B, whose level is set by D6.

18

Counters

The sources of the following circuits are contained in the Sources section beginning on page 694. The figure number contained in the box of each circuit correlates to the source entry in the Sources section.

8-DIGIT UP/DOWN COUNTER

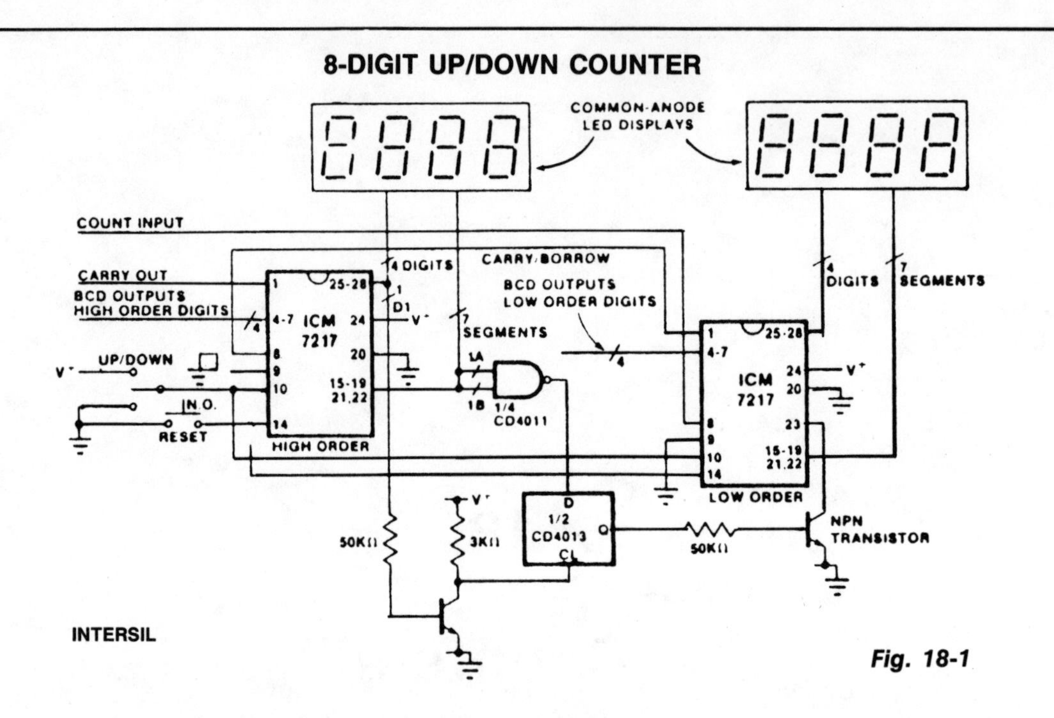

INTERSIL

Fig. 18-1

Circuit Notes

This circuit shows how to cascade counters and retain correct leading zero blanking. The NAND gate detects whether a digit is active since one of the two segments a or b is active on any unblanked number. The flip flop is clocked by the least significant digit of the high order counter, and if this digit is not blanked, the Q output of the flip flop goes high and turns on the npn transistor, thereby inhibiting leading zero blanking on the low order counter.

RING COUNTER WITH VARIABLE TIMING

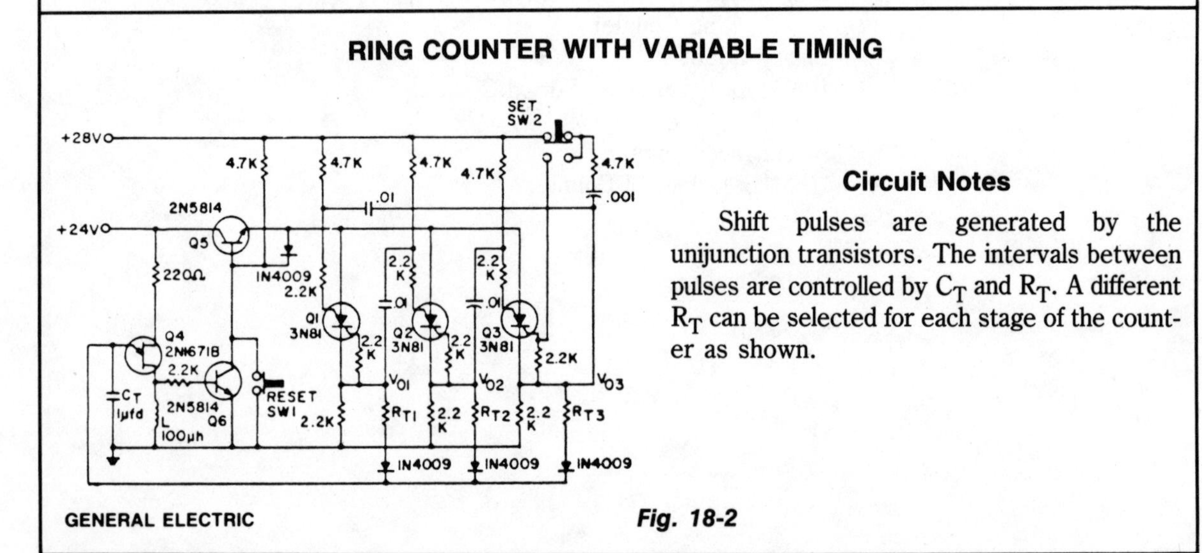

GENERAL ELECTRIC

Fig. 18-2

Circuit Notes

Shift pulses are generated by the unijunction transistors. The intervals between pulses are controlled by C_T and R_T. A different R_T can be selected for each stage of the counter as shown.

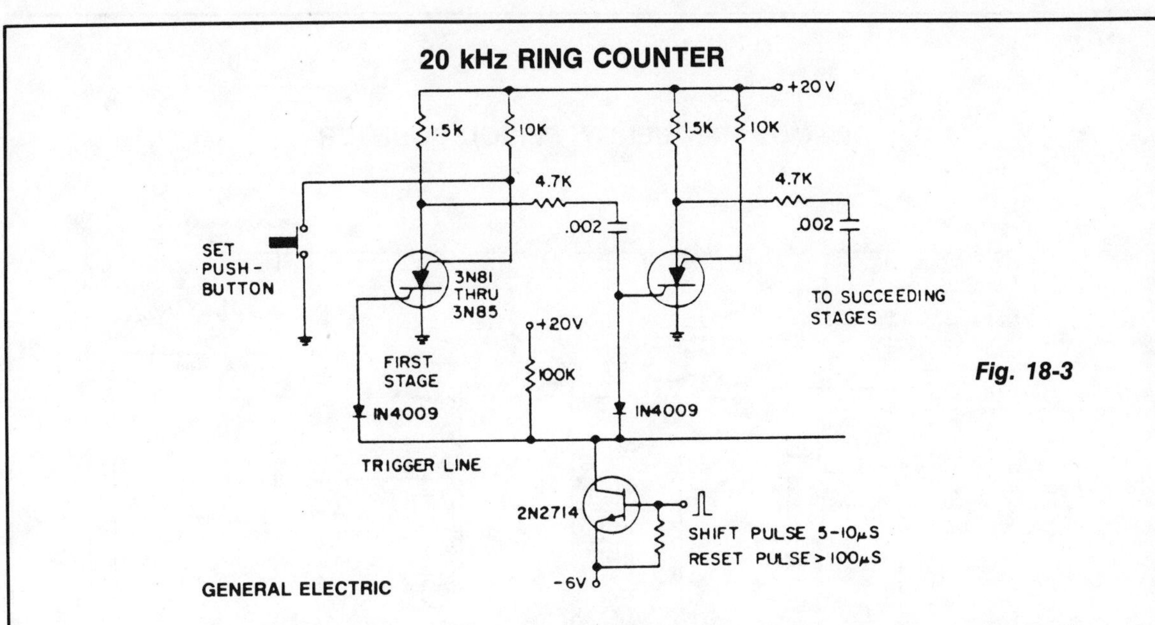

Fig. 18-3

Circuit Notes

The shift pulse turns off the conducting scs by reverse biasing the cathode gate. The charge stored on the coupling capacitor then triggers the next stage. An excessively long shift pulse charges up all the capacitors, turning off all stages. Grounding an anode gate will "set" that stage.

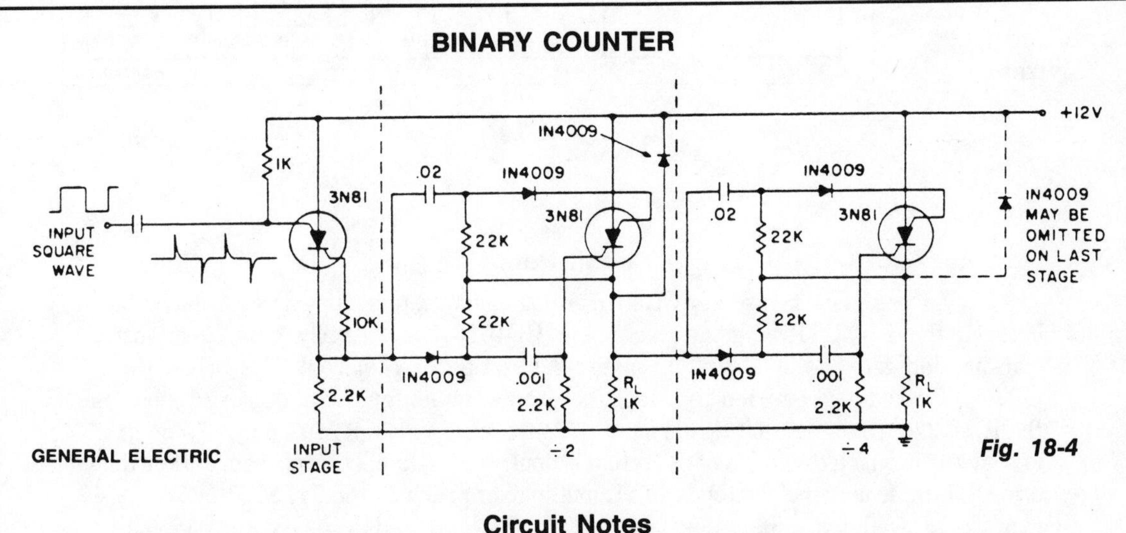

Fig. 18-4

Circuit Notes

Stages are triggered by the positive going edge. The scs is turned on at the cathode gate; turned off at the anode gate. The anode-to-cathode IN4009 suppresses positive transients while the scs is recovering. The input stage generates fast positive edges to trigger the counter.

100 MHz FREQUENCY, PERIOD COUNTER

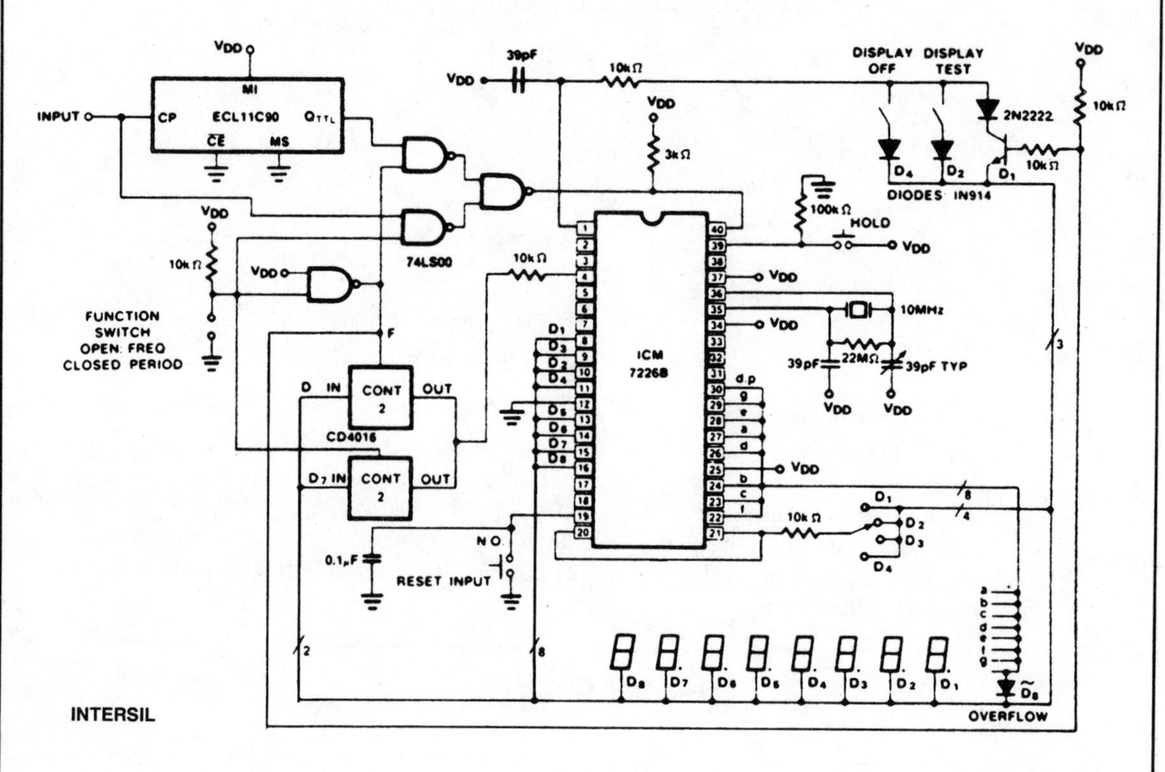

Fig. 18-5

Circuit Notes

The figure shows the use of a CD4016 analog multiplex to multiplex the digital outputs back to the FUNCTION input. Since the CD4016 is a digitally controlled analog transmission gate, no level shifting of the digit output is required. The CD4051's or CD4052's could also be used to select the proper inputs for the multiplexed input on the ICM7226 from 2 or 3 bit digital inputs. These analog multiplexers may also be used in systems in which the mode of operation is controlled by a microprocessor rather than directly from front panel switches. TTL multiplexers such as the 74LS153 or 74LS251 may also be used, but some additional circuitry will be required to convert the digit output to TTL compatible logic levels.

ANALOG COUNTER CIRCUIT

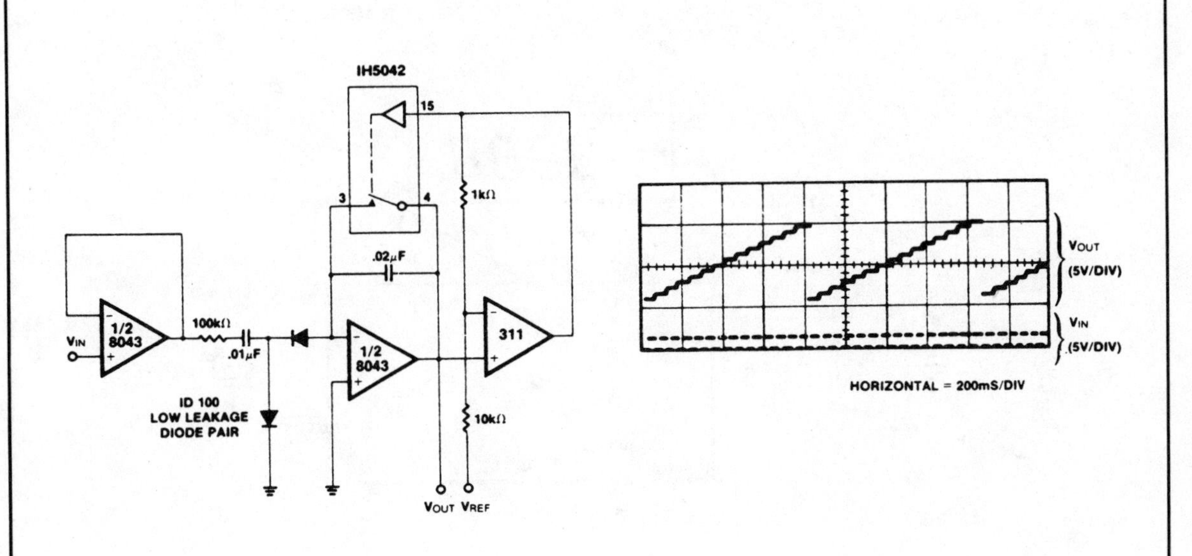

INTERSIL

Fig. 18-6

Circuit Notes

A straightforward circuit using a LM311 for the level detector and a CMOS analog gate to discharge the capacitor is shown. An important property of this type of counter is the ease with which the count can be changed; it is only necessary to change the voltage at which the comparator trips. A low cost A-D converter can also be designed using the same principle since the digital count between reset periods is directly proportional to the analog voltage used as a reference for the comparator. A considerable amount of hysteresis is used in the comparator. This ensures that the capacitor is completely discharged during the reset period. In a more sophisticated circuit, a dual comparator ''window detector'' could be used, the lower trip point is set close to ground to ensure complete discharge. The upper trip point could then be adjusted independently to determine the pulse count.

ATTENDANCE COUNTER

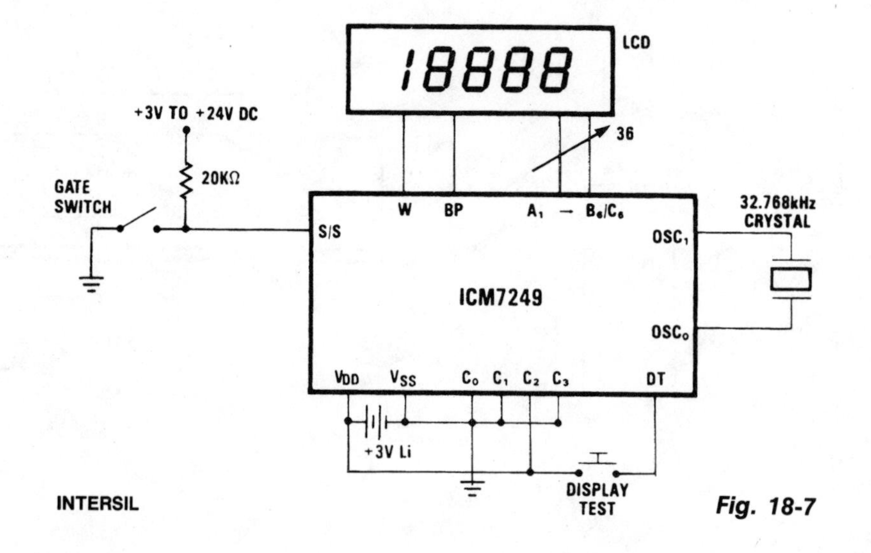

INTERSIL

Fig. 18-7

Circuit Notes

The display shows each increment. By using mode 2, external debouncing of the gate switch is unnecessary, provided the switch bounce is less than 35ms. The 3 V lithium battery can be replaced without disturbing operation if a suitable capacitor is connected in parallel with it. The display should be disconnected, if possible, during the procedure to minimize current drain. The capacitor should be large enough to store charge for the amount of time needed to physically replace the battery ($t = VC/1$). A 100 μF capacitor initially charged to 3 V will supply a current of 1.0 μA for 50 seconds before its voltage drops to 2.5 V, which is the minimum operating voltage for the ICM7249.

Before the battery is removed, the capacitor should be placed in parallel, across the V_{DD} and GND terminals. After the battery is replaced, the capacitor can be removed and the display reconnected.

10 MHz UNIVERSAL COUNTER

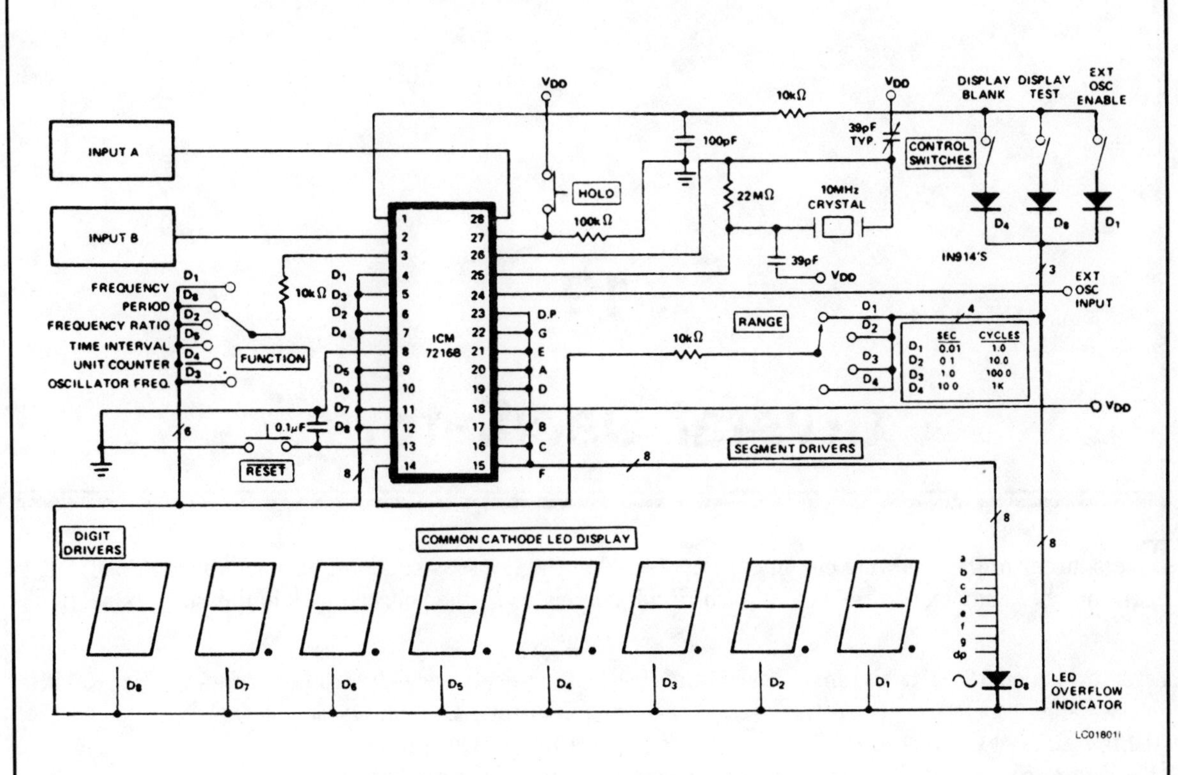

INTERSIL

Fig. 18-8

Circuit Notes

The ICM7216A or B can be used as a minimum component complete Universal Counter. This circuit can use input frequencies up to 10 MHz at INPUT A and 2 MHz at INPUT B. If the signal at INPUT A has a very low duty cycle it may be necessary to use a 74121 monostable multivibrator or similar circuit to stretch the input pulse width to be able to guarantee that it is at least 50 ns in duration.

19

Crystal Oscillators

The sources of the following circuits are contained in the Sources section beginning on page 694. The figure number contained in the box of each circuit correlates to the source entry in the Sources section.

VARACTOR-TUNED 10 MHz CERAMIC RESONATOR OSCILLATOR

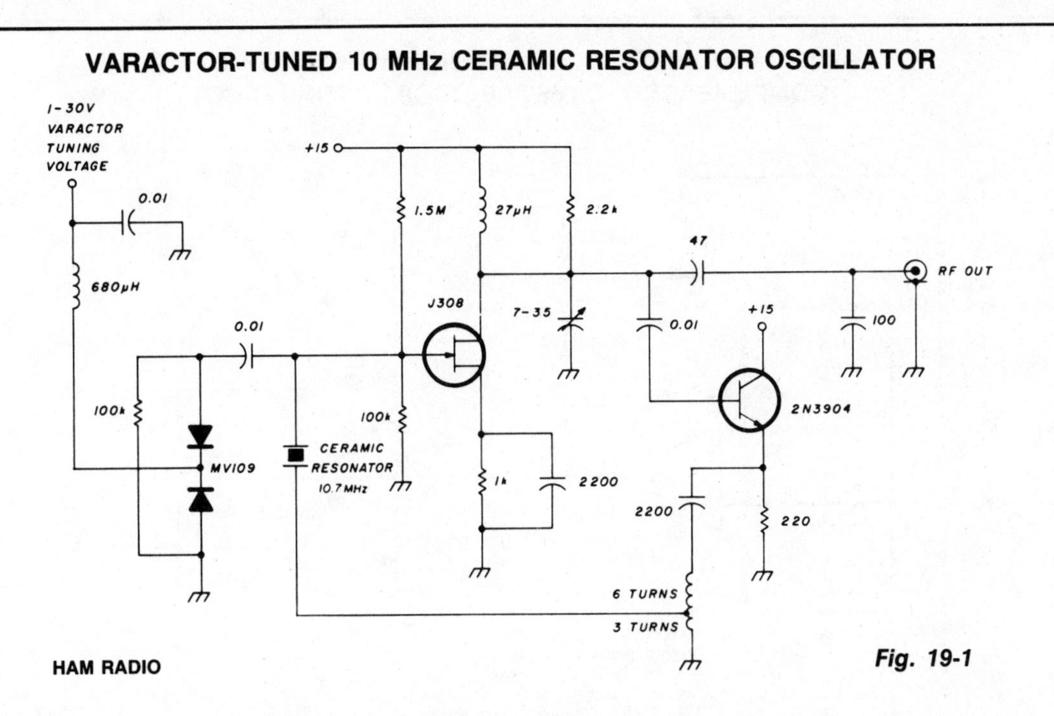

HAM RADIO

Fig. 19-1

Circuit Notes

The FET input amplifier has fixed bias with source feedback. This provides a very high input impedance with very low capacitance. The FET amplifier drives an emitter follower which, in spite of the fact that it has a low output impedance, feeds a transformer with a 3:1 turns ratio for a nine-fold impedance reduction. The result is an impedance at the ceramic resonator of a few ohms maximum. The varactor-tuned ceramic resonator oscillator has a significant frequency-temperature coefficient. The tuning range of the VCO is approximately 232 kHz, with a temperature coefficient of 350 Hz per degree centigrade. When using this circuit as a VCO, the entire 232 kHz range cannot be used because some of the tuning range must be sacrificed for the temperature dependence. If the required tuning range were 200 kHz, leaving 32 kHz for temperature variation, the resulting temperature variation would be more than 90°C.

10 MHz CRYSTAL OSCILLATOR

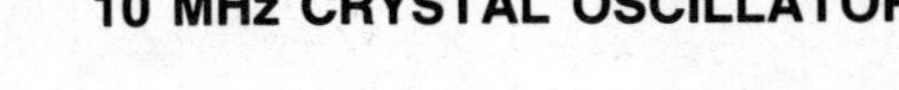

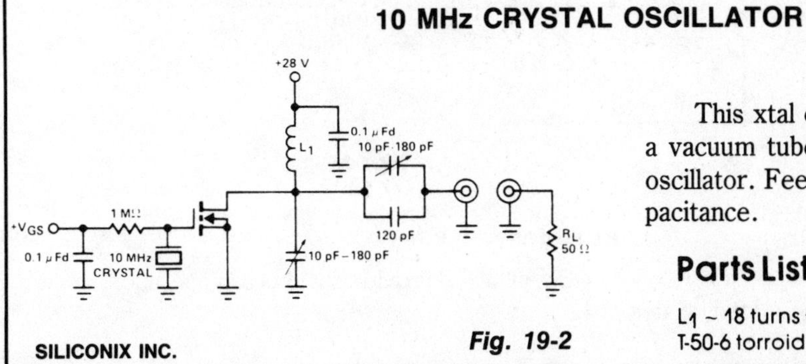

SILICONIX INC.

Fig. 19-2

Circuit Notes

This xtal oscillator is a FET equivalent of a vacuum tube tuned to plate-tuned grid xtal oscillator. Feedback is via the drain to gate capacitance.

Parts List

L_1 ~ 18 turns #22 enameled wire on micrometals T-50-6 torroid core. ≈ 1.0 μH.

LOW POWER, 5 V DRIVEN, TEMPERATURE COMPENSATED CRYSTAL OSCILLATOR (TXCO)

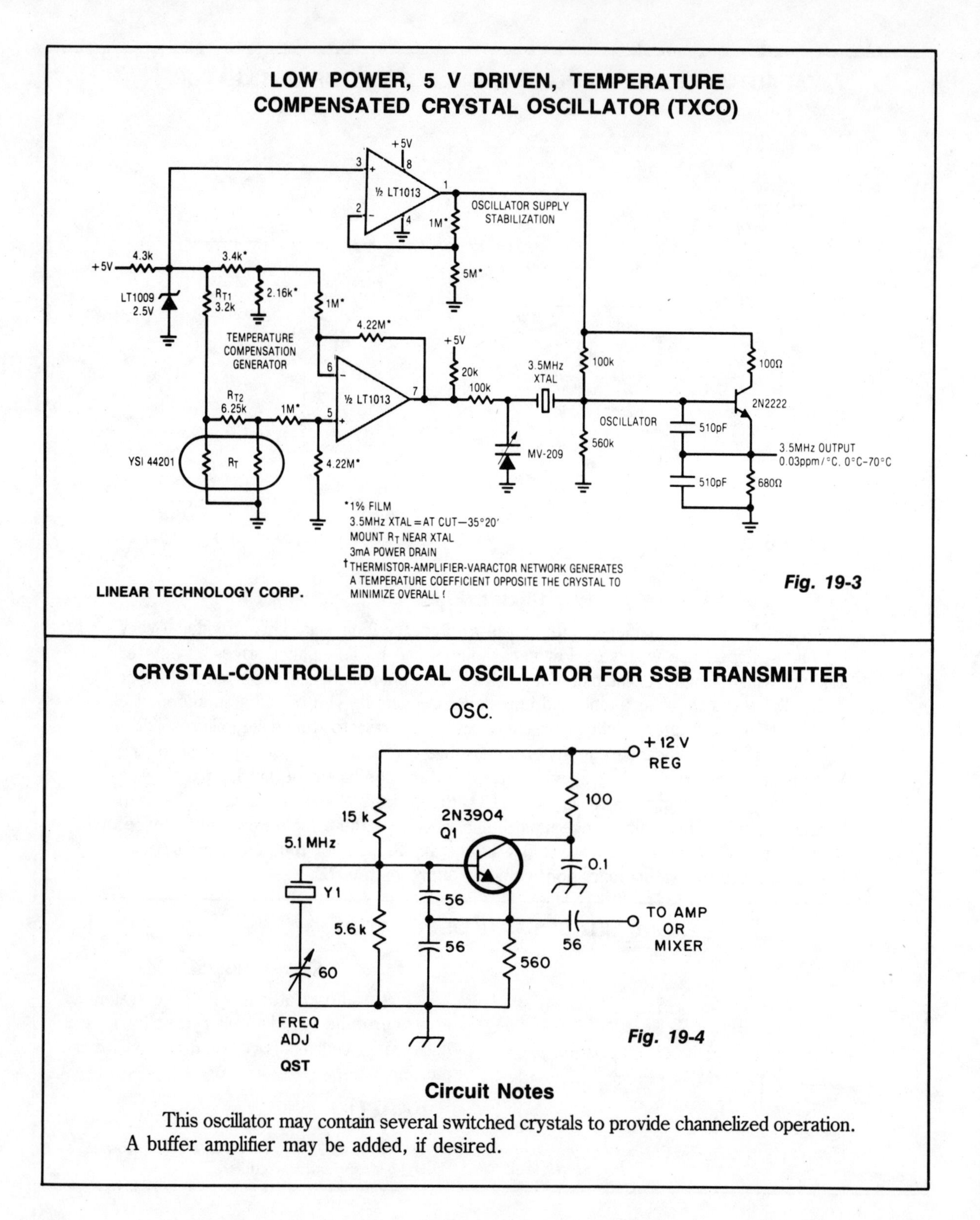

LINEAR TECHNOLOGY CORP.

Fig. 19-3

CRYSTAL-CONTROLLED LOCAL OSCILLATOR FOR SSB TRANSMITTER

QST

Fig. 19-4

Circuit Notes

This oscillator may contain several switched crystals to provide channelized operation. A buffer amplifier may be added, if desired.

CRYSTAL OSCILLATOR

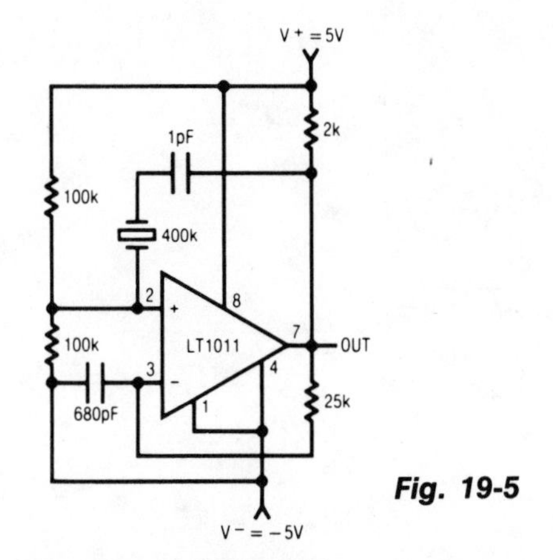

Fig. 19-5

Circuit Notes

This circuit uses an LT1011 comparator biased in its linear mode and a crystal to establish its resonant frequency. This circuit can achieve a few hundred kHz, temperature independent clock frequency with nearly 50% duty cycle.

CRYSTAL-CONTROLLED SIGNAL SOURCE

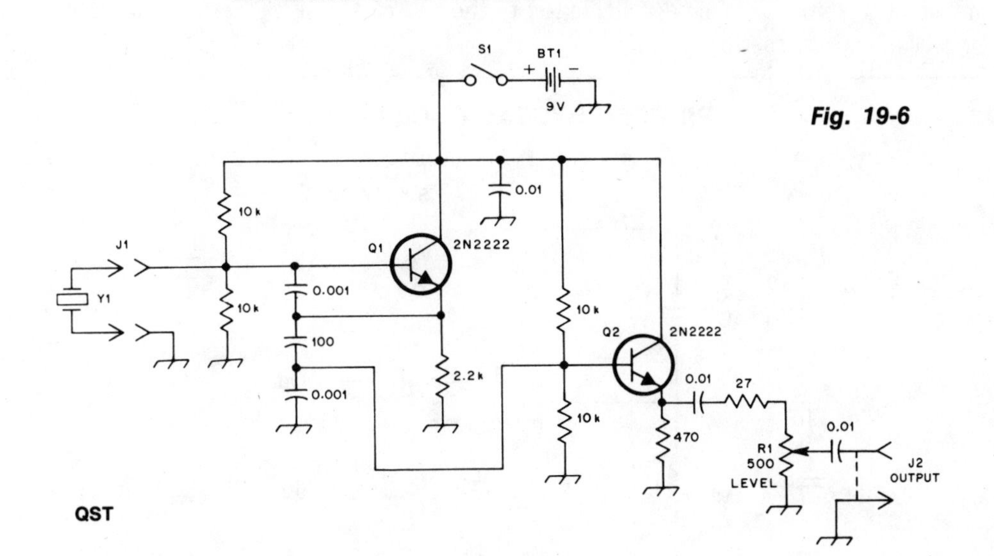

Fig. 19-6

Circuit Notes

This general purpose signal source serves very well in signal-tracing applications. The output level is variable to more than 1 Vrms into a 50 Ω load. Almost any crystal in the 1 to 15 MHz range can be used. Q1 forms a Colpitts oscillator with the output taken from the emitter. A capacitive voltage divider (across the 2.2 K emitter resistor) reduces the voltage applied to the buffer amplifier, Q2. The buffer and emitter follower, provides the low input impedance necessary to drive 50 Ω loads.

1 MHz FET CRYSTAL OSCILLATOR

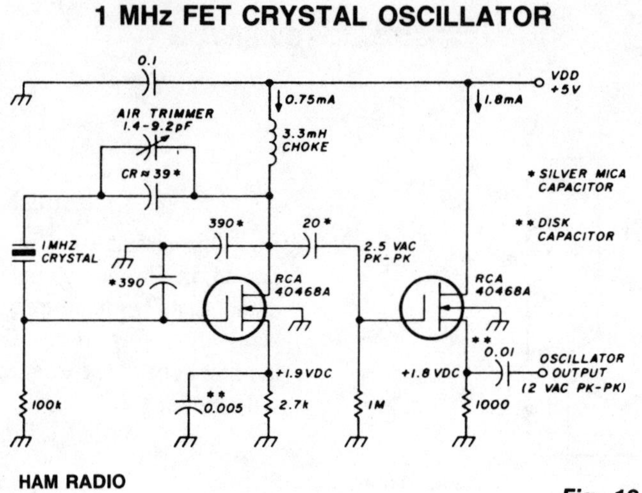

HAM RADIO

Fig. 19-7

Circuit Notes

This stable oscillator circuit exhibits less than 1 Hz frequency change over a V_{DD} range of 3–9 volts. Stability is attributed to the use of MOSFET devices and the use of stable capacitors.

PIERCE CRYSTAL OSCILLATOR

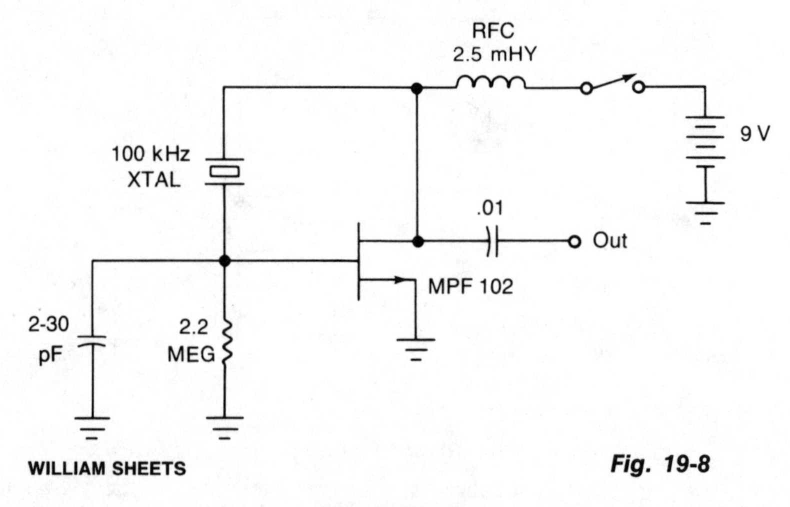

WILLIAM SHEETS

Fig. 19-8

Circuit Notes

The JFET Pierce oscillator is stable and simple. It can be the clock of a microprocessor, a digital timepiece or a calculator. With a probe at the output, it can be used as a precise injection oscillator for troubleshooting. Attach a small length of wire at the output and this circuit becomes a micropower transmitter.

IC-COMPATIBLE CRYSTAL OSCILLATOR

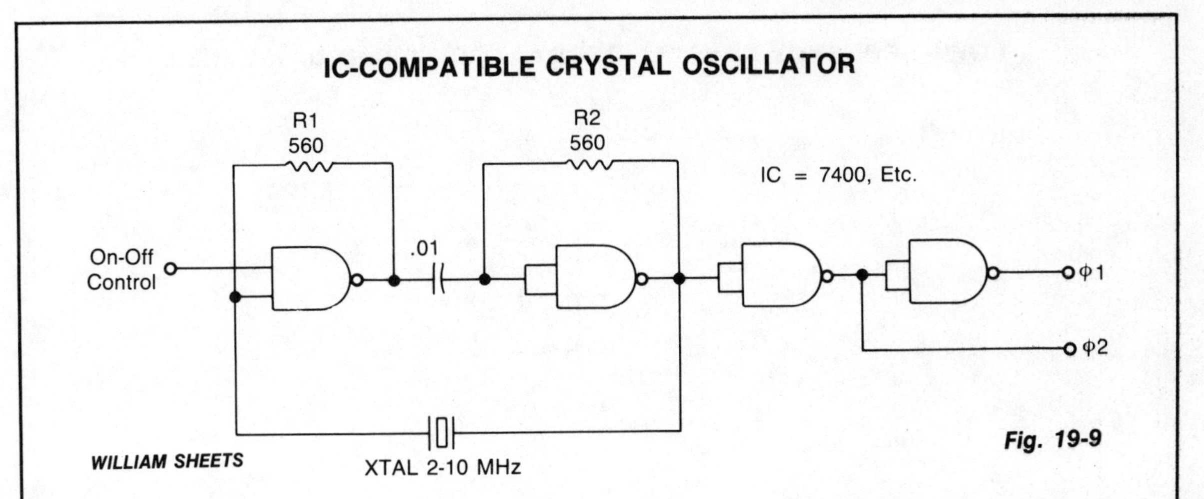

WILLIAM SHEETS

Fig. 19-9

Circuit Notes

Resistors R1 and R2 temperature-stabilize the NAND gates; they also ensure that the gates are in a linear region for starting. Capacitor C1 is a dc block; it must have less than 1/10 ohm impedance at the operating frequency. The crystal runs in a series-resonant mode. Its series resistance must be low; AT-cut crystals for the 1- to 10-MHz range work well. The output waveshape has nearly a 50% duty cycle, with chip-limited rise times. The circuit starts well from 0° to 70°C.

CRYSTAL OSCILLATOR PROVIDES LOW NOISE

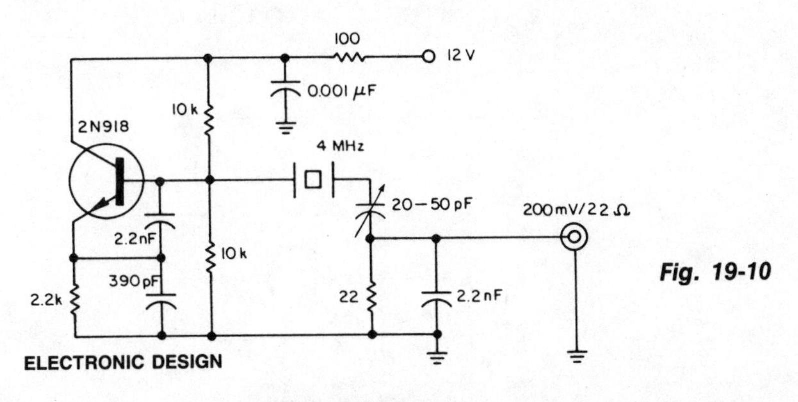

ELECTRONIC DESIGN

Fig. 19-10

Circuit Notes

The oscillator delivers an output of high spectral purity without any substantial sacrifice of the usual stability of a crystal oscillator. The crystal in addition to determining the oscillator's frequency, is used also as a low-pass filter for the unwanted harmonics and as a bandpass filter for the sideband noise. The noise bandwidth is limited to less than 100 Hz. All higher harmonics are substantially suppressed—60 dB down for the third harmonic of the 4-MHz fundamental oscillator frequency.

LOW-FREQUENCY CRYSTAL OSCILLATOR—10 kHz to 150 kHz

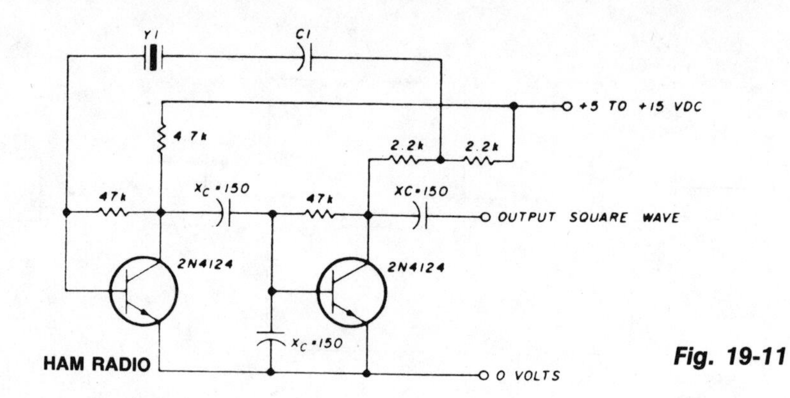

HAM RADIO

Fig. 19-11

Circuit Notes

C1 in series with the crystal may be used to adjust the oscillator output frequency. Value may range between 20 pF and 0.01 μF, or may be a trimmer capacitor and will approximately equal the crystal load capacitance. X values are approximate and can vary for most circuits and frequencies; this is also true for resistance values. Adequate power supply decoupling is required; local decoupling capacitors near the oscillator are recommended. All leads should be extremely short in high frequency circuits.

OVERTONE CRYSTAL OSCILLATOR

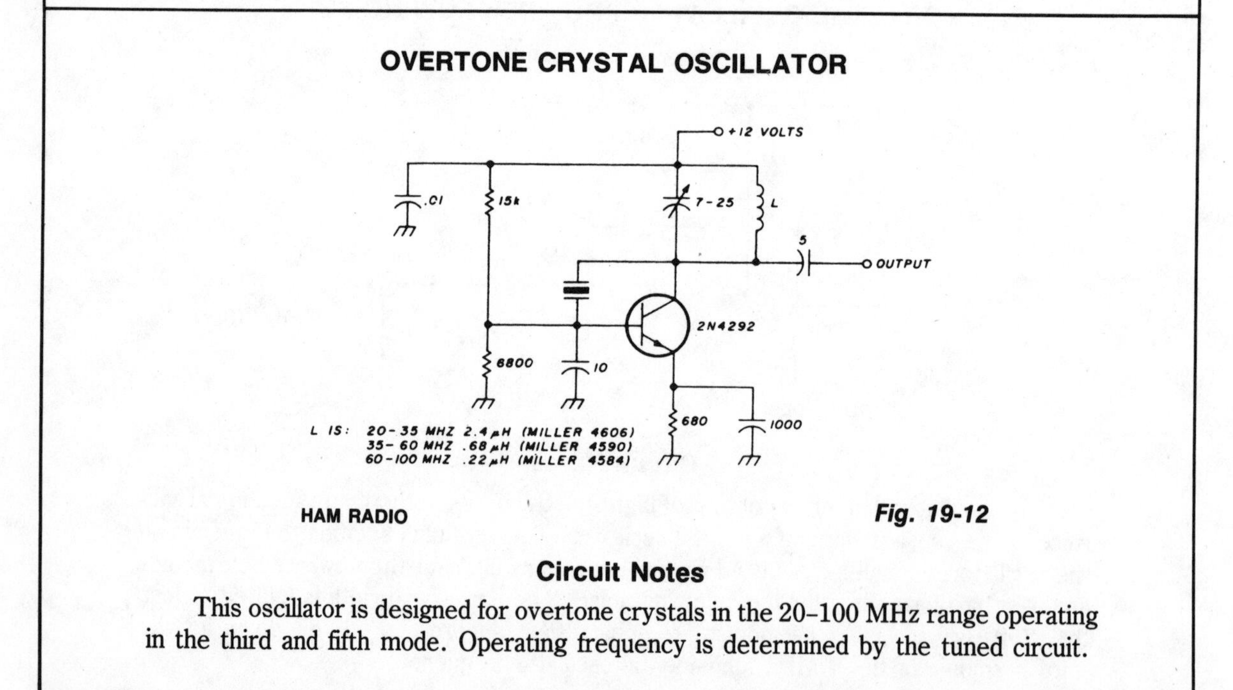

HAM RADIO

Fig. 19-12

Circuit Notes

This oscillator is designed for overtone crystals in the 20–100 MHz range operating in the third and fifth mode. Operating frequency is determined by the tuned circuit.

COLPITTS OSCILLATOR

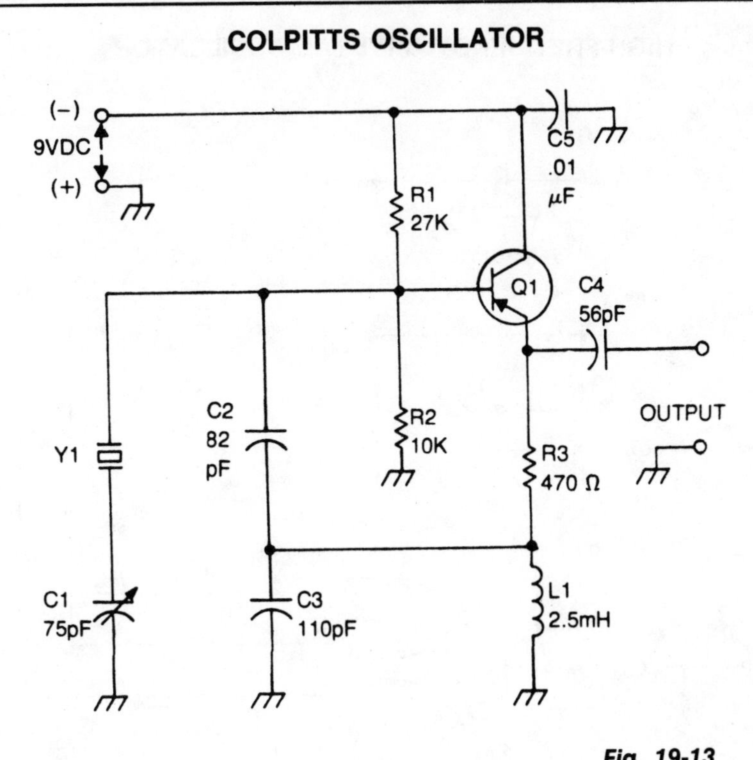

TAB BOOKS, INC.

Fig. 19-13

Circuit Notes

Bias for the pnp bipolar transistor is provided by resistor voltage divider network R1/R2. The collector of the oscillator transistor is kept at ac ground by capacitor C5, placed close to the transistor. Feedback is provided by capacitor voltage divider C2/C3.

CRYSTAL-CONTROLLED OSCILLATOR

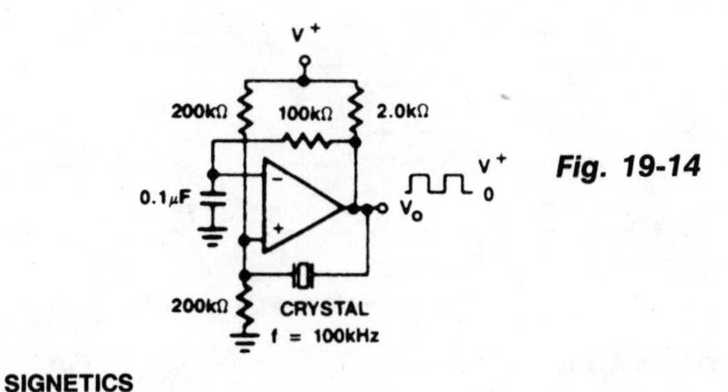

Fig. 19-14

SIGNETICS

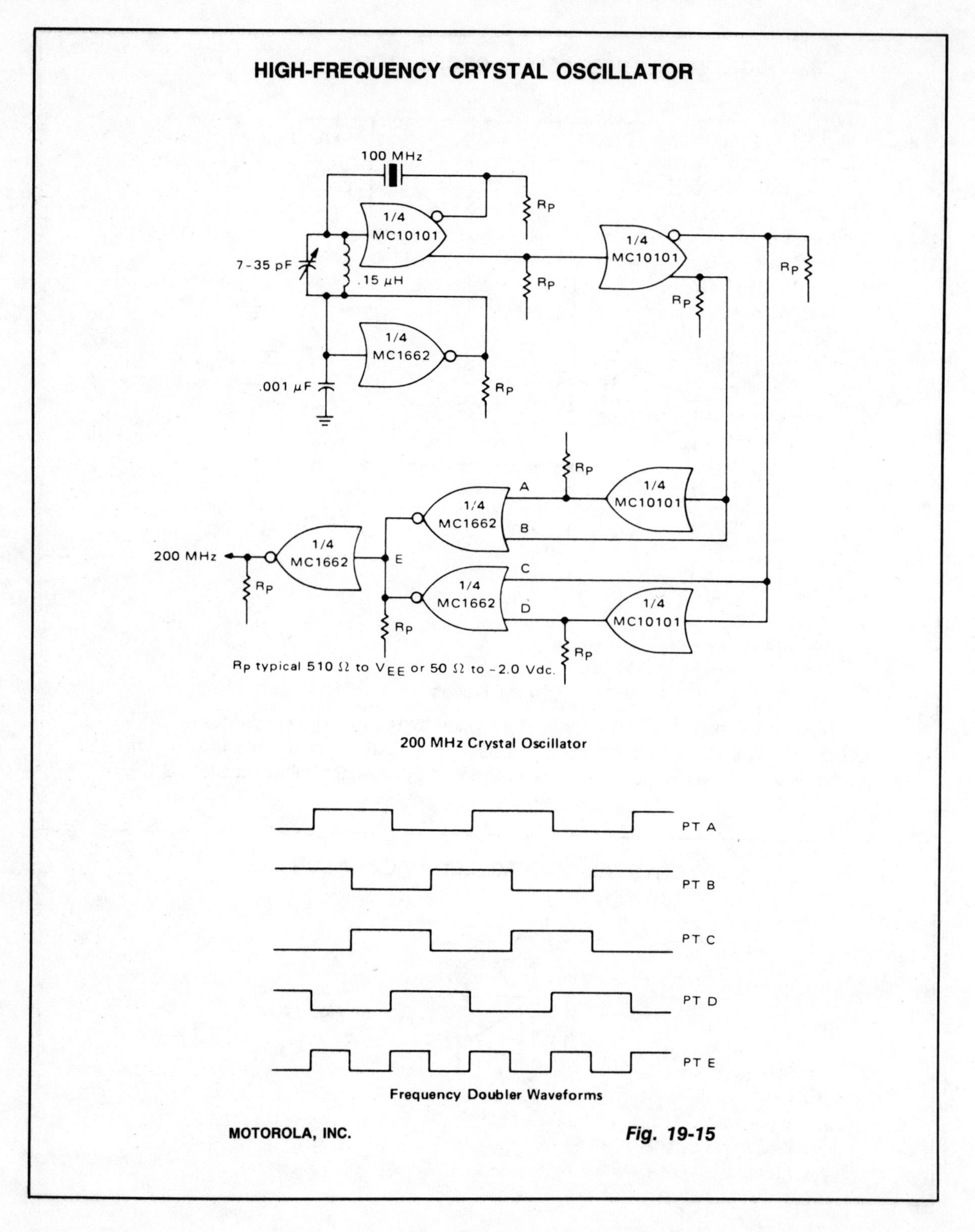

200 MHz Crystal Oscillator

Frequency Doubler Waveforms

Fig. 19-15

HIGH-FREQUENCY CRYSTAL OSCILLATOR, Continued.

Circuit Notes

A high speed oscillator is possible by combining an MECL 10 K crystal oscillator with an MECL III frequency doubler as shown. One section of the MC10101 is connected as a 100 MHz crystal oscillator with the crystal in series with the feedback loop. The LC tank circuit tunes the 100 MHz harmonic of the crystal and may be used to calibrate the circuit to the exact frequency. A second section of the MC10101 buffers the crystal oscillator and gives complementary 100 MHz signals. The frequency doubler consists of two MC10101 gates as phase shifters and two MC1662 NOR gates. For a 50% duty cycle at the output, the delay to the true and complement 100 MHz signals should be 90°. This may be built precisely with 2.5 ns delay lines for the 200 MHz output or approximated by the two MC10101 gates. The gates are easier to incorporate and cause only a slight skew in output signal duty cycle. The MC1662 gates combine the 4 phase 100 MHz signals as shown in Figure B. The outputs of the MC1662's are wire-OR connected to give the 200 MHz signal. MECL III gates are used because of the bandwidth required for 200 MHz signals. One of the remaining MC1662 gates is used as a V_{BB} bias generator for the oscillator. By connecting the NOR output to the input, the circuit stays in the center of the logic swing or at V_{BB}. A 0.001 μF capacitor ensures the V_{BB} circuit does not oscillate.

CRYSTAL-CONTROLLED OSCILLATOR OPERATES FROM ONE MERCURY CELL

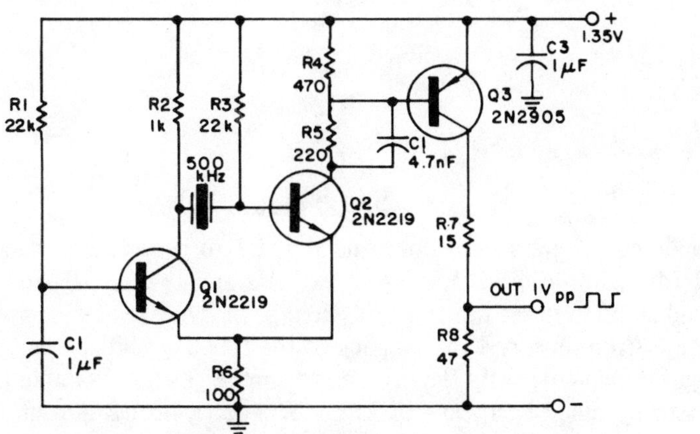

Fig. 19-16

ELECTRONIC DESIGN

Inexpensive crystal controlled oscillator operates from a 1.35-volt source.

Circuit Notes

The circuit is powered by a single 1.35 V mercury cell and provides a 1 V square-wave output. As shown, the crystal is a tuned circuit between transistors Q1 and Q2, which are connected in the common-emitter configuration. Positive feedback provided by means of R permits oscillation. The signal at the collector of Q2 is squared by Q3, which switches between cutoff and saturation. R7 permits short-circuit-proof operation.

HIGH FREQUENCY SIGNAL GENERATOR

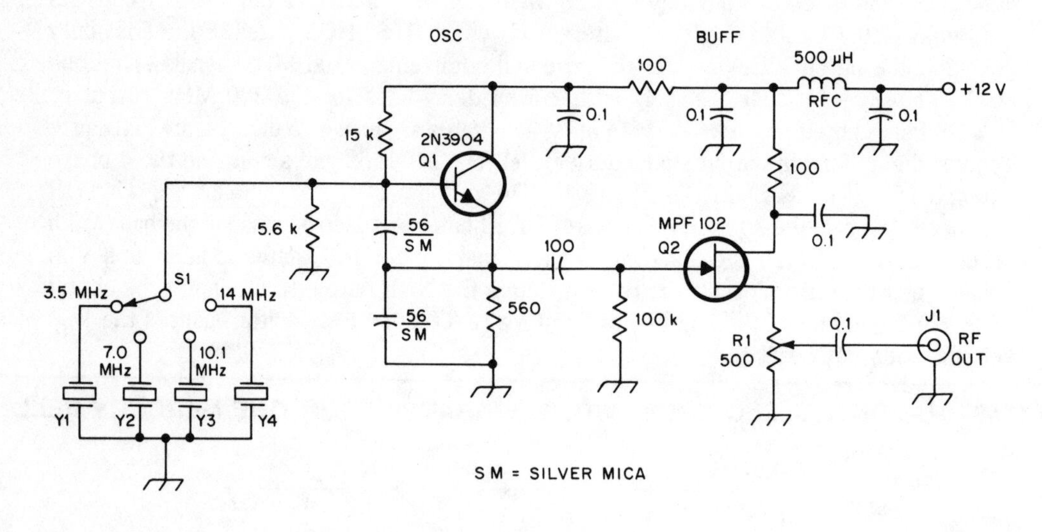

QST

Fig. 19-17

Circuit Notes

A tapped-coil Colpitts oscillator is used at Q1 to provide four tuning ranges from 1.7 to 3.1 MHz, 3.0 to 5.6 MHz, 5.0 to 12 MHz and 11.5 to 31 MHz. A Zener diode (D2) is used at Q1 to lower the operating voltage of the oscillator. A small value capacitor is used at C5 to ensure light coupling to the tuned circuit. Q2 is a source-follower buffer stage. It helps to isolate the oscillator from the generator-output load. The source of Q2 is broadly tuned by means of RFC1. Energy from Q2 is routed to a fed-back, broadband class-A amplifier. A 2 dB attenuator is used at the output of T1 to provide a 50 ohm termination for Q3 and to set the generator-output impedance at 50 ohms. C16, C17 and RFC2 form a brute-force RF decoupling network to keep the generator energy from radiating outside the box on the 12 V supply.

CRYSTAL TESTER

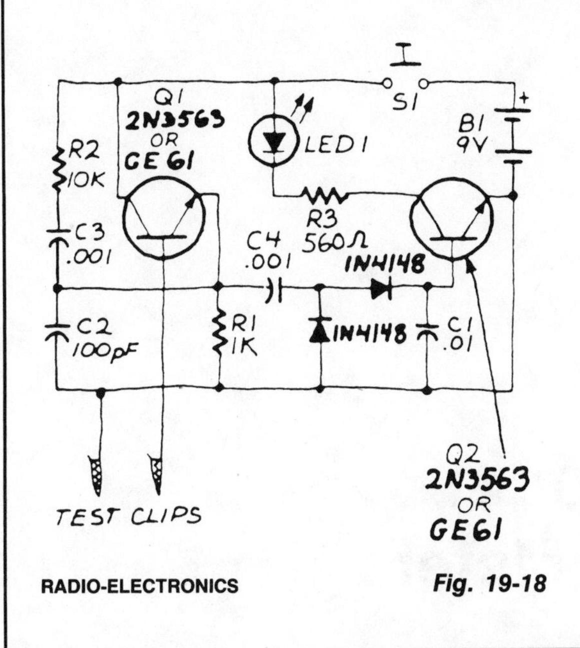

RADIO-ELECTRONICS

Fig. 19-18

Circuit Notes

Transistor Q1, a 2N3563, and its associated components form an oscillator circuit that will oscillate if, and only if, a good crystal is connected to the test clips. The output from the oscillator is then rectified by the two 1N4148 diodes and filtered by C1, a .01 μF capacitor. The positive voltage developed across the capacitor is applied to the base of Q2, another 2N3563, causing it to conduct. When that happens, current flows through LED1, causing it to glow. Since only a good crystal will oscillate, a glowing LED indicates that the crystal is indeed OK. The circuit is powered by a standard nine-volt transistor-radio battery and the SPST pushbutton power-switch is included to prolong battery life.

CRYSTAL-STABILIZED IC TIMER CAN PROVIDE SUBHARMONIC FREQUENCIES

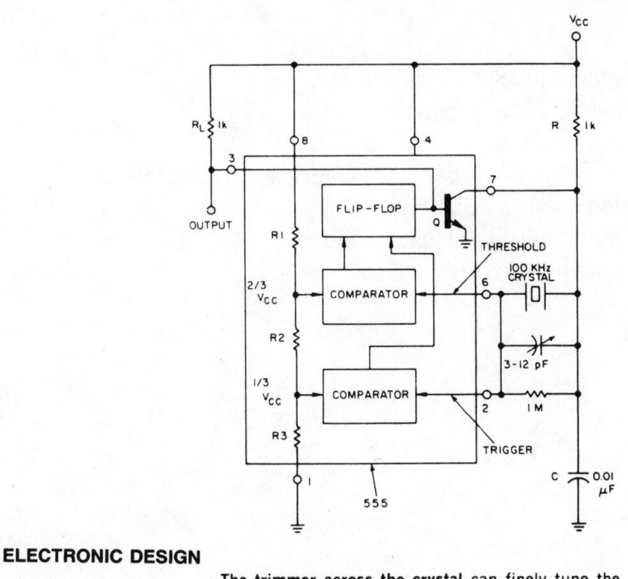

ELECTRONIC DESIGN

Fig. 19-19

The trimmer across the crystal can finely tune the circuit's oscillating frequency.

20
Current Meters

The sources of the following circuits are contained in the Sources section beginning on page 694. The figure number contained in the box of each circuit correlates to the source entry in the Sources section.

AMMETER WITH SIX DECADE RANGE

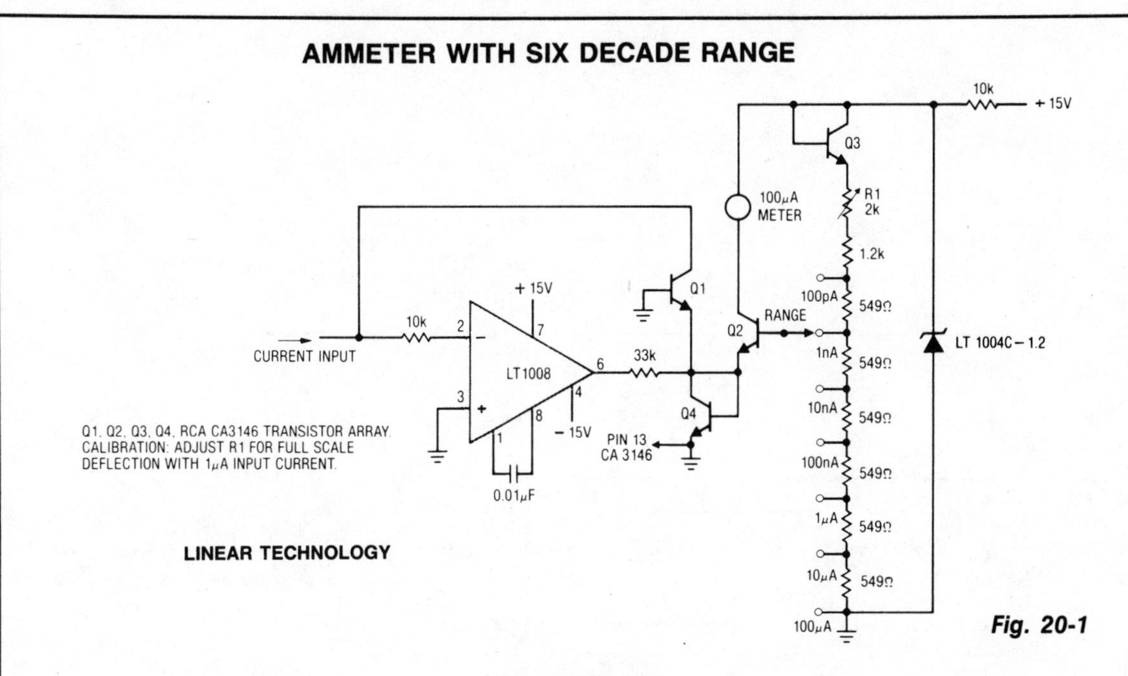

LINEAR TECHNOLOGY

Fig. 20-1

Circuit Notes

The Ammeter measures currents from 100 pA to 100 μA without the use of expensive high value resistors. Accuracy at 100 μA is limited by the offset voltage between Q1 and Q2 and, at 100 pA, by the inverting bias current of the LT1008.

CURRENT SENSING IN SUPPLY RAILS

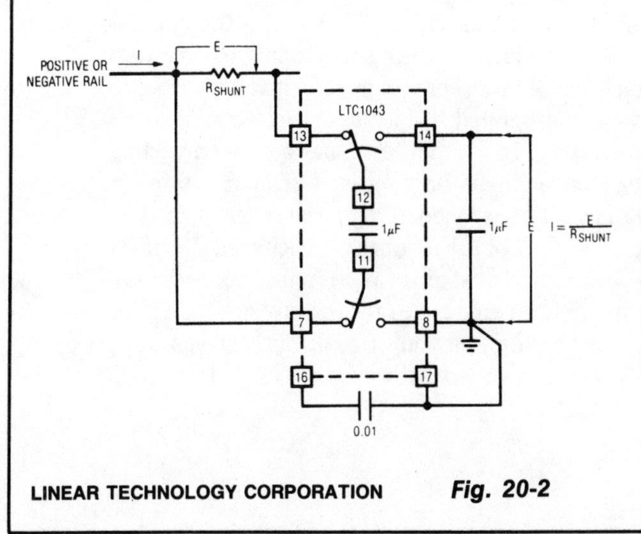

LINEAR TECHNOLOGY CORPORATION

Fig. 20-2

Circuit Notes

The LTC1043 can sense current through a shunt in either of its supply rails. This capability has wide application in battery and solar-powered systems. If the ground-referred voltage output is unloaded by an amplifier, the shunt can operate with very little voltage drop across it, minimizing losses.

PICO AMMETER

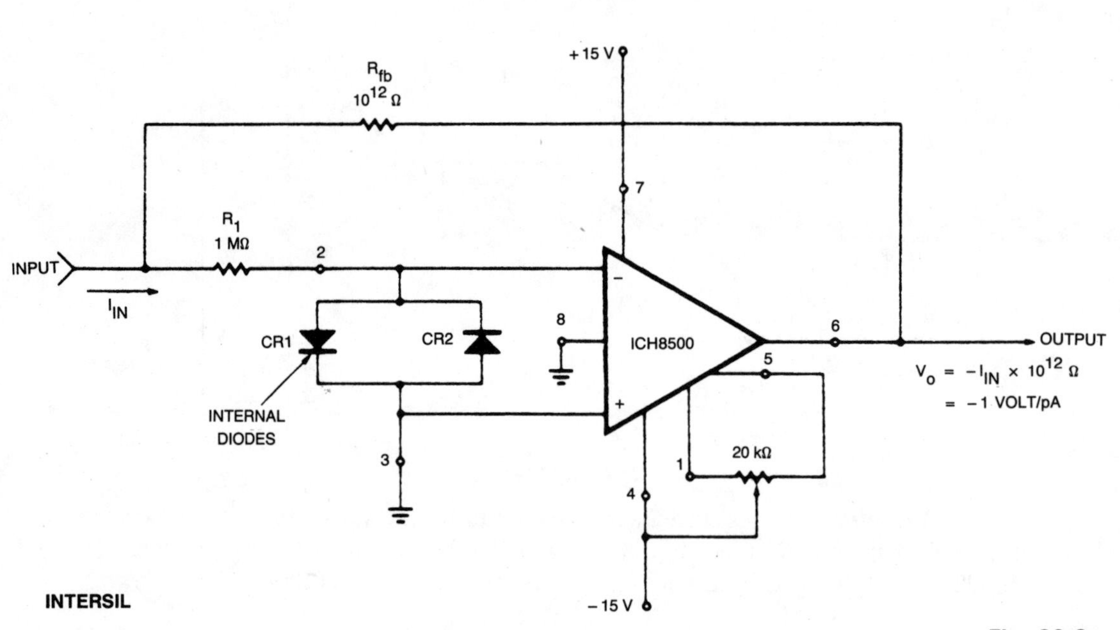

Fig. 20-3

Circuit Notes

Care must be taken to eliminate any stray currents from flowing into the current summing node. This can be accomplished by forcing all points surrounding the input to the same potential as the input. In this case the potential of the input is at virtual ground, or OV. Therefore, the case of the device is grounded to intercept any stray leakage currents that may otherwise exist between the ±15 V input terminals and the inverting input summing junctions. Feedback capacitance should be kept to a minimum in order to maximize the response time of the circuit to step function input currents. The time constant of the circuit is approximately the produce of the feedback capacitance C_{fb} times the feedback resistor R_{fb}. For instance, the time constant of the circuit is 1 sec if C_{fb} = 1pF. Thus, it takes approximately 5 sec (5 time constants) for the circuit to stabilize to within 1% of its final output voltage after a step function of input current has been applied. C_{fb} of less than 0.2 to 0.3 pF can be achieved with proper circuit layout.

ELECTROMETER AMPLIFIER WITH OVERLOAD PROTECTION

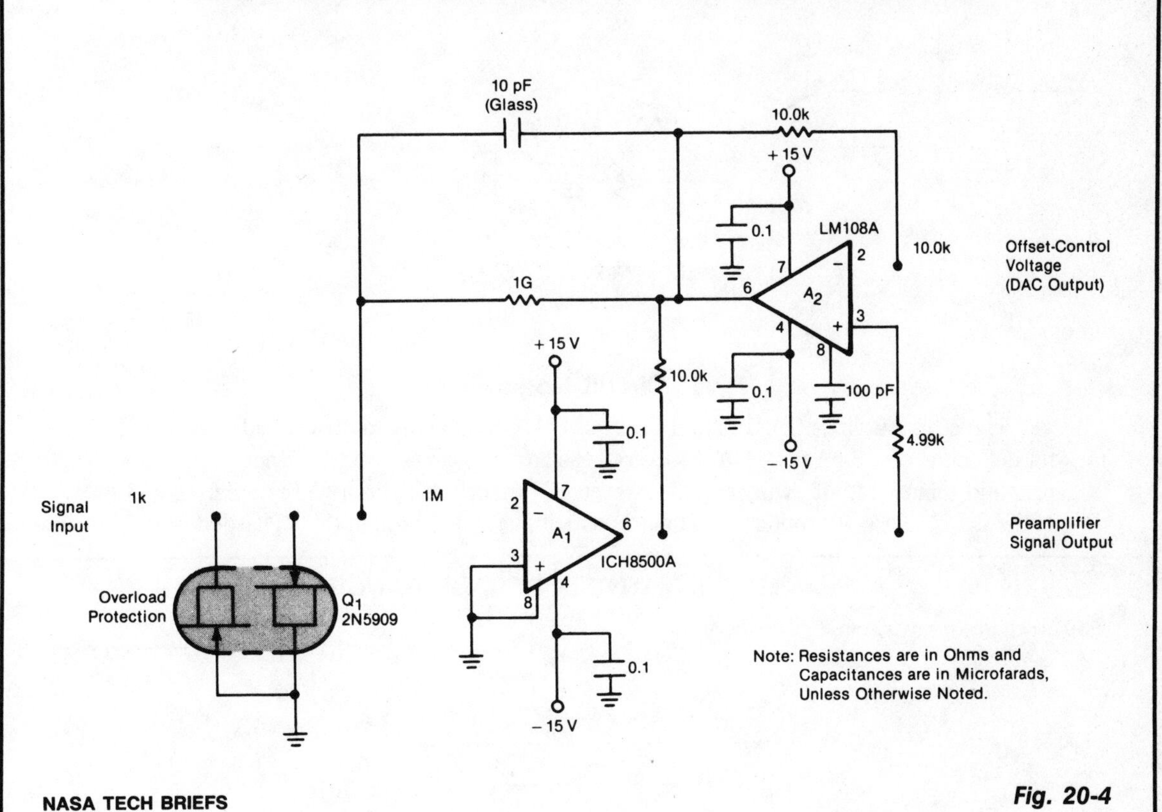

NASA TECH BRIEFS *Fig. 20-4*

Circuit Notes

The preamplifier is protected from excessive input signals of either polarity by the 2N5909 junction field-effect transistor. A nulling circuit makes it possible to set the preamplifier output voltage to zero at a fixed low level (up to $\pm 10^{-8}$A) of the input current. (This level is called the standing current and corresponds to the zero-signal level of the instrumentation.) The opposing (offset) current is generated in the 10^9 feedback resistor to buck the standing current. Different current ranges are reached by feeding the preamplifier output to low and high gain amplifier chains. To reduce noise, each chain includes a 1.5 Hz corner active filter.

GUARDED INPUT PICOAMMETER CIRCUIT

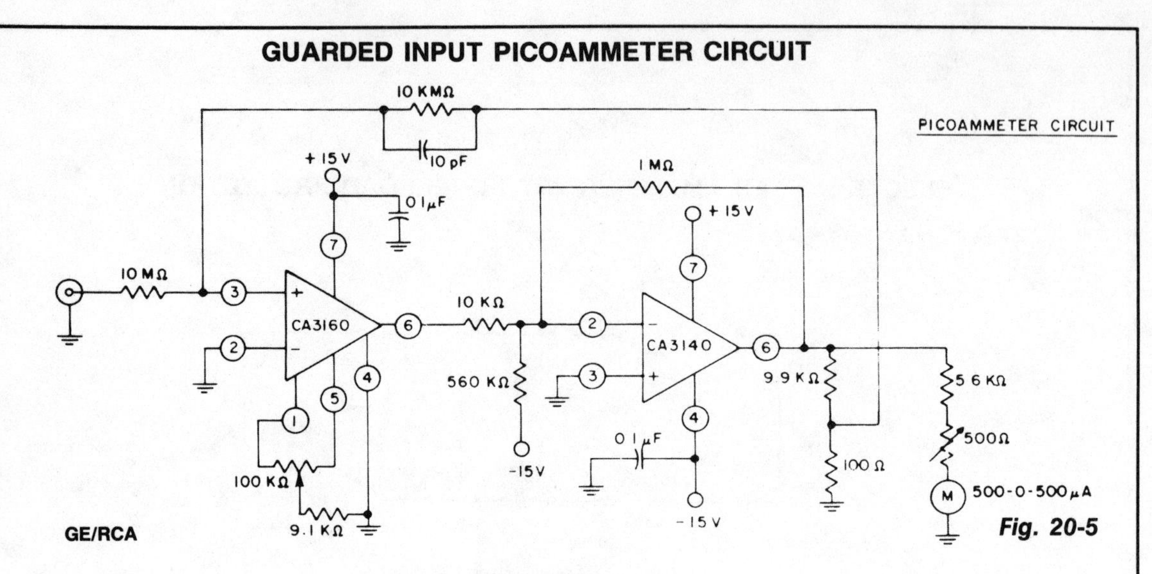

Fig. 20-5

Circuit Notes

The circuit utilizes CA3160 and CA3140 BiMOS op amps to provide a full-scale meter deflection of ±3 pA. The CA3140 serves as an X100 gain stage to provide the required plus and minus output swing for the meter and feedback network. Terminals 2 and 4 of the CA3160 are at ground potential, thus its input is operated in the "guarded mode."

AMMETER WITH SIX DECADE RANGE

LINEAR TECHNOLOGY CORP.

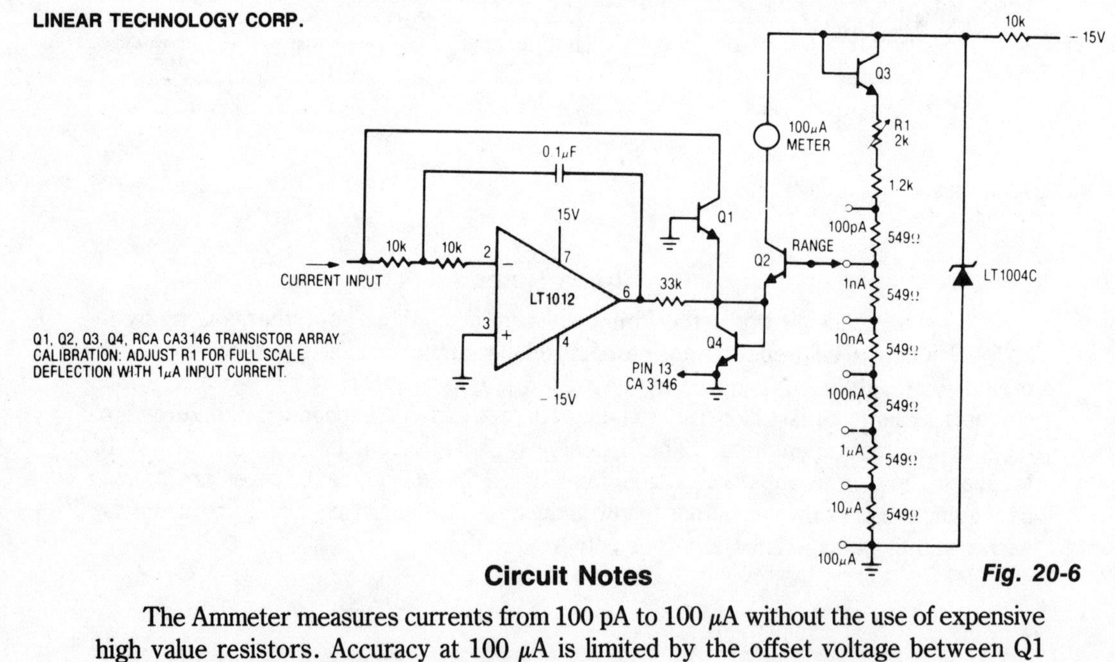

Circuit Notes

Fig. 20-6

The Ammeter measures currents from 100 pA to 100 μA without the use of expensive high value resistors. Accuracy at 100 μA is limited by the offset voltage between Q1 and Q2 and, at 100 pA, by the inverting bias current of the LT1008.

PICOAMMETER CIRCUIT

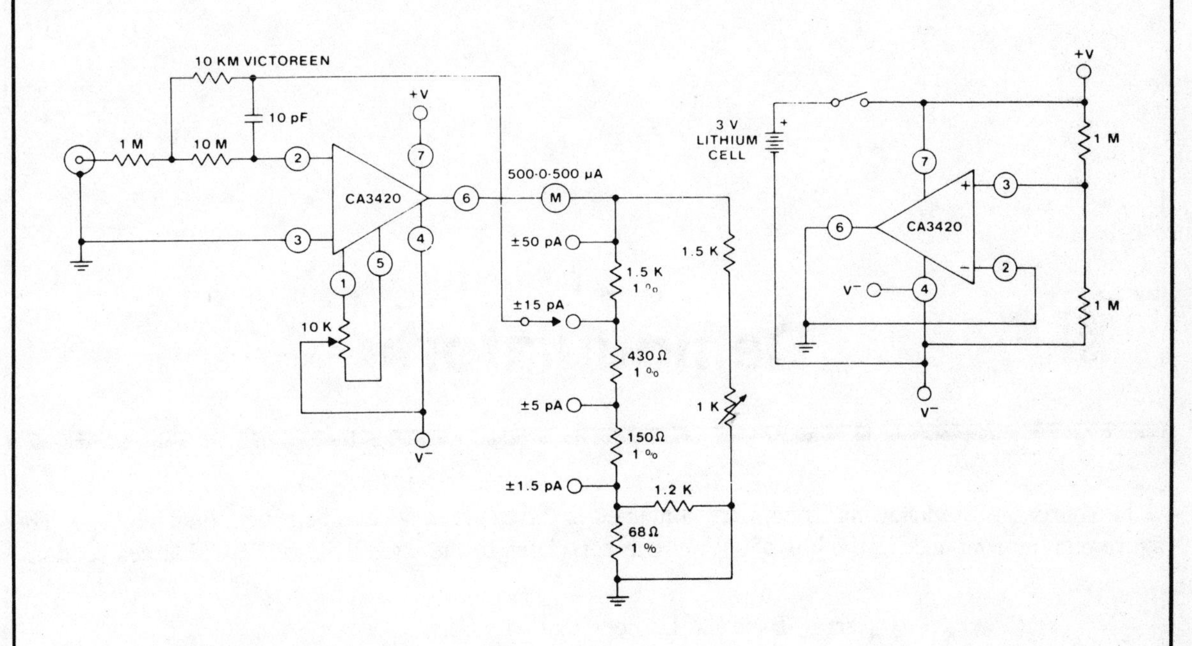

GENERAL ELECTRIC /RCA

Fig. 20-7

Circuit Notes

The circuit uses the exceptionally low input current (0.1pA) of the CA3420 BiMOS op amp. With only a single 10 megohm resistor, the circuit covers the range from ±50 pA maximum to a full-scale sensitivity of ±1.5 pA. Using an additional CA3420, a low-resistance center tap is obtained from a single 3-volt lithium battery.

21
Demodulators

The sources of the following circuits are contained in the Sources section beginning on page 694. The figure number contained in the box of each circuit correlates to the source entry in the Sources section.

Narrow Band FM Demodulator with Carrier Detect
Stereo Demodulator
AM Demodulator
FM Demodulator

NARROW BAND FM DEMODULATOR WITH CARRIER DETECT

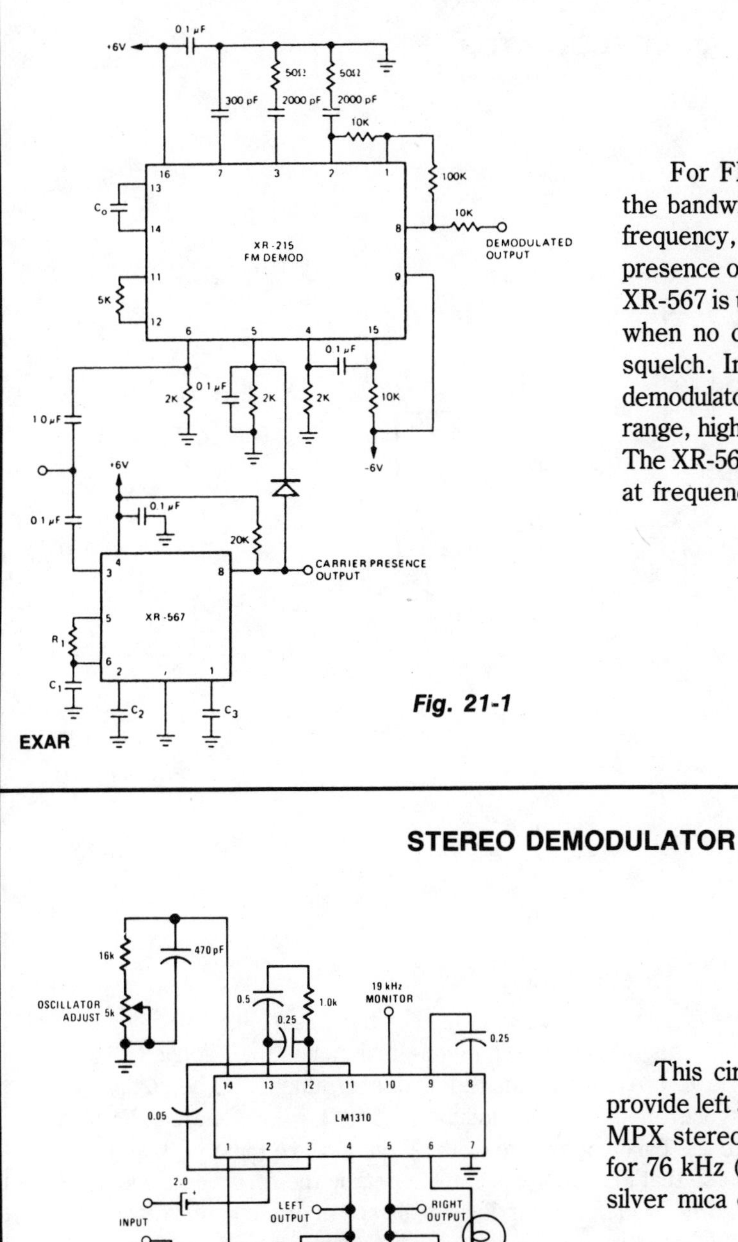

EXAR

Fig. 21-1

Circuit Notes

For FM demodulation applications where the bandwidth is less than 10% of the carrier frequency, an XR-567 can be used to detect the presence of the carrier signal. The output of the XR-567 is used to turn off the FM demodulator when no carrier is present, thus acting as a squelch. In the circuit shown, an XR-215 FM demodulator is used because of its wide dynamic range, high signal/noise ratio and low distortion. The XR-567 will detect the presence of a carrier at frequencies up to 500 kHz.

STEREO DEMODULATOR

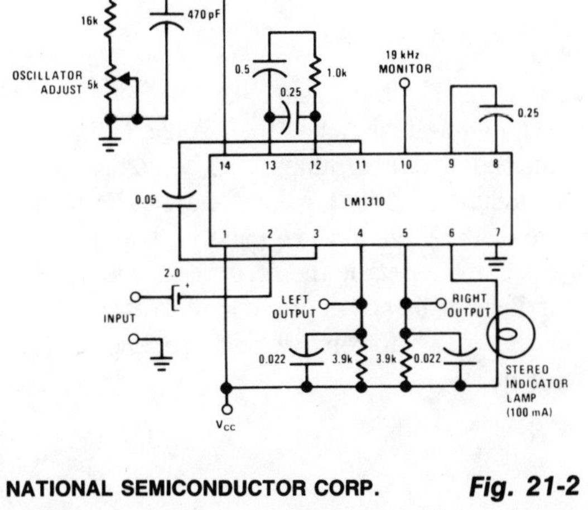

NATIONAL SEMICONDUCTOR CORP.

Fig. 21-2

Circuit Notes

This circuit uses a single IC LM1310 to provide left and right outputs from a composite MPX stereo signal. Oscillator adjust R1 is set for 76 kHz (19 kHz at pin 10). C1 should be a silver mica or NPO ceramic capacitor.

AM DEMODULATOR

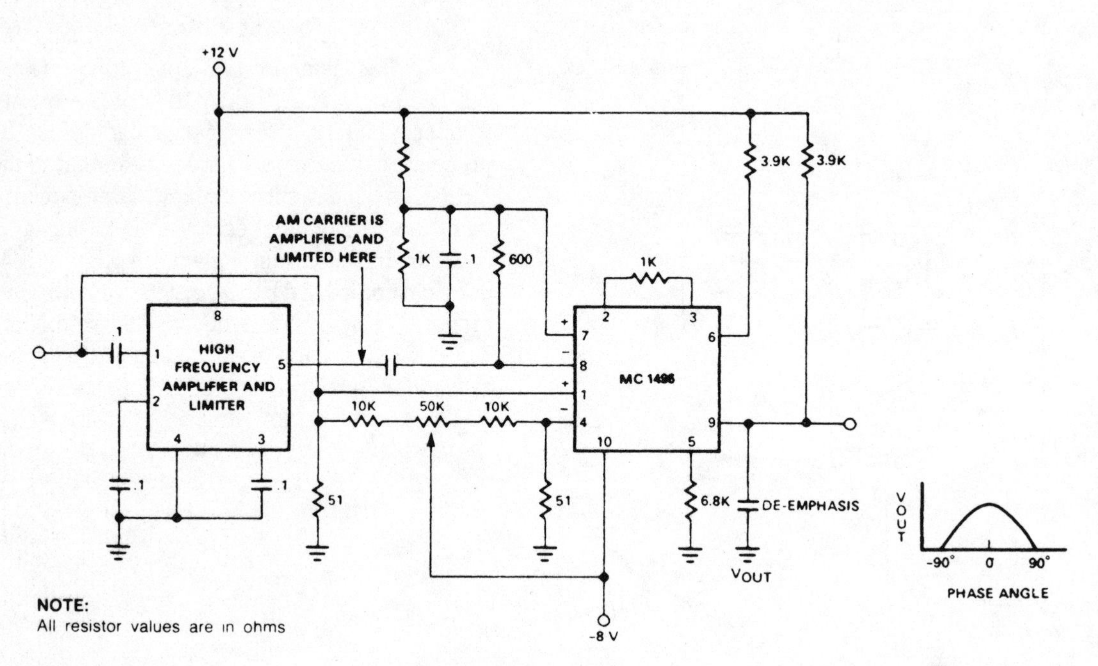

NOTE:
All resistor values are in ohms

Fig. 21-3

SIGNETICS

Circuit Notes

Amplifying and limiting of the AM carrier is accomplished by the if gain block providing 55 dB of gain or higher with a limiting of 40 μV. The limited carrier is then applied to the detector at the carrier ports to provide the desired switching function. The signal is then demodulated by the synchronous AM demodulator (1496) where the carrier frequency is attentuated due to the balanced nature of the device. Care must be taken not to overdrive the signal input so that distortion does not appear in the recorded audio. Maximum conversion gain is reached when the carrier signals are in phase as indicated by the phase-gain relationship. Output filtering is also necessary to remove high frequency sum components of the carrier from the audio signal.

FM DEMODULATOR

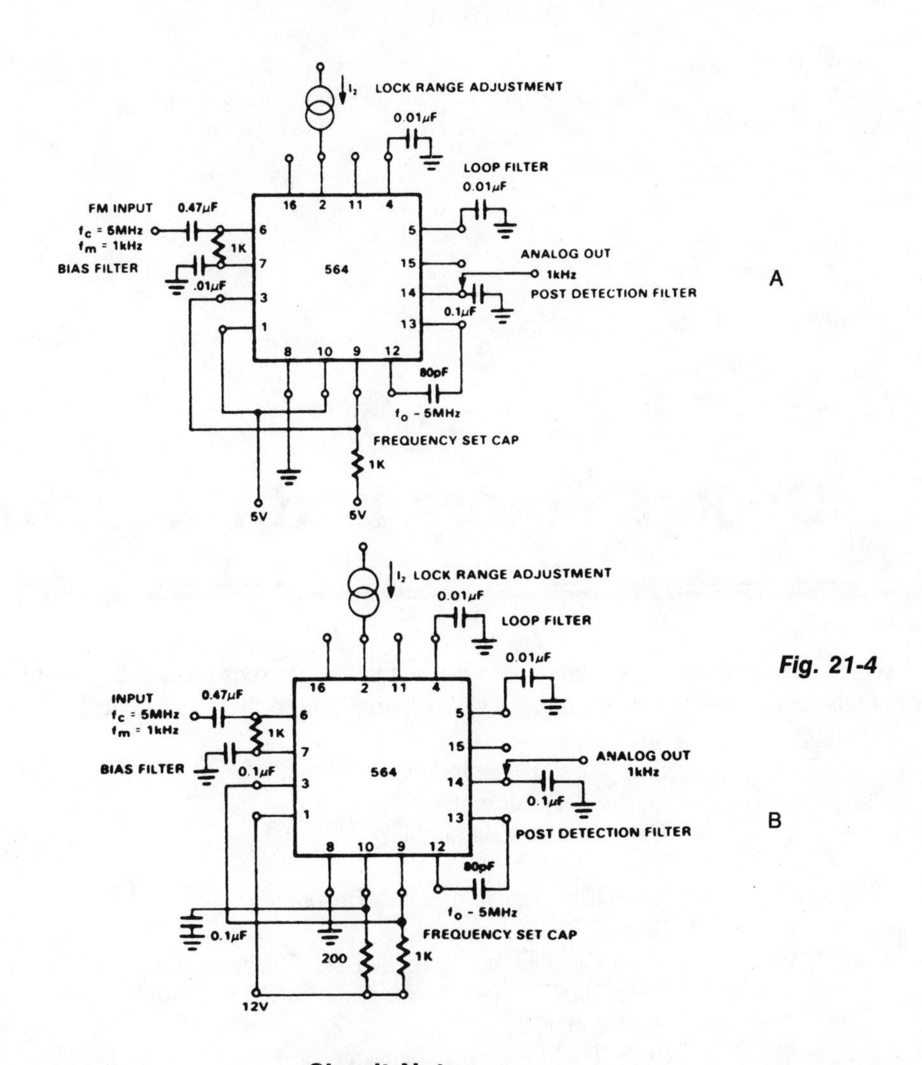

Fig. 21-4

SIGNETICS

Circuit Notes

The NE564 is used as an FM demodulator. The connections for operation at 5 V and 12 V are shown in Figures 21-4A and 21-4B. The input signal is ac coupled with the output signal being extracted at Pin 14. Loop filtering is provided by the capacitors at Pins 4 and 5 with additional filtering being provided by the capacitor at Pin 14. Since the conversion gain of the VCO is not very high, to obtain sufficient demodulated output signal the frequency deviation in the input signal should be 1% or higher.

22
Descramblers and Decoders

The sources of the following circuits are contained in the Sources section beginning on page 694. The figure number contained in the box of each circuit correlates to the source entry in the Sources section.

Sine Wave Descrambler
Outband Descrambler
Gated Pulse Descrambler
SCA Decoder
Dual Time Constant Tone-Decoder
Stereo TV Decoder
Time Division Multiplex (TDM) Stereo Decoder
Frequency Division Multiplex (FDM) Stereo Decoder
SCA (Background Music) Decoder

SINE WAVE DESCRAMBLER

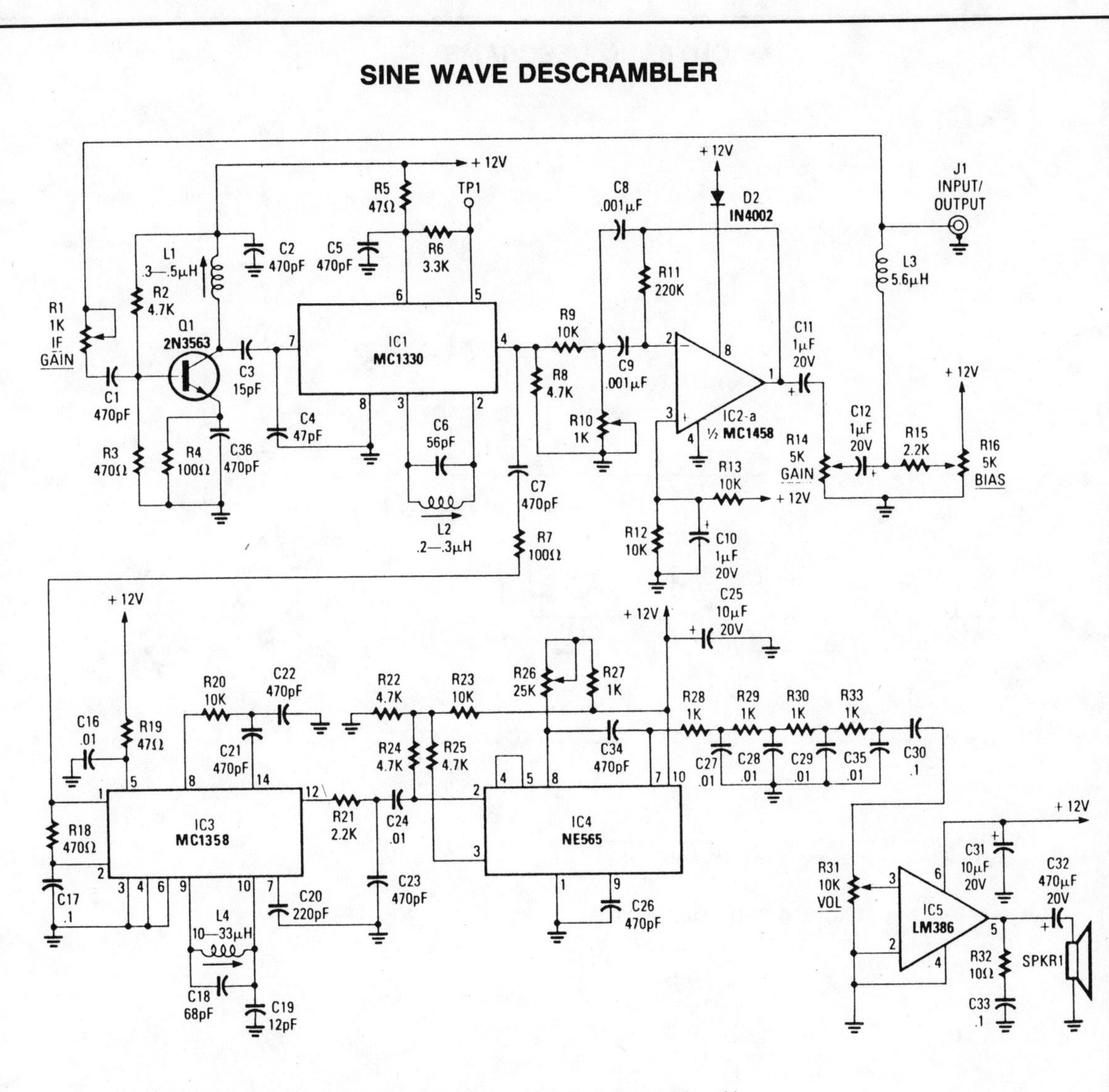

—A COMPLETE SINEWAVE DESCRAMBLER. Easy to build, and relatively easy to align, this circuit completely removes the 15.75-kHz scrambling sinewave.

RADIO-ELECTRONICS *Fig. 22-1*

Circuit Notes

This decoder features a sine wave recovery channel and uses a PIN diode attenuator driven by the sine wave recovery system to cancel out the sine wave sync suppression signal. Kit available from North Country Radio, P.O. Box 53, Wykagyl Station, New York 10804.

OUTBAND DESCRAMBLER

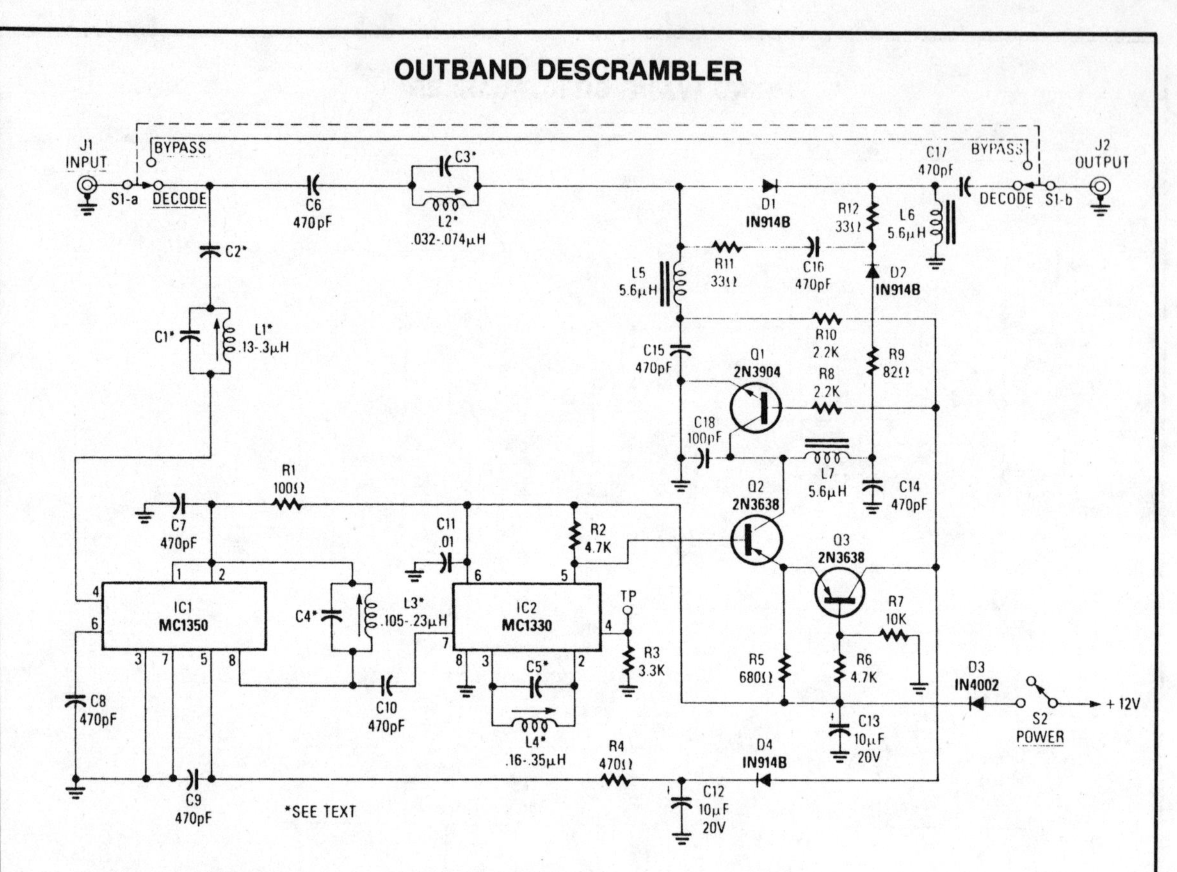

—FOR THE OUTBAND DECODER shown here to work, the cable company must provide at least a 1 millivolt signal. Values for C1–C5 and L1–L4 are found in Table 1.

TABLE 1—CAPACITOR AND COIL VALUES

	50 MHz	90–114 MHz
C1	5 pF	5 pF
C2	47 pF	12 pF
C3	200 pF	82 pF
C4	56 pF	12 pF
C5	56 pF	10 pF
L1	0.2 μH	0.2 μH
L2	0.05 μH	0.03 μH
L3	0.175 μH	0.2 μH
L4	0.175 μH	0.24 μH

RADIO-ELECTRONICS

Fig. 22-2

Circuit Notes

This circuit consists of an amplifier for the synch channel and a video detector which controls an attenuator so that the gain of the systems is increased during synch intervals. Kit available from North Country Radio, P.O. Box 53, Wykagyl Station, New York 10804.

GATED PULSE DESCRAMBLER

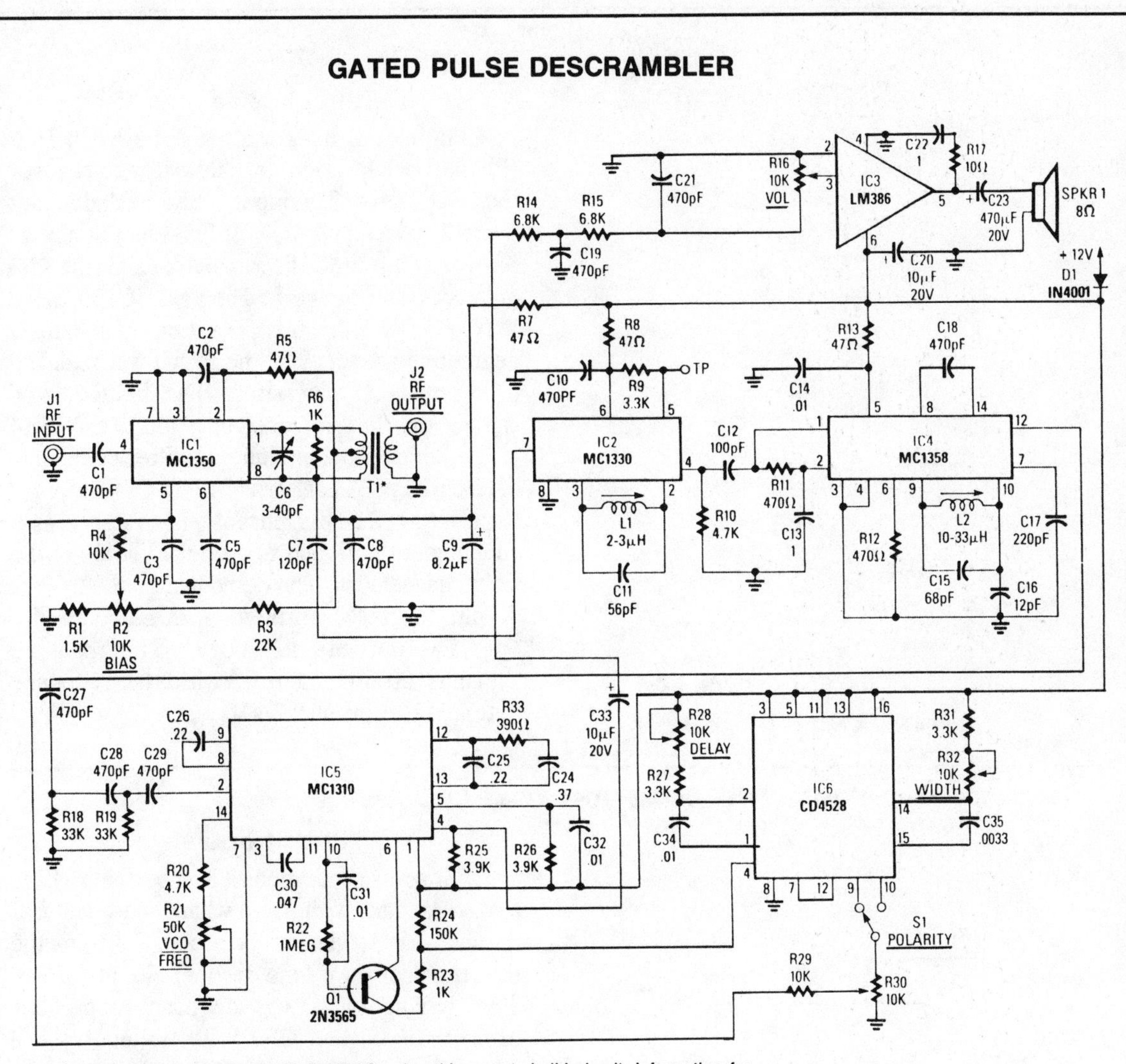

—DESCRAMBLE GATED-PULSE SIGNALS using this easy-to-build circuit. Information for winding transformer T1 and coil L1 can be found in the text.

RADIO-ELECTRONICS

Fig. 22-3

Circuit Notes

This circuit consists of an amplifier and video detector with a second subcarrier detector for synch recovery purposes. A pulse-former circuit modulates the gain of the main channel increasing it during synch intervals. Provision for subcarrier audio descrambling is also provided. Kit available from North Country Radio, P.O. Box 53, Wykagyl Station, New York 10804.

SCA DECODER

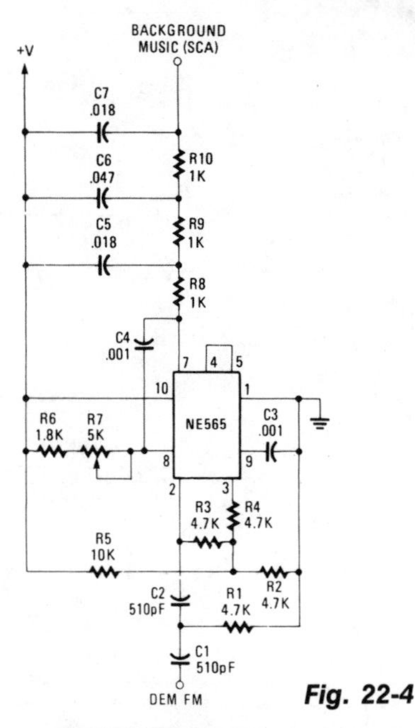

Fig. 22-4

RADIO-ELECTRONICS

Circuit Notes

The circuit uses a Signetics NE565 PLL (Phase-Locked Loop) as a detector to recover the SCA signal. The input to the SCA decoder circuit is connected to an FM receiver at a point between the FM discriminator and the de-emphasis filter network. The PLL, IC1, is tuned to 67 kHz by R7, a 5 K potentiometer. Tuning need not be exact since the circuit will seek and lock onto the subcarrier. The demodulated signal from the FM receiver is fed to the input of the 565 through a high-pass filter consisting of two 510 pF capacitors (C1 and C2) and a 4.7 K resistor (R1). Its purpose is to serve as a coupling network and to attenuate some of the main channel spill. The demodulated SCA signal at pin 7 passes through a three-stage de-emphasis network as shown. The resulting signal is around 50 mV, with the response extending to around 7 kHz.

DUAL TIME CONSTANT TONE DECODER

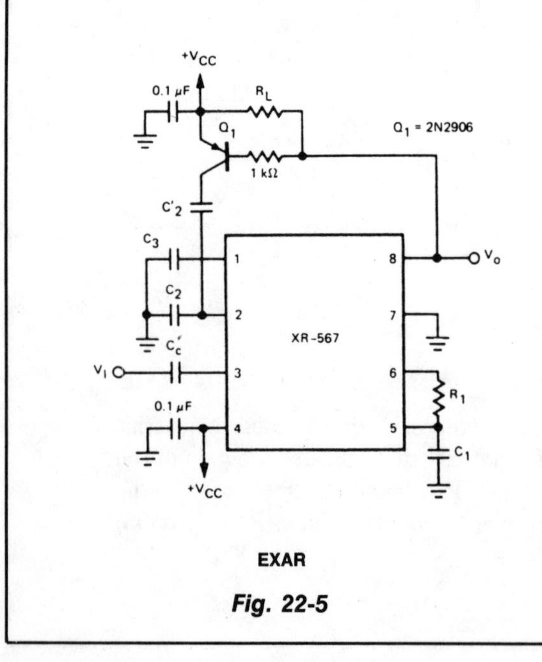

EXAR

Fig. 22-5

Circuit Notes

For some applications it is important to have a tone decoder with narrow bandwidth and fast response time. This can be accomplished by the dual time constant tone decoder circuit shown. The circuit has two low-pass loop filter capacitors, C_2 and C'_2. With no input signal present, the output at pin 8 is high, transistor Q_1 is off, and C'_2 is switched out of the circuit. Thus, the loop low-pass filter is comprised of C_2, which can be kept as small as possible for minimum response time. When an in-band signal is detected, the output at pin 8 will go low, Q_1 will turn on, and capacitor C'_2 will be switched in parallel with capacitor C_2. The low-pass filter capacitance will then be $C_2 + C'_2$. The value of C'_2 can be quite large in order to achieve narrow bandwidth. During the time that no input signal is being received, the bandwidth is determined by capacitor C_2.

STEREO TV DECODER

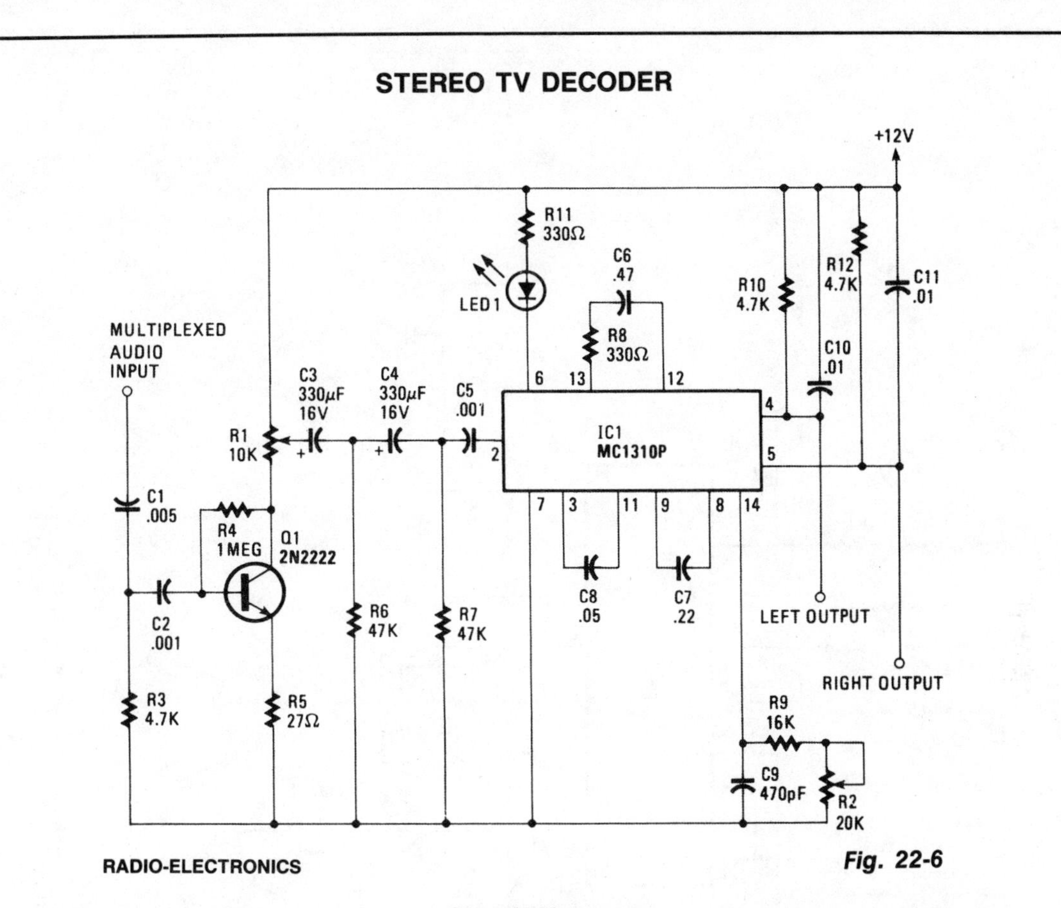

RADIO-ELECTRONICS

Fig. 22-6

Circuit Notes

The composite input signal is preamplified by transistor Q1 and is then coupled to the high-pass filter composed of C3, C4, R6, and R7. The filtered audio is then passed to IC1, an MC1310P "Coilless Stereo Demodulator." That IC is normally used to demodulate broadcast-band FM signals, but by changing the frequency of its on-board VCO (Voltage Controlled Oscillator) slightly (from 19 kHz to 15.734 kHz), we can use that IC to detect stereo-TV signals. A block diagram of the MC1310P is shown in Fig. 22-5. Notice that the components connected to pin 14 control the VCO's frequency, hence the pilot-detect and carrier frequencies. For use in an FM receiver, the VCO would run at four times the 19 kHz pilot frequency (76 kHz), but for our application, it will run at four times the 15.734 kHz pilot frequency of stereo TV, or 62.936 kHz. The MC1310P divides the master VCO signal by two in order to supply the 31.468 kHz carrier that is used to detect the L − R audio signal. The L − R signal undergoes normal FM detection, and at that point we've got two audio signals: L + R and L − R. The decoder block in the IC performs the addition and subtraction to produce the separate left and right signals. R10 and C10 form a de-emphasis network that compensates for the 75 μs pre-emphasis that the left channel underwent; R12 and C11 perform the same function for the right channel.

TIME DIVISION MULTIPLEX (TDM) STEREO DECODER

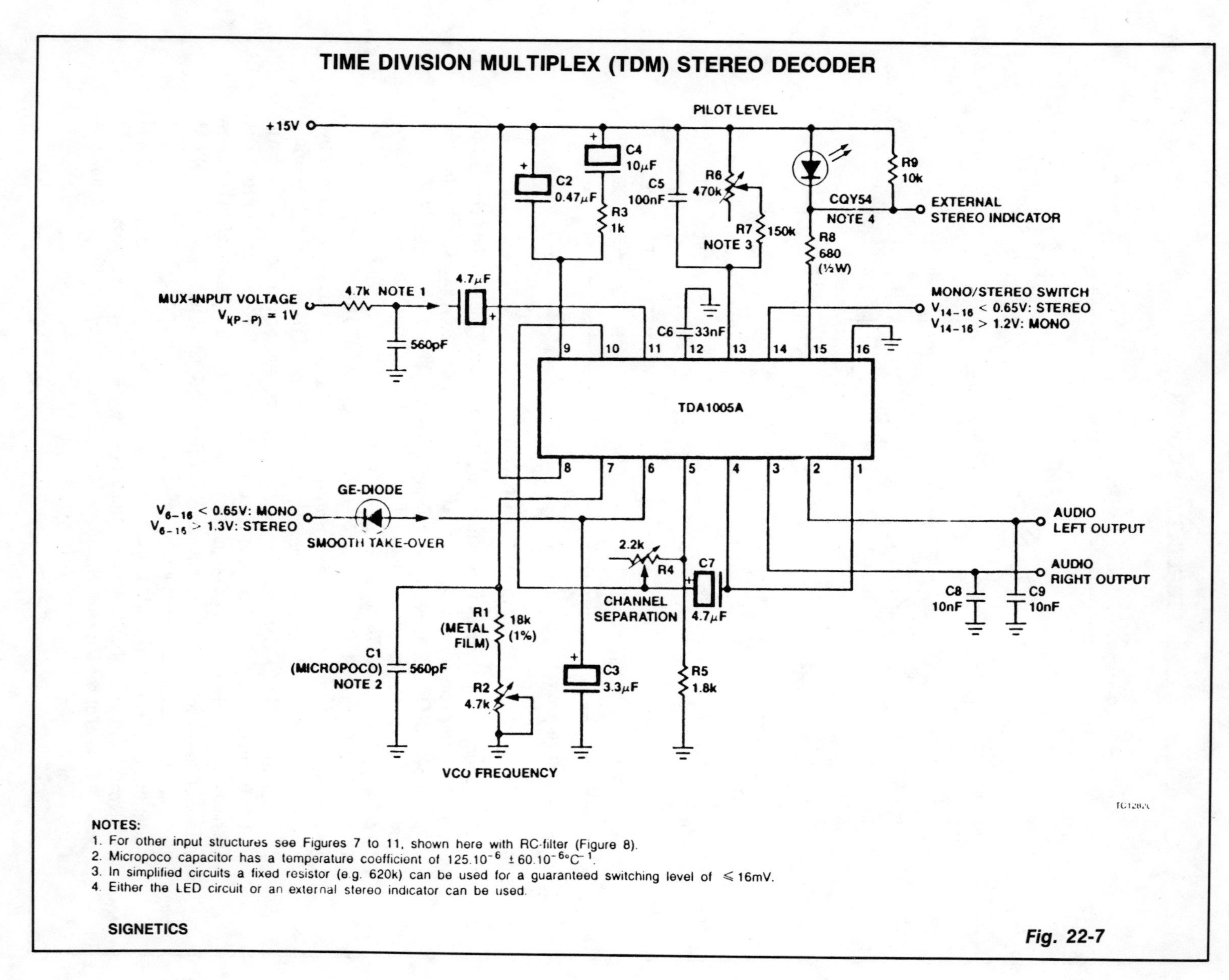

NOTES:
1. For other input structures see Figures 7 to 11, shown here with RC-filter (Figure 8).
2. Micropoco capacitor has a temperature coefficient of $125.10^{-6} \pm 60.10^{-6}$°$C^{-1}$.
3. In simplified circuits a fixed resistor (e.g. 620k) can be used for a guaranteed switching level of ≤ 16mV.
4. Either the LED circuit or an external stereo indicator can be used.

SIGNETICS

Fig. 22-7

FREQUENCY DIVISION MULTIPLEX (FDM) STEREO DECODER

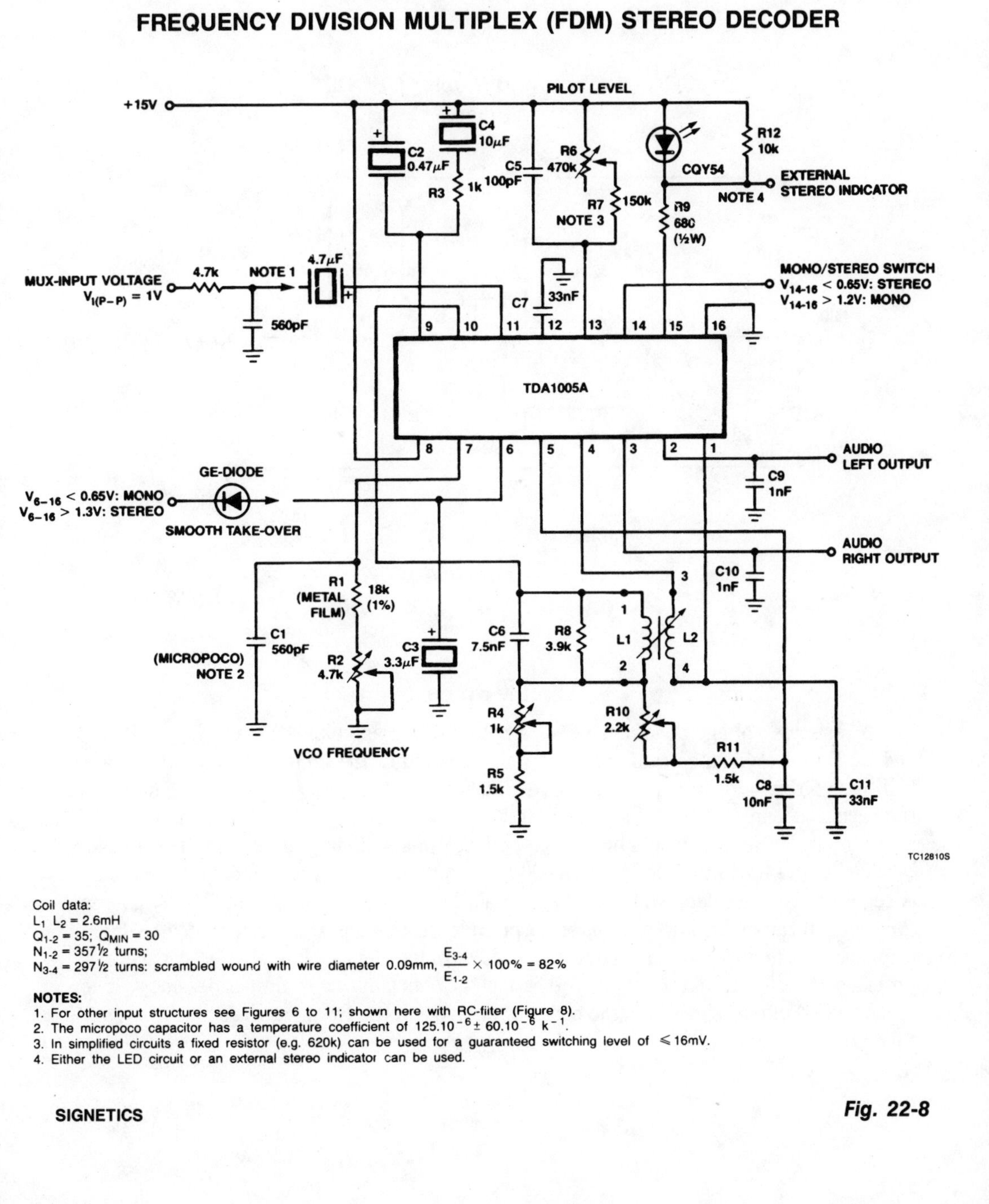

Coil data:
L_1 L_2 = 2.6mH
$Q_{1\text{-}2}$ = 35; Q_{MIN} = 30
$N_{1\text{-}2}$ = 357½ turns;
$N_{3\text{-}4}$ = 297½ turns: scrambled wound with wire diameter 0.09mm, $\frac{E_{3\text{-}4}}{E_{1\text{-}2}} \times 100\% = 82\%$

NOTES:

1. For other input structures see Figures 6 to 11; shown here with RC-filter (Figure 8).
2. The micropoco capacitor has a temperature coefficient of $125.10^{-6} \pm 60.10^{-6}\ k^{-1}$.
3. In simplified circuits a fixed resistor (e.g. 620k) can be used for a guaranteed switching level of ⩽ 16mV.
4. Either the LED circuit or an external stereo indicator can be used.

SIGNETICS

Fig. 22-8

SCA (Background Music) DECODER

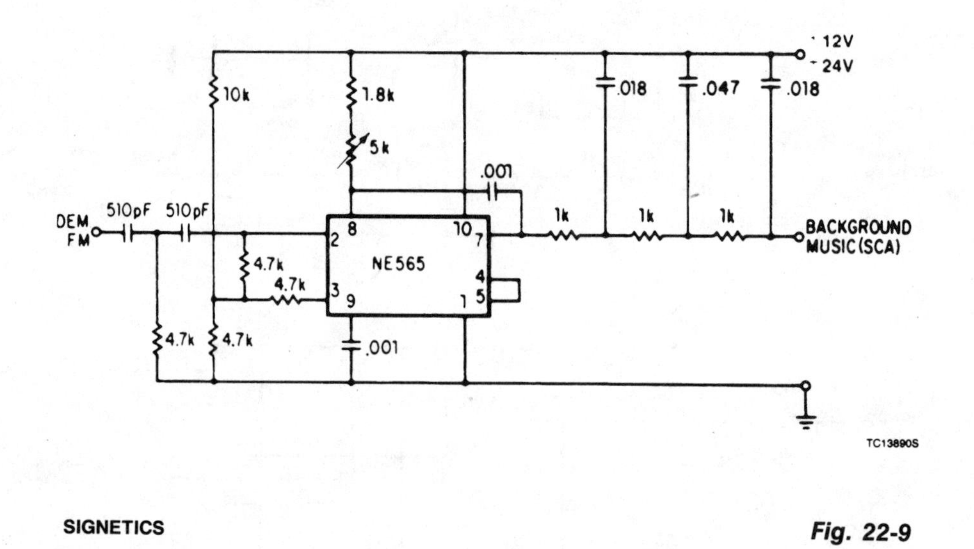

SIGNETICS

Fig. 22-9

Circuit Notes

A resistive voltage divider is used to establish a bias voltage for the input (Pins 2 and 3). The demodulated (multiplex) FM signal is fed to the input through a two-stage high-pass filter, both to effect capacitive coupling and to attenuate the strong signal of the regular channel. A total signal amplitude, between 80 mV and 300 mV, is required at the input. Its source should have an impedance of less than 10,000 ohm. The Phase-Locked Loop is tuned to 67 kHz with a 5000 ohm potentiometer, only approximate tuning is required since the loop will seek the signal. The demodulated output (Pin 7) passes through a three-stage low-pass filter to provide de-emphasis and attenuate the high-frequency noise which often accompanies SCA transmission. Note that no capacitor is provided directly at Pin 7; thus, the circuit is operating as a first-order loop. The demodulated output signal is in the order of 50 mV and the frequency response extends to 7 kHz.

23

Detectors

The sources of the following circuits are contained in the Sources section beginning on page 694. The figure number contained in the box of each circuit correlates to the source entry in the Sources section.

PULSE SEQUENCE DETECTOR

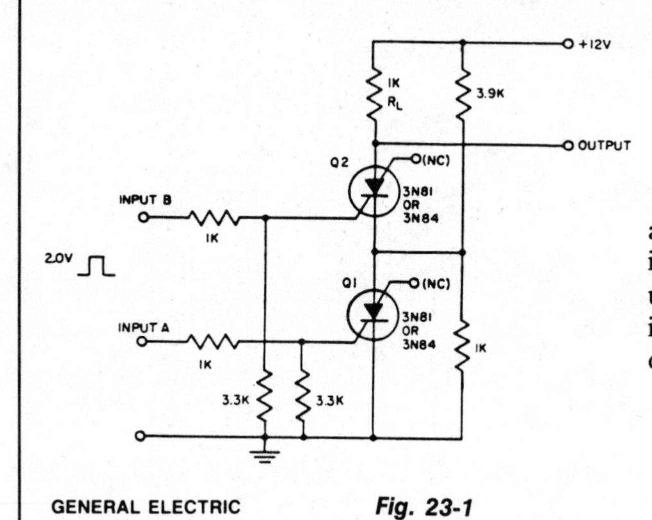

GENERAL ELECTRIC

Fig. 23-1

Circuit Notes

The resistor divider connected between Q1 and Q2 supplies I_H to Q1 after input A triggers it. It also prevents input B from triggering Q2 until Q1 conducts. Consequently, the first B input pulse after input A is applied will supply current to R_L.

VOLTAGE LEVEL DETECTOR

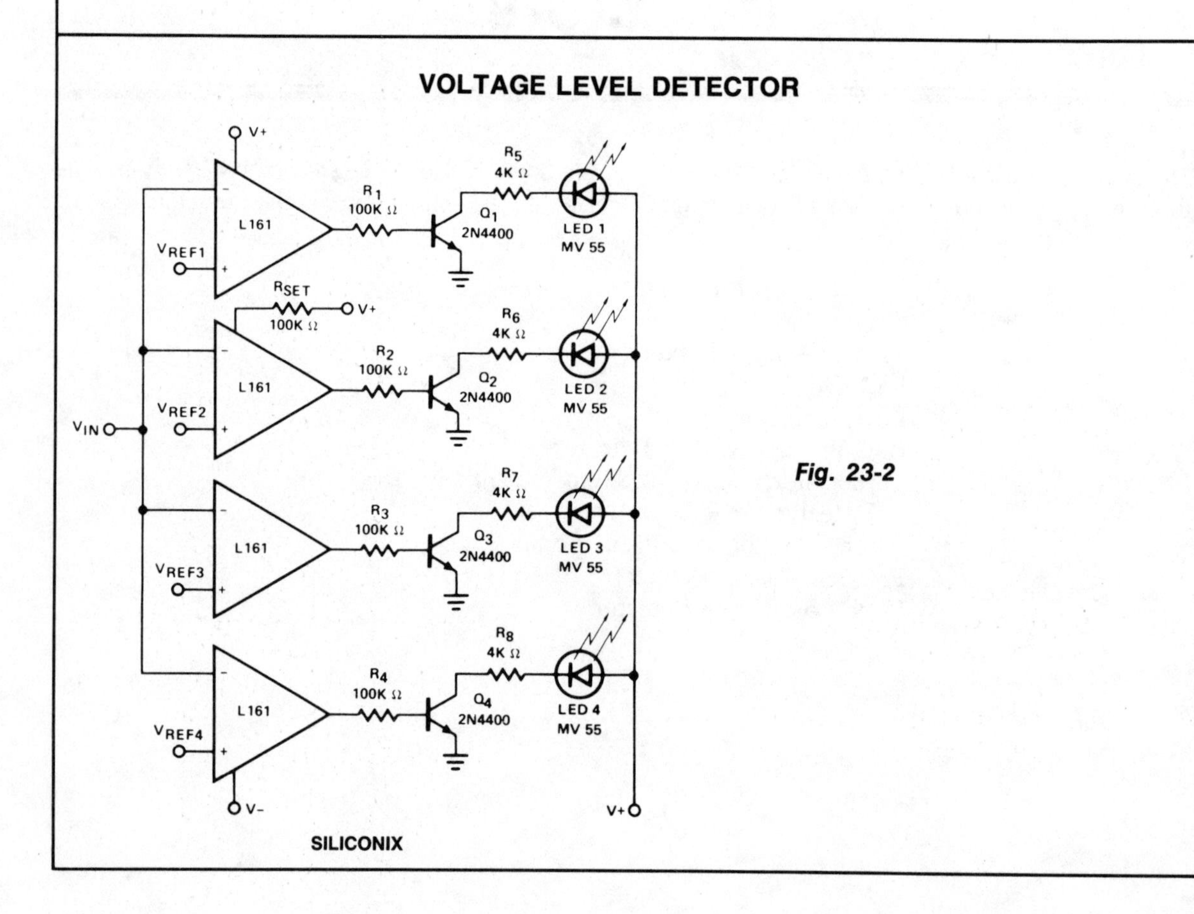

SILICONIX

Fig. 23-2

ZERO-CROSSING DETECTOR

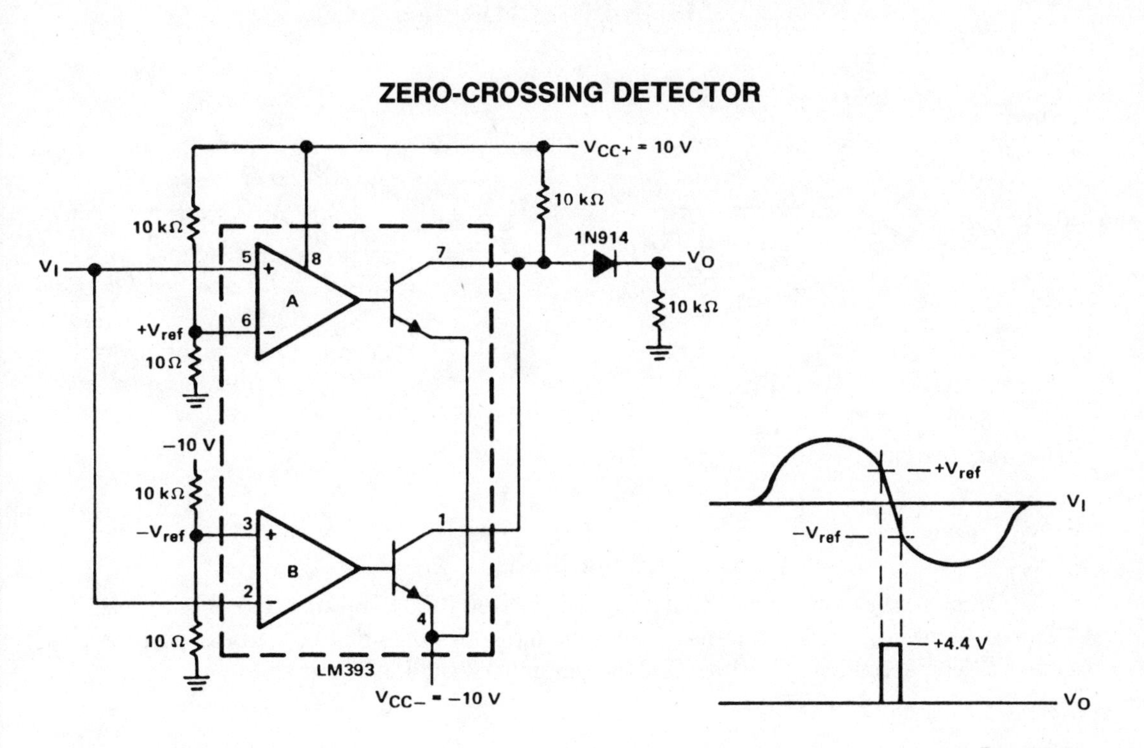

TEXAS INSTRUMENTS

Fig. 23-3

Circuit Notes

This zero-crossing detector uses a dual LM393 comparator, and easily controls hysteresis by the reference levels which are set on the comparator inputs. The circuit illustrated is powered by ±10-V power supplies. The input signal can be an ac signal level up to +8 V. The output will be a positive going pulse of about 4.4 V at the zero-crossover point. These parameters are compatible with TTL logic levels.

The input signal is simultaneously applied to the non-inverting input of comparator A and the inverting input of comparator B. The inverting input of comparator A has a +10 mV reference with respect to ground, while the non-inverting input of comparator B has a −10 mV reference with respect to ground. As the input signal swings positive (greater than +10 mV), the output of comparator "A" will be low while comparator "B" will have a high output. When the input signal swings negative (less than −10 mV), the reverse is true. The result of the combined outputs will be low in either case. On the other hand, when the input signal is between the threshold points (±10 mV around zero crossover), the output of both comparators will be high. If more hysteresis is needed, the ±10 mV window may be made wider by increasing the reference voltages.

PEAK DETECTOR

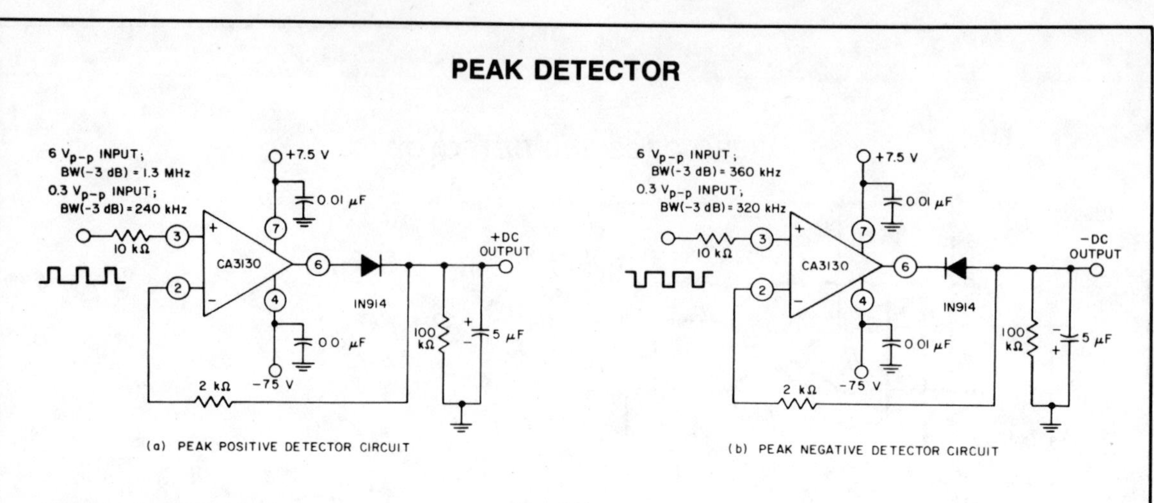

(a) PEAK POSITIVE DETECTOR CIRCUIT

(b) PEAK NEGATIVE DETECTOR CIRCUIT

GENERAL ELECTRIC/RCA

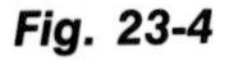

Fig. 23-4

Circuit Notes

Circuits are easily implemented using the CA3130 BiMOS op amp. For large-signal inputs the bandwidth of the peak-negative circuit is less than that of the peak-positive circuit. The second stage of the CA3130 limits bandwidth in this case.

LEVEL DETECTOR

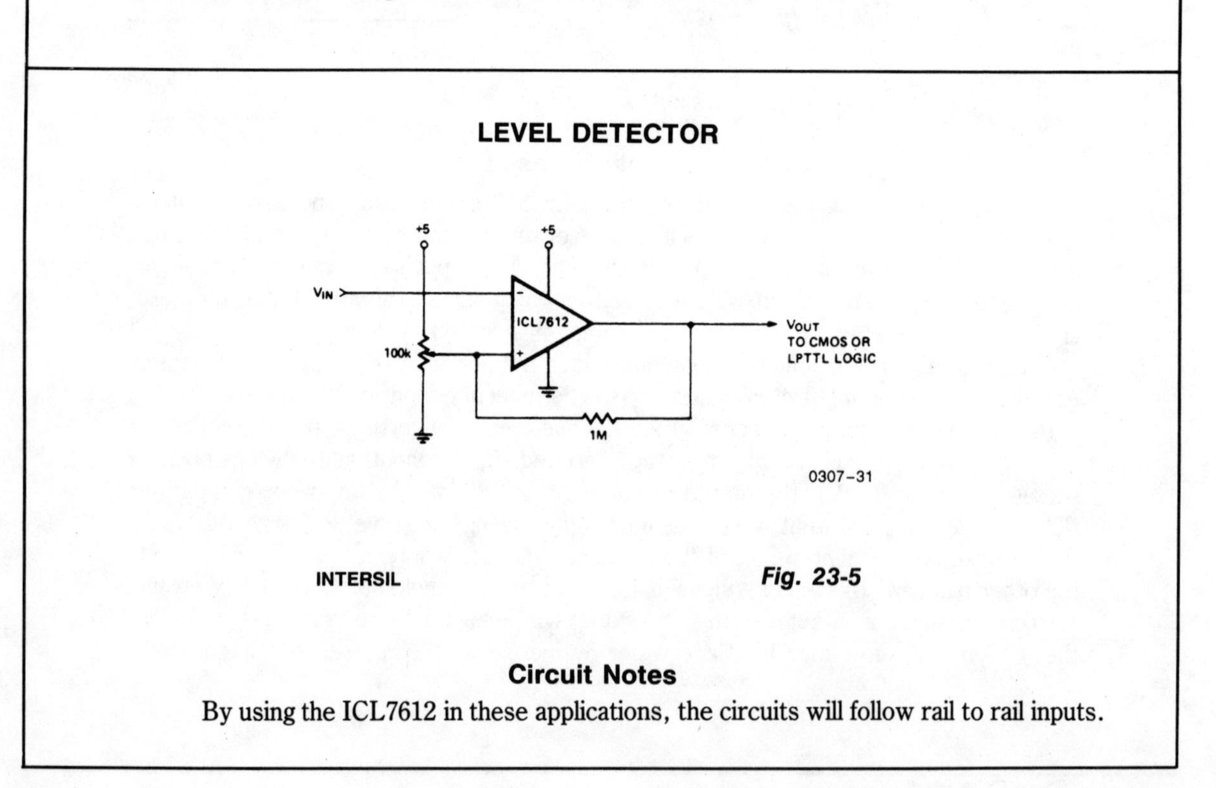

INTERSIL

Fig. 23-5

Circuit Notes

By using the ICL7612 in these applications, the circuits will follow rail to rail inputs.

HIGH FREQUENCY PEAK DETECTOR

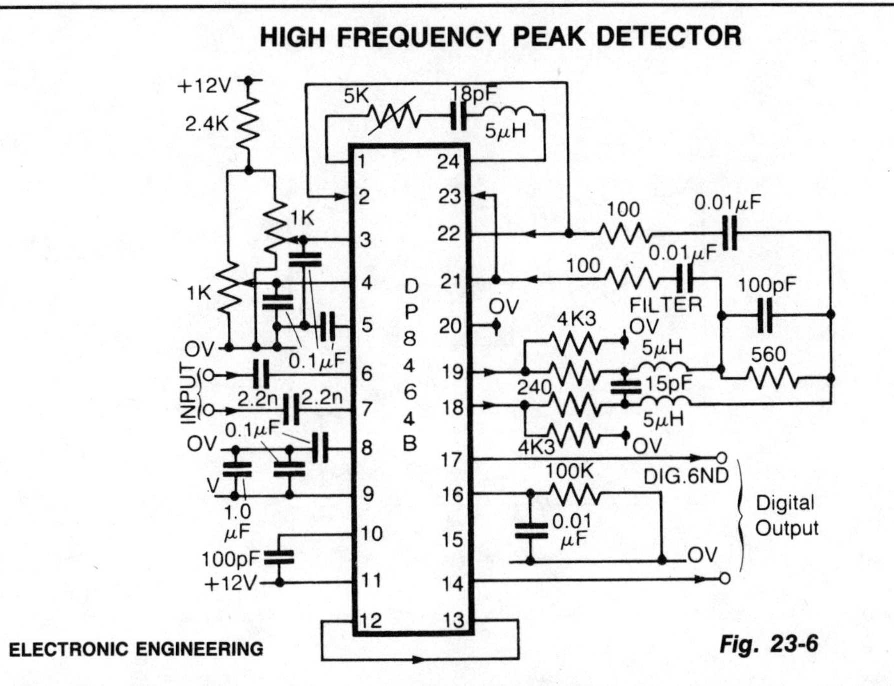

ELECTRONIC ENGINEERING

Fig. 23-6

Circuit Notes

National Semiconductor's DP8464B is primarily intended for use in disk systems as a pulse detector. However it can be easily used as a general purpose peak detector for analogue signals up to 5 MHz. The chip can handle signals between 20 and 66 mV peak-to-peak. The circuit includes a filter with constant group delay characteristics to band limit the signal. Typically the −3 dB point for this filter will be at about 1.5 times the highest frequency of interest. This differentiator network between pins 1 and 24 can be as simple as a capacitor, or can be more complex to band limit the differentiator response.

TACHOMETER, SINGLE PULSE GENERATOR, POWER LOSS DETECTOR, PEAK DETECTOR

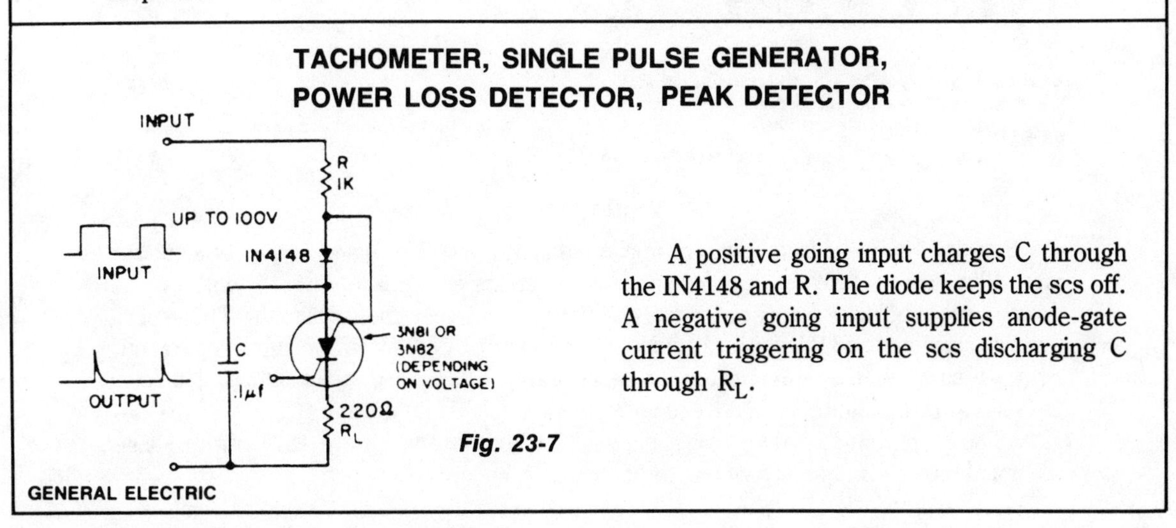

GENERAL ELECTRIC

Fig. 23-7

A positive going input charges C through the IN4148 and R. The diode keeps the scs off. A negative going input supplies anode-gate current triggering on the scs discharging C through R_L.

PHASE DETECTOR WITH 10-BIT ACCURACY

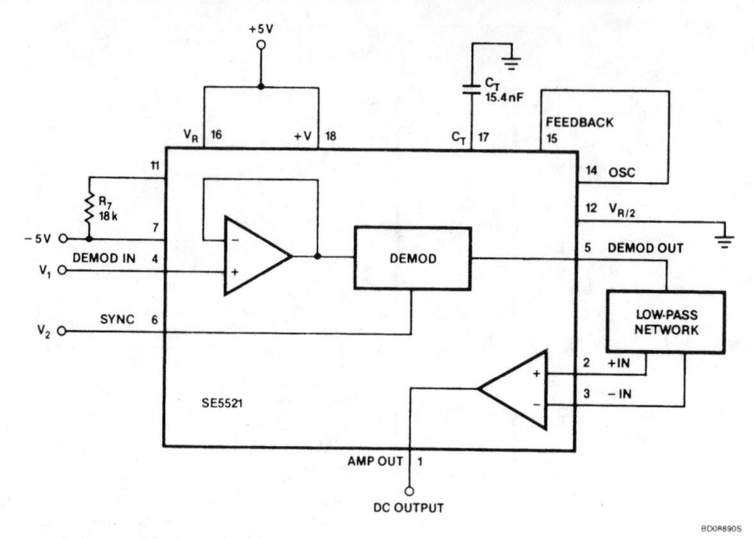

a. Phase Detector Measures Phase Difference Between Signals V_1 and V_2 and Provides dc Output at Pin 1

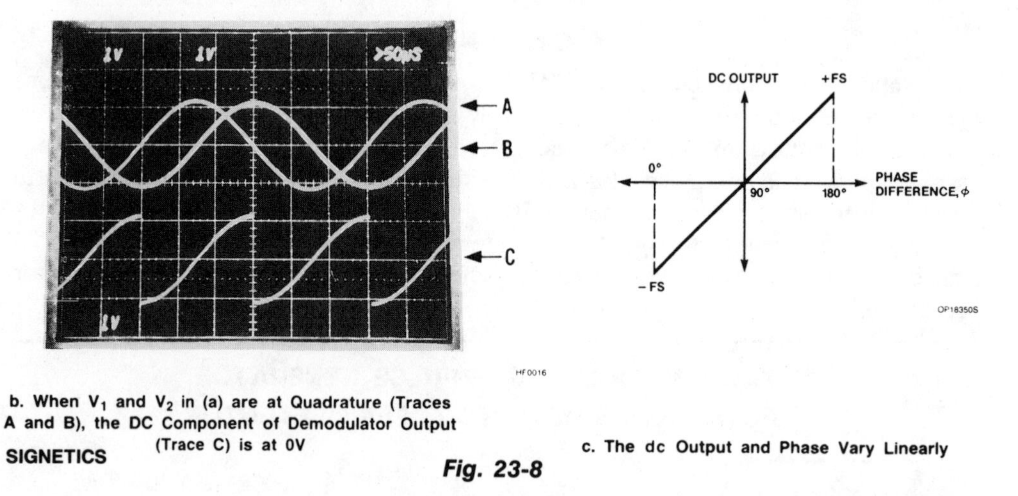

b. When V_1 and V_2 in (a) are at Quadrature (Traces A and B), the DC Component of Demodulator Output (Trace C) is at 0V

c. The dc Output and Phase Vary Linearly

SIGNETICS

Fig. 23-8

Circuit Notes

Signals of identical frequency are applied to sync input (Pin 6) and to the demodulator input (Pin 4), respectively, the demodulator functions as a phase detector with output dc component being proportional to phase difference between the two inputs. The signals must be referenced to 0 V for dual supply operation or to $V_R/2$ for single supply operation. At ±5-V supplies, the demodulator can easily handle 7-V peak-to-peak signals. The low-pass network configured with the uncommitted amplifier dc output at Pin 1 of the device. The dc output is maximum (+ full-scale) when V_1 and V_2 are 180° out of phase and minimum (− full-scale) when the signals are in phase.

FREQUENCY LIMIT DETECTOR

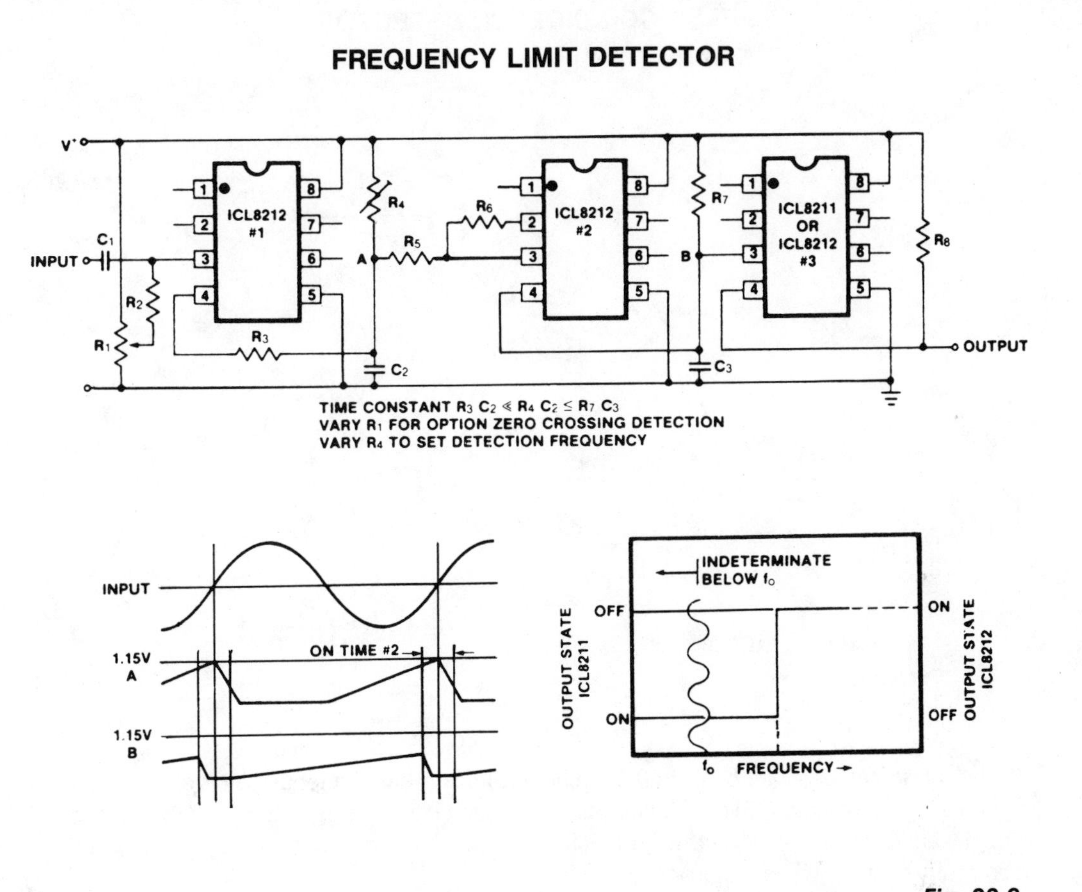

INTERSIL

Fig. 23-9

Circuit Notes

Simple frequency limit detectors providing a GO/NO-GO output for use with varying amplitude input signals may be conveniently implemented with the ICL8211/8212. In the application shown, the first ICL8212 is used as a zero-crossing detector. The output circuit consisting of R_3, R_4 and C_2 results in a slow output positive ramp. The negative range is much faster than the positive range. R_5 and R_6 provide hysteresis so that under all circumstances the second ICL8212 is turned on for sufficient time to discharge C_3. The time constant of R_7C_3 is much greater than R_4C_2. Depending upon the desired output polarities for low and high input frequencies, either an ICL8211 or an ICL8212 may be used as the output driver.

The circuit is sensitive to supply voltage variations and should be used with a stabilized power supply. At very low frequencies the output will switch at the input frequency.

PULSE COINCIDENCE DETECTOR

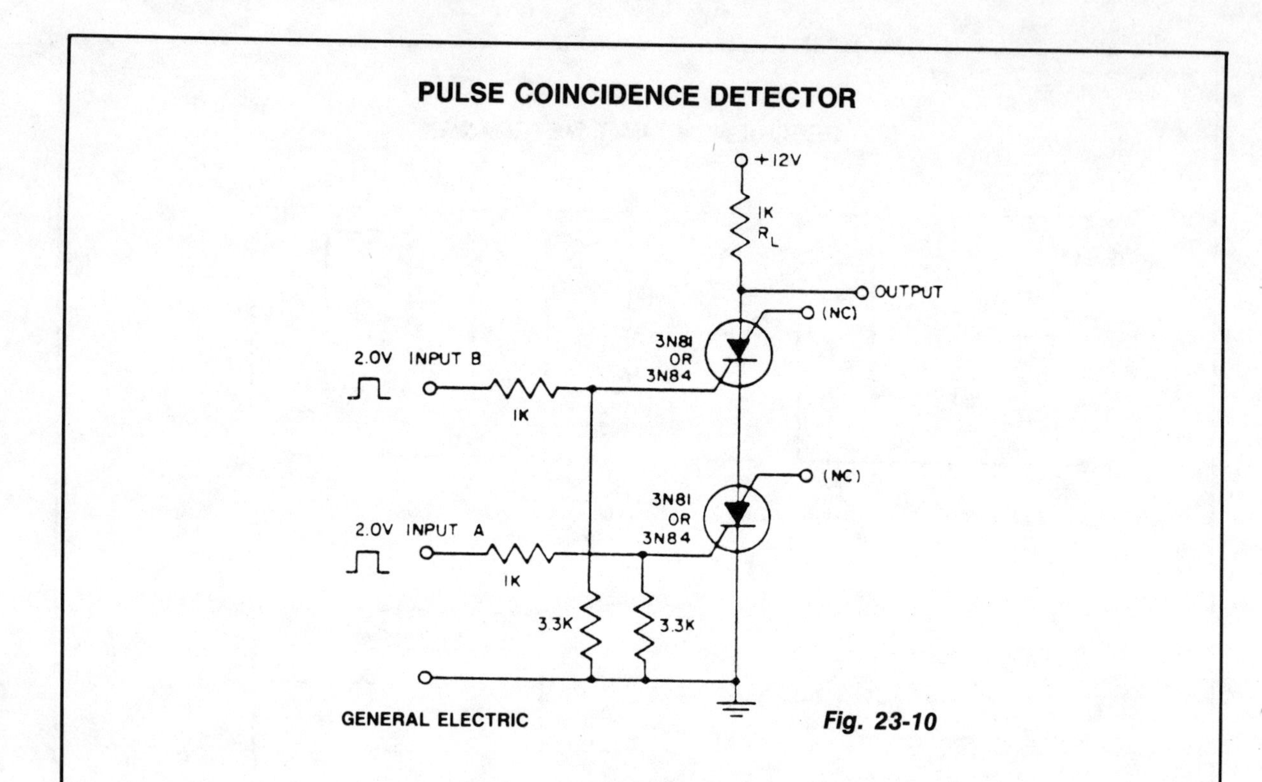

Fig. 23-10

Circuit Notes

Unless inputs A and B (2- to 3-V amplitude) occur simultaneously no voltage exists across R_L. Less than 1 microsecond overlap is sufficient to trigger the scs. Coincidence of negative inputs is detected with gates G_A instead of G_C by using the scs in a complementary SCR configuration.

24
Digital-to-Analog Converters

The sources of the following circuits are contained in the Sources section beginning on page 694. The figure number contained in the box of each circuit correlates to the source entry in the Sources section.

TWO 8-BIT DACS MAKE A 12-BIT DAC

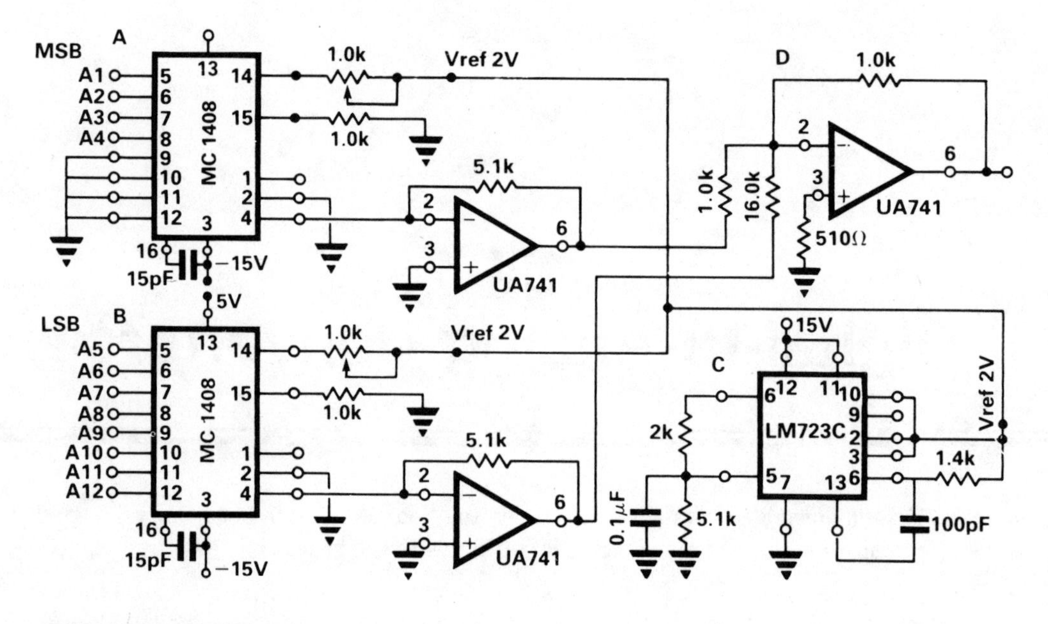

ELECTRONIC ENGINEERING

Fig. 24-1

Circuit Notes

Two MC1408—8-bit D/A converters, A and B in the circuit diagram, are used. The four least significant bits of A are tied to zero. The four most significant bits of the 12-bit data are connected to the remaining four input pins. The eight least significant bits of the 12-bit data are connected to the eight input pins of B. The four most significant bits of the 12-bit data together have a weight of 16 relative to the remaining eight bits. Hence, the output from B is reduced by a factor of 16 and summed with the output from A using the summing op-amp configuration D. Voltage regulator chip, LM7236, is used to provide an accurate reference voltage, 2 V, for the MC1408. The full-scale voltage of the converter is $1/16 \times 9.9609 + 1 \times (9.375)$ or 9.9976 V. The step size of the converter is 2.4 mV.

12-BIT DAC WITH VARIABLE STEP SIZE

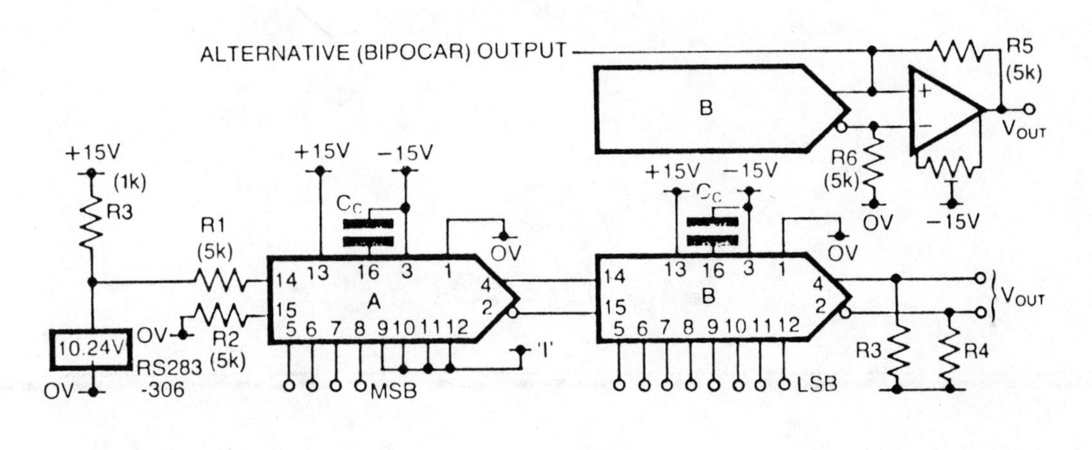

ELECTRONIC ENGINEERING *Fig. 24-2*

Circuit Notes

The step size of the converter is variable by selection of the high order data bits. The first DAC, A, has a stable reference current supplied via the 10.24 V reference IC and R1. R2 provides bias cancellation. As shown, only the first 4 MSB inputs are used, giving a step size of 225/256 × 2.048/16 = 0.127 mA. This current supplies the reference for DAC B whose step size is then 0.1275/256 = 0.498 μA. Complementary voltage outputs are available for unipolar output and using R3 = R4 = 10 K, V_{out} is ±10.2 V approximately, with a step size (1 LSB) of approximately 5 mV. If desired an op amp can be added to the output to provide a low impedance output with bipolar output symmetrical about ground, if R5 = R6 within 0.05%. Note that offset null is required, and all resistors except R2 and R3 should be 1% high stability types.

By using lower order address lines than illustrated for DAC A, a smaller step size (and therefore full-scale output) can be obtained. Unused high order bits can be manipulated high or low to change the relative position of the full-scale output.

25
Dip Meters

The sources of the following circuits are contained in the Sources section beginning on page 694. The figure number contained in the box of each circuit correlates to the source entry in the Sources section.

LITTLE DIPPER

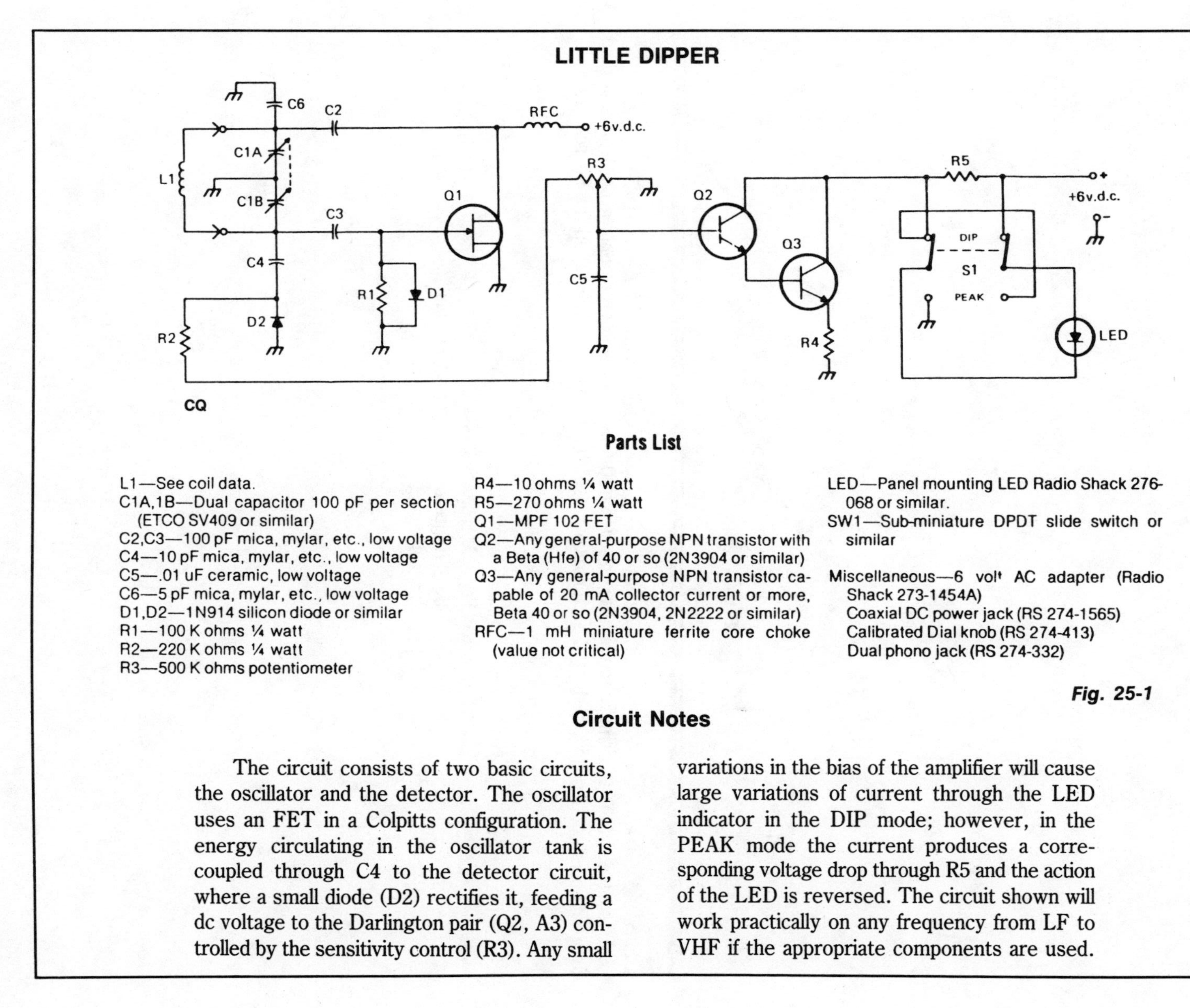

Parts List

L1—See coil data.
C1A,1B—Dual capacitor 100 pF per section (ETCO SV409 or similar)
C2,C3—100 pF mica, mylar, etc., low voltage
C4—10 pF mica, mylar, etc., low voltage
C5—.01 uF ceramic, low voltage
C6—5 pF mica, mylar, etc., low voltage
D1,D2—1N914 silicon diode or similar
R1—100 K ohms ¼ watt
R2—220 K ohms ¼ watt
R3—500 K ohms potentiometer
R4—10 ohms ¼ watt
R5—270 ohms ¼ watt
Q1—MPF 102 FET
Q2—Any general-purpose NPN transistor with a Beta (Hfe) of 40 or so (2N3904 or similar)
Q3—Any general-purpose NPN transistor capable of 20 mA collector current or more, Beta 40 or so (2N3904, 2N2222 or similar)
RFC—1 mH miniature ferrite core choke (value not critical)
LED—Panel mounting LED Radio Shack 276-068 or similar.
SW1—Sub-miniature DPDT slide switch or similar

Miscellaneous—6 volt AC adapter (Radio Shack 273-1454A)
Coaxial DC power jack (RS 274-1565)
Calibrated Dial knob (RS 274-413)
Dual phono jack (RS 274-332)

Fig. 25-1

Circuit Notes

The circuit consists of two basic circuits, the oscillator and the detector. The oscillator uses an FET in a Colpitts configuration. The energy circulating in the oscillator tank is coupled through C4 to the detector circuit, where a small diode (D2) rectifies it, feeding a dc voltage to the Darlington pair (Q2, A3) controlled by the sensitivity control (R3). Any small variations in the bias of the amplifier will cause large variations of current through the LED indicator in the DIP mode; however, in the PEAK mode the current produces a corresponding voltage drop through R5 and the action of the LED is reversed. The circuit shown will work practically on any frequency from LF to VHF if the appropriate components are used.

26
Display Circuits

The sources of the following circuits are contained in the Sources section beginning on page 694. The figure number contained in the box of each circuit correlates to the source entry in the Sources section.

VACUUM FLUORESCENT DISPLAY

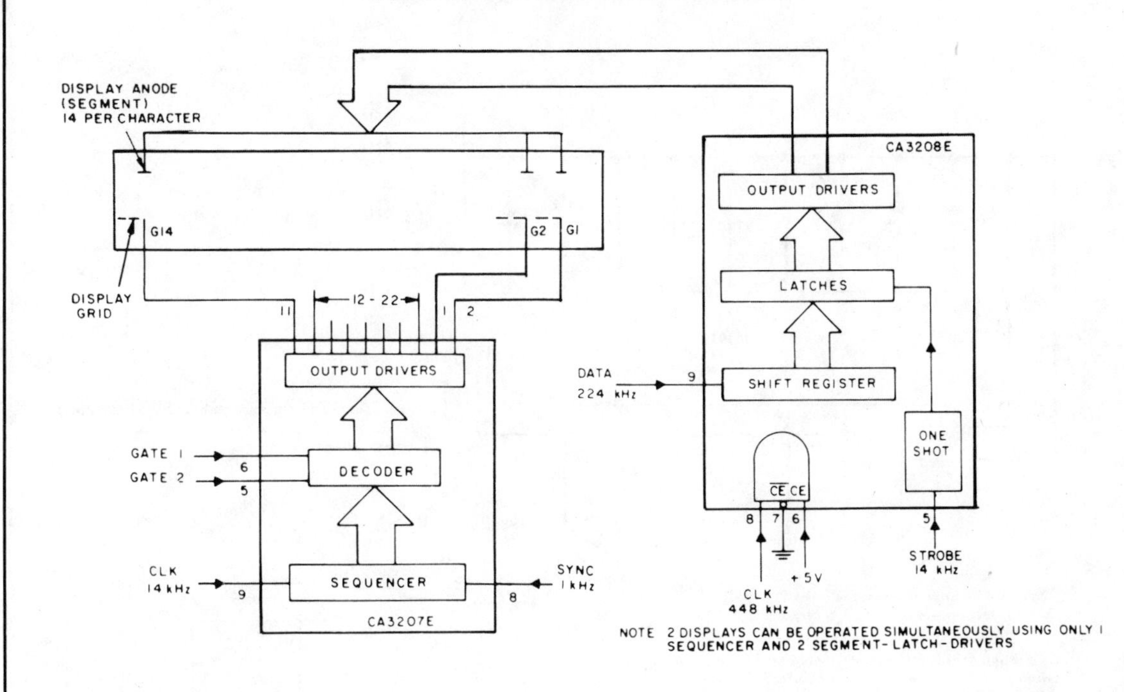

GENERAL ELECTRIC/RCA ***Fig. 26-1***

This circuit uses the CA3207 sequence driver and CA3208 segment latch-driver in combination to drive display devices of up to 14 segments with up to 14 characters of display. The CA3207 selects the digit or character to be displayed in sequence, CA3208 turns on the required alphanumeric segments.

EXPANDED SCALE METER, DOT OR BAR

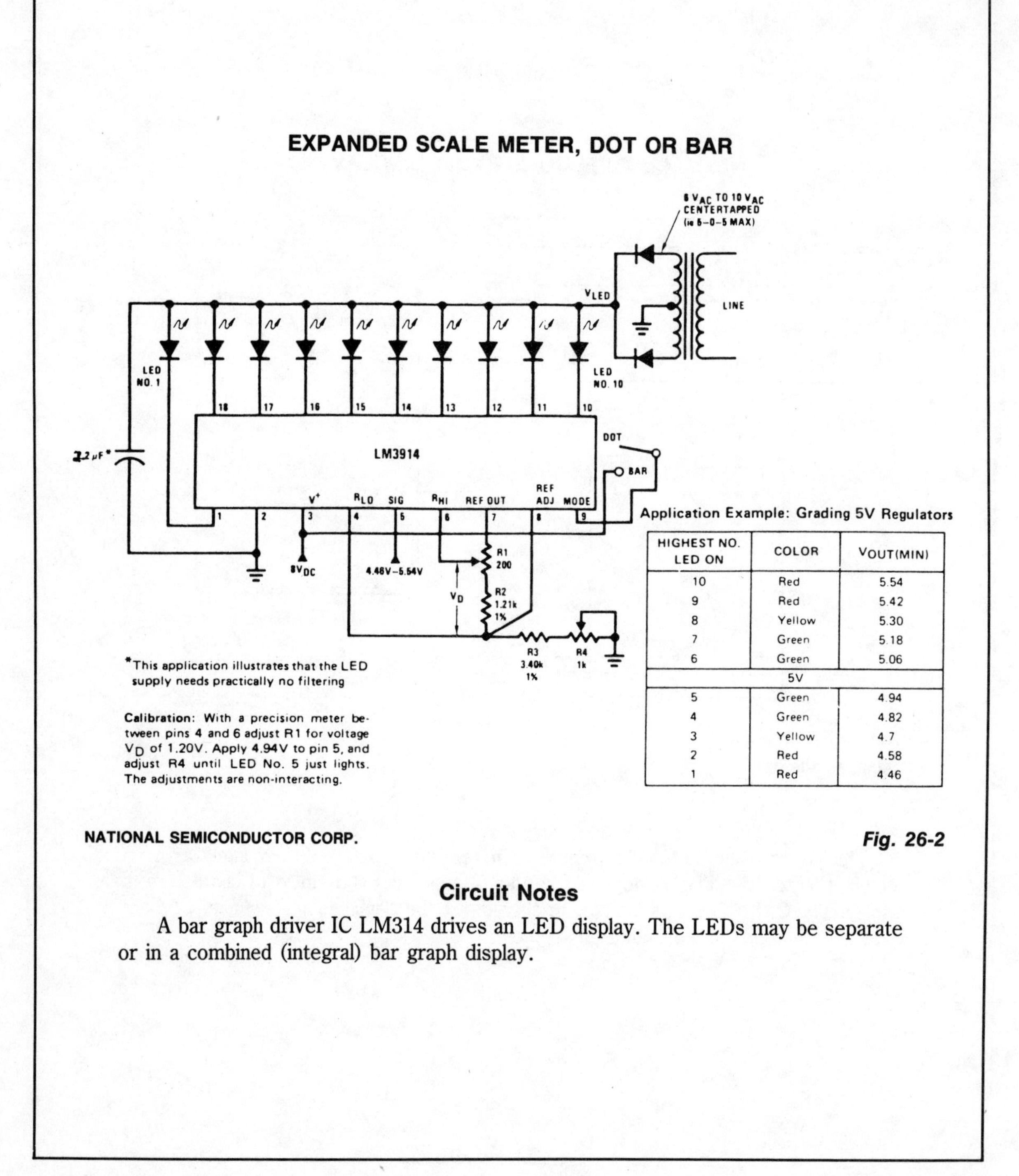

Application Example: Grading 5V Regulators

HIGHEST NO. LED ON	COLOR	$V_{OUT(MIN)}$
10	Red	5.54
9	Red	5.42
8	Yellow	5.30
7	Green	5.18
6	Green	5.06
	5V	
5	Green	4.94
4	Green	4.82
3	Yellow	4.7
2	Red	4.58
1	Red	4.46

*This application illustrates that the LED supply needs practically no filtering

Calibration: With a precision meter between pins 4 and 6 adjust R1 for voltage V_D of 1.20V. Apply 4.94V to pin 5, and adjust R4 until LED No. 5 just lights. The adjustments are non-interacting.

NATIONAL SEMICONDUCTOR CORP.

Fig. 26-2

Circuit Notes

A bar graph driver IC LM314 drives an LED display. The LEDs may be separate or in a combined (integral) bar graph display.

LOW-COST BAR-GRAPH INDICATOR FOR AC SIGNALS

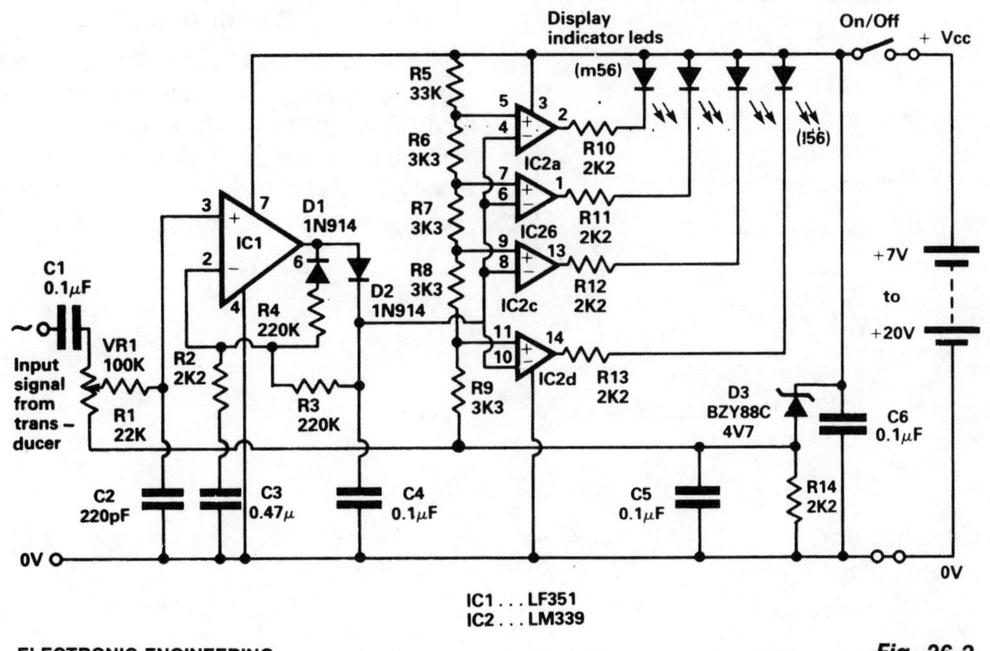

ELECTRONIC ENGINEERING

Fig. 26-3

Circuit Notes

Indicator was designed for displaying the peak level of small ac signals from a variety of transducers including microphones, strain gauges and photodiodes. The circuit responds to input signals contained within the audio frequency spectrum, i.e., 30 Hz to 20 kHz, although a reduced response extends up to 40 kHz. Maximum sensitivity for the component values shown, with VR1 fully clockwise, is 30 mV peak-to-peak. The indicator can be calibrated by setting VR1 when an appropriate input signal is applied.

LED BAR-GRAPH DRIVER

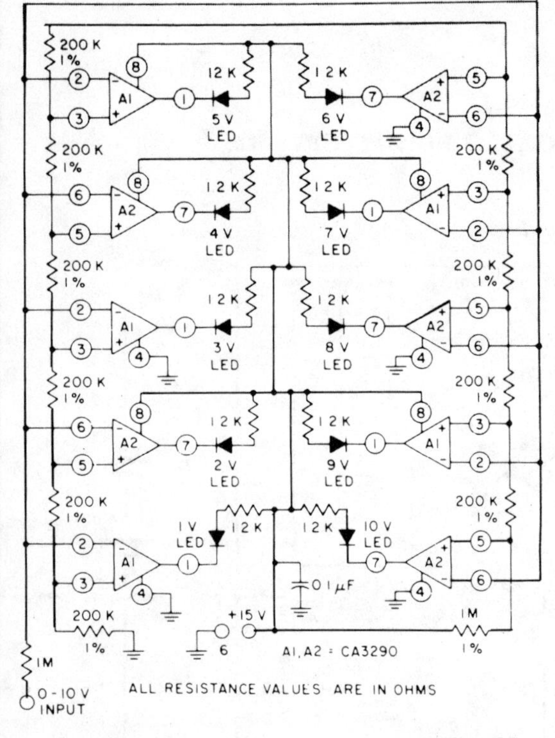

GENERAL ELECTRIC/RCA

Fig. 26-4

Circuit Notes

The circuit uses CA3290 BiMOS dual voltage comparators. Non-inverting inputs of A1 and A2 are tied to voltage divider reference. The input signal is applied to the inverting inputs. LEDs are turned "on" when input voltage the reaches the voltage on the reference divider.

27

Drive Circuits

The sources of the following circuits are contained in the Sources section beginning on page 694. The figure number contained in the box of each circuit correlates to the source entry in the Sources section.

LINE DRIVER PROVIDES FULL RAIL EXCURSIONS

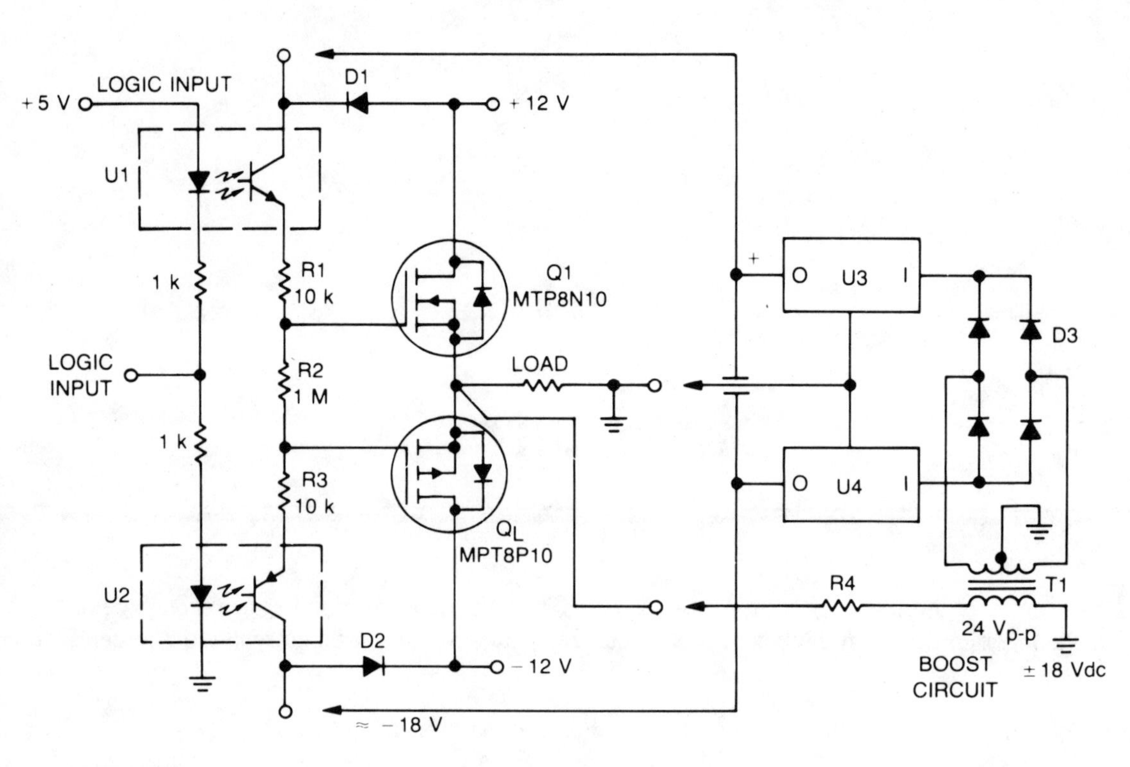

MOTOROLA *Fig. 27-1*

Circuit Notes

The logic input is applied to opto-isolators U1 and U2 with, respectively, npn and pnp emitter follower outputs. Dc balance is adjusted by potentiometer R2. The emitter followers drive the gates of Q1 and Q2, the complementary TMOS pairs. With a ±12 V supply, the swing at the common source output point is about 12 V peak-to-peak.

By adding a ±18-V boost circuit, as shown, the output swing can approach the rail swing. This circuit applies the output to transformer T1, which is rectified by diode bridge D3, regulated by U3 and U4, and then applied to the collectors of U1 and U2. Diodes D1 and D2 are forward-biased when 12-V supplies are used, but they are back-biased when the 18-V boost is used.

FIVE-TRANSISTOR AMPLIFIER BOOSTS FAST PULSES INTO 50-OHM COAXIAL CABLE

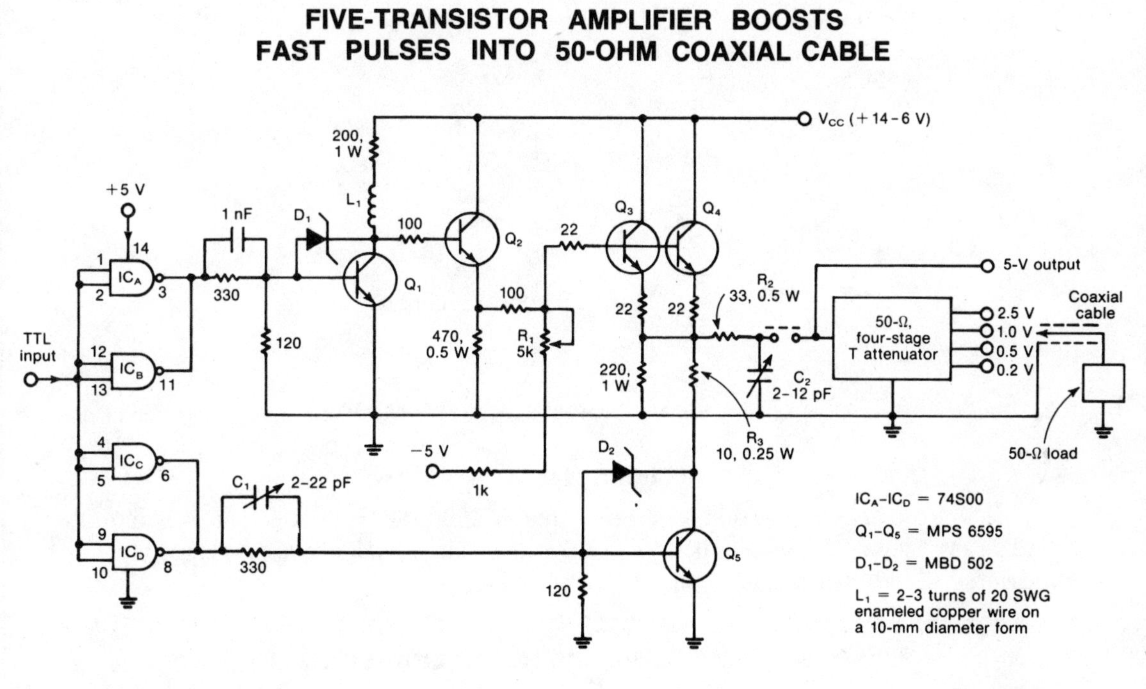

ELECTRONIC DESIGN

Fig. 27-2

Circuit Notes

The circuit works from dc to 50 MHz and will deliver pulses as short as 10 ns. It is driven by a TTL signal through a 740S00 quad Schottky NAND gate, IC_A through IC_D. Transistor Q1, wired as a common-emitter amplifier, drives transistor Q2, a simple emitter follower. Transistors Q3 and Q4, wired in parallel, also form an emitter follower and drive the output. When Q3 and Q4 are both turned off, transistor Q5 works as a low-impedance sink. Schottky diodes D1 and D2 prevent Q1 and Q5 from becoming saturated. To adjust the circuit, potentiometer R1 is set to optimize the output pulse's fall time. Inductor L1, a peaking coil, should be adjusted to improve the rise time to within a permissible 5% overshoot. Likewise, capacitor C1 can be varied to control preshooting. Further output pulse shaping is accomplished with the help of capacitor C2. Resistors R2 and R3 ensure a proper 50-ohm impedance at the amplifier's output when the pulse is on or off, respectively.

50-OHM TRANSMISSION LINE DRIVER

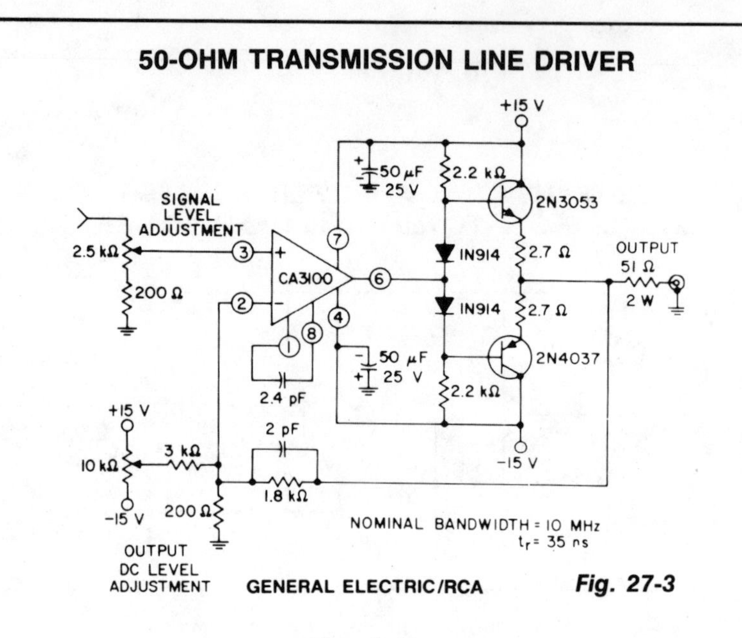

Fig. 27-3

Circuit Notes

This circuit uses a wideband, high slew rate CA3100 BiMOS op amp. The slew rate for this amplifier is 28 V/μs. Output swing is 9 volts peak-to-peak into a terminated line, measured at the termination.

600-OHM BALANCED DRIVER FOR LINE SIGNALS

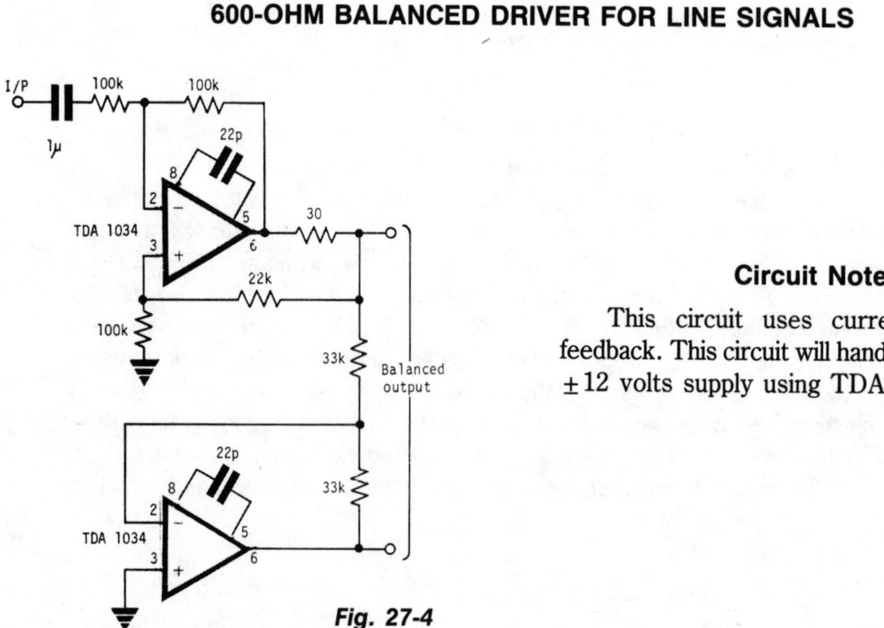

Fig. 27-4

ELECTRONIC ENGINEERING

Circuit Notes

This circuit uses current and voltage feedback. This circuit will handle +24 dBm with ±12 volts supply using TDA 1034s.

HIGH OUTPUT 600-OHM LINE DRIVER

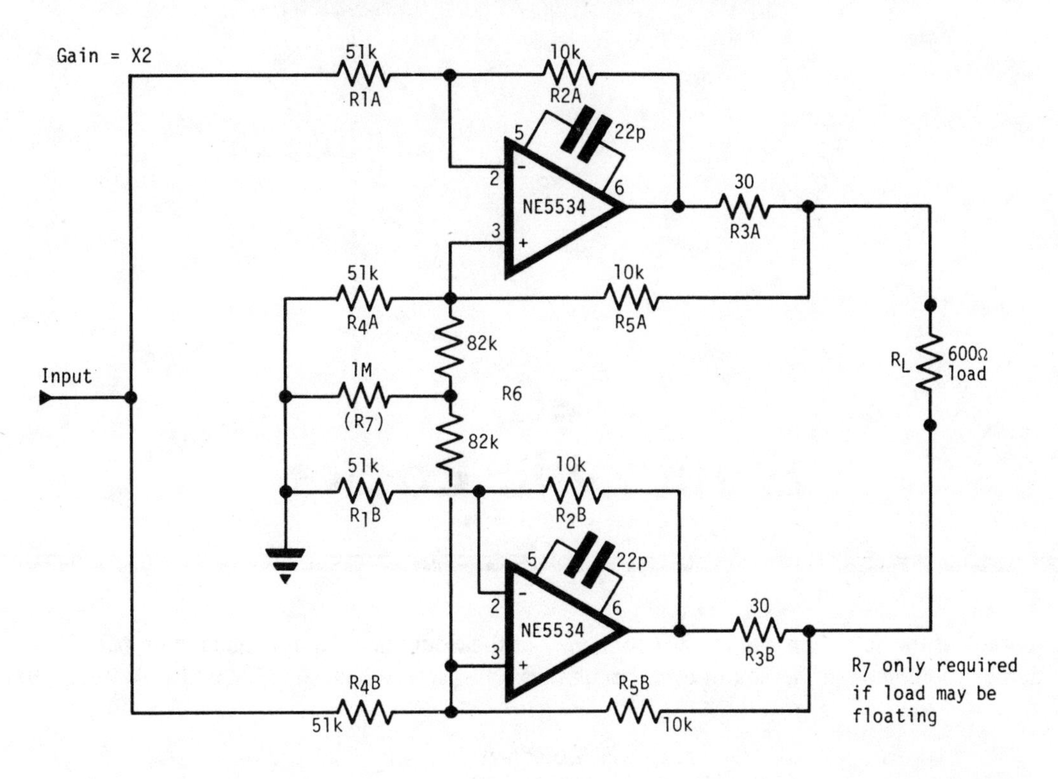

ELECTRONIC ENGINEERING

Fig. 27-5

Circuit Notes

The circuit has a "floating" output, i.e., it behaves like an isolated transformer winding, with the output amplitude remaining unchanged whether the center or either end of the load is grounded. This is achieved by making Z-out, common mode, infinite. The circuit consists of two current-sources in push-pull. Since each has infinite output Z, the common mode output impedance is also infinite. Connecting a resistor between the non-inverting terminals of the op amps reduces the differential Z-out without affecting the Z-common-mode. Since the output is floating, if the load is also floating there is no output ground reference, which results in malfunction. This can be corrected by reducing the common-mode Z slightly. R7 fulfills this function. All resistors should be of close tolerance to give a good balance. The line driver provides +24 dB from ±12 V or +16 dB from ±6 V supplies.

28
Electronic Locks

The sources of the following circuits are contained in the Sources section beginning on page 694. The figure number contained in the box of each circuit correlates to the source entry in the Sources section.

THREE-DIAL COMBINATION LOCK

C1—500-uF, 25-VDC electrolytic capacitor
D1, D2—1N4002 diode
K1—relay with 6-volt coil rated @ 250-ohms, with SPST contacts
Q1—2N5050 SCR
R1, R2—4,700-ohm, ½-watt resistor, 5%
S1, S2, S3—single pole, 10-position rotary or thumbwheel switches
S4—normally closed SPST pushbutton switch
T1—120-VAC to 6.3-VAC @ 300mA power transformer

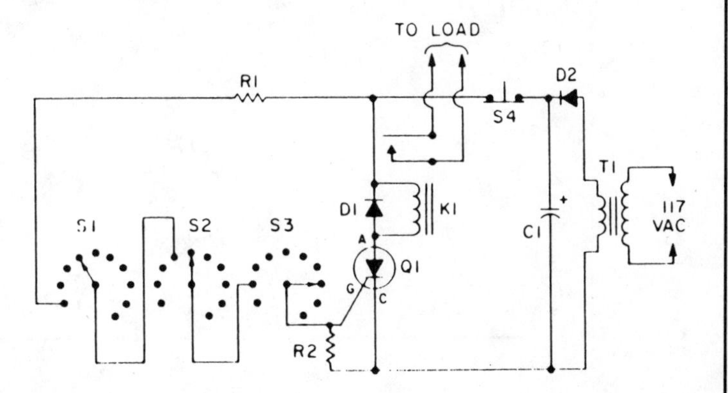

TAB BOOKS, INC.

Fig. 28-1

Circuit Notes

Here's an effective little combination lock that you can put together in one evening's time. To open the lock, simply dial in the correct combination on the three rotary or thumbwheel switches. With the correct combination entered, current flows through R1 into Q1's gate terminal, causing the SCR to latch in a conductive state. This sends a current through relay K1, which responds by closing its contacts and actuating whatever load is attached. After opening the lock, twirl the dials of S1 through S3 away from the correct combination so that nobody gets a look at it. The lock will remain open and your load will remain on because the SCR is latched on. To lock things up, it's only necessary to interrupt the flow of anode current through the SCR by pressing pushbutton S4.

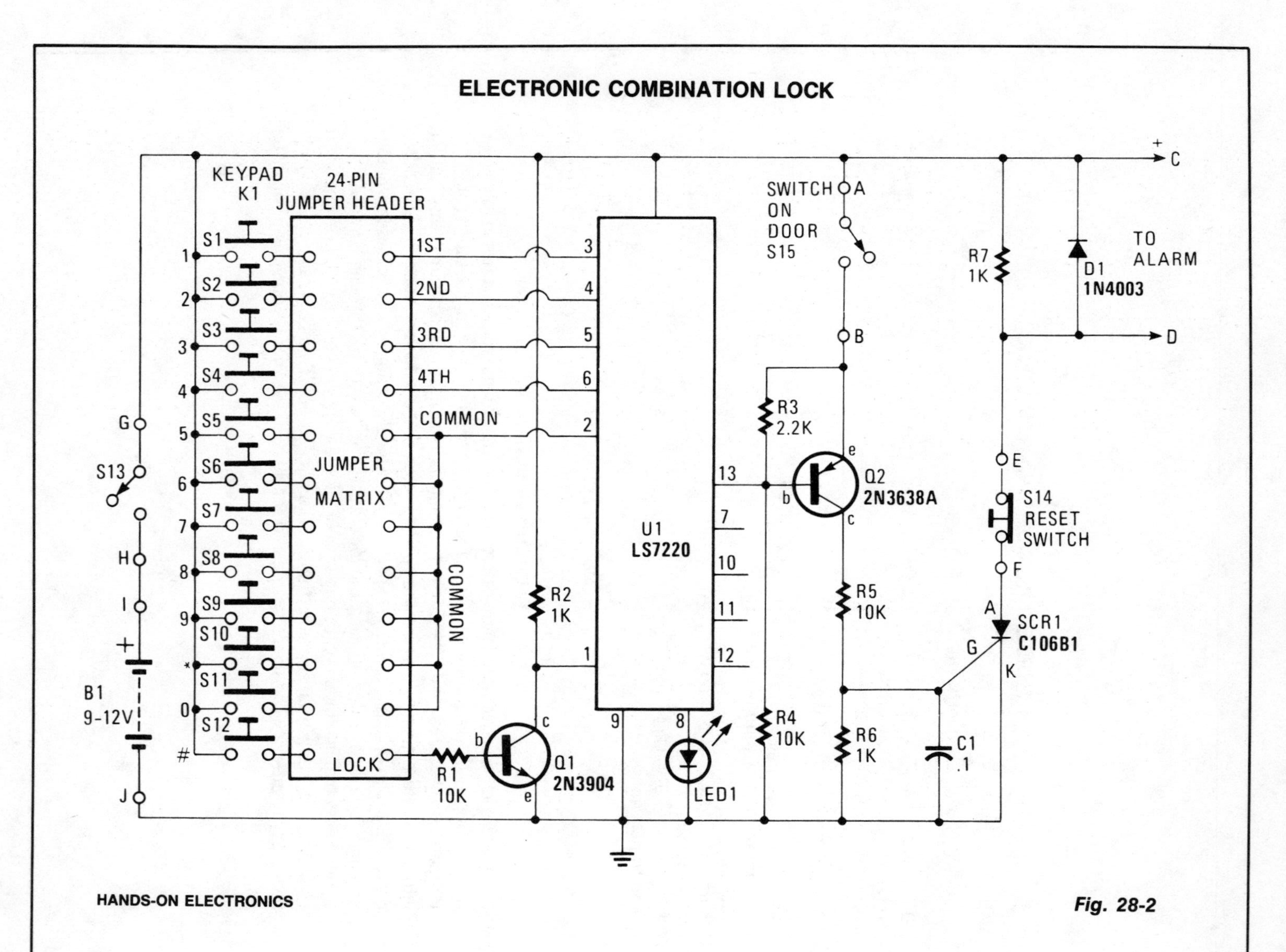
ELECTRONIC COMBINATION LOCK
KEYPAD
K1
24-PIN
JUMPER HEADER
JUMPER
MATRIX
1ST
2ND
3RD
4TH
COMMON
COMMON
LOCK
S1
S2
S3
S4
S5
S6
S7
S8
S9
S10
S11
S12
1
2
3
4
5
6
7
8
9
*
0
#
G
S13
H
I
+
B1
9-12V
J
R1
10K
Q1
2N3904
R2
1K
U1
LS7220
LED1
SWITCH
ON
DOOR
S15
A
B
R3
2.2K
R4
10K
Q2
2N3638A
R5
10K
R6
1K
C1
.1
R7
1K
D1
1N4003
TO
ALARM
C
D
E
S14
RESET
SWITCH
F
SCR1
C106B1
HANDS-ON ELECTRONICS

Fig. 28-2

Circuit Notes

When button S12 (#) is pressed, a positive voltage fed through R1 appears at the base of transistor Q1, turning it on. When Q1 is conducting, pin 1 of U1 is brought to ground (low) or the battery's negative terminal. With pin 1 low, two things occur: Pin 8 of U1 goes high (+9 volts dc), turning on LED 1—indicating that the circuit has been armed—and pin 13 goes from high to low. Transistor Q2 requires a low signal or negative voltage on its base in order to conduct. It also needs a positive voltage on its emitter and a negative voltage on the collector. As long as the door switch (S15) remains open (with the door itself closed), Q2's emitter will not receive the necessary positive voltage. If, however, an unauthorized person opens the door, thus closing switch S15 and placing a positive voltage on the emitter of Q1, the following sequence occurs:

1. Transistor Q2 conducts, receiving the necessary biasing current through a current-divider network consisting of resistors R3 and R4.
2. As Q2 conducts, a voltage drop is developed across the voltage dividers made up of resistors R5 and R6. With R5 at 10,000 ohms and R6 at 1000 ohms, approximately one volt appears at the gate of SCR1. That's enough voltage to trigger the SCR's gate.

29
Emulator Circuits

The sources of the following circuits are contained in the Sources section beginning on page 694. The figure number contained in the box of each circuit correlates to the source entry in the Sources section.

SIMULATED INDUCTOR

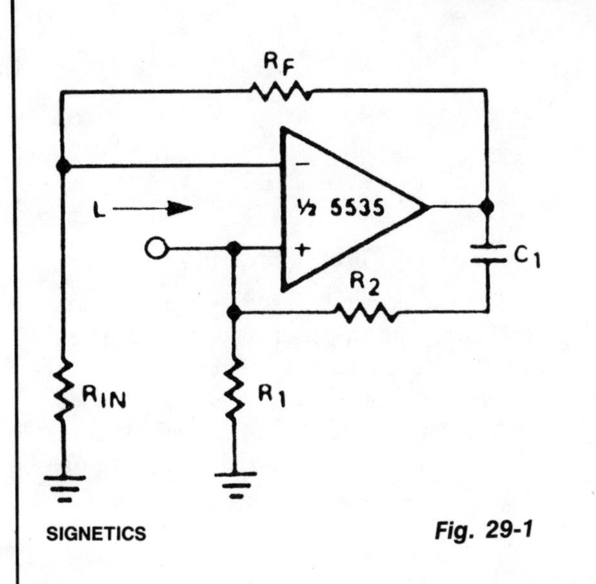

Fig. 29-1

Circuit Notes

With a constant current excitation, the voltage dropped across an inductance increases with frequency. Thus, an active device whose output increases with frequency can be characterized as an inductance. The circuit yields such a response with the effective inductance being equal to: $L = R_1R_2C$. The Q of this inductance depends upon R_1 being equal to R_2. At the same time, however, the positive and negative feedback paths of the amplifier are equal leading to the distinct possibility of instability at high frequencies. R_1 should therefore always be slightly smaller than R_2 to assure stable operation.

RESISTOR MULTIPLIER

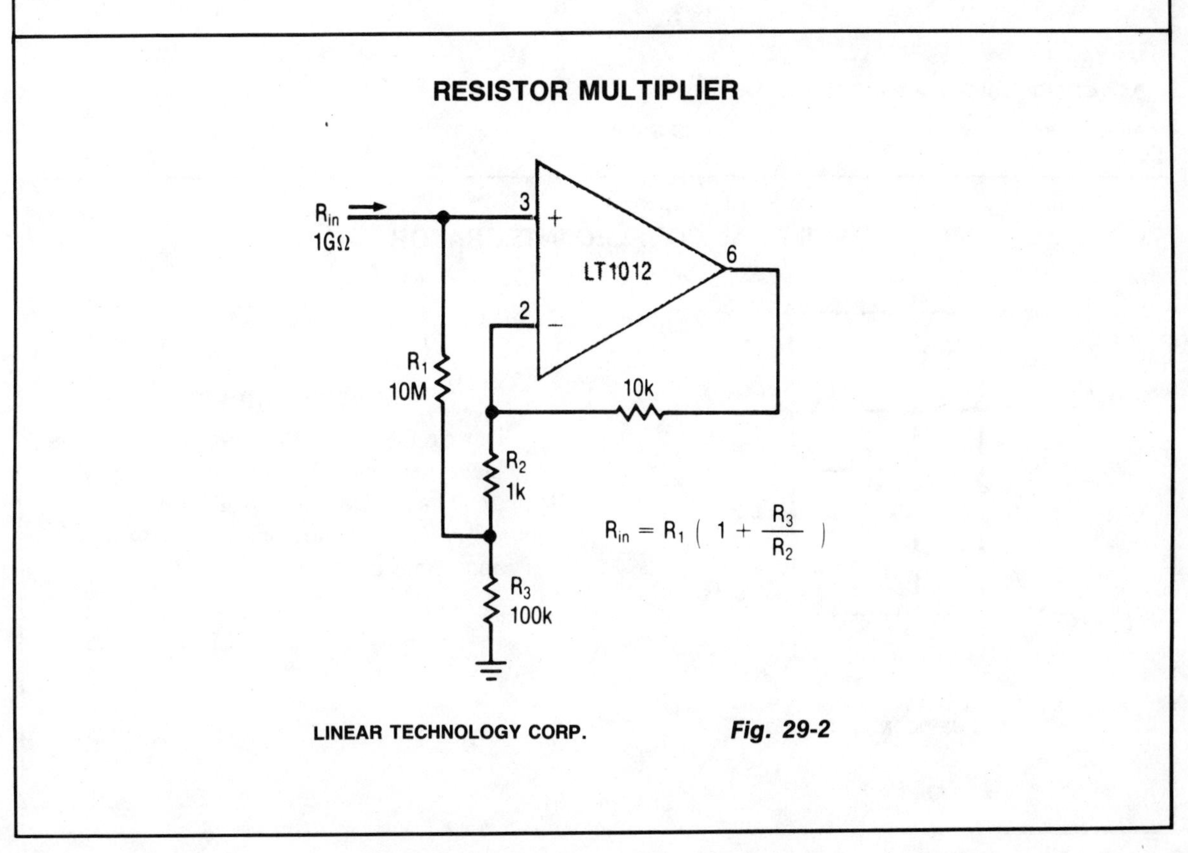

Fig. 29-2

CAPACITANCE MULTIPLIER

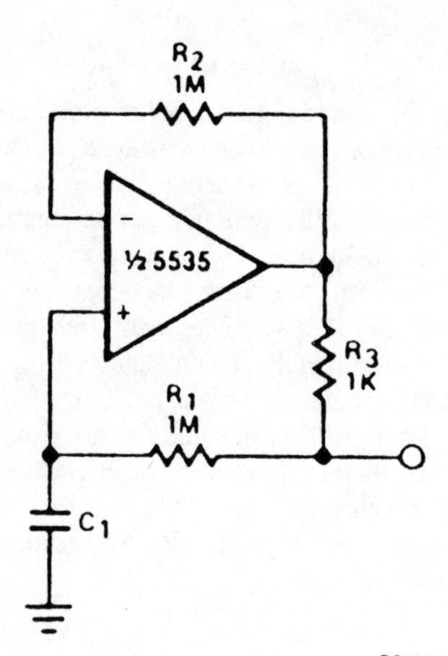

NOTE:
All resistor values are in ohms.

SIGNETICS

Fig. 29-3

Circuit Notes

The circuit can be used to simulate large capacitances using small value components. With the values shown and C - 10 μF, an effective capacitance of 10,000 μF was obtained. The Q available is limited by the effective series resistance. So R1 should be as large as practical.

JFET ac COUPLED INTEGRATOR

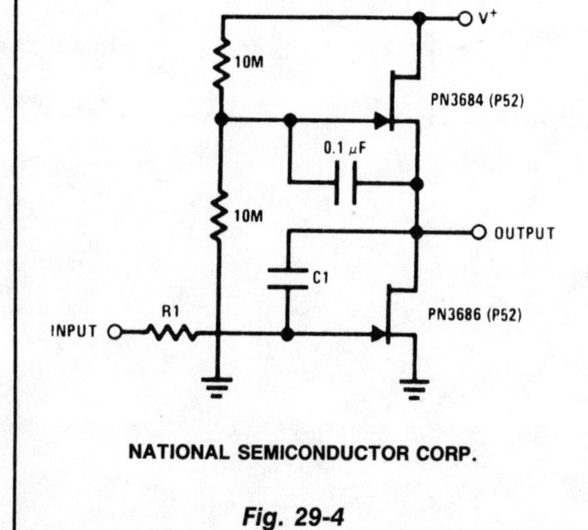

NATIONAL SEMICONDUCTOR CORP.

Fig. 29-4

Circuit Notes

This circuit utilizes the "μ-amp" technique to achieve very high voltage gain. Using C1 in the circuit as a Miller integrator, or capacitance multiplier, allows this simple circuit to handle very long time constants.

30
Fence Chargers

The sources of the following circuits are contained in the Sources section beginning on page 694. The figure number contained in the box of each circuit correlates to the source entry in the Sources section.

BATTERY-POWERED FENCE CHARGER

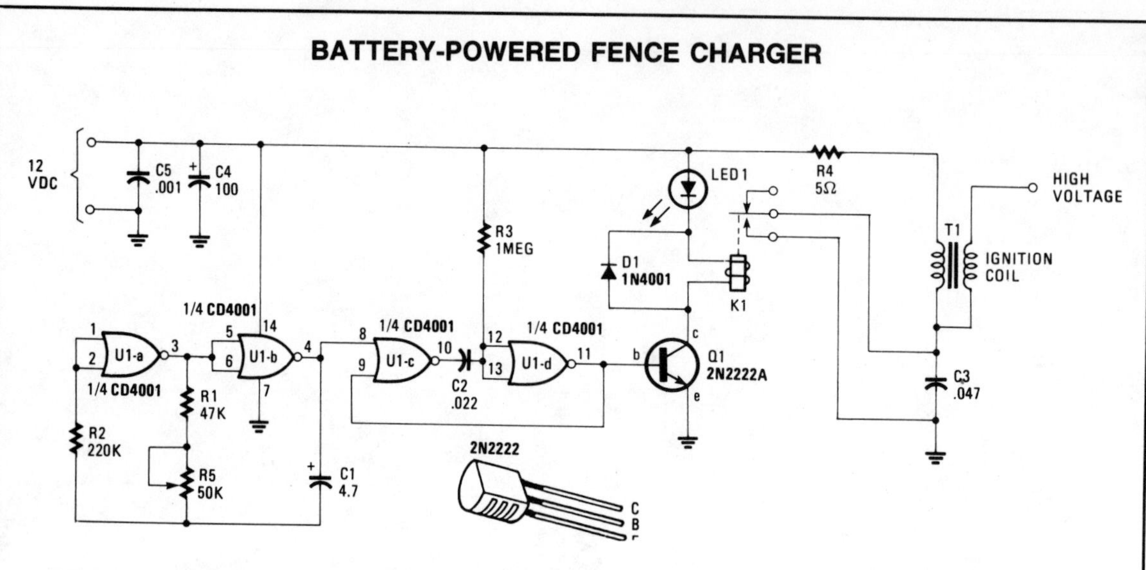

HANDS-ON ELECTRONICS

Fig. 30-1

Circuit Notes

In essence, the circuit is nothing more than an auto ignition coil and a set of points which accomplishes the same thing. A pulsing circuit made from a single CMOS NOR integrated circuit (U1), opens and closes the relay contacts to simulate the action of the original breaker points. The relay pulser is divided into two clocking functions. The first circuit is a free-running squarewave generator that determines the rate or frequency of the pulses that activate the relay. It is essentially a pair of NOR gates connected as inverters and placed in a feedback loop, they are U1-b. The oscillating period of the feedback loop is determined by timing components C1, R1, and variable resistor R5.

ELECTRIC FENCE CHARGER

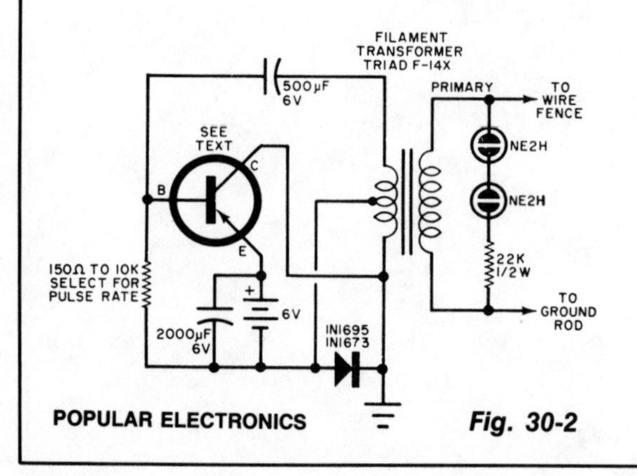

POPULAR ELECTRONICS

Fig. 30-2

Circuit Notes

Any good power transistor can be used in this circuit. The base resistor should be adjusted to obtain a pulse rate of about 50 pulses per minute. The range of values shown can go from 10 pulses to 100 pulses per minute. The single fence wire must be insulated at each supporting pole and should be mounted low enough to prevent an animal from crawling under the wire. The two neon lamps indicate when the unit is operating.

SOLID-STATE ELECTRIC FENCE CHARGER

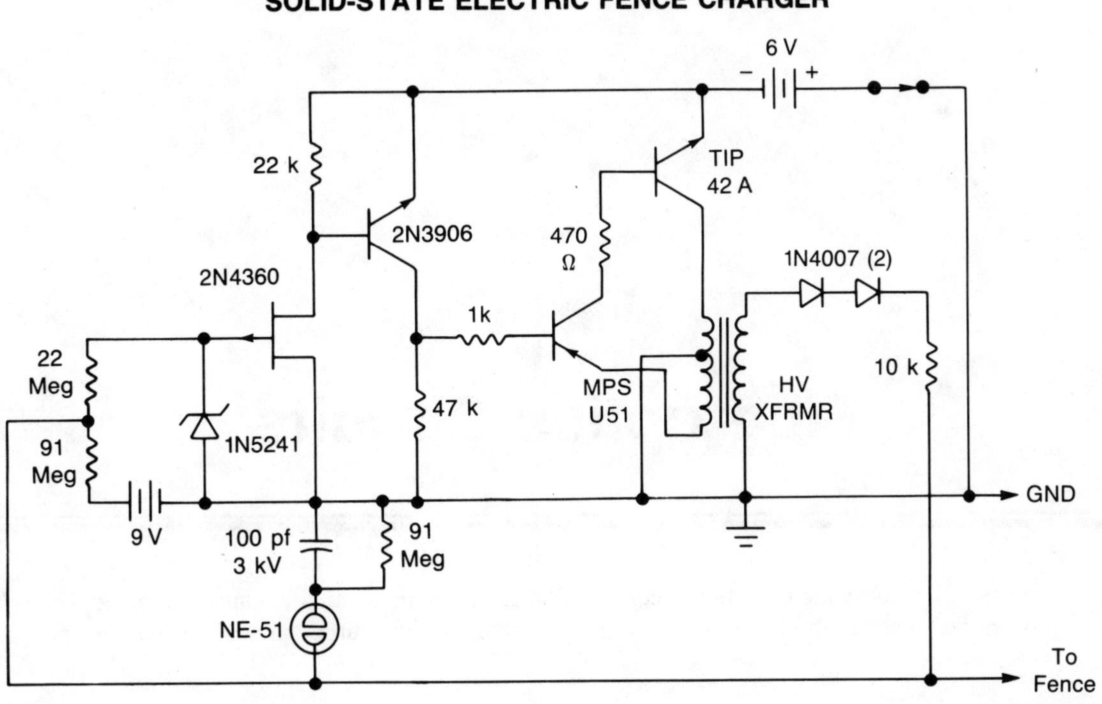

WILLIAM SHEETS

Fig. 30-3

Circuit Notes

A touch-sensing circuit keeps the high-voltage generator cut off until something touches the fence wire. Contact with the fence sensing circuit wire starts the high-voltage generator which applies a series of 500 microsecond pulses at approximately 300 volts to the fence wire. Pulse repetition rate is determined by the intruder's resistance to earth ground. The lower the resistance, the higher the pulse rate. A ground rod is inserted several inches into the ground near the fence wire. In the sensing mode the neon lamp should not flicker or light. If it does, it indicates leakage between the fence wire and ground. If sensitivity is too great, it can be reduced by changing the 91 Meg resistor to 47 or 22 Meg as required.

31
Fiberoptics Circuits

The sources of the following circuits are contained in the Sources section beginning on page 694. The figure number contained in the box of each circuit correlates to the source entry in the Sources section.

10 MHz FIBEROPTIC RECEIVER

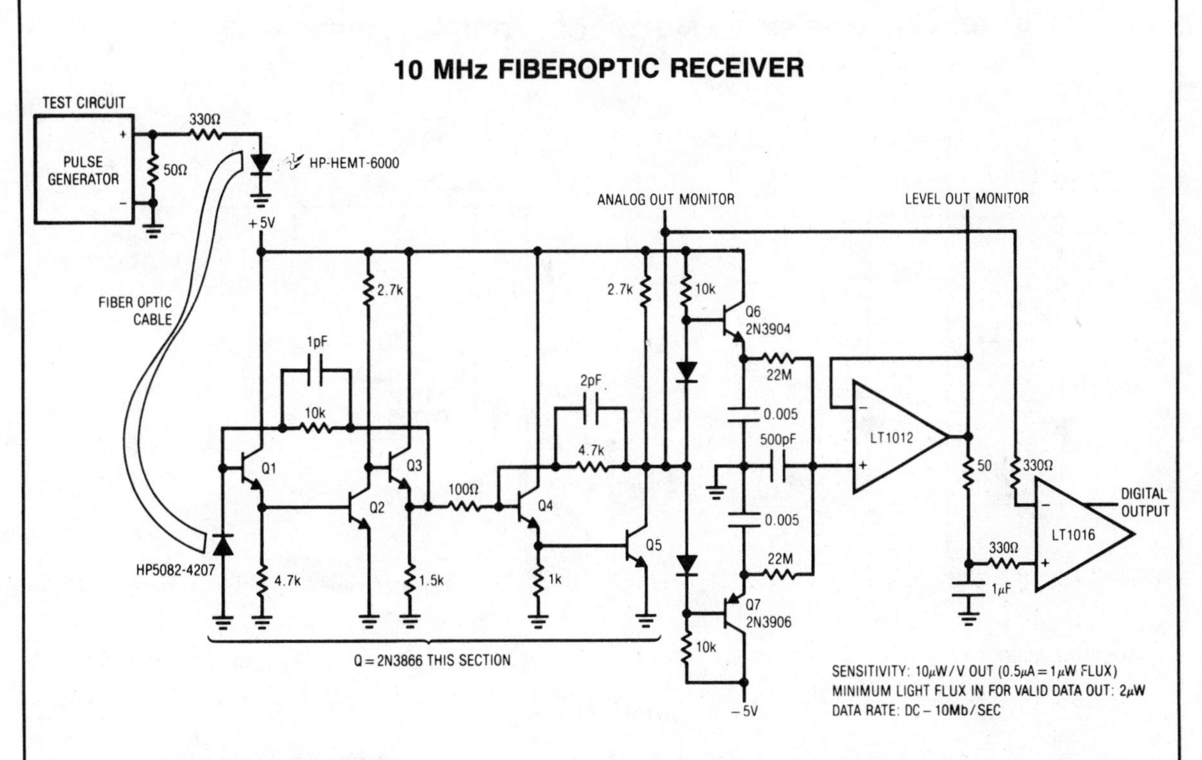

LINEAR TECHNOLOGY CORPORATION *Fig. 31-1*

Circuit Notes

The receiver will accurately condition a wide range of light inputs at up to 10 MHz data rates. The optical signal is detected by the PIN photodiode and amplified by a broadband fed-back stage, Q1-Q3. A second, similar, stage gives further amplification. The output of this stage (Q5's collector) biases a 2-way peak detector (Q6-Q7). The maximum peak is stored in Q6's emitter capacitor while the minimum excursion is retained in Q7's emitter capacitor. The dc value of Q5's output signal's mid-point appears at the junction of the 0.005 μF capacitor and the 22 M ohm unit. This point will always sit midway between the signal's excursions, regardless of absolute amplitude. This signal-adaptive voltage is buffered by the low bias LT1012 to set the trigger voltage at the LT1016's positive input. The LT1016's negative input is biased directly from Q5's collector.

DC VARIABLE SPEED MOTOR CONTROL VIA FIBEROPTICS

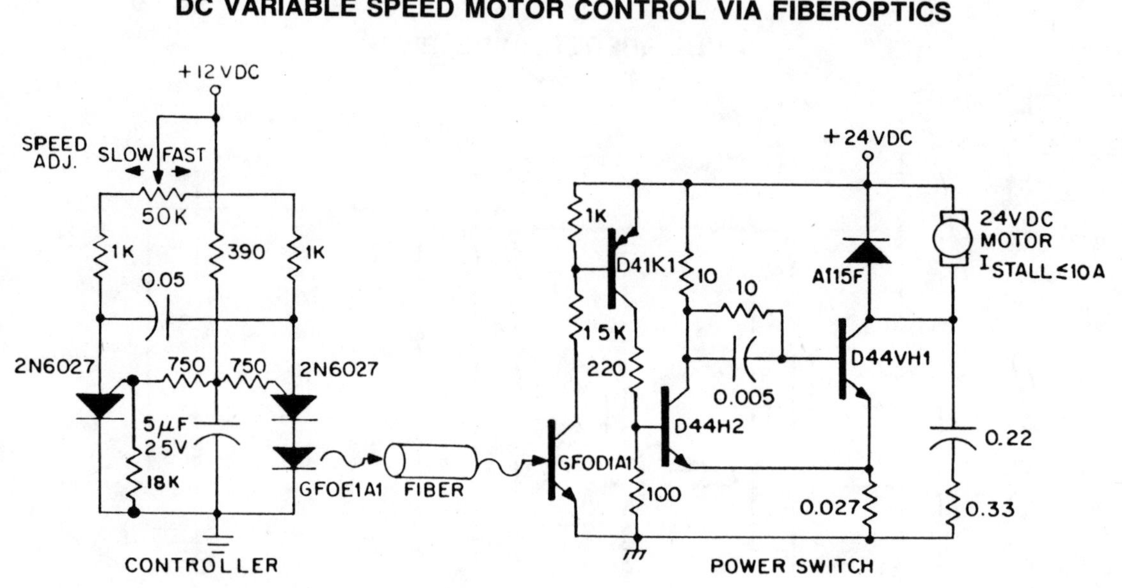

GENERAL ELECTRIC

Fig. 31-2

Circuit Notes

Dc power can also be controlled via fiberoptics. The circuit provides an insulated speed control path for a small dc actuator motor (≤ 1/12 hp). Control logic is a self-contained module requiring about 300 mW at 12 V, which can be battery powered. The control module furnishes infrared pulses, at a rate of 160 Hz, with a duty cycle determined by the position of the speed adjust potentiometer. The programmable unijunction multivibrator provides approximately 10 mA pulses to the GFOE1A1 at duty cycles adjustable over a range of 1% to 99%. The infrared pulses are detected by the GFOD1A1, amplified by the D39C1 pnp Darlington, and supplied to the power drive switch, which is connected in a Schmitt trigger configuration to supply the motor voltage pulses during the infrared pulses. Thus, the motor's average supply voltage is pulse width modulated to the desired speed, while its current is maintained between pulses by the A115F free-wheeling diode. The snubber network connected in parallel with the power switch minimizes peak power dissipation in the output transistor, and enhancing reliability. Larger hp motors can be driven by adding another stage of current gain, while longer fiber range lengths can be obtained with an amplifier transistor driving the GFOE1A1.

FIBEROPTIC INTERFACE

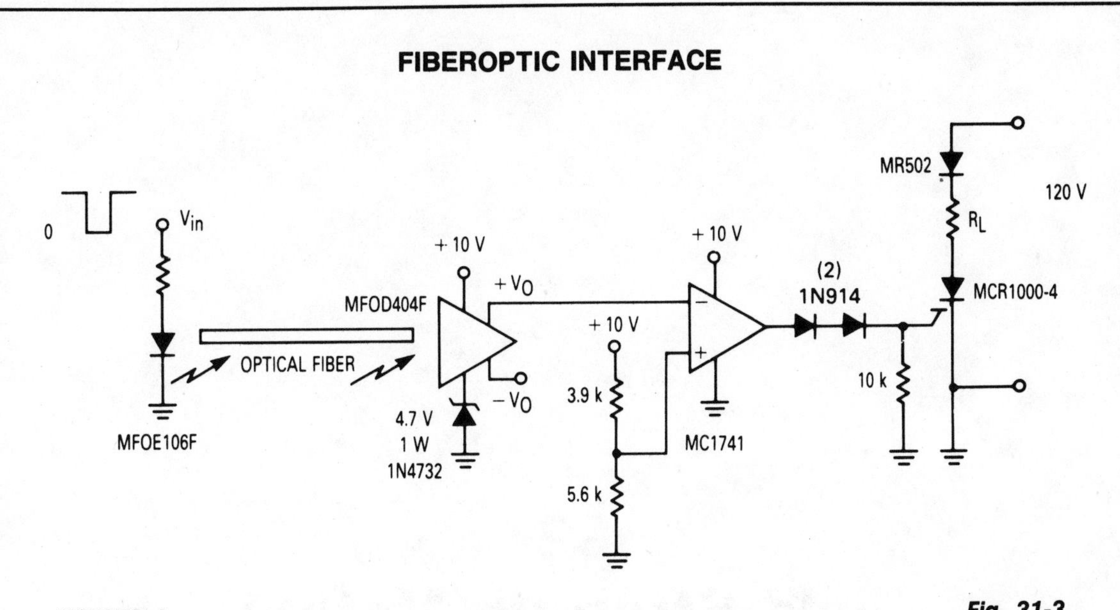

MOTOROLA

Fig. 31-3

Circuit Notes

An op amp is used to interface between a fiberoptic system and the MOS SCR to multi-cycle, half-wave control of a load. This receiver has two complementary outputs, one at a quiescent level of about 0.6 V and the second at 3 V. By adding a 4.7 V zener in series with the return bus, the effective V_{CC} becomes 5.3 V and also the 0.6 V output level is translated up to about 5.3 V. This level is compatible with the reference input (5.9 V) of the single-ended powered op-amp acting as a comparator.

32
Field Strength Meters

The sources of the following circuits are contained in the Sources section beginning on page 694. The figure number contained in the box of each circuit correlates to the source entry in the Sources section.

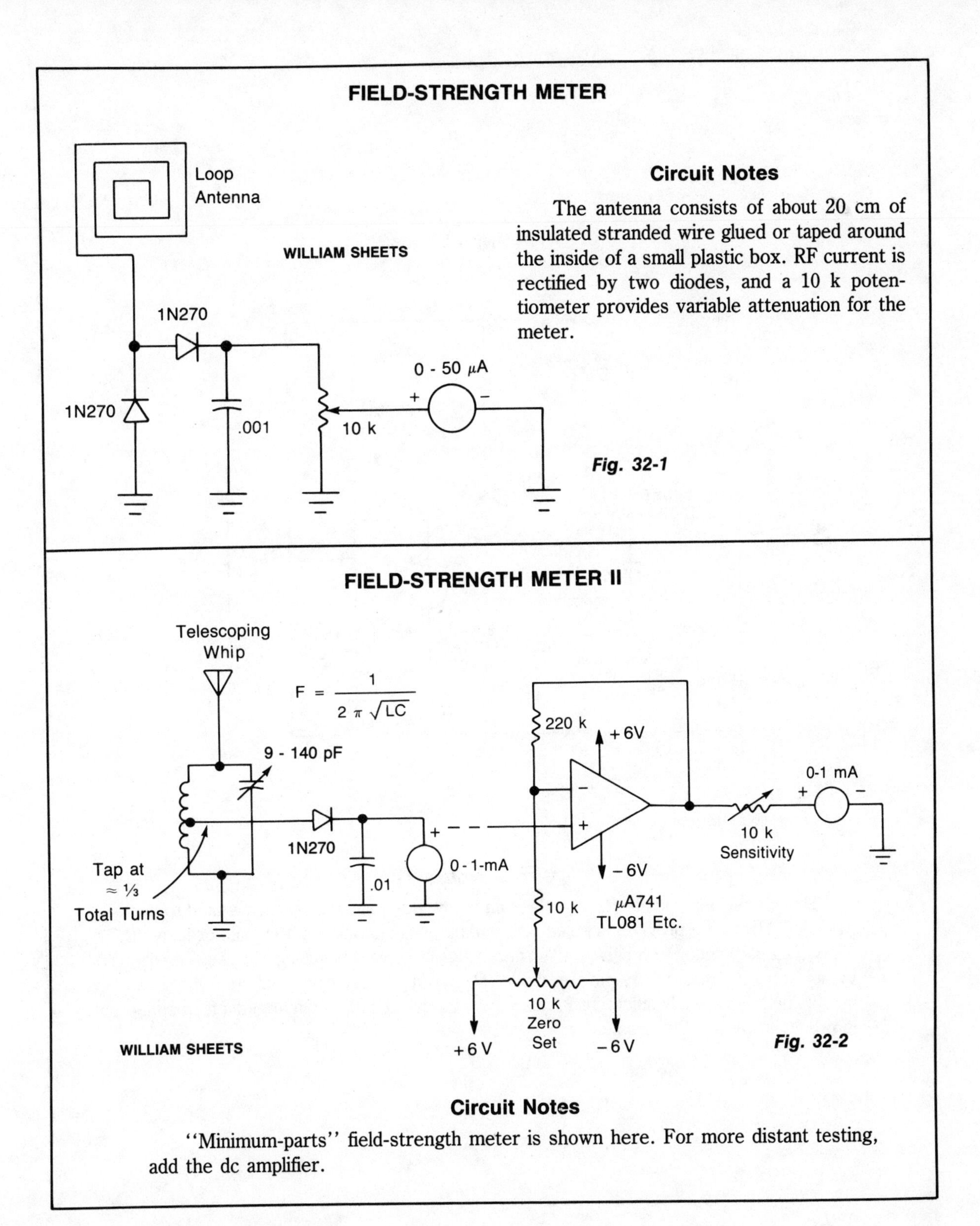

FIELD-STRENGTH METER

Circuit Notes

The antenna consists of about 20 cm of insulated stranded wire glued or taped around the inside of a small plastic box. RF current is rectified by two diodes, and a 10 k potentiometer provides variable attenuation for the meter.

Fig. 32-1

FIELD-STRENGTH METER II

Fig. 32-2

Circuit Notes

"Minimum-parts" field-strength meter is shown here. For more distant testing, add the dc amplifier.

RF SNIFFER

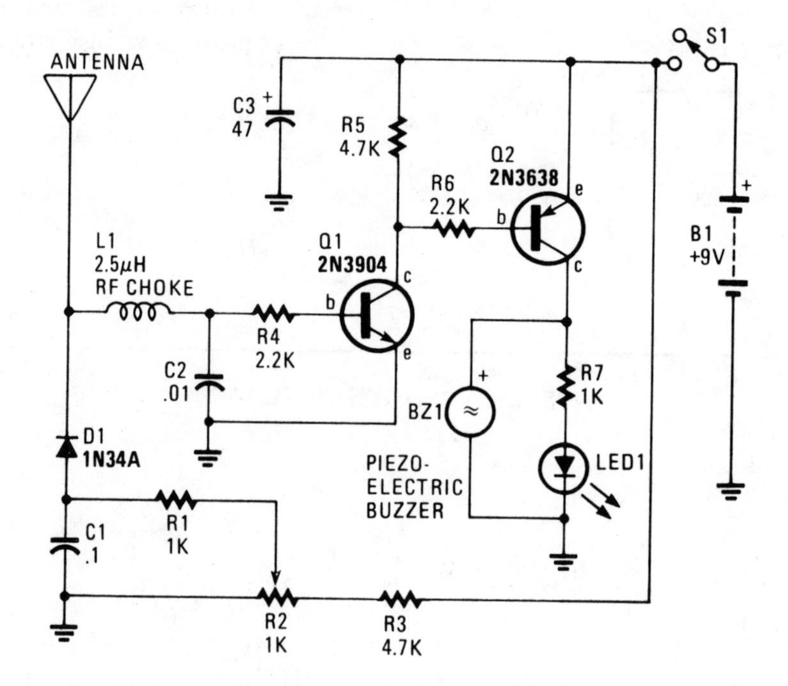

HANDS-ON ELECTRONICS *Fig. 32-3*

Circuit Notes

This circuit responds to RF signals from below the standard broadcast band to well over 500 MHz, and provides a visual and audible indication when a signal is received. The circuit is designed to receive low-powered signals as well as strong sources of energy by adjusting the bias on the pick-up diode, D1, with R2. A very sensitive setting can be obtained by carefully adjusting R2 until the LED just begins to light and a faint sound is produced by the Piezo sounder.

HIGH-SENSITIVITY FIELD STRENGTH METER

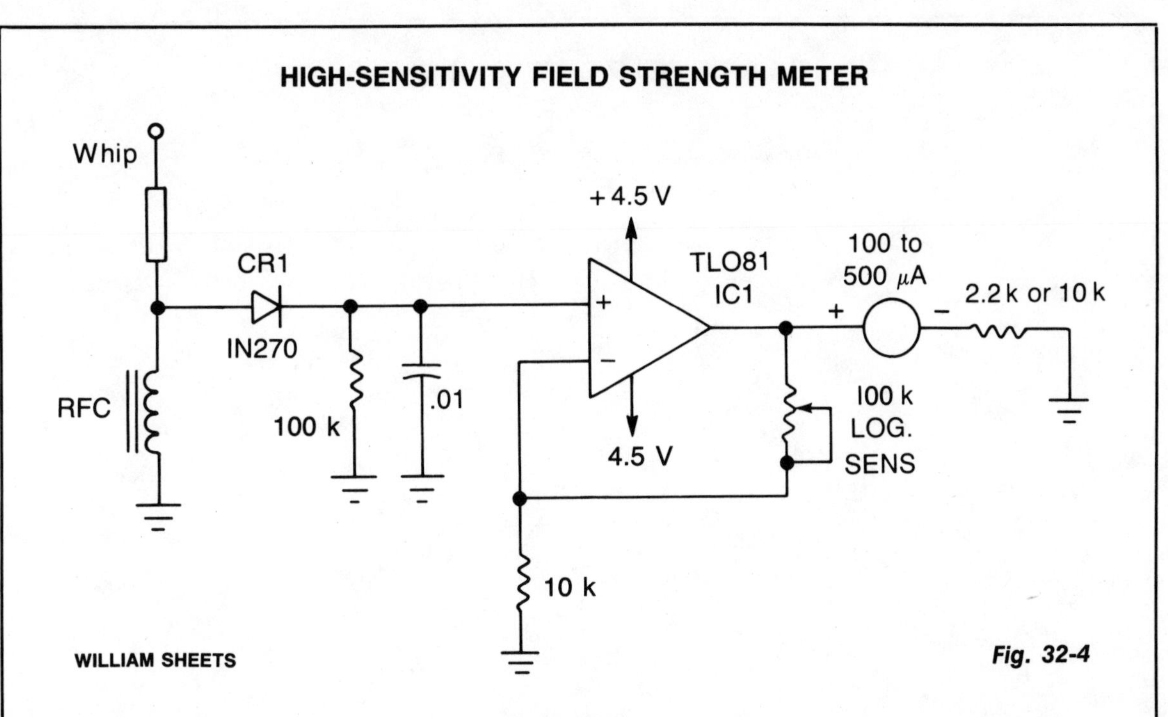

WILLIAM SHEETS

Fig. 32-4

Circuit Notes

A TL081 (IC1 op amp is used to increase sensitivity. RF signal is detected by CR1 and is then amplified by IC1. Full-scale sensitivity is set with the 100 K potentiometer.

TRANSMISSION INDICATOR

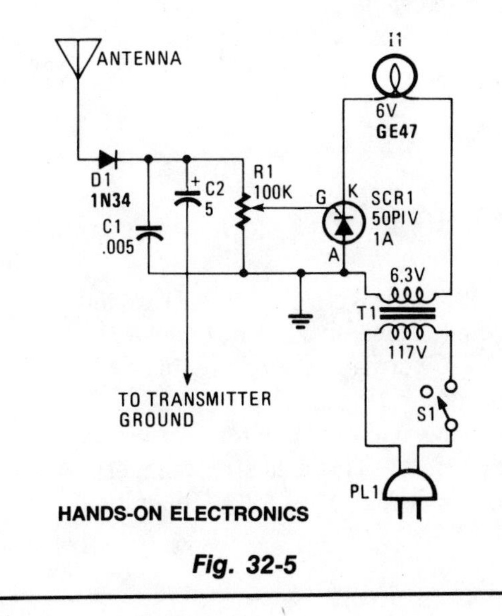

HANDS-ON ELECTRONICS

Fig. 32-5

Circuit Notes

Everytime the push-to-talk button is closed the light will go on. The antenna samples the output RF from the transmitter. That signal is then rectified (detected) by germanium diode D1, and used to charge capacitor C2. The dc output is used to trigger a small silicon-controlled rectifier (SCR1), which permits the current to flow through the small pilot lamp. For lower-power applications, such as CB radio, the antenna will have to be close-coupled to the antenna.

LF OR HF FIELD STRENGTH METER

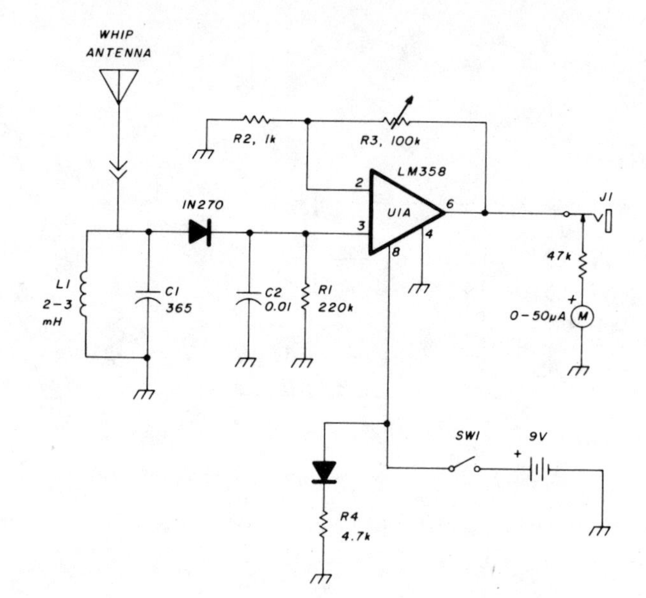

Table 1.

L1	C1 (variable)	Frequency Range	Ham Band
50 μH	30-365 pF	1- 4 MHz	160, 80 meters
3 μH	30-365 pF	5-16 MHz	40, 30, 20 meters
0.9 μH	30-365 pF	9-30 MHz	30, 20, 15, 12, 10 meters
2.5 mH	—	Broadband at reduced gain	

HAM RADIO

Fig. 32-6

Circuit Notes

C1 and L1 resonate on the 1750 meter band, with coverage from 150 kHz to 500 kHz. L1 can be slug-tuned for 160-to-190 kHz coverage alone or a 2.5 mH choke can be used for L1, if desired, using C1 for tuning. A 1N270 germanium diode rectifies the RF signal and C2 is charged at the peak RF level. This dc level is amplified by an LM358. The gain is determined by R2 and R3, 1 100-kilohm linear potentiometer that varies the dc gain from 1 to 100, driving the 50 microampere meter. This field strength meter need not be limited to LF use. The Table shows the L1 and C1 values for HF operation and broadband operation.

33
Filter Circuits

The sources of the following circuits are contained in the Sources section beginning on page 694. The figure number contained in the box of each circuit correlates to the source entry in the Sources section.

LOW COST UNIVERSAL ACTIVE FILTER

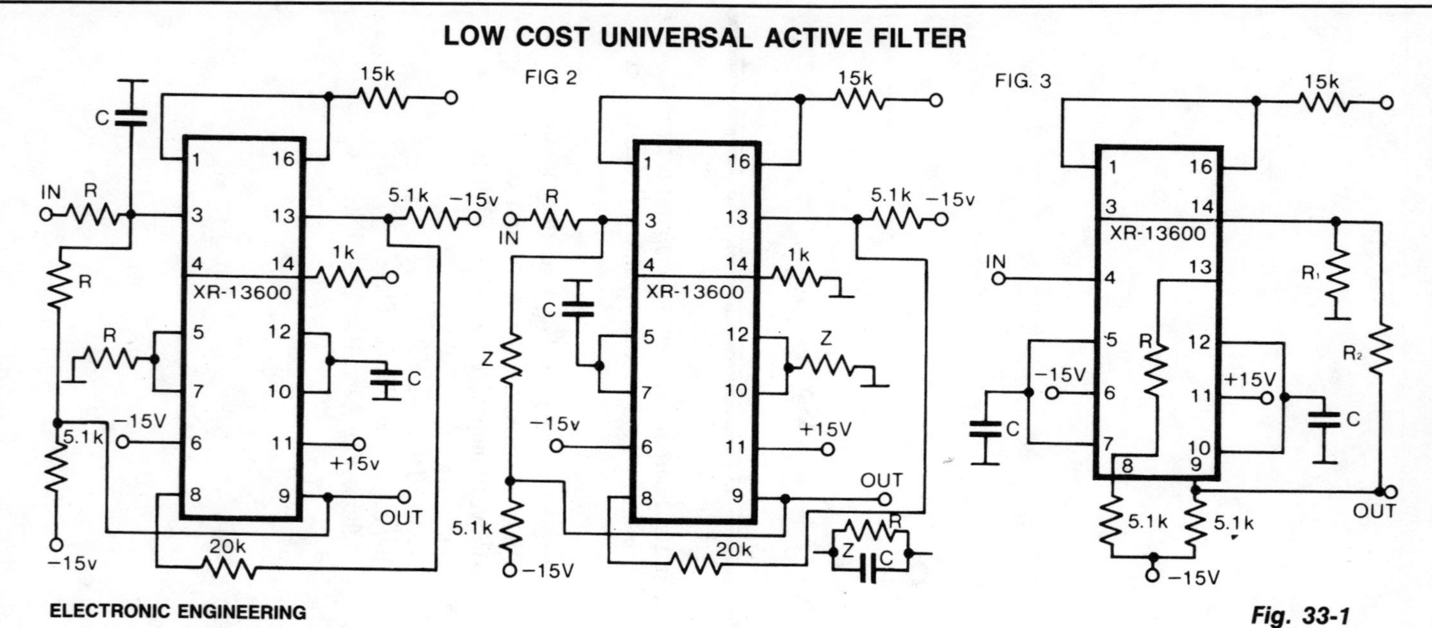

Fig. 33-1

Circuit Notes

The circuit as shown in Fig. 1 gives the bandpass operation the transfer function calculated from

$$F_{BP}(s) = \frac{S/\omega_o}{K}$$

where $K = 1 + s/Q\omega_o + s^2/\omega_o^2$.
The cut-off frequency, ω_o, and the Q-factor are given by

$$\omega_o = g/C \text{ and } Q = gR/2$$

where g is the transconductance at room temperature.

Interchanging the capacitor C with the resistor R at the input of the circuit high-pass operation is obtained. A low-pass filter is obtained by applying two parallel connections of R and C as shown in Fig. 2.

The low-pass operation may be much improved with the circuit as given in Fig. 3. Here the gain and Q may be set up separately with respect to the cut-off frequency according to the equations

$$Q = 1/fB = 1 + R_2/R_1,$$
$$A = Q^2 \text{ and } \omega_o = g\,fB/C.$$

STATE VARIABLE FILTER

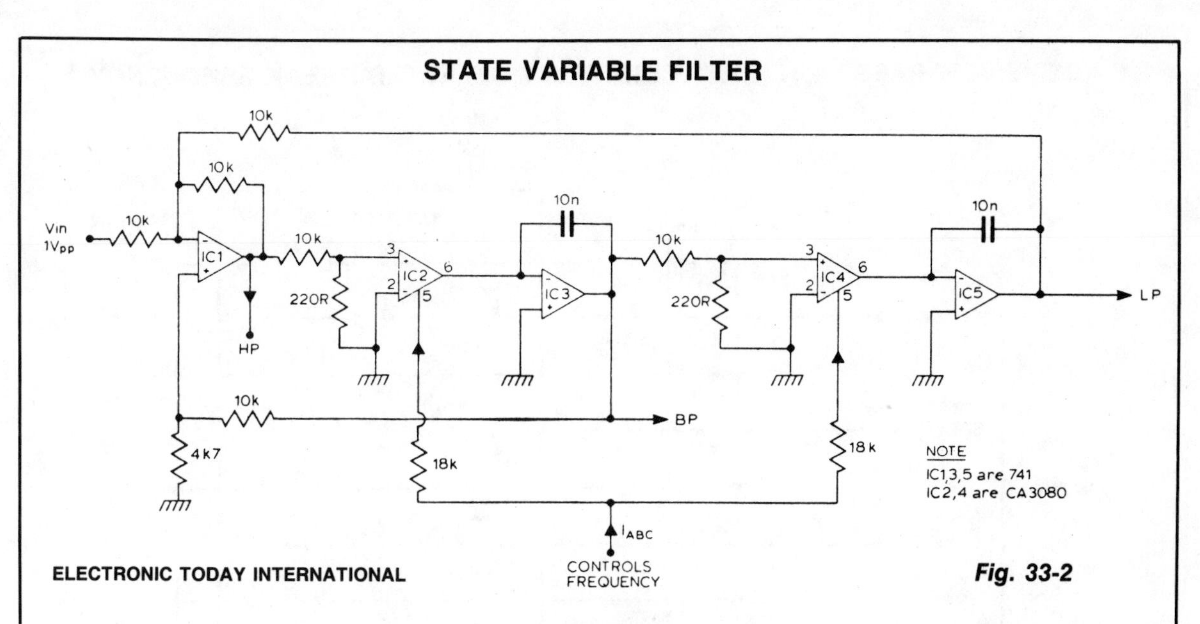

ELECTRONIC TODAY INTERNATIONAL

Fig. 33-2

Circuit Notes

The filter produces three outputs: high-pass, bandpass, and low-pass. Frequency is linearly proportional to the gain of the two integrators. Two CA3080's, (IC2, 4) provide the variable gain, the resonant frequency being proportional to the current I_{ABC}. Using 741 op amps for IC3 a control range of 100 to 1, (resonant frequency) can be obtained. If CA3140's are used instead of 741's then this range can be extended to nearly 10,000 to 1.

WIDEBAND TWO-POLE HIGH-PASS FILTER

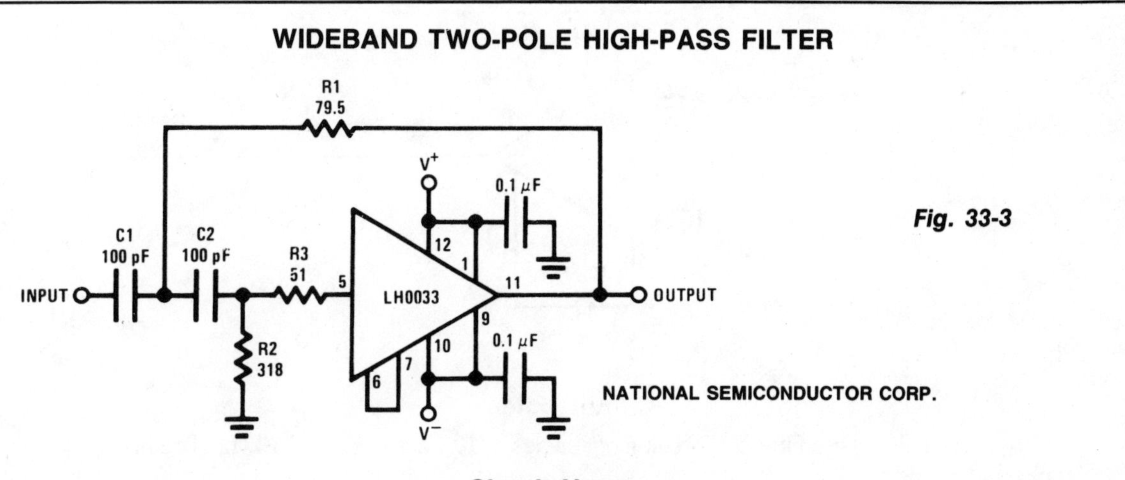

Fig. 33-3

NATIONAL SEMICONDUCTOR CORP.

Circuit Notes

The circuit provides a 10MHz cutoff frequency. Resistor R3 ensures that the input capacitance of the amplifier does not interact with the filter response at the frequency of interest. An equivalent low pass filter is similarly obtained by capacitance and resistance transformation.

ACTIVE LOW-PASS FILTER WITH DIGITALLY SELECTED BREAK FREQUENCY

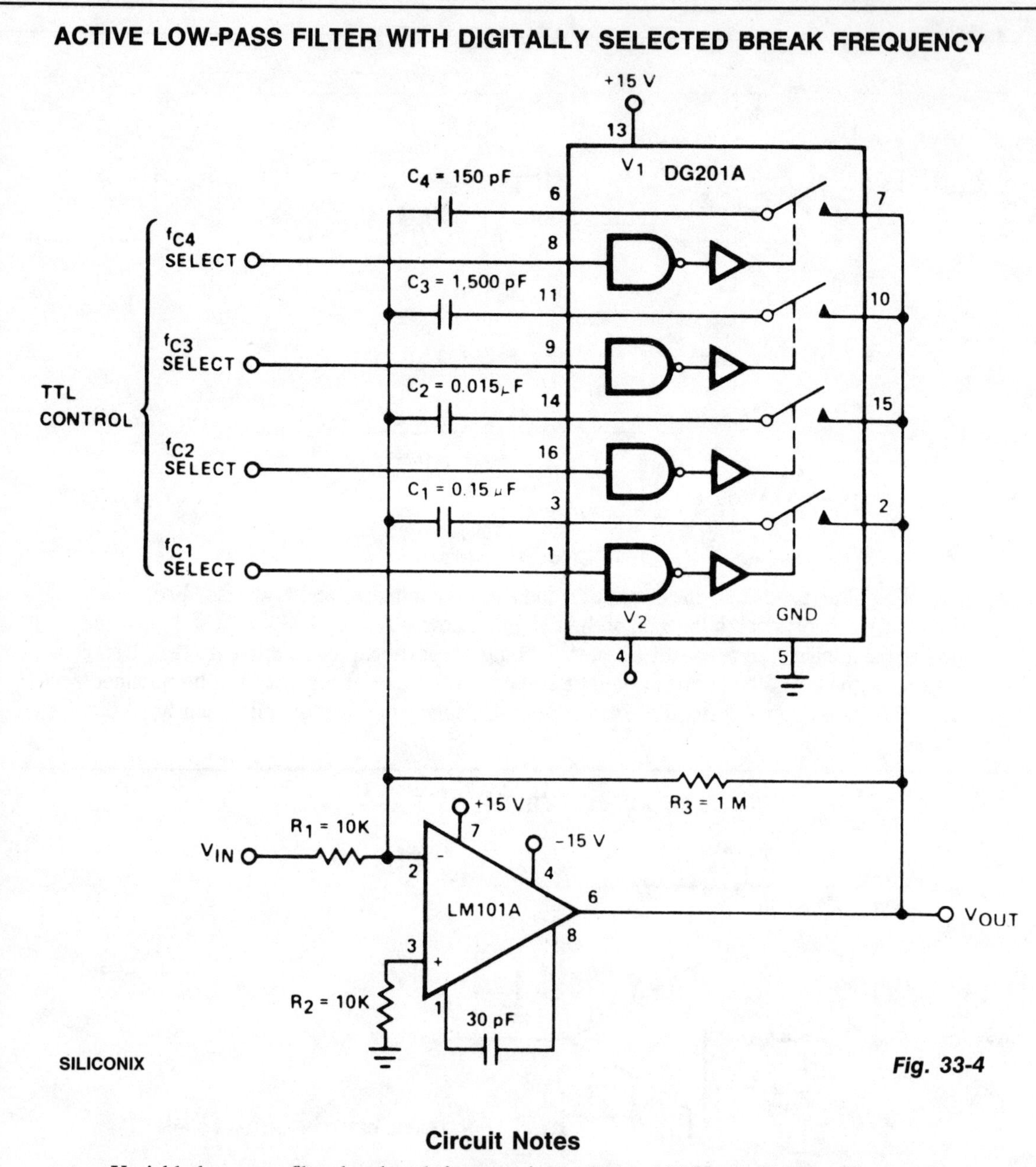

SILICONIX

Fig. 33-4

Circuit Notes

Variable low-pass filter has break frequencies at 1, 10, 100 Hz and 1 kHz. The break frequency is

$$1.\quad fc = \frac{1}{2\,\pi\,R_3\,C_X}$$

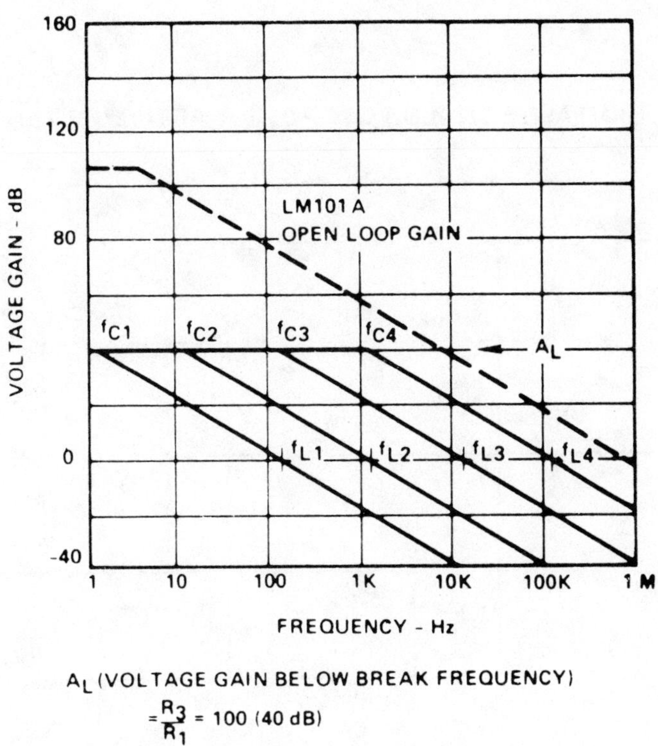

f_c (BREAK FREQUENCY) = $\dfrac{1}{2\pi R_3 C_X}$

f_L (UNITY GAIN FREQUENCY) = $\dfrac{1}{2\pi R_1 C_X}$

MAX ATTENUATION = $\dfrac{r_{DS(on)}}{10K} \approx -40$ dB

The low frequency gain is

$$2. \quad A_L = \frac{R_3}{R_3} = 100 \text{ (40 dB)}$$

A second break frequency (a zero) is introduced by $r_{DS(on)}$ of the DG201A, causing the minimum gain to be

$$3. \quad A_{MIN} = \frac{r_{DS(on)}}{R_1} \approx \frac{100}{10K} = .01,$$

a maximum attenuation of 40 dB (80 dB relative to the low frequency gain).

DIGITALLY TUNED LOW POWER ACTIVE FILTER

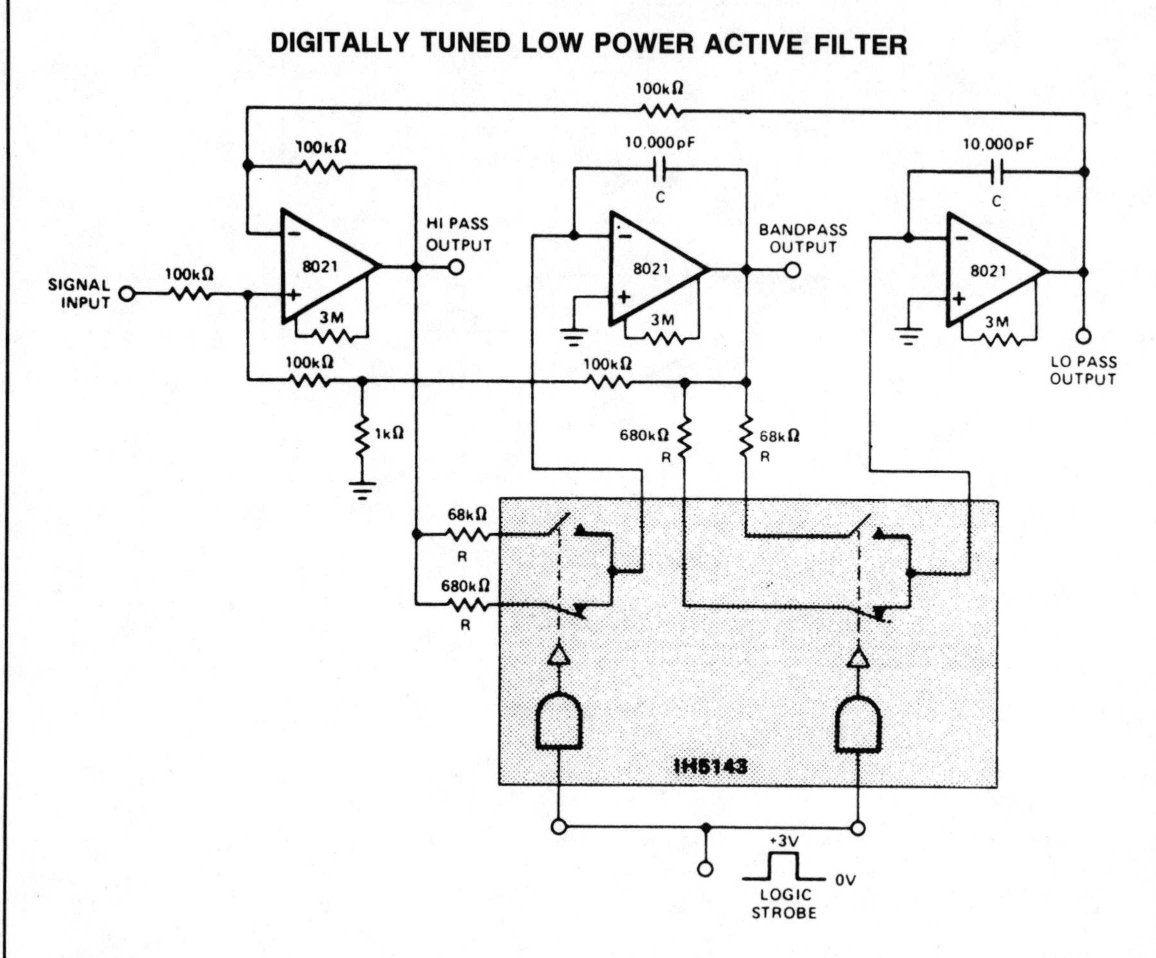

INTERSIL

Fig. 33-5

Circuit Notes

This constant gain, constant Q, variable frequency filter provides simultaneous low-pass, bandpass, and high-pass outputs with the component values shown, the center frequency will be 235 Hz and 23.5 Hz for high and low logic inputs respectively, Q = 100, and gain = 100.

RAZOR-SHARP CW FILTER

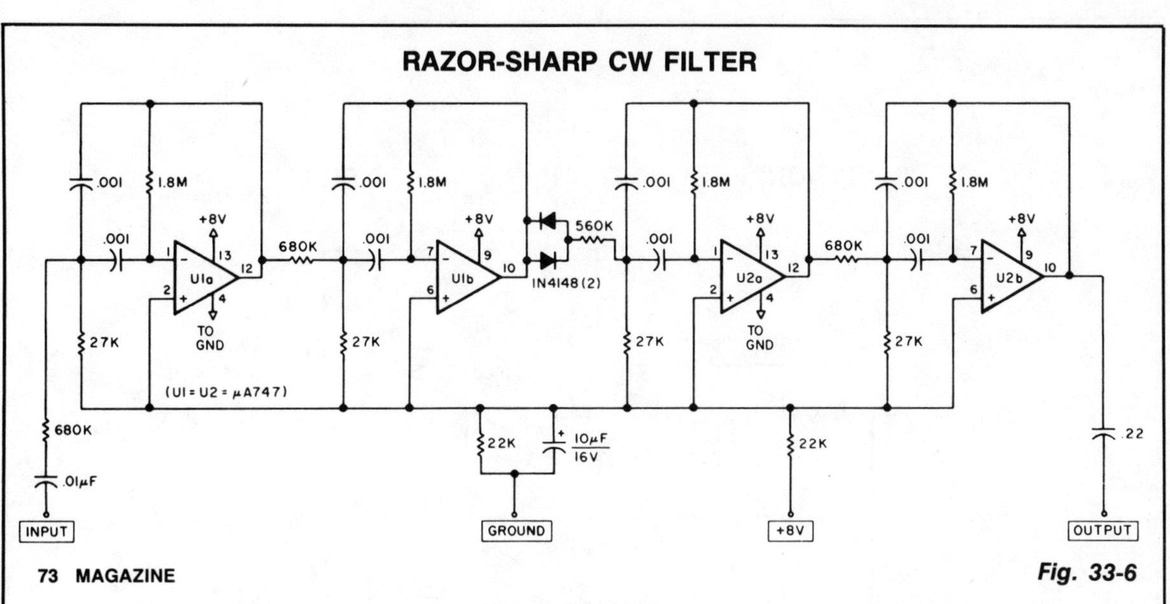

73 MAGAZINE

Fig. 33-6

Circuit Notes

The circuit consists of four stages of active bandpass filtering provided by two type-μA747 integrated-circuit dual op amps and includes a simple threshold detector (diodes D1 and D2) between stages 2 and 3 to reduce low-level background noise. Each of the four filter stages acts as a narrow bandpass filter with an audio bandpass centered at 750 Hz. The actual measured 3-dB bandwidth is only 80 Hz wide.

FIFTH ORDER CHEBYSHEV MULTIPLE FEEDBACK LOW PASS FILTER

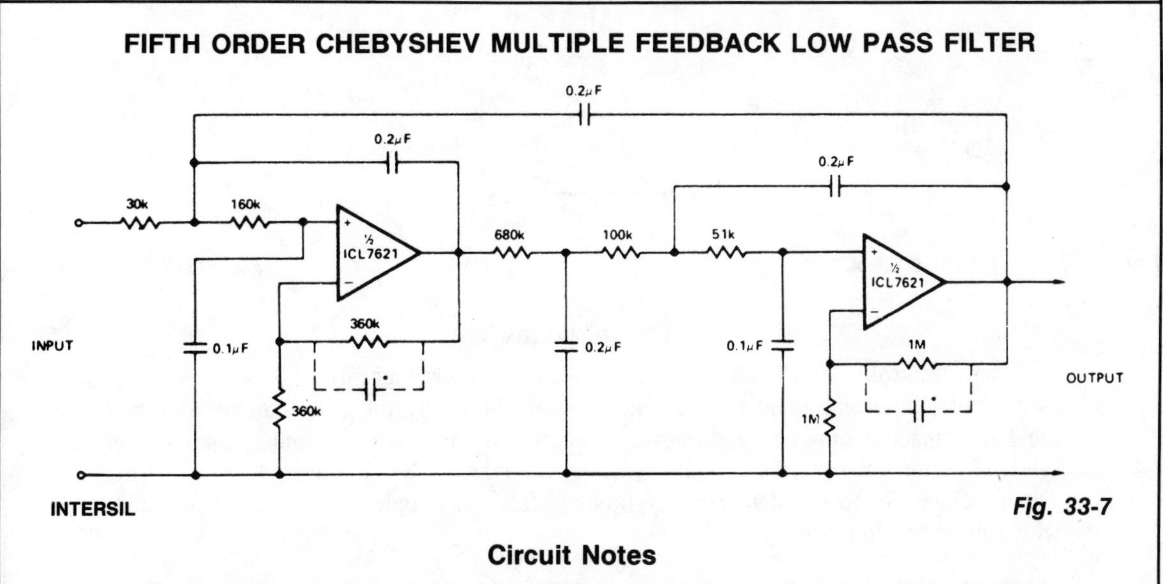

INTERSIL

Fig. 33-7

Circuit Notes

The low bias currents permit high resistance and low capacitance values to be used to achieve low frequency cutoff. f_c = 10 Hz, A_{VCL} = 4, Passband ripple = 0.1 dB. Note that small capacitors (25-50 pF) may be needed for stability in some cases.

PRECISION, FAST SETTLING, LOW-PASS FILTER

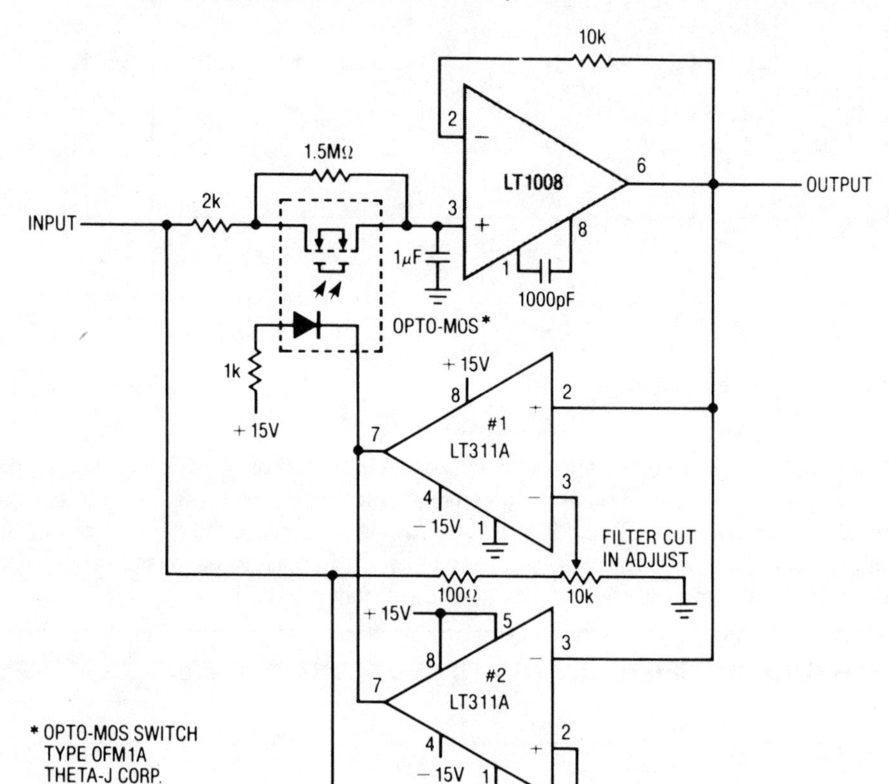

LINEAR TECHNOLOGY *Fig. 33-8*

Circuit Notes

This circuit is useful where fast signal acquisition and high precision are required, as in electronic scales. The filter's time constant is set by the 2 K ohm resistor and the 1 μF capacitor until comparator No. 1 switches. The time constant is then set by the 1.5 M ohm resistor and the 1 μF capacitor. Comparator No. 2 provides a quick reset. The circuit settles to a final value three times as fast as a simple 1.5 M ohm—1 μF filter, with almost no dc error.

PROGRAMMABLE BANDPASS USING TWIN-T BRIDGE

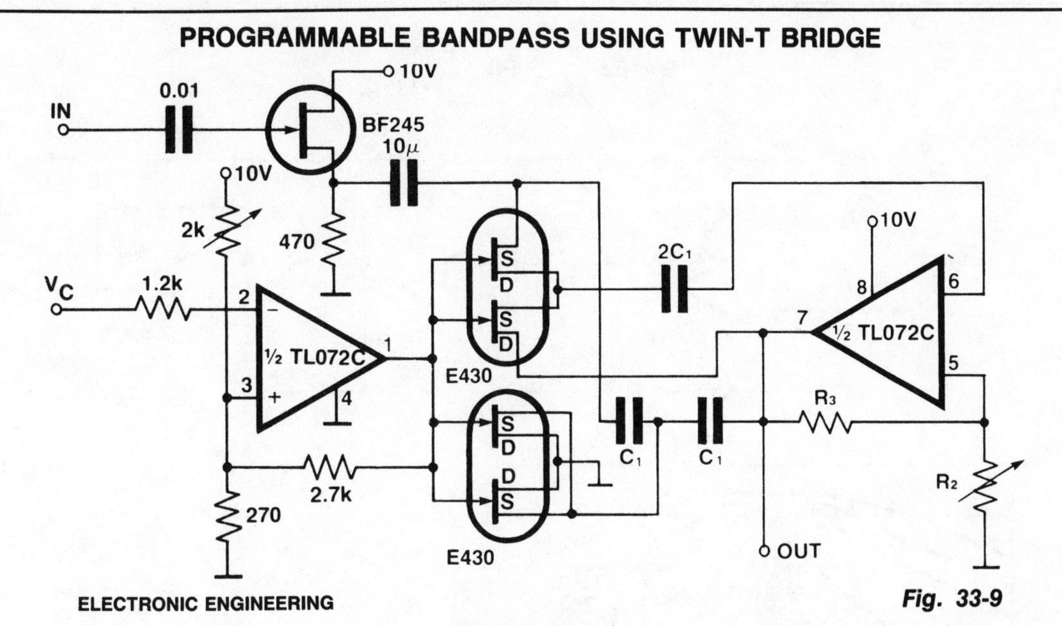

ELECTRONIC ENGINEERING

Fig. 33-9

Circuit Notes

The circuit gives a programmable bandpass where both the cut-over frequency and the gain, A, are controlled independently. In the twin-T bridge the resistors R and R/2 are replaced by two double FETs, E 430, the channel resistance of the first one in the series, the channel resistances of the second one are in parallel as to stimulate the resistance R/2. Both these resistors are controlled by V_C which ranges from 0 V to about 1 V. The gain of the circuit is set by means of the resistors R2 and R3.

ACTIVE BANDPASS FILTER (f0 = 1000 Hz)

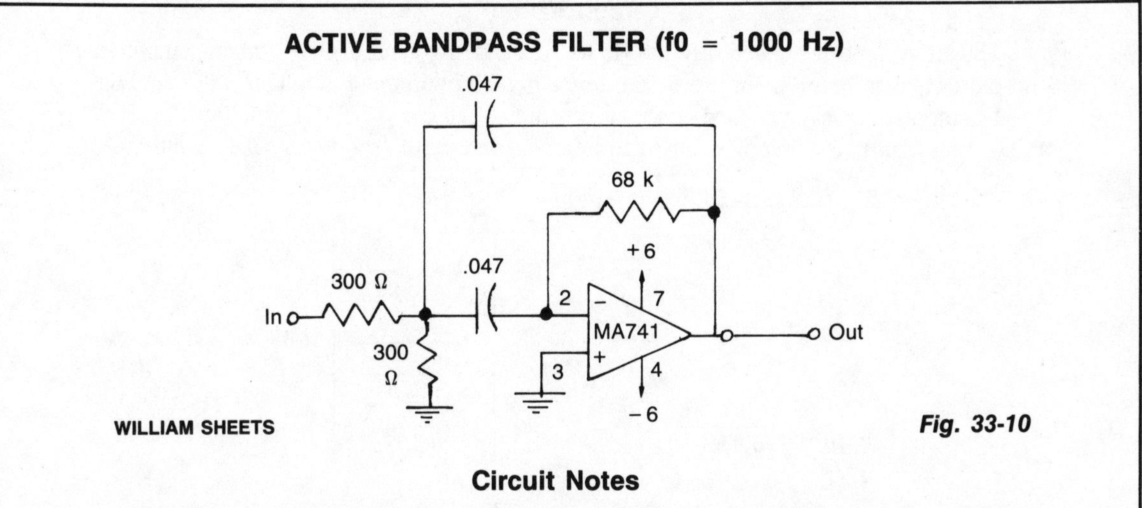

WILLIAM SHEETS

Fig. 33-10

Circuit Notes

This filter has a bandpass centered around 1kHz, for applications such as bridge amplifiers, null detectors, etc.

The circuit uses a µA741 IC and standard 5% tolerance components.

BANDPASS FILTER

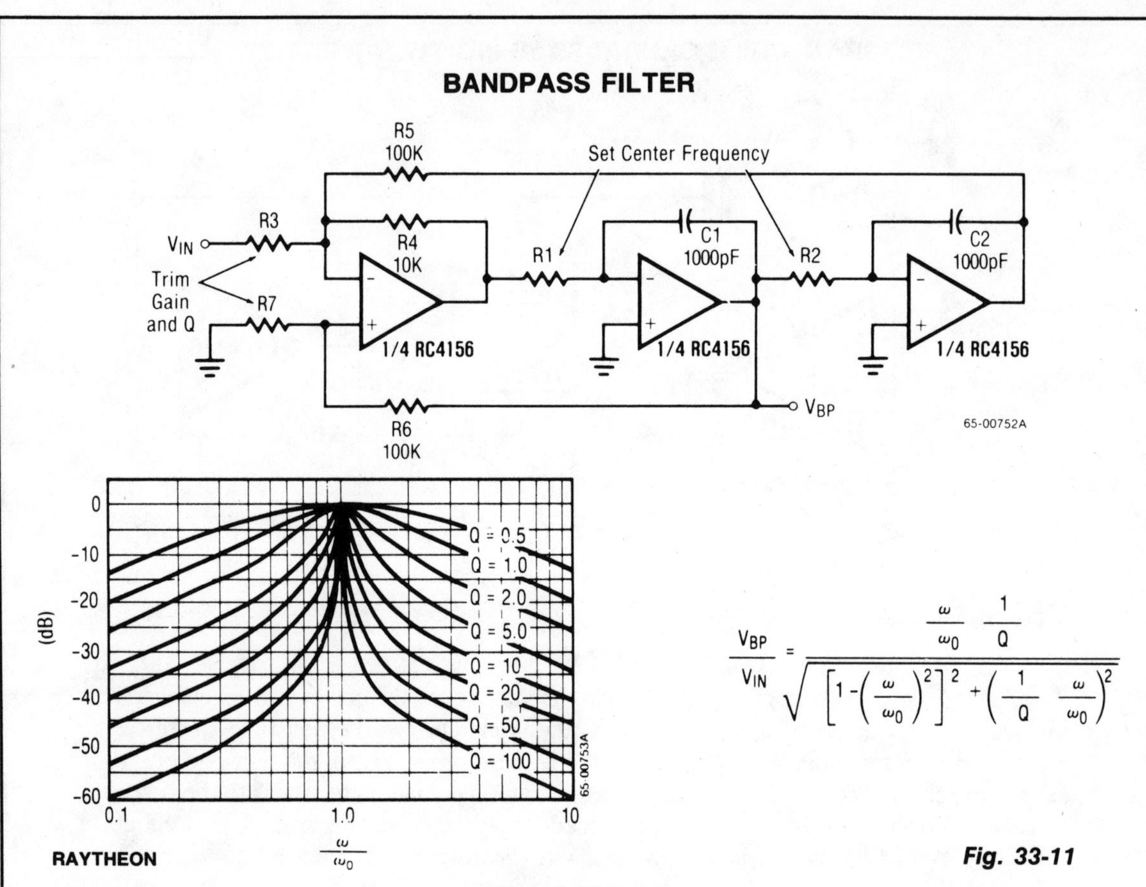

RAYTHEON

Fig. 33-11

Circuit Notes

The input signal is applied through R3 to the inverting input of the summing amplifier and the output is taken from the first integrator. The summing amplifier will maintain equal voltage at the inverting and non-inverting inputs. Defining 1/R1C1 as ω_1 and 1/R2C2 as ω_2, this is now a convenient form to look at the center-frequency ω_0 and filter Q.

$$\omega_0 = \sqrt{0.1\ \omega_1\ \omega_2}$$

$$= 10^{-9}\sqrt{0.1 R1 R2}$$

$$\text{and } Q = \left[\frac{1 + \dfrac{10^5}{R7}}{1.1 + \dfrac{10^4}{R3}}\right]\omega_0$$

The frequency response for various values of Q is shown.

ACTIVE BANDPASS FILTER

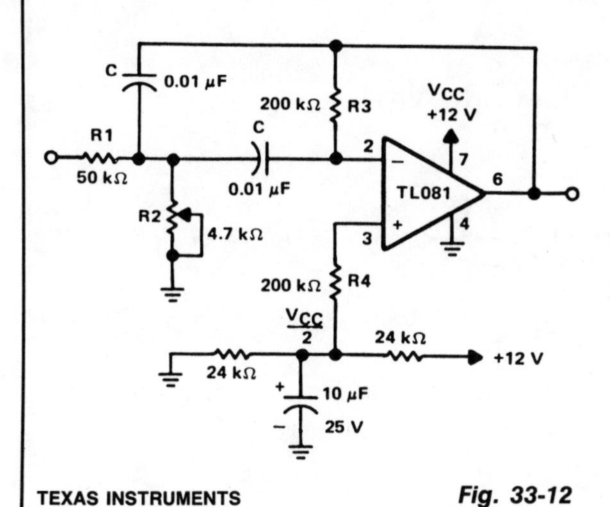

TEXAS INSTRUMENTS

Fig. 33-12

Circuit Notes

The circuit is a two-pole active filter using a TL081 op amp. This type of circuit is usable only for Qs less than 10. The component values for this filter are calculated from the following equations.

$$R1 = \frac{Q}{2\ fGC} \qquad R3 = \frac{2Q}{2fC}$$

$$R2 = \frac{Q}{(2Q^2 - G)2fC} \qquad R4 = R3$$

The values shown are for a center frequency of 800 Hz.

BANDPASS AND NOTCH FILTER

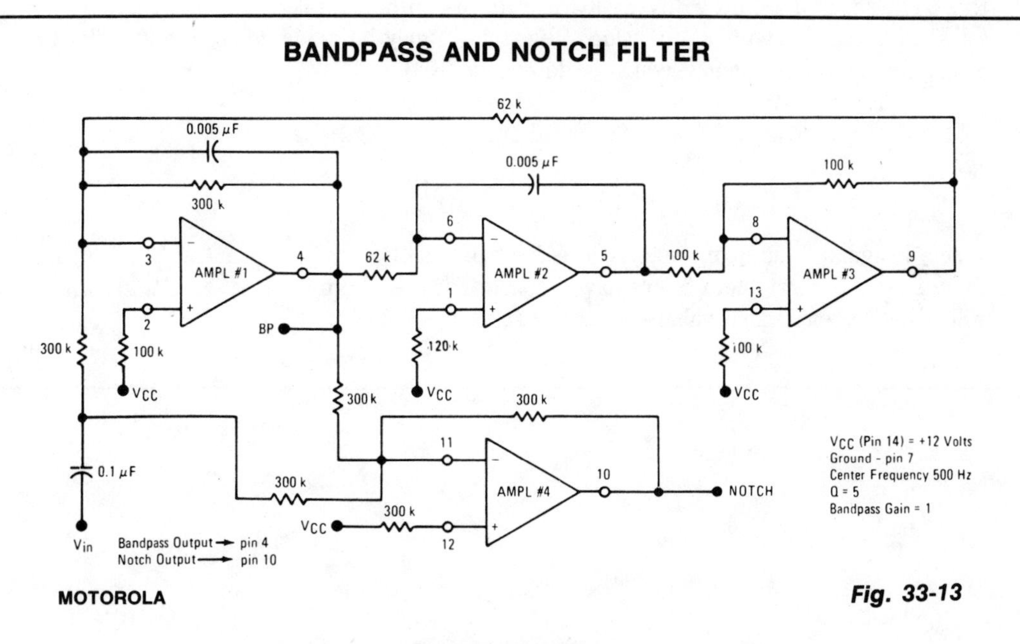

MOTOROLA

Fig. 33-13

Circuit Notes

The Quad op amp MC4301 is used to configure a filter that will notch out a given frequency and produce that notched-out frequency at the BP terminal, useful in communications or measurement setups. By proper component selection any frequency filter up to a few tens of kilohertz can be obtained.

MULTIPLE-FEEDBACK BANDPASS FILTER

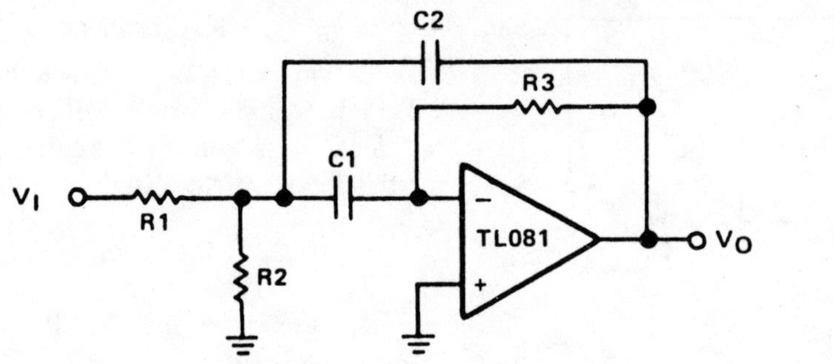

Fig. 33-14

TEXAS INSTRUMENTS

Circuit Notes

The op amp is connected in the inverting mode. Resistor R3 from the output to the inverting input sets the gain and current through the frequency-determining capacitor, C1. Capacitor C2 provides feedback from the output to the junction of R1 and R2. C1 and C2 are always equal in value. Resistor R2 may be made adjustable in order to adjust the center frequency which is determined from:

$$fo = \frac{1}{2\pi C} \quad \frac{1}{R3} \times \frac{R1 + R2\,½}{R1R2}$$

When designing a filter of this type it is best to select a value for C1 and C2, keeping them equal. Typical audio filters have capacitor values from 0.01 μF to 0.1 μF which will result in reasonable values for the resistors.

34

Flashers and Blinkers

The sources of the following circuits are contained in the Sources section beginning on page 694. The figure number contained in the box of each circuit correlates to the source entry in the Sources section.

LOW VOLTAGE LAMP FLASHER

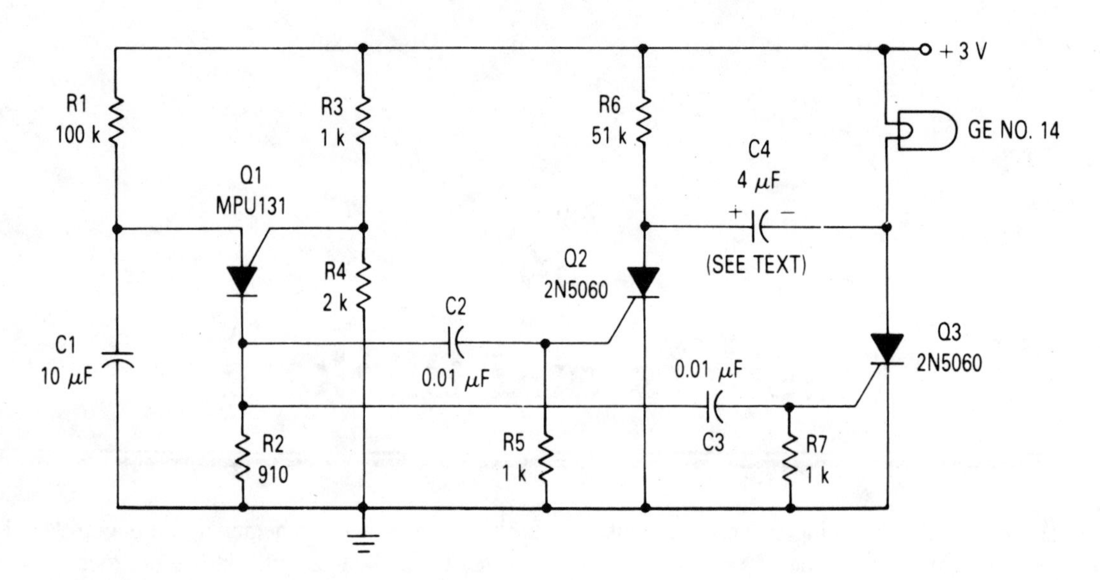

MOTOROLA

Fig. 34-1

Circuit Notes

The circuit is composed of a relaxation oscillator formed by Q1 and an SCR flip-flop formed by Q2 and Q3. With the supply voltage applied to the circuit, the timing capacitor C1 charges to the firing point of the PUT, 2 volts plus a diode drop. The output of the PUT is coupled through two 0.02 μF capacitors to the gate of Q2 and Q3. To clarify operation, assume that Q3 is on and capacitor C4 is charged plus to minus as shown in the figure. The next pulse from the PUT oscillator turns Q2 on. This places the voltage on C4 across Q3 which momentarily reverse biases Q3. This reverse voltage turns Q3 off. After discharging, C4 then charges with its polarity reversed to that shown. The next pulse from Q1 turns Q3 on and Q2 off. Note that C4 is a non-polarized capacitor. For the component values shown, the lamp is on for about ½ second and off the same amount of time.

MINIATURE TRANSISTORIZED LIGHT FLASHER

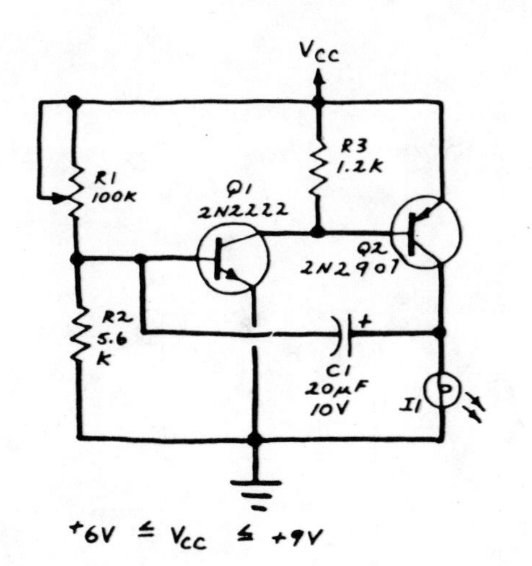

POPULAR ELECTRONICS

Fig. 34-2

Circuit Notes

R1 adjusts the flash rate. The lamp should be a No. 122, No. 222 or other similar, miniature incandescent lamp.

ALTERNATING FLASHER

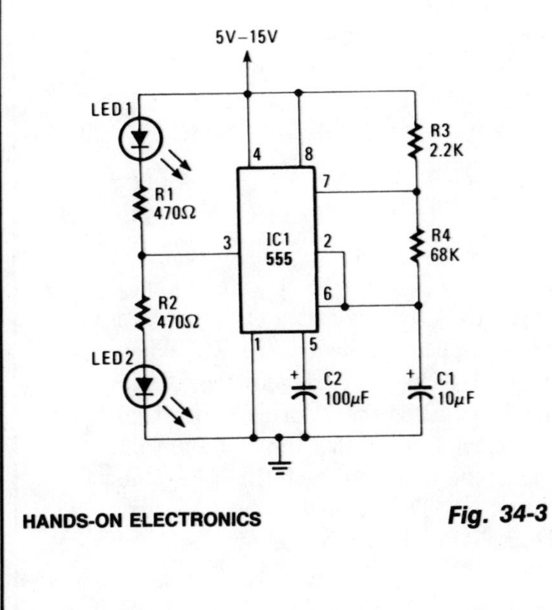

HANDS-ON ELECTRONICS

Fig. 34-3

Circuit Notes

The LED's flash alternately. The flash rate is determined by C1 and R4.

ELECTRONIC LIGHT FLASHER

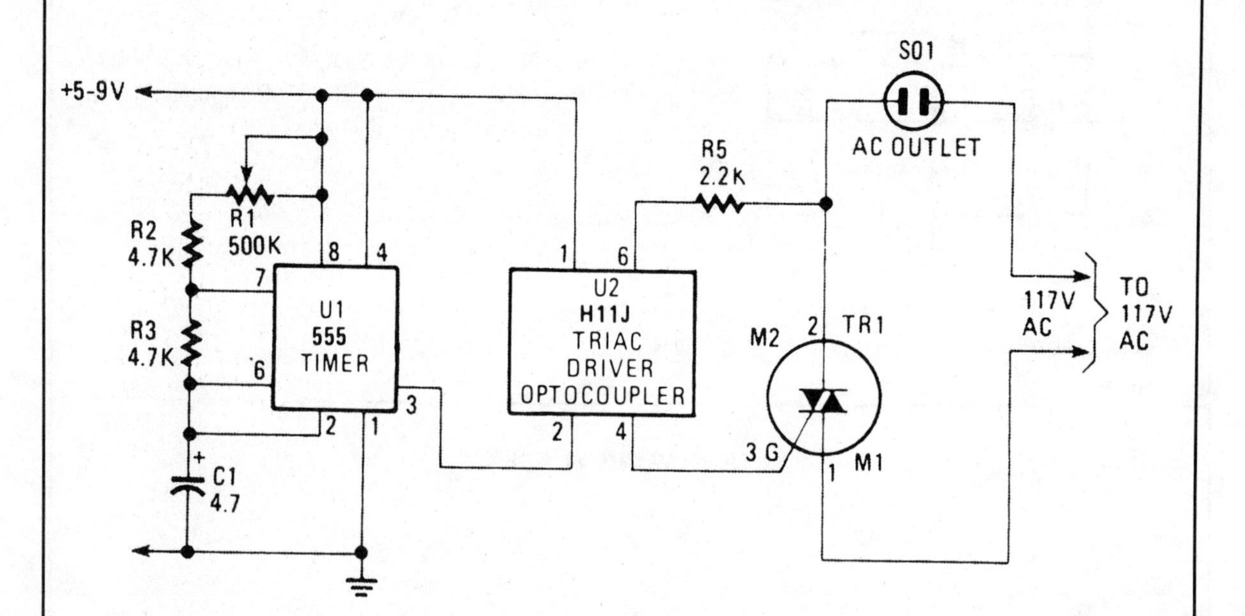

HANDS-ON ELECTRONICS

Fig. 34-4

Circuit Notes

The blinking or flashing rate is determined by U1, a 555 timer integrated circuit. Its output, at pin 3, feeds U2, a HIIJ triac driver. That driver consists of an infrared LED that is coupled internally to a light-activated silicon bilateral switch (DIAC). When the LED internal to U2 is turned on by the timer, U1, its light triggers the DIAC; effectively closing the circuit between pins 4 and 6, and fires the Triac, TR1 through its gate circuit. When the Triac is firing, it acts as a closed circuit that turns on the light (or other device it may be controlling via S01). When the timer turns off, the LED, the DIAC and Triac stop conducting and the light turns off. The sequence then repeats. The flashing rate can be varied by means of R1, a 500,000 ohm potentiometer.

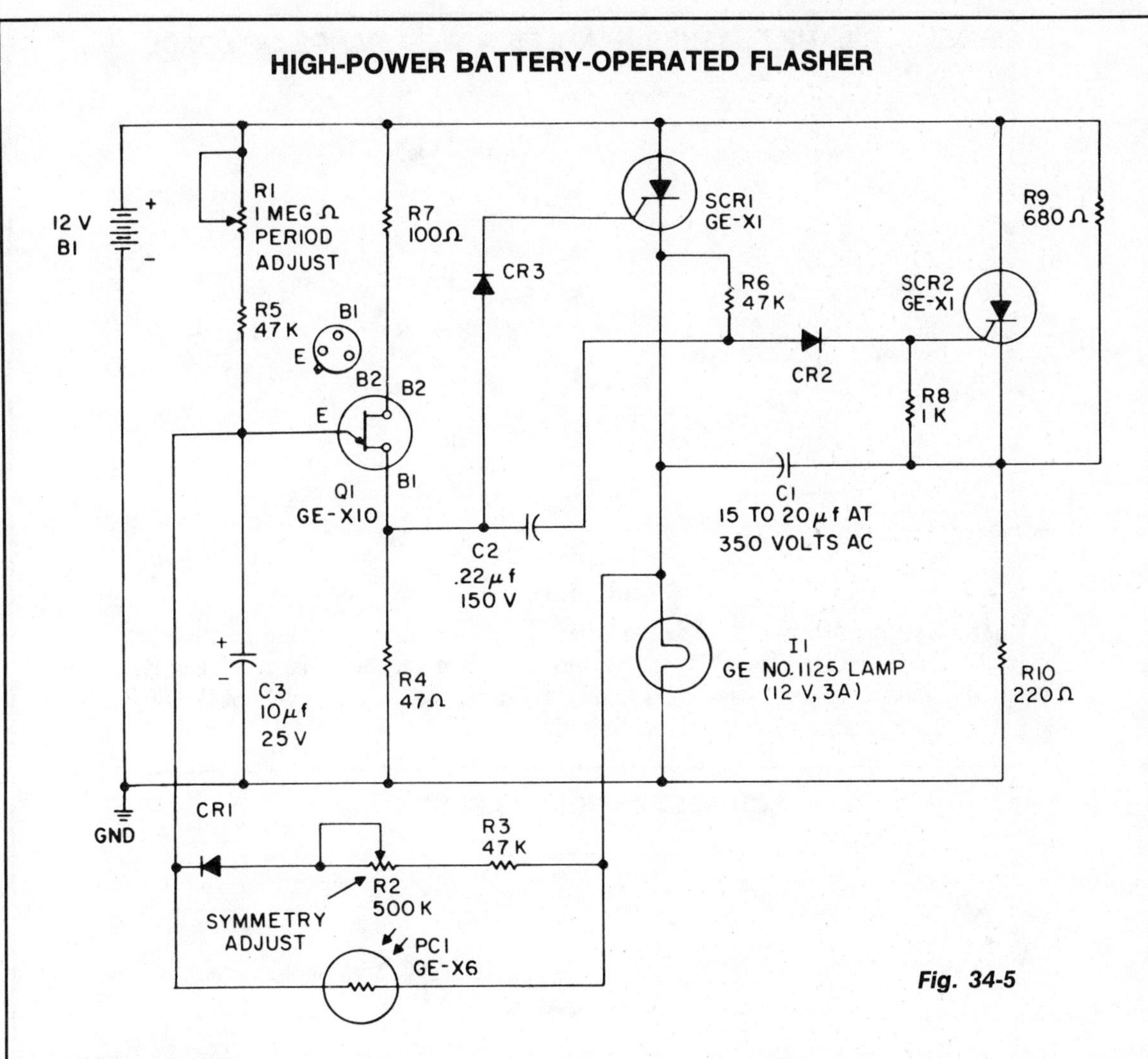

Fig. 34-5

GENERAL ELECTRIC

Circuit Notes

This flasher operates from a 12-volt car or boat battery. It offers 36 to 40-watts output, variable flash rate (up to 60 flashes per minute), independent control of both on and off cycles and photoelectric night and day control that turns the flasher on at night and shuts it off during the day for automatic operation. SCR1 and SCR2 form a basic dc flip-flop. The lamp load is the cathode leg of one SCR so that the other side of the load may be at ground (negative) potential (required in some applications). The flip-flop timing is controlled by a conventional UJT oscillator arrangement (Q1, R1, C3, etc.). Potentiometer R2 and diode CR1 provide on/off timing independence. Photoconductor PC1 locks out the UJT firing circuit during the daylight hours.

SERIES SCR LAMP FLASHER HANDLES A WIDE RANGE OF LOADS

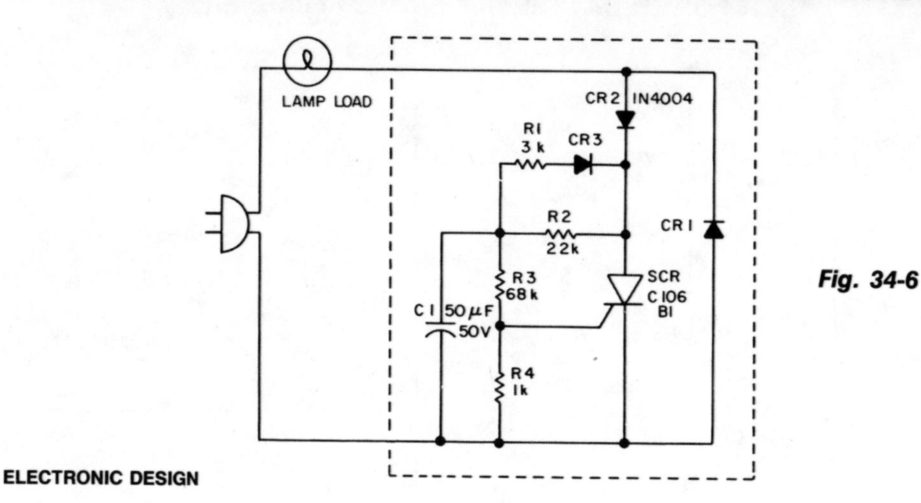

Fig. 34-6

ELECTRONIC DESIGN

Circuit Notes

Brief full-power flashes are obtained when the SCR conducts during positive half cycles of the line voltage. The SCR fires when the voltage at the divider, R3 and R4, reaches the gate-firing level. Diode D1 conducts during the reverse cycle of the SCR and provides preheating current to the lamp filaments.

SCR RELAXATION FLASHER

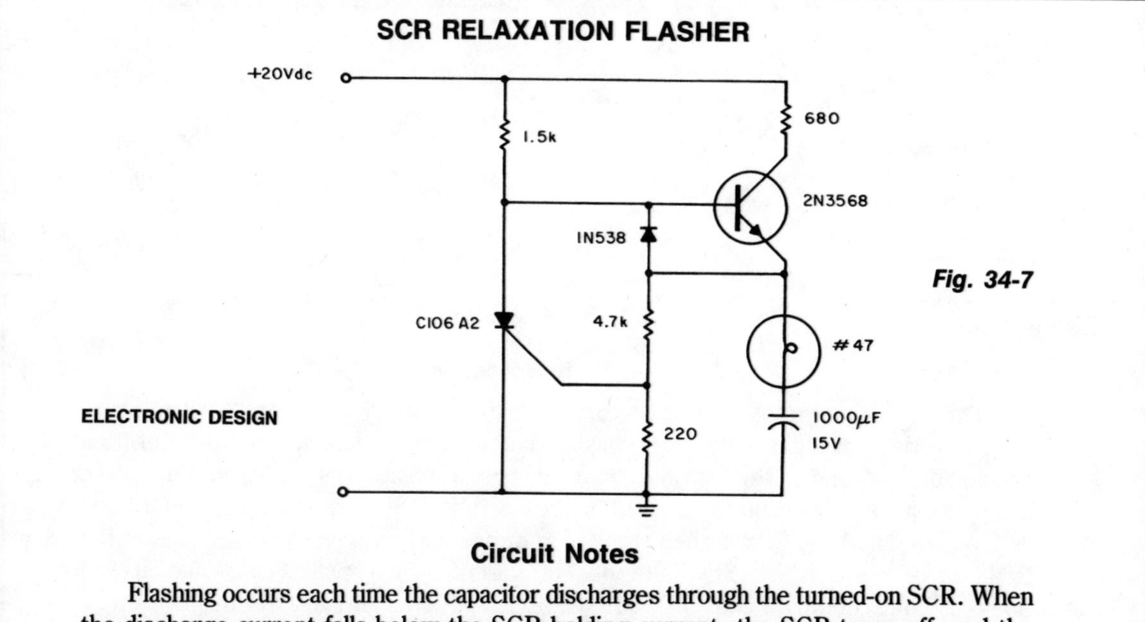

Fig. 34-7

ELECTRONIC DESIGN

Circuit Notes

Flashing occurs each time the capacitor discharges through the turned-on SCR. When the discharge current falls below the SCR holding current, the SCR turns off, and the capacitor begins charging for another cycle. The circuit will maintain a slower but good flashing capability even after considerable battery degradation.

LOW CURRENT CONSUMPTION LAMP FLASHER

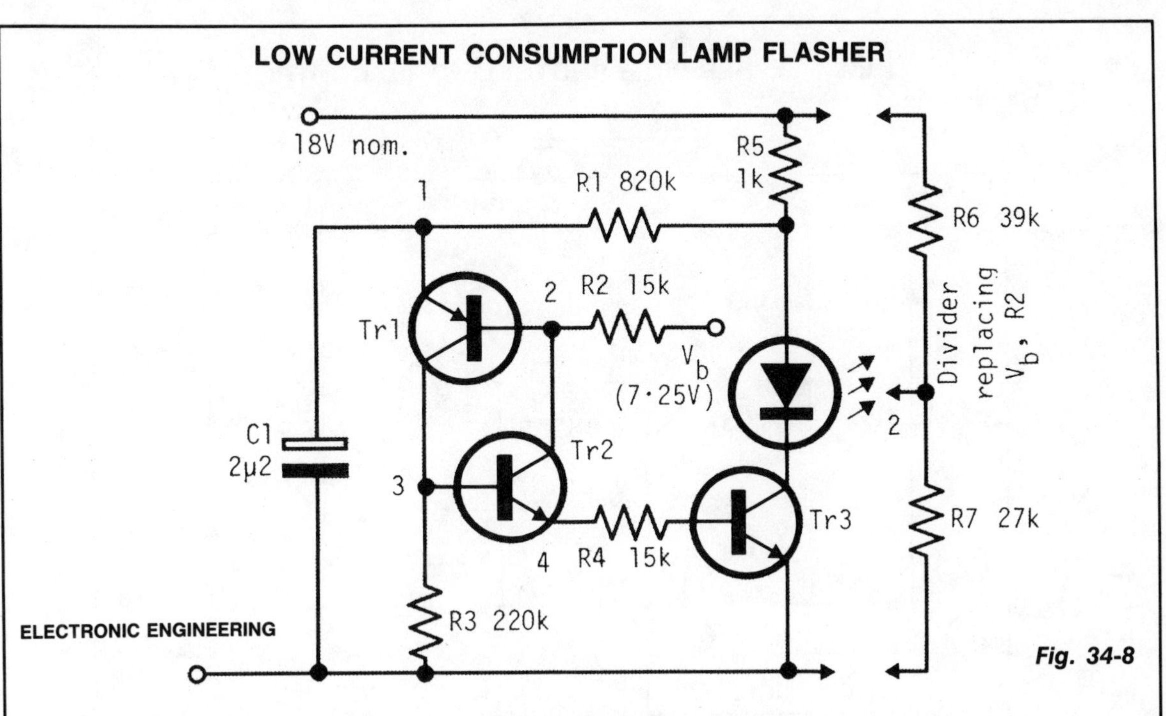

Fig. 34-8

Circuit Notes

The circuit is economical in components, and will work with virtually any transistors and is reliably self-starting. The voltage Vb can be taken from a divider, as shown at the right. If taken from a fixed source, flashing becomes slower as battery voltage falls. The lowest drive current into the base of Tr3 is about (Vb—0,6 V)/(R2 + R4). Resistor R4 limits the initial current from C1 and, as shown, R2 and R4 can be roughly equal when a divider is used for Vb. Resistor R2 equals R6R7/(R6 + R7). With the voltages shown, and with R2 = R4, the on-time is about 1.1 C1R2 and the off-time about 0.28 C1R1. Using the component values shown the period is about 0.55 sec. with a duty cycle of about 7% and a mean battery current including the Vb divider, about 1.5 mA.

LED FLASHER USES UJT

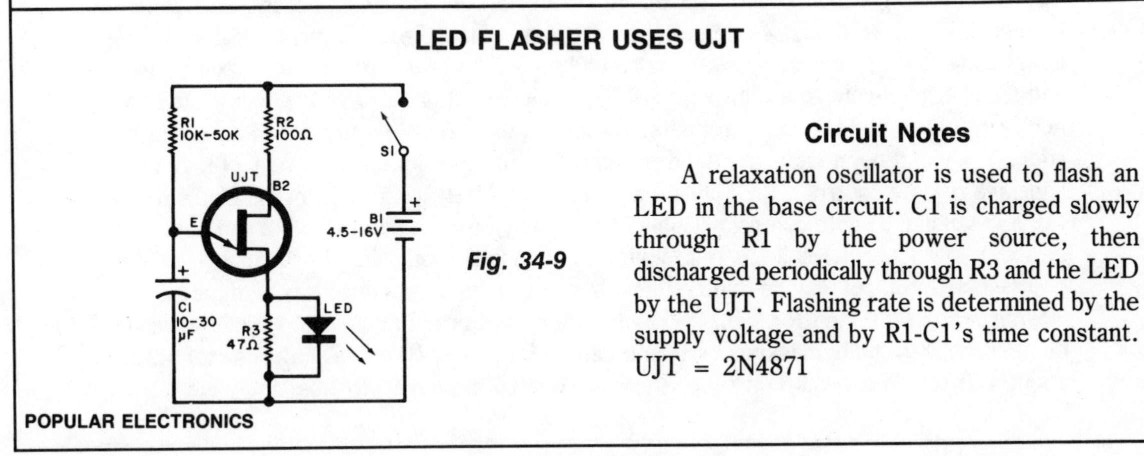

Fig. 34-9

Circuit Notes

A relaxation oscillator is used to flash an LED in the base circuit. C1 is charged slowly through R1 by the power source, then discharged periodically through R3 and the LED by the UJT. Flashing rate is determined by the supply voltage and by R1-C1's time constant. UJT = 2N4871

2 kW FLASHER WITH PHOTOELECTRIC CONTROL

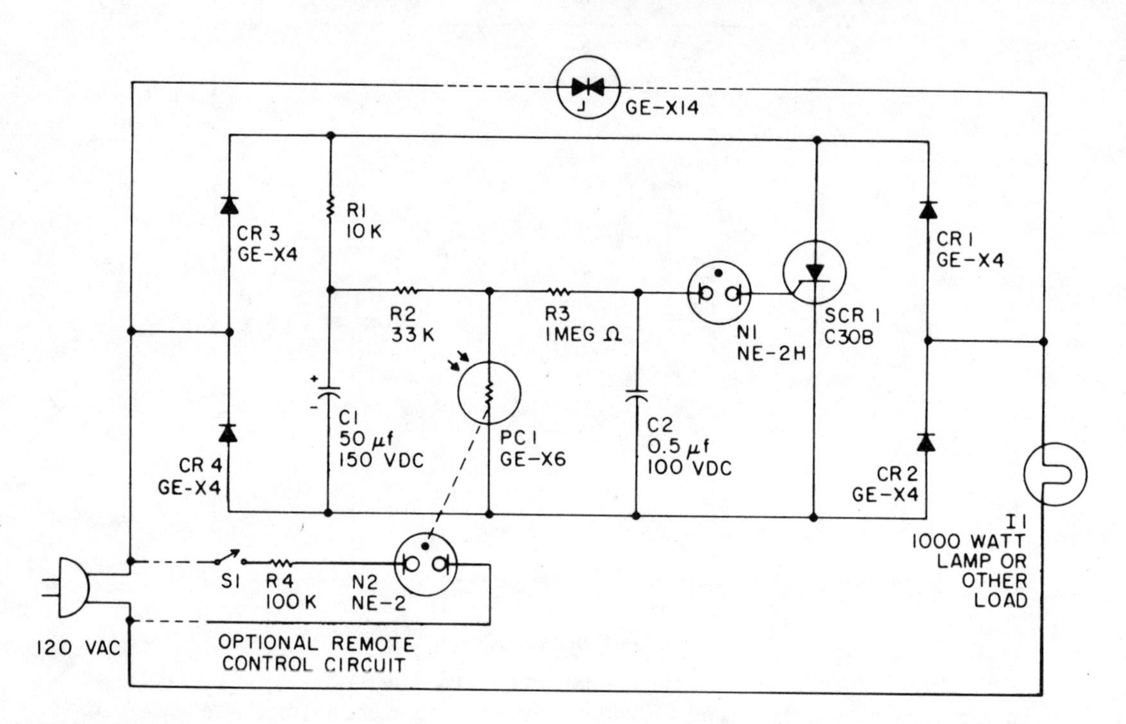

GENERAL ELECTRIC *Fig. 34-10*

Circuit Notes

CR1, CR2, CR3, and CR4 form a bridge circuit with the SCR across the dc legs. With light on the photoconductor PC1, C1 charges through R1 to about 150 Vdc. The resistance of PC1 is low when illuminated, so very little voltage appears across it or C2. At about 90 volts C1 starts discharging through R1 and the SCR, but the SCR cannot turn off until C1 is almost completely discharged. When the SCR turns off during the interval line voltage is near zero, the full supply voltage again appears across the bridge, and C1 charges again to a high voltage. The voltage on C2 also starts rising until the neon lamp fires and the cycle repeats. An alternative remote control can be made by adding a second neon lamp, N2, and masking the photocell so it sees only N2. A very sensitive remote control is thus obtained that is completely isolated from the load circuit. For low-voltage remote control a flashlight lamp may be used instead of N2 and operated at about ½ its normal voltage thus giving exceptionally long life. Performance of the photoelectric control may be inverted (flash when the photoconductor is illuminated) by interchanging PC1, and R2. Sensitivity in either the normal or inverted modes can be decreased by partially masking PC1, and can be increased by increasing resistor R2 to about 470 K. To increase on time, increase C1; to increase off time, increase R3.

SEQUENTIAL FLASHER

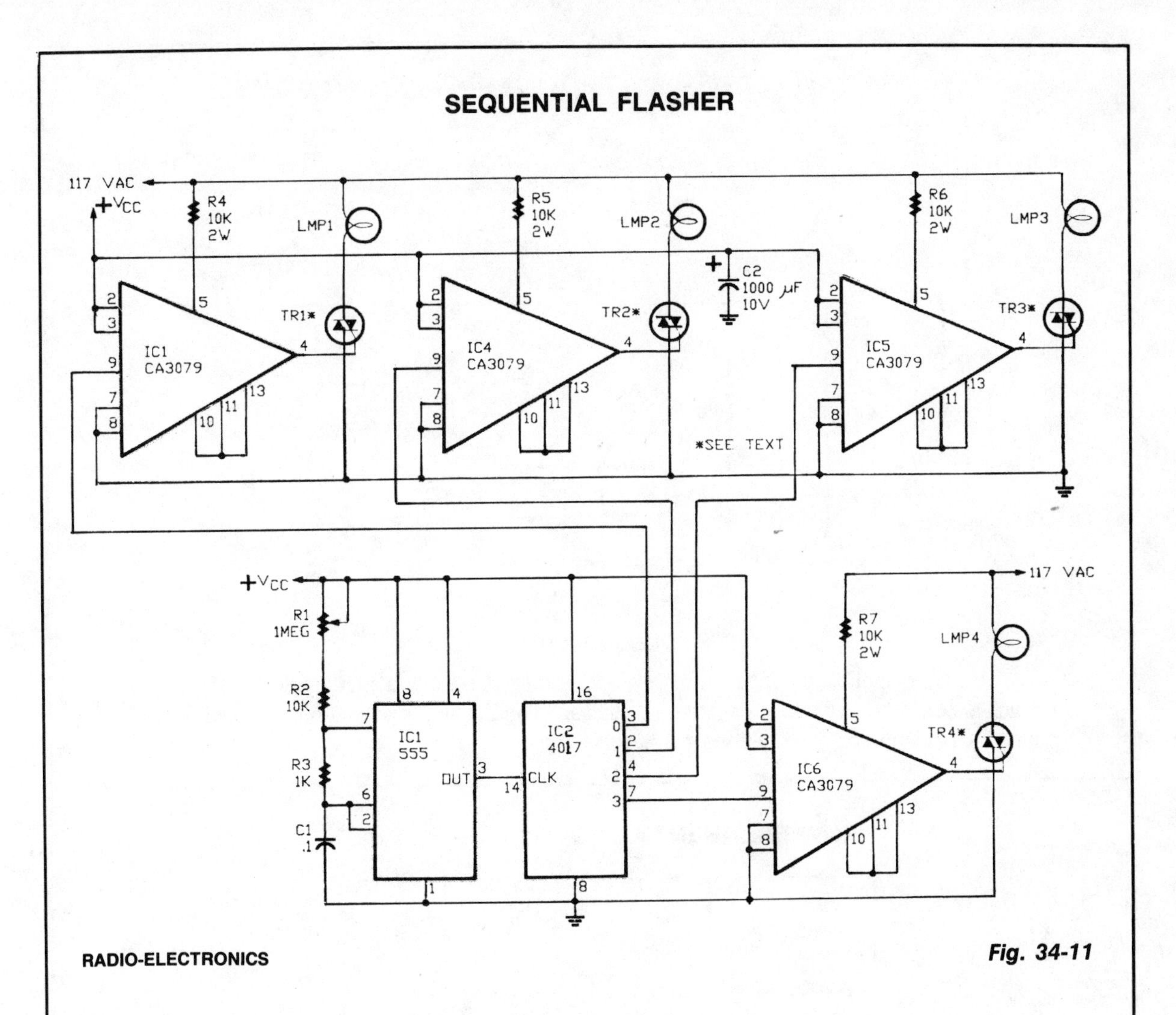

RADIO-ELECTRONICS

Fig. 34-11

Circuit Notes

A 555 timer, IC1, drives a 4017 CMOS decade counter. Each of the 4017's first four outputs drives a CA3079 zero-voltage switch. Pin 9 of the CA3079 is used to inhibit output from pin 4, thereby disabling the string of pulses that IC normally delivers. Those pulses occur every 8.3 ms, i.e., at a rate of 120 Hz. Each pulse has a width of 120 µs. Due to the action of the CA3079, the lamps connected to the TRI-AC's turn on and off near the zero crossing of the ac waveform. Switching at that point increases lamp life by reducing the inrush of current that would happen if the lamp were turned on near the high point of the ac waveform. In addition, switching at the zero crossing reduces Radio-Frequency Interference (RFI) considerably.

CAUTION: The CA3079's are driven directly from the 117-volt ac power line, so use care.

1 kW FLIP-FLOP FLASHER CIRCUIT

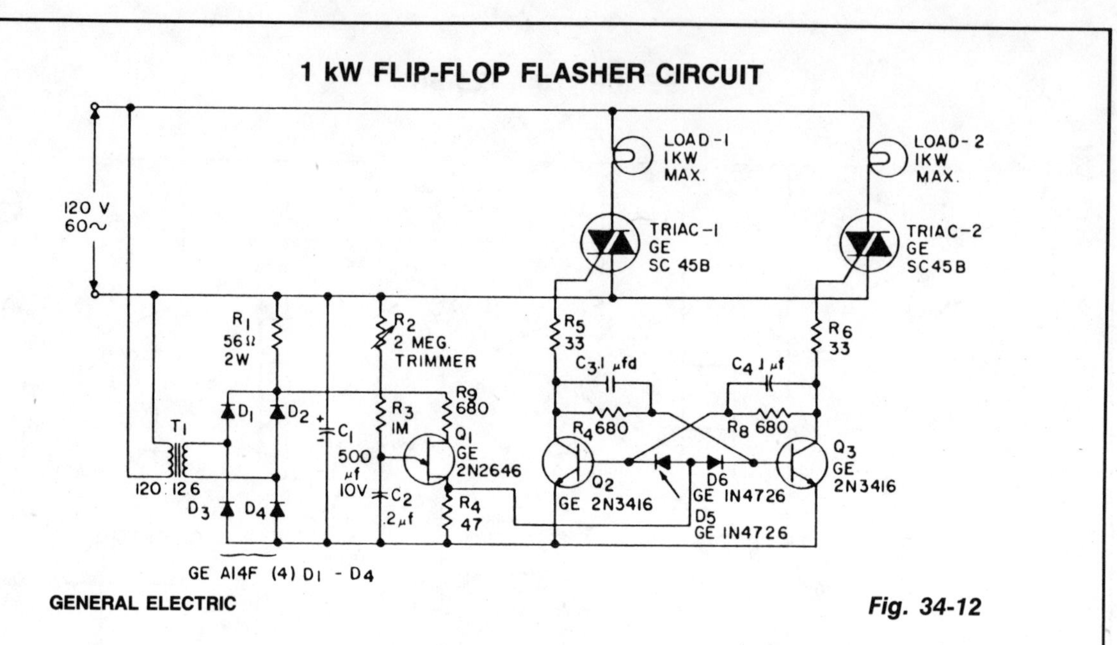

GENERAL ELECTRIC

Fig. 34-12

Circuit Notes

This is an application of the static switch circuit where the control logic is a flip-flop which is controlled by the unijunction transistor. The flashing rate can be adjusted from about 0.1 second to a 10 second cycle time.

LOW FREQUENCY OSCILLATOR-FLASHER

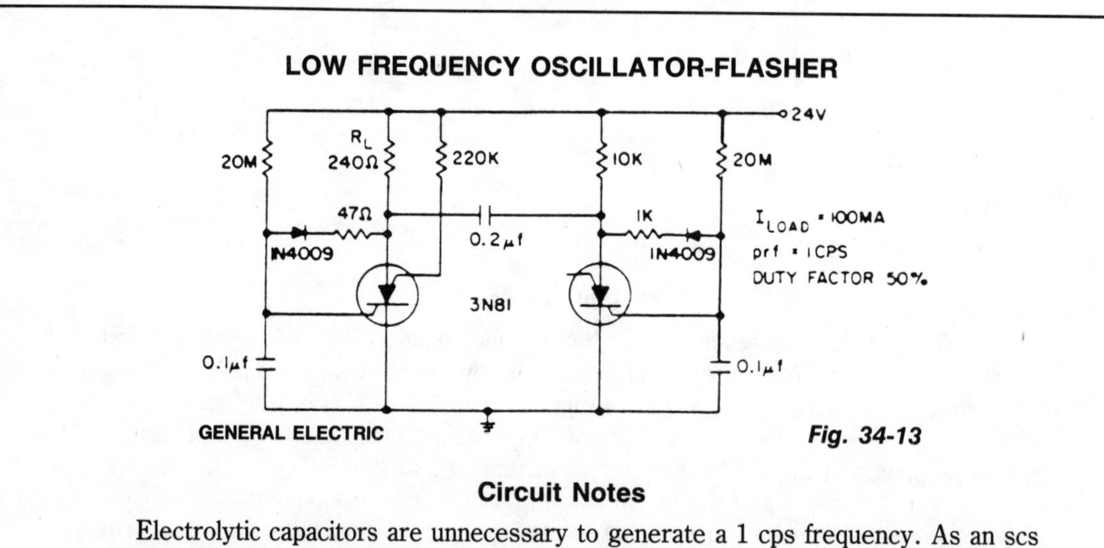

GENERAL ELECTRIC

Fig. 34-13

Circuit Notes

Electrolytic capacitors are unnecessary to generate a 1 cps frequency. As an scs triggers on, the 0.2 μF commutating capacitor turns off the other one and charges its gate capacitor to a negative potential. The gate capacitor charges towards 24 volts through 20 M retriggering its scs. Battery power is delivered to the load with 88% efficiency. The 20 M resistors can be varied to change prf or duty factor.

HIGH DRIVE OSCILLATOR/FLASHER

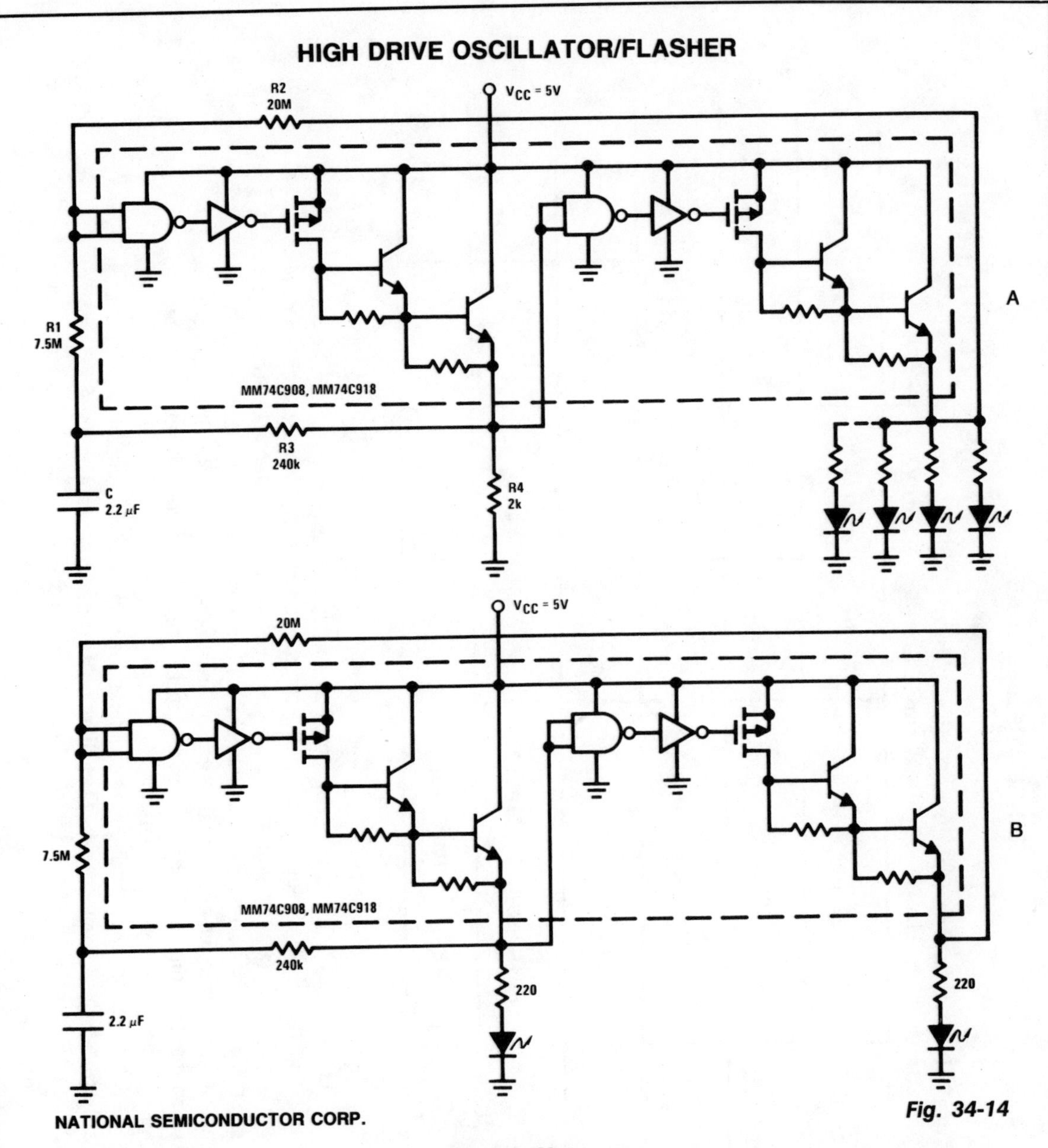

NATIONAL SEMICONDUCTOR CORP.

Fig. 34-14

Circuit Notes

The driver in the package is connected as a Schmitt trigger oscillator (A) where R1 and R2 are used to generate hysteresis. R3 and C are the inverting feedback timing elements and R4 is the pull-down load for the first driver. Because of its current capability, the circuit can be used to drive an array of LEDs or lamps. If resistor R4 is replaced by an LED (plus a current limiting resistor), the circuit becomes a double flasher with the 2 LEDs flashing out of phase (B).

TRANSISTORIZED FLASHERS

Flasher circuit performance as a function of temperature using limit sample transistors.			AMBIENT TEMPERATURE -40°F	77°F	212°F
CIRUIT #1	Flashes per Minute	min.	58.0	60.0	59.6
		max.	56.7	59.6	58.7
	Flash Duration in %	min.	14.5	14.6	14.3
		max.	15.1	14.9	15.3
CIRCUIT #2	Flashes per Minute	min.	58.6	60.0	59.0
		max.	55.6	58.1	56.4
	Flash Duration in %	min.	26.3	27.5	22.2
		max.	28.8	29.1	29.6
CIRCUIT #3	Flashes per Minute	min.	59.0	60.0	61.5
		max.	55.4	57.7	55.2
	Flash Duration in %	min.	45.2	46.0	48.2
		max.	45.6	46.2	47.0

Circuit Notes

Transistors Q1 and Q2 are connected as a free running multivibrator. The output, at the emitter of Q2, drives the base of the common emitter amplifier Q3, which controls the lamp. This circuit configuration permits the flash duration, the interval between flashes, and the lamp type to be varied independently. Flash duration is proportional to the product of R2C2 while the off interval is proportional to the product of R3C1. Consequently, when the flash timing must be accurately maintained, these component tolerances will have to be held to similar limits.

All three circuits described are designed for barricade warning flasher lights such as used in highway construction. They differ only in flash duration which normally is 15%, 25%, or 50% of the flash rate. Performance has been checked at ambient temperatures of −40°F, 77°F, and 212°F. A GE 5 volt, 90 milliampere type No. 1850 lamp is used.

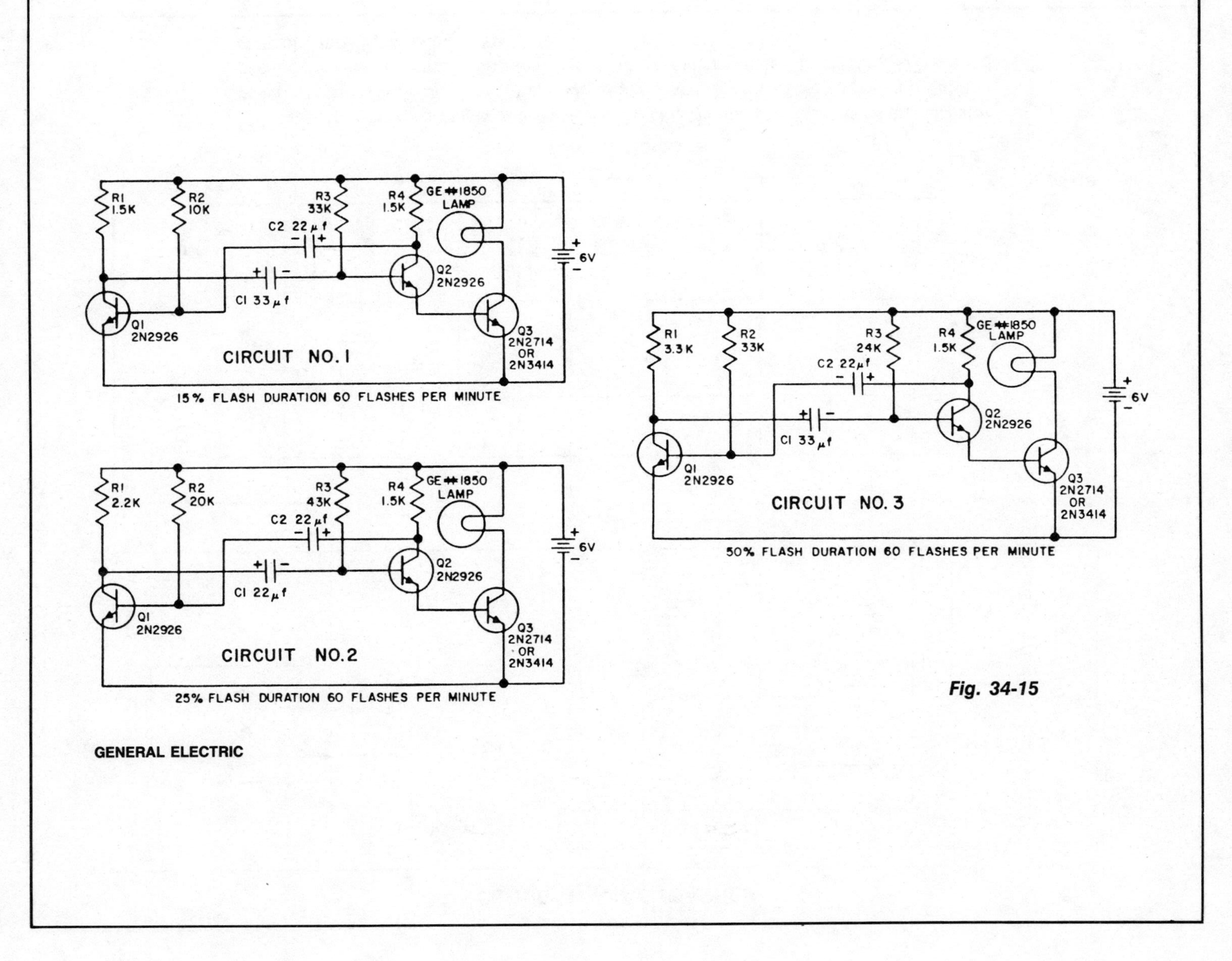

Fig. 34-15

GENERAL ELECTRIC

SEQUENTIAL AC FLASHER

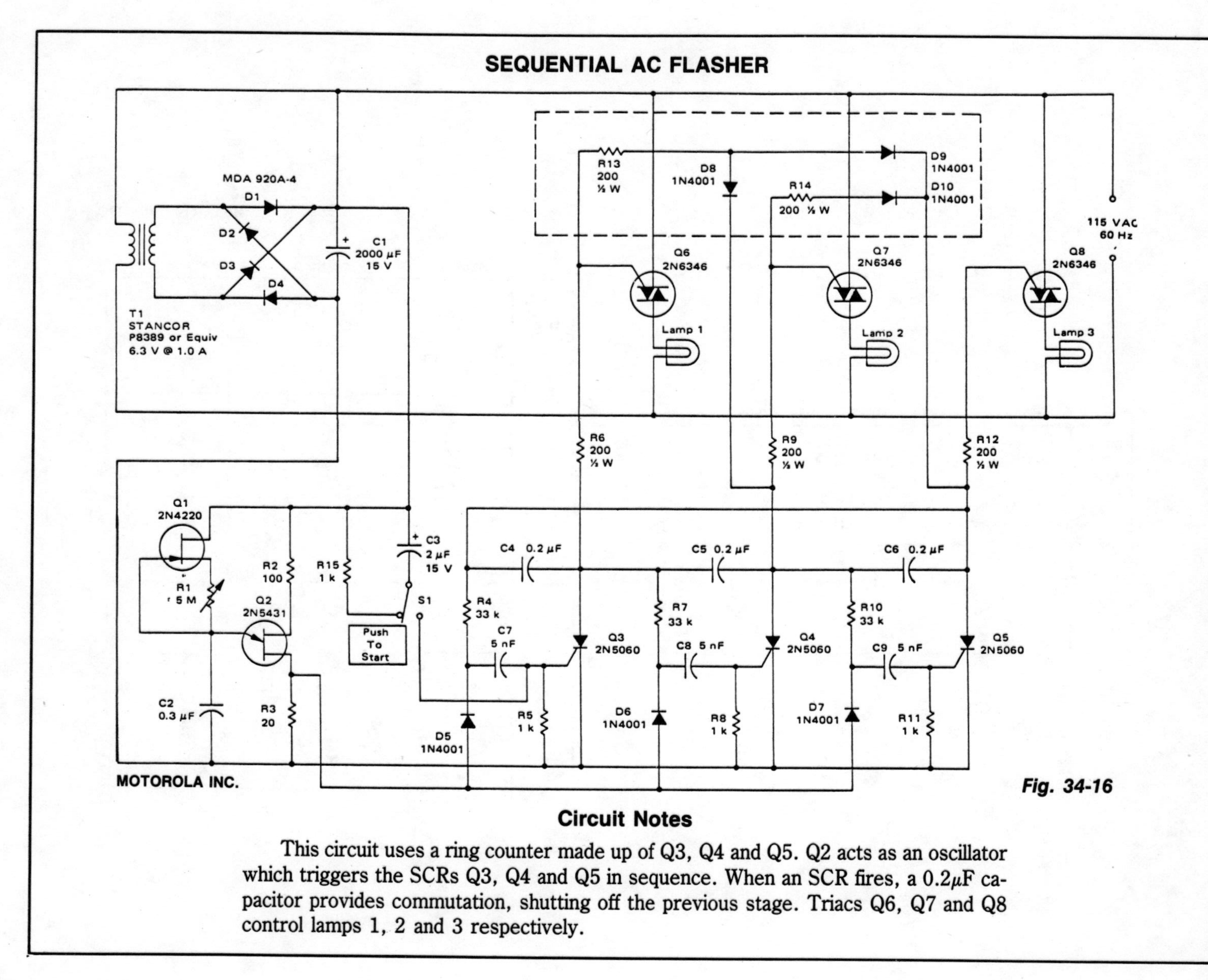

Fig. 34-16

Circuit Notes

This circuit uses a ring counter made up of Q3, Q4 and Q5. Q2 acts as an oscillator which triggers the SCRs Q3, Q4 and Q5 in sequence. When an SCR fires, a 0.2μF capacitor provides commutation, shutting off the previous stage. Triacs Q6, Q7 and Q8 control lamps 1, 2 and 3 respectively.

ASTABLE FLIP-FLOP WITH STARTER

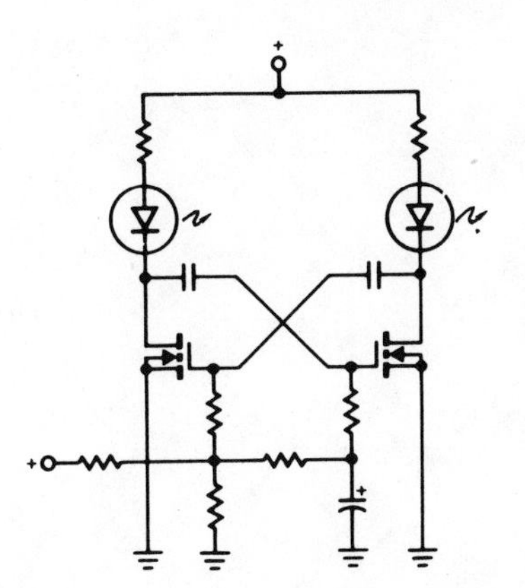

Circuit Notes

A pair of non-zenered MOSPOWER transistors, a pair of LEDs and a simple RC circuit make an easy sequential flasher with almost unlimited sequencing time—from momentary to several seconds. The infinite input resistance of the MOSFET gate allows for very long sequencing times that are impossible when using bipolars. One precaution, though, don't wire your circuit using phenolic or printed circuit boards when looking for slow sequencing (they exhibit too much leakage!).

SILICONIX, INC. *Fig. 34-17*

LED FLASHER USES PUT

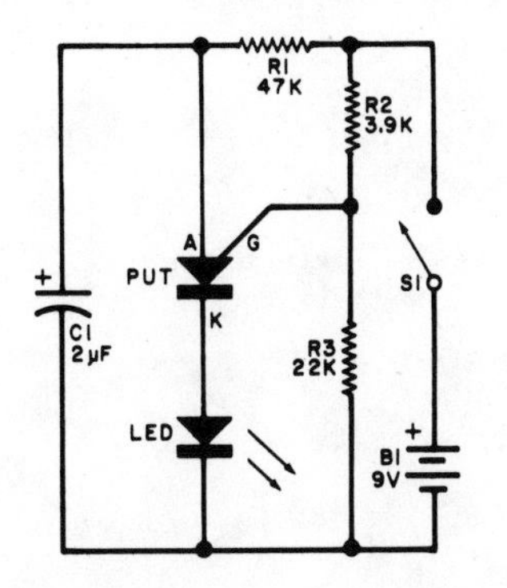

Circuit Notes

This flasher circuit operates as a relaxation oscillator with C1 discharged periodically through the LED as the PUT switches on. The flashing rate is about 100/minute with the component values listed.

POPULAR ELECTRONICS *Fig. 34-18*

35

Flow Detectors

The sources of the following circuits are contained in the Sources section beginning on page 694. The figure number contained in the box of each circuit correlates to the source entry in the Sources section.

THERMALLY BASED ANEMOMETER (AIR FLOWMETER)

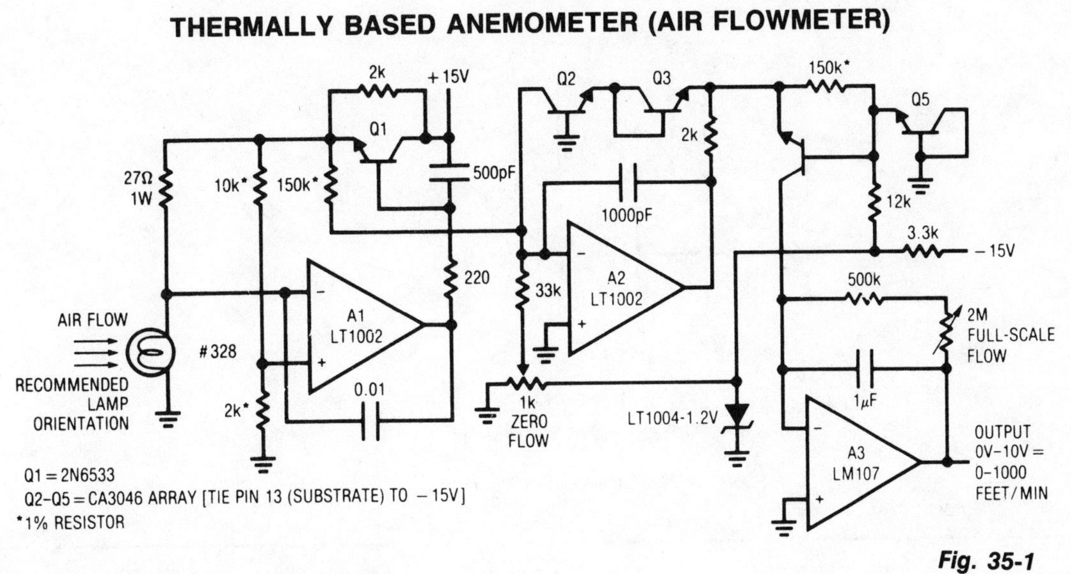

Fig. 35-1

LINEAR TECHNOLOGY CORPORATION

Circuit Notes

This design used to measure air or gas flow works by measuring the energy required to maintain a heated resistance wire at constant temperature. The positive temperature coefficient of a small lamp, in combination with its ready availability, makes it a good sensor. A type 328 lamp is modified for this circuit by removing its glass envelope. The lamp is placed in a bridge which is monitored by A1. A1's output is current amplified by Q1 and fed back to drive the bridge. When power is applied, the lamp is at a low resistance and Q1's emitter tries to come full on. As current flows through the lamp, its temperature quickly rises, forcing its resistance to increase. This action increases A1's negative input potential. Q1's emitter voltage decreases and the circuit finds a stable operating point. To keep the bridge balanced, A1 acts to force the lamp's resistance, hence its temperature, constant. The 20 k - 2 k bridge values have been chosen so that the lamp operates just below the incandescence point.

To use this circuit, place the lamp in the air flow so that its filament is at a 90° angle to the flow. Next, either shut off the air flow or shield the lamp from it and adjust the zero flow potentiometer for a circuit output of 0 V. Then, expose the lamp to air flow of 1000 feet/minute and trim the full flow potentiometer for 10 V output. Repeat these adjustments until both points are fixed. With this procedure completed, the air flowmeter is accurate within 3% over the entire 0-1000 foot/minute range.

AIR FLOW DETECTOR

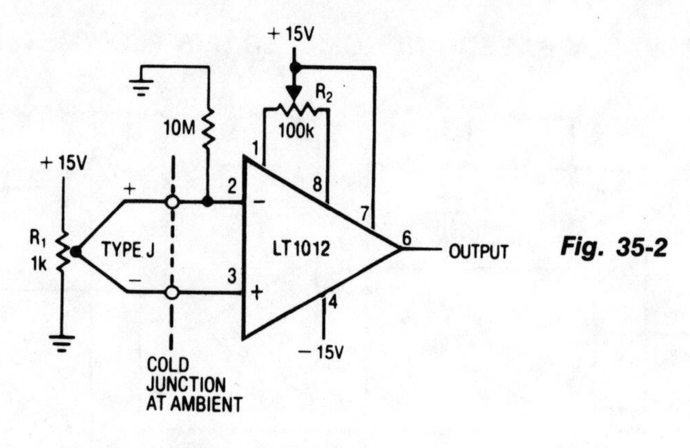

Fig. 35-2

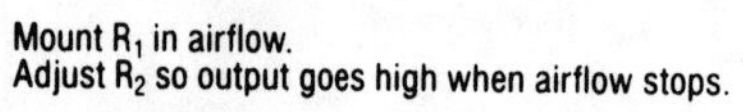

Mount R_1 in airflow.
Adjust R_2 so output goes high when airflow stops.

LINEAR TECHNOLOGY CORP.

36

Fluid and Moisture Detectors

The sources of the following circuits are contained in the Sources section beginning on page 694. The figure number contained in the box of each circuit correlates to the source entry in the Sources section.

RAIN WARNING BLEEPER

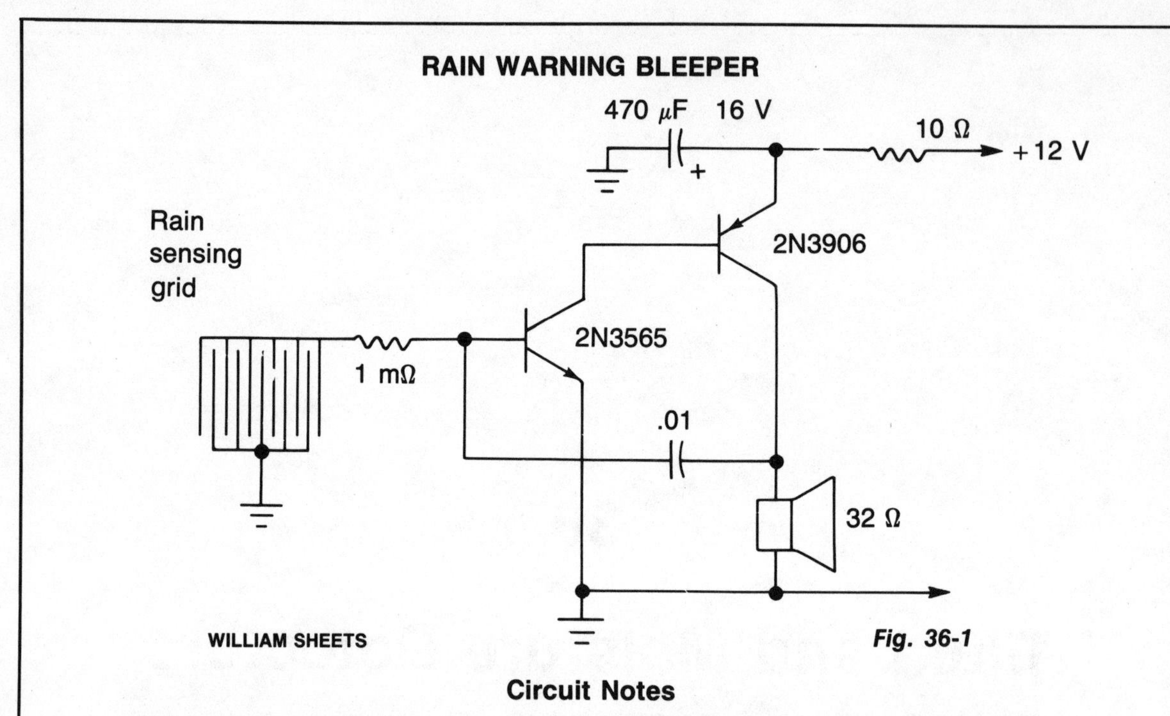

Fig. 36-1

Circuit Notes

One small spot of rain on the sense pad of this bleeper will start this audio warning. It can also be operated by rising water. The circuit has two transistors, with feedback via capacitor C1, but Trl cannot operate as long as the moisture sense pad is dry. When the pad conducts, Tr1 and Tr2 form an audio oscillatory circuit, the pitch depends somewhat on the resistance.

WATER-LEVEL INDICATOR

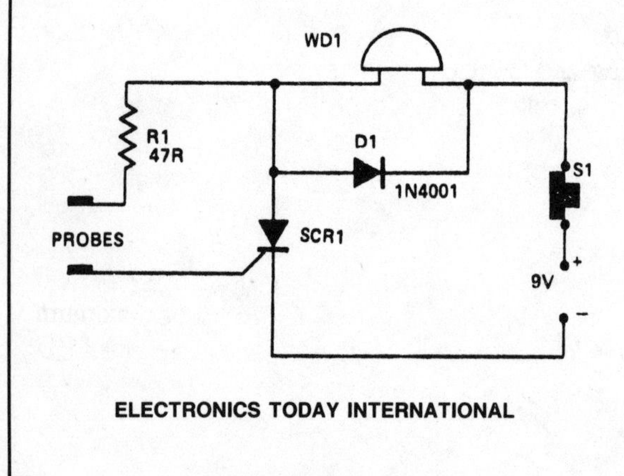

ELECTRONICS TODAY INTERNATIONAL

Fig. 36-2

Circuit Notes

In this a warning device WD1 is in series with SCR1. When the liquid level causes a conductive path between the probes, the SCR conducts sounding WD1. The warning device may be a Sonalert (TM), a lamp or a buzzer. D1 acts as a transient suppressor. Press S1 to reset the circuit.

ACID RAIN MONITOR

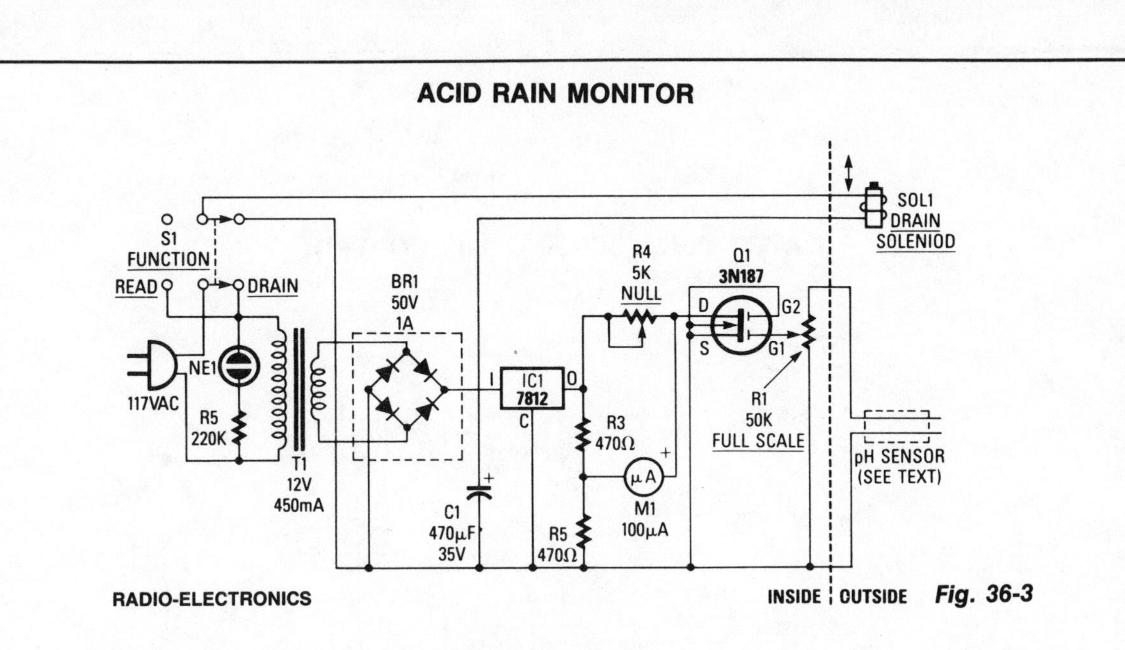

RADIO-ELECTRONICS *Fig. 36-3*

Circuit Notes

A bridge rectifier and 12-volt regulator powers the MOSFET sensing circuit. The unregulated output of the bridge rectifier operates the drain solenoid via switch S1. The sensor itself is built from two electrodes, one made of copper, the other of lead. In combination with the liquid trapped by the sensor, they form a miniature lead-acid cell whose output is amplified by MOSFET Q1. The maximum output produced by our prototype cell was about 50 μA. MOSFET Q1 serves as the fourth leg of a Wheatstone bridge. When acidity causes the sensor to generate a voltage, Q1 turns on slightly, so its drain-to-source resistance decreases. That resistance variation causes an imbalance in the bridge, and that imbalance is indicated by meter M1.

PLANT-WATER MONITOR

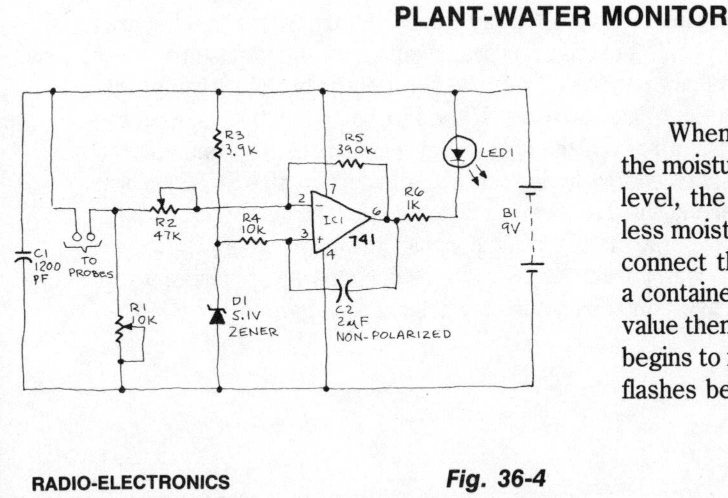

Circuit Notes

When the soil is moist, the LED glows. If the moisture falls below a certain predetermined level, the LED begins to flash. If there is still less moisture, the LED turns off. To calibrate, connect the battery and insert the probe into a container of dry soil. Set R1 to its maximum value then reduce that resistance until the LED begins to flash. The range over which the LED flashes before going out is adjusted using R2.

RADIO-ELECTRONICS *Fig. 36-4*

WATER-LEVEL SENSING AND CONTROL

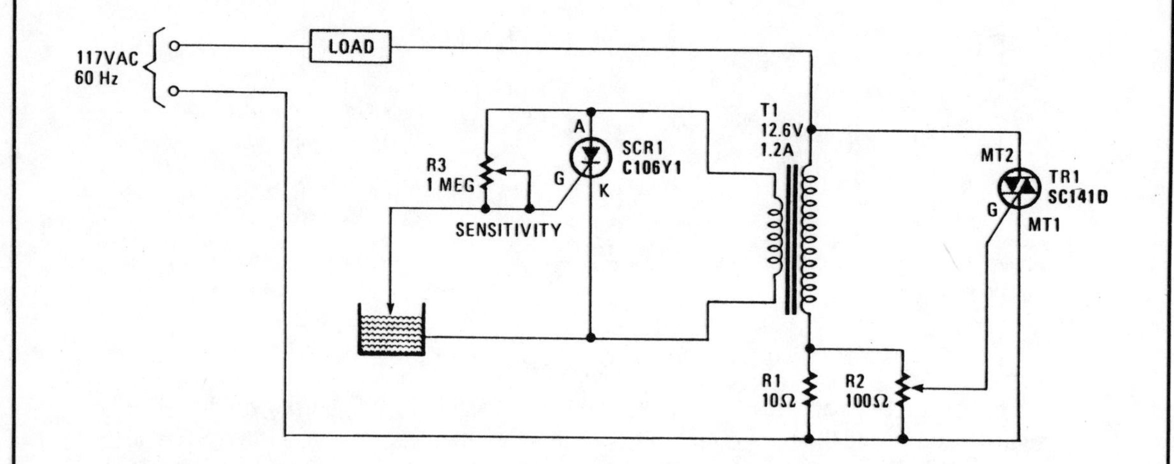

HANDS-ON ELECTRONICS *Fig. 36-5*

Circuit Notes

The operation of the circuit is based on the difference in the primary impedance of a transformer when its secondary is loaded and when it is open-circuit. The impedance of the primary of T1 and resistor R1 are in series with the load. The triac's gate-control voltage is developed across parallel resistors R1 and R2. When the water level is low, the probe is out of the water and SCR1 is triggered on. It conducts and imposes a heavy load on transformer T1's secondary winding. That load is reflected back into the primary, gating triac TR1 on, which energizes the load. If the load is an electric value in the water-supply line, it will open and remain open until the water rises and touches the probe, which shorts SCR1's gate and cathode, thereby turning off the SCR1, which effectively open-circuits the secondary. The open-circuit condition—when reflected back to the primary winding—removes the triac's trigger signal, thereby turning the water off. The load may range from a water valve, a relay controlling a pump supplying water for irrigation, or a solenoid valve controlling the water level in a garden lily pond.

SINGLE CHIP PUMP CONTROLLER

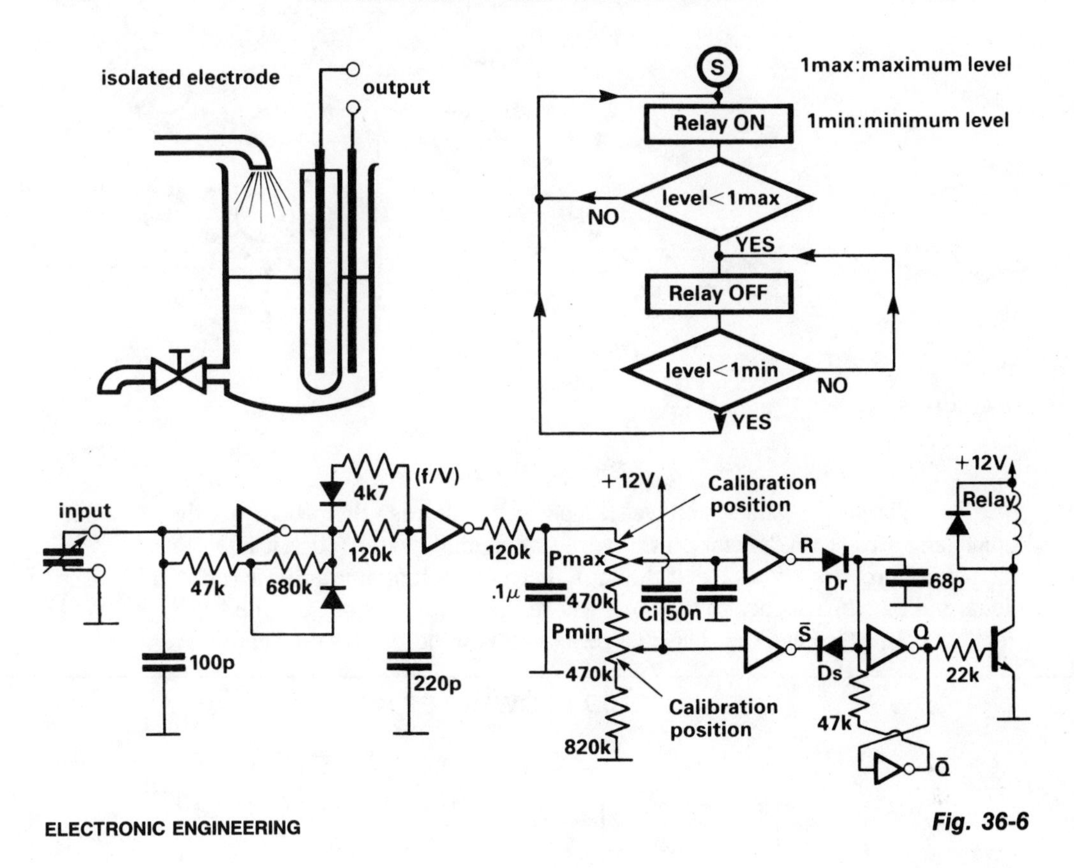

ELECTRONIC ENGINEERING

Fig. 36-6

Circuit Notes

This circuit controls the level of a tank using a bang-bang controlled electrical pump. The actual level of liquid is measured by a capacitive level-meter. The first inverter performs as a capacitance to frequency converter. It is a Schmitt oscillator and its frequency output decreases as the capacitance increases. The second inverter is a monostable which performs as a frequency to voltage converter (f/V). Its output is applied to the maximum and minimum level comparator inputs. Maximum and minimum liquid levels may be set by the potentiometers. The maximum level (1 max) may be preset between the limits: 65 pF less than C (1 max) less than 120 pF. The minimum level is presetable and the limits are: 0 less than C (1 min) less than 25 pF.

PLANT WATER GAUGE

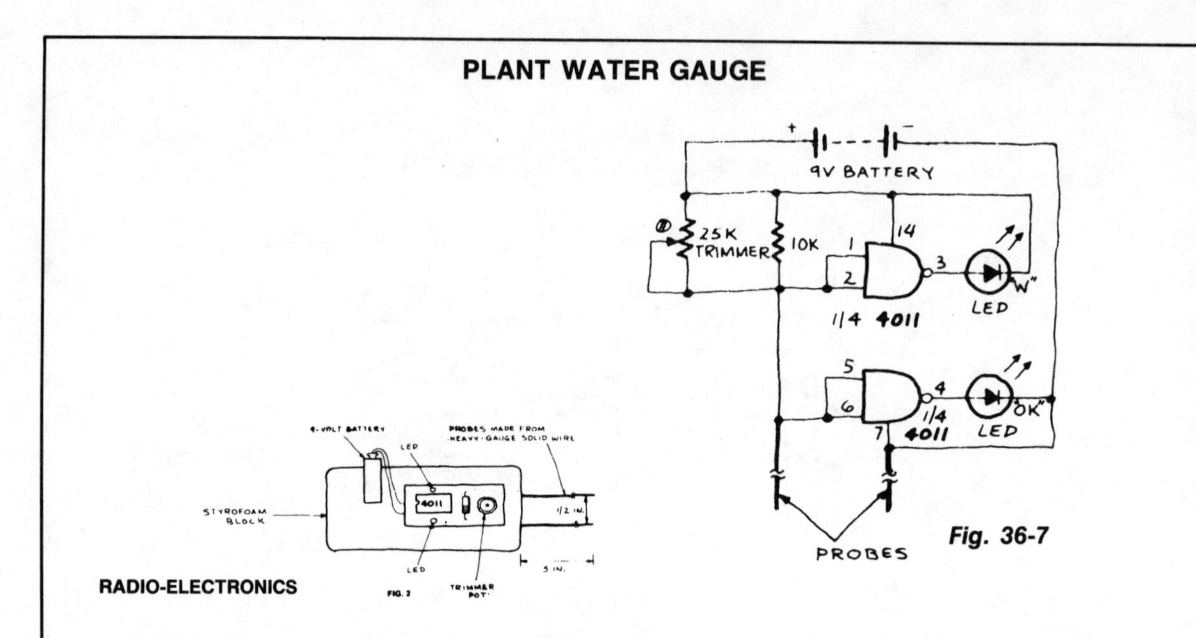

RADIO-ELECTRONICS

Fig. 36-7

Circuit Notes

To calibrate the gauge, connect the battery and press the probes gently into a pot containing a plant that is just on the verge of needing water (stick it in so that only an inch of the probe is left visible at the top). Turn the potentiometer until the "OK" LED lights and then turn it back to the point where that LED goes out and the "W", or "Water", LED just comes on. The device should now be properly adjusted.

LIQUID FLOWMETER

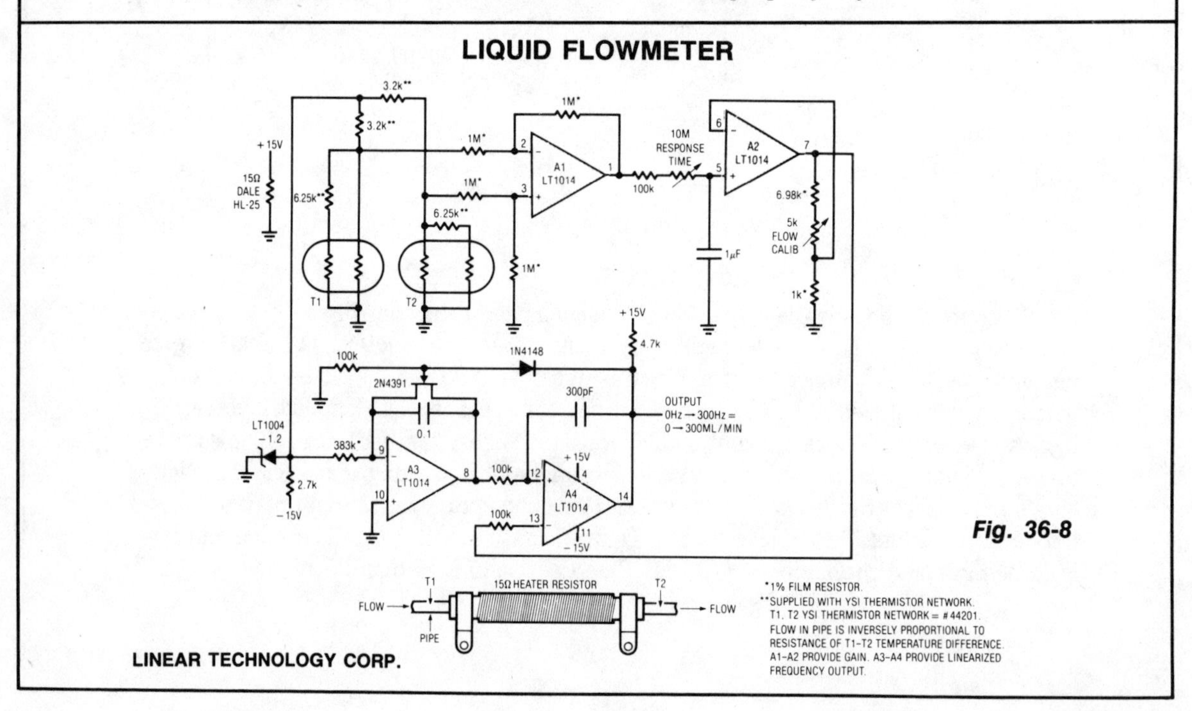

LINEAR TECHNOLOGY CORP.

Fig. 36-8

37
Frequency Meters

The sources of the following circuits are contained in the Sources section beginning on page 694. The figure number contained in the box of each circuit correlates to the source entry in the Sources section.

POWER-FREQUENCY METER

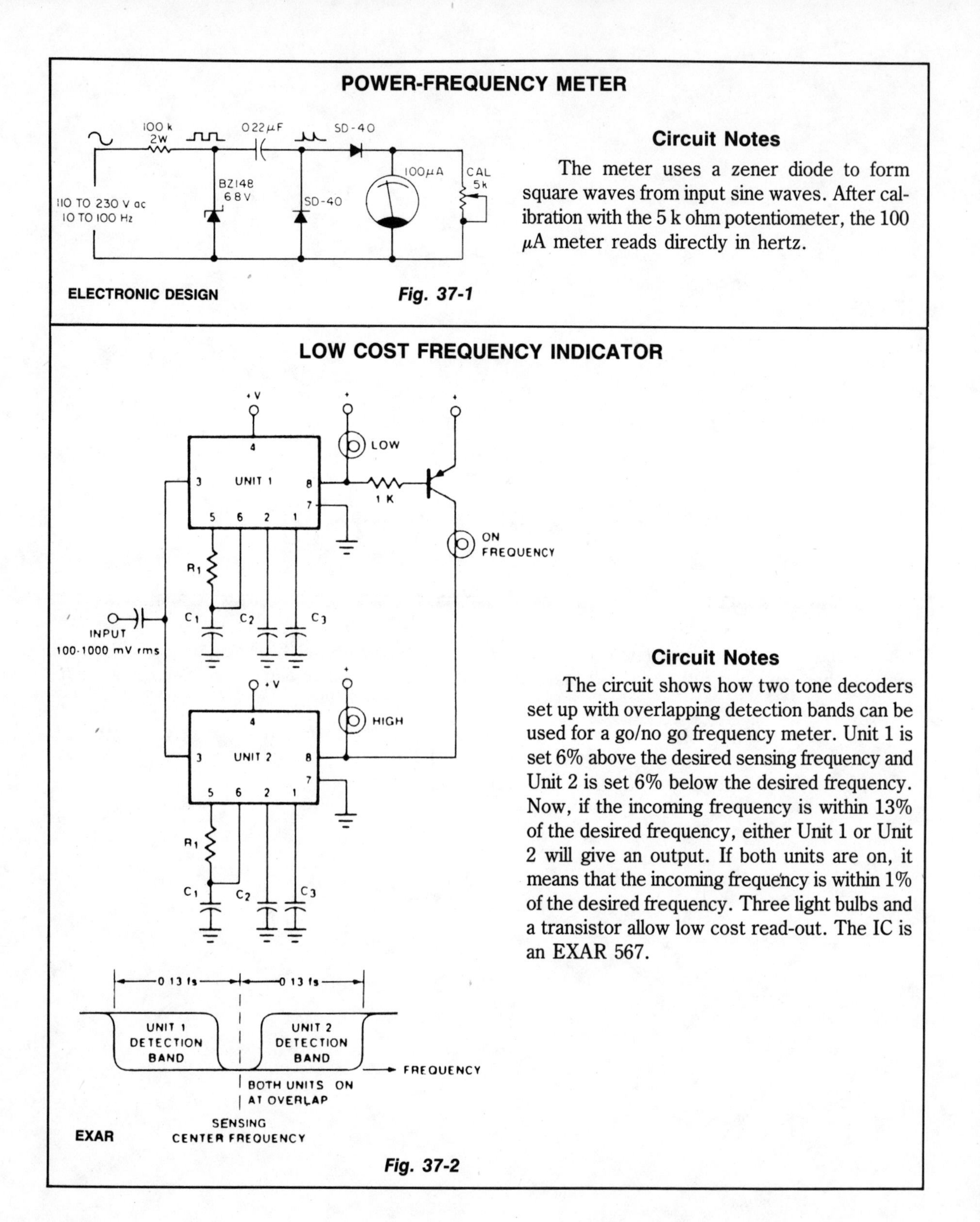

Circuit Notes

The meter uses a zener diode to form square waves from input sine waves. After calibration with the 5 k ohm potentiometer, the 100 μA meter reads directly in hertz.

ELECTRONIC DESIGN

Fig. 37-1

LOW COST FREQUENCY INDICATOR

Circuit Notes

The circuit shows how two tone decoders set up with overlapping detection bands can be used for a go/no go frequency meter. Unit 1 is set 6% above the desired sensing frequency and Unit 2 is set 6% below the desired frequency. Now, if the incoming frequency is within 13% of the desired frequency, either Unit 1 or Unit 2 will give an output. If both units are on, it means that the incoming frequency is within 1% of the desired frequency. Three light bulbs and a transistor allow low cost read-out. The IC is an EXAR 567.

EXAR

Fig. 37-2

38

Frequency Multiplier and Divider Circuits

The sources of the following circuits are contained in the Sources section beginning on page 694. The figure number contained in the box of each circuit correlates to the source entry in the Sources section.

NONSELECTIVE FREQUENCY TRIPLER USES TRANSISTOR SATURATION CHARACTERISTICS

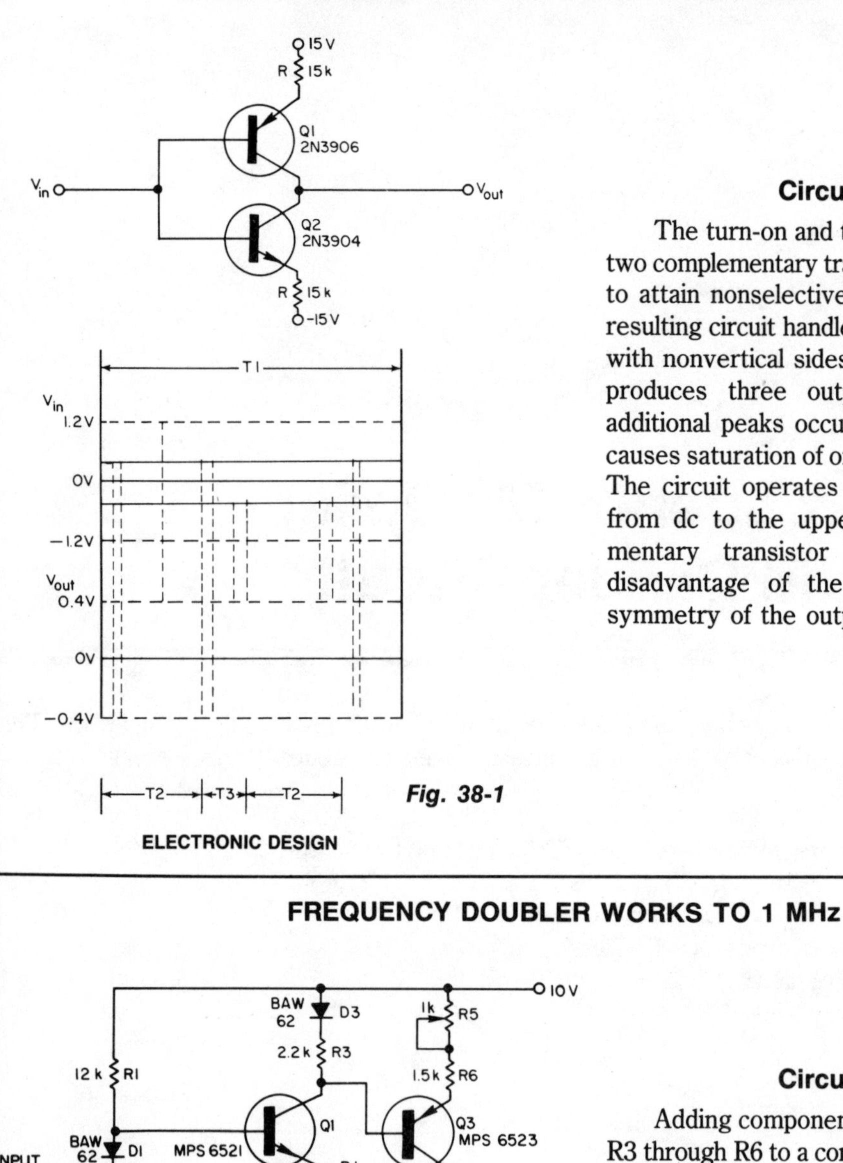

Fig. 38-1

ELECTRONIC DESIGN

Circuit Notes

The turn-on and turn-off characteristics of two complementary transistors can be combined to attain nonselective frequency tripling. The resulting circuit handles any periodic waveform with nonvertical sides. Each input signal peak produces three output signal peaks. The additional peaks occur where the input signal causes saturation of one of the two transistors. The circuit operates over a frequency range from dc to the upper limits of the complementary transistor pair. About the only disadvantage of the circuit is the lack of symmetry of the output signal peaks.

FREQUENCY DOUBLER WORKS TO 1 MHz

ELECTRONIC DESIGN

Fig. 38-2

Circuit Notes

Adding components Q3, D3, and resistors R3 through R6 to a conventional complementary symmetry class AB buffer can double the frequency of an input sine wave.

LOW FREQUENCY DIVIDER

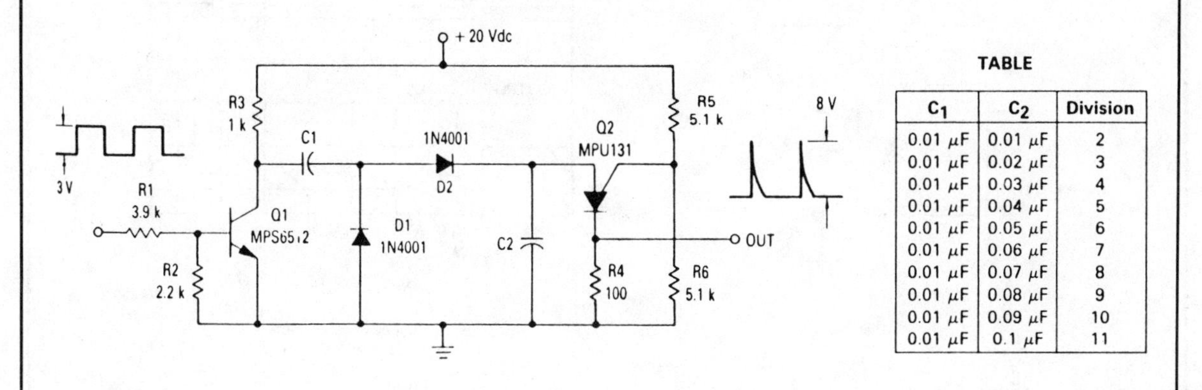

TABLE

C_1	C_2	Division
0.01 μF	0.01 μF	2
0.01 μF	0.02 μF	3
0.01 μF	0.03 μF	4
0.01 μF	0.04 μF	5
0.01 μF	0.05 μF	6
0.01 μF	0.06 μF	7
0.01 μF	0.07 μF	8
0.01 μF	0.08 μF	9
0.01 μF	0.09 μF	10
0.01 μF	0.1 μF	11

MOTOROLA *Fig. 38-3*

Circuit Notes

The ratio of capacitors C1 and C2 determines division. With a positive pulse applied to the base of Q1, assume that C1 = C2 and that C1 and C2 are discharged. When Q1 turns off, both C1 and C2 charge to 10 volts each through R3. On the next pulse to the base of Q1, C1 is again discharged but C2 remains charged to 10 volts. As Q1 turns off this time, C1 and C2 again charge. This time C2 charges to the peak point firing voltage of the PUT causing it to fire. This discharges capacitor C2 and allows capacitor C1 to charge to the line voltage. As soon as C2 discharges and C1 charges, the PUT turns off. The next cycle begins with another positive pulse on the base of Q1 which again discharges C1. The input and output frequency can be approximated by the equation

$$f_{in} = \frac{(C1 = C2)}{C1} f_{out}$$

For a 10 kHz input frequency with an amplitude of 3 volts, the table shows the values for C1 and C2 needed to divide by 2 to 11.

FREQUENCY DIVIDER

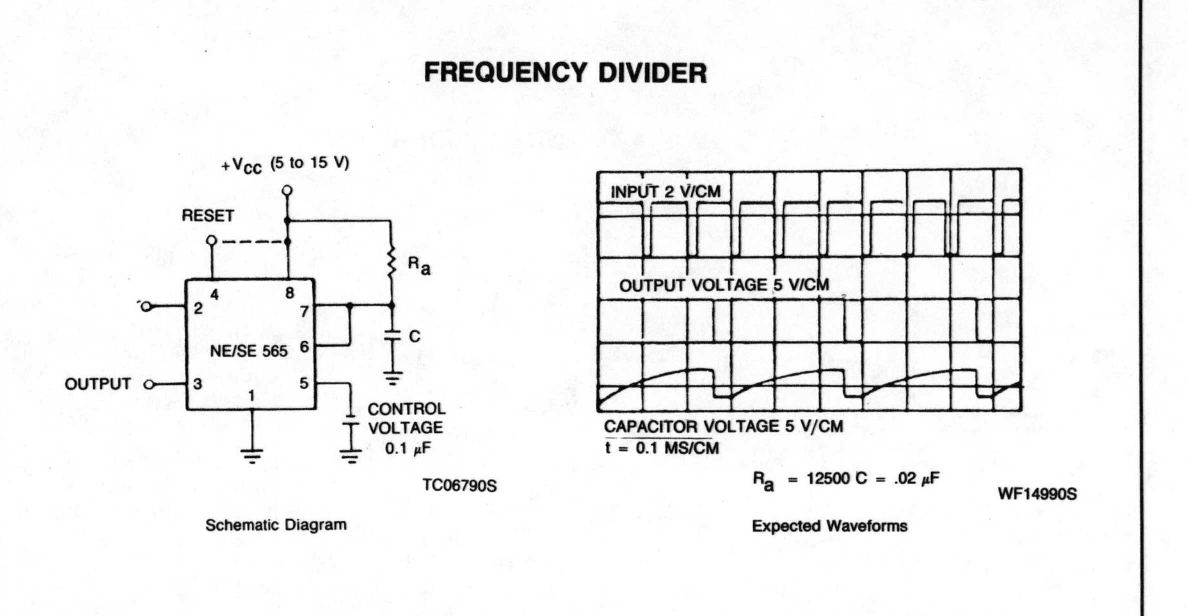

SIGNETICS

Fig. 38-4

Circuit Notes

If the input frequency is known, the timer can easily be used as a frequency divider by adjusting the length of the timeing cycle. Figure shows the waveforms of the timer when used as a divide-by-three circuit. This application makes use of the fact that this circuit cannot be retriggered during the timing cycle.

39

Frequency-to-Voltage Converters

The sources of the following circuits are contained in the Sources section beginning on page 694. The figure number contained in the box of each circuit correlates to the source entry in the Sources section.

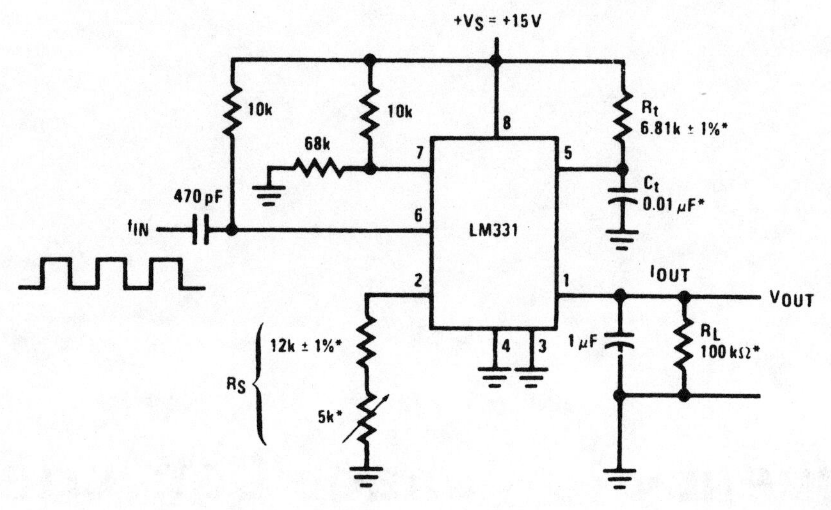

TL/H/5680-7

$$V_{OUT} = f_{IN} \times 2.09V \times \frac{R_L}{R_S} \times (R_tC_t)$$

*Use stable components with low temperature coefficients.

Simple Frequency-to-Voltage Converter, 10 kHz Full-Scale, ±0.06% Non-Linearity

NATIONAL SEMICONDUCTOR CORP. *Fig. 39-1*

Circuit Notes

In these applications, a pulse input at f_{IN} is differentiated by a C-R network and the negative-going edge at pin 6 causes the input comparator to trigger the timer circuit. Just as with a V-to-F converter, the average current flowing out of pin 1 is $I_{AVERAGE} = i \times (1.1\ R_1C_1) \times f$. In this simple circuit, this current is filtered in the network $R_L = 100$ k ohm and 1 μF. The ripple will be less than 10 mV peak, but the response will be slow, with a 0.1 second time constant, and settling of 0.7 second to 0.1%

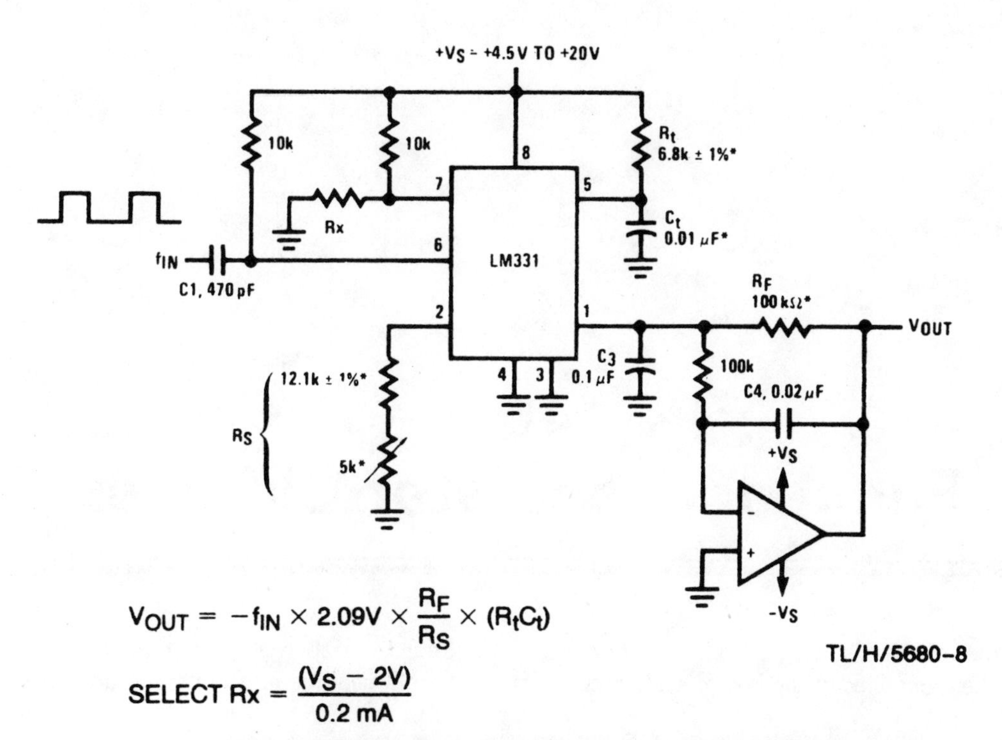

$$V_{OUT} = -f_{IN} \times 2.09V \times \frac{R_F}{R_S} \times (R_tC_t)$$

$$\text{SELECT } R_X = \frac{(V_S - 2V)}{0.2\ mA}$$

*Use stable components with low temperature coefficients.

Precision Frequency-to-Voltage Converter, 10 kHz Full-Scale with 2-Pole Filter, ±0.01% Non-Linearity Maximum

accuracy. In the precision circuit, an operational amplifier provides a buffered output and also acts as a 2-pole filter. The ripple will be less than 5 mV peak for all frequencies above 1 kHz, and the response time will be much quicker than in Part 1. However, for input frequencies below 200 Hz, this circuit will have worse ripple than the figure. The engineering of the filter time-constants to get adequate response and small enough ripple simply requires a study of the compromises to be made. Inherently, V-to-F converter response can be fast, but F-to-V response cannot.

40
Function Generator Circuits

The sources of the following circuits are contained in the Sources section beginning on page 694. The figure number contained in the box of each circuit correlates to the source entry in the Sources section.

QUAD OP AMP GENERATES FOUR DIFFERENT SYNCHRONIZED WAVEFORMS SIMULTANEOUSLY

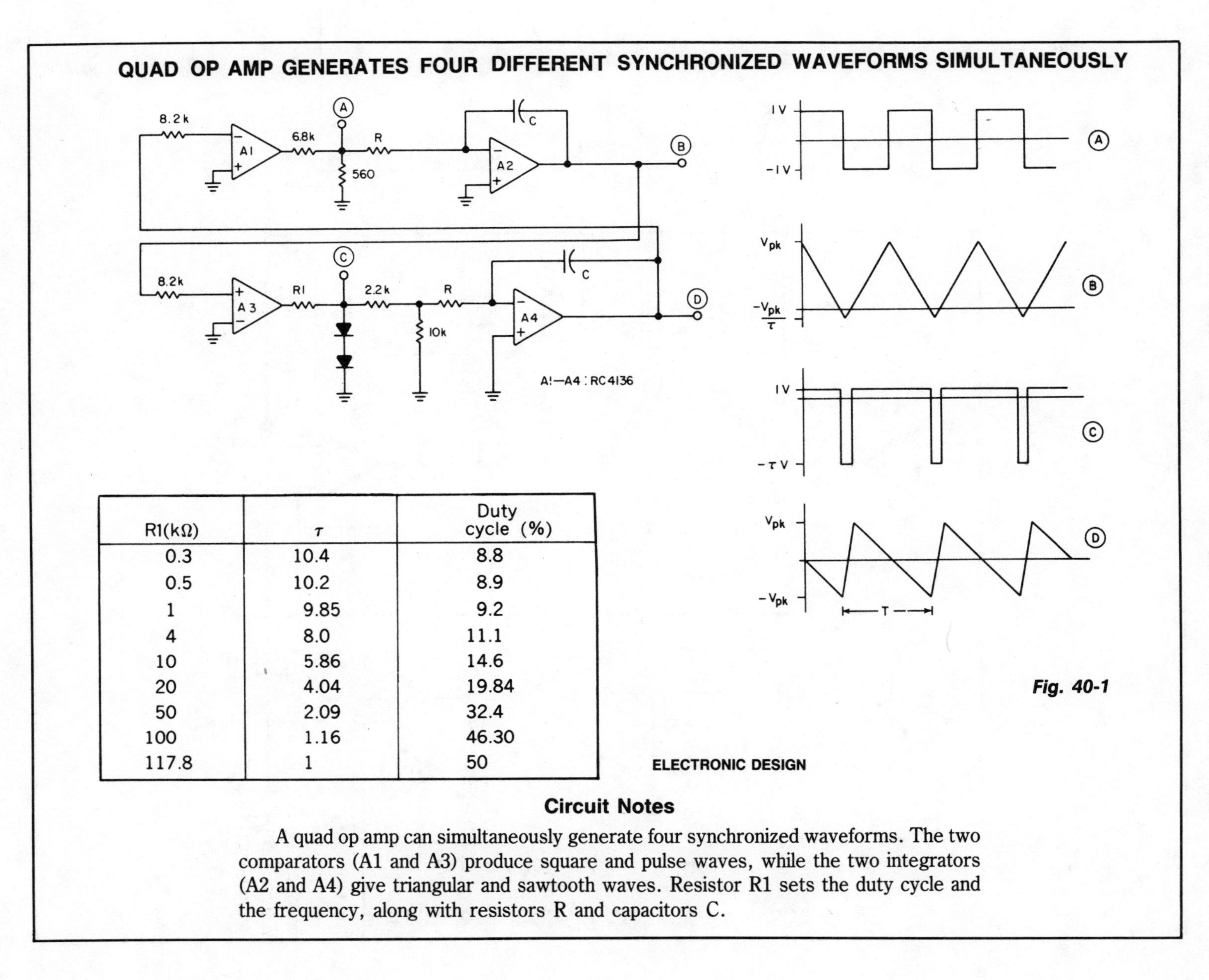

R1(kΩ)	τ	Duty cycle (%)
0.3	10.4	8.8
0.5	10.2	8.9
1	9.85	9.2
4	8.0	11.1
10	5.86	14.6
20	4.04	19.84
50	2.09	32.4
100	1.16	46.30
117.8	1	50

Fig. 40-1

ELECTRONIC DESIGN

Circuit Notes

A quad op amp can simultaneously generate four synchronized waveforms. The two comparators (A1 and A3) produce square and pulse waves, while the two integrators (A2 and A4) give triangular and sawtooth waves. Resistor R1 sets the duty cycle and the frequency, along with resistors R and capacitors C.

A SINE/COSINE GENERATOR FOR 0.1 - 10 kHz

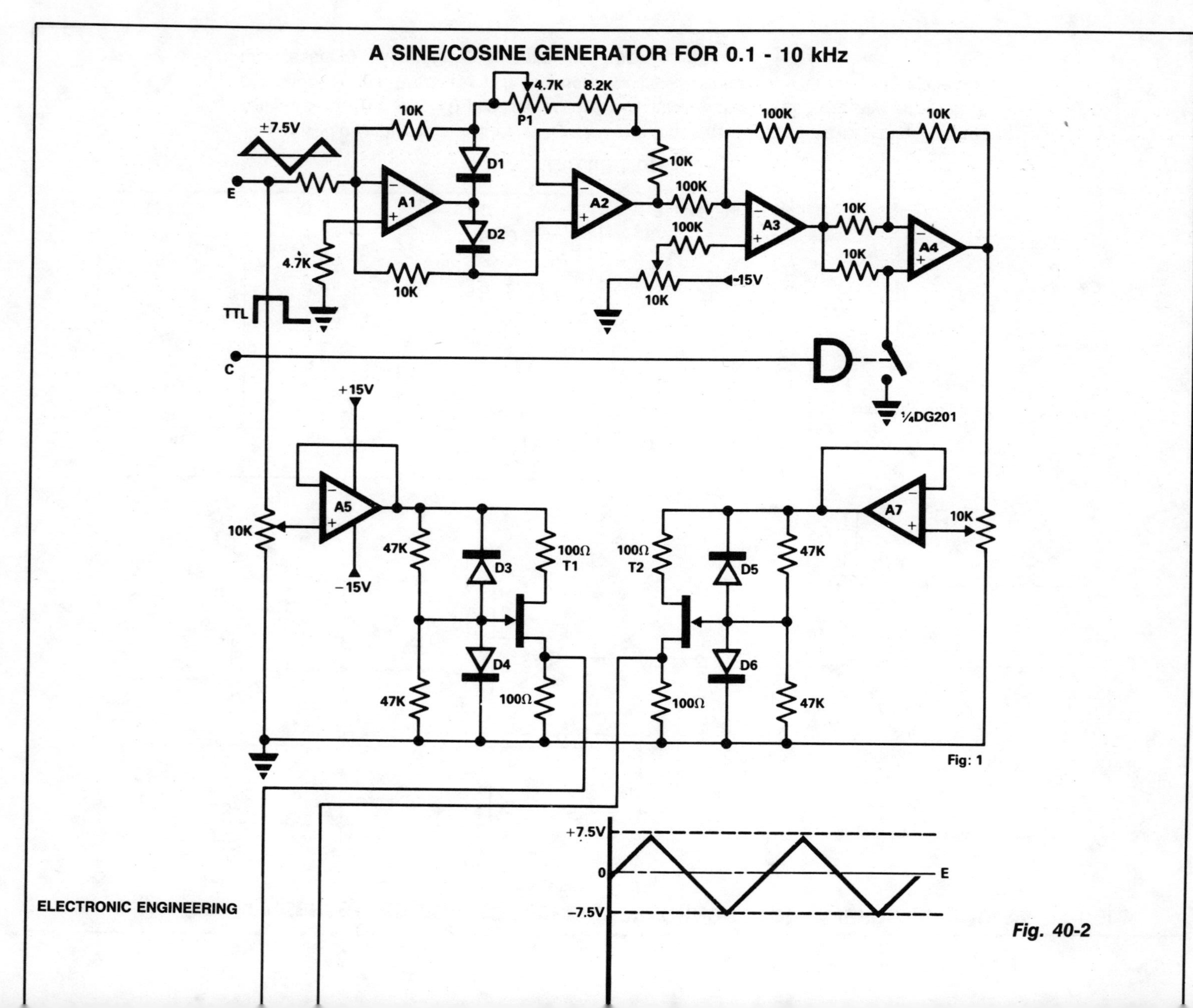

ELECTRONIC ENGINEERING

Fig. 40-2

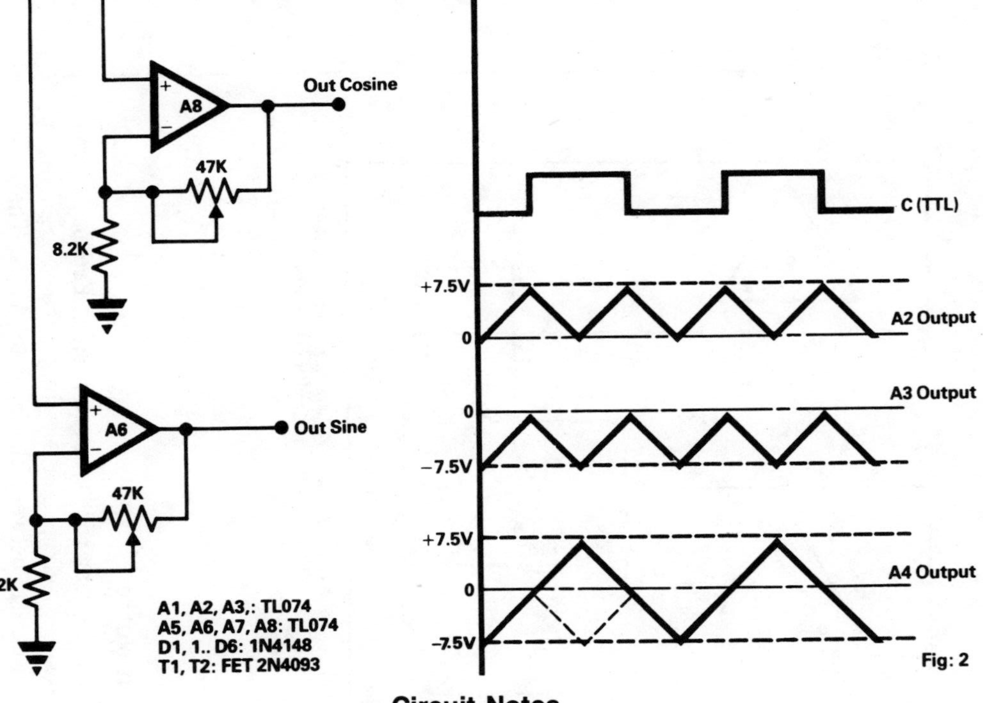

Fig: 2

Circuit Notes

The scheme presented delivers waveforms from any function generator producing a triangular output and a synchronized TTL square wave. A1 and A2 act as a two-phase current rectifier by inverting the negative voltage appearing at the input of A1.

Positive input: Both A1 and A2 work as unity gain followers, D1 and D2 being in the off-state.

Negative input: A1 has a $-2/3$ gain (D1 off and D2 on), A2 has a $+3/2$ gain and the total voltage transfer is -1 between output and input. P1 allows a fine trimming of the -1 gain for the negative input signals. A3 adds a continuous voltage to the rectified positive signal in order to attack A4 which acts as a $\pm$ multiplier commanded by the TTL input through the analog switch. The signal polarity is reconstructed and the output of A4 delivers a triangular waveform shifted by 90° with respect to the input signal, Fig. 2. The original and the shifted voltages are fed into the triangle to sine converters through A5 and A7 working as impedance converters. Over the frequency dynamic ranges from 0.1 Hz to 10 kHz, the phase shift is constant and the distortion on the sine voltage is less than 1%.

OSCILLATOR OR AMPLIFIER WITH WIDE FREQUENCY RANGE

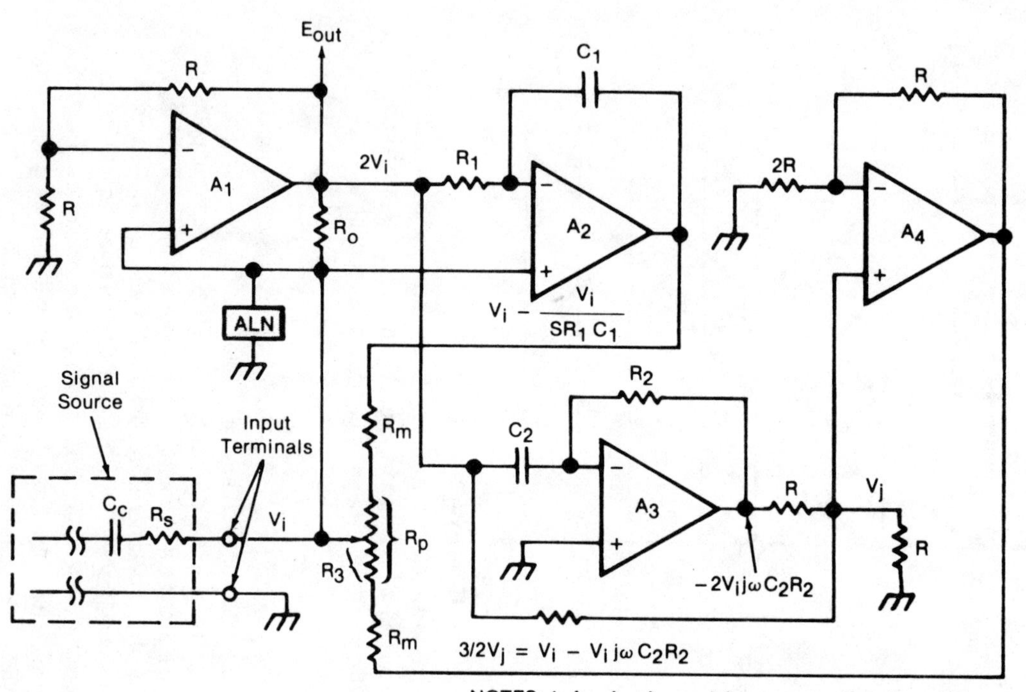

Fig. 40-3

Circuit Notes

An oscillator/amplifier is resistively tunable over a wide frequency range. Feedback circuits containing operational amplifiers, resistors, and capacitors synthesize the electrical effects of an inductance and capacitance in parallel between the input terminals. The synthetic inductance and capacitance, and, therefore, the resonant frequency of the input admittance, are adjusted by changing a potentiometer setting. The input signal is introduced in parallel to the noninverting input terminals of operational amplifiers A_1 and A_2 and to the potentiometer cursor. The voltages produced by the feedback circuits in response to input voltage V_1 are indicated at the various circuit nodes.

LINEAR TRIANGLE/SQUARE WAVE VCO

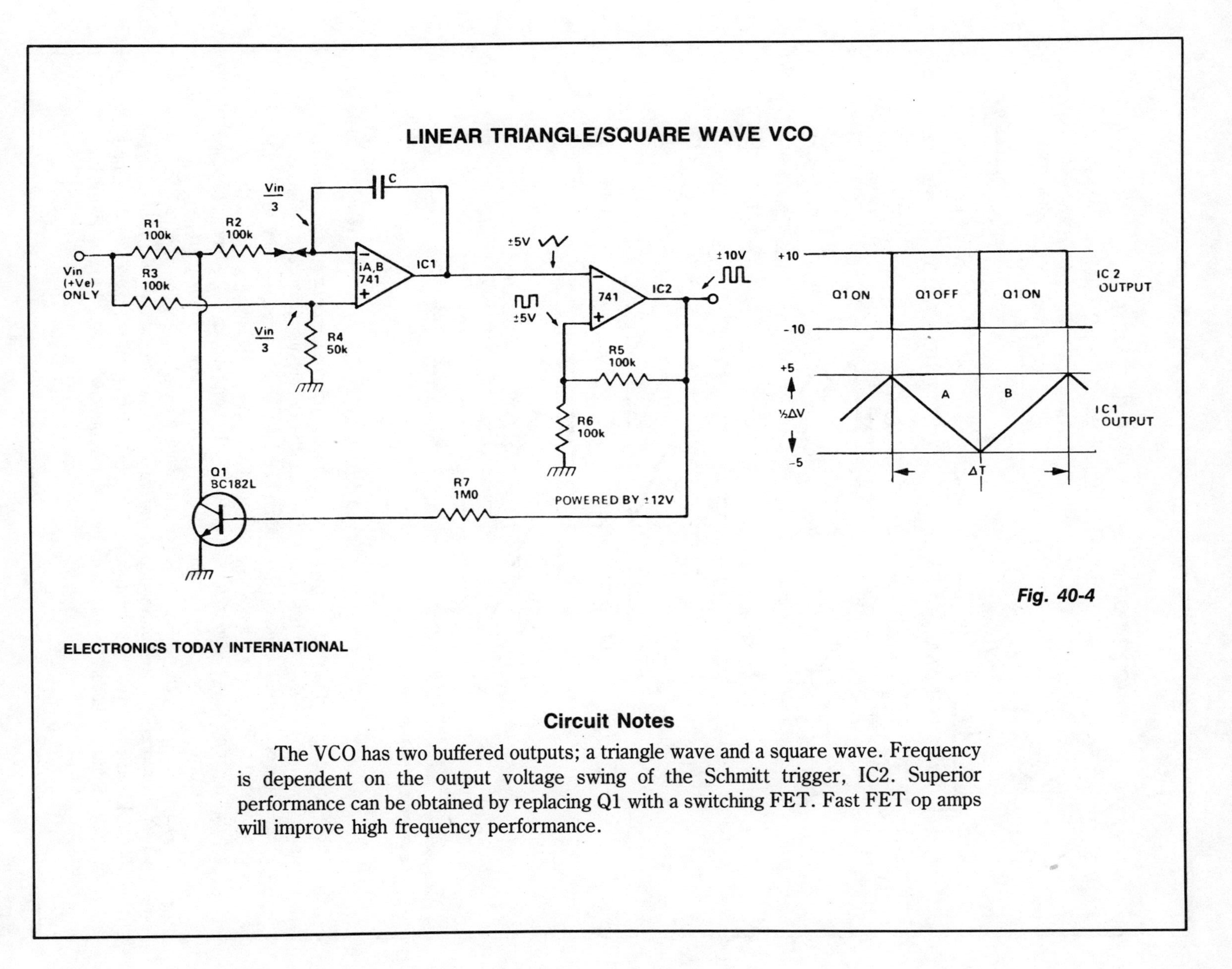

Fig. 40-4

ELECTRONICS TODAY INTERNATIONAL

Circuit Notes

The VCO has two buffered outputs; a triangle wave and a square wave. Frequency is dependent on the output voltage swing of the Schmitt trigger, IC2. Superior performance can be obtained by replacing Q1 with a switching FET. Fast FET op amps will improve high frequency performance.

CIRCUIT FOR MULTIPLYING PULSE WIDTHS

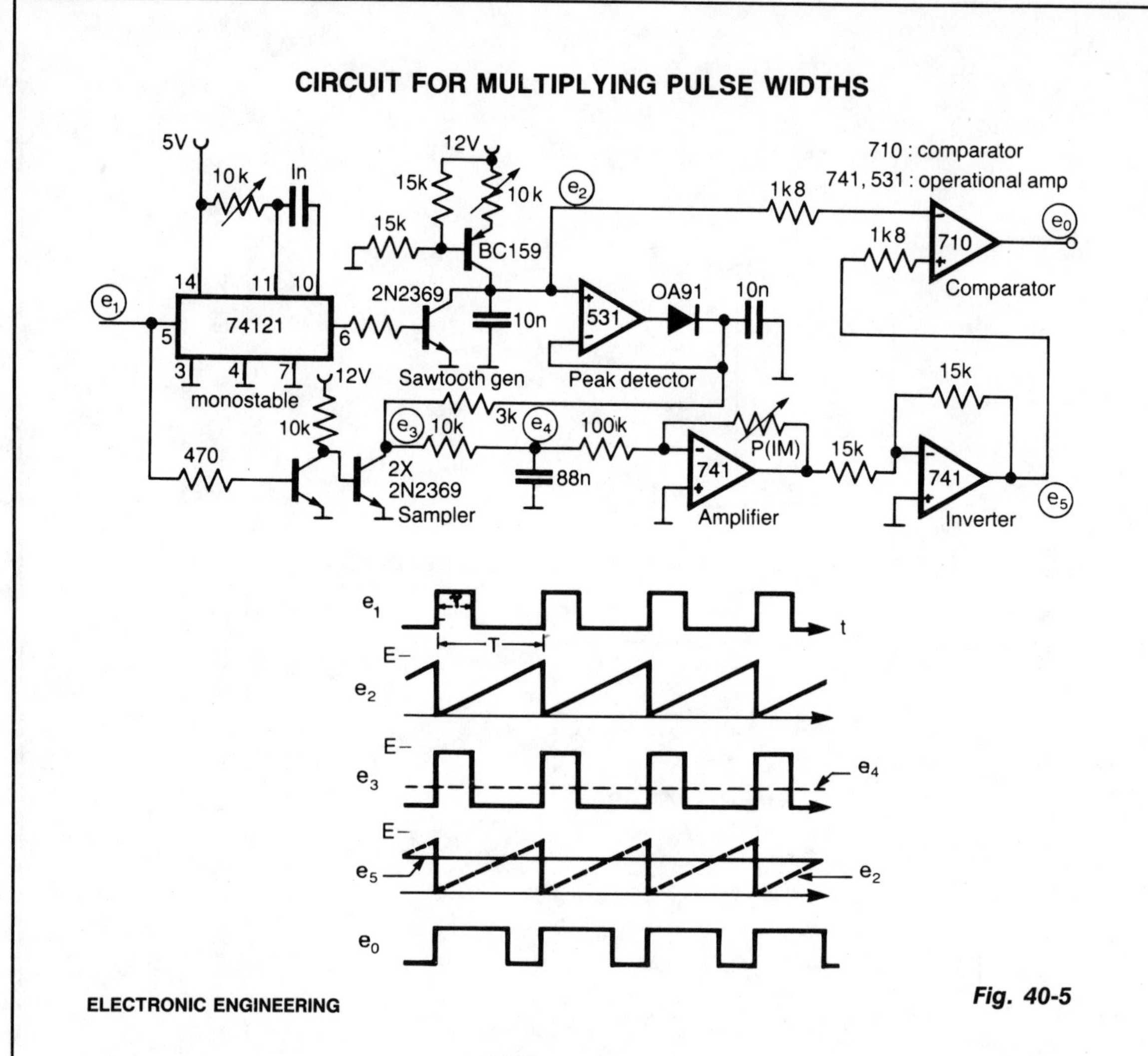

ELECTRONIC ENGINEERING

Fig. 40-5

Circuit Notes

A circuit for multiplying the width of incoming pulses by a factor greater or less than unity is simple to build and has the feature that the multiplying factor can be selected by adjusting one potentiometer only. The multiplying factor is determined by setting the potentiometer P in the feedback of a 741 amplifier. The input pulses e_1 of width τ and repetition period T is used to trigger a sawtooth generator at its rising edges to produce the waveform e_2 having a peak value of (E) volt. This peak value is then sampled by the input pulses to generate the pulse train e_3 having an average value of $e_4(= E\ E\tau/T)$ which is proportional to τ and independent on T. The dc voltage e_4 is amplified by a factor k and compared with sawtooth waveform e_2 giving output pulses of duration k τ. The circuit is capable of operating over the frequency range 10 kHz - 100 kHz.

PROGRAMMABLE VOLTAGE CONTROLLED FREQUENCY SYNTHESIZER

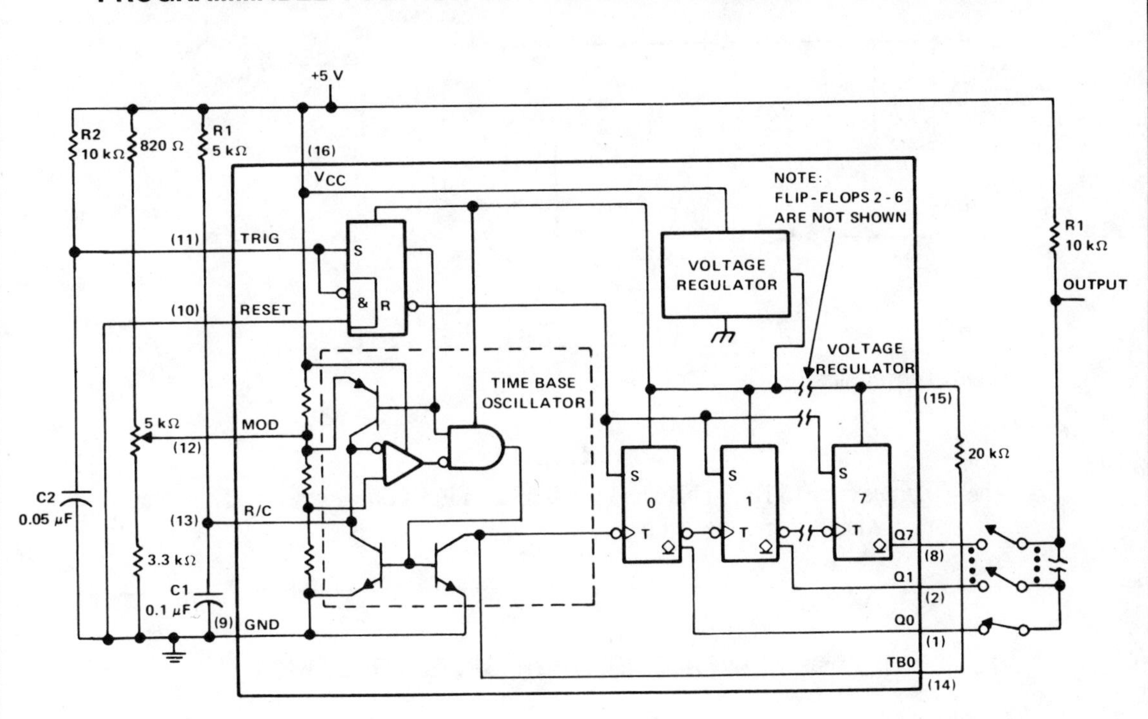

TEXAS INSTRUMENTS

Fig. 40-6

Circuit Notes

The µA2240 consists of four basic circuit elements: (1) a time-base oscillator, (2) an eight-bit counter, (3) a control flip-flop, and (4) a voltage regulator. The basic frequency of the time-base oscillator (TBO) is set by the external time constant determined by the values of R1 and C1 (1?R1C1 = 2 kHz). The open-collector output of the TBO is connected to the regulator output via a 20 k ohm pull-up resistor, and drives the input to the eight-bit counter. At power-up, a positive trigger pulse is detected across C2 which starts the TBO and sets all counter outputs to a low state. Once the µA2240 is initially triggered, any further trigger inputs are ignored until it is reset. In this astable operation, the µA2240 will free-run from the time it is triggered until it receives an external reset signal. Up to 255 discrete frequencies can be synthesized by connecting different counter outputs.

EMITTER-COUPLED RC OSCILLATOR

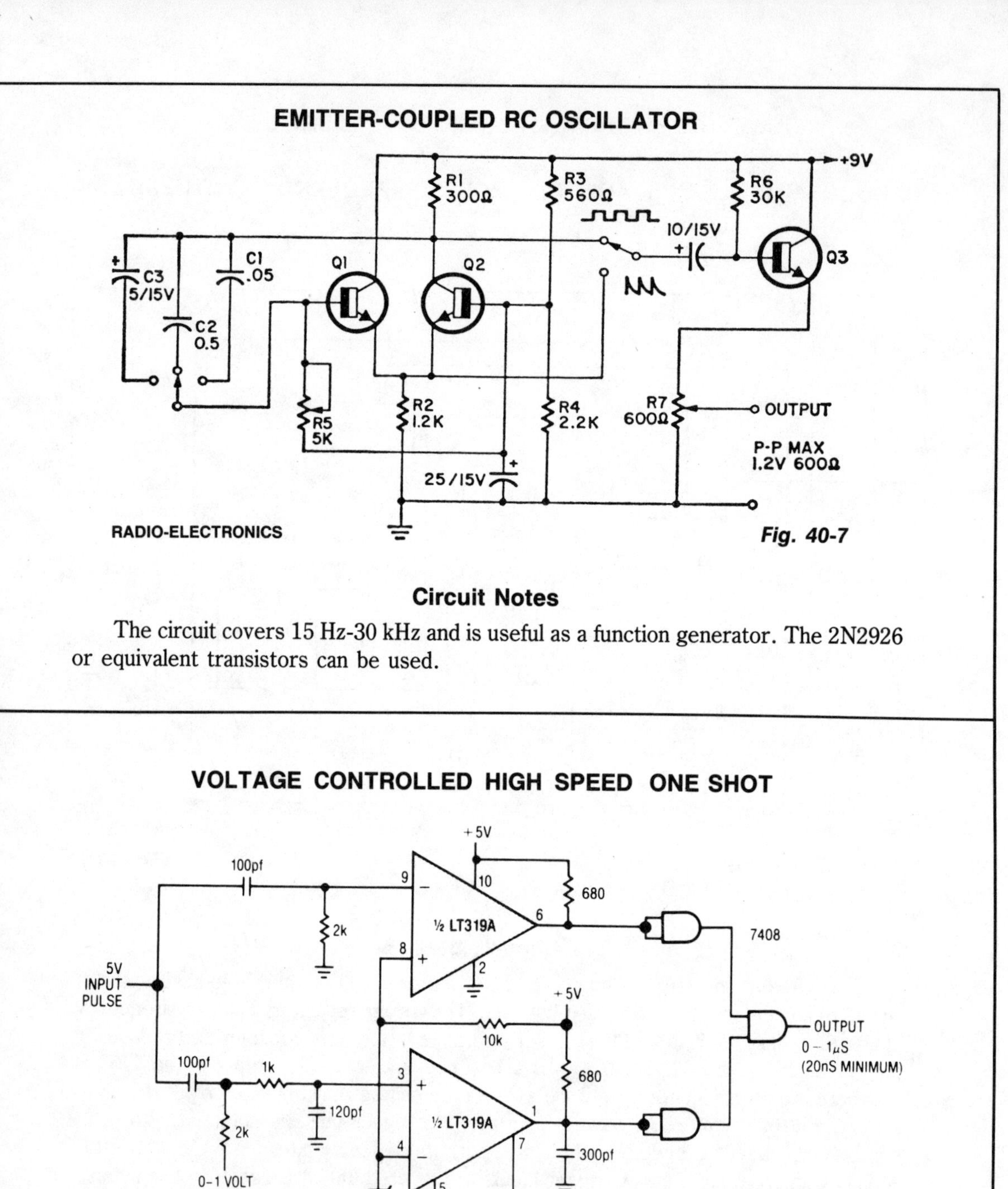

Fig. 40-7

Circuit Notes

The circuit covers 15 Hz-30 kHz and is useful as a function generator. The 2N2926 or equivalent transistors can be used.

VOLTAGE CONTROLLED HIGH SPEED ONE SHOT

Fig. 40-8

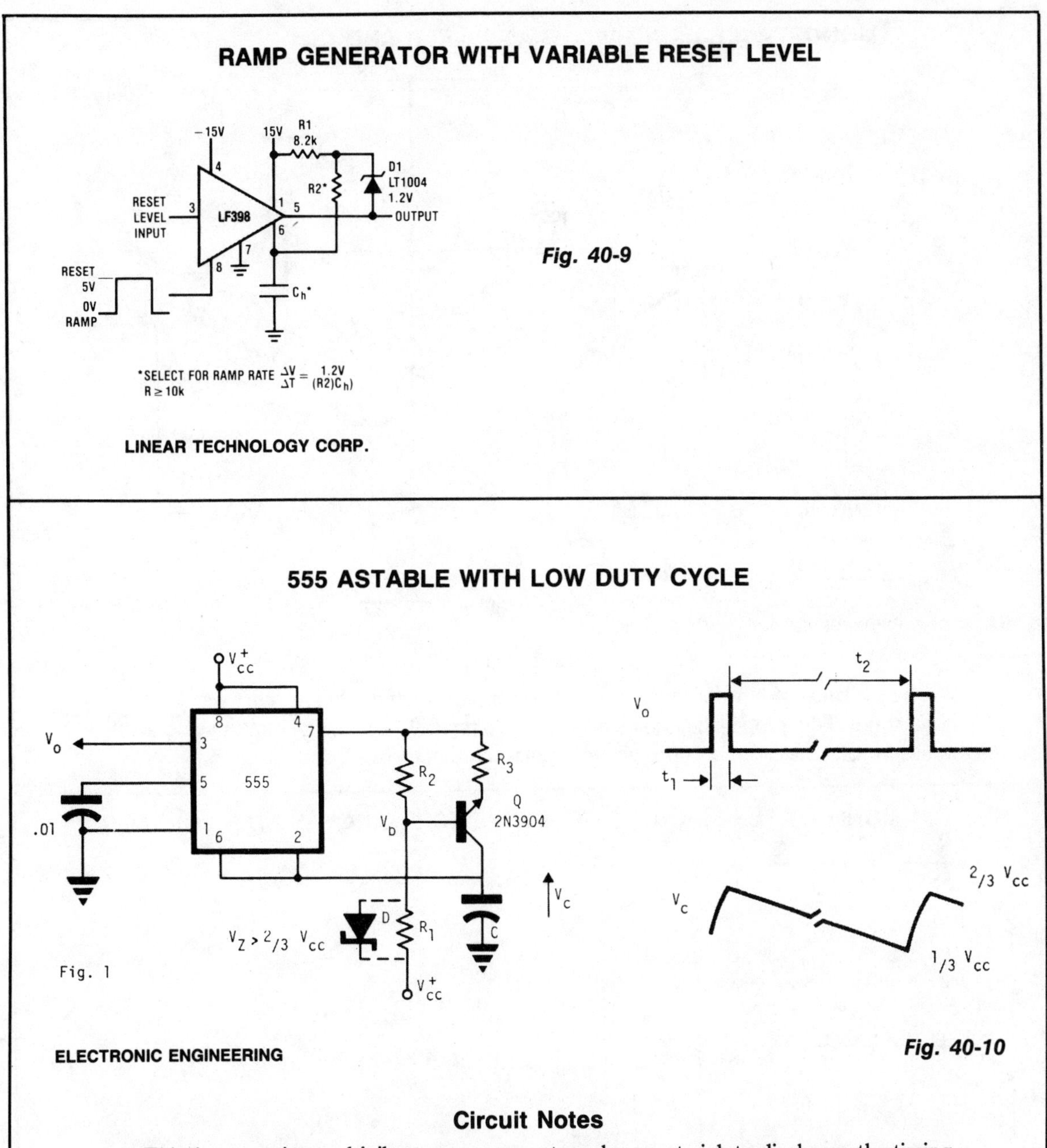

Fig. 40-9

Fig. 40-10

Circuit Notes

This free-running multivibrator uses an external current sink to discharge the timing capacitor, C. Therefore, interval t_2 may easily be 1000 × the pulse duration, t_1, which defines a positive output. Capacitor voltage, V_C, is a negative going ramp with exponential rise during the pulse output periods.

MONOSTABLE USING VIDEO AMPLIFIER AND COMPARATOR

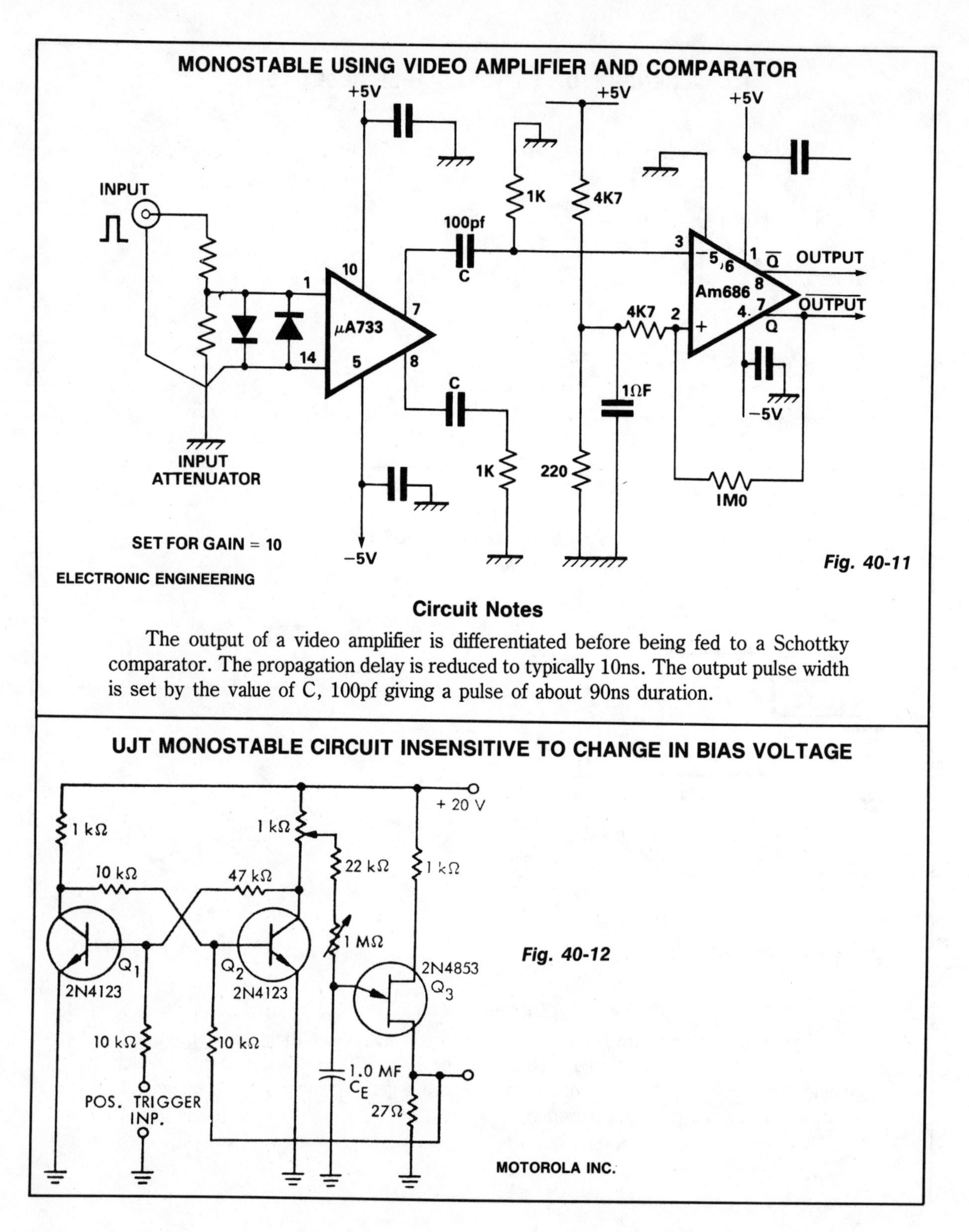

Fig. 40-11

Circuit Notes

The output of a video amplifier is differentiated before being fed to a Schottky comparator. The propagation delay is reduced to typically 10ns. The output pulse width is set by the value of C, 100pf giving a pulse of about 90ns duration.

UJT MONOSTABLE CIRCUIT INSENSITIVE TO CHANGE IN BIAS VOLTAGE

Fig. 40-12

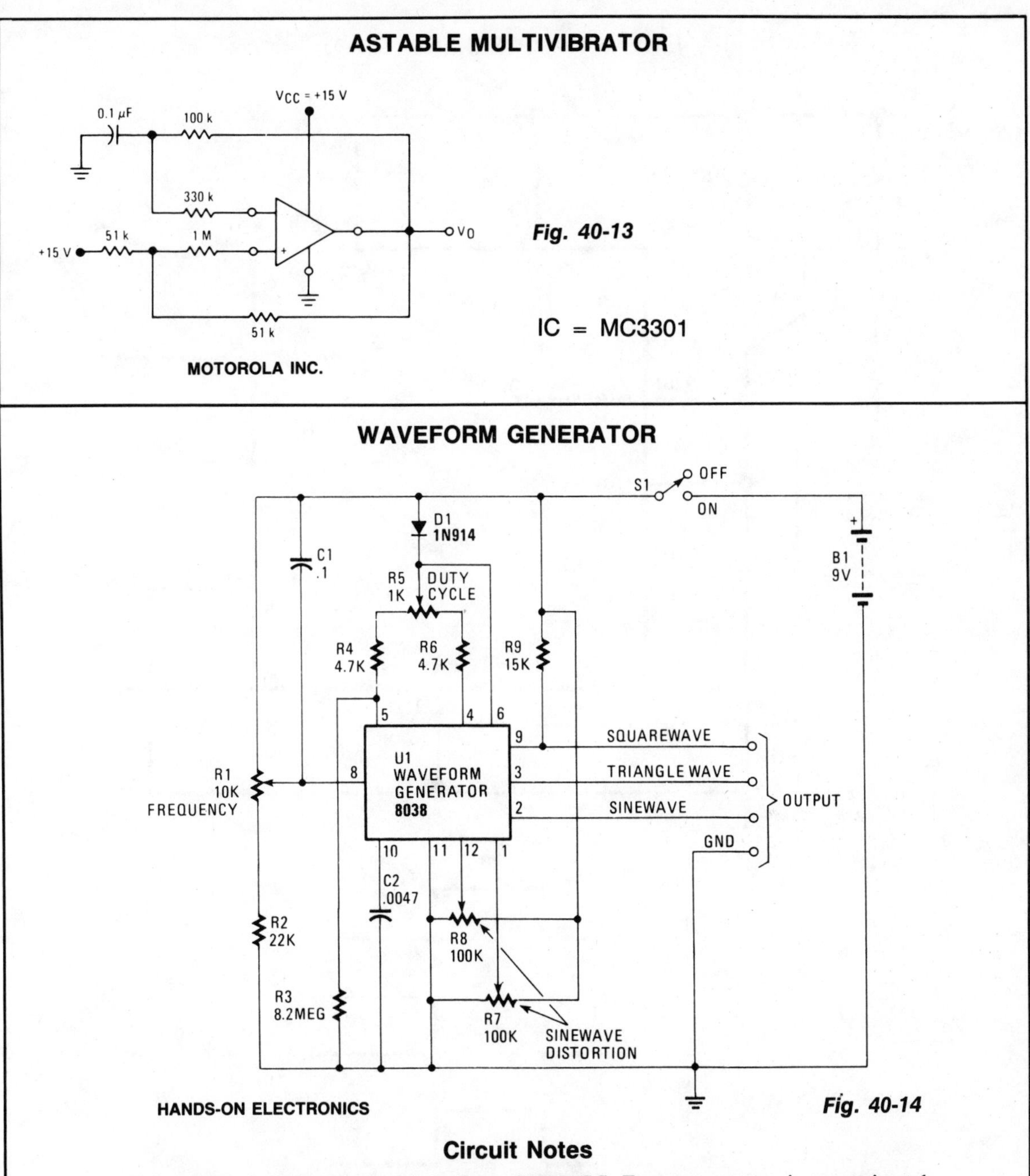

Circuit Notes

The circuit is designed around the Intersil 8038CC. Frequency range is approximately 20 Hz to 20 kHz—a tuning range of 1000:1 with a single control. The output frequency depends on the value of C2 and on the setting of potentiometer R1. Other values of C2 change the frequency range. Increase the value of C2 to lower the frequency. The lowest possible frequency is around .001 Hz and the highest is around 300 kHz.

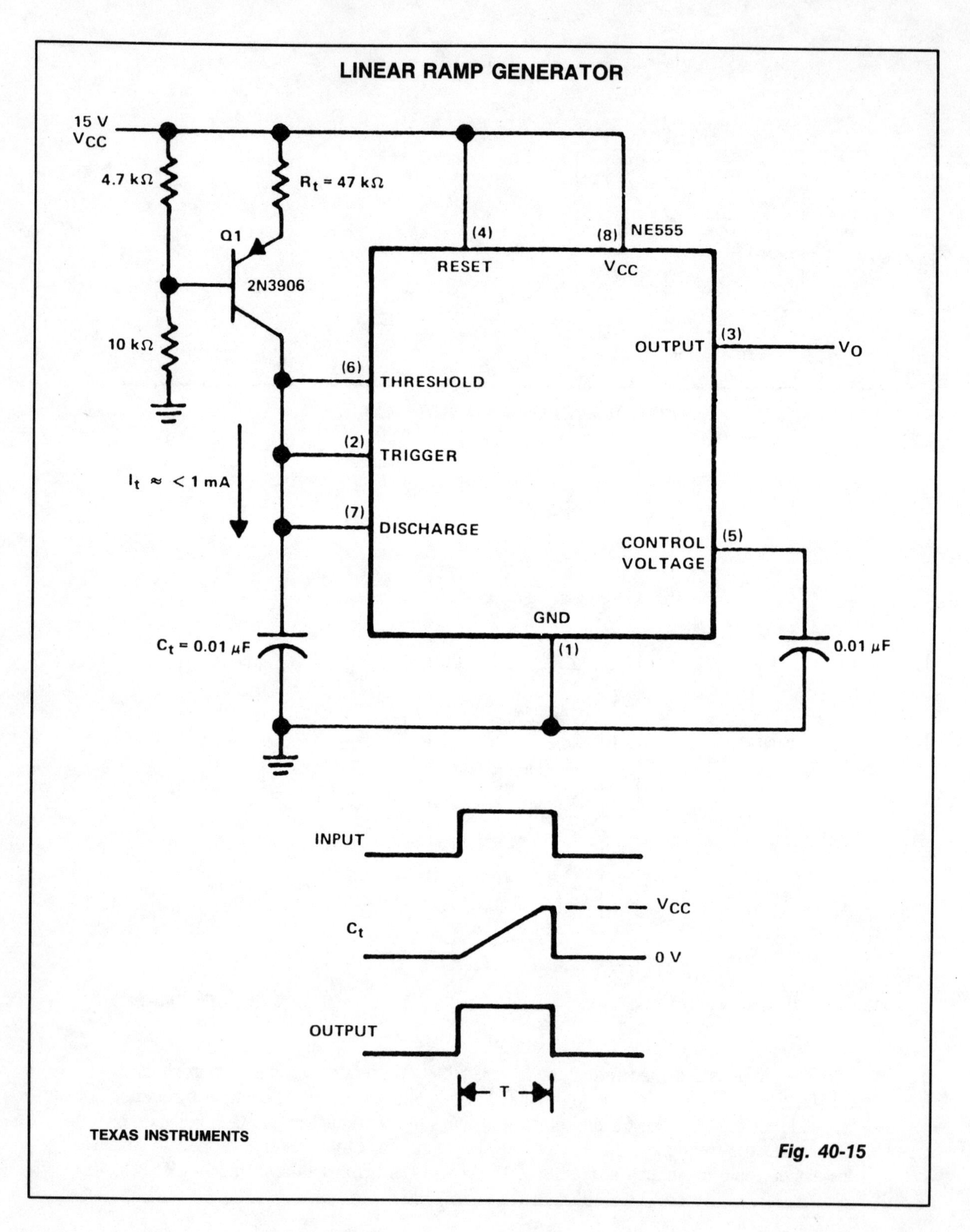
LINEAR RAMP GENERATOR
15 V
VCC
4.7 kΩ
Rt = 47 kΩ
Q1
2N3906
10 kΩ
(4)
(8)
NE555
RESET
VCC
(3)
OUTPUT
VO
(6)
THRESHOLD
(2)
TRIGGER
It ≈ < 1 mA
(7)
DISCHARGE
CONTROL
VOLTAGE
(5)
GND
(1)
Ct = 0.01 μF
0.01 μF
INPUT
Ct
VCC
0 V
OUTPUT
T
TEXAS INSTRUMENTS

Fig. 40-15

LINEAR RAMP GENERATOR, Continued.

Circuit Notes

The linear charging ramp is most useful where linear control of voltage is required. Some possible applications are a long period voltage controlled timer, a voltage to pulse width converter, or a linear pulse width modulator. Q1 is the current source transistor, supplying constant current to the timing capacitor C_t. When the timer is triggered, the clamp on C_t is removed and C_t charges linearly toward V_{CC} by virtue of the constant current supplied by Q1. The threshold at pin 6 is ⅔ V_{CC}; here, it is termed V_C. When the voltage across C_t reaches V_C volts, the timing cycle ends. The timing expression for output pulse with T is:

In general, I_t should be 1 mA value compatible with the NE555.

FUNCTION GENERATOR

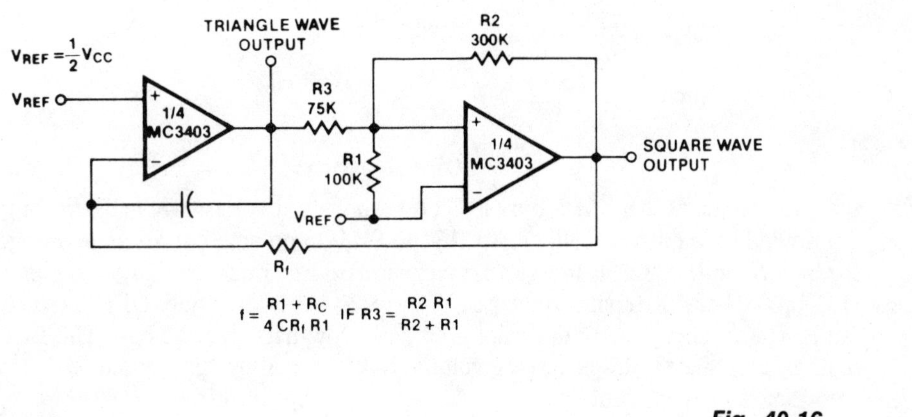

SIGNETICS

Fig. 40-16

WAVEFORM GENERATOR

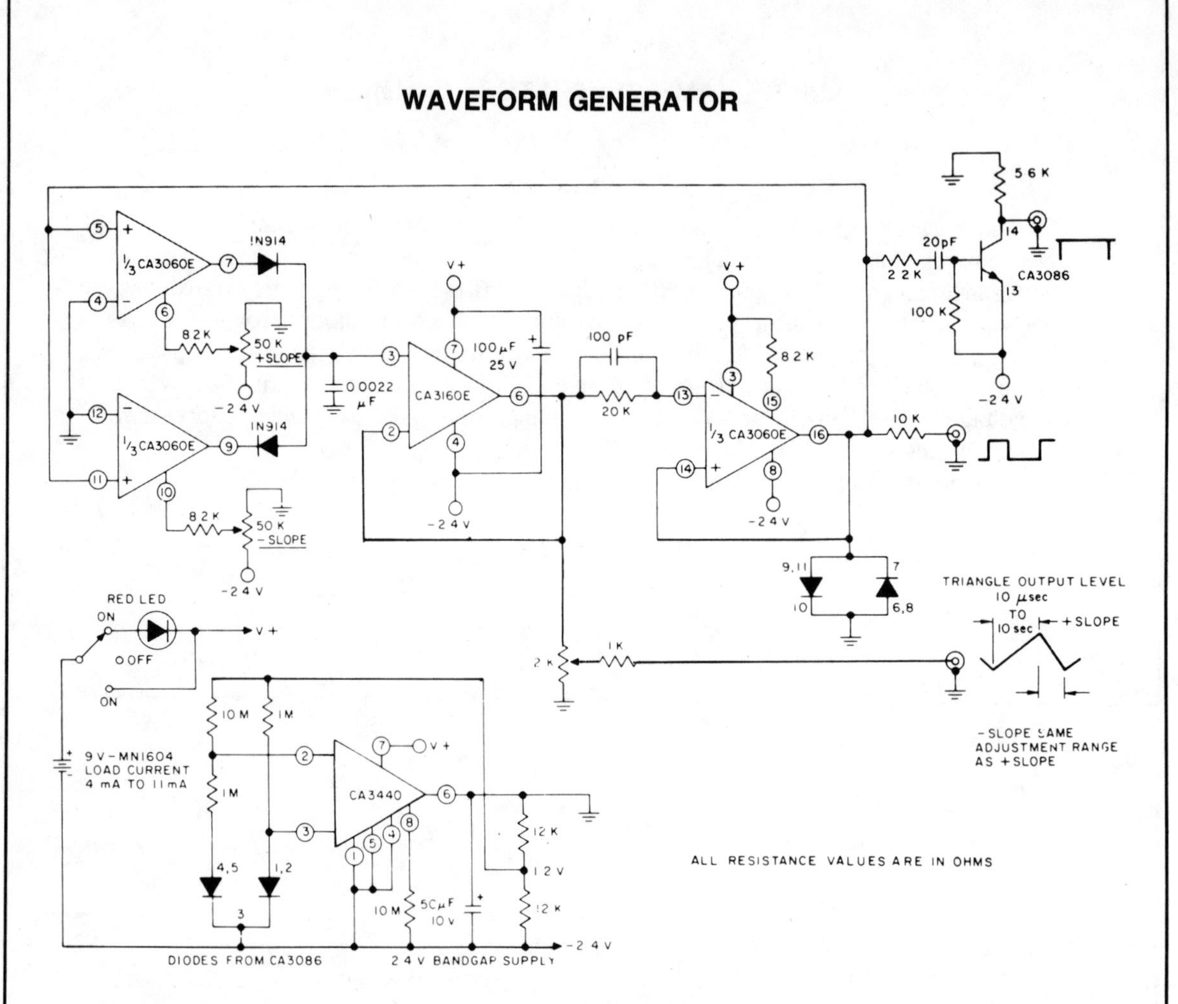

GENERAL ELECTRIC/RCA *Fig. 40-17*

Circuit Notes

The circuit uses a CA3060 triple OTA (two units serve as switched current generators controlled by a third amplifier). A CA3160 BiMOS op amp serves as a voltage follower to buffer the 0.0022 μF integrating capacitor. The circuit has an adjustment range of 1,000,000:1 and a timing range of 20 μs to 20 sec. The "ON-OFF" switch actuates an LED that serves as both a pilot light and a low-battery indicator. The LED extends battery life, since it drops battery voltage to the circuit by approximately 1.2 volts, thus reducing supply current.

SINGLE SUPPLY FUNCTION GENERATOR

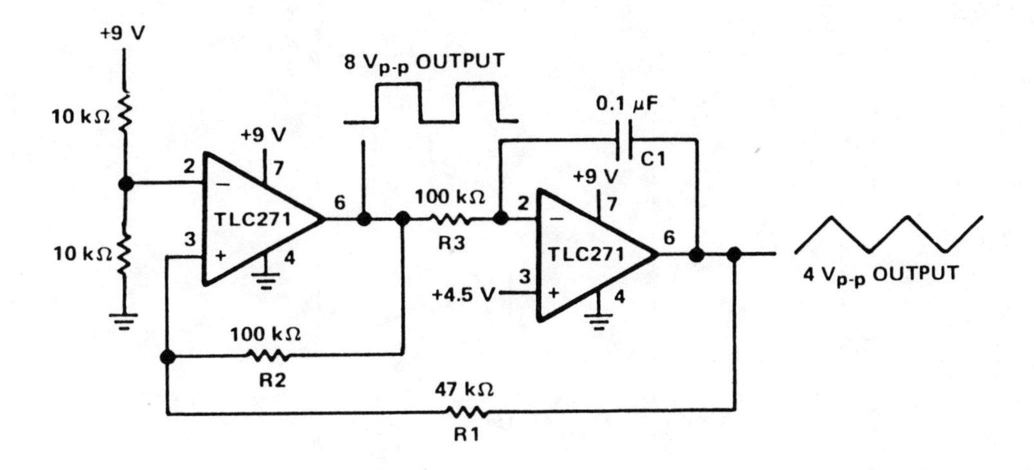

TEXAS INSTRUMENTS

Fig. 40-18

Circuit Notes

The circuit has both square-wave and triangle-wave output. The left section is similar in function to a comparator circuit that uses positive feedback for hysteresis. The inverting input is biased at one-half the V_{CC} voltage by resistors R4 and R5. The output is fed back to the non-inverting input of the first stage to control the frequency. The amplitude of the square wave is the output swing of the first stage, which is 8 V peak-to-peak. The second stage is basically an op amp integrator. The resistor R3 is the input element and capacitor C1 is the feedback element. The ratio R1/R2 sets the amplitude of the triangle wave, as referenced to the square-wave output. For both waveforms, the frequency of oscillation can be determined by the equation:

$$fo = \frac{1}{4R3C1} \quad \frac{R2}{R1}$$

The output frequency is approximately 50 Hz with the given components.

PRECISE WAVE GENERATOR

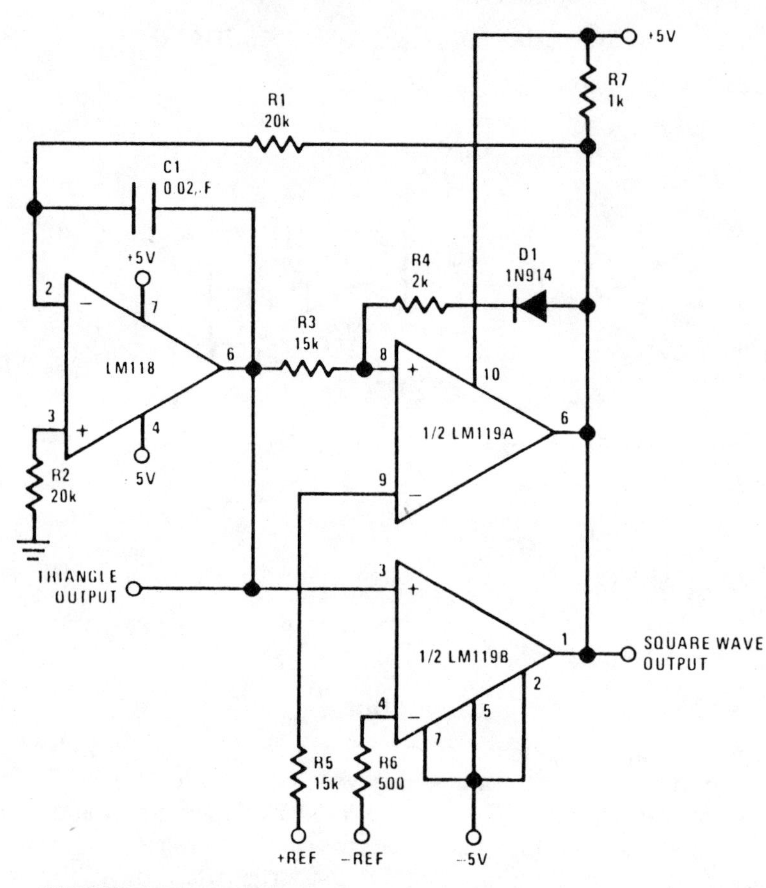

NATIONAL SEMICONDUCTOR

Fig. 40-19

Circuit Notes

The positive and negative peak amplitude is controllable to an accuracy of about ± 0.01 V by a dc input. Also, the output frequency and symmetry are easily adjustable. The oscillator consists of an integrator and two comparators—one comparator sets the positive peak and the other the negative peak of the triangle wave. If R1 is replaced by a potentiometer, the frequency can be varied over at least a 10 to 1 range without affecting amplitude. Symmetry is also adjustable by connecting a 50 kΩ resistor from the inverting input of the LM118 to the arm of the 1 kΩ potentiometer. The ends of the potentiometer are connected across the supplies. Current for the resistor either adds or subtracts from the current through R1, changing the ramp time.

41
Games

The sources of the following circuits are contained in the Sources section beginning on page 694. The figure number contained in the box of each circuit correlates to the source entry in the Sources section.

ELECTRONIC ROULETTE

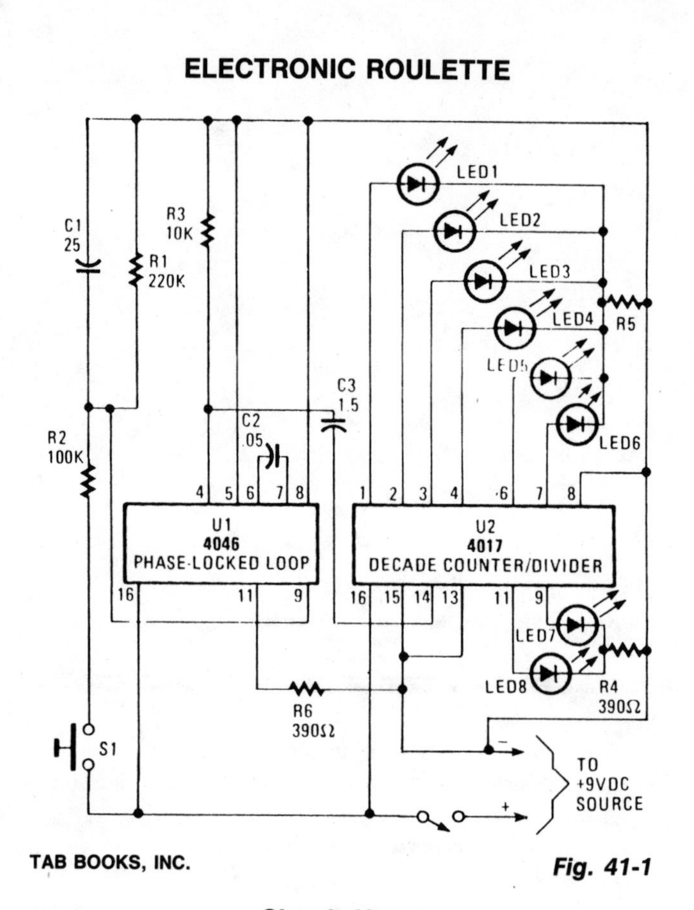

TAB BOOKS, INC.

Fig. 41-1

Circuit Notes

U1 (a 4046 PLL containing a voltage controlled oscillator or VCO, two phase comparators, a source follower, and a Zener diode) is used to produce a low-frequency, pulsed output of about 40 Hz. The VCO's frequency range is determined by R6 and C2, which can be altered by varying the voltage at pin 9. The rising voltage causes the frequency to rise from zero to threshold and remain at that frequency as long as S1 is closed. When S1 is opened, C1 discharges slowly through R1 to ground and the voltage falls toward zero. That produces a decreasing pulse rate. The output of U1 at pin 4 is connected to the clock input of U2 (a 4017 decade decoder/driver) at pin 14 via C3. U2 sequentially advances through each of its ten outputs (0 to 9)—pins 1 to 7, and 9 to 11—with each input pulse. As each output goes high, its associated LED is lighted, and extinguished when it returns to the low state. Only eight outputs are used in the circuit, giving two numbers to the spinner of the house. The circuit can be set up so that the LED's lights sequence or you can use some staggered combination; the LEDs grouped in a straight line or a circle.

LIE DETECTOR

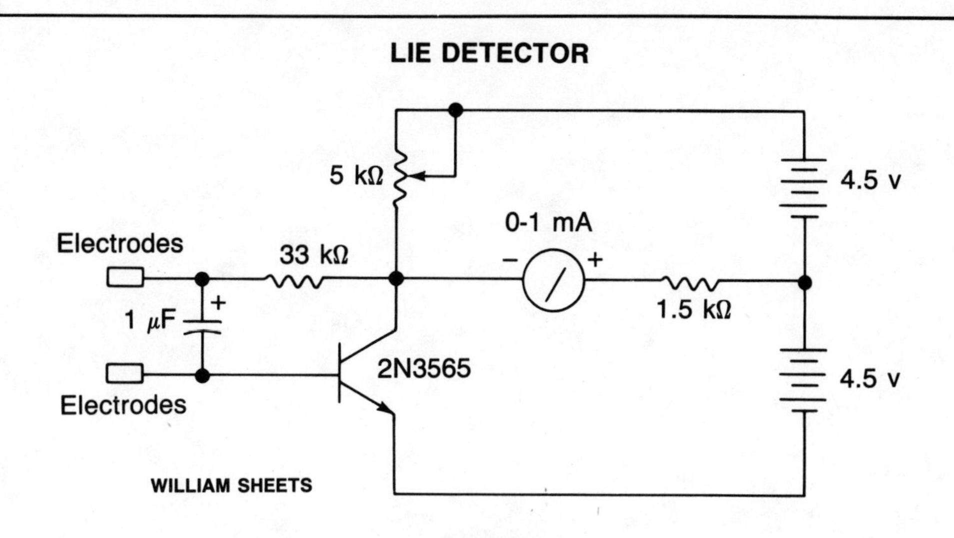

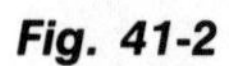

Fig. 41-2

Circuit Notes

The two probes shown are held in the hands and the skin resistance applies bias to the transistor. The 5 k ohm pot is set for zero deflection on the meter. When the "subject" is embarrassed or lies, sweating on the hands takes place, increasing the bias to the transistor and upsetting the bridge balance.

42
Gas and Smoke Detectors

The sources of the following circuits are contained in the Sources section beginning on page 694. The figure number contained in the box of each circuit correlates to the source entry in the Sources section.

GAS AND VAPOR DETECTOR

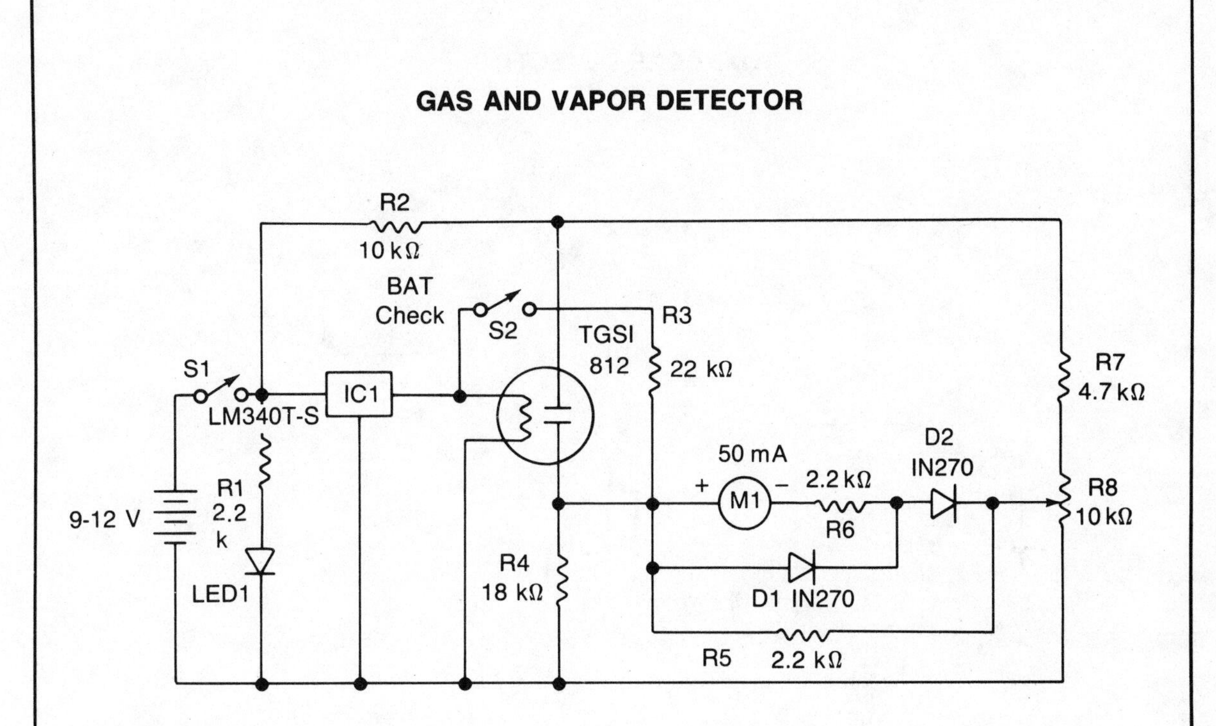

WILLIAM SHEETS *Fig. 42-1*

Circuit Notes

The power drain is approximately 150 mA. IC1 provides a regulated 5-volt supply for the filament heater of the sensor. The gas sensitive element is connected as one arm of a resistance bridge consisting of R4, R7, R8 and the meter M1 with its associated resistors R5 and R6. The bridge can be balanced by adjusting R8 so that no current flows through the meter. A change in the sensor's resistance, caused by detection of noxious gases, will unbalance the circuit and deflect the meter. Diodes D1, D2 and resistor R5 protect the meter from overload while R6 determines overall sensitivity. R2 limits the current through the sensor; R1 and LED1 indicate that the circuit is working, so that you do not drain the battery leaving the unit on inadvertently; R3 and S2 give a battery level check.

TOXIC GAS DETECTOR

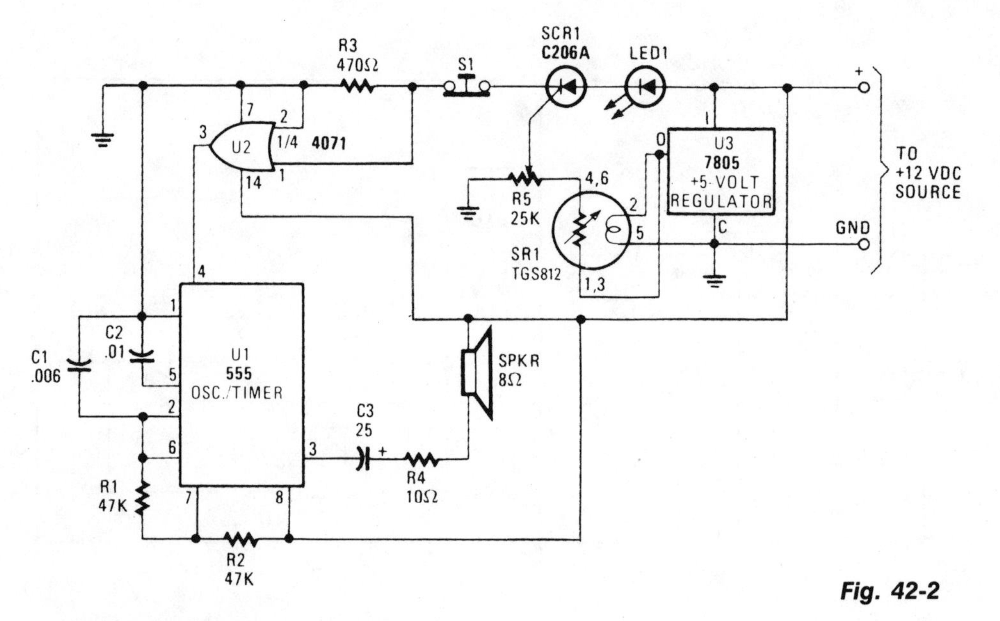

HANDS-ON ELECTRONICS

Fig. 42-2

Circuit Notes

The major device in the circuit is SR1 (a TGS812 toxic-gas sensor manufactured by Figaro Engineering Inc.) The gas-sensitive semiconductor (acting like a variable resistor in the presence of toxic gas) decreases in electrical resistance when gaseous toxins are absorbed from the sensor surface. A 25,000 ohm potentiometer (R5) connected to the sensor serves as a load, voltage-dividing network, and sensitivity control and has its center tap connected to the gate of SCR1. When toxic fumes come in contact with the sensor, decreasing its electrical resistance, current flows through the load (potentiometer R5). The voltage developed across the wiper of R5, which is connected to the gate of SCR1, triggers the SCR into conduction. With SCR1 now conducting, pin 1-volt supply for the semiconductor elements of the TGS812 in spite of the suggested 10 volts, thus reducing the standby current. A 7805 regulator is used to meet the 5-volt requirement for the heater and semiconductor elements.

GAS ANALYZER

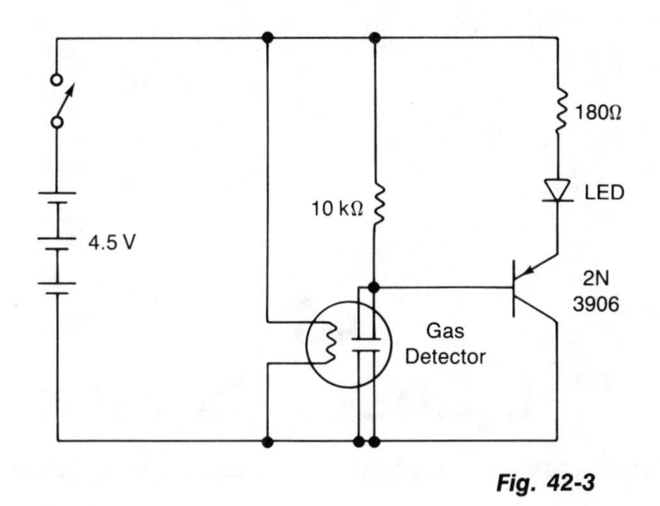

Fig. 42-3

WILLIAM SHEETS

Circuit Notes

The circuit shows a simple yes/no gas detector. Three 1.5-V D cells are used as a power supply, with S1 acting as an on/off switch. The heater is energized directly from the battery, while the electrodes are in series with a 10 k resistor. The voltage across this resistor is monitored by a pnp transistor. When the sensor is in clean air, the resistance between the electrodes is about 40 k, so that only about 0.9 V is dropped across the 10 k resistor. This is insufficient to turn on the transistor, because of the extra 1.6 V required to forward bias the light emitting diode (LED) in series with the emitter. When the sensor comes in contact with contaminated air, the resistance starts to fall, increasing the voltage dropped across the 10 k resistor. When the sensor resistance falls to about 10 k or less, the transistor starts to turn on, current passes through the LED, causing it to emit. The 180 ohm resistor limits the current through the LED to a safe value.

43
Hall Effect Circuits

The sources of the following circuits are contained in the Sources section beginning on page 694. The figure number contained in the box of each circuit correlates to the source entry in the Sources section.

ANGLE OF ROTATION DETECTOR

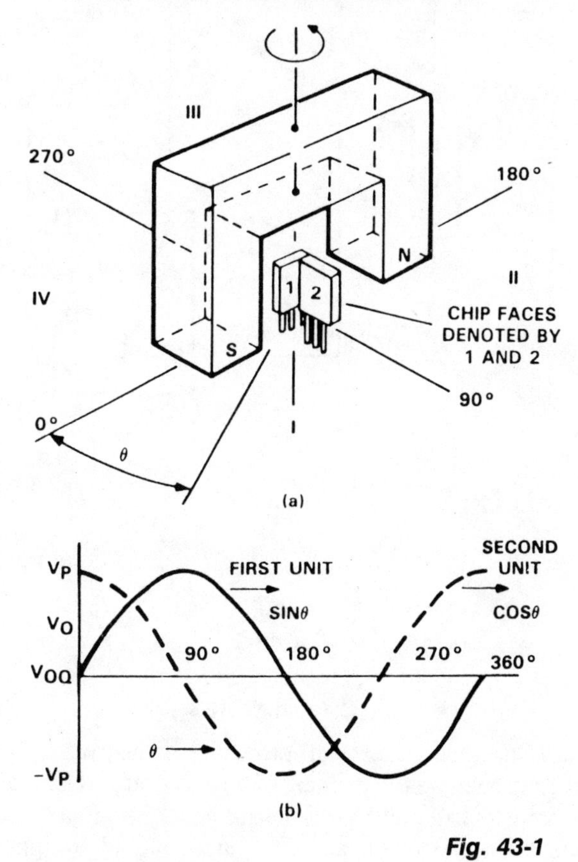

Fig. 43-1

TEXAS INSTRUMENTS

Circuit Notes

The figure shows two TL3103 linear Hall-effect devices used for detecting the angle of rotation. The TL3103s are centered in the gap of a U-shaped permanent magnet. The angle that the south pole makes with the chip face of unit #1 is defined as angle 0. Angle 0 is set to 0° when the chip face of unit #1 is perpendicular to the south pole of the magnet. As the south pole of the magnet sweeps through a 0° to 90° angle, the output of the sensor increases from 0°. Sensor unit #2 decreases from its peak value of +Vp at 0° to a value V_{OQ} at 90°. So, the output of sensor unit #1 is a sine function of 0 and the output of unit #2 is a cosine function of 0 as shown. Thus, the first sensor yields the angle of rotation and the second sensor indicates the quadrant location.

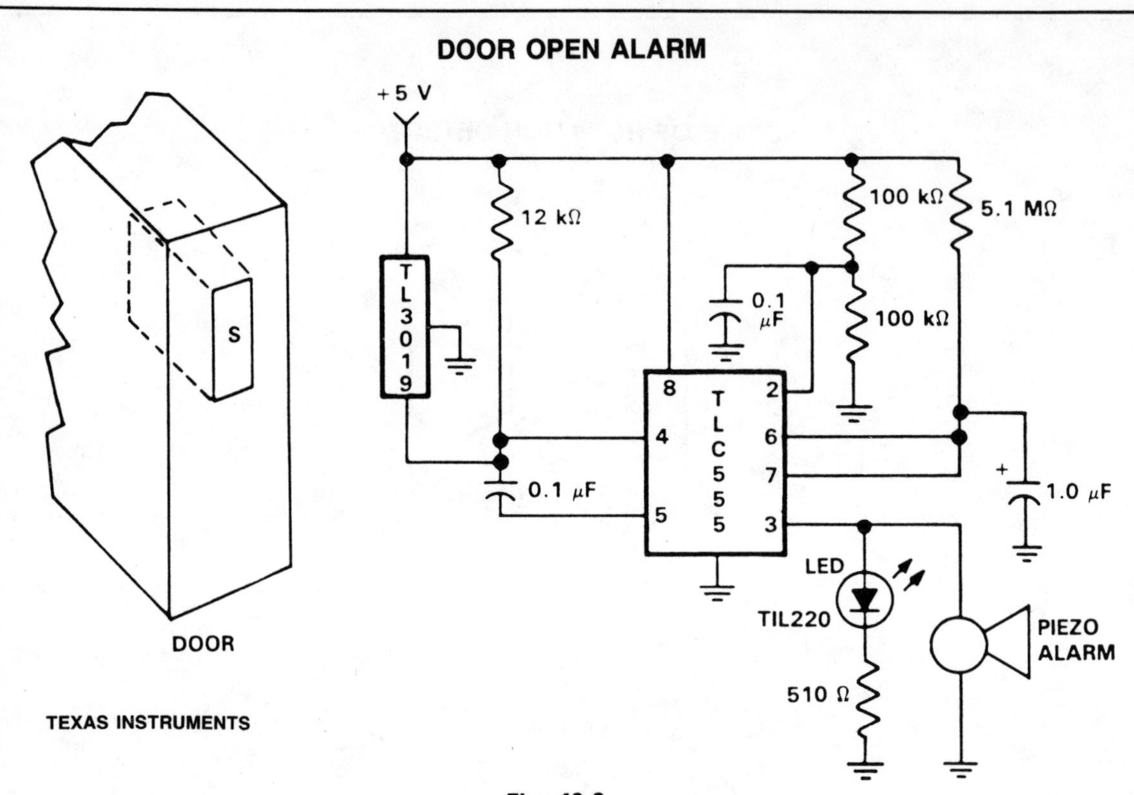

Fig. 43-2

Circuit Notes

Door open alarms are used chiefly in automotive, industrial, and appliance applications. This type of circuit can sense the opening of a refrigerator door. When the door opens, a triac could be activated to control the inside light. The figure shows a door position alarm. When the door is opened, an LED turns on and the piezo alarm sounds for approximately 5 seconds. This circuit uses a TL3019 Hall-effect device for the door sensor. This normally open switch is located in the door frame. The magnet is mounted in the door. When the door is in the closed position, the TL3019 output goes to logic low, and remains low until the door is opened. This design consists of a TLC555 monostable timer circuit. The 1 μF capacitor and 5.1 M ohm resistor on pins 6 and 7 set the monostable RC time constant. These values allow the LED and piezo alarm to remain on about 5 seconds when triggered. One unusual aspect of this circuit is the method of triggering. Usually a 555 timer circuit is triggered by taking the trigger, pin 2, low which produces a high at the output, pin 3. In this configuration with the door in the closed position, the TL3019 output is held low. The trigger, pin 2, is connected to ½ the supply voltage V_{CC}. When the door opens, a positive high pulse is applied to control pin 5 through a 0.1 μF capacitor and also to reset pin 4. This starts the timing cycle. Both the piezo alarm and the LED visual indicator are activated.

44
Humidity Sensors

The sources of the following circuits are contained in the Sources section beginning on page 694. The figure number contained in the box of each circuit correlates to the source entry in the Sources section.

Relative Humidity Sensor Signal Conditioner

RELATIVE HUMIDITY SENSOR SIGNAL CONDITIONER

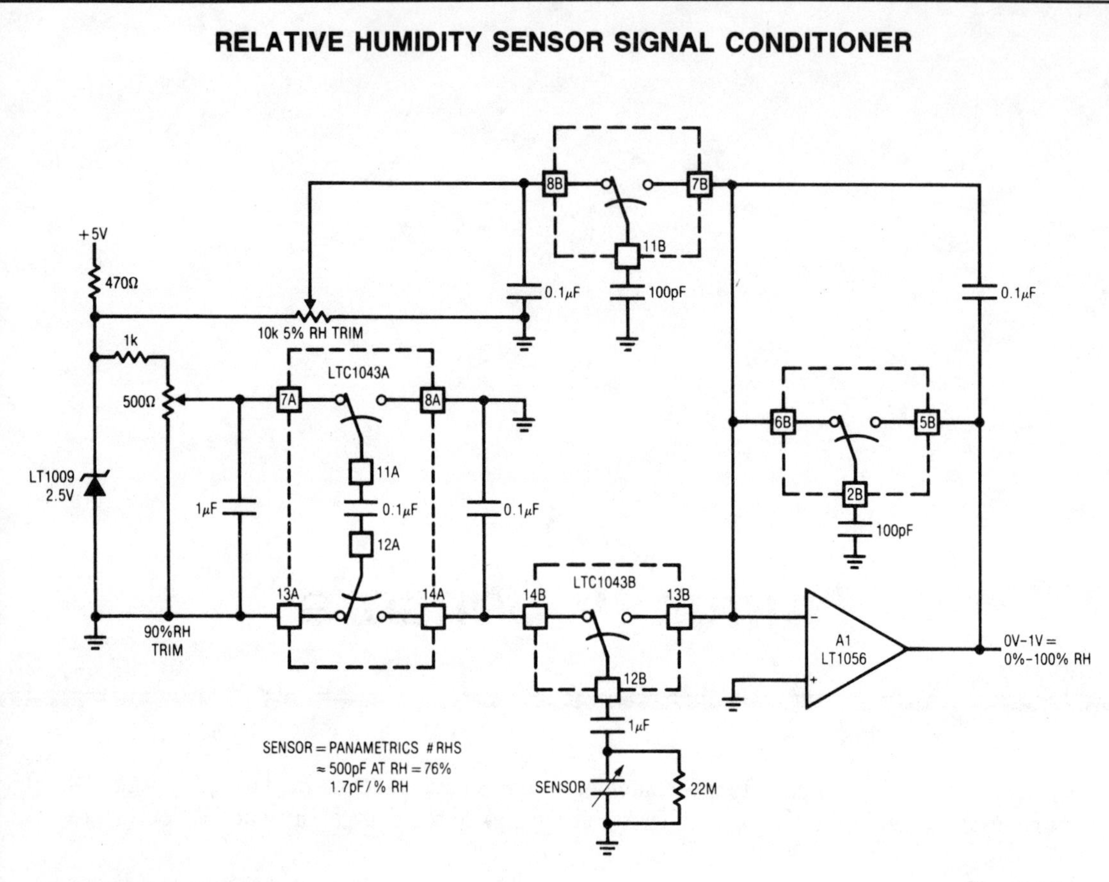

LINEAR TECHNOLOGY CORPORATION *Fig. 44-1*

Circuit Notes

This circuit combines two LTC1043s with a based humidity transducer in a simple charge-pump based circuit. The sensor specified has a nominal 400 pF capacitance at RH = 76%, with a slope of 1.7 pF/% RH. The average voltage across this device must be zero. This provision prevents deleterious electrochemical migration in the sensor. The LTC1043A inverts a resistively scaled portion of the LT1009 reference, generating a negative potential at pin 14A. The LTC1043B alternately charges and discharges the humidity sensor via pins 12B, 13B, and 14B. With 14B and 12B connected, the sensor charges via the 1 μF unit to the negative potential at pin 14A. When the 14B-12B pair

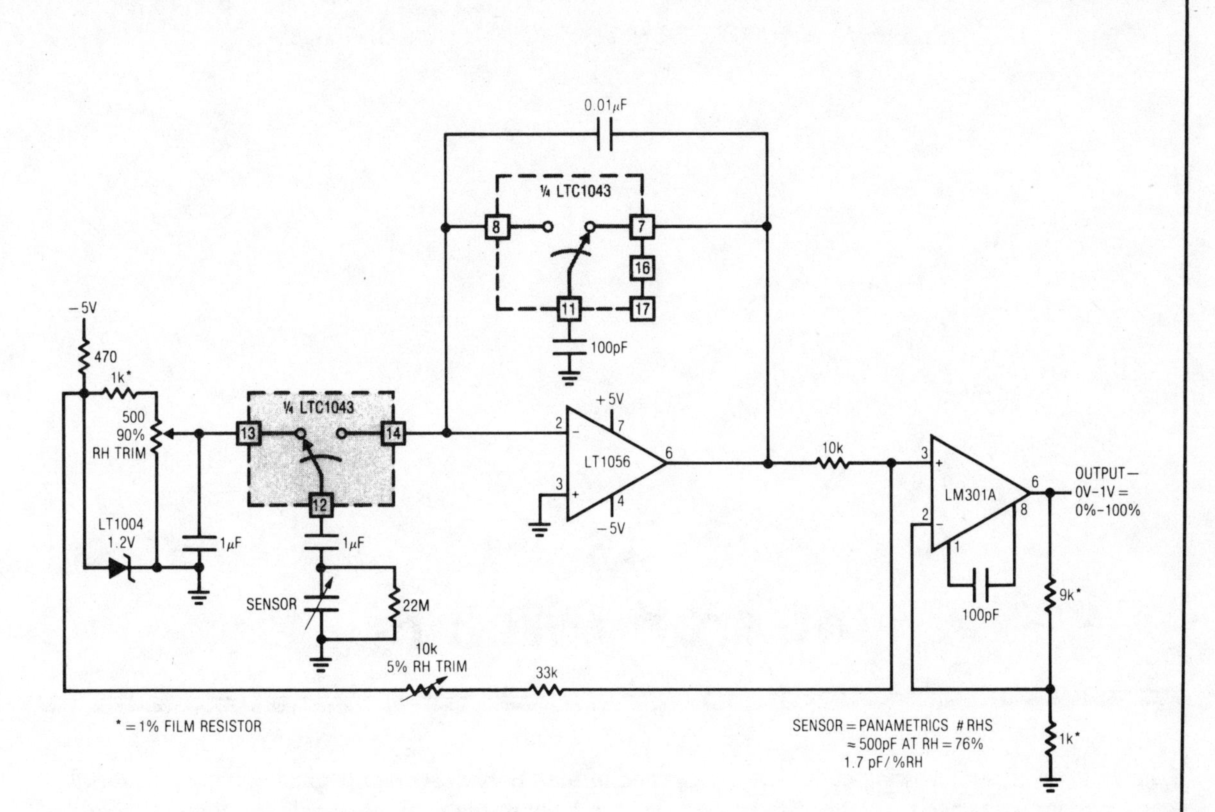

opens, 12B is connected to A1's summing point via 13B. The sensor now discharges into the summing point through the 1 μF capacitor. Since the charge voltage is fixed, the average current into the summing point is determined by the sensor's humidity related value. The 1 μF value ac couples the sensor to the charge-discharge path, maintaining the required zero average voltage across the device. The 22M resistor prevents accumulation of charge, which would stop current flow. The average current into A1's summing point is balanced by packets of charge delivered by the switched-capacitor gives A1 an integrator-like response, and its output is dc.

45
Infrared Circuits

The sources of the following circuits are contained in the Sources section beginning on page 694. The figure number contained in the box of each circuit correlates to the source entry in the Sources section.

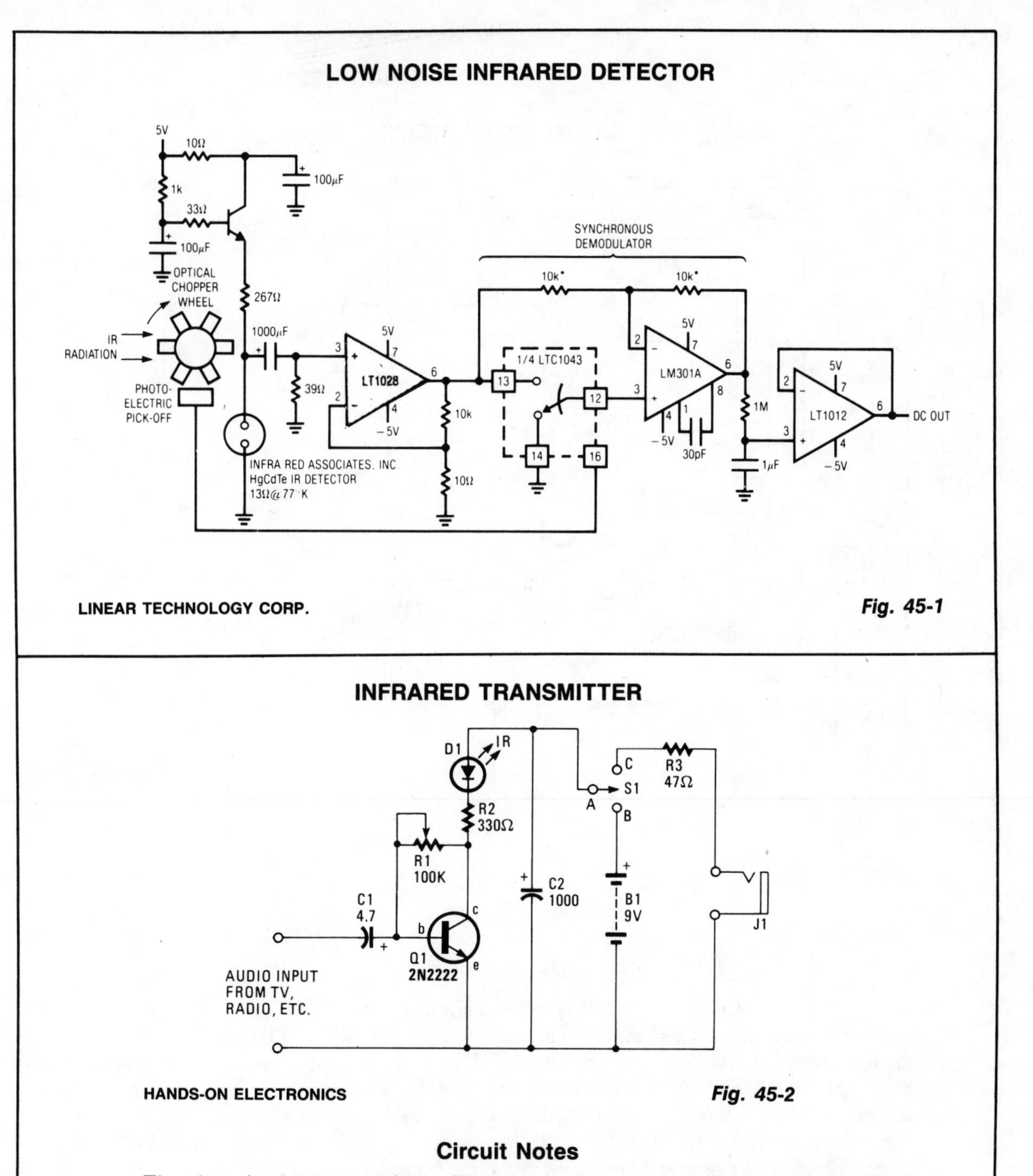

LINEAR TECHNOLOGY CORP. *Fig. 45-1*

HANDS-ON ELECTRONICS *Fig. 45-2*

Circuit Notes

The ultra-simple one-transistor, IR transmitter shown is designed to transmit the sound from any 8 or 16 ohm audio source, such as a TV, radio, or tape recorder on an infrared beam of light.

INFRARED TRANSMITTER

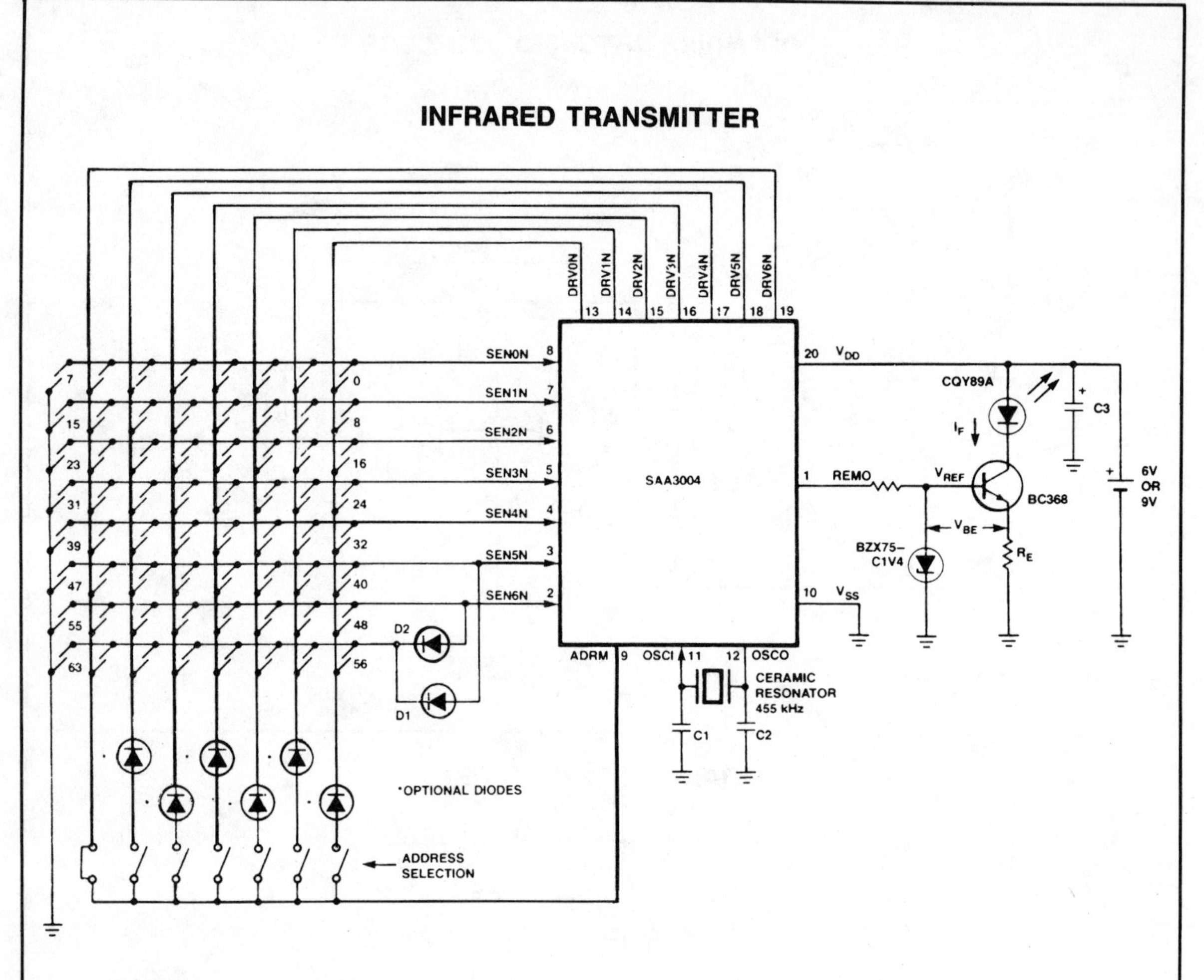

SIGNETICS

Fig. 45-3

Circuit Notes

The transmitter keyboard is arranged as a scanned matrix. The matrix consists of 7 driver outputs and 7 sense inputs. The driver outputs DRVON to DRV6N are open-drain n-channel transistors and they are conductive in the stand-by mode. The 7 sense inputs (SENON to SEN6N) enable the generation of 56 command codes. With 2 external diodes all 64 commands are addressable. The sense inputs have p-channel pull-up transistors, so that they are HIGH until they are pulled LOW by connecting them to an output via a key depression to initiate a code transmission.

INVISIBLE INFRARED PULSED LASER RIFLE

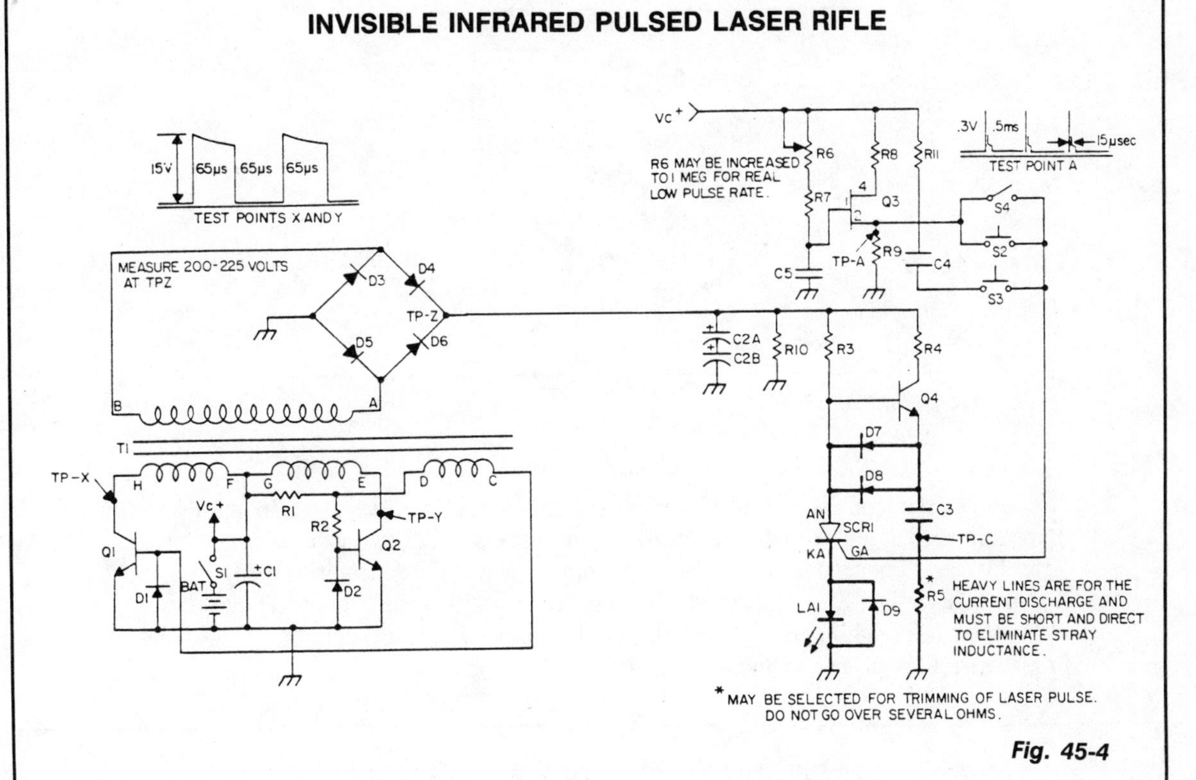

Fig. 45-4

TAB BOOKS INC.

Circuit Notes

The device generates an adjustable frequency of low to medium powered IR pulses of invisible energy and must be treated with care.

The portable battery pack is stepped up to 200 to 300 volts by the inverter circuit consisting of Q1, Q2, and T1. Q1 conducts until saturated, at which time, the base no longer can sustain it in an "on" state and Q1 turns "off," causing the magnetic field in its collector winding to collapse thus producing a voltage or proper phase in the base drive winding that turns on Q2 until saturated, repeating the above sequence of events in an "on/off" action. The diodes connected at the bases provide a return path for the base drive current. The stepped up squarewave voltage on the secondary of T1 is rectified and integrated on C2.

INFRARED RECEIVER

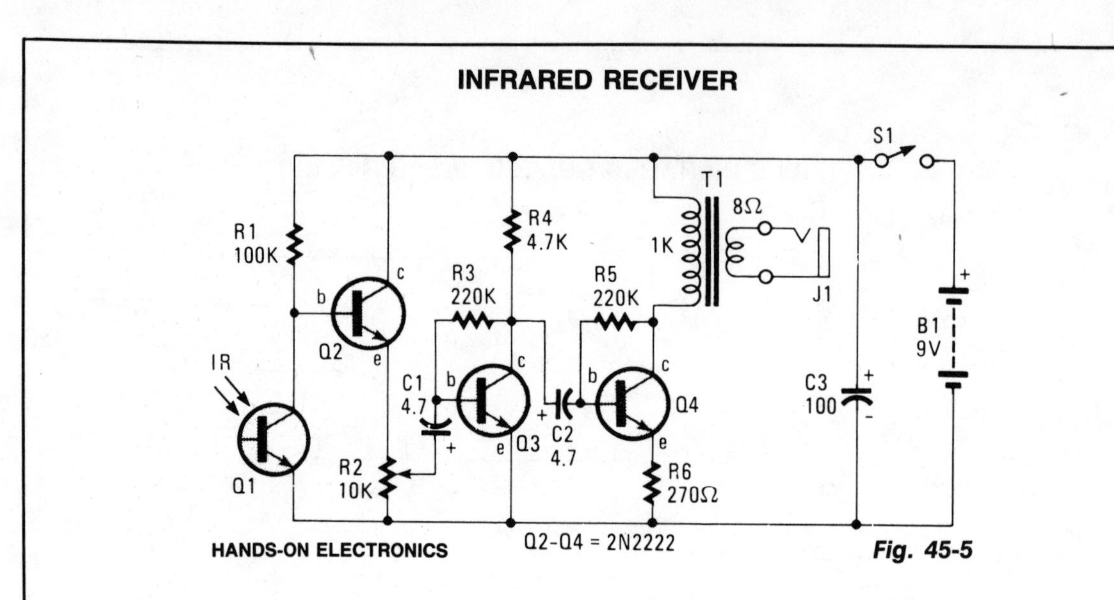

HANDS-ON ELECTRONICS

Fig. 45-5

Circuit Notes

The circuit consists of Q1—a phototransistor that responds to an intensity of amplitude-modulated IR light source—and a three-stage, high-gain audio amplifier. Transformer T1 is used to match the output impedance of the receiver to today's popular low-impedance (low-Z) headphones; but if a set of 1000-2000 ohm, magnetic (not crystal), high-impedance (high-Z) phones are to be used, remove T1 and connect the high-Z phones in place of T1's primary winding—the 1000-ohm winding.

PULSED INFRARED DIODE EMITTER DRIVE

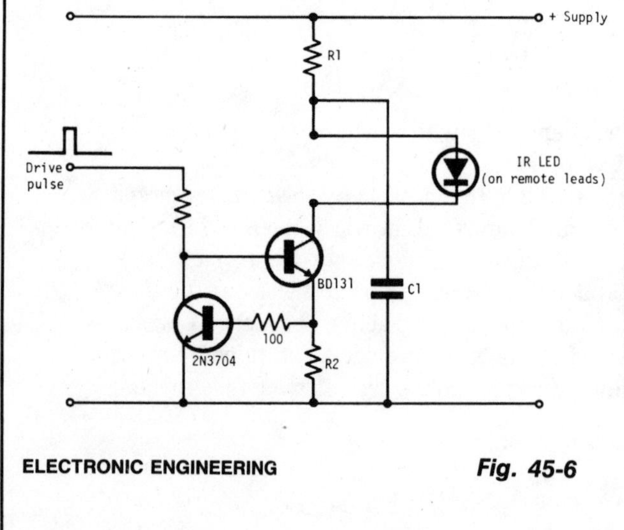

Circuit Notes

Q1 and Q2 form a constant current drive defined by R2. (I approximates to the reciprocal of R2 in the circuit shown for values of I greater than 1 amp). The pulse current is drawn from C1 which is recharged during the time between pulses via R1. The value of C1 is determined from the duration and magnitude of the peak current required, and the time constant R1 C1 is determined from the duration between pulses.

ELECTRONIC ENGINEERING

Fig. 45-6

46

Instrumentation Amplifiers

The sources of the following circuits are contained in the Sources section beginning on page 694. The figure number contained in the box of each circuit correlates to the source entry in the Sources section.

INSTRUMENT PREAMP

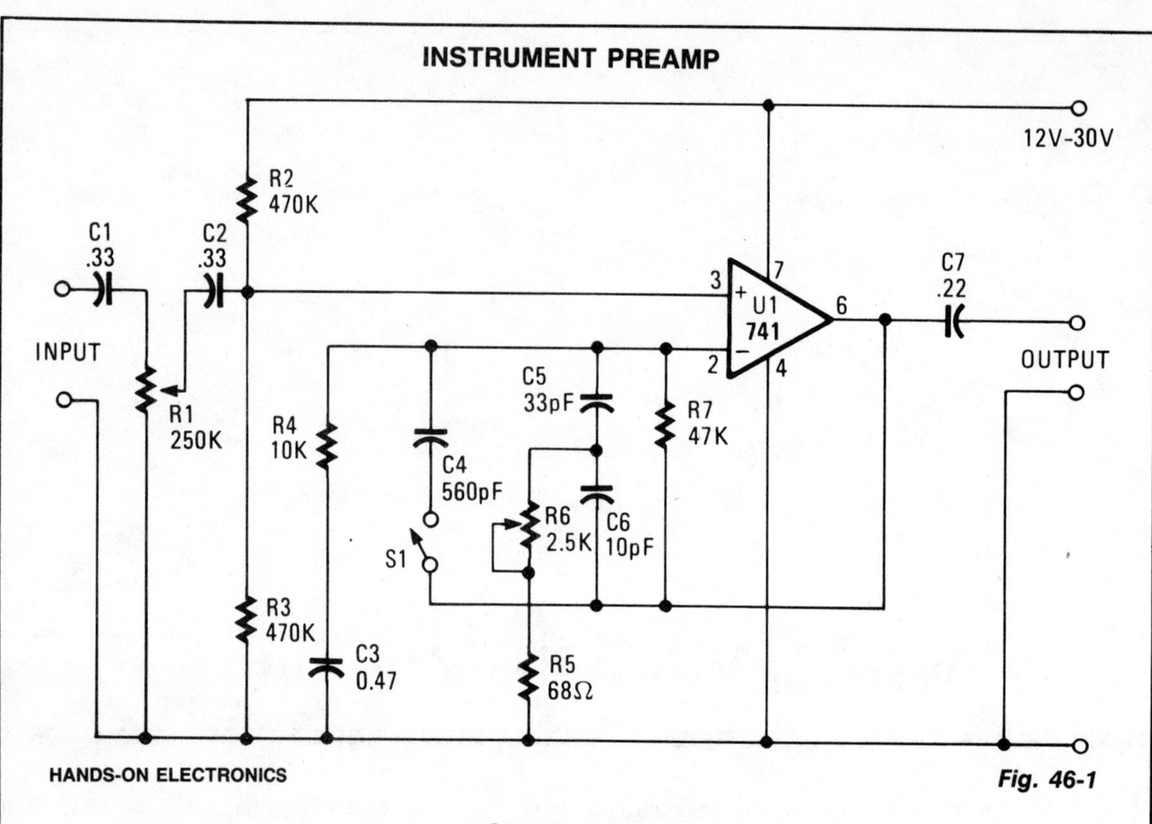

HANDS-ON ELECTRONICS

Fig. 46-1

Circuit Notes

The input impedance is the value of potentiometer R1. If your instrument has extra-deep bass, change capacitor C1 to 0.5 μF. What appears to be an extra part in the feedback loop is a brightening tone control. The basic feedback from the op amp's output (pin 6) to the inverting input (pin 2) consists of resistor R7, and the series connection of resistor R4 and capacitor C3, which produce a voltage gain of almost 5 (almost 14 dB). That should be more extra oomph than usually needed. If the circuit is somewhat short on bass response, increase the value of capacitor C3 to 1 to 10 μF. Start with 1 μF and increase the value until you get the bass effect you want.

INSTRUMENTATION AMPLIFIER WITH ±100 VOLT COMMON MODE RANGE

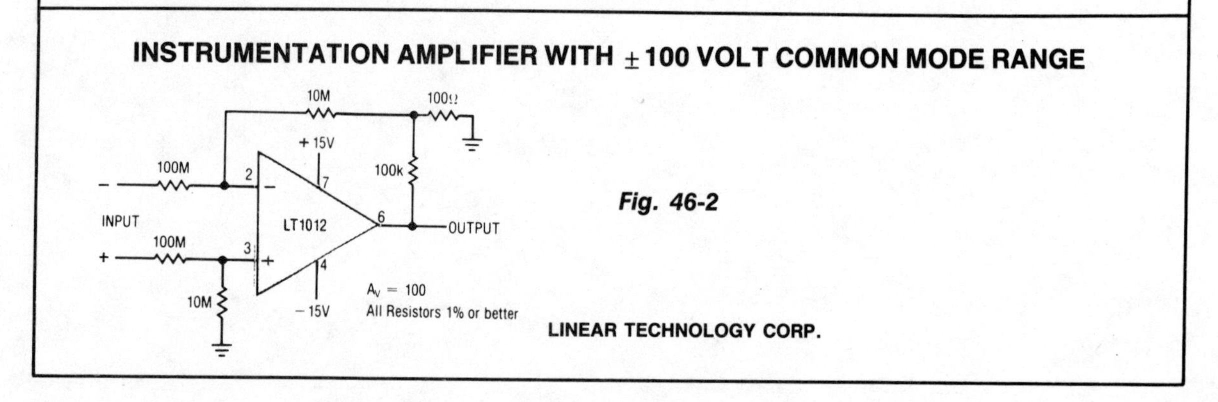

Fig. 46-2

LINEAR TECHNOLOGY CORP.

CURRENT-COLLECTOR HEAD-AMPLIFIER

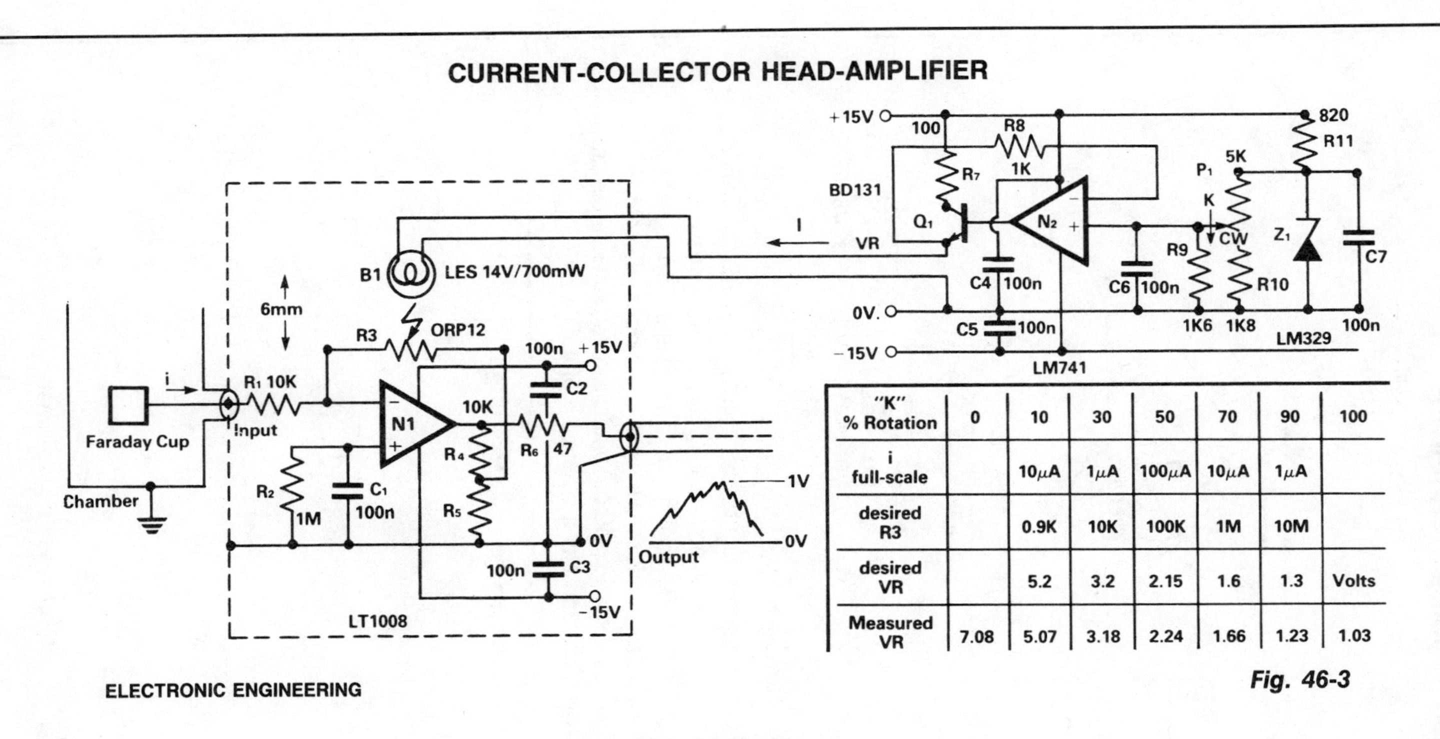

"K" % Rotation	0	10	30	50	70	90	100
i full-scale		10μA	1μA	100μA	10μA	1μA	
desired R3		0.9K	10K	100K	1M	10M	
desired VR		5.2	3.2	2.15	1.6	1.3	Volts
Measured VR	7.08	5.07	3.18	2.24	1.66	1.23	1.03

ELECTRONIC ENGINEERING

Fig. 46-3

Circuit Notes

To amplify small current signals such as from an electron-collector inside a vacuum chamber, it is convenient for reasons of noise and bandwidth to have a "head-amplifier" attached to the chamber. The op amp N1 is a precision bipolar device with extremely low bias current and offset voltage (1) as well as low noise, which allows the 100:1 feedback attenuator R4:R5. The resistance of R3 can be varied from above 10M to below 1R, and so the nominal 0 to 1 V-peak output signal corresponds to input current ranges of 1 nA to 10 μA; this current i enters via the protective resistor R1. Light from the bulb B1 shines on R3, and the filament current I is controlled by the op amp N2.

The reference voltage VR is "shaped" by the resistors R9R10 so as to tailor the bulb and LDR characteristics to the desired current ranges. Thus, rotation of the calibrated knob K gives the appropriate resistance to R3 for the peak-current scale shown.

INSTRUMENTATION METER DRIVER

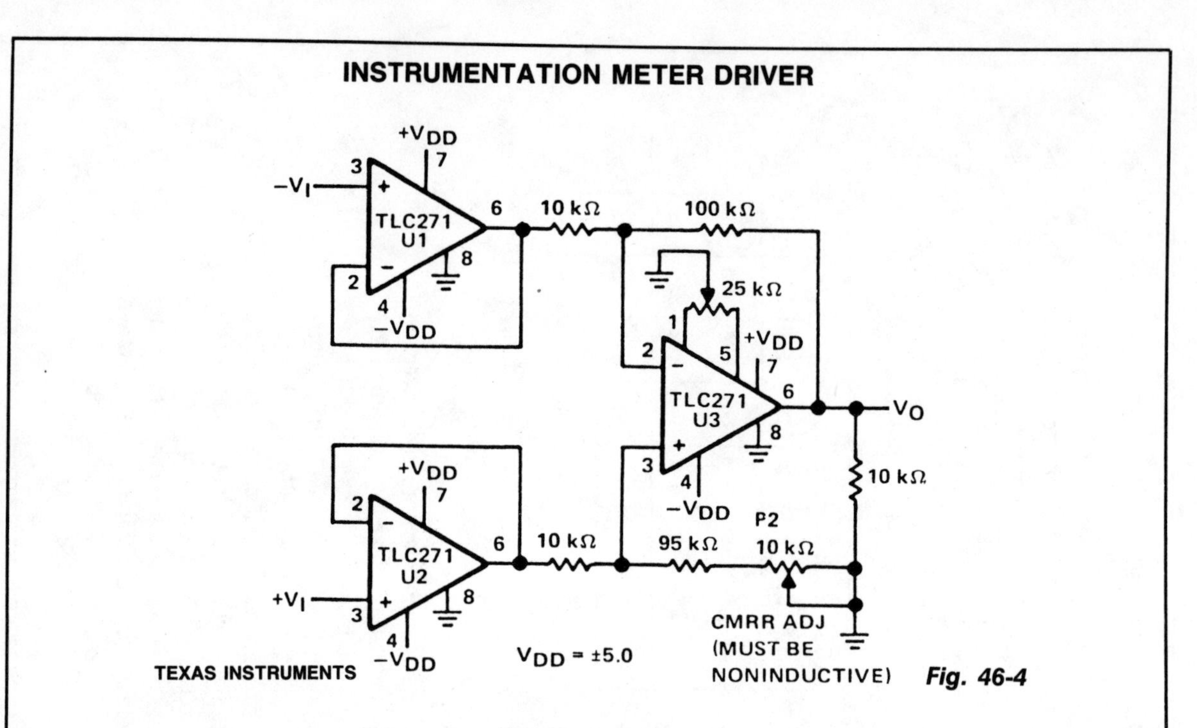

TEXAS INSTRUMENTS *Fig. 46-4*

Circuit Notes

Three op amps U1, U2, and U3 are connected in the basic instrumentation amplifier configuration. Operating from ±5 V, pin 8 of each op amp is connected directly to ground and provides the ac performance desired in this application (high bias mode). P1 is for offset error correction and P2 allows adjustment of the input common mode rejection ratio. The high input impedance allows megohms without loading. The resulting circuit frequency response is 200 kHz at −3 dB and has a slew rate of 4.5 V/μs.

SATURATED STANDARD CELL AMPLIFIER

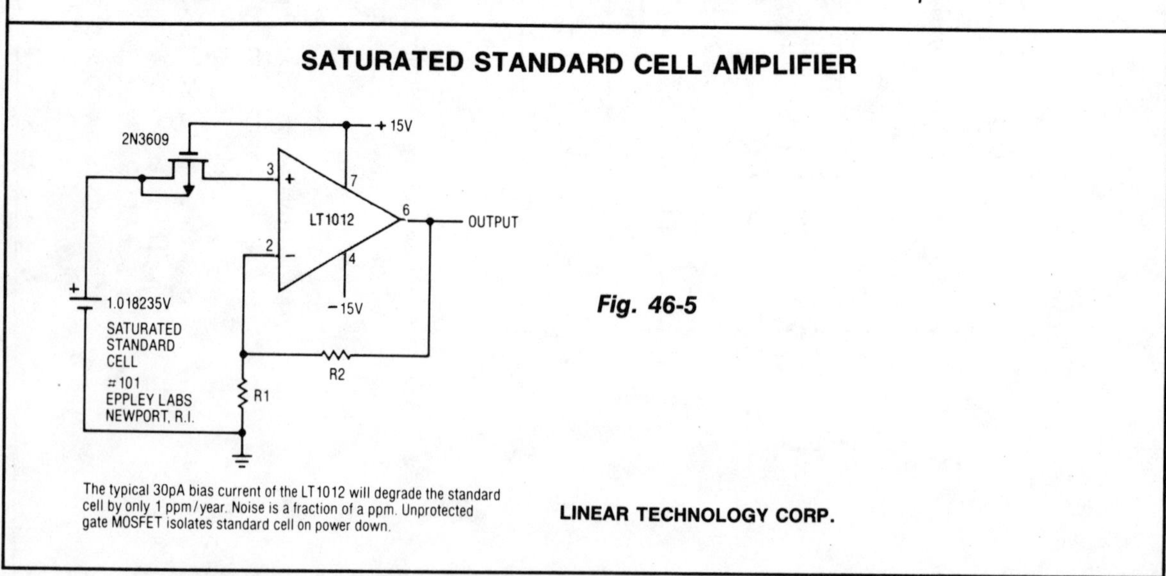

The typical 30pA bias current of the LT1012 will degrade the standard cell by only 1 ppm/year. Noise is a fraction of a ppm. Unprotected gate MOSFET isolates standard cell on power down.

Fig. 46-5

LINEAR TECHNOLOGY CORP.

47

Integrator Circuits

The sources of the following circuits are contained in the Sources section beginning on page 694. The figure number contained in the box of each circuit correlates to the source entry in the Sources section.

IMPROVED NON-INVERTING INTEGRATOR

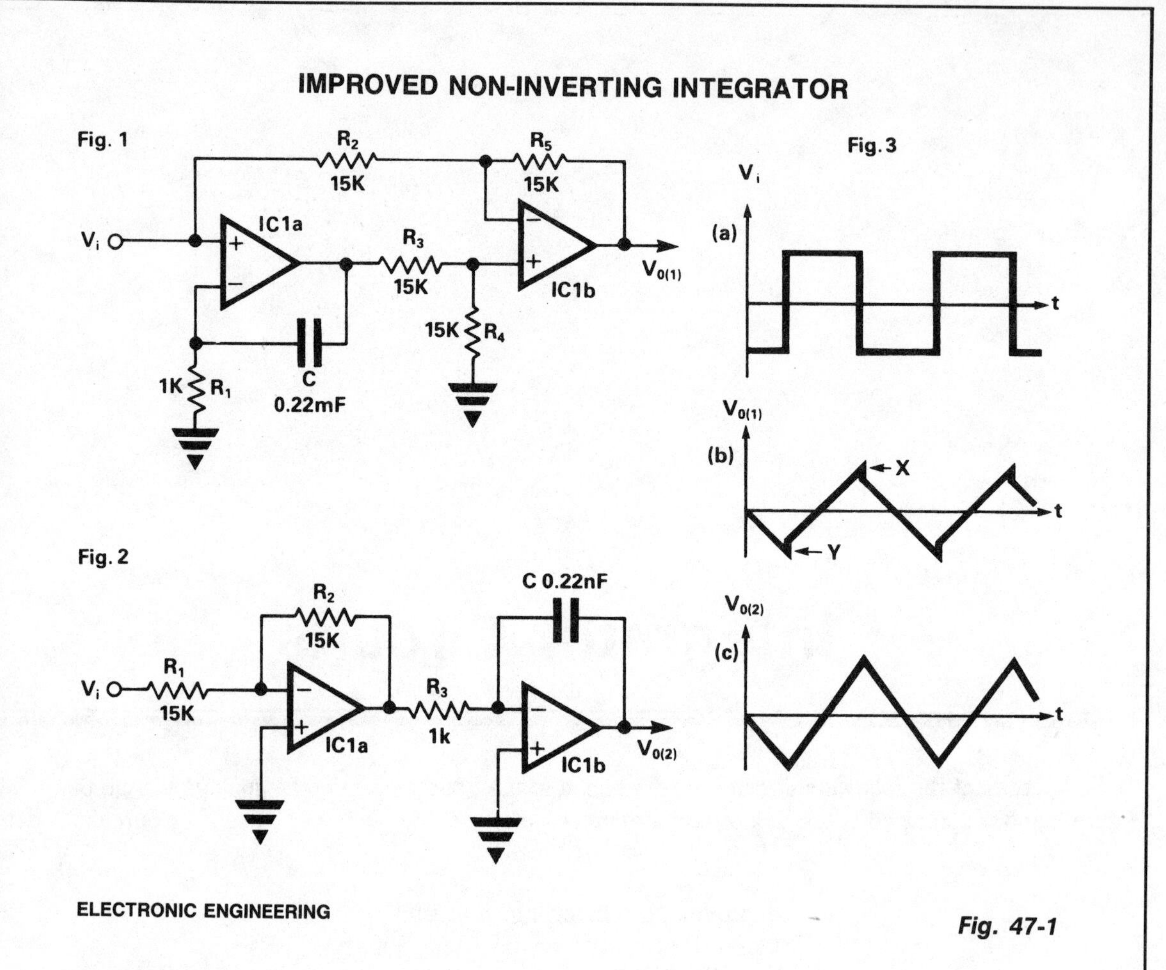

ELECTRONIC ENGINEERING

Fig. 47-1

Circuit Notes

In the circuit in Fig. 1, IC1a produces the integral term required but also has the side effect of producing a proportional term not required, so this term is subtracted by IC1b leaving a pure integral. If the ratio R2/R5 does not exactly match the ratio of R3/R4, the subtraction will not be complete and a small amount of the proportional term will reach the output. The result of this with a squarewave input is shown in Fig. 3a as small steps in the output waveform at points X and Y.

This effect can be completely removed by using the simplified circuit shown in Fig. 2. Here the signal is pre-inverted by IC1a, then fed to a standard inverting integrator IC1b. The result is a non-inverting integrator with the advantage that the unwanted proportional term is never produced, so it does not need to be subtracted.

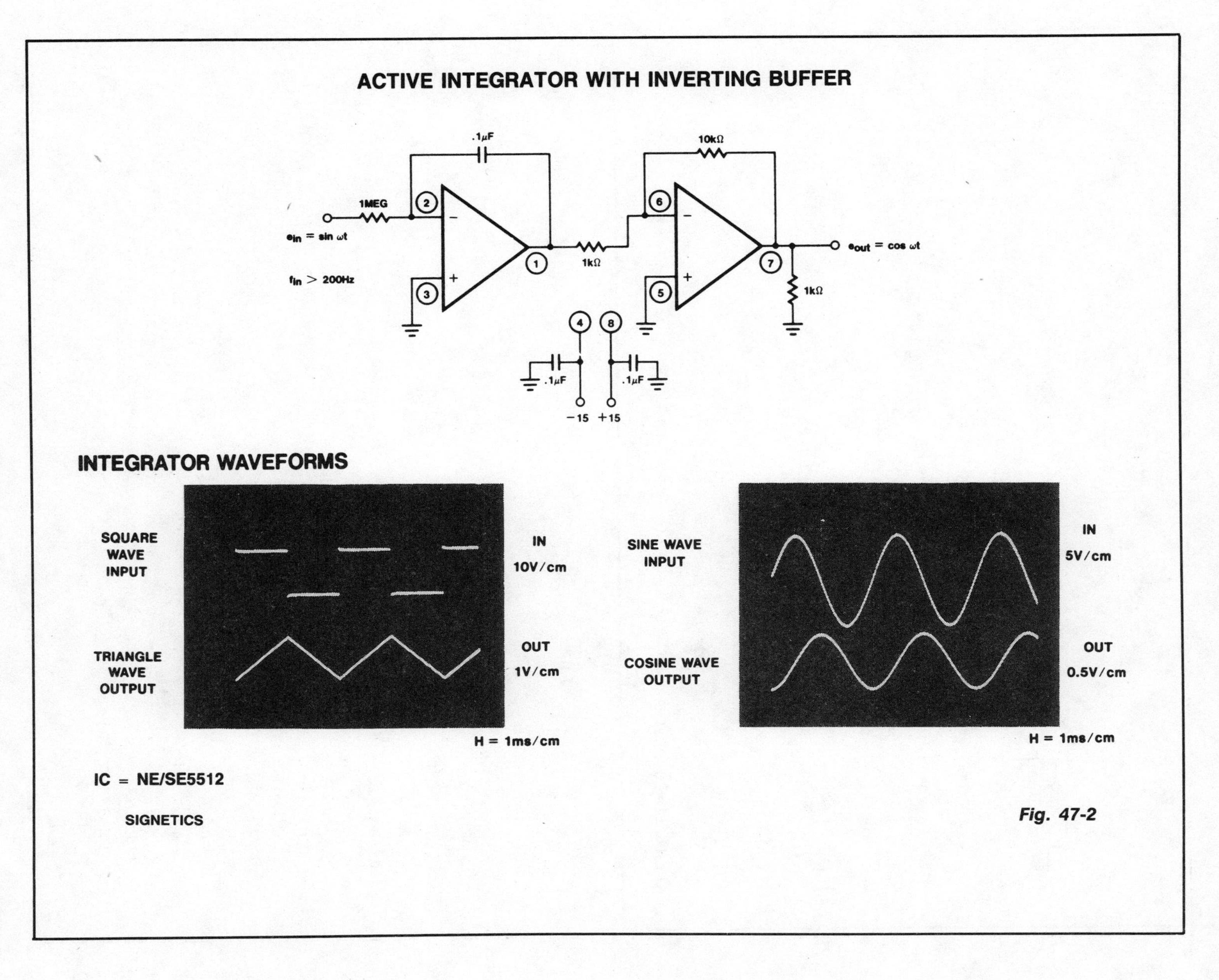
ACTIVE INTEGRATOR WITH INVERTING BUFFER
.1μF
10kΩ
1MEG
e_{in} = sin ωt
f_{in} > 200Hz
1kΩ
e_{out} = cos ωt
1kΩ
.1μF
.1μF
−15 +15
INTEGRATOR WAVEFORMS
SQUARE WAVE INPUT
TRIANGLE WAVE OUTPUT
IN
10V/cm
OUT
1V/cm
H = 1ms/cm
SINE WAVE INPUT
COSINE WAVE OUTPUT
IN
5V/cm
OUT
0.5V/cm
H = 1ms/cm
IC = NE/SE5512
SIGNETICS

Fig. 47-2

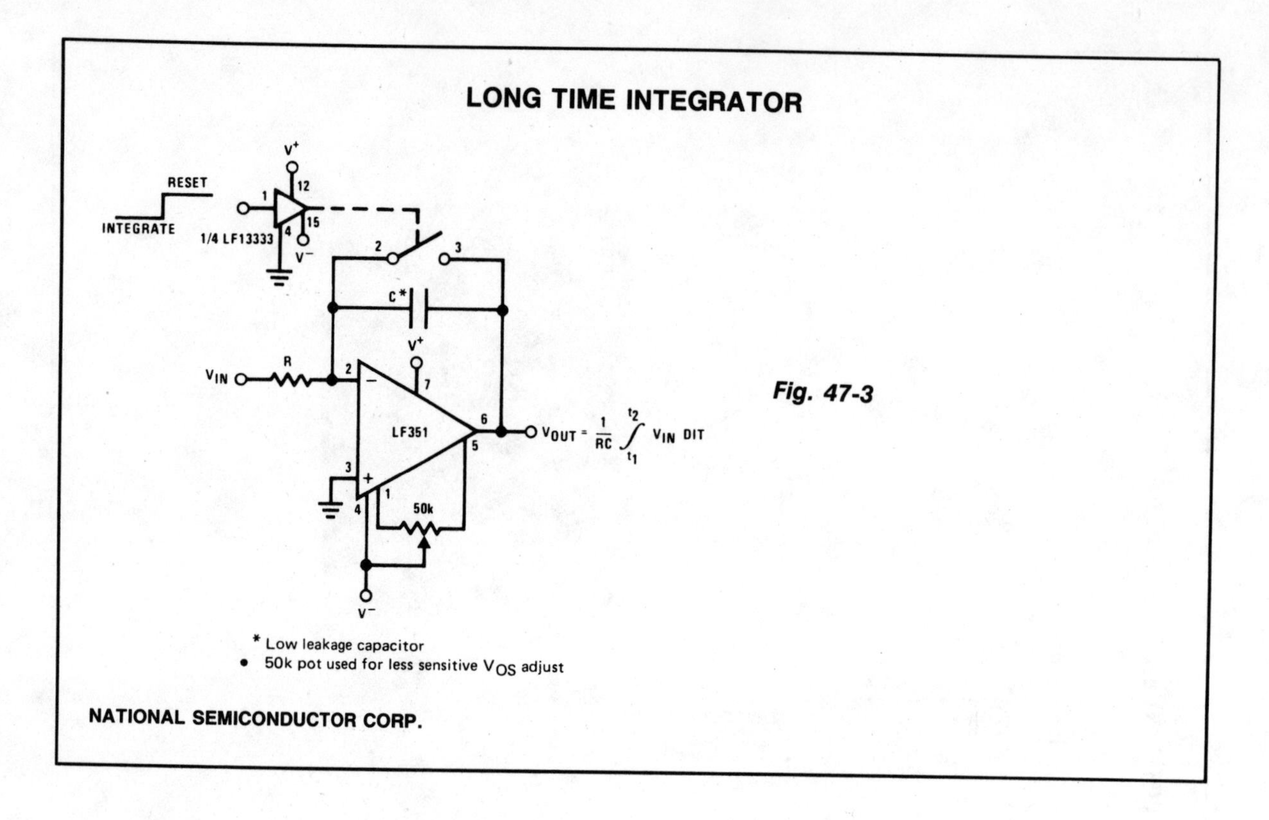
LONG TIME INTEGRATOR
V+
RESET
INTEGRATE
1/4 LF13333
1
12
15
4
V−
2
3
C*
V+
VIN
R
2
7
LF351
6
5
3
1
4
50k
V−
VOUT = 1/RC ∫ VIN DIT
t2
t1
* Low leakage capacitor
• 50k pot used for less sensitive VOS adjust
NATIONAL SEMICONDUCTOR CORP.
Fig. 47-3

48
Intercom Circuits

The sources of the following circuits are contained in the Sources section beginning on page 694. The figure number contained in the box of each circuit correlates to the source entry in the Sources section.

INTERCOM

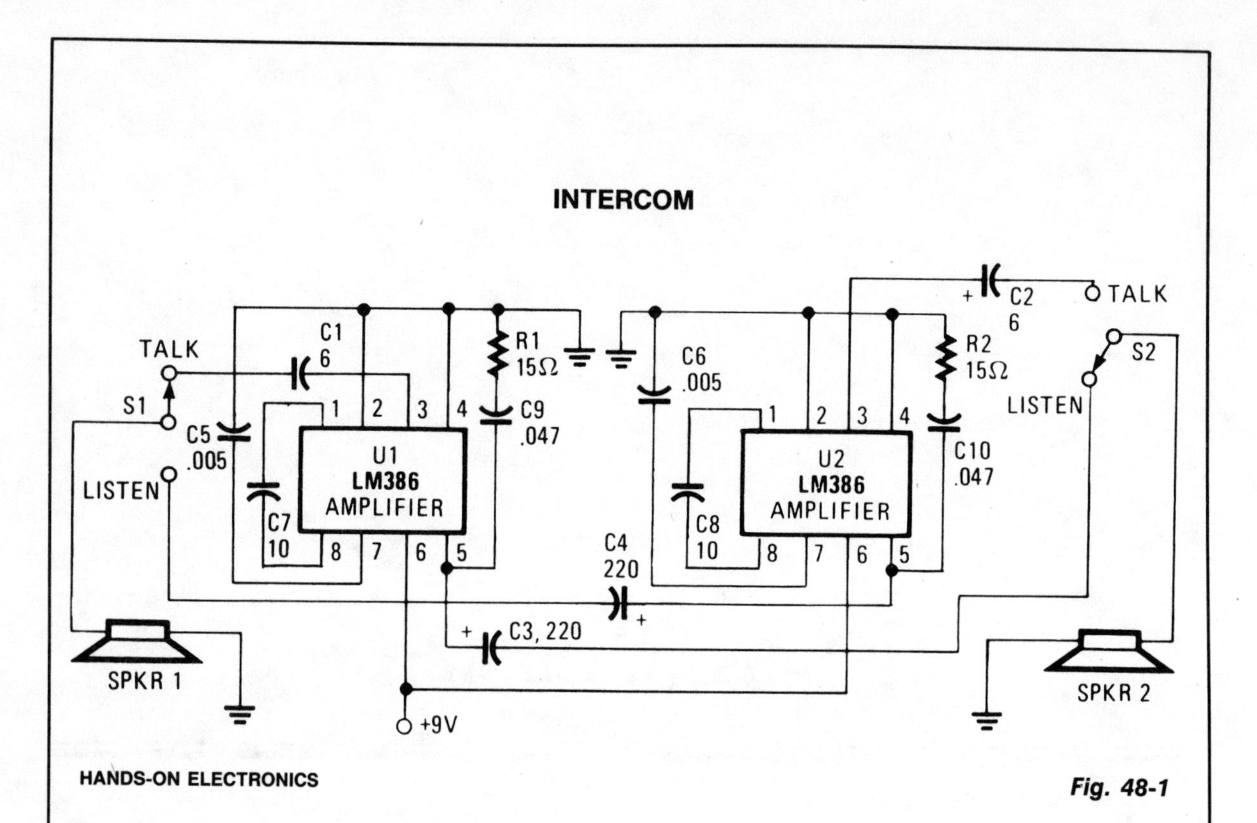

HANDS-ON ELECTRONICS

Fig. 48-1

Circuit Notes

The circuit consists of separate amplifiers—one for each station—rather than a single amplifier and a time sharing arrangement. U1 and U2 are low-voltage audio amplifiers, each of which operates as separate entities with switches at either station controlling which will transmit or receive. With capacitors C7 and C8 included in the circuit, the amplifiers have a gain of 200. Omitting those two components drops the gain to about 20. Other gain levels are available with the addition of a series-connected R/C combination connected between pin 1 and pin 8—for example, a 1000 ohm resistor and 10 μF capacitor for a gain of about 150.

PARTY-LINE INTERCOM

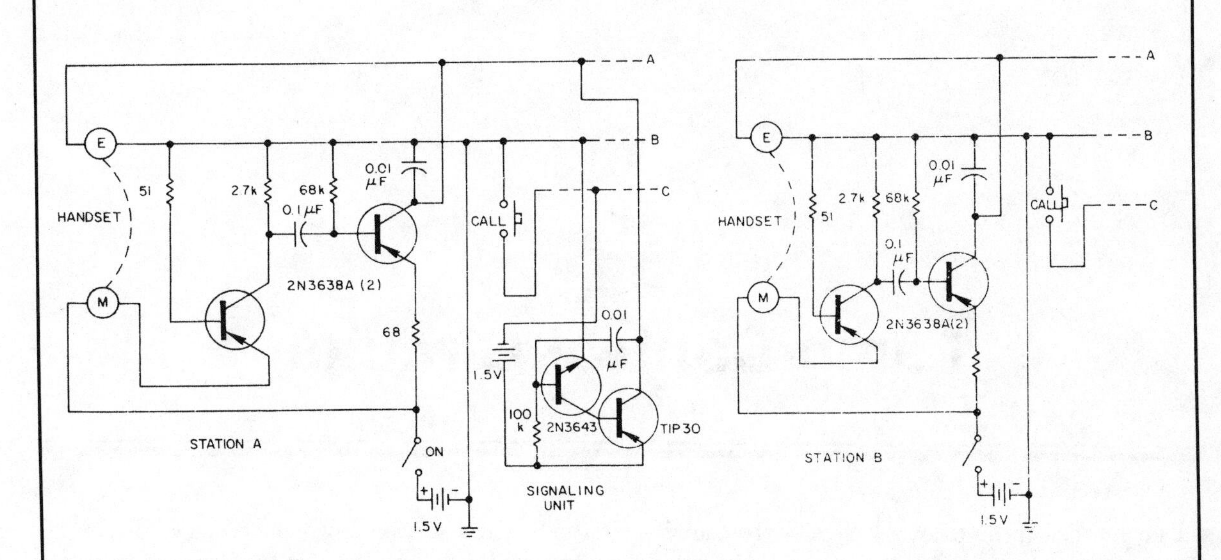

ELECTRONIC DESIGN

Fig. 48-2

Circuit Notes

A large number of intercom stations can be tied together. All units are connected in parallel, and the entire system is buzzed by only one signaling circuit. Each unit is powered individually from 1.5-V cells for redundancy. For greater signal volume, 3-V sources can be used for the supplies without changing any other parts of the system. The carbon microphone of a standard telephone handset at each station feeds into a common-base amplifier, and a tandem high-gain common-emitter stage drives the intercom line. All phone earpieces are in parallel across the line. The signaling circuit, also connected across the line, is a simple oscillator that drives all the earpieces.

49
Lamp-Control Circuits

The sources of the following circuits are contained in the Sources section beginning on page 694. The figure number contained in the box of each circuit correlates to the source entry in the Sources section.

VOLTAGE REGULATOR FOR A PROJECTION LAMP

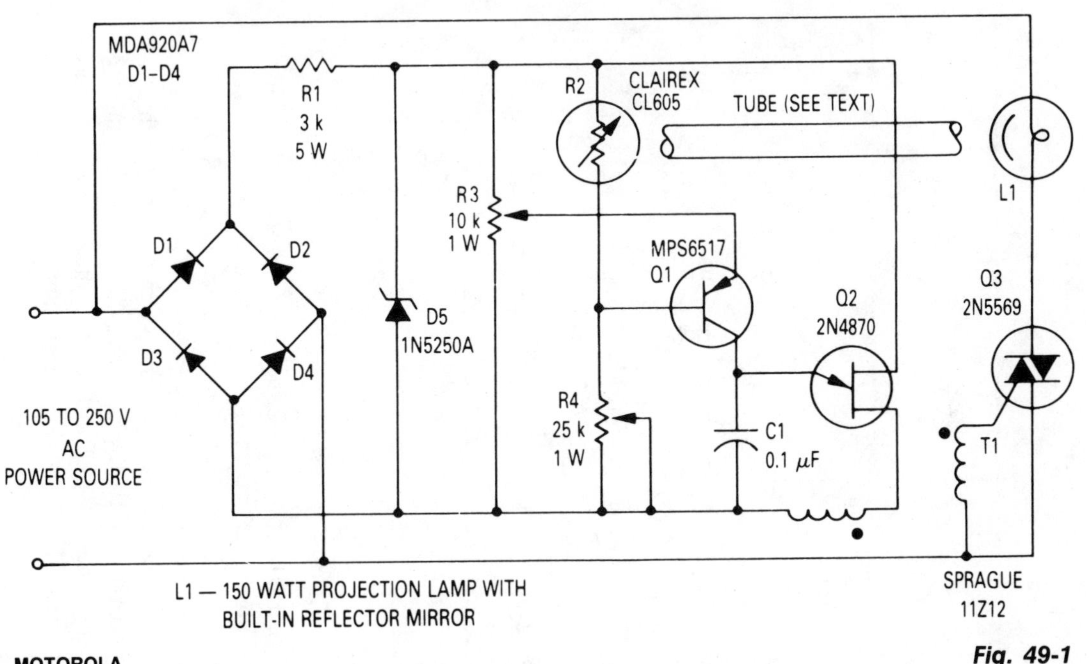

MOTOROLA

Fig. 49-1

Circuit Notes

The circuit will regulate the rms output voltage across the load (a projection lamp) to 100 volts ±2% for an input voltage between 105 and 250 volts ac. This is accomplished by indirectly sensing the light output of lamp L1 and applying this feedback signal to the firing circuit (Q1 and Q2) which controls the conduction angle of TRIAC Q3. The lamp voltage is provided by TRIAC Q3, whose conduction angle is set by the firing circuit for unijunction transistor Q2. The circuit is synchronized with the line through the full-wave bridge rectifier. The voltage to the firing circuit is limited by zener diode D5. Phase control of the supply voltage is set by the charging rate of capacitor C1. Q2 will fire when the voltage on C1 reaches approximately 0.65 times the zener voltage. The charging rate of C1 is set by the conduction of Q1, which is controlled by the resistance of photocell R2. Potentiometers R3 and R4 are used to set the lamp voltage to 100 volts when the line voltage is 105 volts and 250 volts, respectively.

MACHINE VISION ILLUMINATION STABILIZER

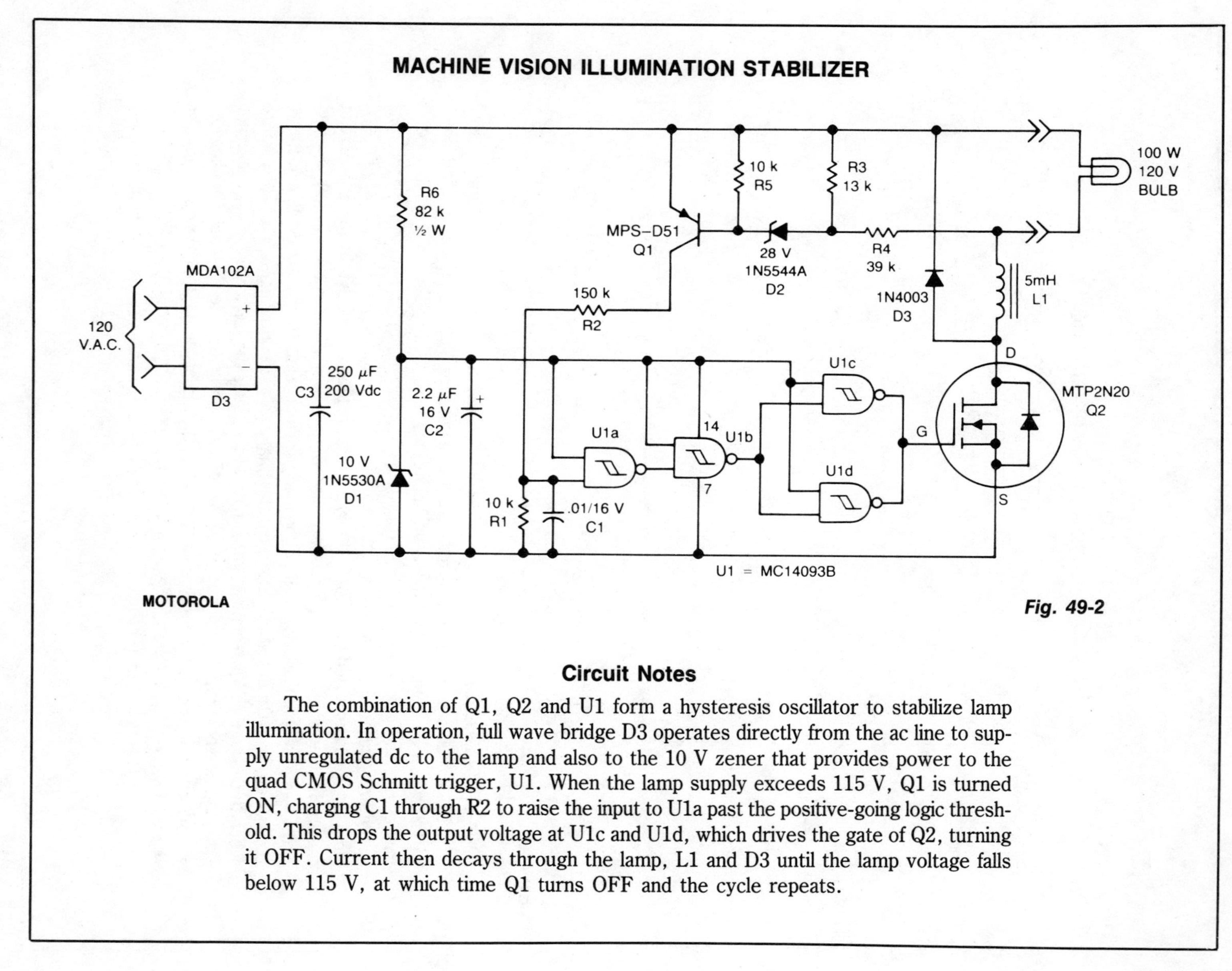

Fig. 49-2

Circuit Notes

The combination of Q1, Q2 and U1 form a hysteresis oscillator to stabilize lamp illumination. In operation, full wave bridge D3 operates directly from the ac line to supply unregulated dc to the lamp and also to the 10 V zener that provides power to the quad CMOS Schmitt trigger, U1. When the lamp supply exceeds 115 V, Q1 is turned ON, charging C1 through R2 to raise the input to U1a past the positive-going logic threshold. This drops the output voltage at U1c and U1d, which drives the gate of Q2, turning it OFF. Current then decays through the lamp, L1 and D3 until the lamp voltage falls below 115 V, at which time Q1 turns OFF and the cycle repeats.

DC LAMP DIMMER

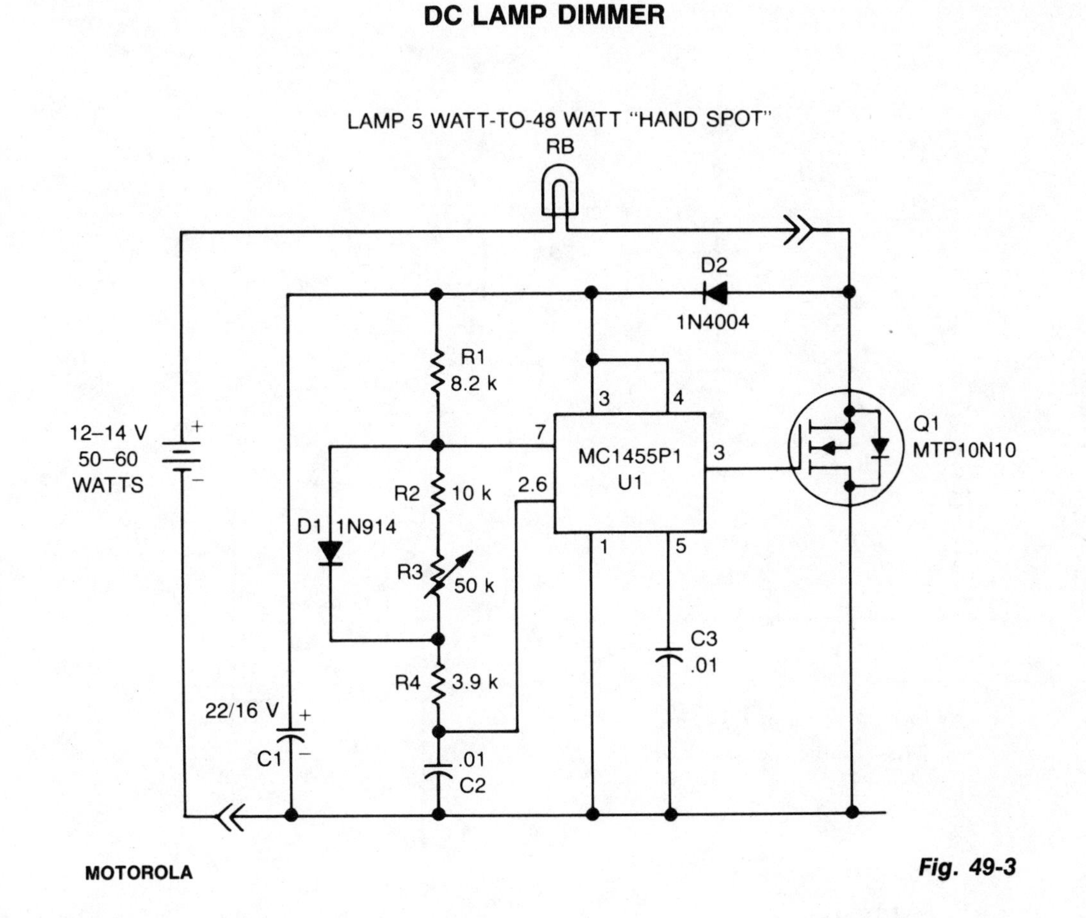

Fig. 49-3

Circuit Notes

A low power, low cost dc lamp dimmer for a two-wire portable "flashlight" can be realized with little or no heatsinking. In addition, a single potentiometer, R3 adjusts lamp brightness.

Battery power is stored in C1 for U1, which is a free-running multivibrator whose frequency is determined by R1, R2, R3, R4, and C2. U1 drives the gate of Q1, turning it and the lamp ON and OFF at a rate proportional to the multivibrator duty cycle.

AUTOMATIC LIGHT CONTROLLER FOR CARPORT

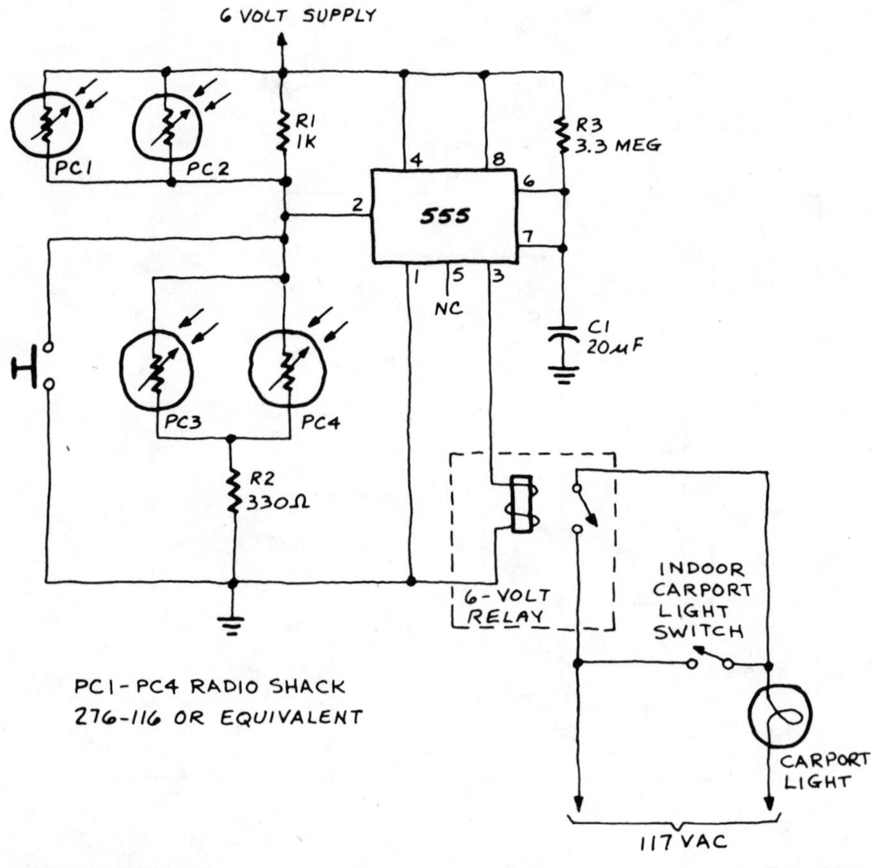

RADIO-ELECTRONICS ***Fig. 49-4***

Circuit Notes

A 555 timer IC, operating in the one-shot mode, is triggered by light striking photoresistors. These normally have a resistance of several megohms but, in the presence of light, that resistance drops to several hundred ohms, permitting current from the six-volt source to flow in the circuit. The R-C combination shown gives an on-time of about two minutes. Photoresistors PC3 and PC4 are mounted at heatlight-height. When headlights illuminate the photoresistor, the timer starts. That actuates a relay, RY1, and the lights are turned on. The lights are automatically turned off when the timer's two minutes are up.

800 W LIGHT-DIMMER

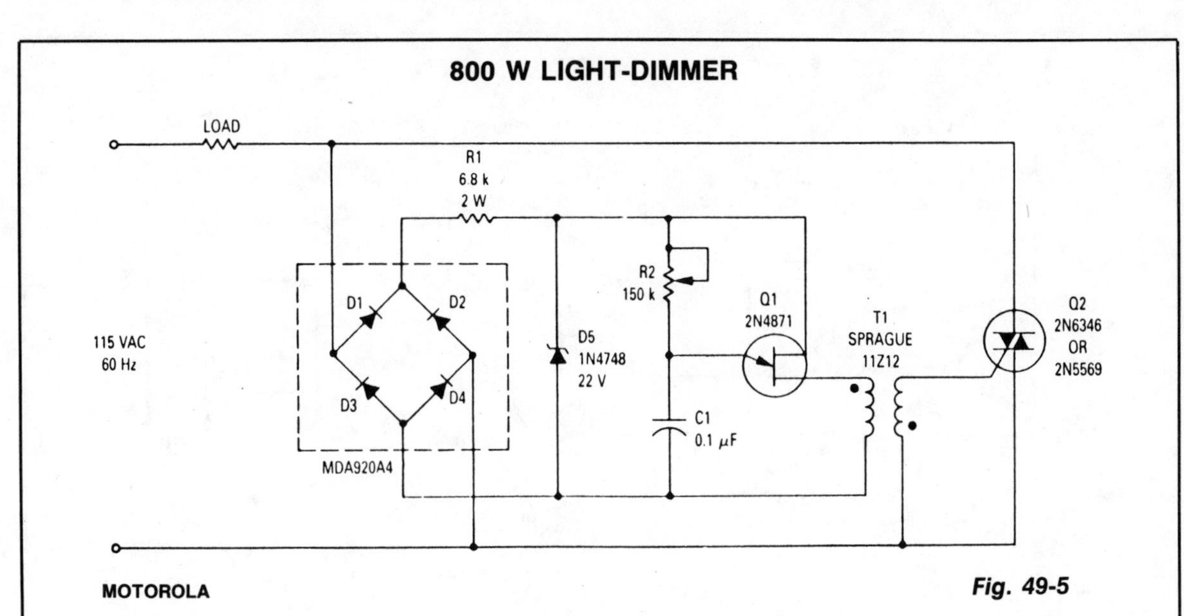

MOTOROLA

Fig. 49-5

Circuit Notes

This wide-range light dimmer circuit uses a unijunction transistor and a pulse transformer to provide phase control for the TRIAC. The circuit operates from a 115 volt, 60 Hz source and can control up to 800 watts of power to incandescent lights. The power to the lights is controlled by varying the conduction angle of the TRIAC from 0° to about 170°. The power available at 170° conduction is better than 97% of that at the full 180°.

LAMP DIMMER

LOAD

20 k

20 k

120 VAC

500 k

2N6343A

0.22 μF

MBS4991

MOTOROLA

Fig. 49-6

Circuit Notes

A full range power controller suitable for lamp dimming and similar applications operate from a 120 volt, 60 Hz ac source, and can control up to 1000 watts of power to incandescent bulbs. The power to the bulbs is varied by controlling the conduction angle of TRIAC Q1. At the end of each positive half-cycle when the applied voltage drops below that of the capacitor, gate current flows out of the SBS and it switches on, discharging the capacitor to near zero volts. The RC network shown across the TRIAC represents a typical snubber circuit that is normally adequate to prevent line transients from accidentally firing the TRIAC.

RUGGED LAMP DRIVER IS SHORT-CIRCUIT PROOF

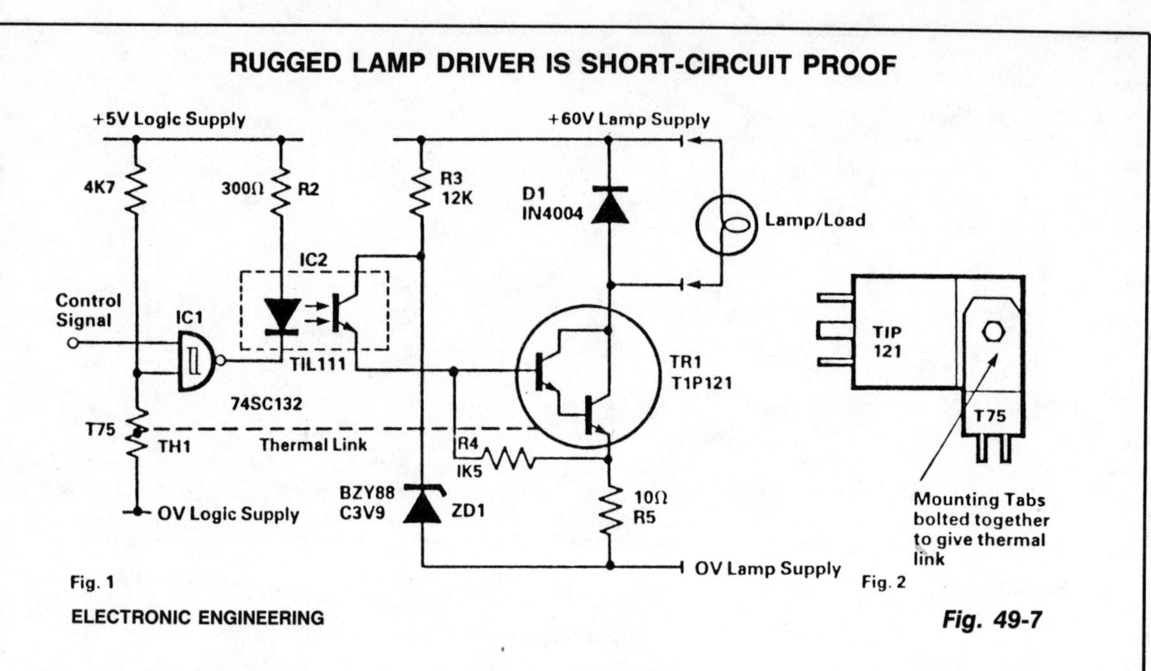

ELECTRONIC ENGINEERING

Fig. 49-7

Circuit Notes

This circuit is capable of driving filament lamps of nominal rating 200 mA at 60 V dc from a CMOS logic signal.

The lamp or load is connected in series with the Darlington transistor TR1 and emitter resistor R5. The Zener diode ZD1 establishes a soft reference voltage on the collector of the optical coupler IC2. When the logic control signal from the processor switches the optocoupler on via IC1, base drive is applied to TR1 and the lamp is switched on.

TRIAC LAMP DIMMER

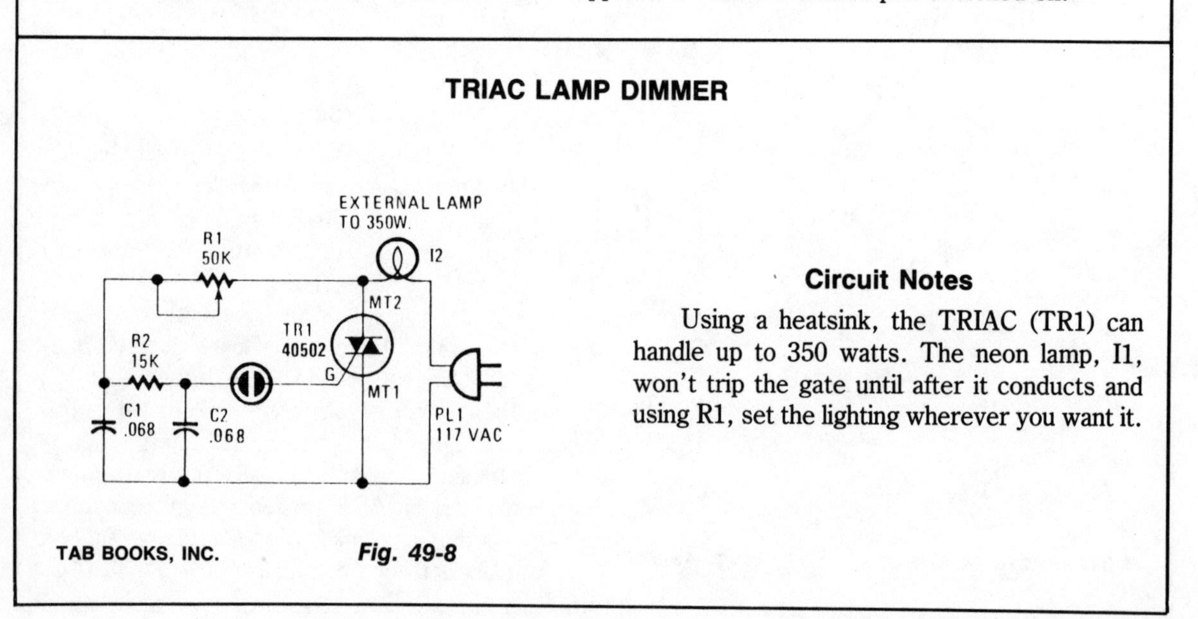

Circuit Notes

Using a heatsink, the TRIAC (TR1) can handle up to 350 watts. The neon lamp, I1, won't trip the gate until after it conducts and using R1, set the lighting wherever you want it.

TAB BOOKS, INC.

Fig. 49-8

TRIAC ZERO-POINT SWITCH

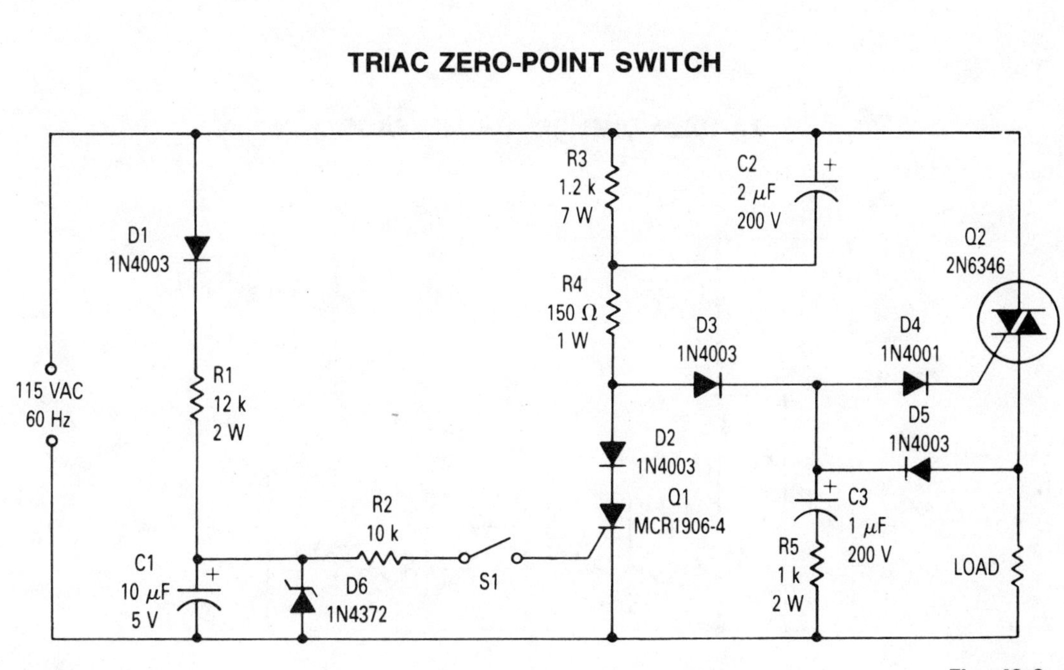

MOTOROLA

Fig. 49-9

Circuit Notes

On the initial part of the positive half cycle, the voltage is changing rapidly from zero causing a large current to flow into capacitor C2. The current through C2 flows through R4, D3, and D4 into the gate of the TRIAC Q2 causing it to turn on very close to zero voltage. Once Q2 turns on, capacitor C3 charges to the peak of the line voltage through D5. When the line voltage passes through the peak, D5 becomes reverse-biased and C3 begins to discharge through D4 and the gate of Q2. At this time the voltage on C3 lags the line voltage. When the line voltage goes through zero there is still some charge on C3 so that when the line voltage starts negative C3 is still discharging into the gate of Q2. Thus Q2 is also turned on near zero on the negative half cycle. This operation continues for each cycle until switch S1 is closed, at which time SCR Q1 is turned on. Q1 shunts the gate current away from Q2 during each positive half cycle keeping Q2 from turning on. Q2 cannot turn on during the negative cycle because C3 cannot charge unless Q2 is on during the positive half cycle.

TANDEM DIMMER (CROSS-FADER)

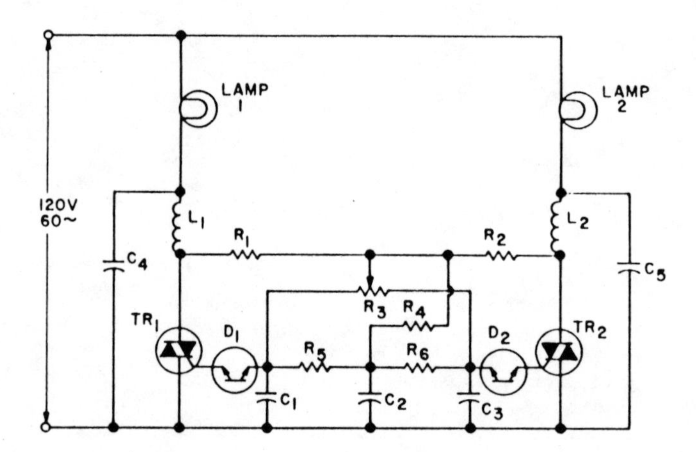

$R_1 = R_2 = 6800\Omega$, 1 WATT
$R_3 = 150K\ \Omega$ LINEAR POT. 1W
$R_5 = R_6 = 22K\Omega$, 1/2 W.
$R_4 = 15K\Omega$, 1/2 W.
$TR_1 = TR_2 =$ TRIAC
$D_1 = D_2 =$ GE ST-2 DIAC

$L_1 = L_2 = 60\mu hy$ (FERRITE CORE)
$C_1 = C_2 = C_3 = 0.1\mu f$ 50V
$C_4 = C_5 = 0.1\mu f$ 200 VOLTS

NOTE: TOTAL LIGHT LEVEL (SUM OF LAMPS 1+2) CONSTANT WITHIN 15%.

GENERAL ELECTRIC

Fig. 49-10

Circuit Notes

This cross fader circuit can be used for fading between two slide projectors. As R_3 is moved to either side of center, one triac is fired earlier in each half cycle, and the other later. The total light output of both lamps stays about the same for any control position.

50

Laser Circuits

The sources of the following circuits are contained in the Sources section beginning on page 694. The figure number contained in the box of each circuit correlates to the source entry in the Sources section.

LASER LIGHT DETECTOR

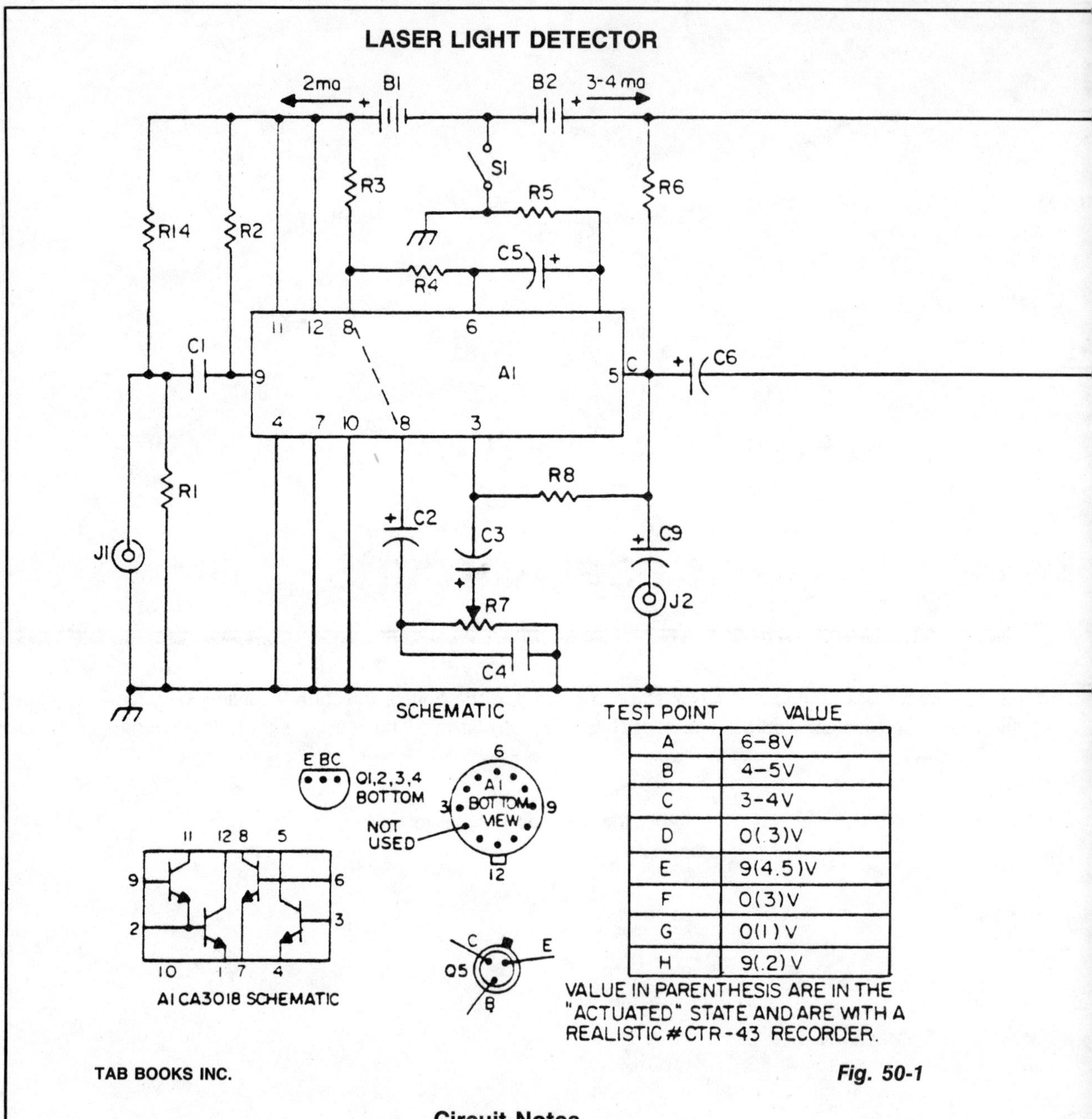

TEST POINT	VALUE
A	6–8V
B	4–5V
C	3–4V
D	0(.3)V
E	9(4.5)V
F	0(3)V
G	0(1) V
H	9(.2) V

VALUE IN PARENTHESIS ARE IN THE "ACTUATED" STATE AND ARE WITH A REALISTIC #CTR-43 RECORDER.

TAB BOOKS INC. *Fig. 50-1*

Circuit Notes

The laser light detector utilizes a sensitive photo transistor (Q5) placed at the focal point of a lens (LE2). The output of Q5 is fed to a sensitive amplifier consisting of array (A1) and is biased via the voltage divider consisting of R14 and R1. The base is not used. Q5 is capacitively coupled to a Darlington pair for impedance transforming and is further fed to a capacitively coupled cascaded pair of common-emitter amplifiers for further signal

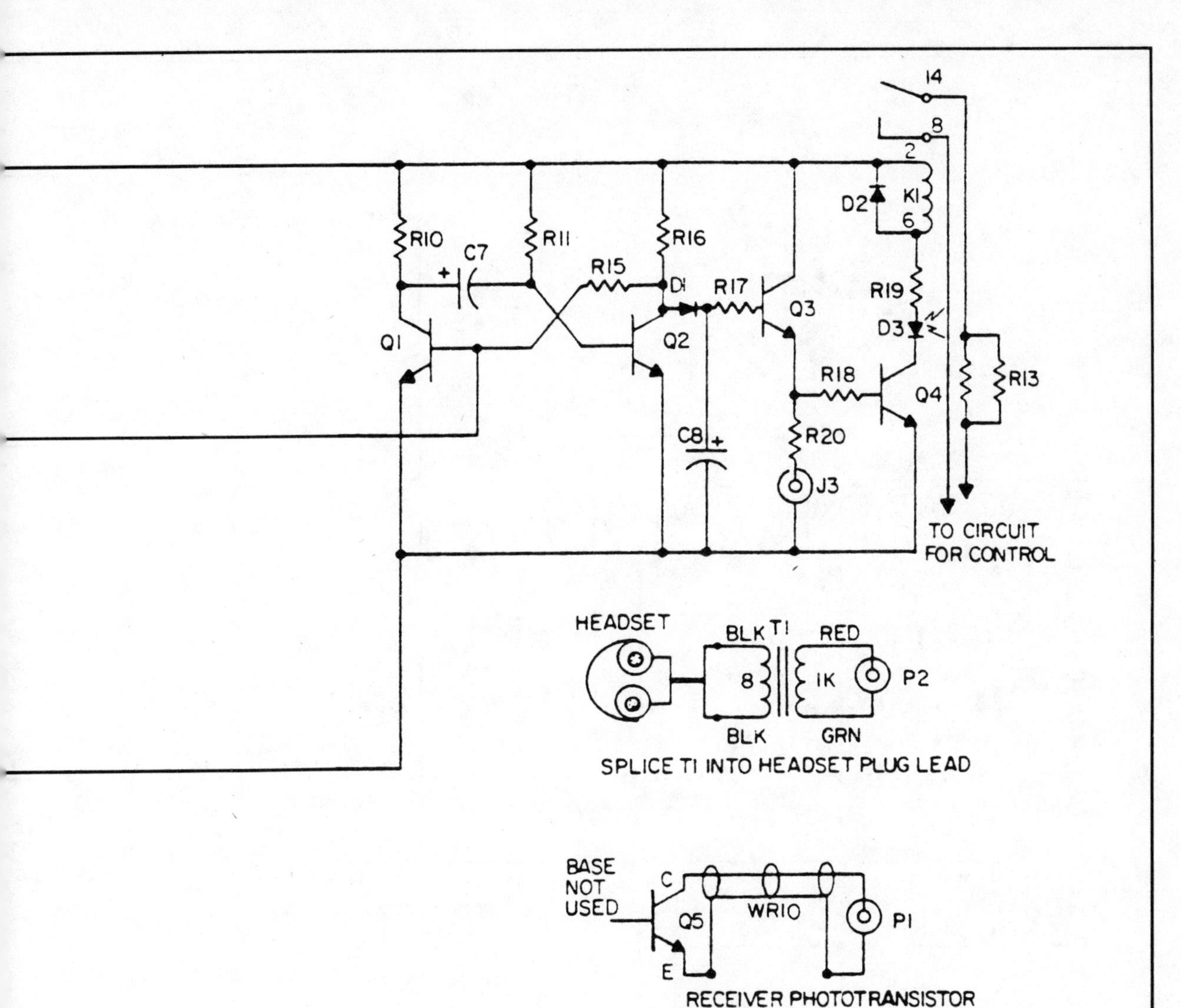

amplification. Sensitivity control (R7) controls base drive to the final transistor of the array and hence controls overall system sensitivity. Output of the amplifier array is capacitively coupled to a one-shot consisting of Q1 and Q2 in turn integrating the output pulses of Q2 onto capacitor C8 through D1. This dc level now drives relay drivers Q3 and Q4 activating K1 along with energizing indicator D3, consequently controlling the desired external circuitry. The contacts of K1 are in series with low ohm resistor R13 to prevent failure when switching capacitive loads. J2 allows "listening" to the intercepted light beam via headsets. This is especially useful when working with pulsed light sources such as GaAs lasers or any other varying periodic light source.

STABILIZING A LASER DISCHARGE CURRENT

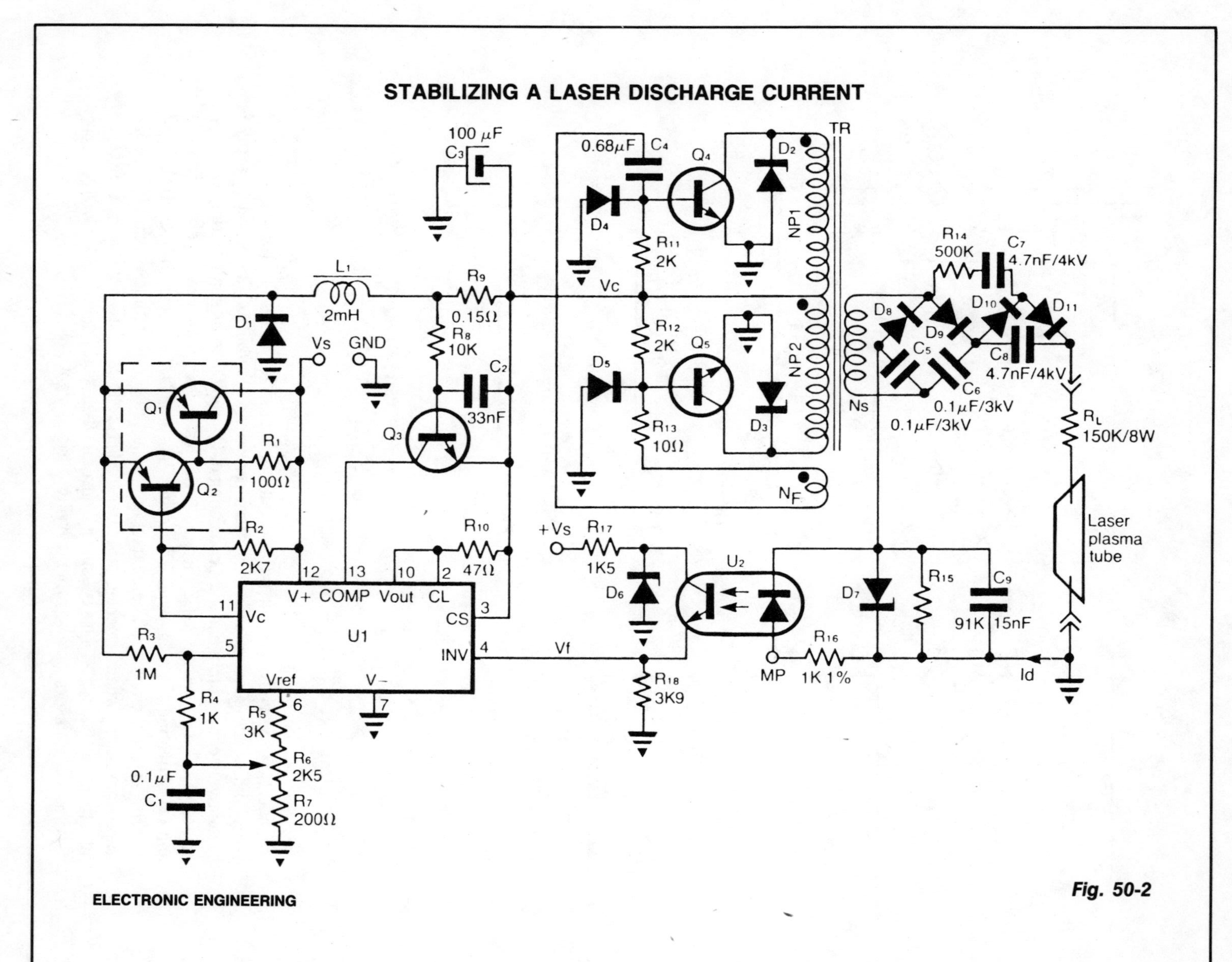

ELECTRONIC ENGINEERING

Fig. 50-2

						D_1	BA159
						D_2; D_3	PL33Z
				Q_1	KD167	D_4; D_5	1N4148
		$Np_1 = Np_2$	8 turns	Q_2	BD140	D_6	PL7V5
U_1	LM723	Ns	1100 turns	Q_3	BC173C	D_7	PL47Z
U_2	MB101	Nf	5 turns	Q_4; Q_5	2N5496	$D_8 \ldots D_{11}$	LA40

Circuit Notes

The circuit uses a free-running push-pull dc to dc high voltage converter to get the necessary voltage for the laser plasma tube supply. The supply voltage V_C of this converter, is adjusted by a switch-mode power supply in order to keep the load current constant, at set value. The linear opto-electronic isolator U2, connected in series with the laser plasma tube, gives a voltage V_F proportional to the discharge current I_D across R18, having the correct polarity to drive directly the inverting input of U1, D7, R15 protects the optoisolator diode against damage produced by the high voltage ignition pulse.

Due to the high operating frequency of the high voltage converter (25 kHz) the ripple of the laser output power is less than 2.10^{-4}. The stability of I_D is better than 10^{-2}, for variations of supply voltage V_s is the range of $\pm 10\%$, and depends on the optoisolator sensitivity.

51

Light-Controlled Circuits

The sources of the following circuits are contained in the Sources section beginning on page 694. The figure number contained in the box of each circuit correlates to the source entry in the Sources section.

PHOTO ALARM

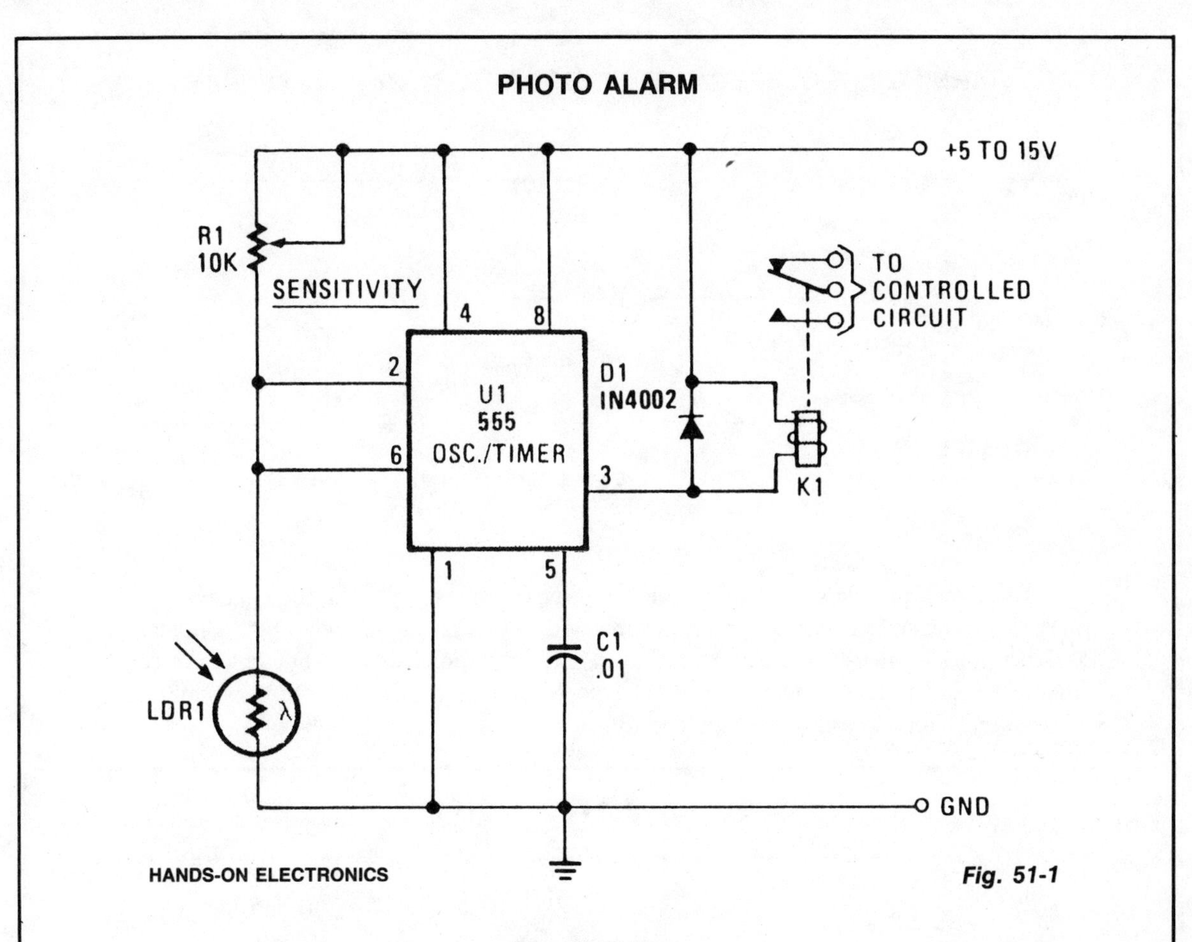

HANDS-ON ELECTRONICS

Fig. 51-1

Circuit Notes

LDR1, a cadmium sulphide (CDS) photoresistive cell is used as the lower leg of a voltage divider between V_{CC} and ground. The timer terminals 2 and 6 are connected to the junction of the photocell and SENSITIVITY control R1. The resistance of the photoresistive cell varies inversely as the light intensity; resistance is high when the illumination level is low; low in bright light. (The Radio Shack CDS cell 276-116 has a typically wide resistance range—about 3 megohms in darkness and 100 ohms in bright light.) When the light is interrupted or falls below a level set by SENSITIVITY control R1, the rise in LDR1's resistance causes the voltage on pins 2 and 6 to rise. If the control is set so the voltage rises above ⅔ V_{CC}, the relay pulls in. The relay drops out when the light level increases and the drop across the photocell falls below ⅓ V_{CC}. (The circuit can be modified by placing relay K1 and diode D1 between pin 3 and ground. In this case, the relay drops out when the voltage on pins 2 and 6 rises above ⅔ V_{CC}, and pulls in when it falls below ⅓ V_{CC}. This modification is valuable when the relay has single-throw contacts.) Opening and closing of the relay contacts occurs at different illumination levels. This ⅓ V_{CC} hysteresis is an advantage that prevents the circuit from hunting and the relay from chattering when there are very small changes in illumination.

WARNING LIGHT OPERATES FROM BATTERY POWER SUPPLY

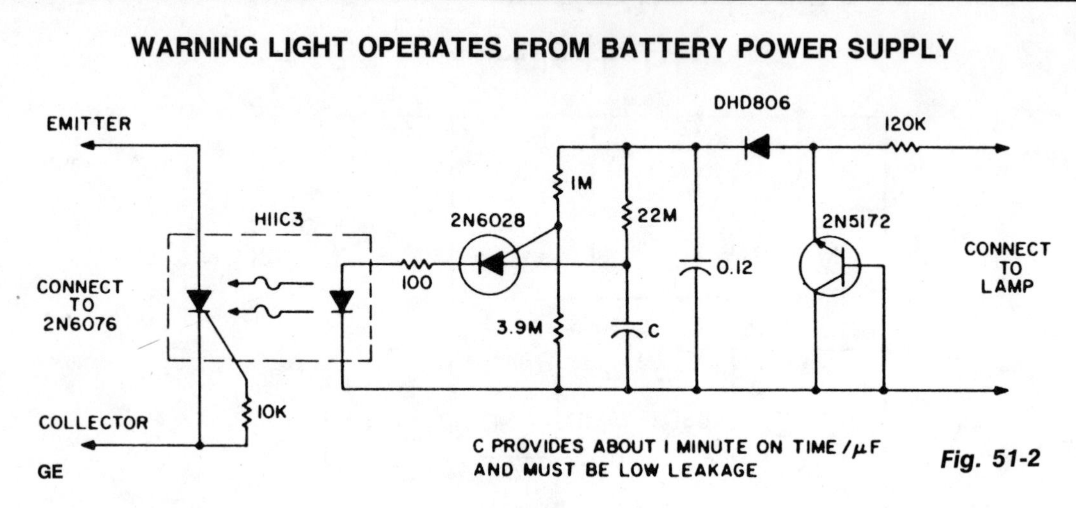

Fig. 51-2

Circuit Notes

The circuit provides illumination when darkness comes. By using the gain available in darlington transistors, this circuit is simplified to use just a photodarlington sensor, a darlington amplifier, and three resistors. The illumination level will be slightly lower than normal, and longer bulb life can be expected, since the D40K saturation voltage lowers the lamp operating voltage slightly.

LIGHT-OPERATED SWITCH

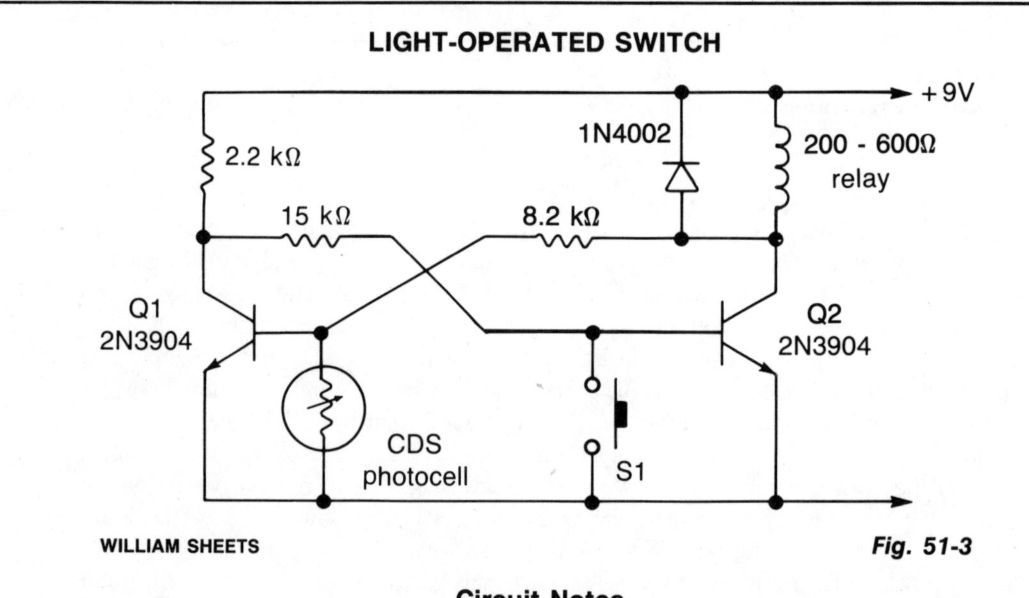

Fig. 51-3

Circuit Notes

This circuit uses a flip-flop arrangement of Q1 and Q2. Normally Q1 is conducting heavily. Light on CDS photocell causes Q1 bias to decrease, cutting it off, turning on Q2, removing the remaining bias from Q1. Reset is accomplished by depressing S1.

PHOTOELECTRIC SWITCH

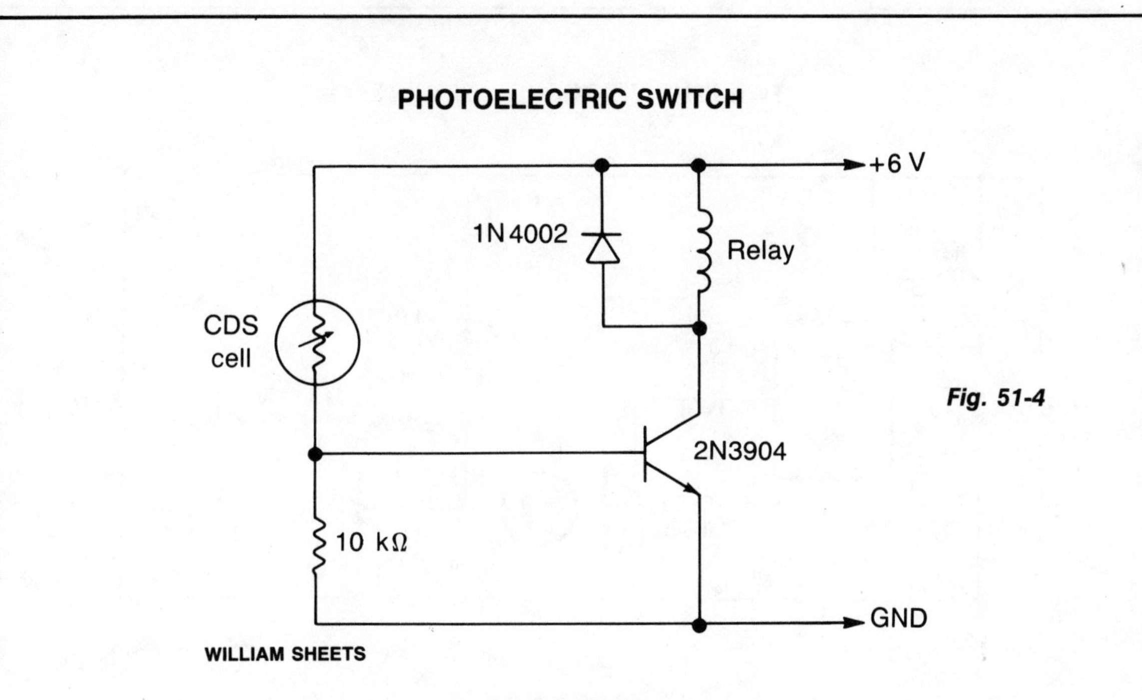

Fig. 51-4

Circuit Notes

The CDS cell resistance decreases in the presence of light, turning on the 2N3904 relay driver.

BACK-BIASED GaAsP LED OPERATES AS LIGHT SENSOR

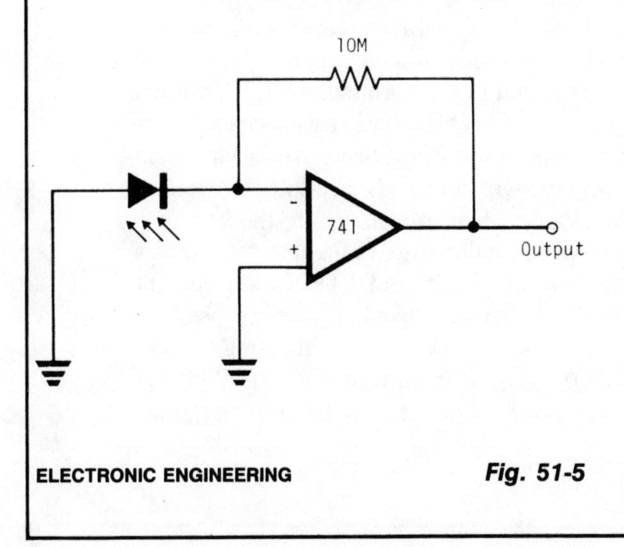

ELECTRONIC ENGINEERING

Fig. 51-5

Circuit Notes

Using a simple 741 amplifier connected as a current-to-voltage converter with the LED as the current source, the voltage at the output is proportional to incident light. The junction is biased only by the difference between the summing node junction potential and ground, preventing the possibility of reverse breakdown. The photon-generated current equals the short-circuit current of the junction, which is linearly related to incident light. The sensor requires a level of incident illumination that depends on the degree of opacity of the LED package.

TWILIGHT-TRIGGERED CIRCUIT

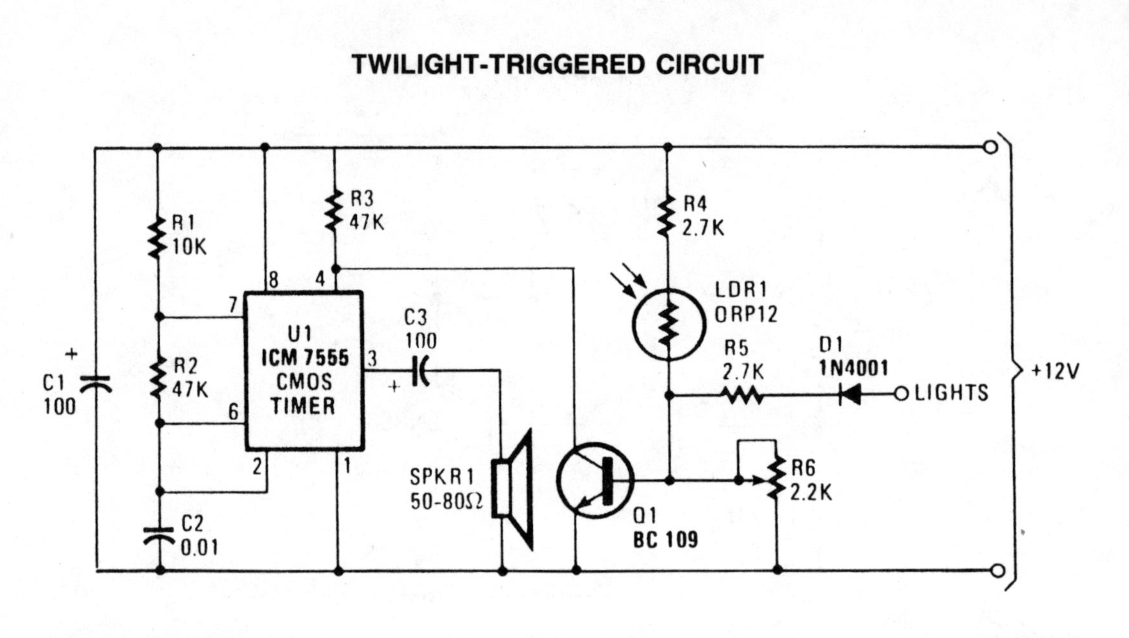

HANDS-ON ELECTRONICS *Fig. 51-6*

Circuit Notes

As dusk begins to fall, the sensor (a cadmium-sulfide light-dependent resistor or LDR) operates a small horn to provide an audible reminder that it's time to turn on your lights. To turn the circuit off—simply turn your headlights on and the noise stops. The base of Q1 is fed through a voltage divider formed by R4, LDR1—a light-dependent resistor with an internal resistor of about 100 ohms under bright-light conditions and about 10 megohms in total darkness—potentiometer R6. Q1's base voltage depends on the light level received by LDR1 and the setting of R6. If LDR1 detects a high light level, its resistance decreases, thereby providing a greater base current for Q1, causing it to conduct. When Q1 conducts, pin 4 of U1 is pulled to near ground potential, muting the oscillator. If, on the other hand, LDR1 detects a low light level, its resistance increases (reducing base currentto Q1), cutting off the transistor and enabling the oscillator. In actual practice, you set R6 so that at a suitable light level (dusk), the oscillator will sound. The anode of diode D1 connects to the light switch, where it connects to the vehicle's parking lights. With the lights switched off, that point is connected to the negative chassis by way of the parking lamp. That has no effect on the circuit, as D1 blocks any current flow to ground from Q1's base via R6 and the sidelight lamps. When the lights are switched on, the anode of D1 is connected to the positive supply via the parking lamp switch, thereby applying a voltage to the base of Q1, biasing it into conduction. With Q1 conducting, pin 4 of U1 is pulled virtually to ground, disabling the oscillator even though LDR1's resistance is not enough to do so.

AUTOMATIC MOORING LIGHT

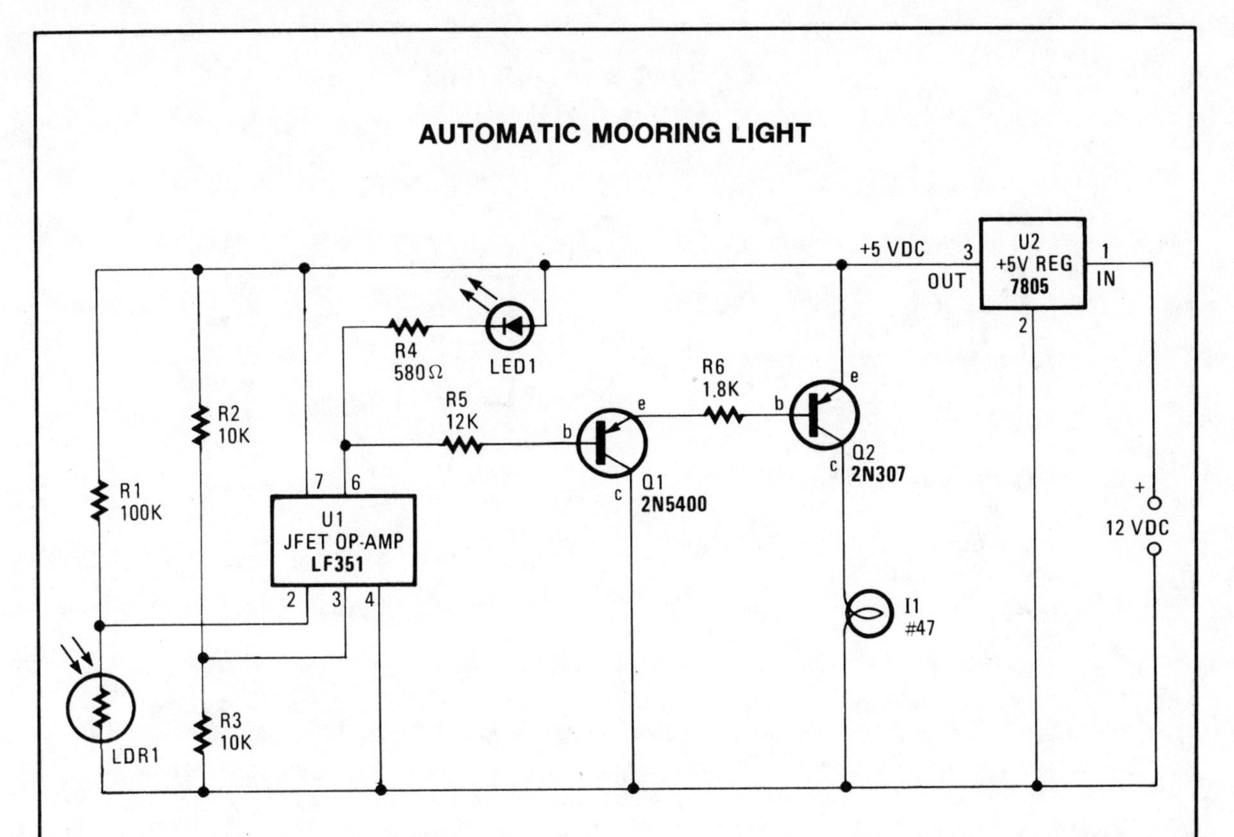

HANDS-ON ELECTRONICS

Fig. 51-7

Circuit Notes

Integrated-circuit U1—an LF351 or 741 op amp—is used as a comparator to control the light. Resistors R2 and R3 provide a reference voltage of about 2.5 volts at pin 3 of U1. When daylight falls on light-dependent resistor LDR1, its resistance is low: about 1000 ohms. In darkness, the LDR's resistance rises to about 1 megohm. Since R1 is 100,000 ohms, and the LDR in daylight is 1000 ohms, the voltage ratio is 100 to 1; the voltage drop across the LDR is less than the 2.5 volt reference voltage and pin 2 of U1 is held at that voltage. In that state, the output at pin 6 of U1 is positive at about 4.5 volts, a value that reverse-biases Q1 to cutoff, which in turn holds Q2 in cutoff, thereby keeping lamp I1 off. When darkness falls, the LDR's resistance rises above R1's value and the voltage at pin 2 of U1 rises above the reference voltage of 2.5 volts. U1's output terminal (pin 6) falls to less than a volt and Q1 is biased on. The base-to-emitter current flow turns Q2 on, which causes current to flow through the lamp. When daylight arrives, the LDR's resistance falls sharply, which causes the lamp to be turned off, ready to repeat the next night/day cycle.

ELECTRONIC WAKE-UP CALL

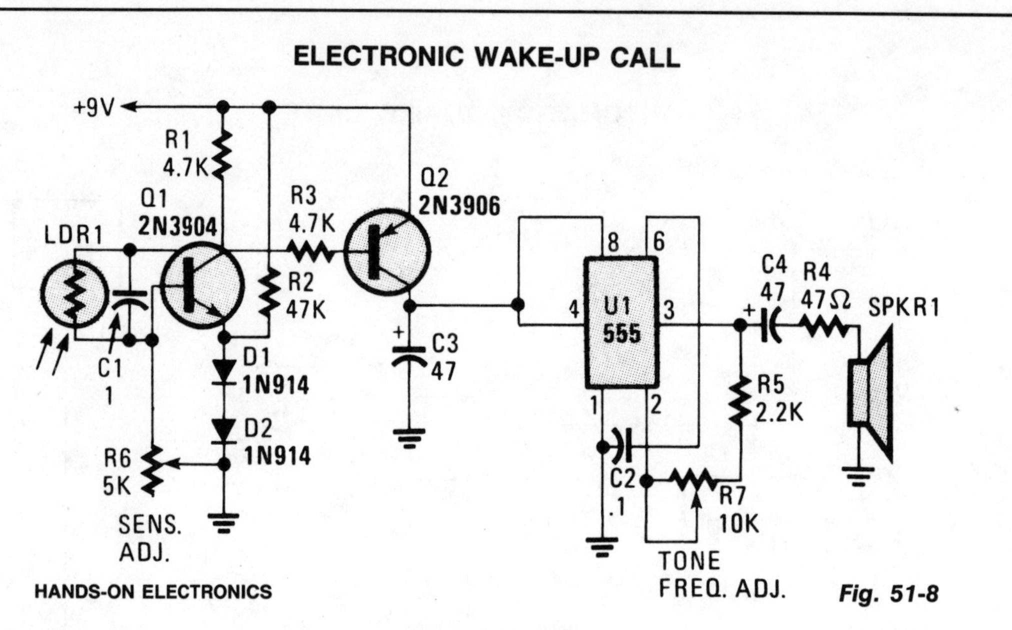

HANDS-ON ELECTRONICS

Fig. 51-8

Circuit Notes

A cadmium sulfide photocell (LDR1, which is a light-dependent resistor) is connected to the base and collector of an npn transistor, Q1. When light hits LDR1, the internal resistance goes from a very high (dark) value to a low (light) value, supplying base current to Q1, turning it on. The voltage across R1 produces a bias that turns Q2 on, which in turn, supplies the positive voltage to U1 at pin 8 (the positive-supply input) and pin 4 (the reset input), to operate the 555 audio oscillator circuit. The circuit's sensitivity to light can be set via R6 (a 50,000 ohm potentiometer). R7 sets the audio tone to the most desirable sound. The squarewave audio tone is fed from U1 pin 3 to a small speaker through coupling capacitor C4 and current limiting resistor R4.

PHOTODIODE SENSOR AMPLIFIER

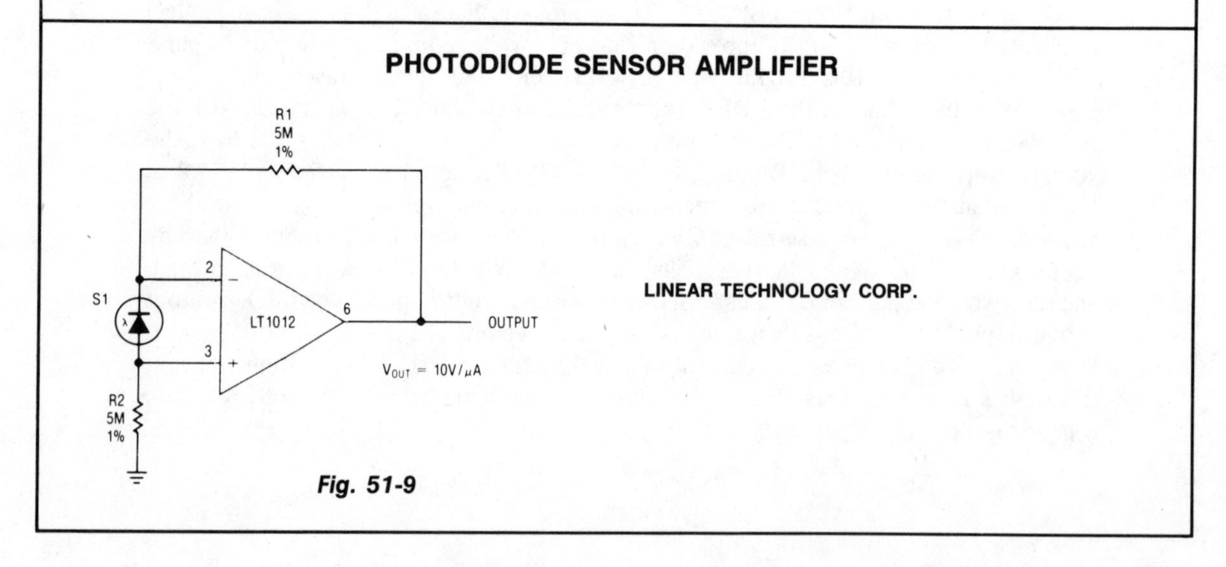

Fig. 51-9

LIGHT-SEEKING ROBOT

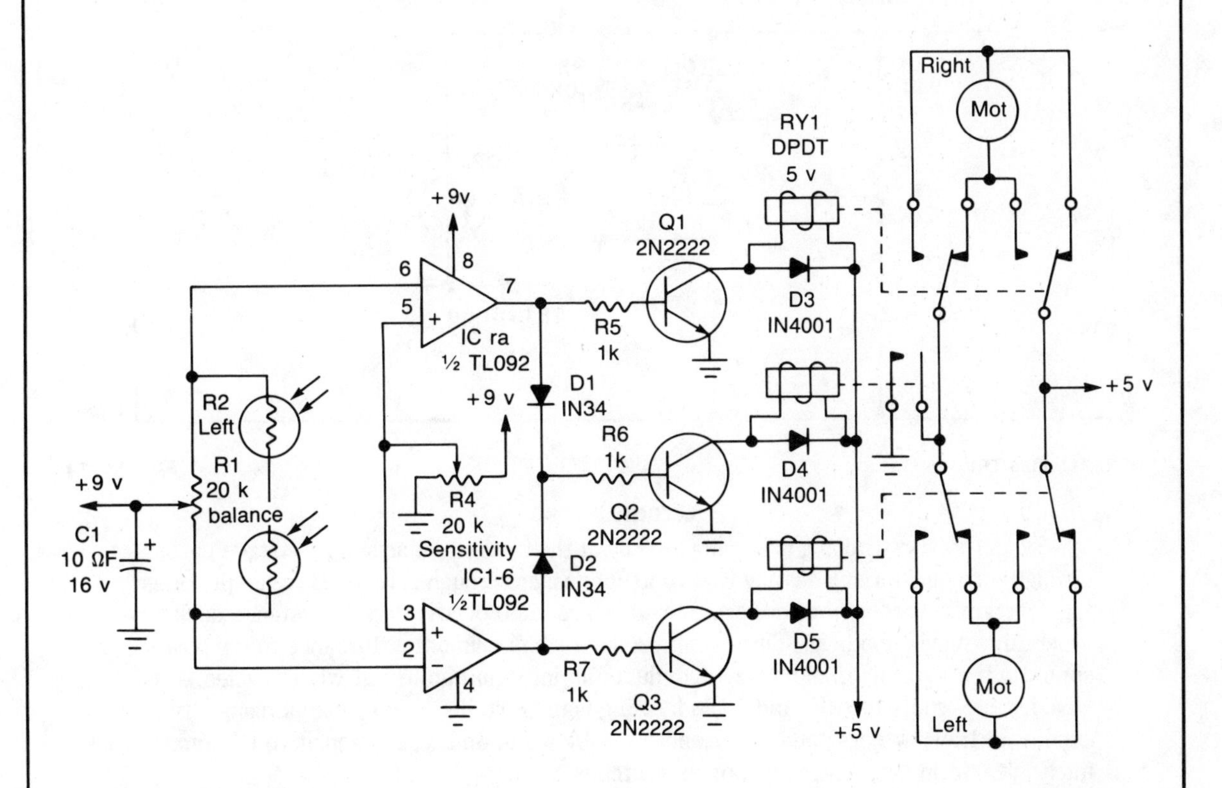

RADIO-ELECTRONICS *Fig. 51-10*

Circuit Notes

The circuit is light seeking; it will follow a flashlight around a darkened room. A pair of photocells determine the direction in which the robot will move. Each photocell is connected to an op amp configured as a comparator. When sufficient light falls on photocell R2, the voltage at the inverting input (pin 6) of IC1-a will fall below the voltage at the non-inverting input (pin 5), so the output of the comparator will go high, and transistors Q1 and Q2 will turn on. That will enable relays RY1 and RY2, and thereby provide power for the right motor. The robot will then turn left. Likewise, when light falling on R3 lowers its resistance, Q2 and Q3 will turn on, the left motor will energize, and the robot will turn right.

SYNCHRONOUS PHOTOELECTRIC SWITCH

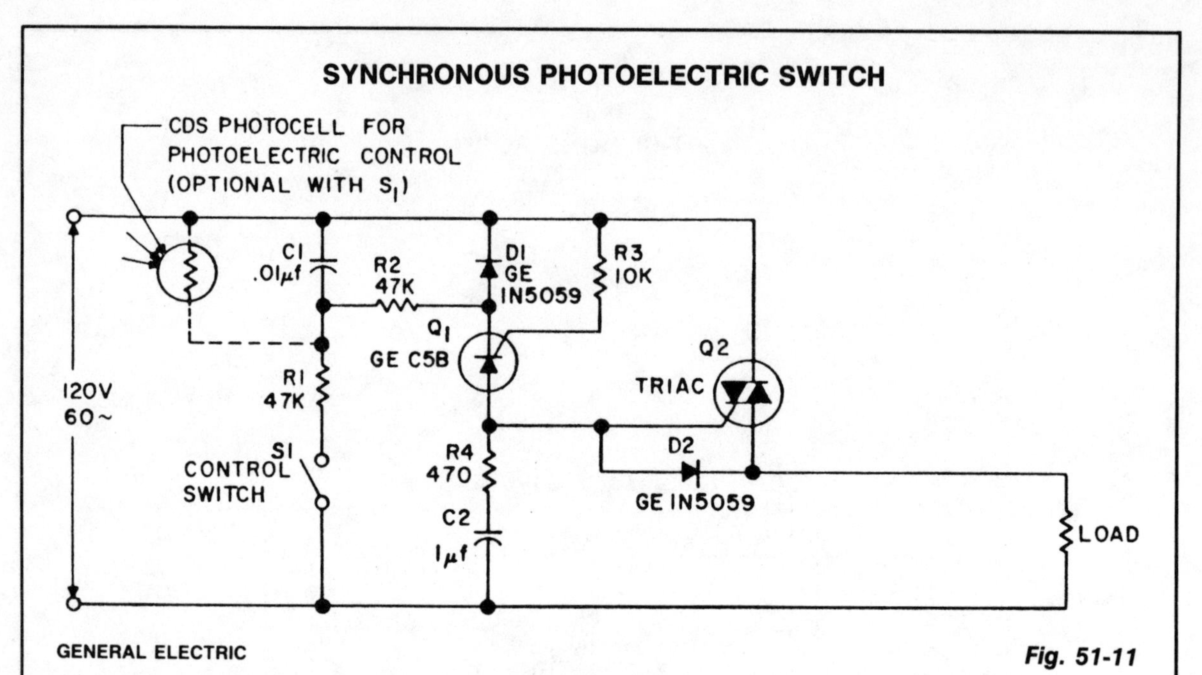

GENERAL ELECTRIC

Fig. 51-11

Circuit Notes

Synchronous switching is turning on only at the instant the ac supply voltage passes through zero, and turning off only when current passes through zero. This circuit provides this function in response to either a mechanical switch or a variable resistance such as a cadmium-sulfide photocell. This circuit produces the minimum disturbance to the power supply when switching, and always conducts an integral number of whole cycles. It is ideal for use wherever RFI and audio filtering is undesirable, where magnetizing inrush current of transformers causes nuisance fuse-blowing, and where sensitive equipment must operate in the vicinity of power switches.

PHOTOCURRENT INTEGRATOR

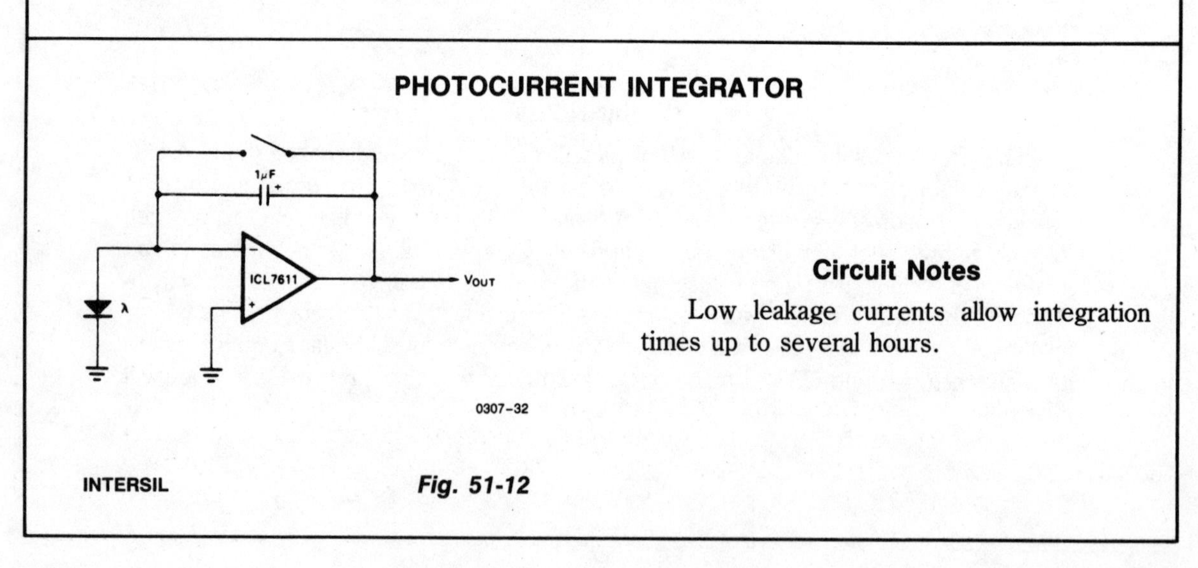

Circuit Notes

Low leakage currents allow integration times up to several hours.

INTERSIL

Fig. 51-12

ROBOT EYES

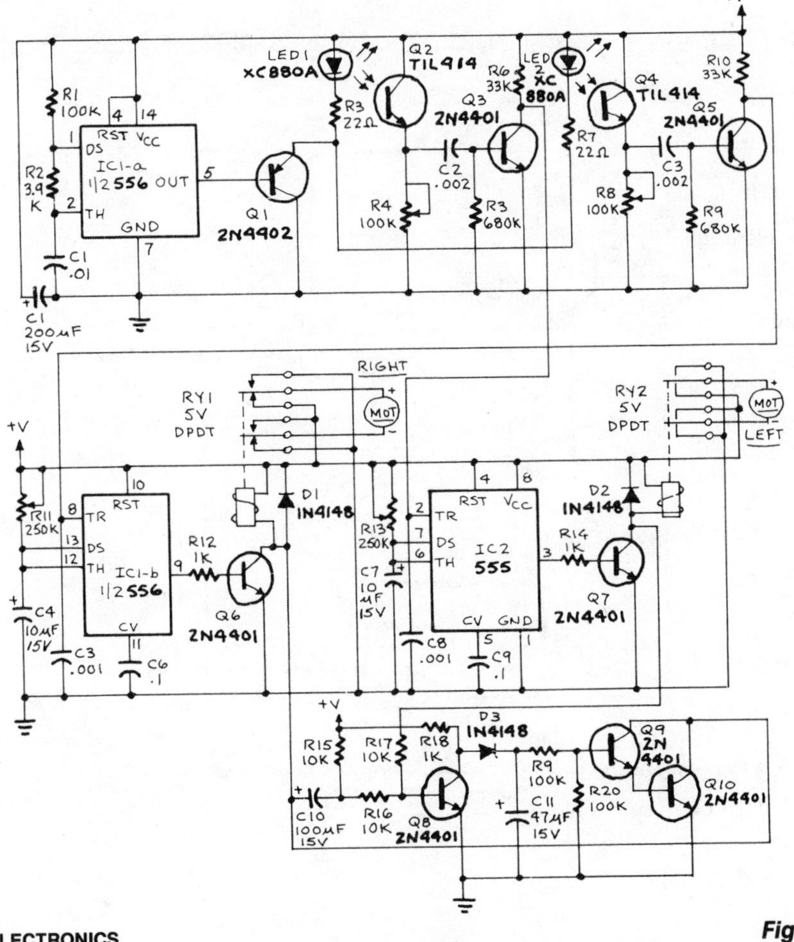

RADIO ELECTRONICS

Fig. 51-13

Circuit Notes

An infrared LED and a phototransistor are used for each eye. Half of a 556 timer IC (IC1-a) functions as an astable multivibrator oscillating at a frequency of about 1 kHz. That IC drives transistor Q1 which in turn drives the two infrared LED's, LED1 and LED2. The right eye is composed of LED1 and Q2. If an obstacle appears in front of the right eye, pulses from LED1 are reflected by the obstacle and detected by Q2. The signal from Q2 is amplified by Q3, which triggers IC2, a 555. That IC operates in the monostable mode, and it provides a pulse output with a width of as much as 2.75 seconds, depending on the setting of R11. That pulse output energizes relay RY1, and that reverses the polarity of the voltage applied to the motor. Corresponding portions of the circuit of the left eye operate in the same fashion, using the unused half of the 556 (IC1-b). That action causes the robot to turn away from an obstacle.

MODULATED LIGHT-BEAM CIRCUIT CANCELS AMBIENT LIGHT EFFECTS

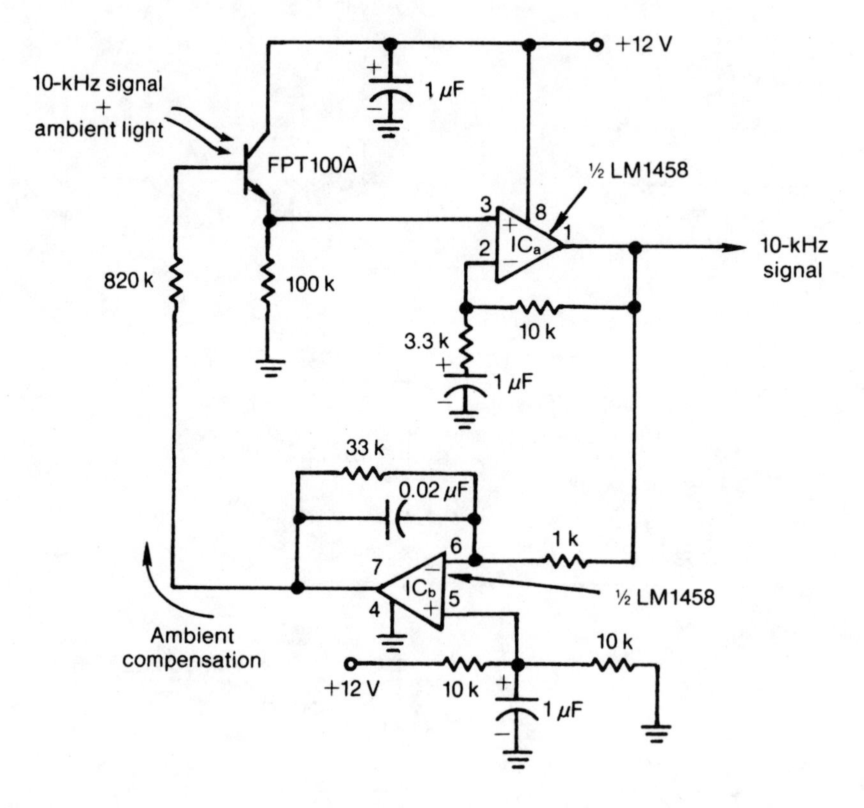

Fig. 51-14

ELECTRONIC DESIGN

Circuit Notes

Feedback control of the phototransistor in this optical detector helps negate the effects of varying ambient light sources. The output of a modulated visible-light LED is detected, amplified, buffered, and fed through a low-pass filter. Ambient light signals below the LED's 10-kHz modulating rate reach the detector's base out of phase with incoming ambient light and cancel the undesired effects.

MONOSTABLE PHOTOCELL CIRCUIT HAS SELF-ADJUSTING TRIGGER LEVEL

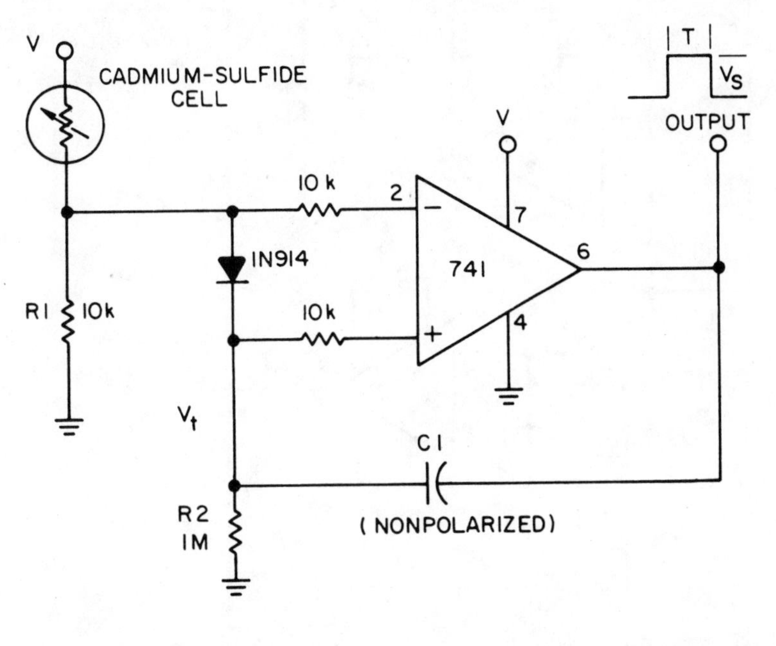

ELECTRONIC DESIGN

Fig. 51-15

Circuit Notes

A photocell circuit provides automatic threshold adjustment. Monostable action prevents undesired retriggering of the output. With only one op amp IC, the circuit offers: Automatic adjustment of its trigger level to accommodate various light sources, changes in ambient light and misalignments; A built-in monostable action to provide only a single output pulse during a preset time; Feedback action to raise the threshold level after triggering and to speed switching. The feedback also eliminates the circuit's tendency to oscillate during switching.

THERMALLY STABILIZED PIN PHOTODIODE SIGNAL CONDITIONER

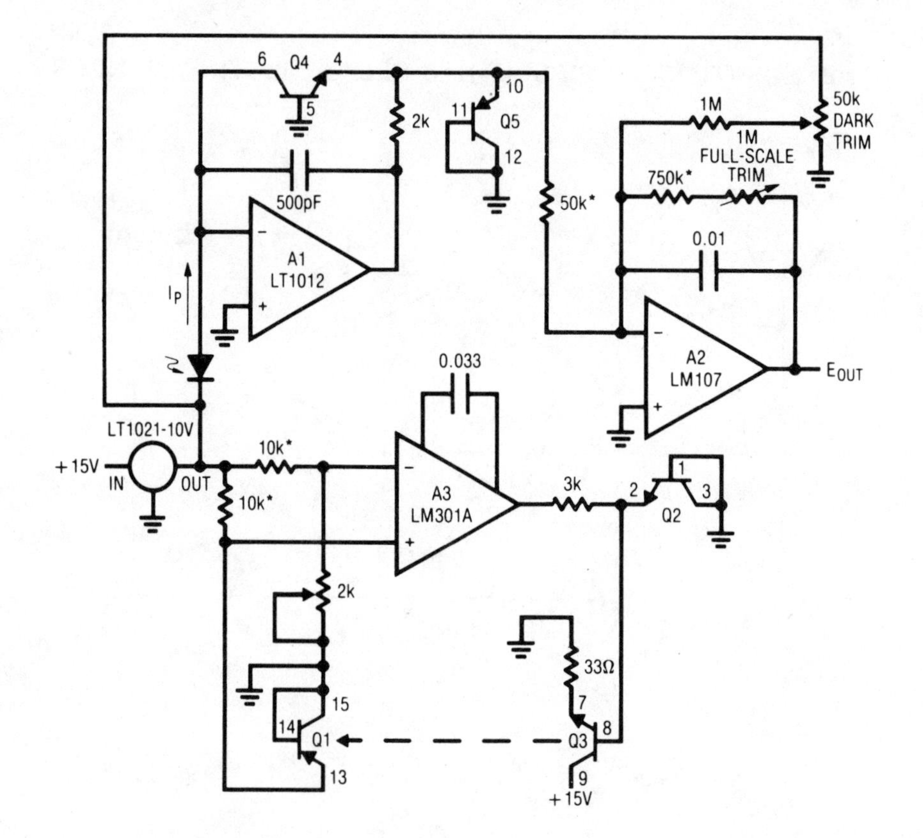

LINEAR TECHNOLOGY CORPORATION

Fig. 51-16

Circuit Notes

The photodiode specified responds linearly to light intensity over a 100 dB range. Digitizing the diodes linearly amplified output would require an A-D converter with 17 bits of range. This requirement can be eliminated by logarithmically compressing the diode's output in the signal conditioning circuity. A1 and Q4 convert the diode's photocurrent to voltage output with a logarithmic transfer function. A2 provides offsetting and additional gain. A3 and its associated components form a temperature control loop

RESPONSE DATA

LIGHT (900μM)	DIODE CURRENT	CIRCUIT OUTPUT
1mW	350μA	10.0V
100μW	35μA	7.85V
10μW	3.5μA	5.70V
1μW	350nA	3.55V
100nW	35nA	1.40V
10nW	3.5nA	−0.75V

= HP-5082-4204 PIN PHOTODIODE.

Q1–Q5 = CA3096.

CONNECT SUBSTRATE OF CA3096 ARRAY TO Q4'S EMITTER.

*1% RESISTOR

which maintains Q4 at constant temperature (all transistors in this circuit are part of a CA3096 monolithic array). The 0.033 μF value at A3's compensation pins gives good loop damping if the circuit is built using the array's transistors in the location shown. Because of the array die's small size, response is quick and clean. A full-scale step requires only 250 ms to settle to final value. To use this circuit, first set the thermal control loop. To do this, ground Q3's base and set the 2 k pot so A3's negative input voltage is 55 mV above its positive input. This places the servo's setpoint at about 50°C (25°C ambient + (2.2 mV/°C × 25°C rise = 55 mV = 50°C). Unground Q3's base and the array will come to temperature. Next, place the photodiode in a completely dark environment and adjust the "dark trim" so A2's output is 0 V. Finally, apply or electrically simulate 1 mW of light and set the "full-scale" trim for 10 V out. Once adjusted, this circuit responds logarithmically to light inputs from 10nW to 1mW with an accuracy limited by the diode's 1% error.

52
Logic Amplifiers

The sources of the following circuits are contained in the Sources section beginning on page 694. The figure number contained in the box of each circuit correlates to the source entry in the Sources section.

LOW POWER INVERTING AMPLIFIER WITH DIGITALLY SELECTABLE GAIN

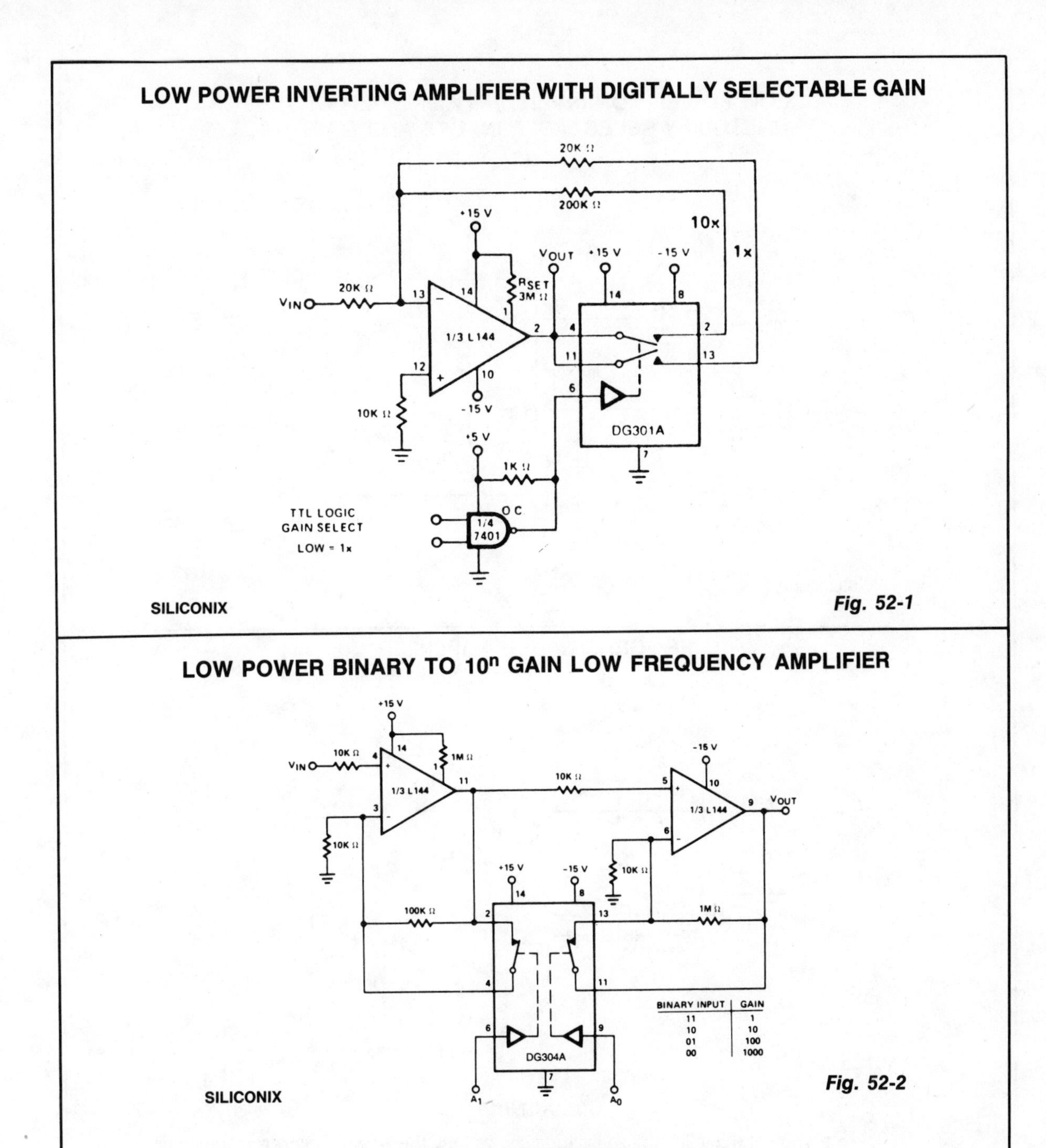

BINARY INPUT	GAIN
11	1
10	10
01	100
00	1000

SILICONIX

Fig. 52-1

LOW POWER BINARY TO 10^n GAIN LOW FREQUENCY AMPLIFIER

SILICONIX

Fig. 52-2

Circuit Notes

Gain increases by decades as the binary input decreases from 1,1 to 0,0. Minimum gain is 1 and maximum gain is 1000. Since the switch is static in this type of amplifier the power dissipation of the switch will be less than a tenth of a milliwatt.

LOW POWER NON-INVERTING AMPLIFIER WITH DIGITALLY SELECTABLE INPUTS AND GAIN

PROGRAMMABLE AMPLIFIER

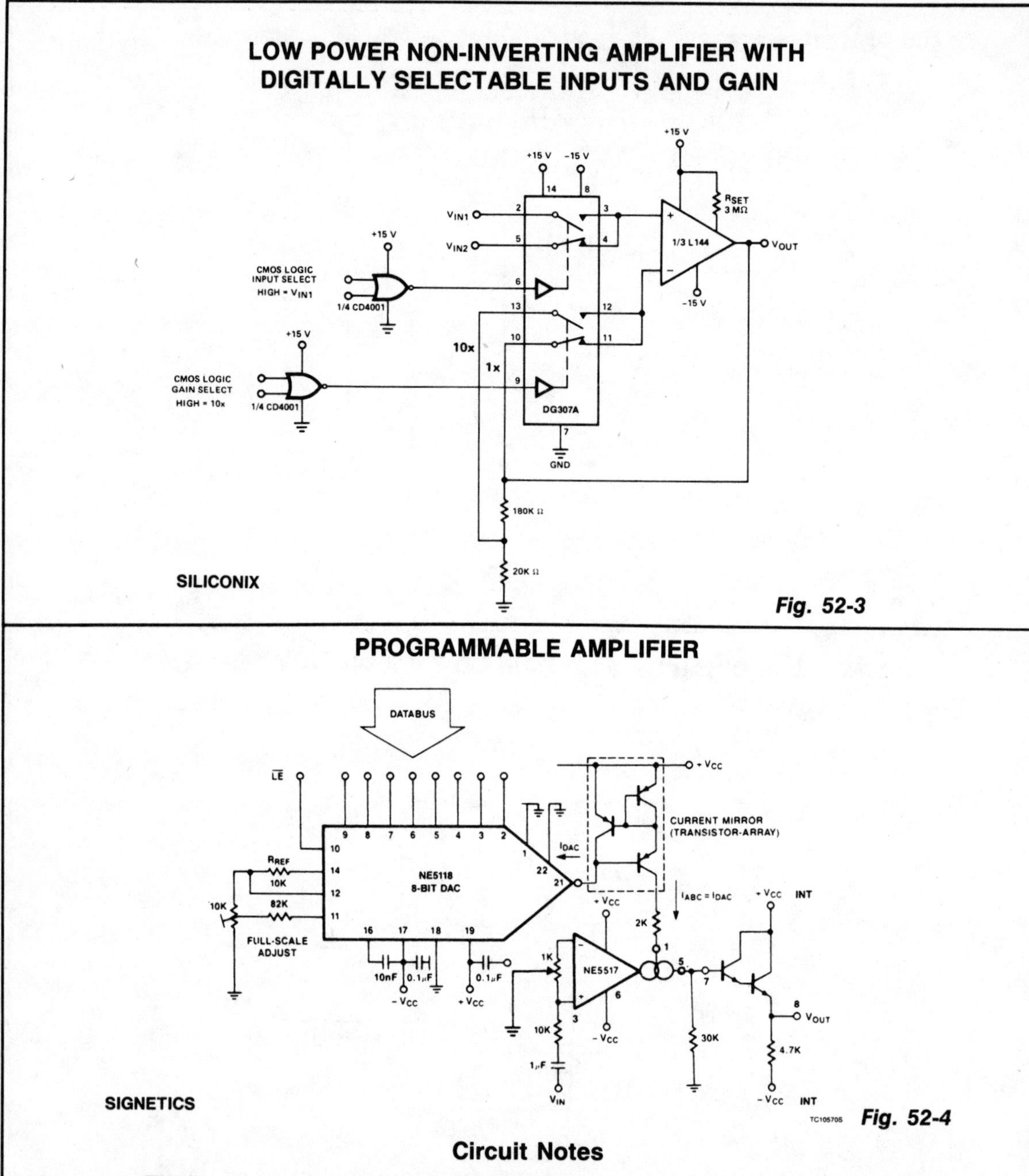

Fig. 52-3

Fig. 52-4

Circuit Notes

The intention of the following application shows how the NE5517 works in connection with a DAC. In the application, the NE5118 is used—an 8-bit DAC with current output—its input register making this device fully μP-compatible. The circuit consists of three functional blocks; the NE5118 which generates a control current equivalent to the applied data byte, a current mirror, and the NE5517.

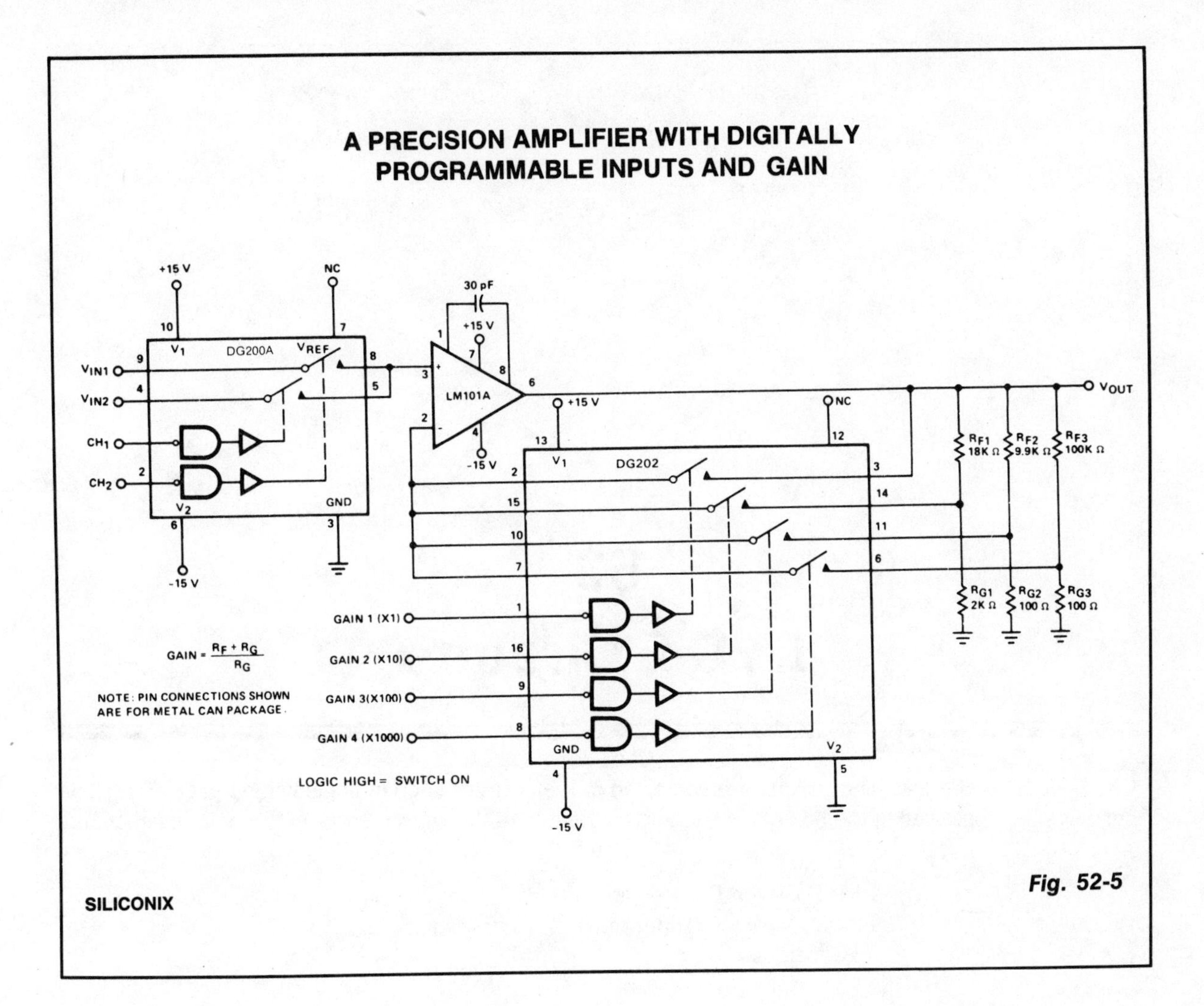
A PRECISION AMPLIFIER WITH DIGITALLY PROGRAMMABLE INPUTS AND GAIN
+15 V
NC
30 pF
+15 V
DG200A
LM101A
-15 V
VIN1
VIN2
CH1
CH2
V1
V2
VREF
GND
-15 V
+15 V
NC
DG202
GND
VOUT
RF1 18K Ω
RF2 9.9K Ω
RF3 100K Ω
RG1 2K Ω
RG2 100 Ω
RG3 100 Ω
GAIN 1 (X1)
GAIN 2 (X10)
GAIN 3(X100)
GAIN 4 (X1000)
GAIN = (RF + RG) / RG
NOTE: PIN CONNECTIONS SHOWN ARE FOR METAL CAN PACKAGE.
LOGIC HIGH = SWITCH ON
-15 V
SILICONIX

Fig. 52-5

53
LVDT Circuits

The sources of the following circuits are contained in the Sources section beginning on page 694. The figure number contained in the box of each circuit correlates to the source entry in the Sources section.

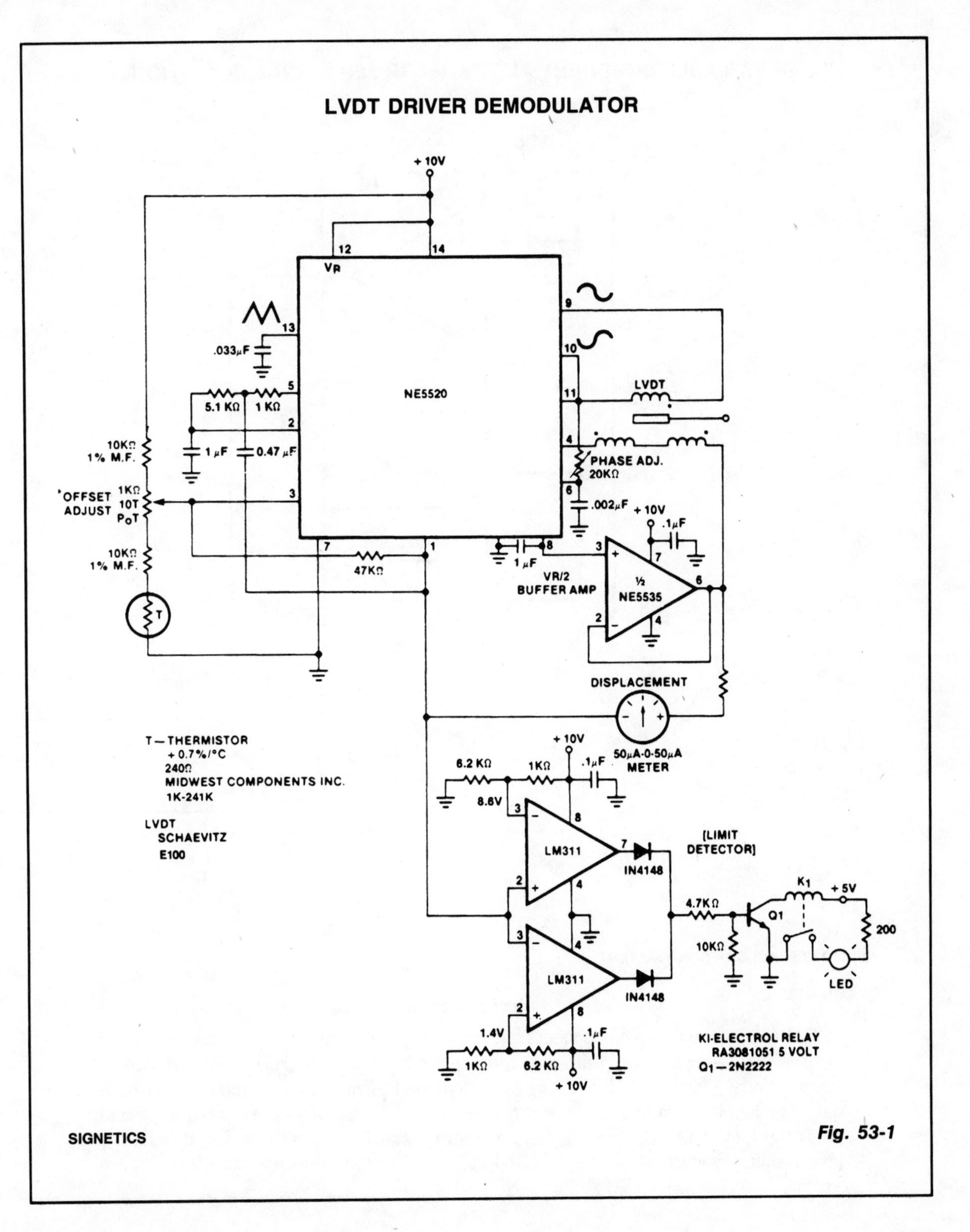
LVDT DRIVER DEMODULATOR
+10V
12
14
VR
NE5520
9
10
11
4
6
13
.033μF
5
5.1 KΩ
1 KΩ
2
1 μF
0.47 μF
3
10KΩ
1% M.F.
OFFSET
ADJUST
1KΩ
10T
PoT
10KΩ
1% M.F.
T
7
1
47KΩ
LVDT
PHASE ADJ.
20KΩ
.002μF
+10V
.1μF
8
1 μF
VR/2
BUFFER AMP
½
NE5535
DISPLACEMENT
50μA-0-50μA
METER
T—THERMISTOR
+0.7%/°C
240Ω
MIDWEST COMPONENTS INC.
1K-241K
LVDT
SCHAEVITZ
E100
6.2 KΩ
1KΩ
.1μF
8.6V
LM311
IN4148
[LIMIT
DETECTOR]
K1
+5V
4.7KΩ
Q1
10KΩ
200
LED
1.4V
1KΩ
6.2 KΩ
KI-ELECTROL RELAY
RA3081051 5 VOLT
Q1—2N2222
SIGNETICS
Fig. 53-1

LINEAR VARIABLE DIFFERENTIAL TRANSFORMER SIGNAL CONDITIONER

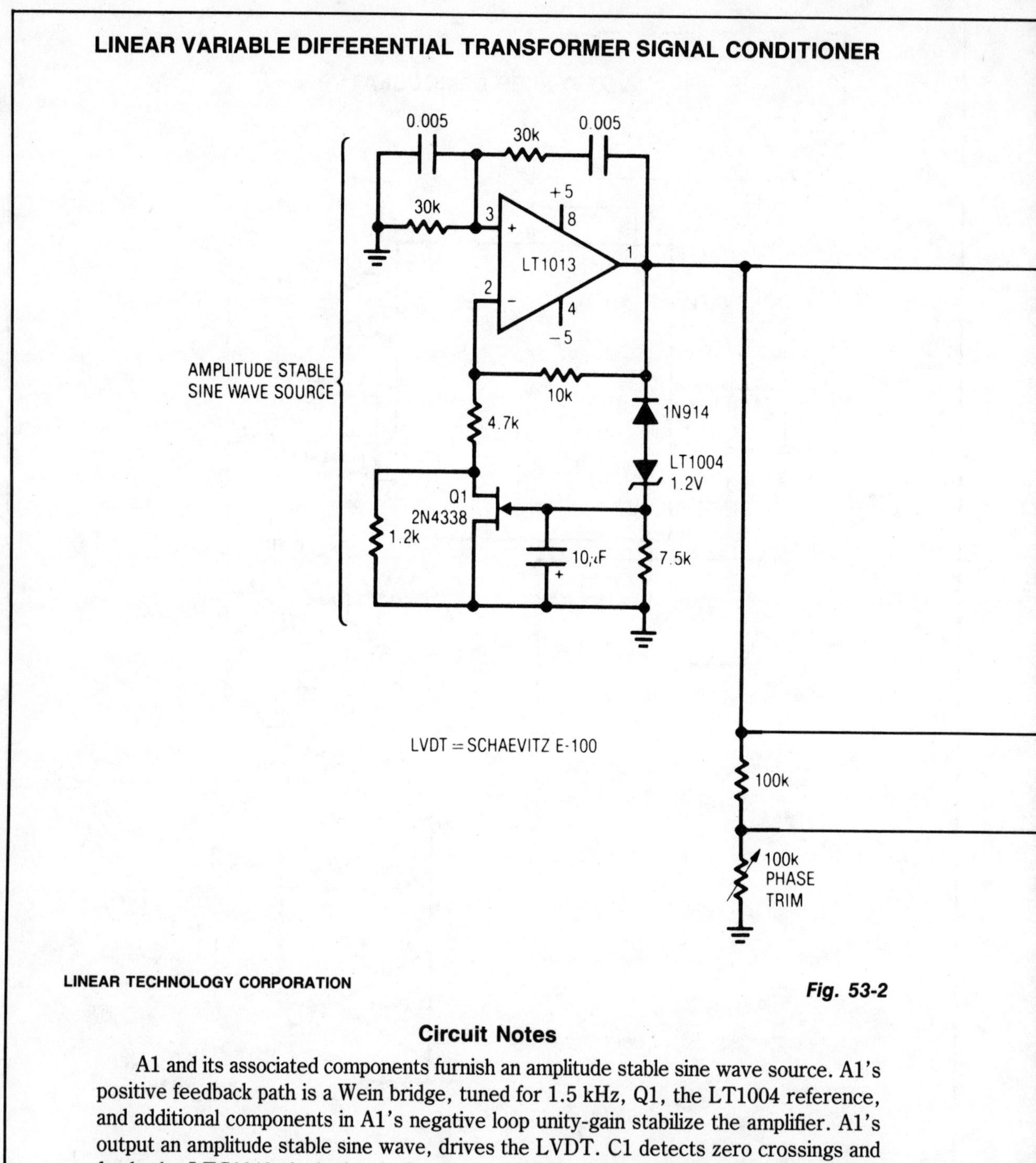

LINEAR TECHNOLOGY CORPORATION

Fig. 53-2

Circuit Notes

A1 and its associated components furnish an amplitude stable sine wave source. A1's positive feedback path is a Wein bridge, tuned for 1.5 kHz, Q1, the LT1004 reference, and additional components in A1's negative loop unity-gain stabilize the amplifier. A1's output an amplitude stable sine wave, drives the LVDT. C1 detects zero crossings and feeds the LTC1043 clock pin. A speed-up network at C1's input compensates LVDT phase shift, synchronizing the LTC1043's clock to the transformer's output zero

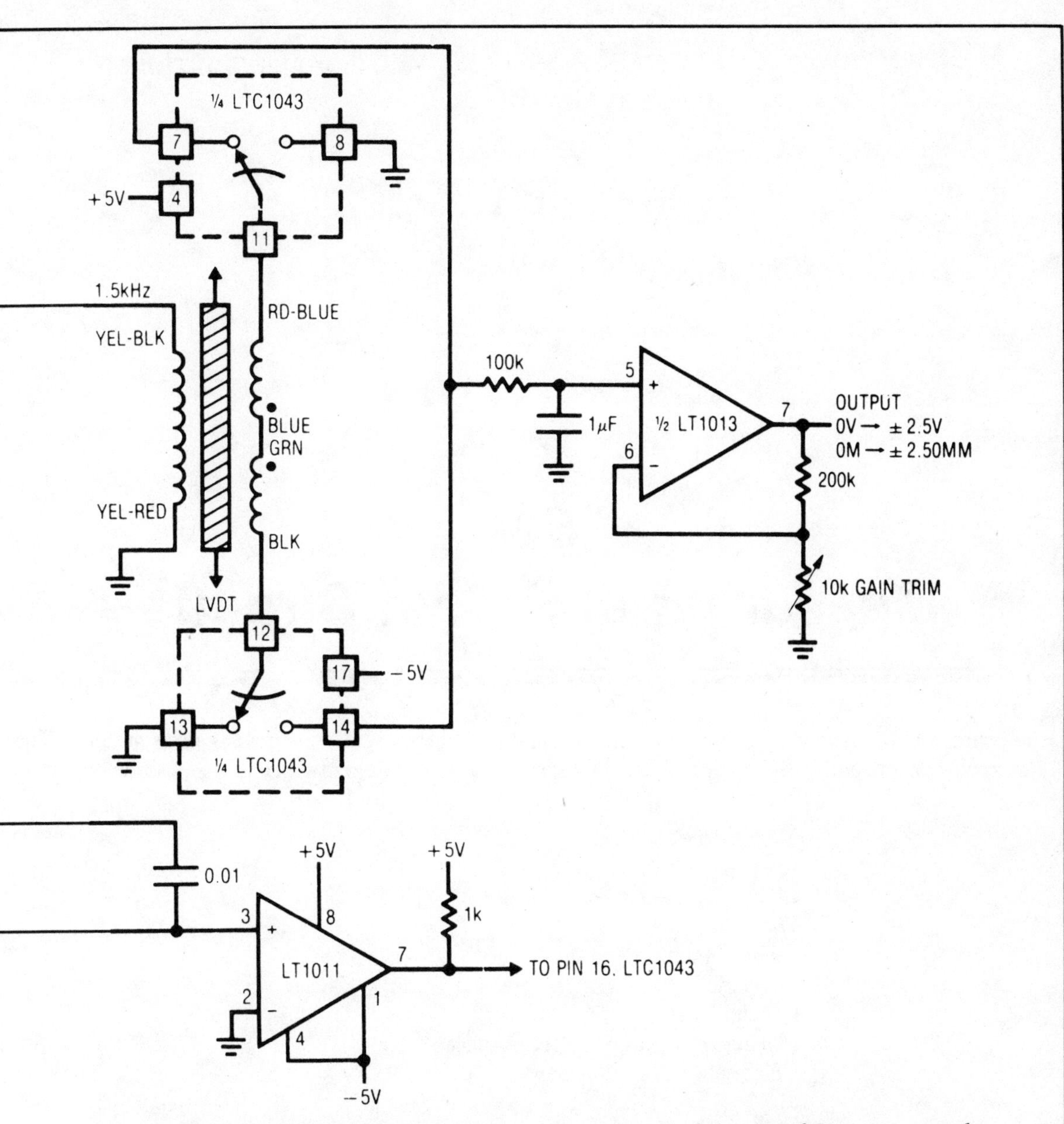

crossings. The LTC1043 alternately connects each end of the transformer to ground, resulting in positive half-wave rectification at pins 7 and 14. These points are summed at a low-pass filter which feeds A2. A2 furnishes gain scaling and the circuit's output. The LTC1043's synchronized clocking means the information presented to the low-pass filter is amplitude and phase sensitive. The circuit output indicates how far the core is from center and on which side. To calibrate this circuit, center the LVDT core in the transformer and adjust the phase trim for 0 V output. Next, move the core to either extreme position and set the gain trim for 2.50 V output.

54
Measuring and Test Circuits

The sources of the following circuits are contained in the Sources section beginning on page 694. The figure number contained in the box of each circuit correlates to the source entry in the Sources section.

MAGNETOMETER

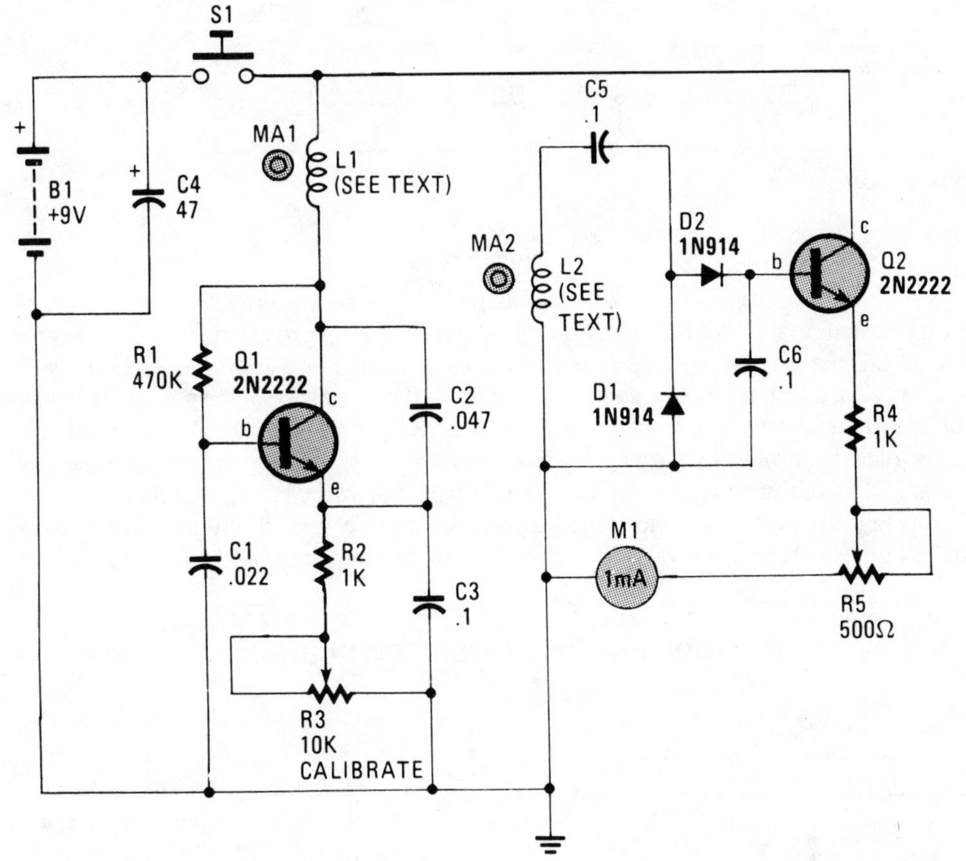

HANDS-ON ELECTRONICS

Fig. 54-1

Circuit Notes

The circuit uses two general-purpose npn transistors, Q1 and Q2, and a special hand-wound, dual-coil probe ferrets out the magnetism. Q1 and its associated components form a simple VLF oscillator circuit, with L1, C2, and C3 setting the frequency. The VLF signal received by the pickup coil, L2, is passed through C5 and rectified by diodes D1 and D2. The small dc signal output from the rectifier is fed to the base of Q2 (configured as an emitter follower), which is then fed to a 0-1 mA meter, M1.

RESISTANCE-RATIO DETECTOR

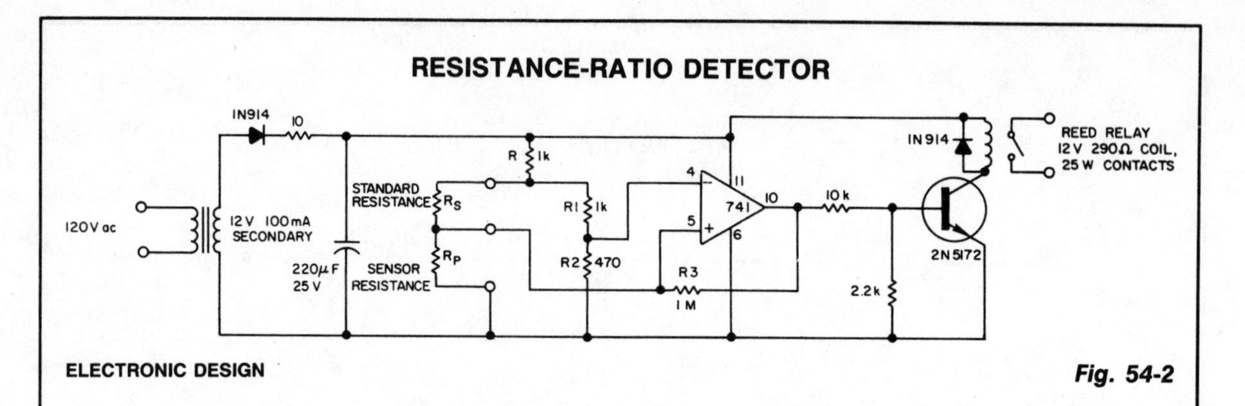

ELECTRONIC DESIGN

Fig. 54-2

Circuit Notes

Applications such as photoelectric control, temperature detection and moisture sensing require a circuit that can accurately detect a given resistance ratio. A simple technique that uses an op amp as a sensing element can provide 0.5% accuracy with low parts cost. The reed-relay contacts close when the resistance of the sensor Rp equals 47% of the standard Rs. Adjusting either R1 or R2 provides a variable threshold; the threshold is controlled by varying R3. For the most part, the type of resistors used for R1 and R2 determines the accuracy and stability of the circuit. With metal-film resistors, less than 0.5% change in ratio sensing occurs over the commercial temperature range (0 to 70 C) with ac input variations from 105 to 135 V.

CONTINUITY TESTER FOR PCB'S

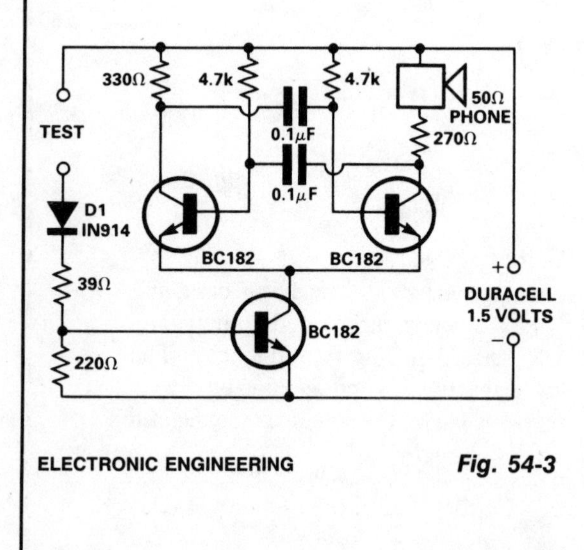

ELECTRONIC ENGINEERING

Fig. 54-3

Circuit Notes

The continuity tester is for tracing wiring on printed circuit boards. It only consumes any appreciable power when the test leads are shorted, so no On/Off switch is used or required. The applied voltage at the test terminals is insufficient to turn on diodes or other semiconductors. Resistors below 50 ohms act as short circuit; above 100 ohms as open circuit. The circuit is a simple multivibrator—T1 and T2, which are switched on by transistor T3. The components in the base of T3 are D1, R1, R2, and the test resistance. With a 1.5 volt supply, there is insufficient voltage to turn on a semiconductor connected to the test terminals. The phone is a telephone earpiece but a 30 ohm speaker would work equally as well.

WIRE TRACER

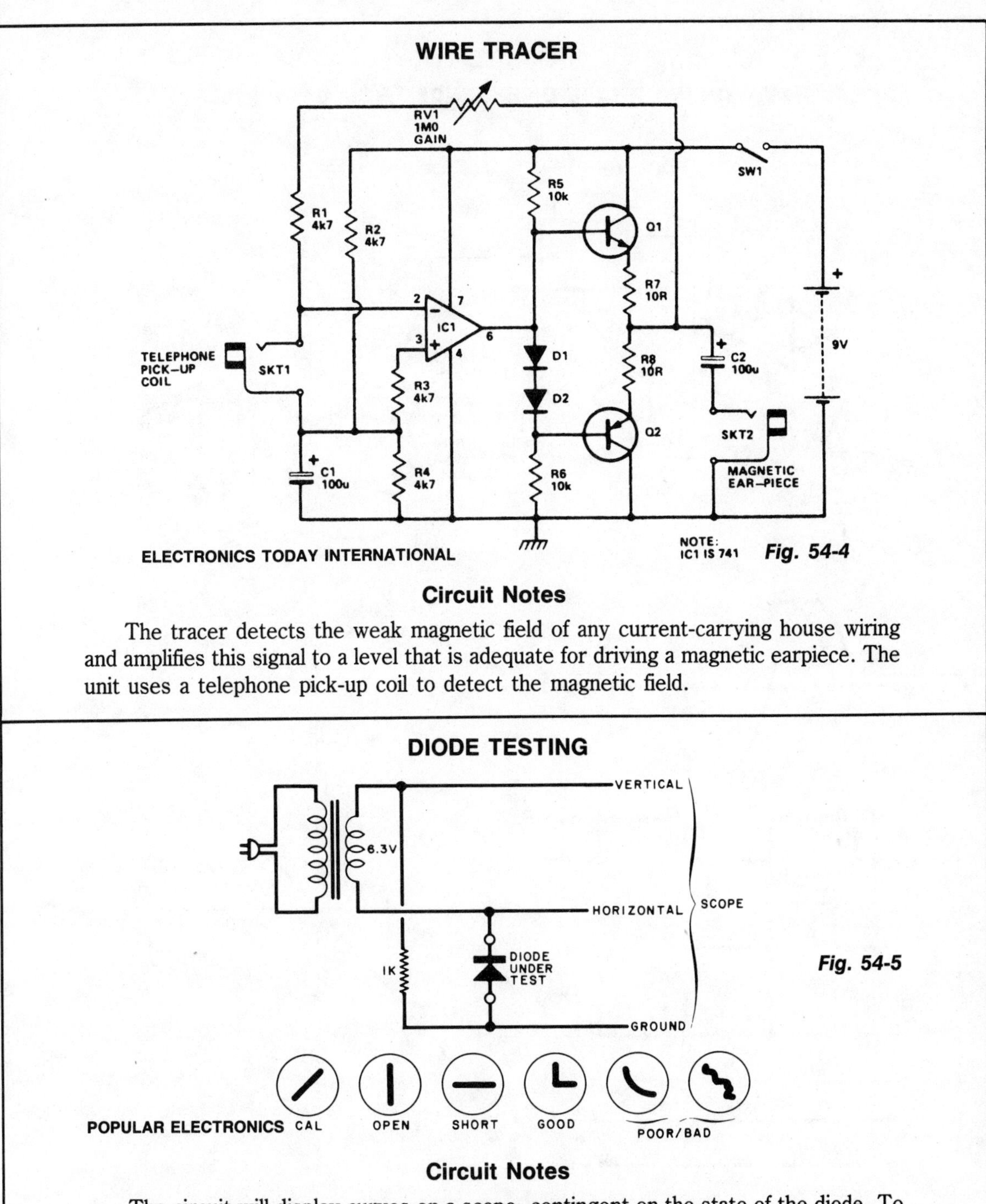

ELECTRONICS TODAY INTERNATIONAL

Fig. 54-4

Circuit Notes

The tracer detects the weak magnetic field of any current-carrying house wiring and amplifies this signal to a level that is adequate for driving a magnetic earpiece. The unit uses a telephone pick-up coil to detect the magnetic field.

DIODE TESTING

Fig. 54-5

POPULAR ELECTRONICS

Circuit Notes

The circuit will display curves on a scope, contingent on the state of the diode. To "calibrate," substitute a 1000-ohm resistor for the diode and adjust the scope gains for a 45-degree line. The drawings show some expected results.

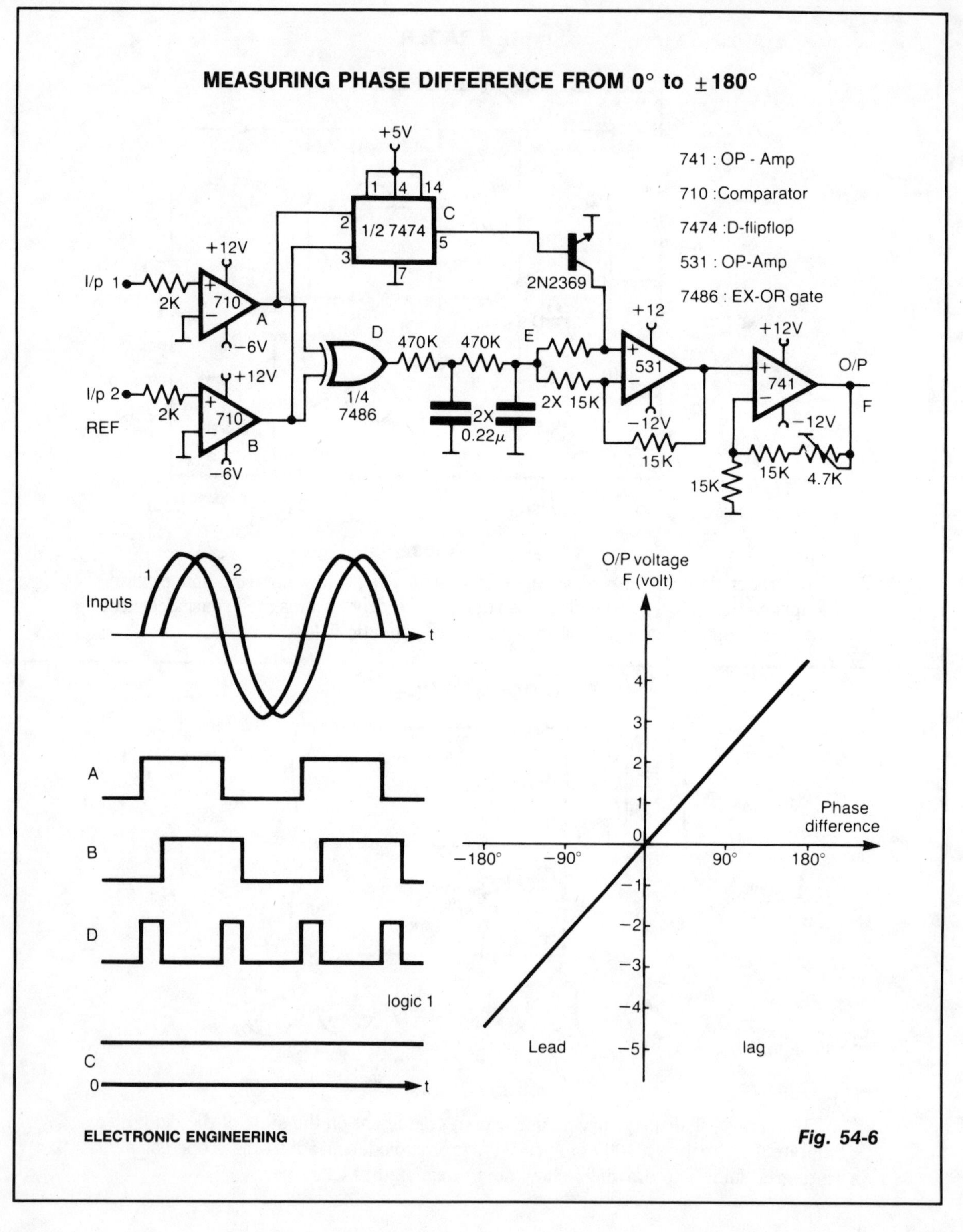
MEASURING PHASE DIFFERENCE FROM 0° to ±180°
+5V
1
4
14
2
1/2 7474
C
5
3
7
741 : OP - Amp
710 :Comparator
7474 :D-flipflop
531 : OP-Amp
7486 : EX-OR gate
+12V
I/p 1
2K
710
A
−6V
2N2369
+12
+12V
D
470K
470K
E
531
741
O/P
I/p 2
+12V
2K
REF
710
B
−6V
1/4
7486
2X
0.22μ
2X 15K
−12V
−12V
F
15K
15K
15K
4.7K
Inputs
1
2
t
A
B
D
logic 1
C
0
t
O/P voltage
F (volt)
4
3
2
1
0
−1
−2
−3
−4
−5
Phase
difference
−180°
−90°
90°
180°
Lead
lag
ELECTRONIC ENGINEERING
Fig. 54-6

MEASURING PHASE DIFFERENCE FROM 0° to ±180°, Continued.

Circuit Notes

This method is capable of measuring phase between 0 to ±180°. The generated square waves A and B are fed to a D flip-flop which gives an output C equal to logic 1 when input 1 leads input 2 and equal to logic 0 in case of lagging. When C = logic 0, the output of the amplifier F will be positive proportional to the average value E of the output of the EX-OR. When C = logic 1, F will be negative and also proportional to E by the same factor. Hence, the output of the meter is positive in case of lagging and negative for leading. The circuit is tested for sinusoidal inputs and indicates a linearity within 1%. Measurements are unaffected by the frequency of the inputs up to 75 kHz.

GROUND TESTER

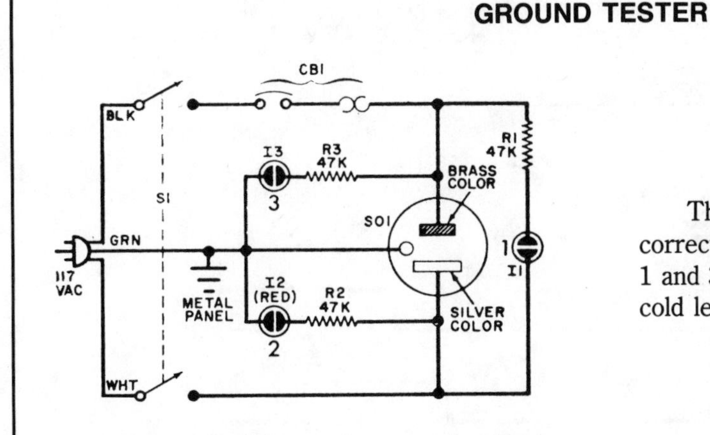

POPULAR ELECTRONICS *Fig. 54-7*

Circuit Notes

The circuit is simple and foolproof if wired correctly. Under normal conditions, only lamps 1 and 3 should be lit. If lamp 2 comes on, the cold lead is 117 volts above ground.

MAKING SLOW LOGIC PULSES AUDIBLE

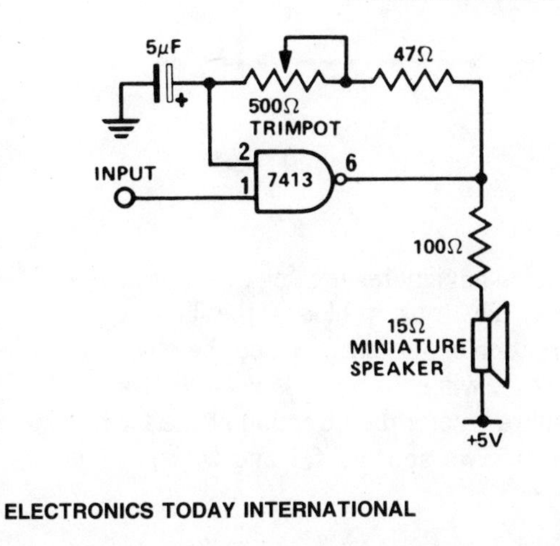

ELECTRONICS TODAY INTERNATIONAL *Fig. 54-8*

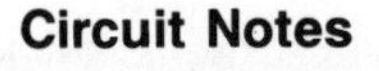

Circuit Notes

This circuit is useful for monitoring slow logic pulses as a keying monitor or digital clock alarm. The Schmitt trigger is connected as an oscillator. The trimpot controls the pitch of the output. When the input goes high, the circuit will oscillate.

UNIDIRECTIONAL MOTION SENSOR

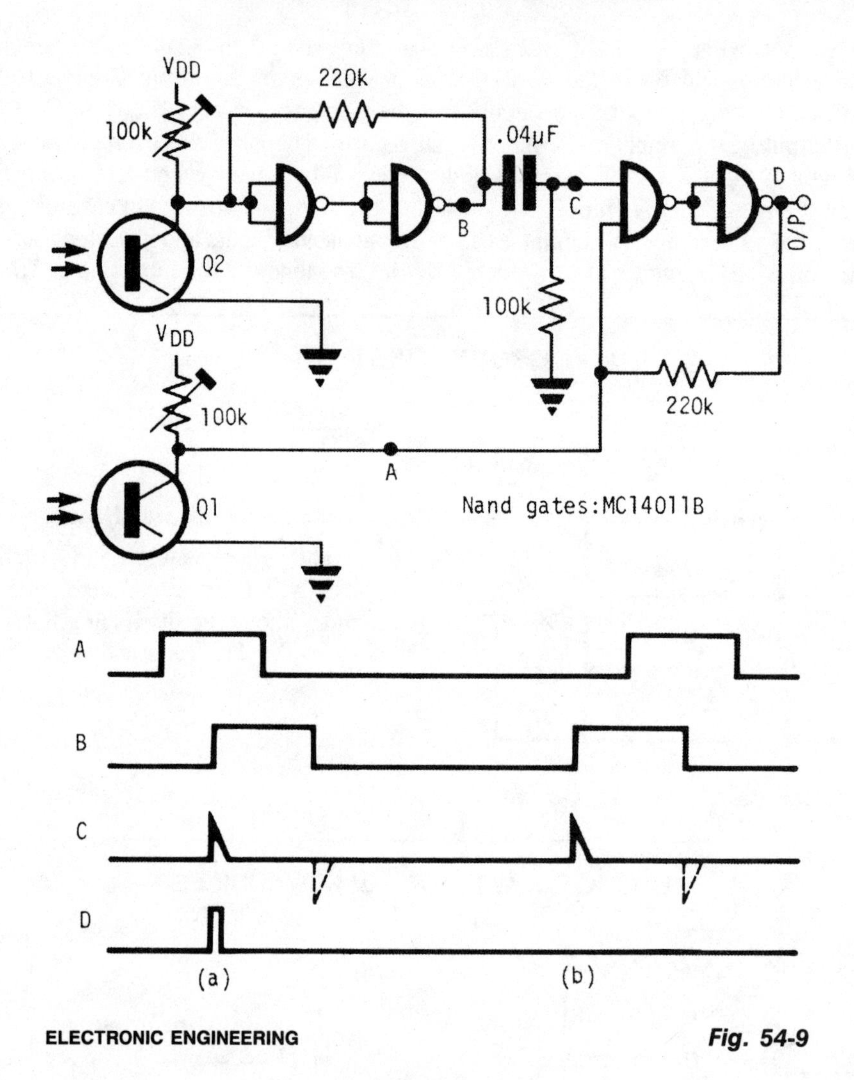

ELECTRONIC ENGINEERING

Fig. 54-9

Circuit Notes

This circuit detects an object passing in one direction but ignores it going the opposite way. Two sensors define the sense of direction. The object blocks the light to phototransistor Q1 or Q2 first dependent on the direction of approach. When the object passes Q1 then Q2, an output pulse is generated at D; while no pulse is seen at D as the object passes Q2 then Q1. Object length (measured along the direction of the two sensors) should be greater than the separation of the two sensors Q1 and Q2.

55
Medical Electronics Circuits

The sources of the following circuits are contained in the Sources section beginning on page 694. The figure number contained in the box of each circuit correlates to the source entry in the Sources section.

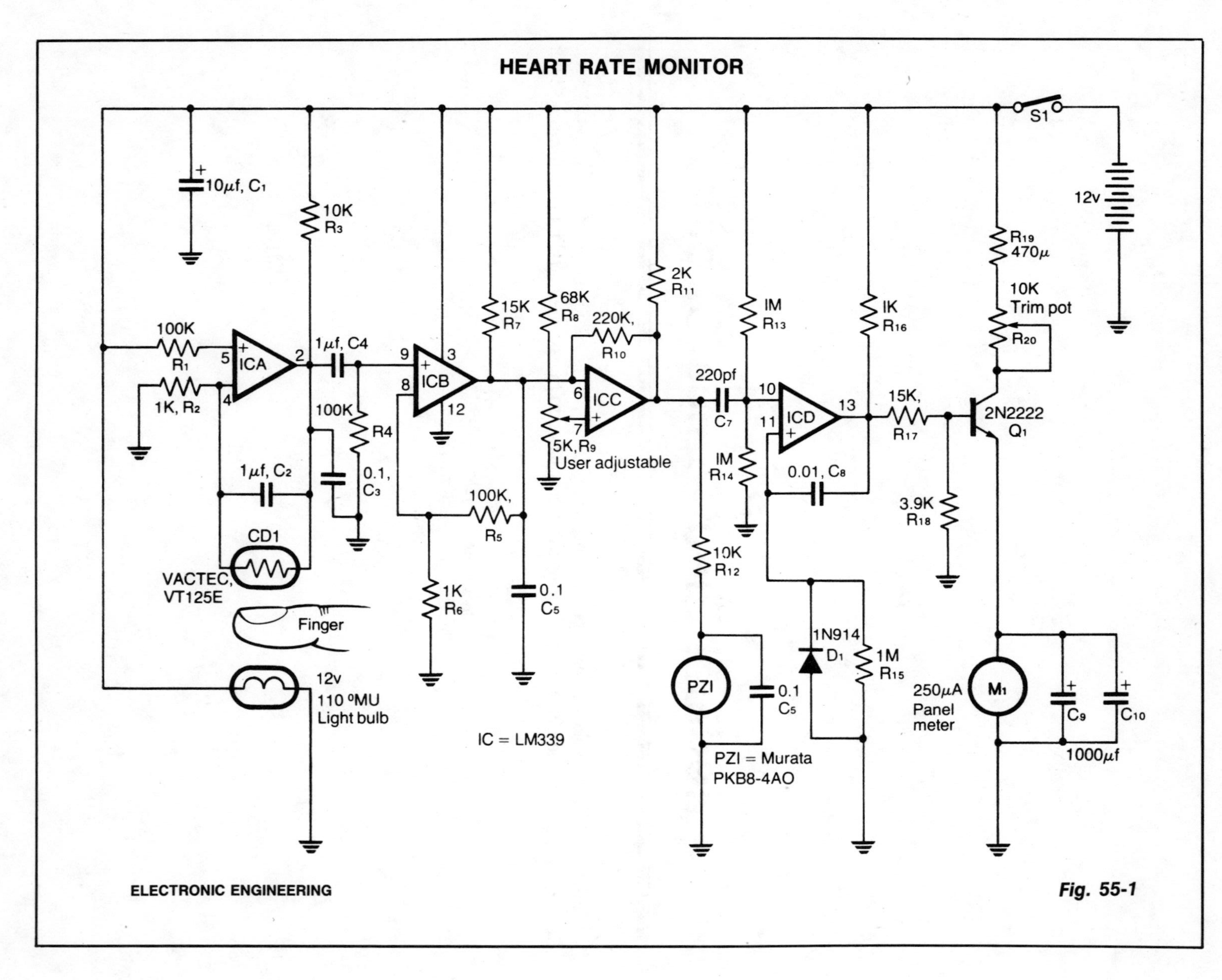
HEART RATE MONITOR
10μf, C1
10K R3
100K R1
1K, R2
ICA
5
4
2
1μf, C4
100K
R4
0.1, C3
1μf, C2
CD1
VACTEC, VT125E
Finger
12v
110 ºMU
Light bulb
ICB
9
8
3
12
15K R7
68K R8
220K, R10
100K, R5
1K R6
0.1 C5
ICC
6
7
5K, R9
User adjustable
2K R11
IC = LM339
220pf
C7
IM R13
IM R14
10K R12
PZI
0.1 C5
PZI = Murata
PKB8-4AO
ICD
10
11
13
0.01, C8
1N914 D1
1M R15
IK R16
15K, R17
3.9K R18
2N2222 Q1
R19 470μ
10K Trim pot
R20
S1
12v
250μA Panel meter
M1
C9
C10
1000μf
ELECTRONIC ENGINEERING
Fig. 55-1

HEART RATE MONITOR, Continued.

Circuit Notes

Light filtering through the finger tip is detected by the cadmium sulphide photoresistor CD1 which forms the feedback network for transducer amplifier section ICA producing a weak signal which is further amplified by ICB. This signal is now compared against a user adjusted threshold, comparator ICD triggers gating on the piezoelectric buzzer PZ1. On each falling edge of the comparator's output signal one-shot multivibrator ICD produces a 2 μs pulse which is inverted by Q1 and averaged by the RC network consisting of M1, C6 and C7. The 10 K trimpot R20 in Q1's collector circuit sets the scale factor for M_1 where full scale is 150 beats per minute.

MEDICAL INSTRUMENT PREAMPLIFIER

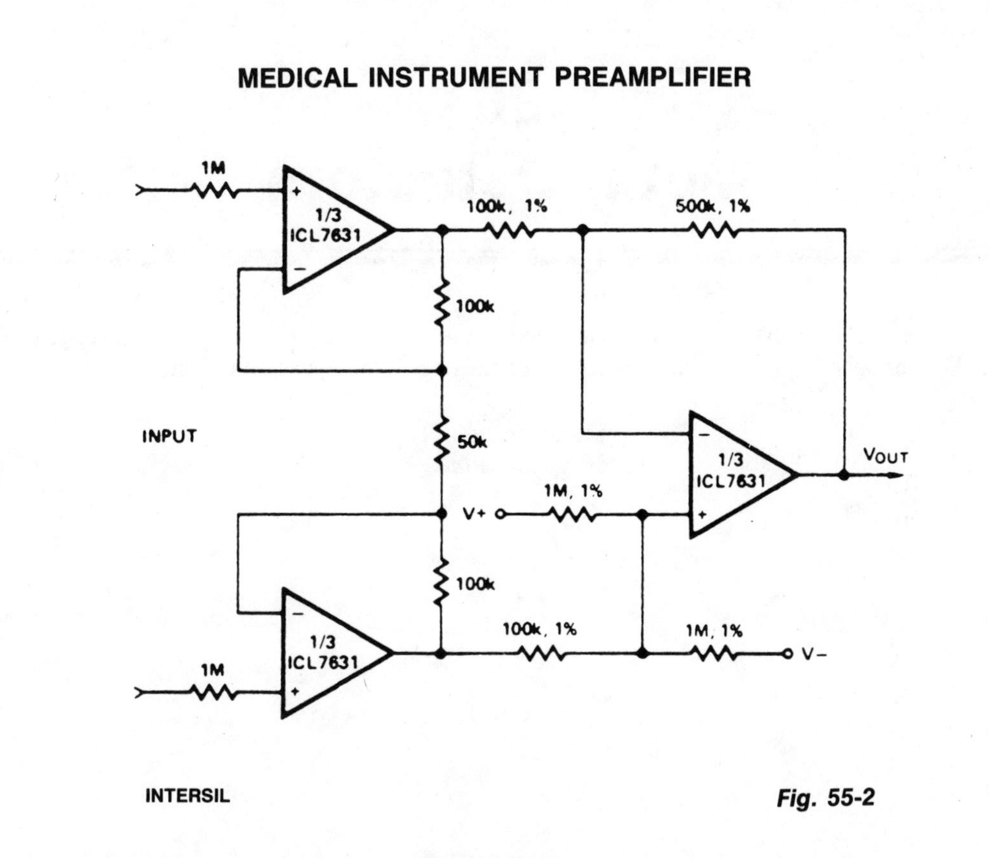

INTERSIL

Fig. 55-2

Circuit Notes

Note that A_{VOL} = 25; single Ni-cad battery operation. Input current (from sensors connected to the patient) is limited to less than 5 μA under fault conditions.

56
Metal Detectors

The sources of the following circuits are contained in the Sources section beginning on page 694. The figure number contained in the box of each circuit correlates to the source entry in the Sources section.

METAL LOCATOR II

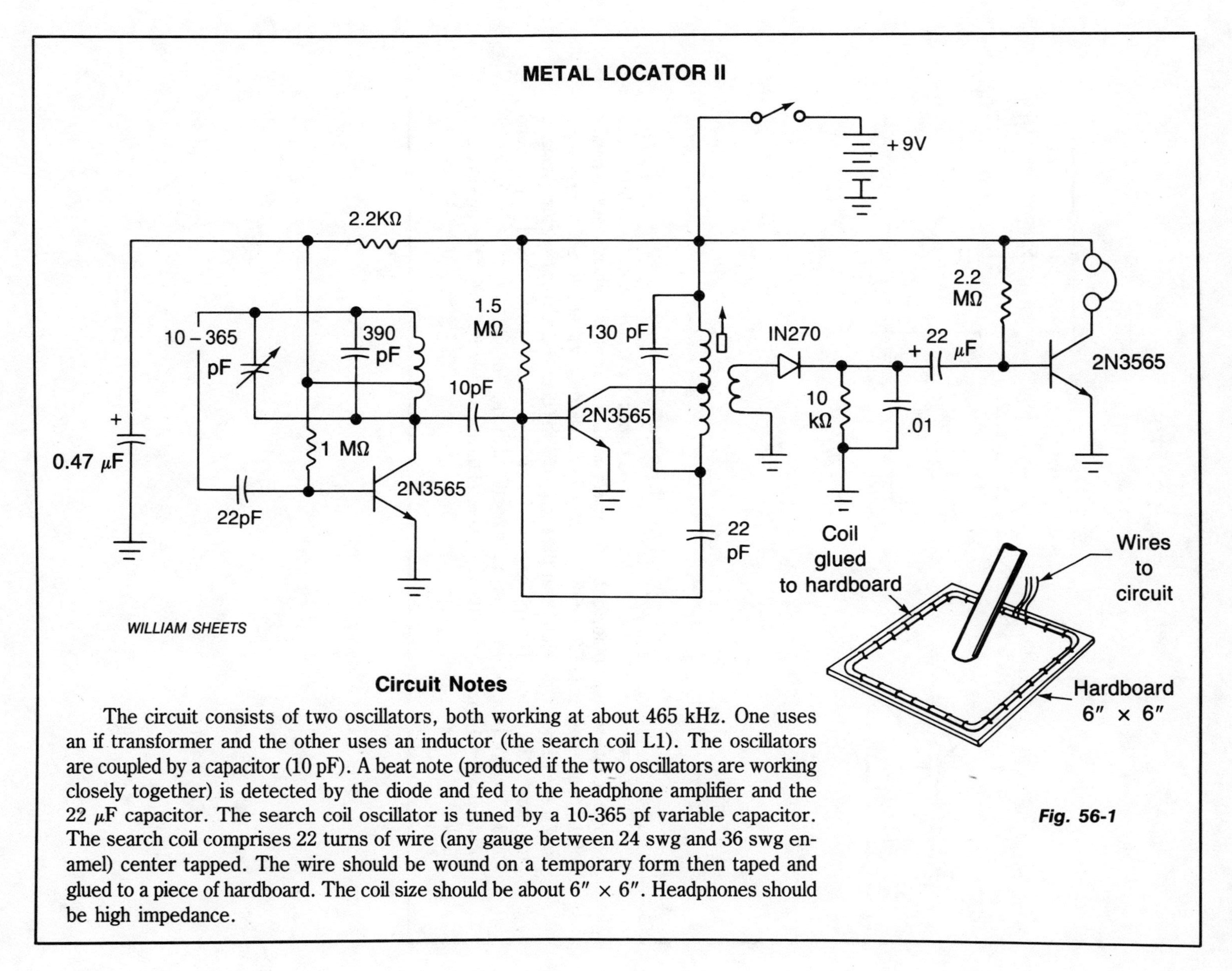

WILLIAM SHEETS

Circuit Notes

The circuit consists of two oscillators, both working at about 465 kHz. One uses an if transformer and the other uses an inductor (the search coil L1). The oscillators are coupled by a capacitor (10 pF). A beat note (produced if the two oscillators are working closely together) is detected by the diode and fed to the headphone amplifier and the 22 μF capacitor. The search coil oscillator is tuned by a 10-365 pf variable capacitor. The search coil comprises 22 turns of wire (any gauge between 24 swg and 36 swg enamel) center tapped. The wire should be wound on a temporary form then taped and glued to a piece of hardboard. The coil size should be about 6″ × 6″. Headphones should be high impedance.

Fig. 56-1

METAL LOCATOR

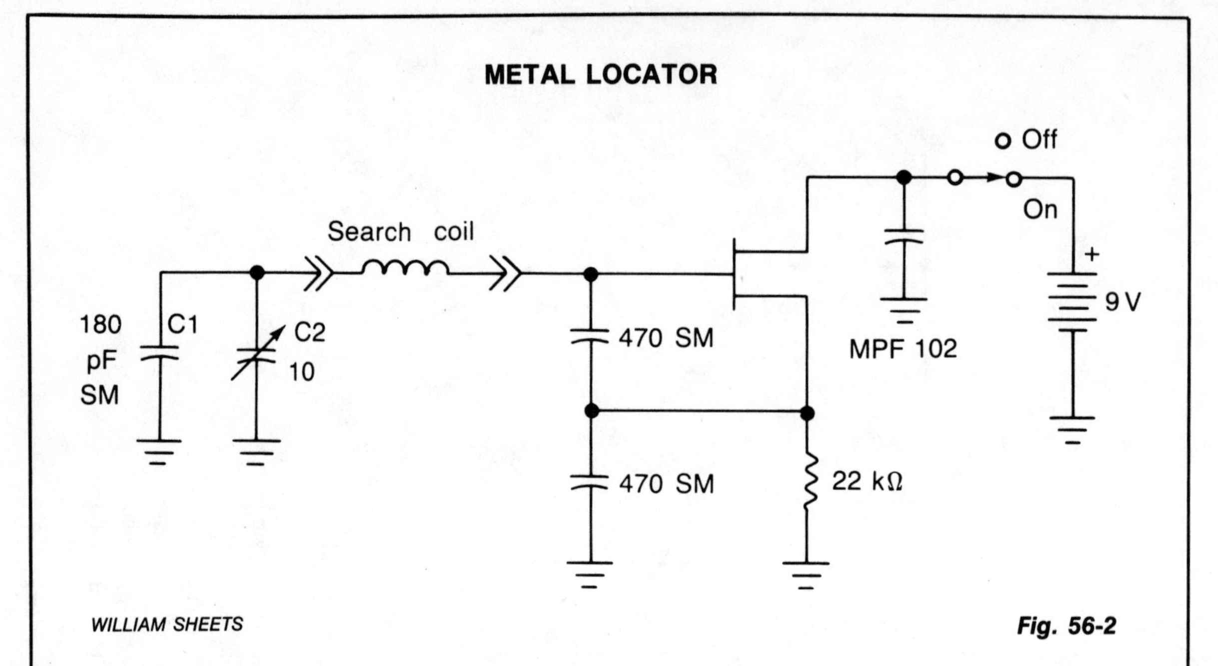

WILLIAM SHEETS

Fig. 56-2

Circuit Notes

The search coil, C1 and C2 form a tuned circuit for the oscillator which is tuned near the center of the broadcast band. Tune a portable radio to a station near the middle of the band, then tune C2 until a squeal is heard as the two signals mix to produce a beat (heterodyne) note. Metal near the search coil will detune the circuit slightly, changing the pitch of the squeal. The search coil is 20 turns of number 30 enameled wire, wound on a 6″ × 8″ wood or plastic form. It is affixed at the end of a 30″ to 40″ wooden or plastic pole, and connected to the remainder of the metal detector circuit through a coaxial cable.

57
Metronomes

The sources of the following circuits are contained in the Sources section beginning on page 694. The figure number contained in the box of each circuit correlates to the source entry in the Sources section.

SIMPLE METRONOME

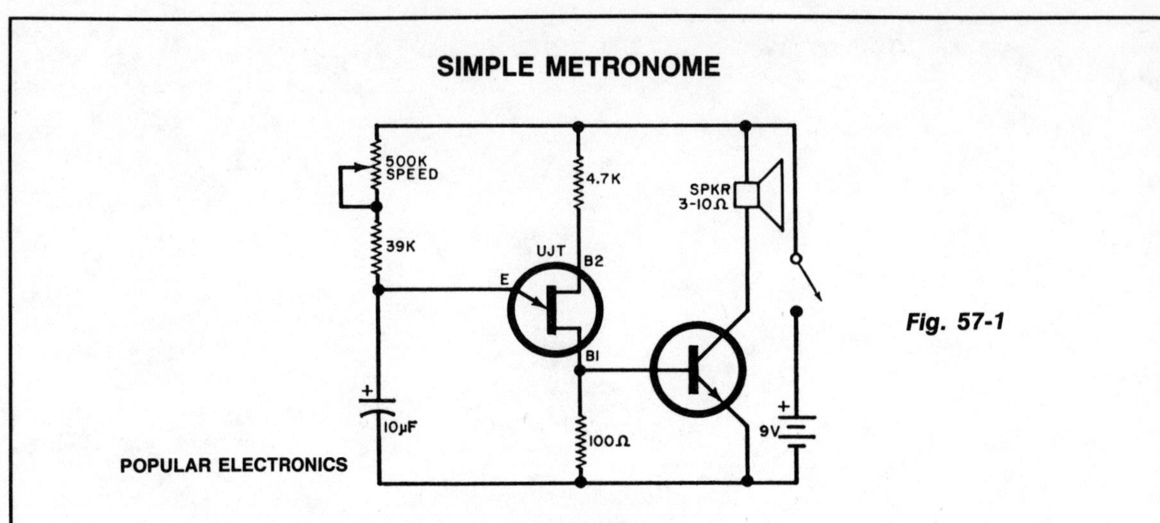

Fig. 57-1

Circuit Notes

Adjustable from 15 to 240 beats per minute. The UJT oscillator output is applied to a general purpose npn transistor which drives the speaker.

UJT = 2N4871
NPNxistor = TIP31

METRONOME I

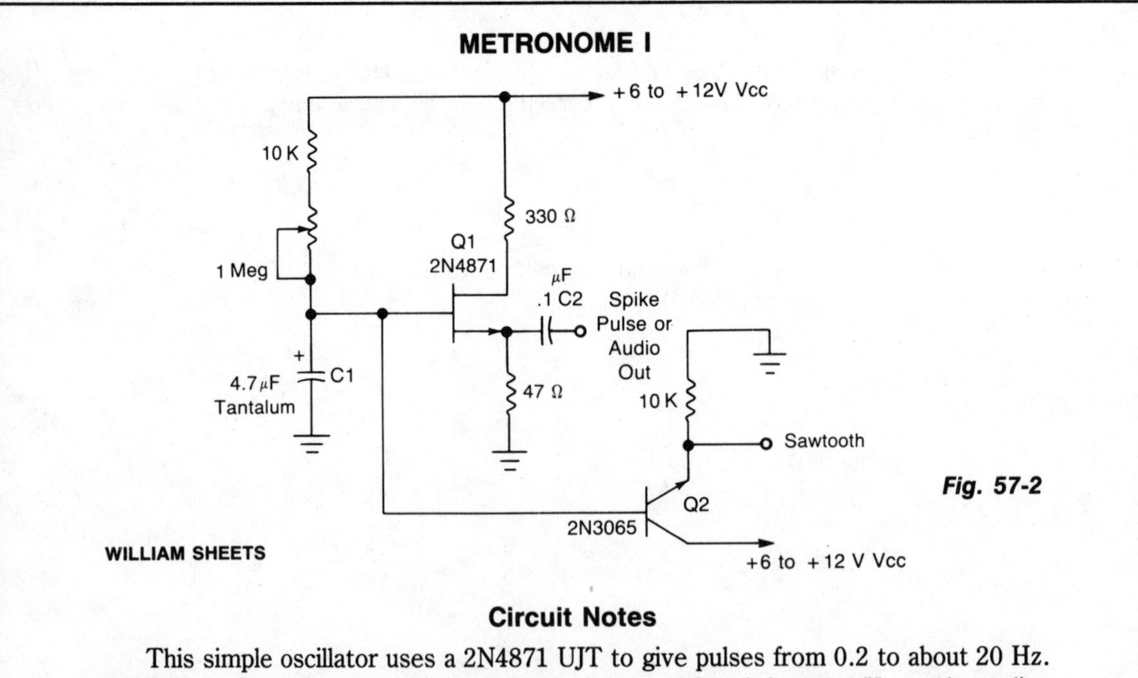

Fig. 57-2

Circuit Notes

This simple oscillator uses a 2N4871 UJT to give pulses from 0.2 to about 20 Hz. A spike is available at C2, a sawtooth at the emitter of Q2 of about 2-3 V p-p, depending on V_{CC}.

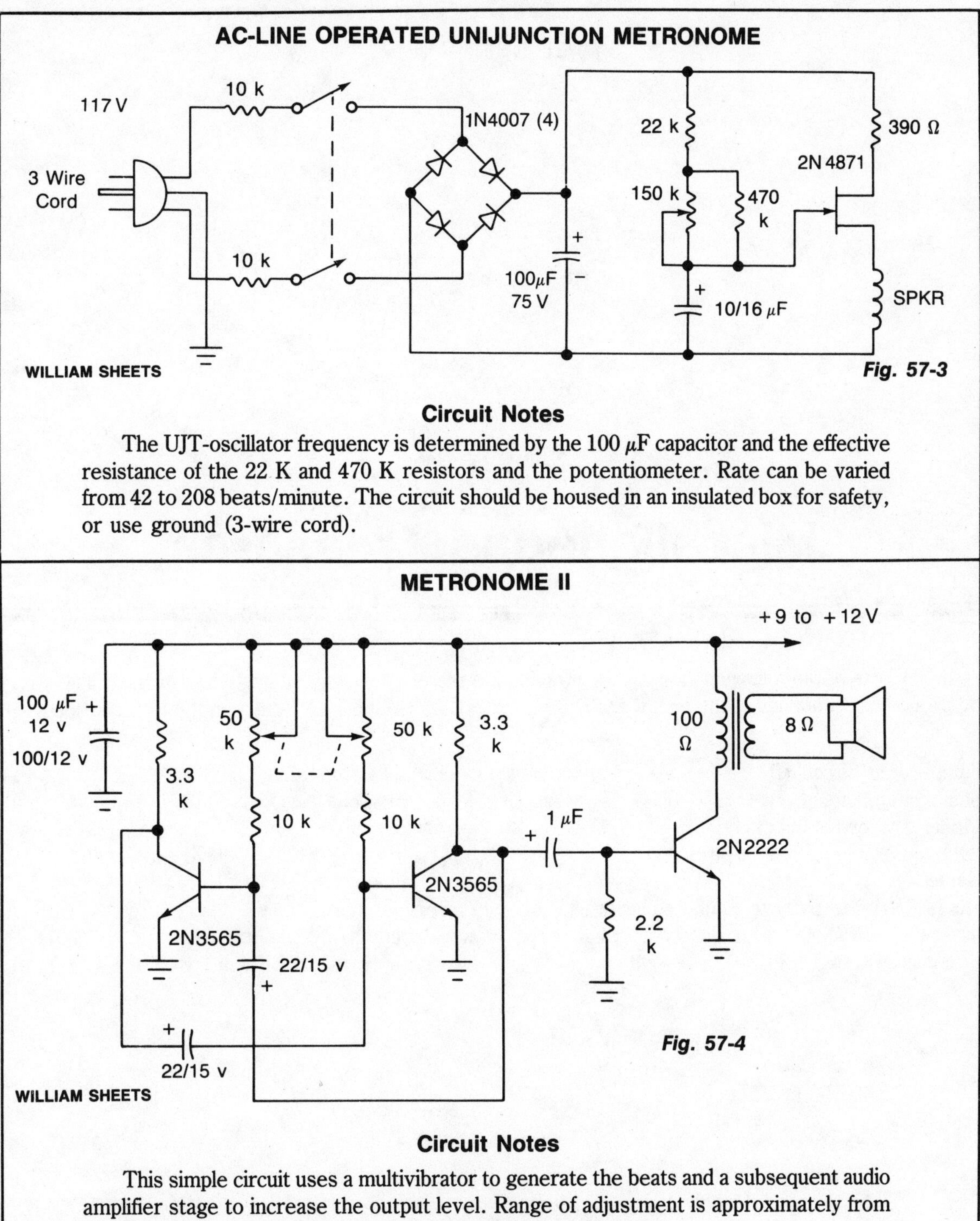

Fig. 57-3

Circuit Notes

The UJT-oscillator frequency is determined by the 100 μF capacitor and the effective resistance of the 22 K and 470 K resistors and the potentiometer. Rate can be varied from 42 to 208 beats/minute. The circuit should be housed in an insulated box for safety, or use ground (3-wire cord).

Fig. 57-4

Circuit Notes

This simple circuit uses a multivibrator to generate the beats and a subsequent audio amplifier stage to increase the output level. Range of adjustment is approximately from 40 to 200 beats per minute set by the gauged potentiometer.

58
Miscellaneous Treasures

The sources of the following circuits are contained in the Sources section beginning on page 694. The figure number contained in the box of each circuit correlates to the source entry in the Sources section.

SQUIB-FIRING CIRCUIT (I)

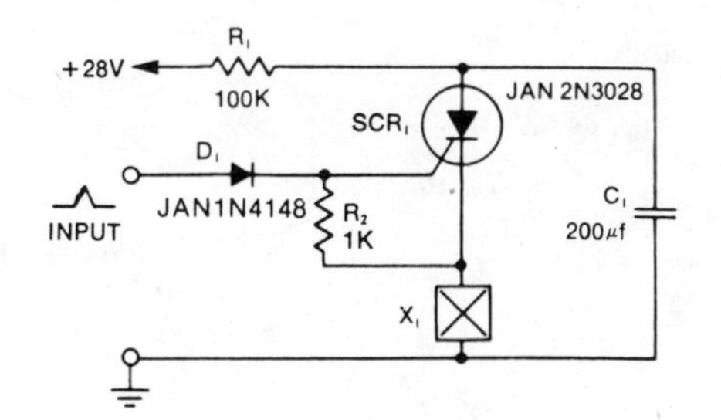

UNITRODE CORP.

Fig. 58-1

Circuit Notes

Capacitor C1 is charged to +28 V through R1 and stores energy for firing the squib. A positive pulse of 1 mA applied to the gate of SCR1 will cause it to conduct, discharging C1 into the squib load X1. With the load in the cathode circuit, the cathode rises immediately to +28 V as soon as the SCR is triggered on. Diode D1 decouples the gate from the gate trigger source, allowing the gate to rise in potential along with the cathode so that the negative gate-to-cathode voltage rating is not exceeded. This circuit will reset itself after test firing, since the available current through R1 is less than the holding current of the SCR. After C1 has been discharged, the SCR automatically turns off—allowing C1 to recharge.

SQUIB-FIRING CIRCUIT (II)

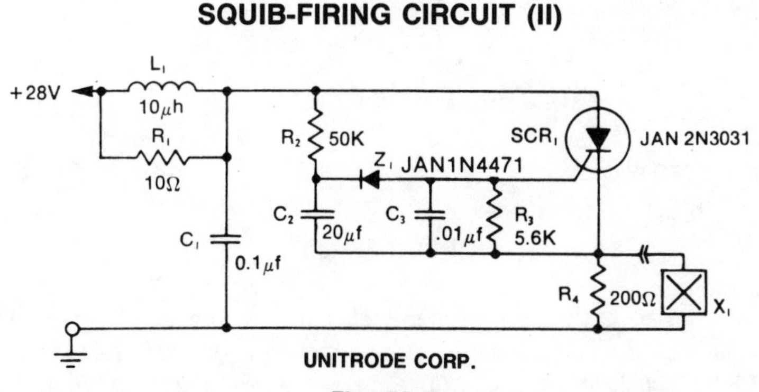

UNITRODE CORP.

Fig. 58-2

Circuit Notes

The LRC input network limits the anode dv/dt to a safe value—below 30 V/µs. R1 provides critical damping to prevent voltage overshoot. While a simple RC filter section could be used, the high current required by the squib would dictate a small value of resistance and a much larger capacitor. Resistor R3 provides dc bias stabilization, while C3 provides stiff gate bias during the transient interval when anode voltage is applied. The SCR is fired one second after arming by means of the simple R2C2Z1 time delay network. R4 provides a load for the SCR for testing the circuit with the squib disconnected—limiting the current to a level well within the continuous rating of the SCR. The circuit can be reset by opening the +28 V supply and then re-arming.

MODEL ROCKET LAUNCHER

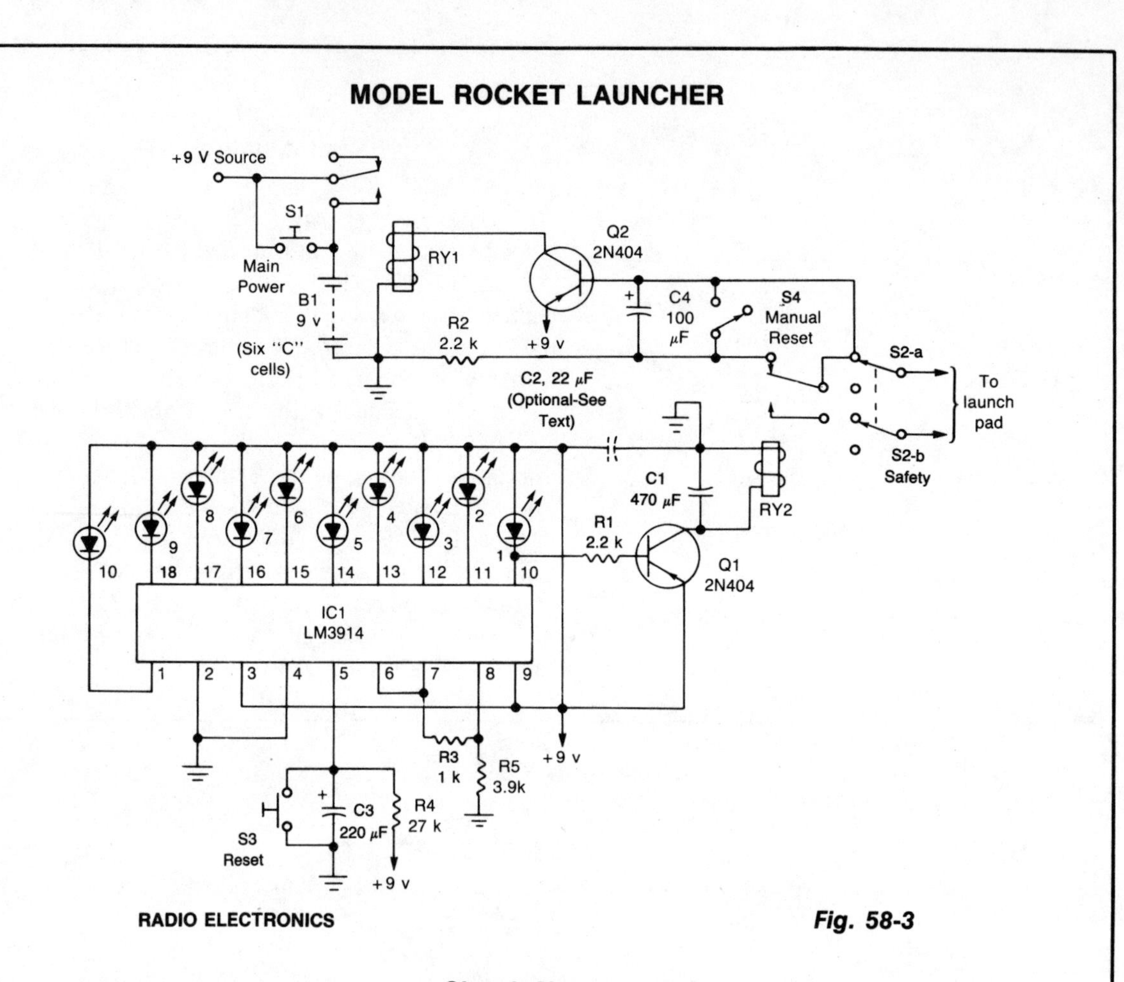

RADIO ELECTRONICS

Fig. 58-3

Circuit Notes

The circuit consists of the launch timer itself and an automatic-off timer. When power is applied to that IC, the countdown LED's sequence is on until they are all lit. When the last one LED1, is fully lit, transistor Q1 saturates, energizing RY2. When that happens, a circuit between the lantern battery at the launch pad and the nickel-chromium wire is completed; the wire heats up as before, and the rocket is launched. Resistor R4 and capacitor C3 determine the countdown timing; with the values shown it should be approximately 10 seconds. Resistors R3 and R5 set the LED brightness. Safety is of the utmost importance. That's the purpose of the second half of the circuit. When RY2 opens, the current flow to Q2 is disrupted. But, because of the presence of R2 and C4, the transistor remains saturated for about 3 seconds. After that, however, the transistor stops conducting and RY1 is de-energized. That cuts off the power to the rest of the circuit, and RY2 de-energizes again, breaking the circuit to the launch pad. Switch S3 is used to reset the countdown. Once that is done, pressing S1 starts the launch sequence; the rest is automatic. Switch S4 is used to latch RY1 manually if needed.

PUSH-ON/PUSH-OFF ELECTRONIC SWITCH

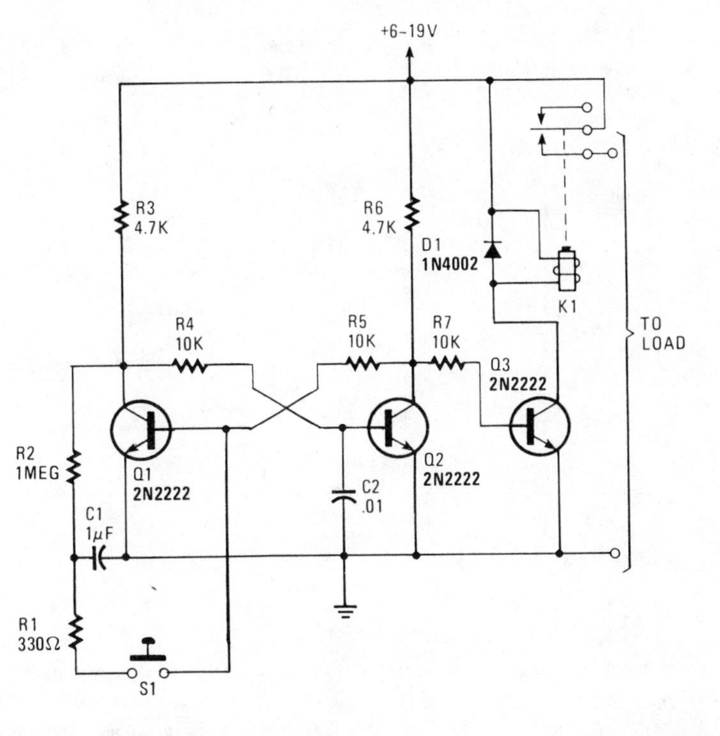

HANDS-ON ELECTRONICS

Fig. 58-4

Circuit Notes

Transistors Q1 and Q2 make up the flip-flop while Q3 drives a reed relay. When power is first applied to the circuit, Q1 and Q3 are conducting and Q2 is cut off. Momentarily closing S1 causes the flip-flop to switch states—Q1 cuts off and Q2 conducts. When Q2 is conducting, its collector drops to around 0.6 volt. That prevents base current from flowing into Q3 so it is cut off, de-energizing relay K1. The flip-flop changes state every time S1 is pressed. Capacitors C1 and C2 ensure that Q1 is always the transistor that turns on when power is first applied to the circuit. When power is first applied to the basic flip-flop, the initial status is random—Q1 and Q2 both try to conduct and, usually, the transistor with the higher gain will take control, reaching full conduction and cutting off the other one. However, differences in the values of the collector and coupling resistors will also influence the initial state at power-on. With C2 in the circuit, it and R4 form an R-C network that slightly delays the rise in Q2's base voltage. That gives Q1 sufficient time to reach saturation and thus take control.

GAME FEEDER CONTROLLER

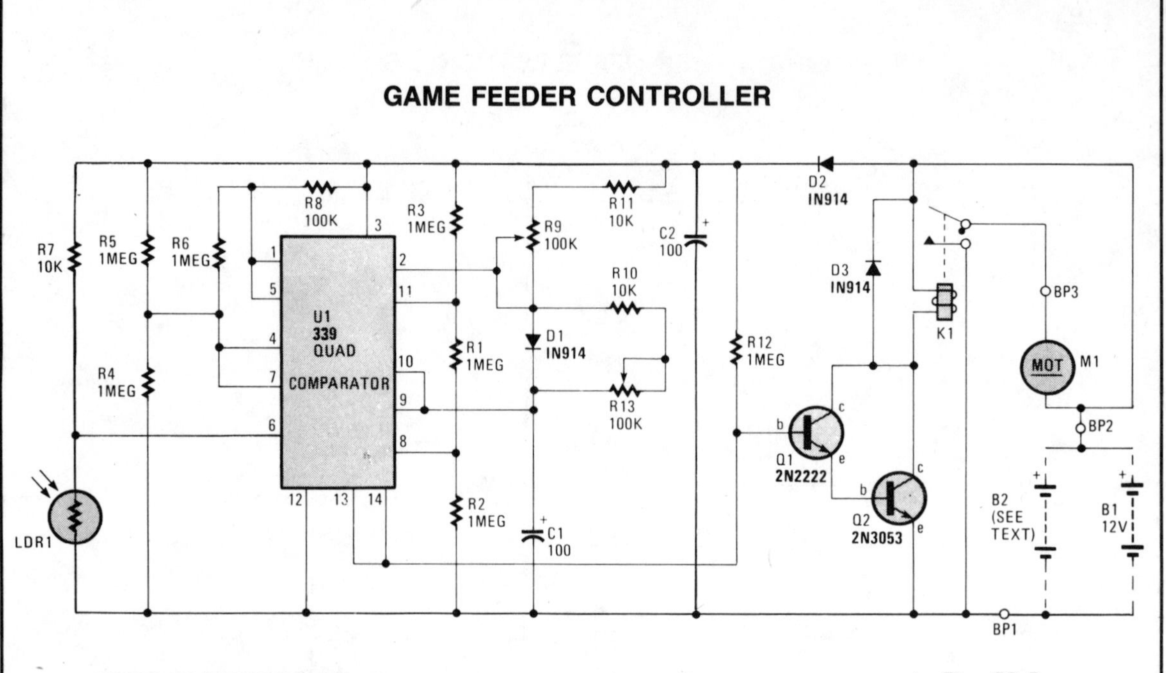

HANDS-ON ELECTRONICS *Fig. 58-5*

Circuit Notes

The circuit is built around an LM339 quad comparator, U1, which forms the basis of a Schmitt trigger, timer circuit, and a window comparator. One comparator within the LM339 (pins 1, 7, 6), plus LDR1, R4, R5, R6 and R8, is used as a Schmitt trigger. The timer circuit (which receives its input from the Schmitt trigger) consists of R9, R10, R11, R13. The last two-fourth's of U1 (pins 8, 9, 10, 11, 13 and 14) are wired as a window comparator. The two inputs to the window comparator are derived from the charge on capacitor C1—which is fed to pins 9 and 10 of U1. The other inputs are picked from two points along a voltage-divider network, consisting of R1, R2, and R3. Diode D1 is used as a blocking diode, forcing capacitor C1 to discharge through R10 and R13. The window comparator looks for any voltage falling between one-third and two-thirds of the supply voltage. When the voltage falls between those two points, the output of the window comparator (pins 13/14) goes high. Transistors Q1, and Q2 are turned on, when the pins 13/14 junction goes high, energizing the relay, K1. The energized relay provides a dc path to ground, activating the motor, M1, which reloads the feeder. The timer circuit also provides immunity from triggering, due to lightning. The on-time of relay K1 is determined by the charge cycle of C1, R11, and R9 or the discharge cycle of C1, R10, and R13. Changing the value of either a resistor or the capacitor, changes the timing cycle.

SINGLE LED CAN INDICATE FOUR LOGIC STATES

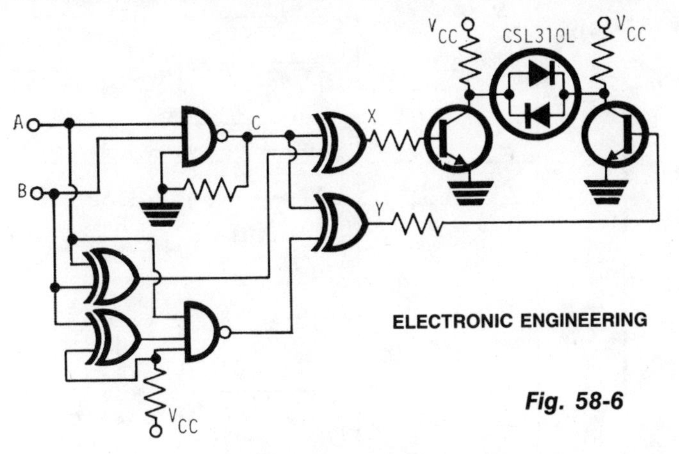

ELECTRONIC ENGINEERING

Fig. 58-6

Circuit Notes

The LED is the CSL310L which contains a red LED and a green LED connected back to back and mounted close together in a single moulding. The LED can emit red or green light by controlling the polarity of the applied voltage and if the polarity is switched at a rate of several hundred Hertz the emitted light appears yellow. The four combinations of inputs A and B can therefore be converted to four LED states—red, green, yellow and off. The truth table shows the LED colors corresponding to the combinations of A and B levels.

Truth Table

A	B	X	Y	LED color
0	0	1	0	red
0	1	0	0	off
1	0	0	1	green
1	1	C	C	yellow

INEXPENSIVE RADIO-CONTROL USES ONLY ONE SCR

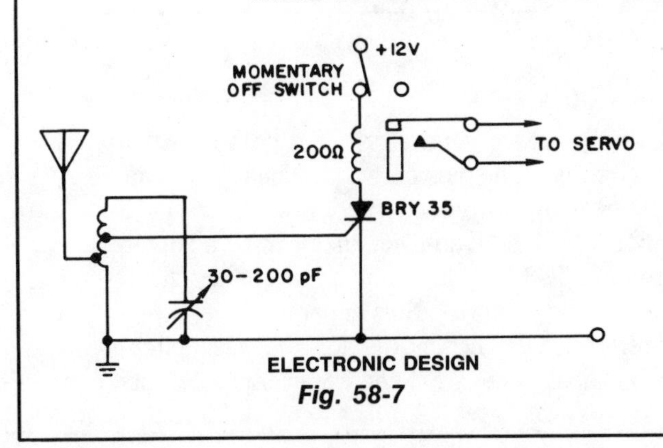

ELECTRONIC DESIGN

Fig. 58-7

Circuit Notes

A simple and effective receiver for actuating garage doors, alarms, warning systems, etc. The SCR, which has a very low trigger current 30 μA is typical—it requires an input power of only 30 μW to activate the relay. A high Q tuned antenna circuit assures rejection of spurious signals. A whip or wire antenna is adequate up to 100 feet from a low power transistor transmitter. A momentary-off switch resets the circuit.

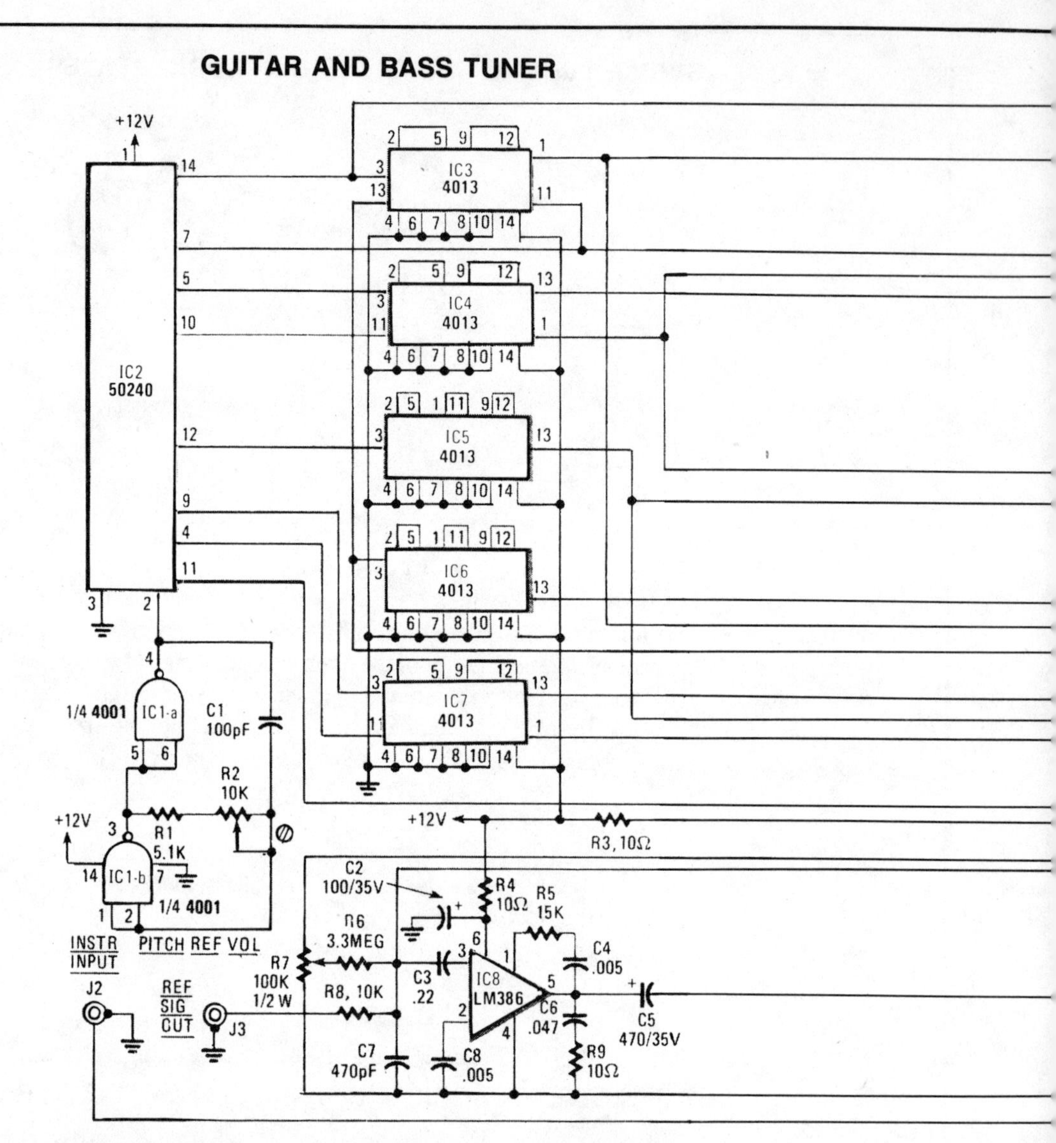

Fig. 58-8

Circuit Notes

The heart of the circuit is IC2, a 50240 top-octave generator. That device uses a single input-frequency to generate all twelve notes of the musical scale. The input signal is provided by IC1, a 4001 quad 2-input NOR gate. Two sections of that IC are used to form an oscillator that runs at approximately 2 MHz. The frequency can be adjusted by trimmer potentiometer R2. Dual D flip-flops, IC3-IC7, are used as frequency dividers. They divide down the upper-octave frequencies from IC2, thus generating the lower-frequency notes required for the pitch references. The chords for the bass pitch-references are composed of three notes each. Those notes are taken from various outputs

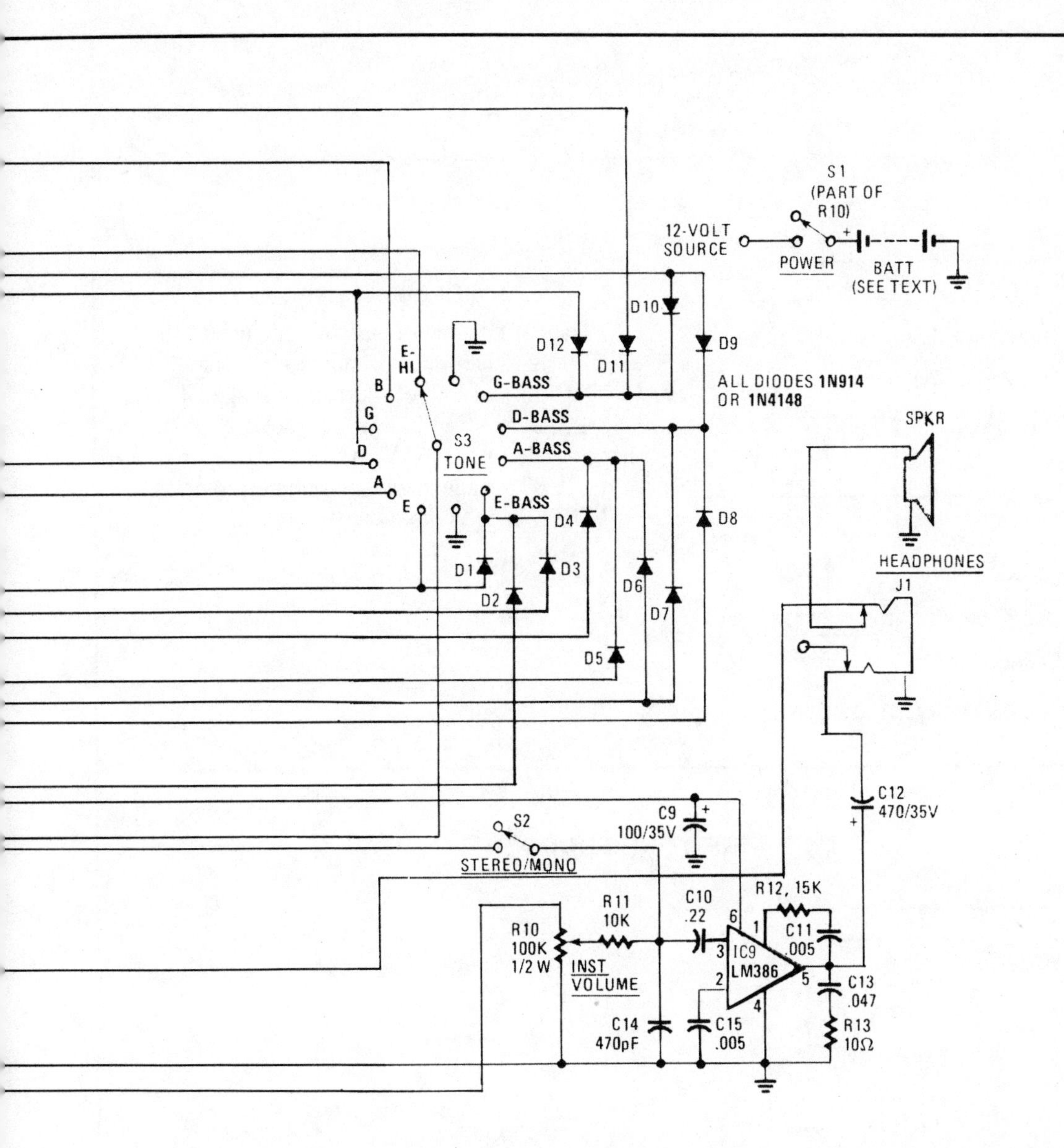

of IC2-IC7 through isolation diodes D1-D12. All signals are routed to the TONE switch, S3. The wiper arm of that switch is connected through R7 to the input of audio power-amplifier IC8, an LM386. The resistor acts as a volume control for the pitch reference. Another LM386, IC9, serves as an amplifier for the instrument being tuned, with R10 acting as its volume control. The outputs of IC8 and IC9 are coupled, through C5 and C12 respectively, to the headphone jack, J1. Switch S2 STEREO/MONO is used to mix the reference and instrument signals at IC9 for mono operation. Power is supplied by eight "AA" cells connected in series.

4-CHANNEL COMMUTATOR

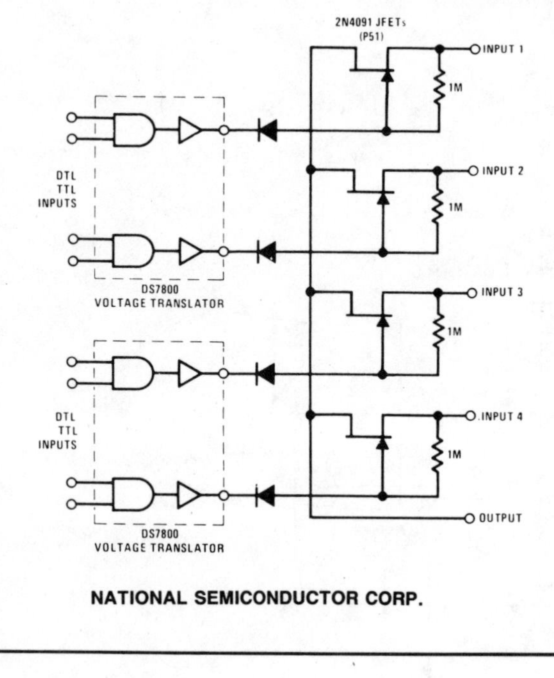

NATIONAL SEMICONDUCTOR CORP.

Circuit Notes

This 4-channel commutator used the 2N4091 to achieve low channel on resistance (<30 ohm) and low off current leakage. The DS7800 voltage translator is a monolithic device that provides from 10 V to −20 V gate drive to the JFETs while at the same time providing DTL/TTL logic compatibility.

Fig. 58-9

TWO-WIRE TONE ENCODER

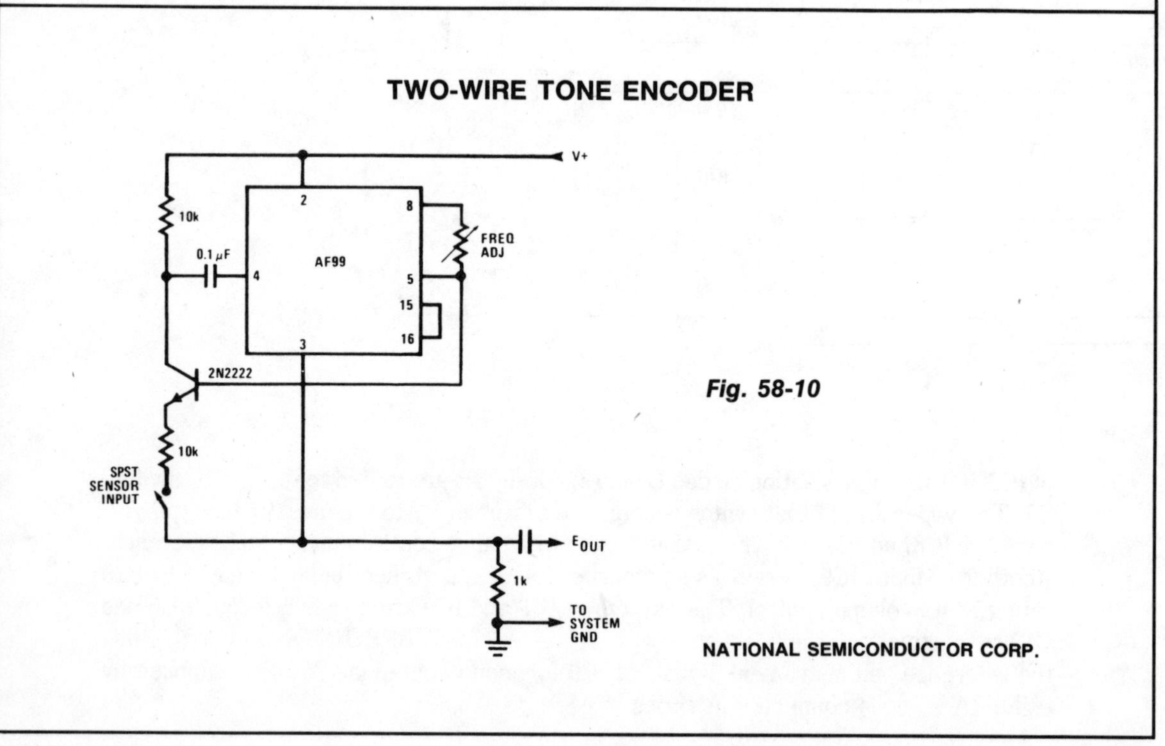

Fig. 58-10

NATIONAL SEMICONDUCTOR CORP.

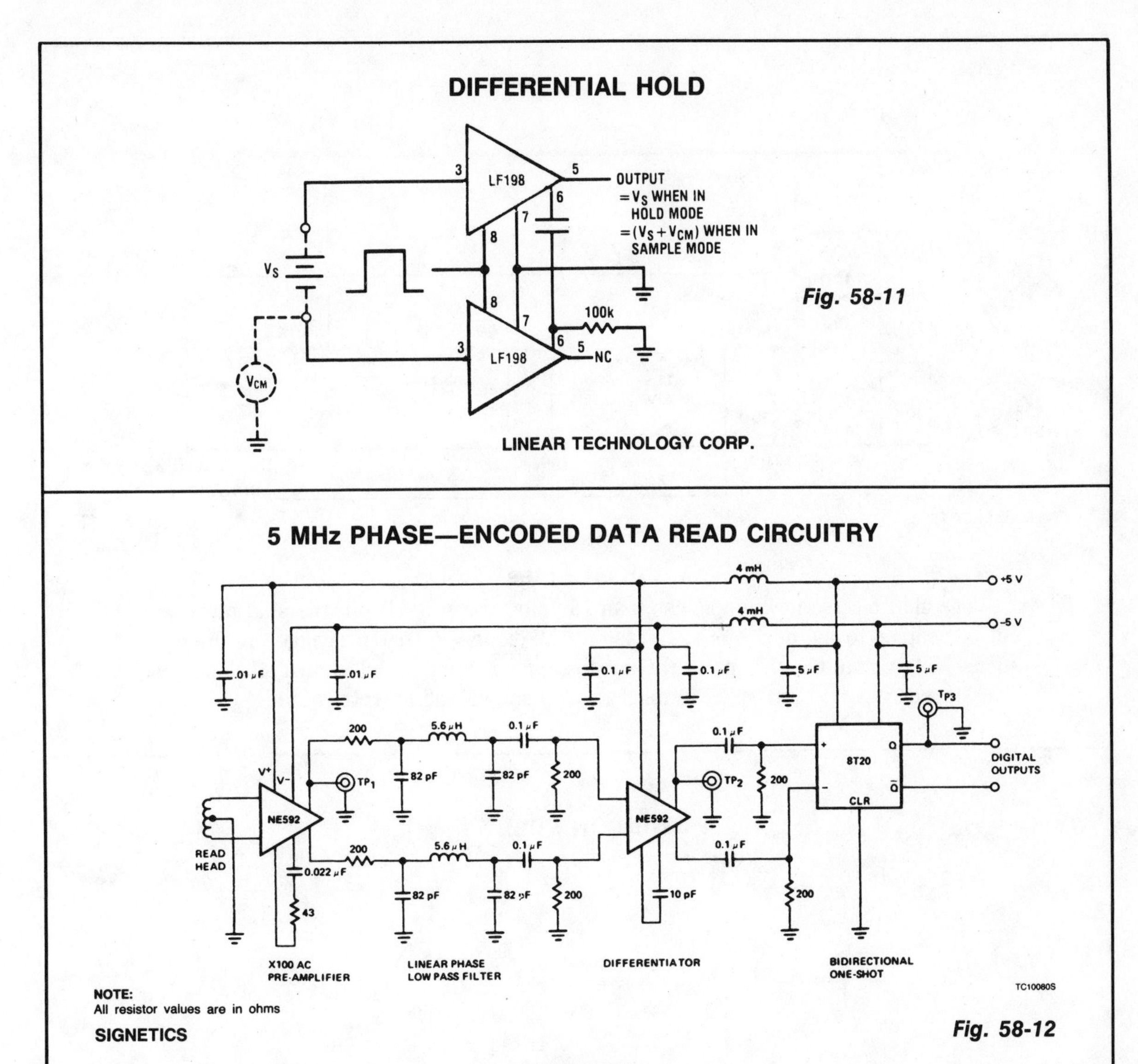

Fig. 58-11

Fig. 58-12

Circuit Notes

Readback data is applied directly to the input of the first NE592. This amplifier functions as a wide-band ac coupled amplifier with a gain of 100. By direct coupling of the readback head to the amplifier, no matched terminating resistors are required and the excellent common-mode rejection ratio of the amplifier is preserved. The dc components are also rejected because the NE592 has no gain at dc due to the capacitance across the gain select terminals. The output of the first stage amplifier is routed to a linear phase shift low-pass filter, with a characteristic impedance of 200 ohms. The second NE592 is utilized as a low noise differentiator/amplifier stage. The output of the differentiator/amplifier is connected to the 8T20 bidirectional monostable unit to provide the proper pulses at the zero-crossing points of the differentiator.

SHIFT REGISTER

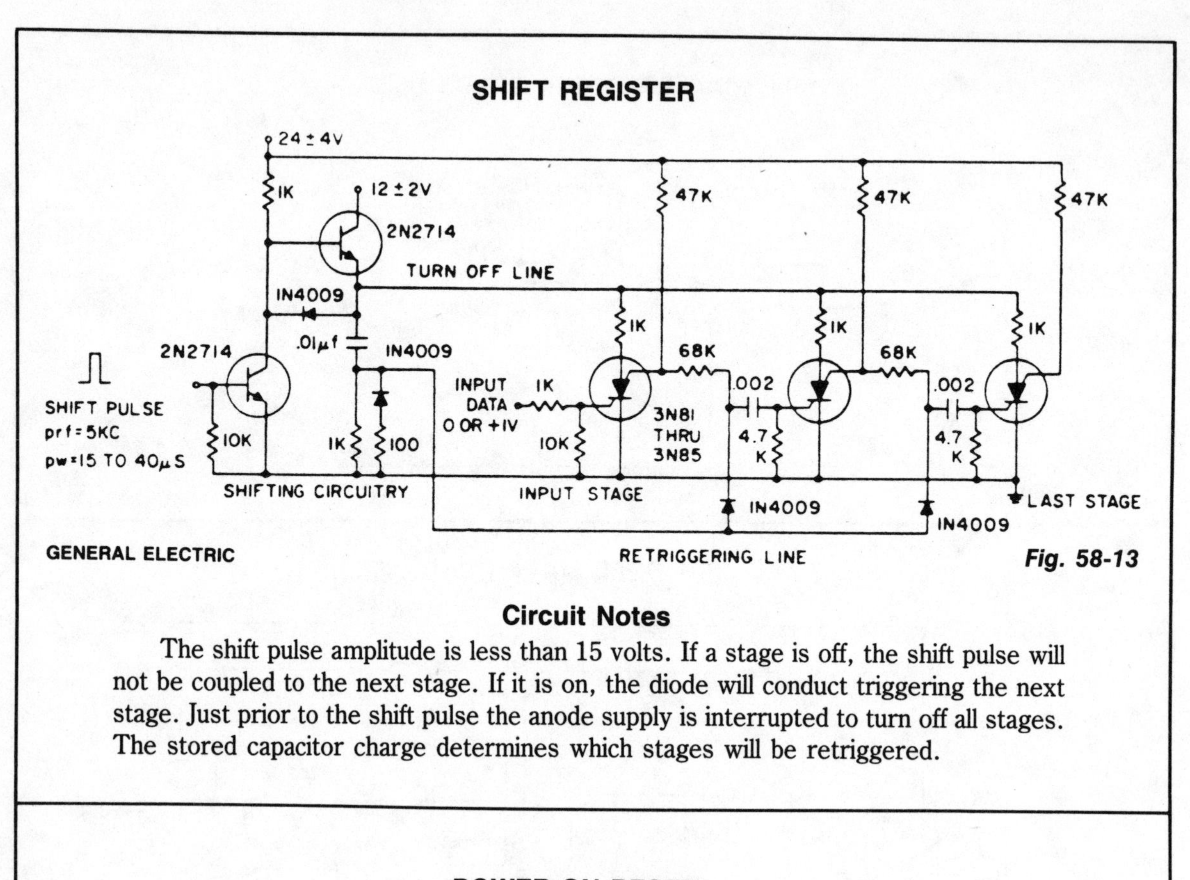

GENERAL ELECTRIC

Fig. 58-13

Circuit Notes

The shift pulse amplitude is less than 15 volts. If a stage is off, the shift pulse will not be coupled to the next stage. If it is on, the diode will conduct triggering the next stage. Just prior to the shift pulse the anode supply is interrupted to turn off all stages. The stored capacitor charge determines which stages will be retriggered.

POWER-ON RESET

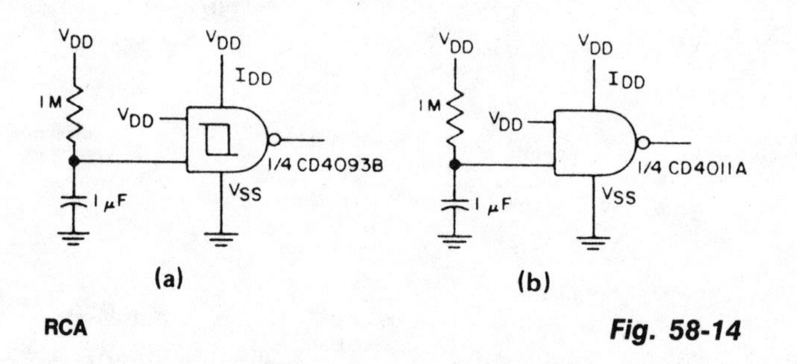

RCA

Fig. 58-14

Circuit Notes

A reset pulse is often required at power-on in a digital system. This type of reset pulse is ideally provided by this circuit. Because of the high input impedance of the Schmitt trigger, long reset pulse times may be achieved without the excess dissipation that results when both output devices are on simultaneously, as in an ordinary gate device (B).

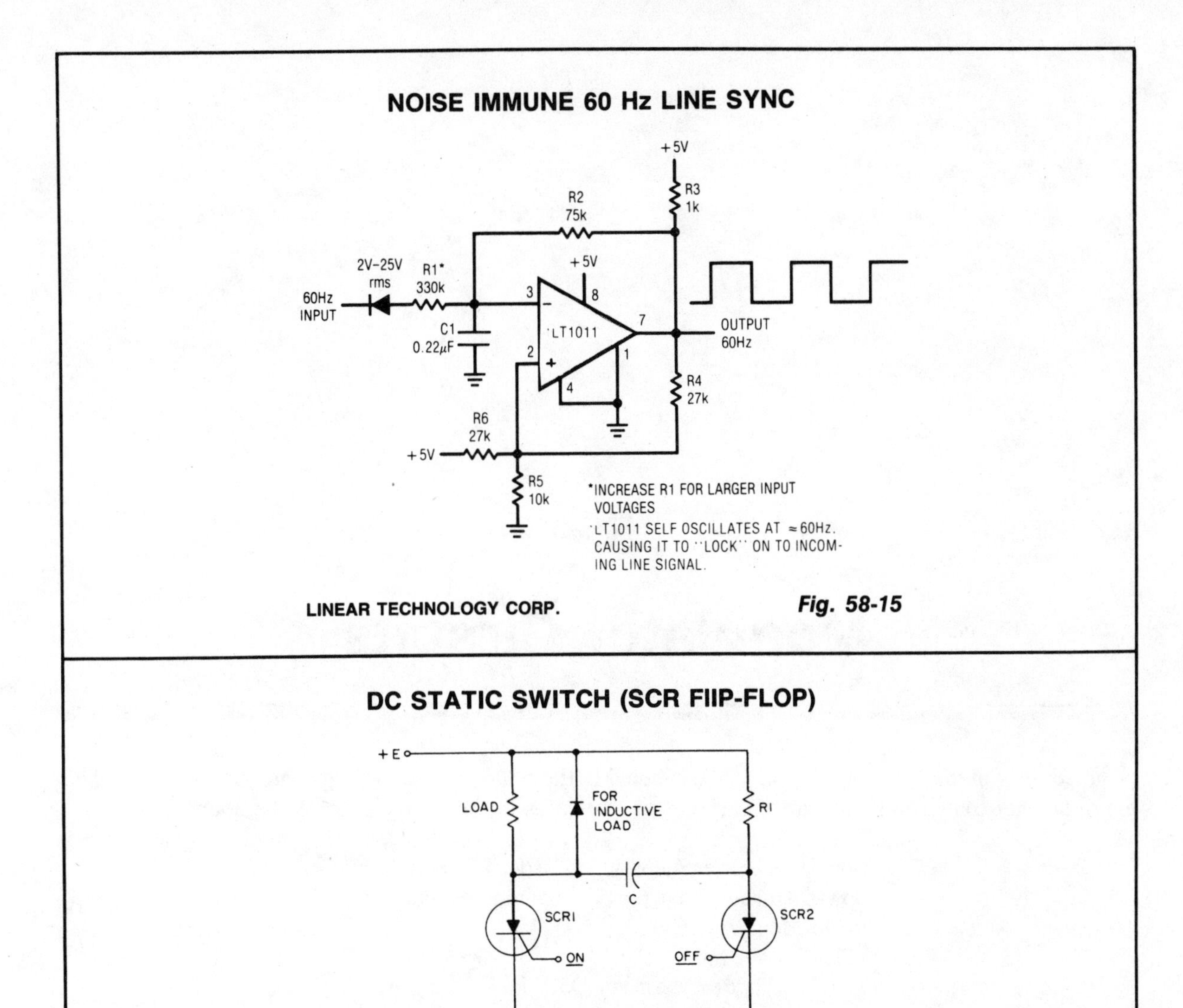

GENERAL ELECTRIC

Fig. 58-16

Circuit Notes

This circuit is a static SCR switch for use in a dc circuit. When a low power signal is applied to the gate of SCR1, this SCR is triggered and voltage is applied to the load. The right hand plate of C charges positively with respect to the left hand plate through R1. When SCR2 is triggered on, capacitor C is connected across SCR1, so that this SCR is momentarily reverse biased between anode and cathode. This reverse voltage turns SCR1 off provided the gate signal is not applied simultaneously to both gates. The current through the load will decrease to zero in an exponential fashion as C becomes charged.

59

Modulator Circuits

The sources of the following circuits are contained in the Sources section beginning on page 694. The figure number contained in the box of each circuit correlates to the source entry in the Sources section.

DOUBLE SIDEBAND, SUPPRESSED CARRIER RF MODULATOR

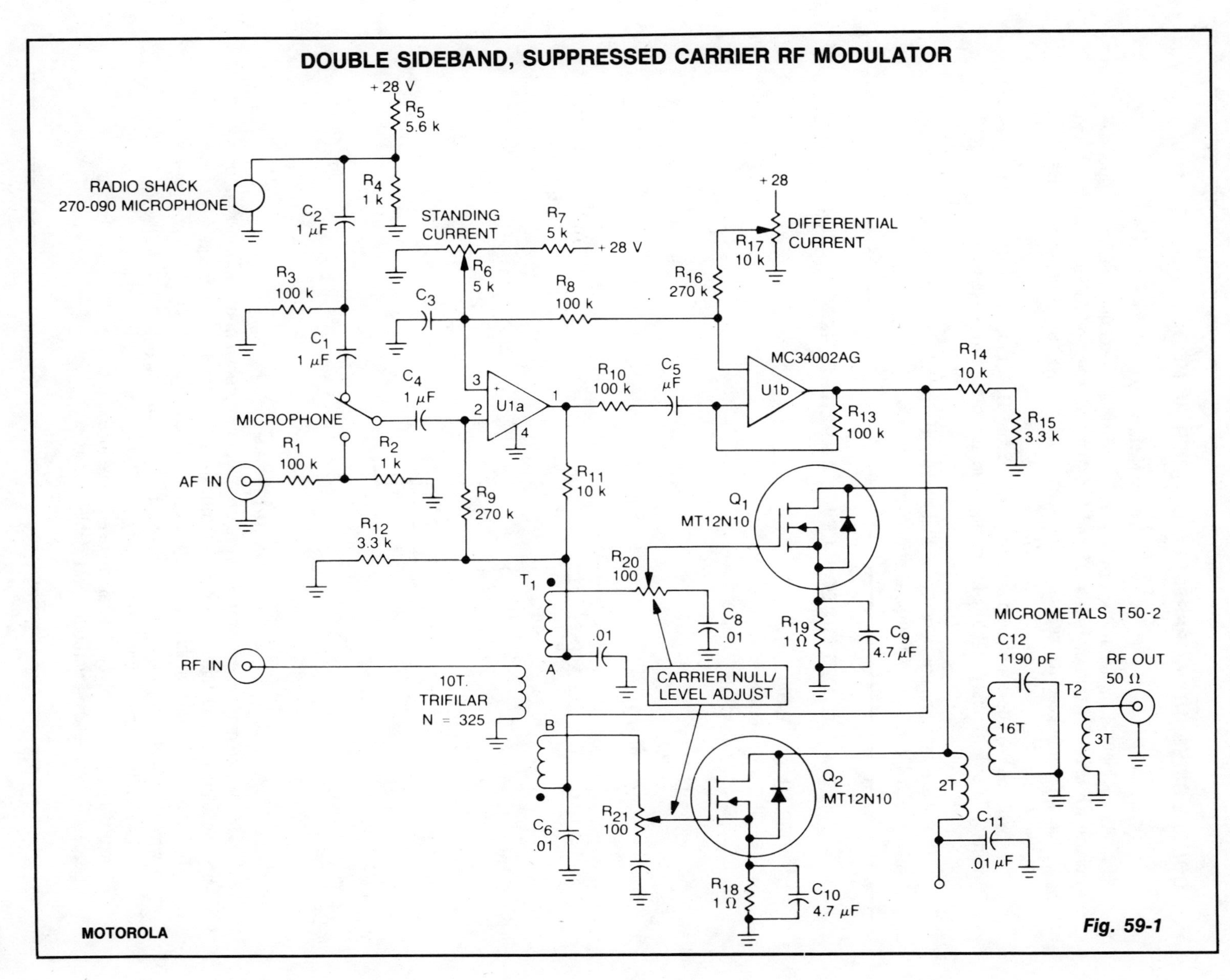

Fig. 59-1

DOUBLE SIDEBAND, SUPPRESSED CARRIER RF MODULATOR, Continued.

Circuit Notes

An RF input is applied to the primary of T1, which applies equal amplitude, opposite phase RF drive for output FETs Q1 and Q2. With no AF modulation at points A and B, the opposite phase RF signals cancel each other and no output appears at the 50 V output connector.

When AF modulation is applied to points A and B, a modulated RF output is obtained. The dc stability and low frequency gain are improved by source resistors R18 and R19.

A phase inverter consisting of a dual op amp (U1a and U1b) produces the out-of-phase, equal amplitude AF modulation signals.

LOW-DISTORTION LOW-LEVEL AMPLITUDE MODULATOR

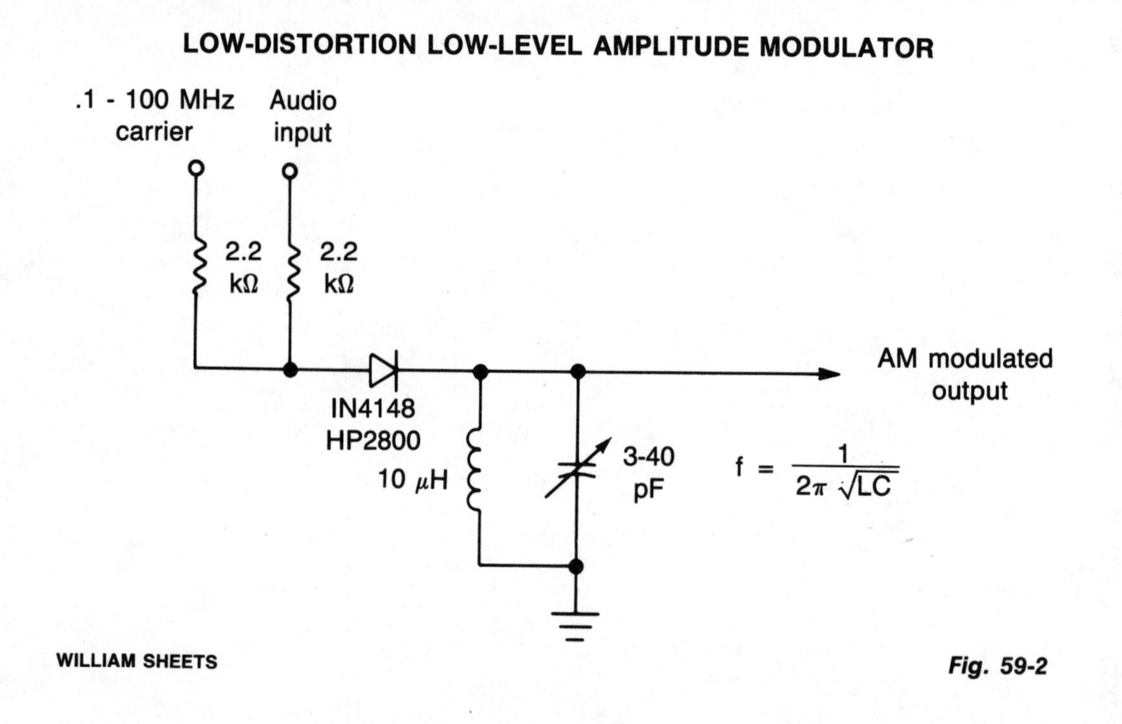

WILLIAM SHEETS

Fig. 59-2

Circuit Notes

This simple diode modulator delivers excellent results when used for high percentage modulation at low signal levels. Constants are shown for a carrier frequency of about 10 MHz, but, with a suitable tank, the circuit will give good results at any frequency at which the diode approximates a good switch. To extend frequency above that for which the IN4148 is suited, a hot-carrier diode (HP2800, etc.) can be substituted. A shunt resistor across the tank circuit can be used to reduce the circuit Q so as to permit high percentage modulation without appreciable distortion.

VIDEO MODULATOR CIRCUIT

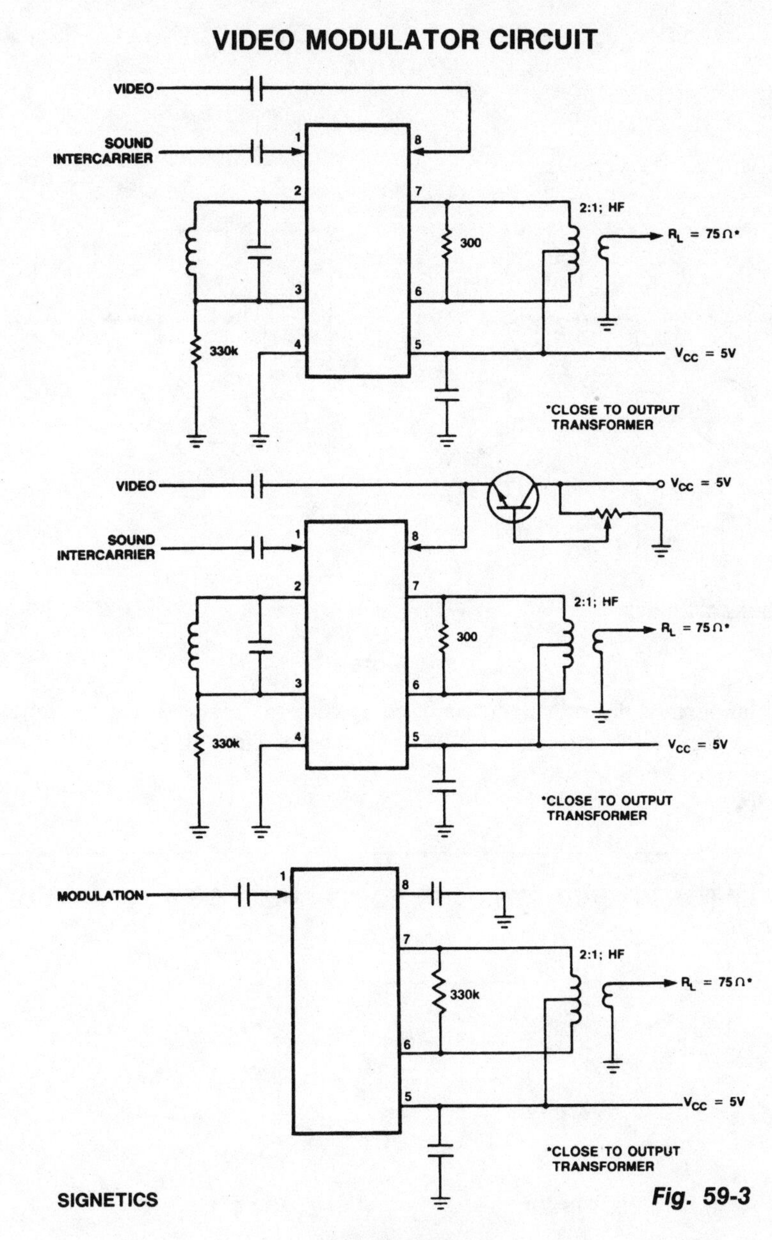

Fig. 59-3

Circuit Notes

These are modulator circuits for modulation of video signals on a VHF/UHF carrier. The circuits require a 5 V power supply and few external components for the negative modulation mode. For positive modulation an external clamp circuit is required. The circuits can be used as general-purpose modulators without additional external components. The IC is TDA6800.

VIDEO MODULATOR

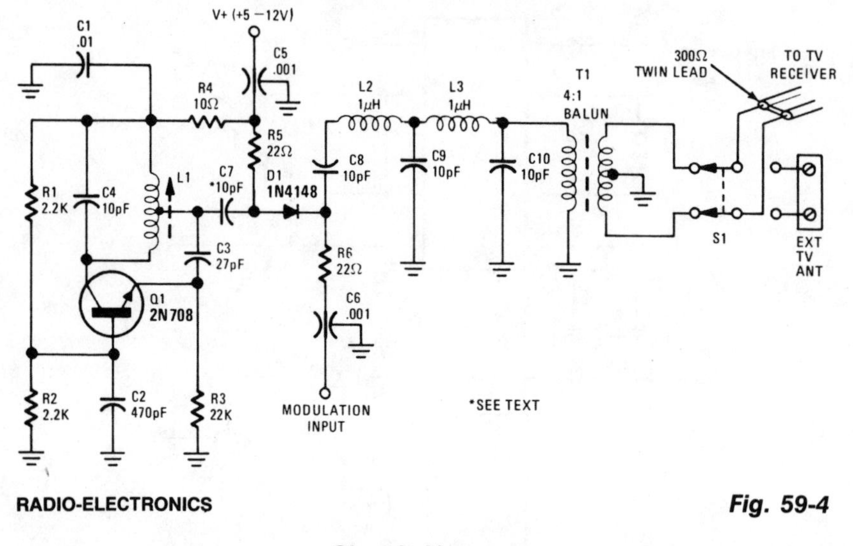

RADIO-ELECTRONICS *Fig. 59-4*

Circuit Notes

This circuit permits direct connection of composite video signals from video games and microcomputers to the antenna terminals of TV sets. The output signal level is controlled by the modulation input.

TTL OSCILLATOR INTERFACES DATA FOR DISPLAY BY A TELEVISION SET

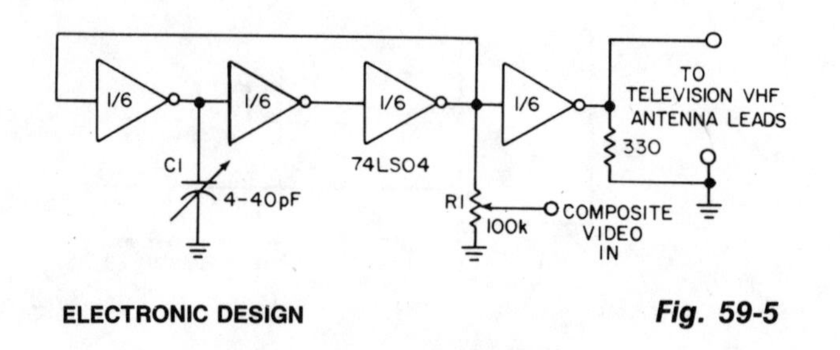

ELECTRONIC DESIGN *Fig. 59-5*

Circuit Notes

Three gates of a 74LS04 form the oscillator circuit. Capacitor C1 allows fine-frequency adjustment to a specific television channel and helps stabilize the circuit. Potentiometer R1 acts as the mixing input and provides adjustment of the contrast ratio for the best viewing. A fourth gate buffers and helps stabilize the oscillator.

60

Motor Control Circuits

The sources of the following circuits are contained in the Sources section beginning on page 694. The figure number contained in the box of each circuit correlates to the source entry in the Sources section.

BI-DIRECTIONAL PROPORTIONAL MOTOR CONTROL

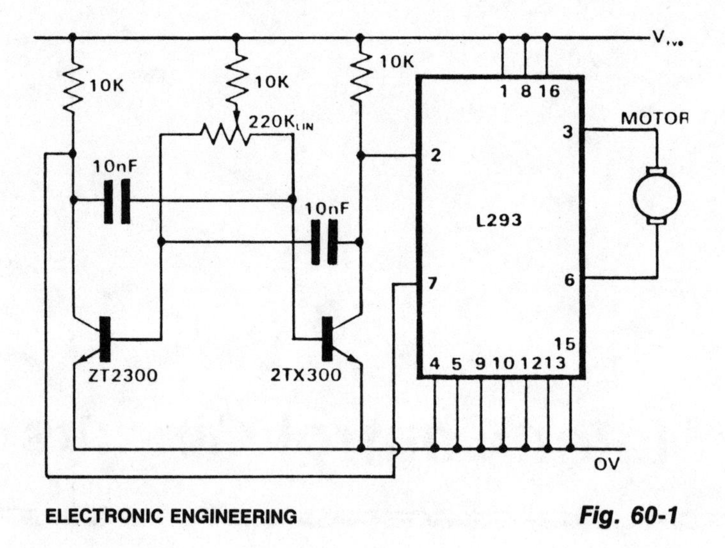

ELECTRONIC ENGINEERING *Fig. 60-1*

Circuit Notes

The control of both direction and of proportional motor speed is achieved by rotation of a single potentiometer. The motor driver is an SGS integrated circuit L293 which will drive up to 1 amp in either direction, depending on the logic state of input 1 and input 2 as per table.

I/P 1	I/P 2	Function
High	Low	Motor turns one way
Low	High	Motor reverses

By applying a variable M/S ratio flip-flop to these inputs, both speed and direction will be controlled. With RV1 in its center position the M/S will be 1:1 whereby the motor will remain stationary due to its inability to track at the flip-flop frequency. Movement of RV1 in either direction will gradually alter the M/S ratio and provide an average voltage bias in one direction proportional to the M/S ratio.

AC MOTOR CONTROL

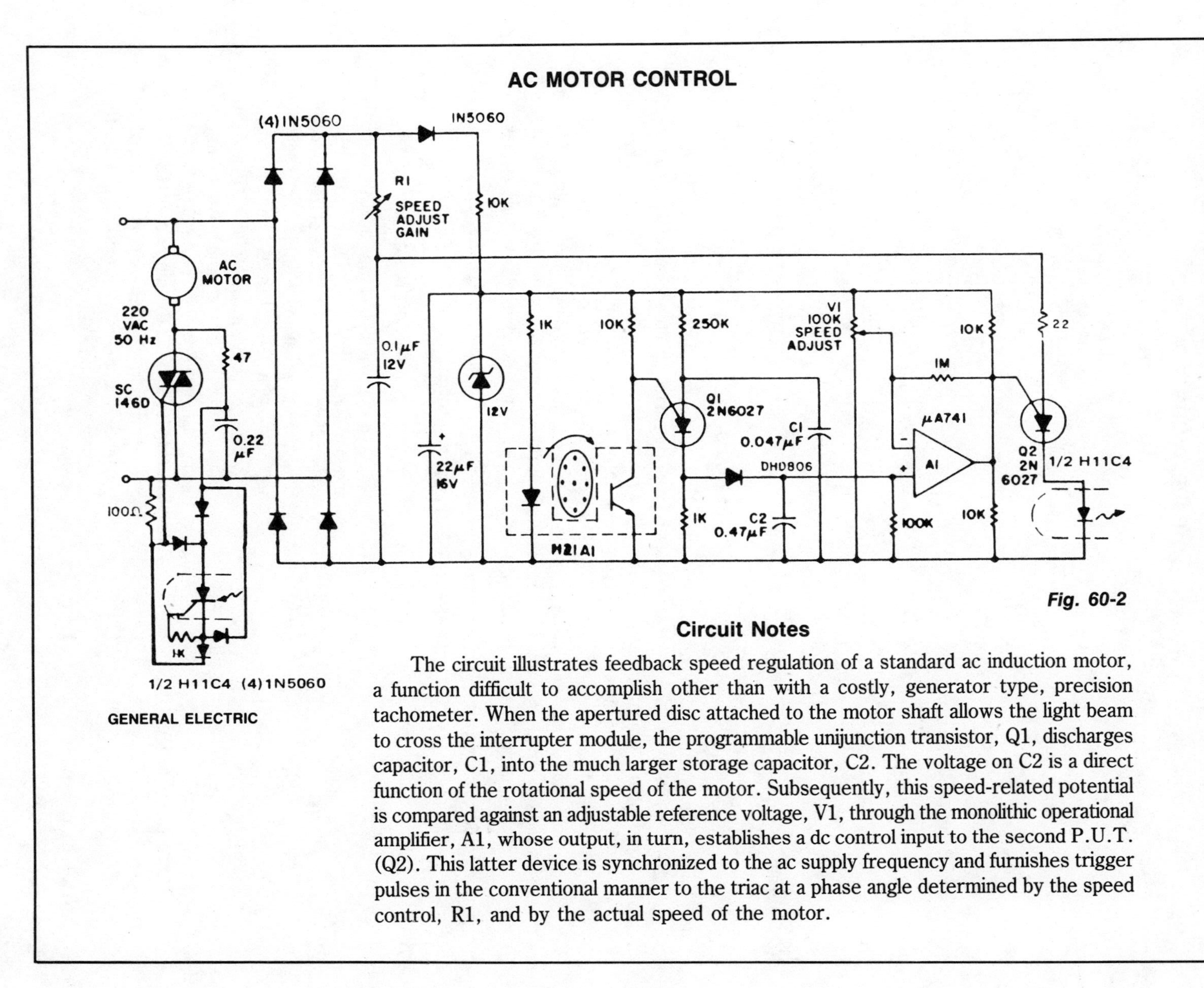

Fig. 60-2

Circuit Notes

The circuit illustrates feedback speed regulation of a standard ac induction motor, a function difficult to accomplish other than with a costly, generator type, precision tachometer. When the apertured disc attached to the motor shaft allows the light beam to cross the interrupter module, the programmable unijunction transistor, Q1, discharges capacitor, C1, into the much larger storage capacitor, C2. The voltage on C2 is a direct function of the rotational speed of the motor. Subsequently, this speed-related potential is compared against an adjustable reference voltage, V1, through the monolithic operational amplifier, A1, whose output, in turn, establishes a dc control input to the second P.U.T. (Q2). This latter device is synchronized to the ac supply frequency and furnishes trigger pulses in the conventional manner to the triac at a phase angle determined by the speed control, R1, and by the actual speed of the motor.

PWM MOTOR SPEED CONTROL

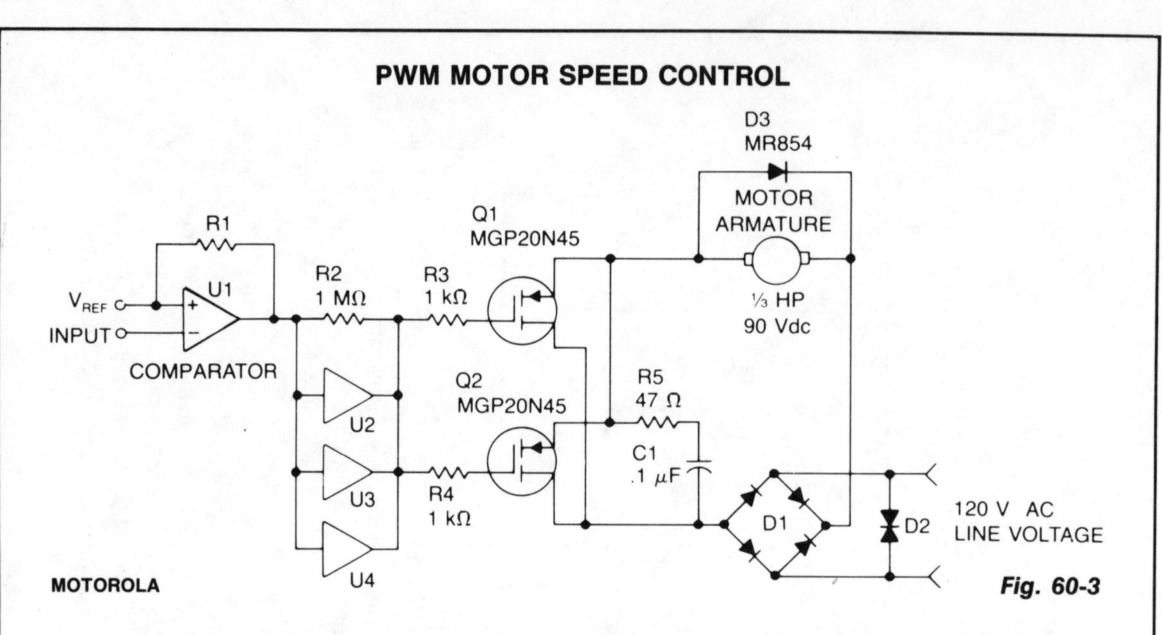

MOTOROLA

Fig. 60-3

Circuit Notes

Speed control is accomplished by pulse width modulating the gates of two MGP20N45 TMOS devices. Therefore, motor speed is proportional to the pulse width of the incoming digital signal, which can be generated by a microprocessor or digital logic.

The incoming signal is applied to comparator U1, then to paralleled inverters U2, U3, and U4 that drive the two TMOS devices, which, in turn, control power applied to the motor armature. Bridge rectifier D1 supplies fullwave power that is filtered by R5 and C1. Free-wheeling diode D3 (MR854) prevents high voltage across Q1 and Q2. A back-to-back zener diode, D2, protects against transients and high voltage surges.

STEPPING MOTOR DRIVER

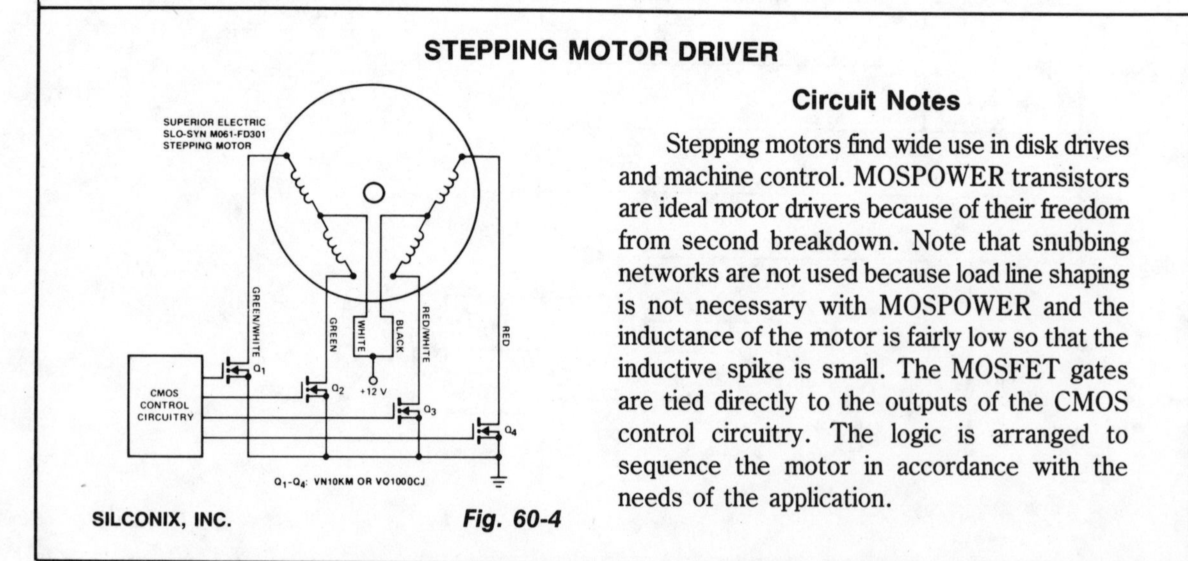

SILCONIX, INC.

Fig. 60-4

Circuit Notes

Stepping motors find wide use in disk drives and machine control. MOSPOWER transistors are ideal motor drivers because of their freedom from second breakdown. Note that snubbing networks are not used because load line shaping is not necessary with MOSPOWER and the inductance of the motor is fairly low so that the inductive spike is small. The MOSFET gates are tied directly to the outputs of the CMOS control circuitry. The logic is arranged to sequence the motor in accordance with the needs of the application.

LOW-COST SPEED REGULATOR FOR DC MOTORS

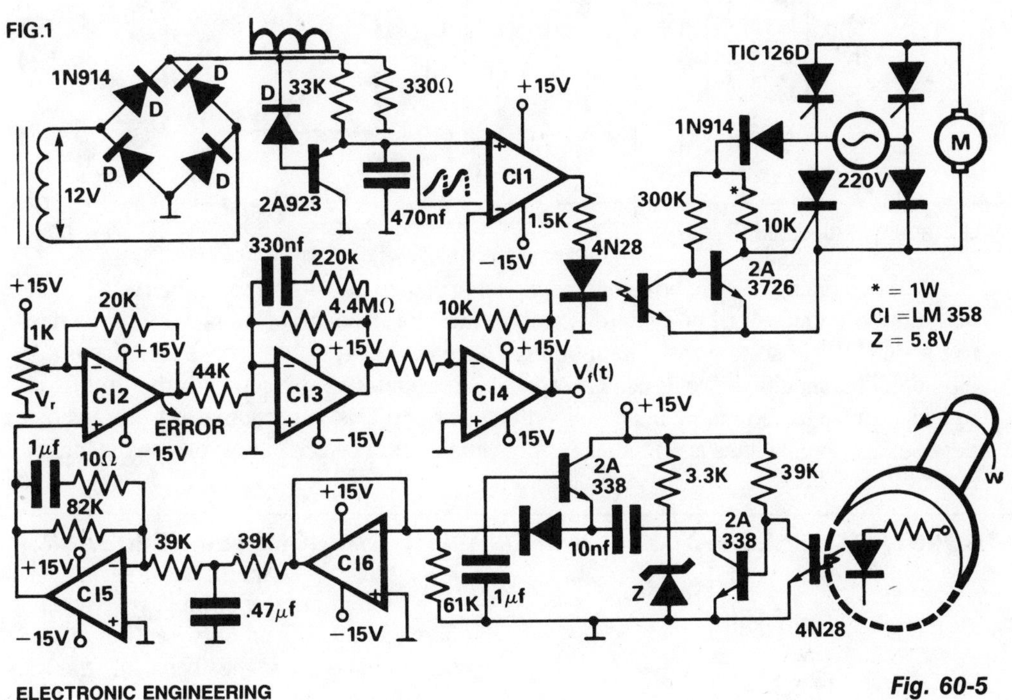

ELECTRONIC ENGINEERING

Fig. 60-5

Circuit Notes

A four thyristor controlled bridge is used for operation in two quadrants of the torque-speed characteristics. In the trigger circuits the usual pulse transformers were replaced by self biased circuits which minimize gate power consumption and increase noise immunity. Electrical isolation is guaranteed by the use of optocouplers. The trigger pulses are generated by the comparison between an error signal, previously processed and amplified, and a line synchronism signal. The converter's output is a dc voltage proportional to the speed, which after being compared with a reference signal, becomes the error signal.

MOTOR SPEED CONTROL CIRCUIT

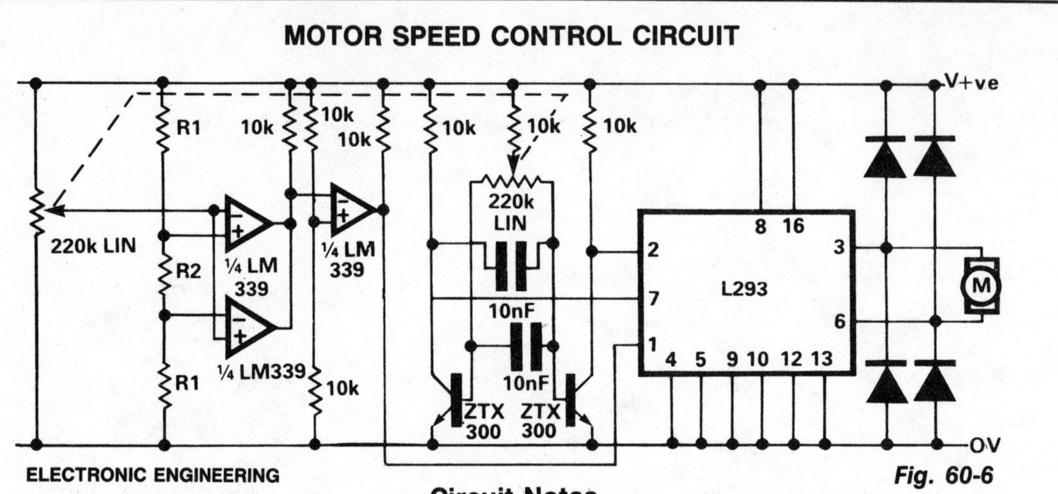

Fig. 60-6

Circuit Notes

A shortcoming of the above bi-directional proportional motor control circuit is that with the potentiometer in its center position the motor does not stop, but creeps due to the difficulty in setting the potentiometer for an exact 1:1 mark-space ratio from the flip-flop. This modified circuit uses a second potentiometer, ganged with the first used to inhibit drive to the motor near the center position. This potentiometer is connected between the supply lines and feeds a window comparator which in turn drives the inhibit input of the L293.

CONSTANT SPEED MOTOR CONTROL USING TACHOMETER FEEDBACK

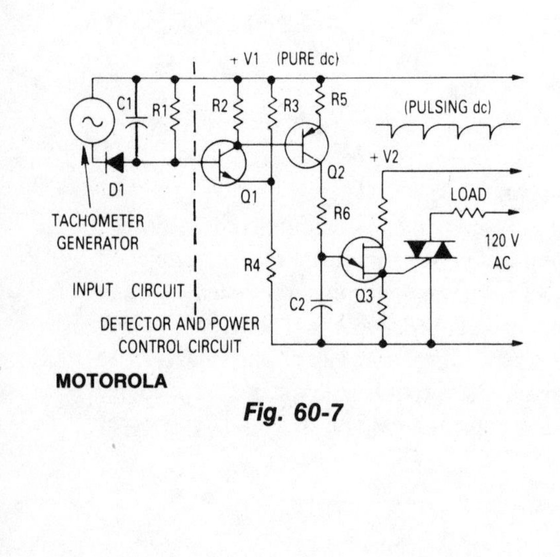

Fig. 60-7

Circuit Notes

The generator output is rectified then filtered and applied between the positive supply voltage and the base of the detector transistor. This provides a negative voltage which reduces the base-voltage when the speed increases. In normal operation, if the tachometer voltage is less than desired, the detector transistor is turned on, then turns on Q2 which causes the timing capacitor for the unijunction transistor to charge quickly. As the tachometer output approaches the voltage desired, the base-emitter voltage is reduced to the point at which Q1 is almost cut off. Thereby, the collector current which charges the unijunction timing capacitor is reduced, causing that capacitor to charge slowly and trigger the thyristor later in the half cycle. In this manner, the average power to the motor is reduced until just enough power to maintain the desired motor speed is allowed to flow.

BACK EMF PM MOTOR SPEED CONTROL

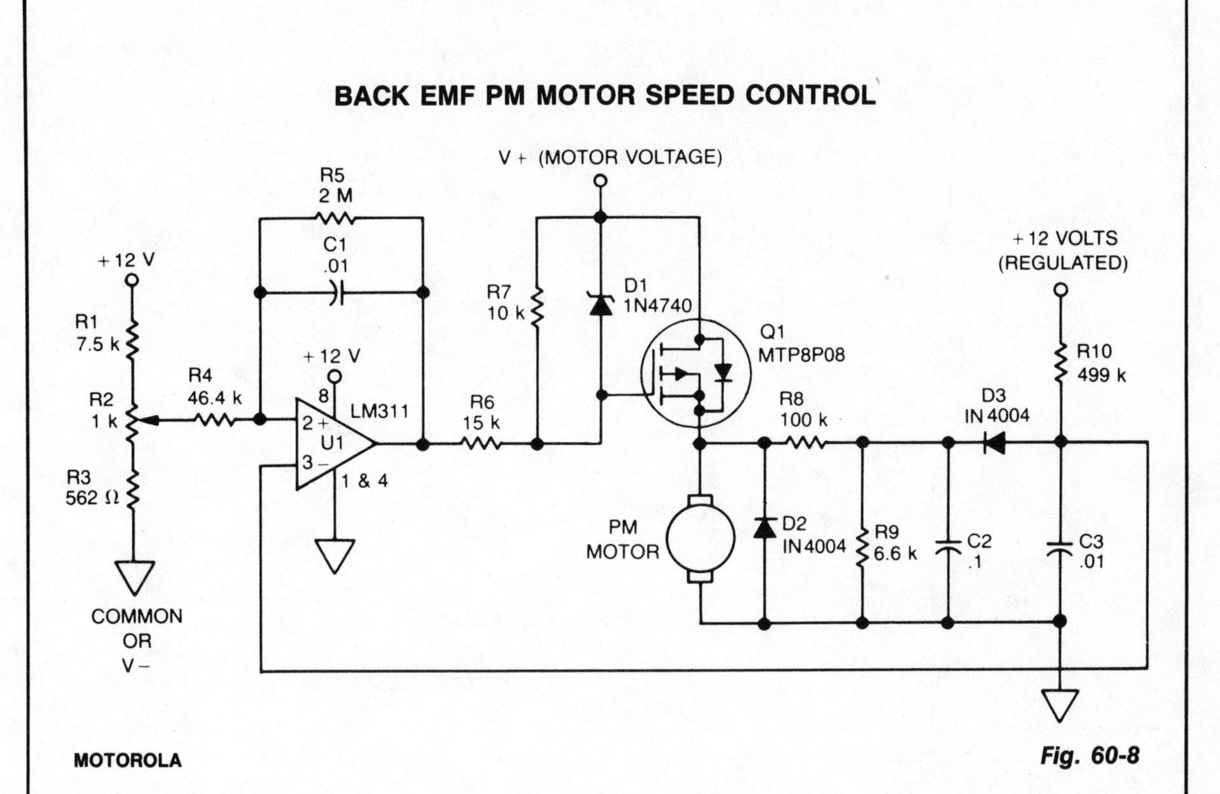

Fig. 60-8

Circuit Notes

The use of power MOSFETs allows a direct interface between logic and motor power, which permits circuit simplicity as well as high efficiency. This speed control circuit can be packaged on a 22-pin, double-sided, 3.5 × 4-in. pc board.

A 12 V control supply and a TRW BL11, 30 V motor are used; with minor changes other motor and control voltages can be accommodated. For example, a single 24 V rail could supply both control and motor voltages. Motor and control voltages are kept separate here because CMOS logic is used to start, stop, reverse and oscillate the motor with a variable delay between motor reversals.

Motor speed is established by potentiometer R2, which applies a corresponding dc voltage to the + input of comparator U1, whose output is then applied to TMOS device MTP8P08 (Q1). Zener diode D1 limits the drive to Q1. The output of Q1 drives the permanent magnet motor.

Back emf is obtained from the motor via the network consisting of R8, R9, R10, C2, C3 and D3; it is applied to—input of comparator U1.

DC MOTOR SPEED CONTROL

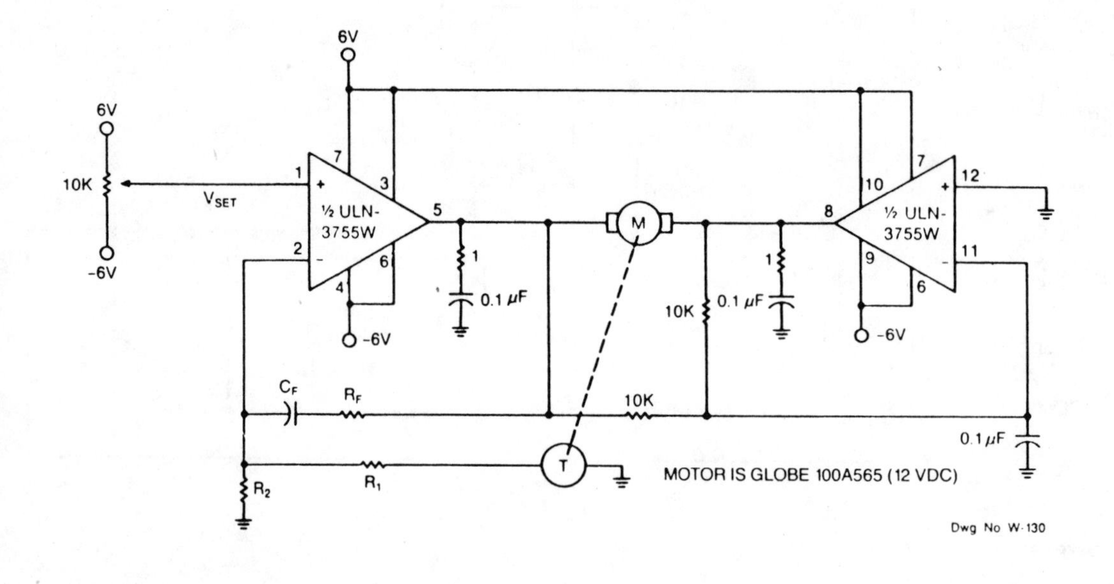

SPRAGUE ELECTRIC CO.

Fig. 60-9

Circuit Notes

Power op amps provide accurate speed control for dc motors. The circuit provides bidirectional speed control. The amplifiers' push-pull configuration ensures a full rail-to-rail voltage swing (minus the output stages' saturation drops) across the motor in either direction. The circuit uses a mechanically-coupled tachometer to provide speed-stabilizing feedback to the first amplifier section. The motor's speed and direction of rotation is set by adjusting the 10 k ohm potentiometer at the amplifier's noninverting input. The RFCF feedback network prevents oscillation by compensating for the inherent dynamic mechanical lag of the motor. Select the RFCF time constant to match the particular motor's characteristics.

REVERSING MOTOR DRIVE, DC CONTROL SIGNAL

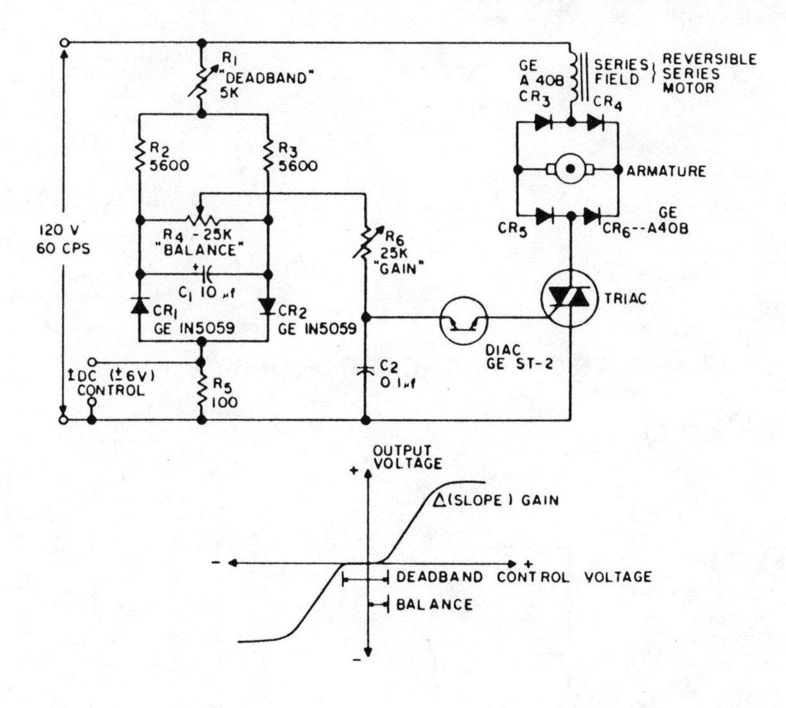

GENERAL ELECTRIC *Fig. 60-10*

Circuit Notes

This is a positioning servo drive featuring adjustment of balance, gain, and deadband. In addition to control from a dc signal, mechanical input can be fed into the balance control, or that control could be replaced by a pair of resistance transducers for control by light or by temperature.

N-PHASE MOTOR DRIVERS

SINGLE-PHASE AC MOTOR DRIVER

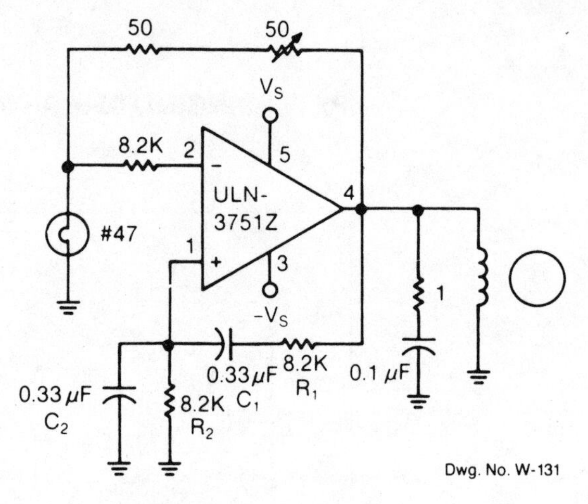

TWO-PHASE AC MOTOR DRIVER

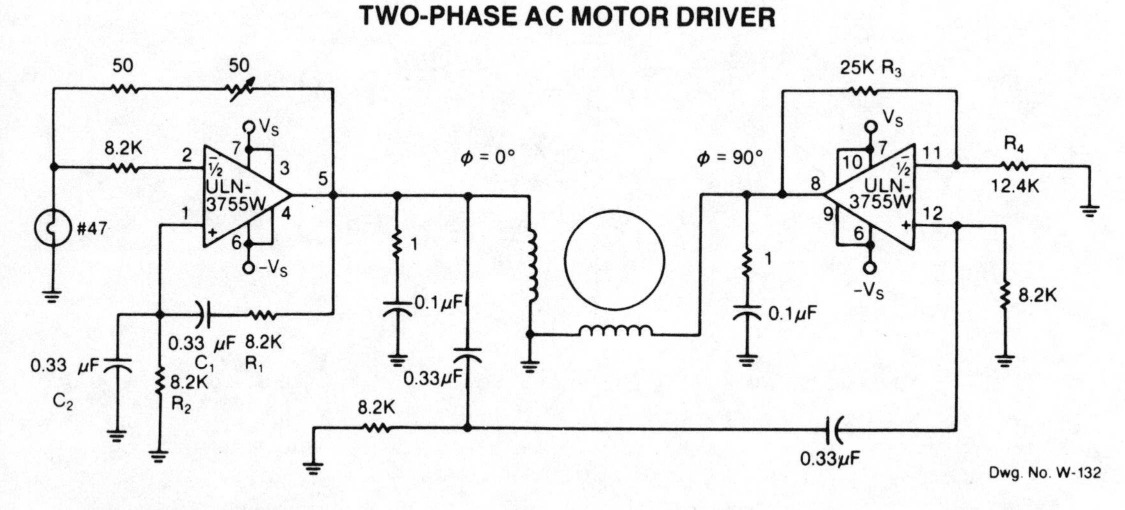

SPRAGUE ELECTRIC CO.

Fig. 60-11

Circuit Notes

Because of its high amplification factor and built-in power-output stage, an integrated power operational amplifier makes a convenient driver for ac motors. One op amp can be configured as an oscillator to generate the required ac signal. The power-output stage, of course, supplies the high-current drive to the motor. The controlling op amp is

THREE-PHASE AC MOTOR DRIVER

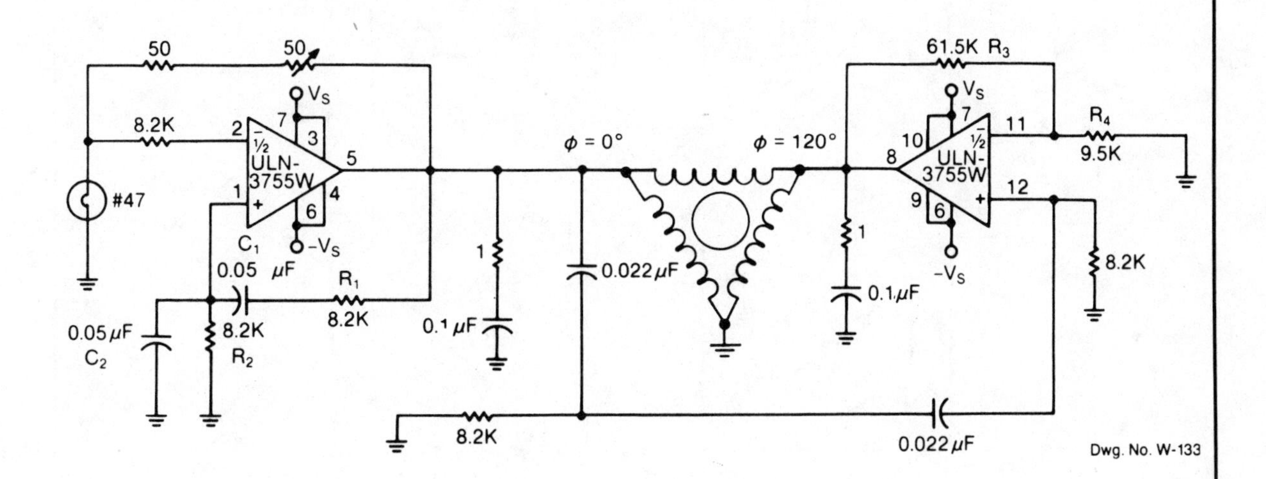

configured as a Wein bridge oscillator. The R1C1, R2C2 feedback networks determine the oscillation frequency, according to the following expression:

$$f_o = \frac{1}{(2\ \pi\sqrt{R_1R_2C_1C_2})}$$

By varying either R1 or R2, the oscillator frequency can be adjusted over a narrow range. The R3/R4 ratio sets the second amplifier's gain to compensate for signal attenuation occurring in the phase shifters. The circuits can be driven from an external source, such as a pulse or square wave, setting the gain of the left-hand amplifier to a level less than that required for oscillation. The RC feedback networks then function as an active filter causing the outputs to be sinusoidal.

SERVO MOTOR DRIVE AMPLIFIER

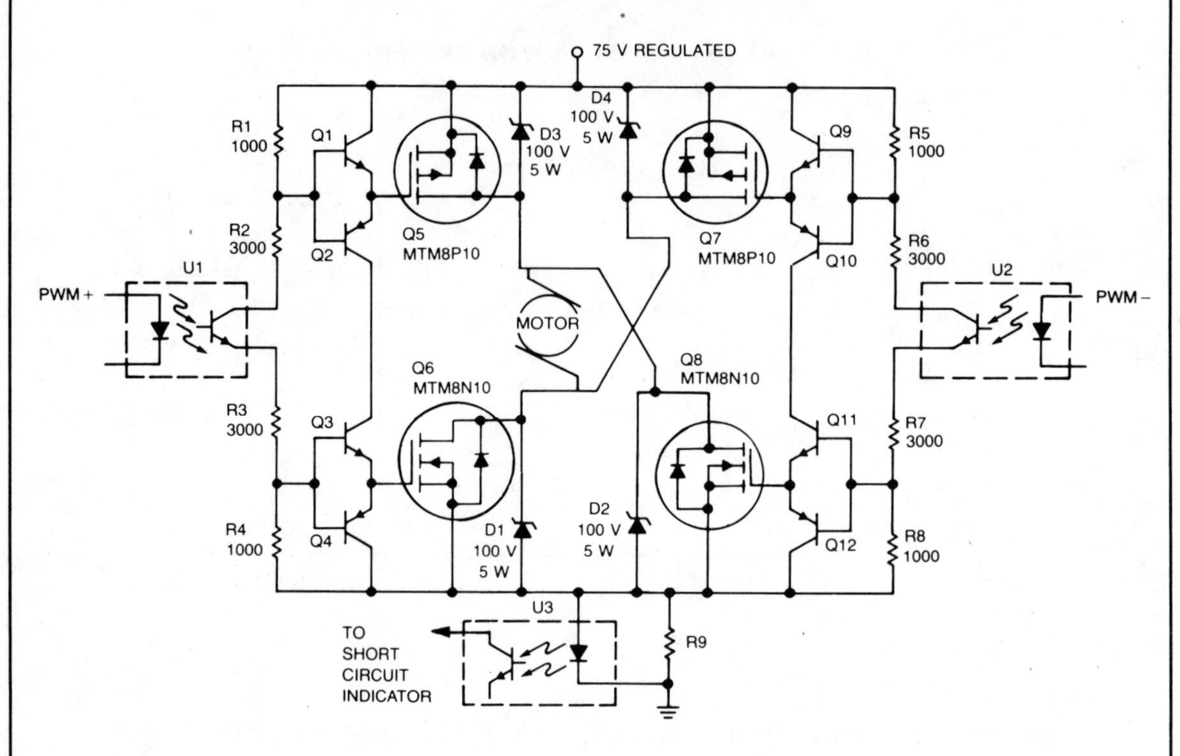

MOTOROLA

Fig. 60-12

Circuit Notes

Digital ICs and opto-isolators provide the drive for this TMOS servo amplifier, resulting in fewer analog circuits and less drift. Fast and consistent turn-on and turn-off characteristics also enable accurate analog output results directly from the digital signal without the need for analog feedback.

An "H" bridge configuration is employed for the servo amplifier, which obtains complementary PWM inputs from digital control circuits. The PWM inputs are applied via opto-isolators, which keep the digital control logic isolated from the 75 V supply used for the amplifier. A short circuit indicator is provided by opto-isolator U3; if there is a short, the drop across R9 increases to a value sufficient to activate the isolator and send a short indication to the digital control logic.

DC SERVO DRIVE EMPLOYS BIPOLAR CONTROL INPUT

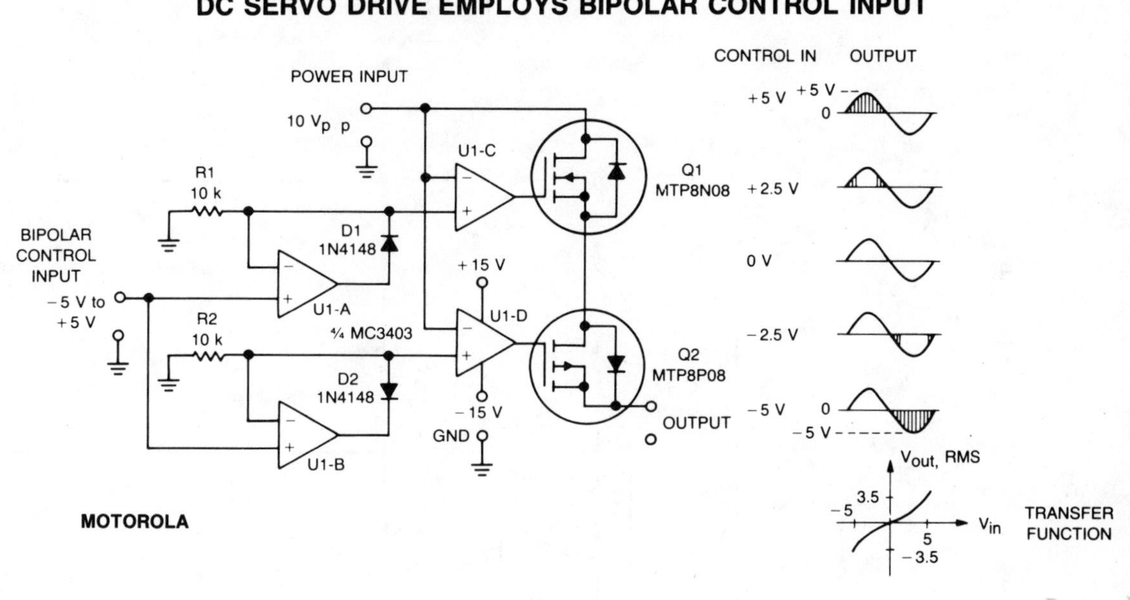

Fig. 60-13

Circuit Notes

This circuit accepts bipolar control inputs of ±5 V and provides a phase-chopped output to a dc load (such as a servo motor) of the same polarity as the input. The rms voltage of the output is closely proportional to the control input voltage.

N-channel and p-channel TMOS devices, Q1 and Q2, are connected in anti-series to form a bidirectional switch through which current can flow in either the forward or reverse direction. Control circuits turn Q1 and Q2 on when they are reverse biased, bypassing their reverse rectifier and increasing circuit efficiency. Each device is allowed to turn off only when forward biased.

The Q1-Q2 switch connects the ac power source to the load when its instantaneous voltage is the same polarity and less than the control voltage. U1a is configured as an ideal positive rectifier whose output follows the control voltage when it is positive, and is zero otherwise. Similarly, U1b is a negative rectifier. U1c turns Q1 on whenever the ac input voltage is lower than the positive rectifier output. For negative control voltages, Q1 is turned on only during the negative half-cycle. For positive control voltages, Q1 is turned on during the end portions of the positive half-cycle. Similarly, U1d turns Q2 on whenever the ac input voltage is higher than the output of the negative rectifier.

400 Hz SERVO AMPLIFIER

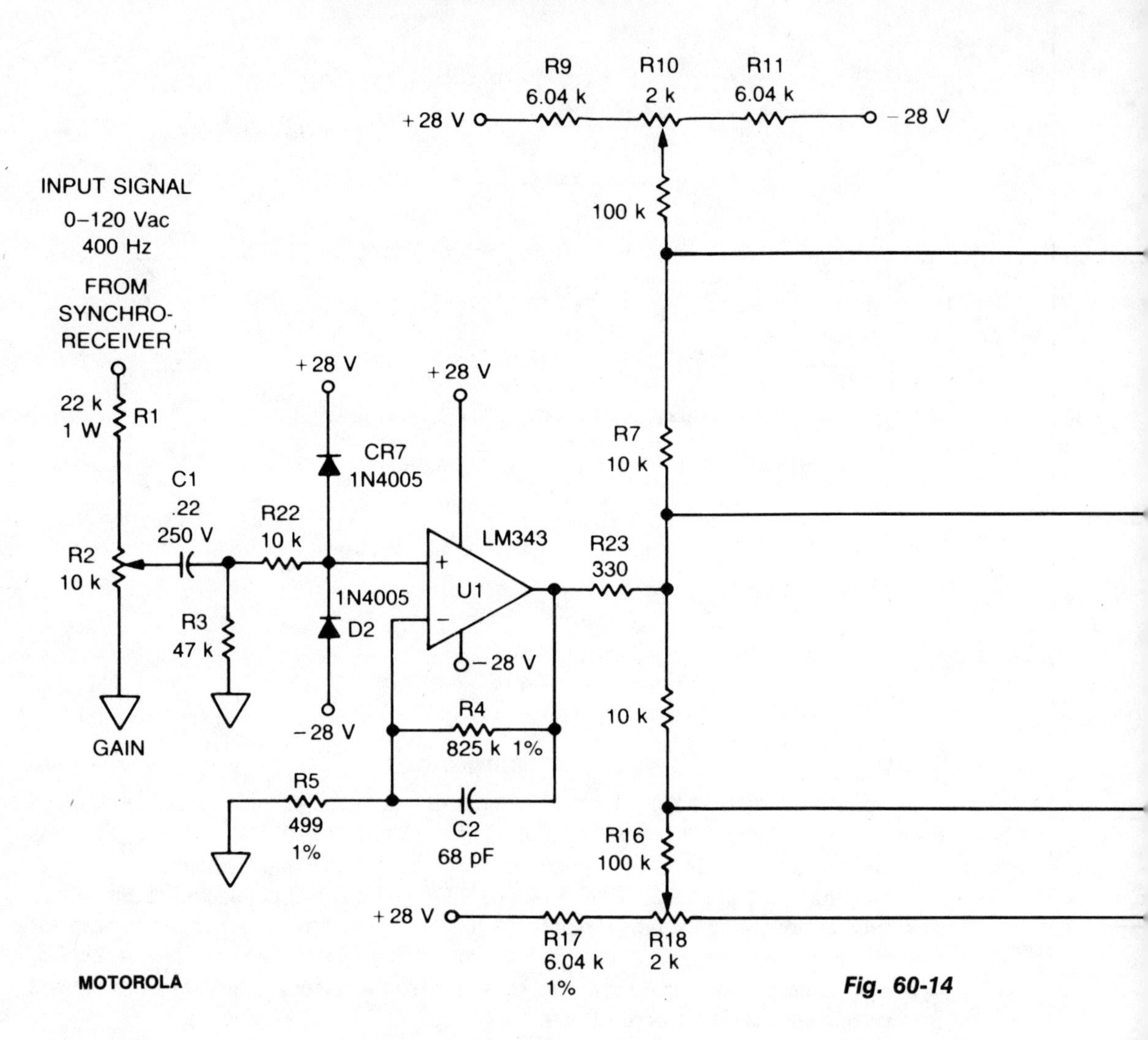

Fig. 60-14

Circuit Notes

The signal from a synchro receiver or a variable resistive cam follower (potentiometer) is boosted by operational amplifier U1, whose output swing is limited by back-to-back zeners D3 and D4. The signal is then applied to operational amplifiers U2 and U3, which drive the gates of Q1 and Q2 respectively. The npn transistor (Q3) is a fast current limiter for the n-channel MTM8N10; a pnp transistor (Q4) performs the same function for the p-channel MTM8P10. Capacitors C3 and C4 eliminate the need for accurate dc offset zeroing. T1 steps up the output voltage to 120 V for the 400 Hz servo motor.

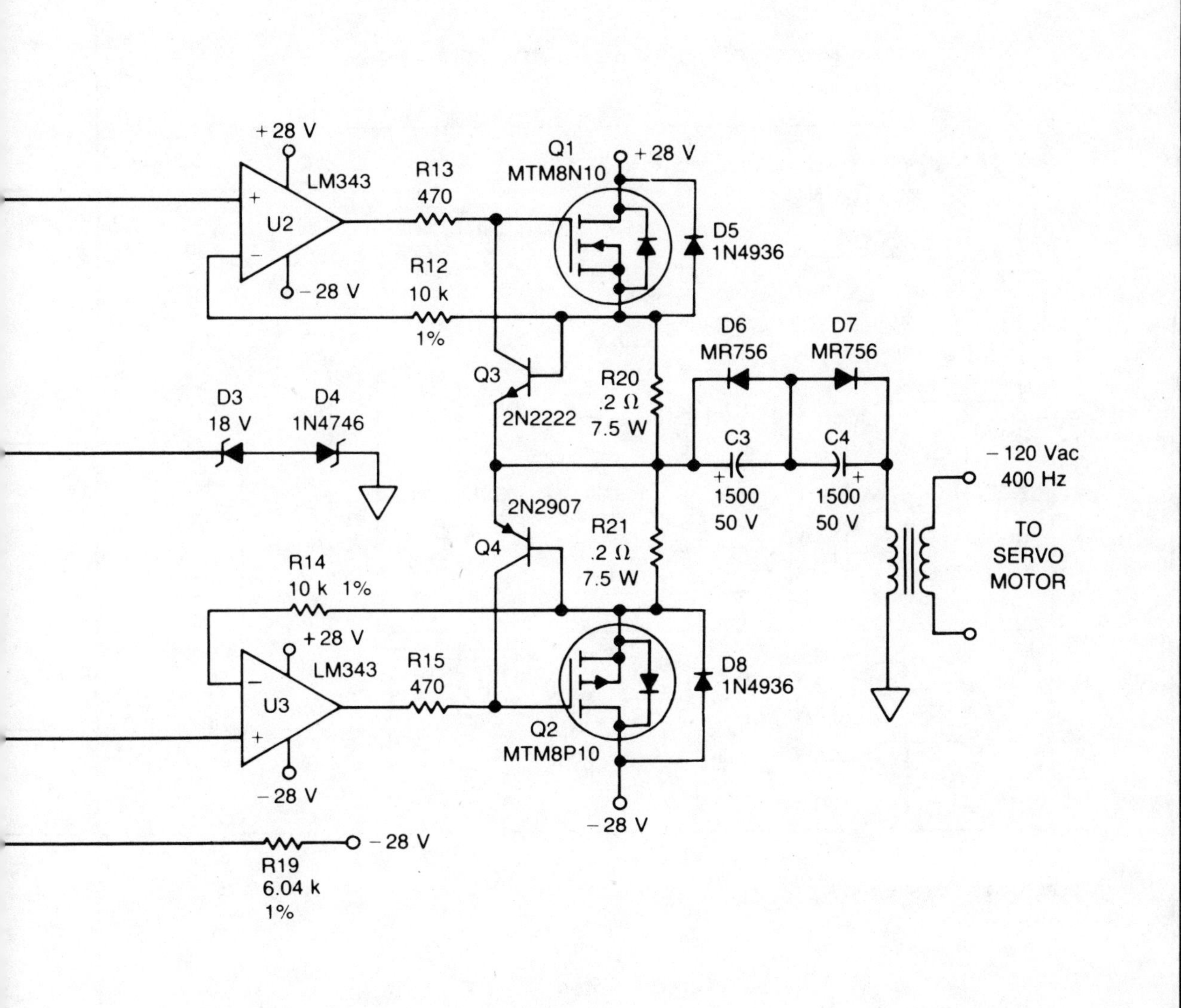
+28 V
LM343
U2
−28 V
R13
470
Q1
MTM8N10
+28 V
D5
1N4936
R12
10 k
1%
Q3
2N2222
R20
.2 Ω
7.5 W
D6
MR756
D7
MR756
D3
18 V
D4
1N4746
C3
1500
50 V
C4
1500
50 V
−120 Vac
400 Hz
TO
SERVO
MOTOR
2N2907
Q4
R21
.2 Ω
7.5 W
R14
10 k 1%
+28 V
LM343
U3
−28 V
R15
470
D8
1N4936
Q2
MTM8P10
−28 V
−28 V
R19
6.04 k
1%

THREE-PHASE POWER-FACTOR CONTROLLER

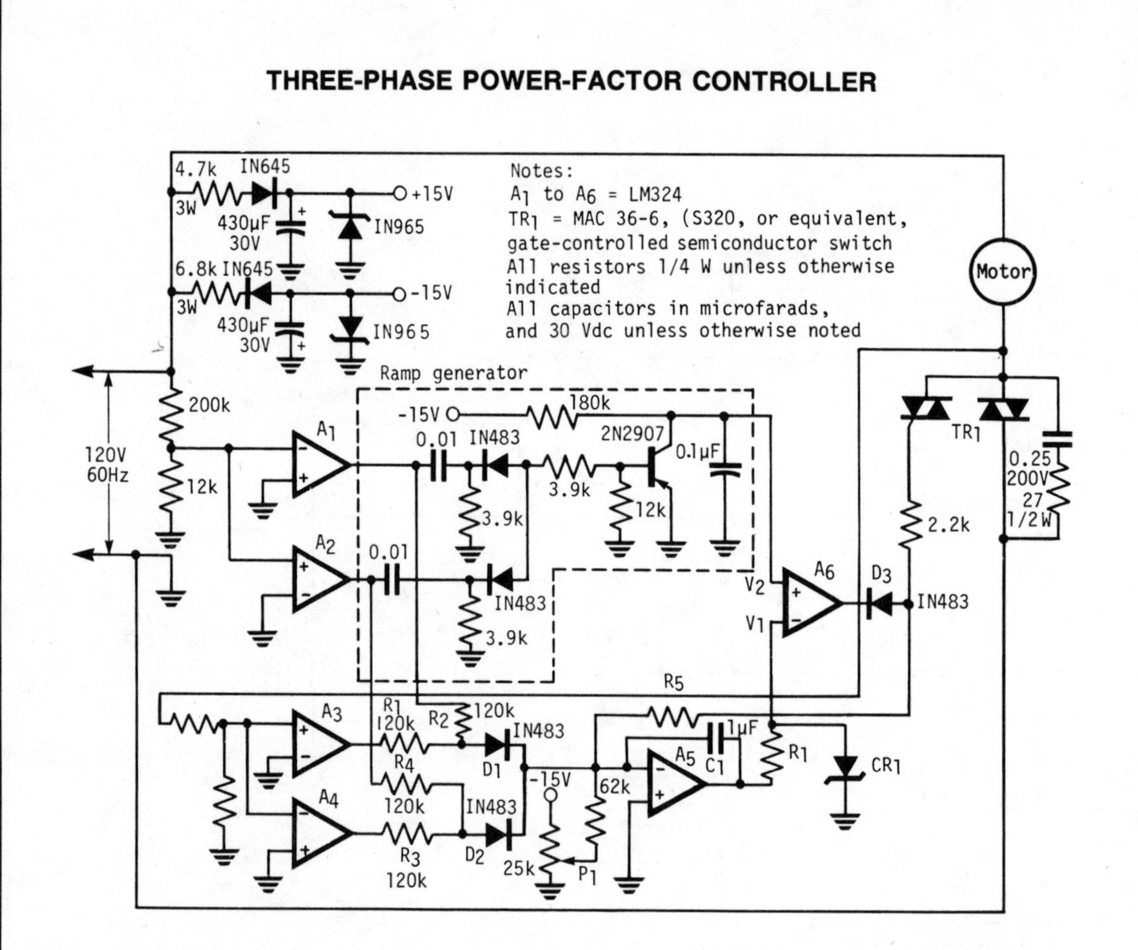

ELECTRONIC ENGINEERING

Fig. 60-15

Circuit Notes

The modified power-factor controller, developed at the Marshall Space Flight Center, employs a phase detector for each of the three phase windings of a delta-connected induction motor. The phase-difference sum is the basis for control. Instabilities of earlier systems are overcome with improved feedback control incorporating a 20Hz bandwidth signal.

MOTOR/TACHOMETER SPEED CONTROL

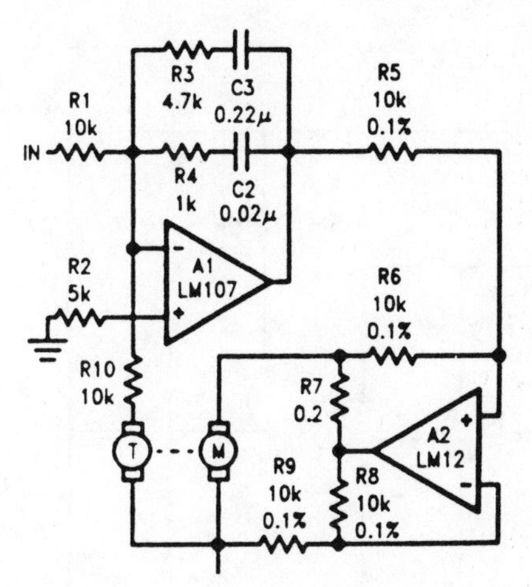

NATIONAL SEMICONDUCTOR CORP.

Fig. 60-16

Circuit Notes

The tachometer, on the same shaft as the dc motor, is simply a generator. It gives a dc output voltage proportional to the speed of the motor. A summing amplifier, A1, controls its output so that the tachometer voltage equals the input voltage, but of opposite sign. With current drive to the motor, phase lag to the tachometer is 90°, before the second order effects come in. Compensation on A1 is designed to give less than 90° phase shift over the range of frequencies where the servo loop goes through unity gain. Should response time be of less concern, a power op amp could be substituted for A1 to drive the motor directly. Lowering break frequencies of the compensation would, of course, be necessary. The circuit could also be used as a position servo. All that is needed is a voltage indicating the sense and magnitude of the motor shaft displacement from a desired position. This error signal is connected to the input, and the servo works to make it zero. The tachometer is still required to develop a phase-correcting rate signal because the error signal lags the motor drive by 180°.

CLOSED LOOP, TACHOMETER FEEDBACK CONTROL

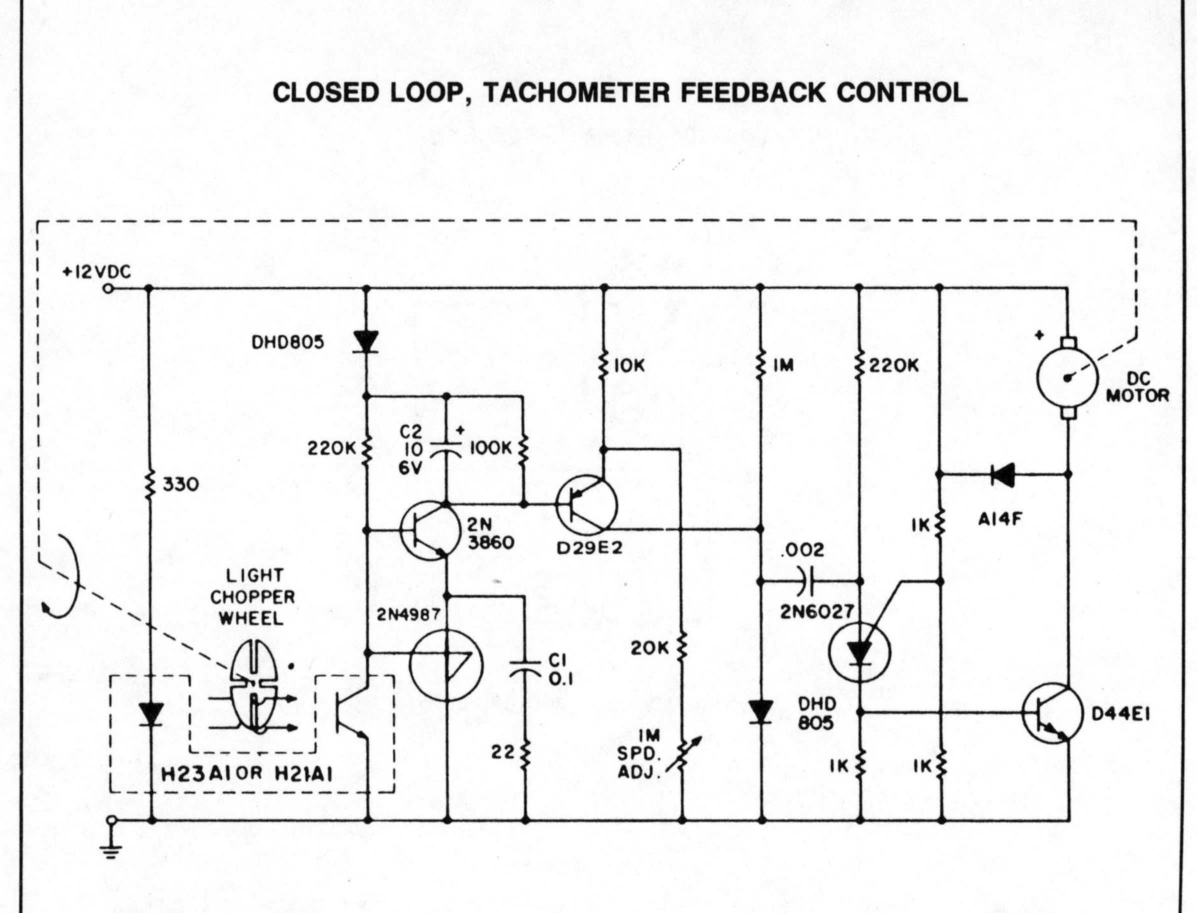

GENERAL ELECTRIC

Fig. 60-17

Circuit Notes

The system utilizes the H21A1 and a chopper disc to provide superior speed regulation when the dynamic characteristics of the motor system and the feedback system are matched to provide stability. The tachometer feedback system illustrated was designed around specific motor/load combinations and may require modification to prevent hunting or oscillation with other combinations. This dc motor control utilizes the optachometer circuit previously shown to control a P.U.T. pulse generator that drives the D44E1 darlington transistor which powers the motor.

61

Multiplier Circuits

The sources of the following circuits are contained in the Sources section beginning on page 694. The figure number contained in the box of each circuit correlates to the source entry in the Sources section.

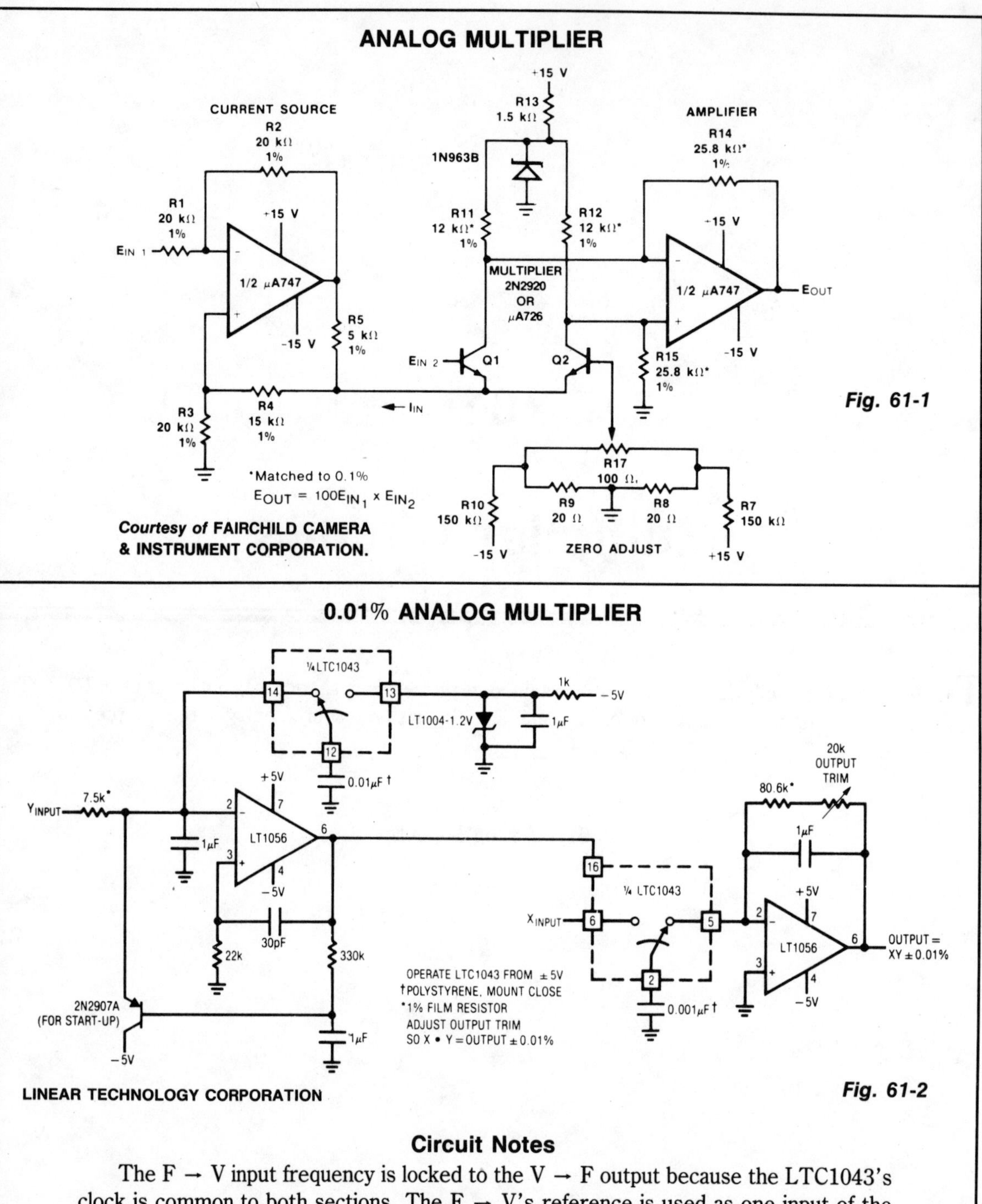

Courtesy of **FAIRCHILD CAMERA & INSTRUMENT CORPORATION.**

Fig. 61-1

LINEAR TECHNOLOGY CORPORATION

Fig. 61-2

Circuit Notes

The F → V input frequency is locked to the V → F output because the LTC1043's clock is common to both sections. The F → V's reference is used as one input of the multiplier, while the V → F furnishes the other. To calibrate, short the X and Y inputs to 1.7320 V and trim for a 3-V output.

62

Noise Reduction Circuits

The sources of the following circuits are contained in the Sources section beginning on page 694. The figure number contained in the box of each circuit correlates to the source entry in the Sources section.

AUDIO SQUELCH CIRCUIT

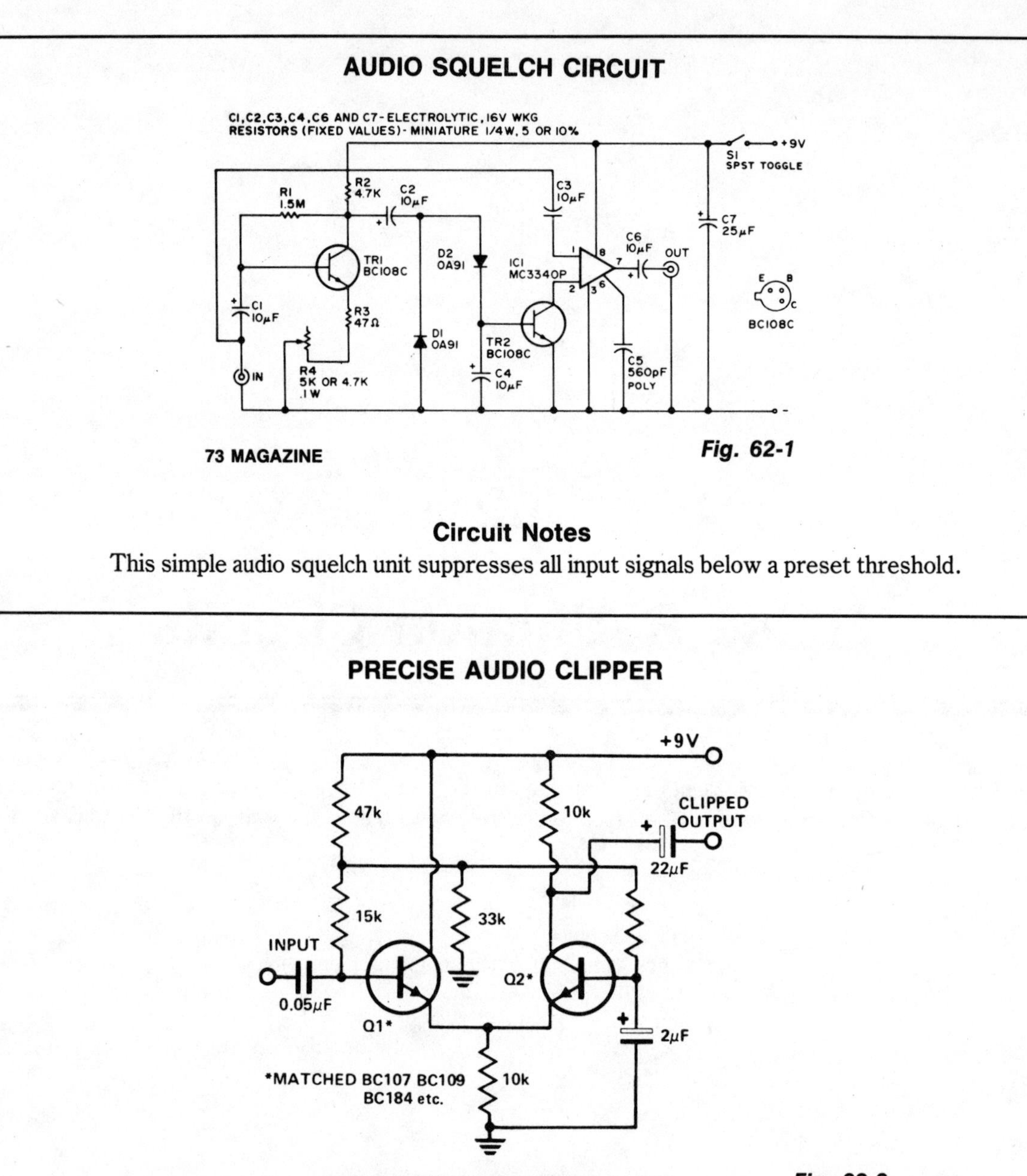

73 MAGAZINE

Fig. 62-1

Circuit Notes

This simple audio squelch unit suppresses all input signals below a preset threshold.

PRECISE AUDIO CLIPPER

ELECTRONICS INTERNATIONAL TODAY

Fig. 62-2

Circuit Notes

A differential amplifier makes an excellent audio clipper and can provide precise, symmetrical clipping. The circuit shown commences clipping at an input of 100 mV. The output commences clipping at ±3 V. Matching Q7 and Q2 is necessary for good symmetrical clipping. (If some asymmetry can be tolerated, this need not be done.)

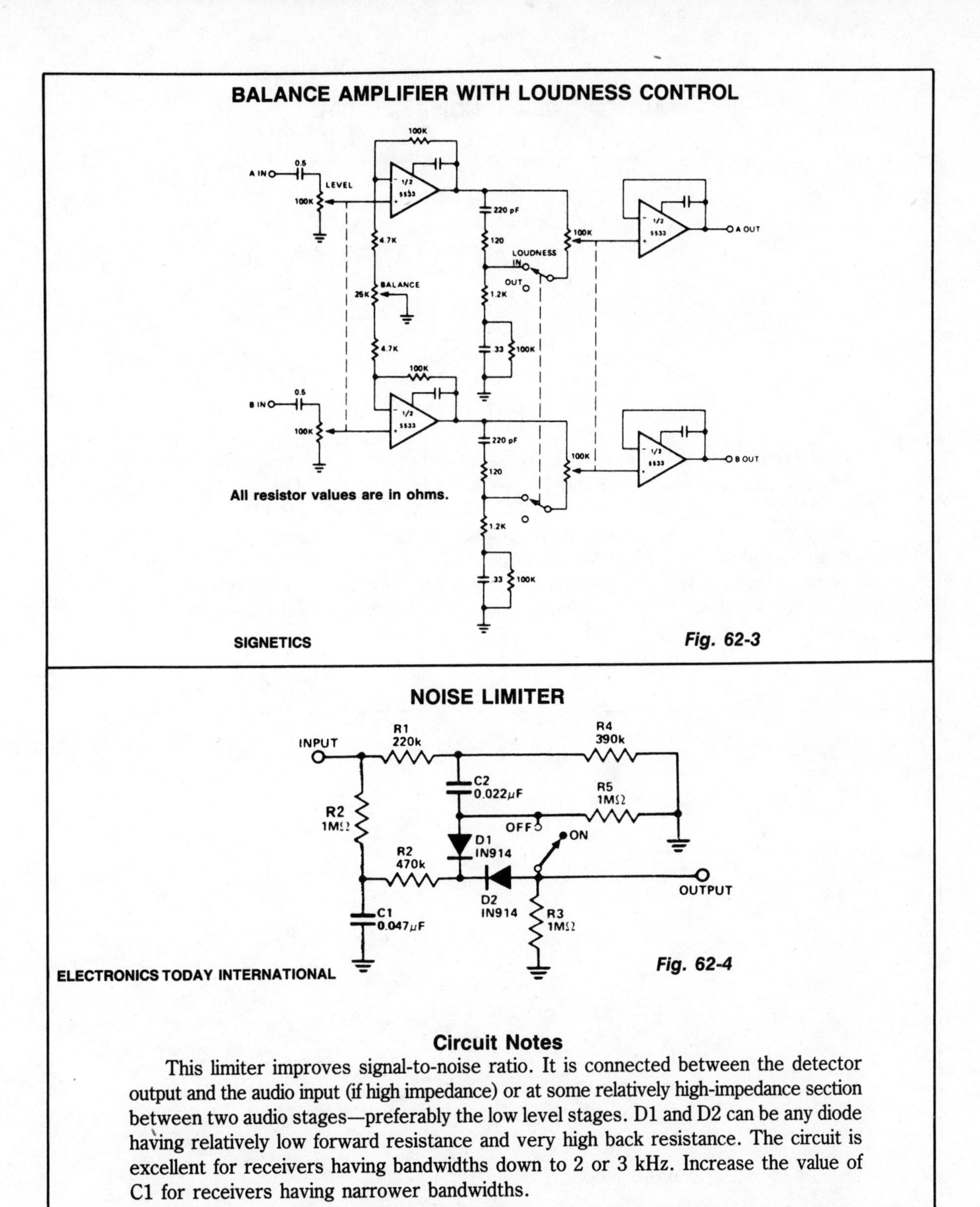

Fig. 62-3

Fig. 62-4

Circuit Notes

This limiter improves signal-to-noise ratio. It is connected between the detector output and the audio input (if high impedance) or at some relatively high-impedance section between two audio stages—preferably the low level stages. D1 and D2 can be any diode having relatively low forward resistance and very high back resistance. The circuit is excellent for receivers having bandwidths down to 2 or 3 kHz. Increase the value of C1 for receivers having narrower bandwidths.

AUDIO-POWERED NOISE CLIPPER

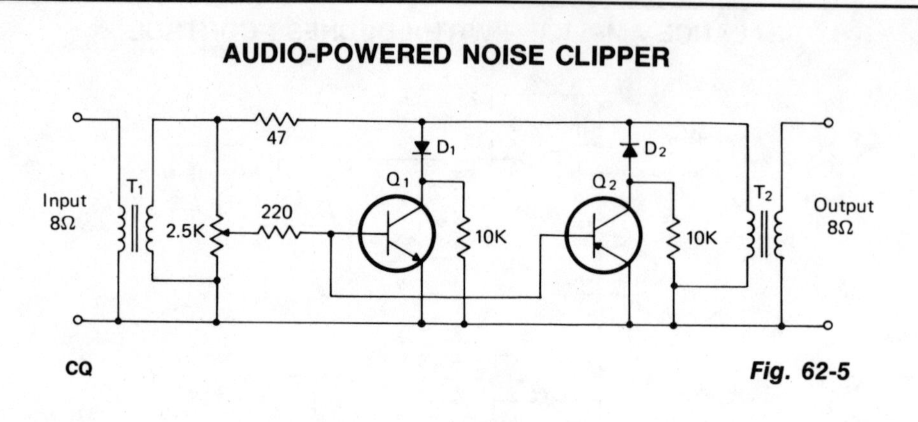

CQ

Fig. 62-5

Circuit Notes

T1 and T2 are 600 to 8 ohm transformers (any transistor radio output transformers with 500 to 4 ohm impedance may be used). Q1 is a 2N2222 npn transistor, and Q2 is a 2N2907 pnp transistor. D1 and D2 1N270 signal diodes (HEP 134 or 135). Two transistors, powered by the audio power contained within the signal, will clip signal peaks which exceed the threshold established by the 2.5 K potentiometer. The diodes isolate the positive and negative clipping circuits represented by the npn and pnp transistors, respectively. A desired audio operating level can be established and the potentiometer needs little or no further adjustment.

63
Notch Filters

The sources of the following circuits are contained in the Sources section beginning on page 694. The figure number contained in the box of each circuit correlates to the source entry in the Sources section.

ADJUSTABLE Q NOTCH FILTER

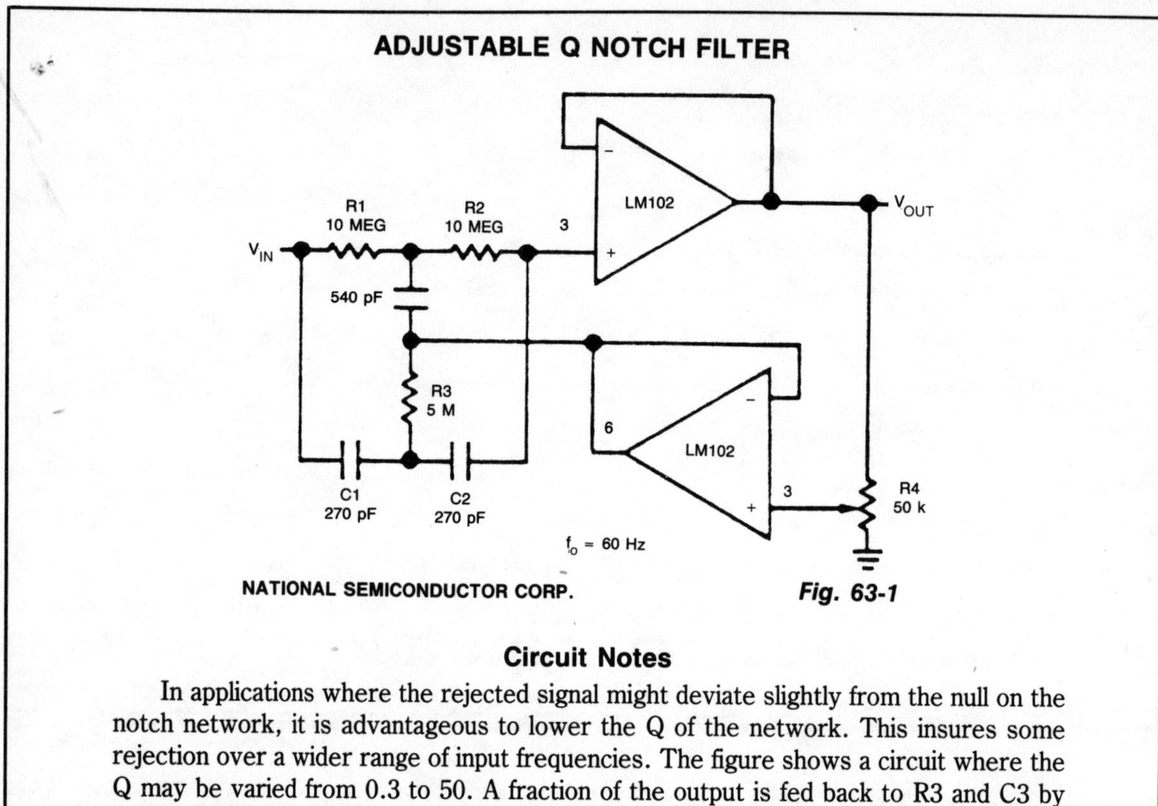

NATIONAL SEMICONDUCTOR CORP.

Fig. 63-1

Circuit Notes

In applications where the rejected signal might deviate slightly from the null on the notch network, it is advantageous to lower the Q of the network. This insures some rejection over a wider range of input frequencies. The figure shows a circuit where the Q may be varied from 0.3 to 50. A fraction of the output is fed back to R3 and C3 by a second voltage follower, and the notch Q is dependent on the amount of signal fed back. A second follower is necessary to drive the twin "T" from a low-resistance source so that the notch frequency and depth will not change with the potentiometer setting.

1800 Hz NOTCH FILTER

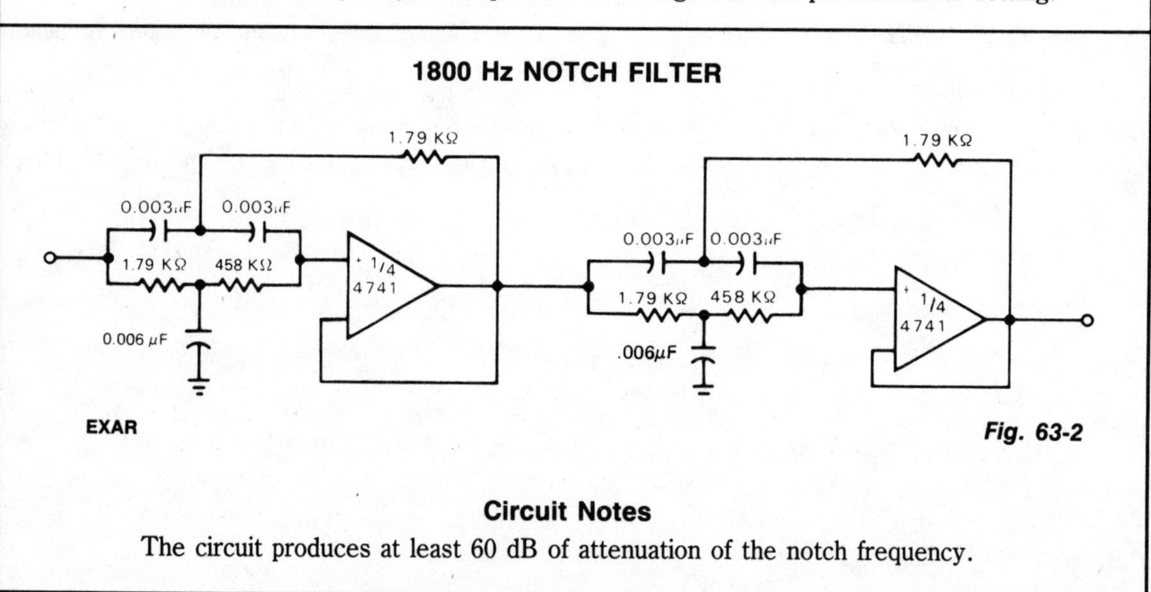

EXAR

Fig. 63-2

Circuit Notes

The circuit produces at least 60 dB of attenuation of the notch frequency.

550 Hz NOTCH FILTER

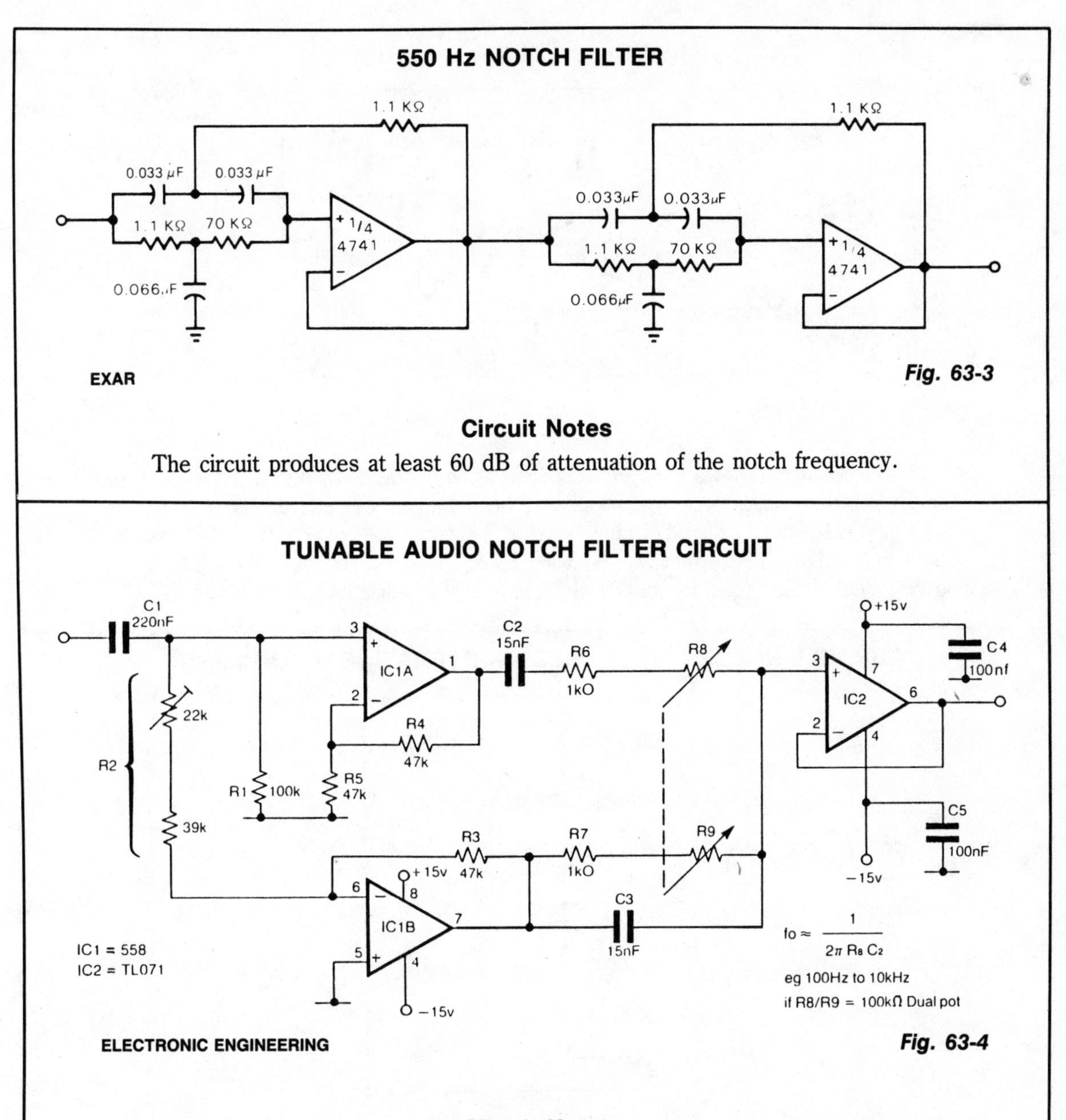

EXAR

Fig. 63-3

Circuit Notes

The circuit produces at least 60 dB of attenuation of the notch frequency.

TUNABLE AUDIO NOTCH FILTER CIRCUIT

ELECTRONIC ENGINEERING

Fig. 63-4

Circuit Notes

The circuit requires only one dual-ganged potentiometer to tune over a wide range; if necessary over the entire audio range in one sweep. The principle used is that of the Wien bridge, fed from anti-phase inputs. The output should be buffered as shown with a FET input op amp, particularly if a high value pot is used. An op amp with differential outputs (eg., MC1445) may be used in place of the driver ICS; R2 may be made trimmable to optimize the notch.

AUDIO NOTCH FILTER

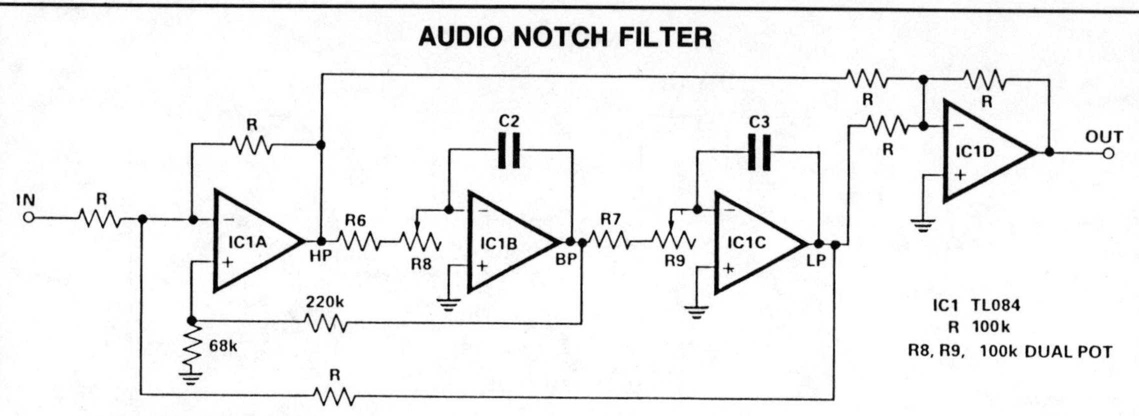

ELECTRONIC ENGINEERING

Fig. 63-5

Circuit Notes

With the circuit shown here the response at one octave off tune is within 10% of the far out response: notch sharpness may be increased or reduced by reducing or increasing respectively the 68 K ohm resistor. Linearity tracking of R8 and R9 has no effect on notch depth. The signals at HP and LP are always in antiphase, notch will always be very deep at the tuned frequency, despite tolerance variations in R6-9 and C2, C3.

TUNABLE NOTCH FILTER USES AN OPERATIONAL AMPLIFIER

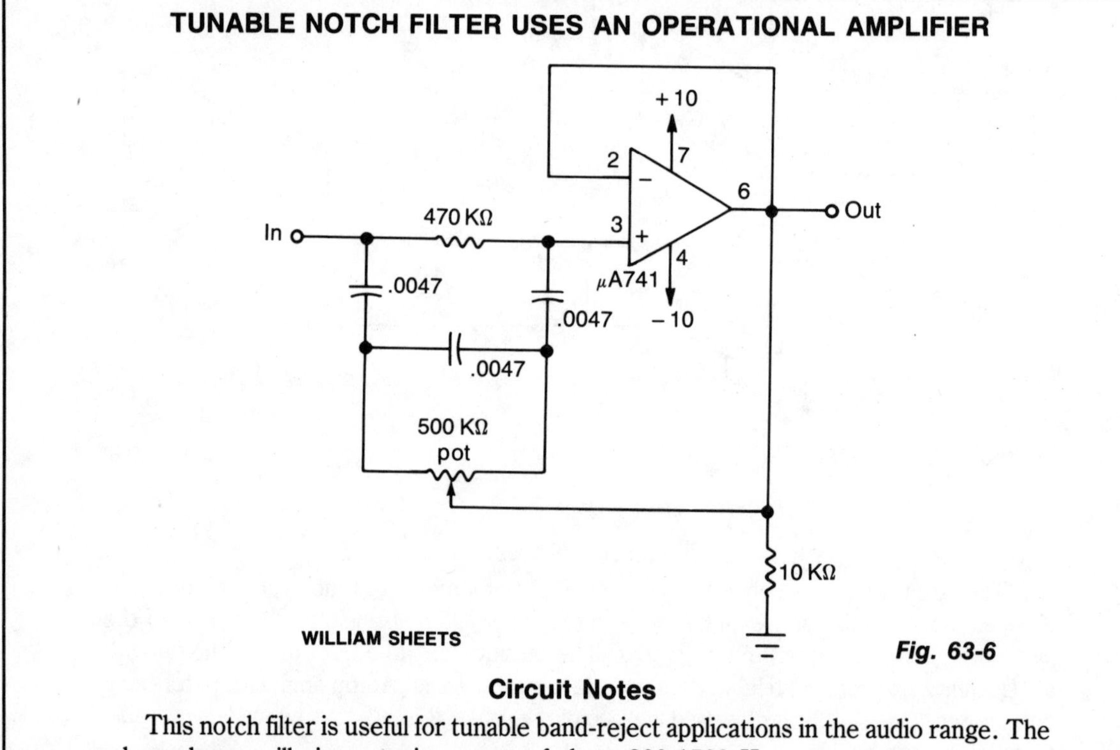

WILLIAM SHEETS

Fig. 63-6

Circuit Notes

This notch filter is useful for tunable band-reject applications in the audio range. The values shown will give a tuning range of about 300-1500 Hz.

ACTIVE BAND-REJECT FILTER

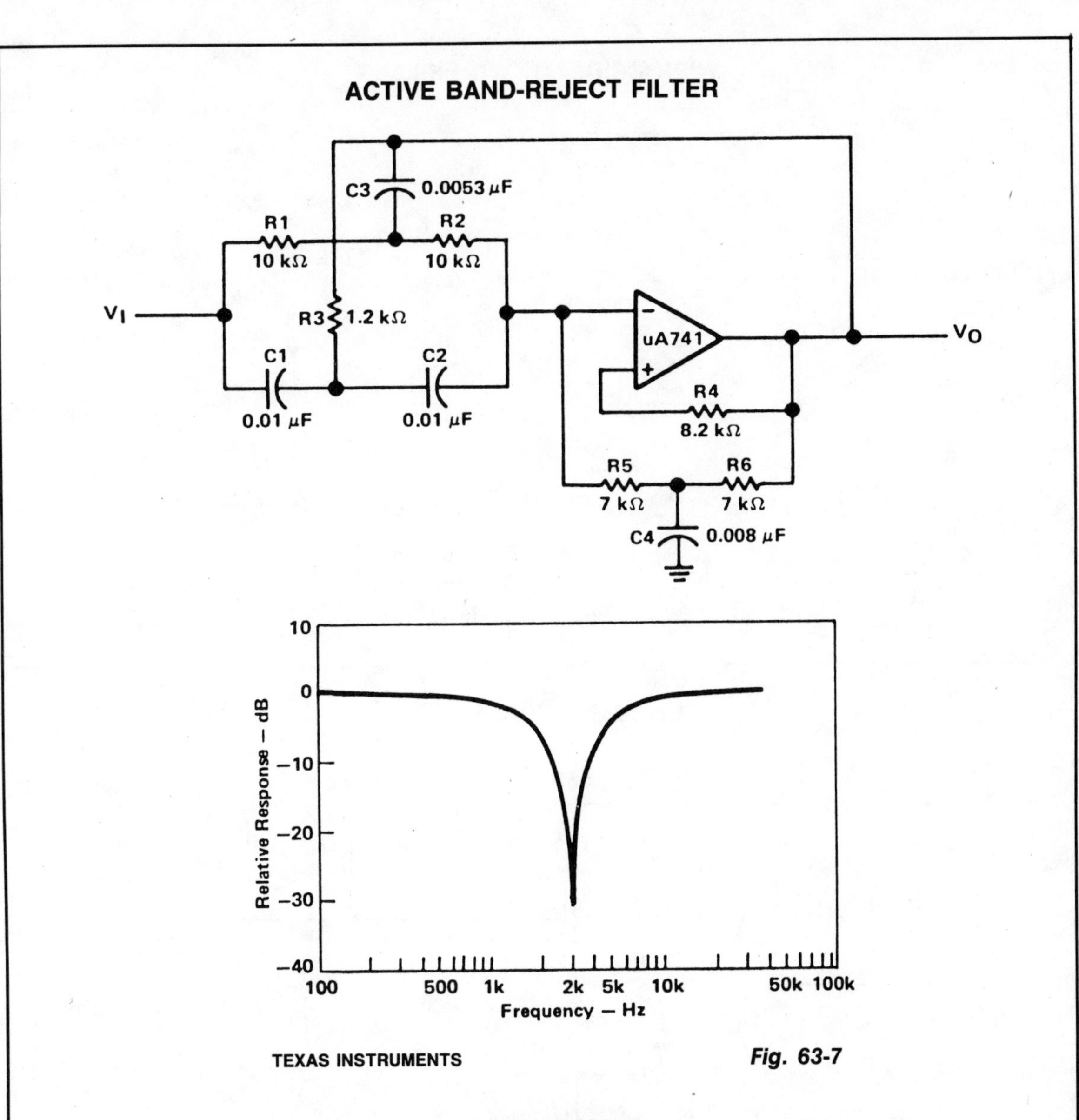

TEXAS INSTRUMENTS

Fig. 63-7

Circuit Notes

A filter with a band-reject characteristic is frequently referred to as a notch filter. A typical circuit using a μA 741 is the unity-gain configuration for this type of active filter shown. The filter response curve shown is a second-order band-reject filter with a notch frequency of 3 kHz. The resulting Q of this filter is about 23, with a notch depth of –31 dB. Although three passive T networks are used in this application, the operational amplifier has become a sharply tuned low-frequency filter without the use of inductors or large-value capacitors.

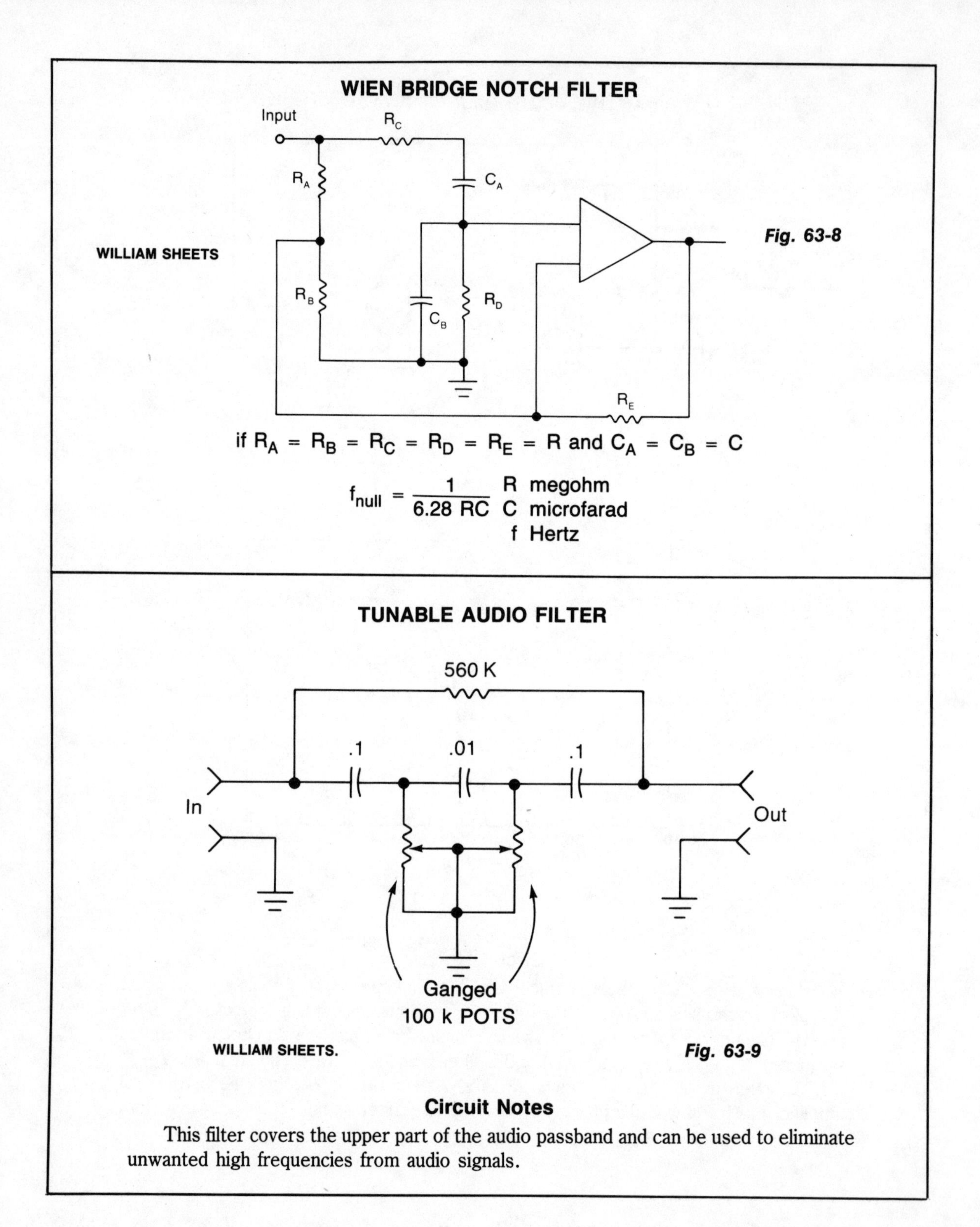

if $R_A = R_B = R_C = R_D = R_E = R$ and $C_A = C_B = C$

$$f_{null} = \frac{1}{6.28\ RC}$$

R megohm
C microfarad
f Hertz

Circuit Notes

This filter covers the upper part of the audio passband and can be used to eliminate unwanted high frequencies from audio signals.

PASSIVE BRIDGED, DIFFERENTIATOR TUNABLE NOTCH FILTER

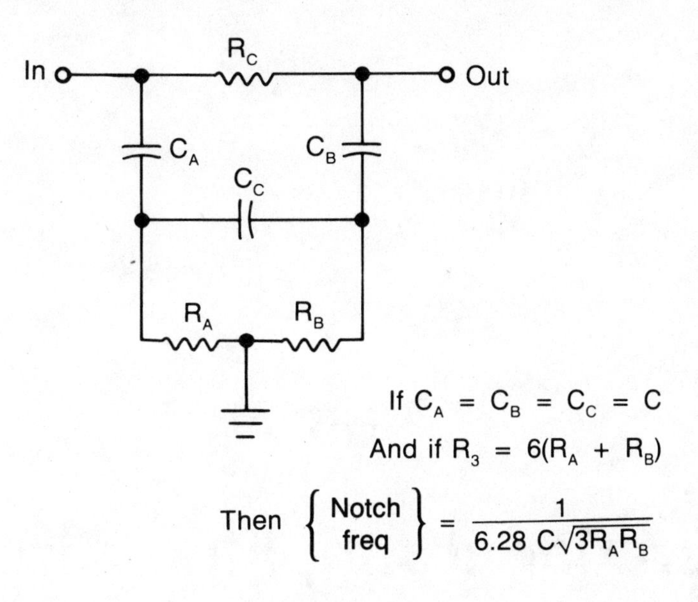

If $C_A = C_B = C_C = C$

And if $R_3 = 6(R_A + R_B)$

$$\text{Then} \left\{ \begin{array}{c} \text{Notch} \\ \text{freq} \end{array} \right\} = \frac{1}{6.28\, C\sqrt{3R_A R_B}}$$

If R_A and R_B is made a potentiometer then the filter can be variable.

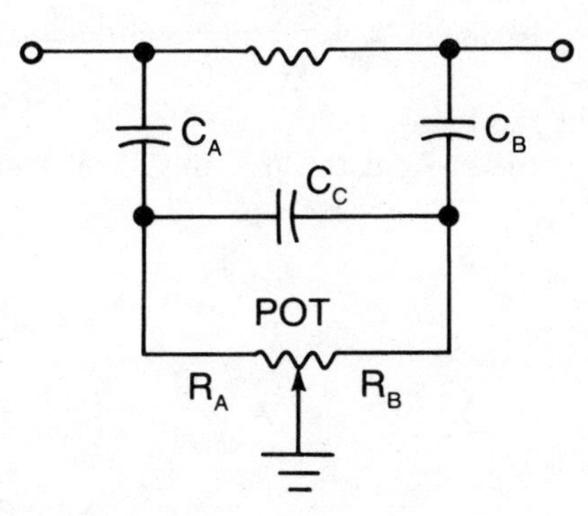

WILLIAM SHEETS

Fig. 63-10

R_A and R_B are sections of potentiometer.

64

Operational Amplifier Circuits

The sources of the following circuits are contained in the Sources section beginning on page 694. The figure number contained in the box of each circuit correlates to the source entry in the Sources section.

VARIABLE GAIN AND SIGN OP AMP CIRCUIT

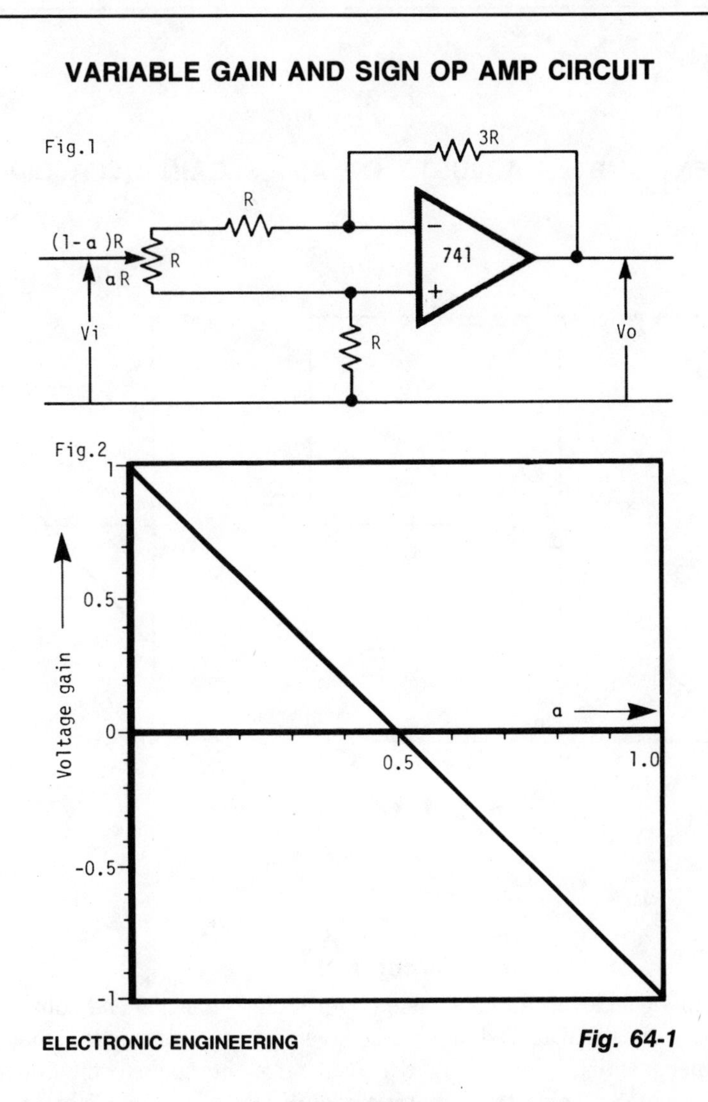

ELECTRONIC ENGINEERING

Fig. 64-1

Circuit Notes

The gain of the amplifier is smoothly controllable between the limits of +1 to −1. It is adjustable over this range using a single potentiometer. The voltage gain of the arrangement is given by:

$$\frac{V_o}{V_i} = \frac{2(1-2\alpha)}{(1+\alpha)(2-\alpha)}$$

Where α represents the fractional rotation of the potentiometer, R.

SINGLE POTENTIOMETER ADJUSTS OP AMP'S GAIN OVER BIPOLAR RANGE

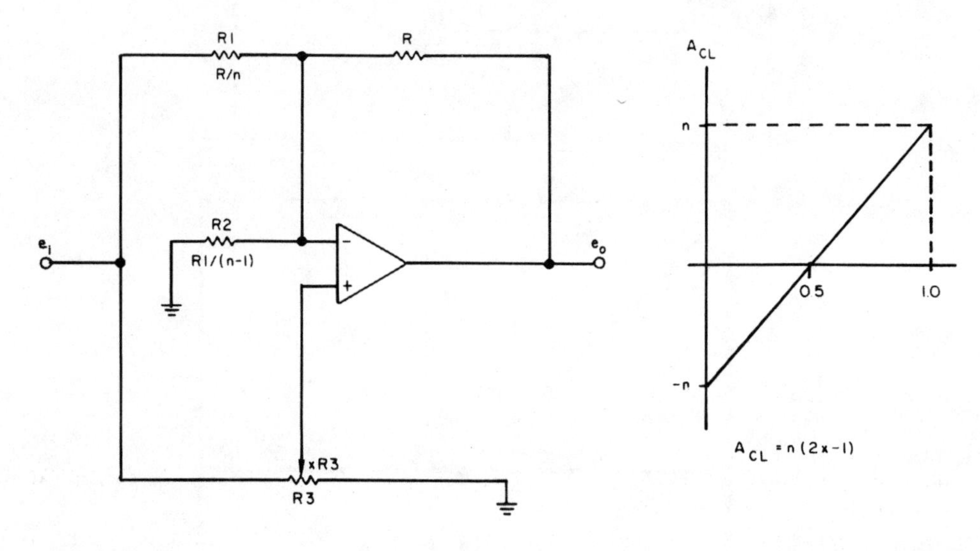

Fig. 64-2

ELECTRONIC DESIGN

Circuit Notes

An op amp's gain level can be adjusted over its full inverting and noninverting gain range. R3 varies the signal applied to both the inverting and noninverting amplifier inputs. When the wiper position (denoted by x) equals zero, the noninverting amplifier input is grounded. This also holds the voltage across R2 at zero, so R2 has no effect on operation. Now only R1 and R carry feedback current, and the amplifier operates at a gain of $-n$. At the other pot extreme, where $x = 1$, the input signal is connected directly to the noninverting input. Since feedback maintains a near-zero voltage between the amplifier inputs, the amplifier's inverting input will also be near the input signal level, thus little voltage is across R1, also now the gain is $+n$. The amplifier should be driven from a low impedance source to minimize source loading error, low offset op amps should be used.

65

Optically-Coupled Circuits

The sources of the following circuits are contained in the Sources section beginning on page 694. The figure number contained in the box of each circuit correlates to the source entry in the Sources section.

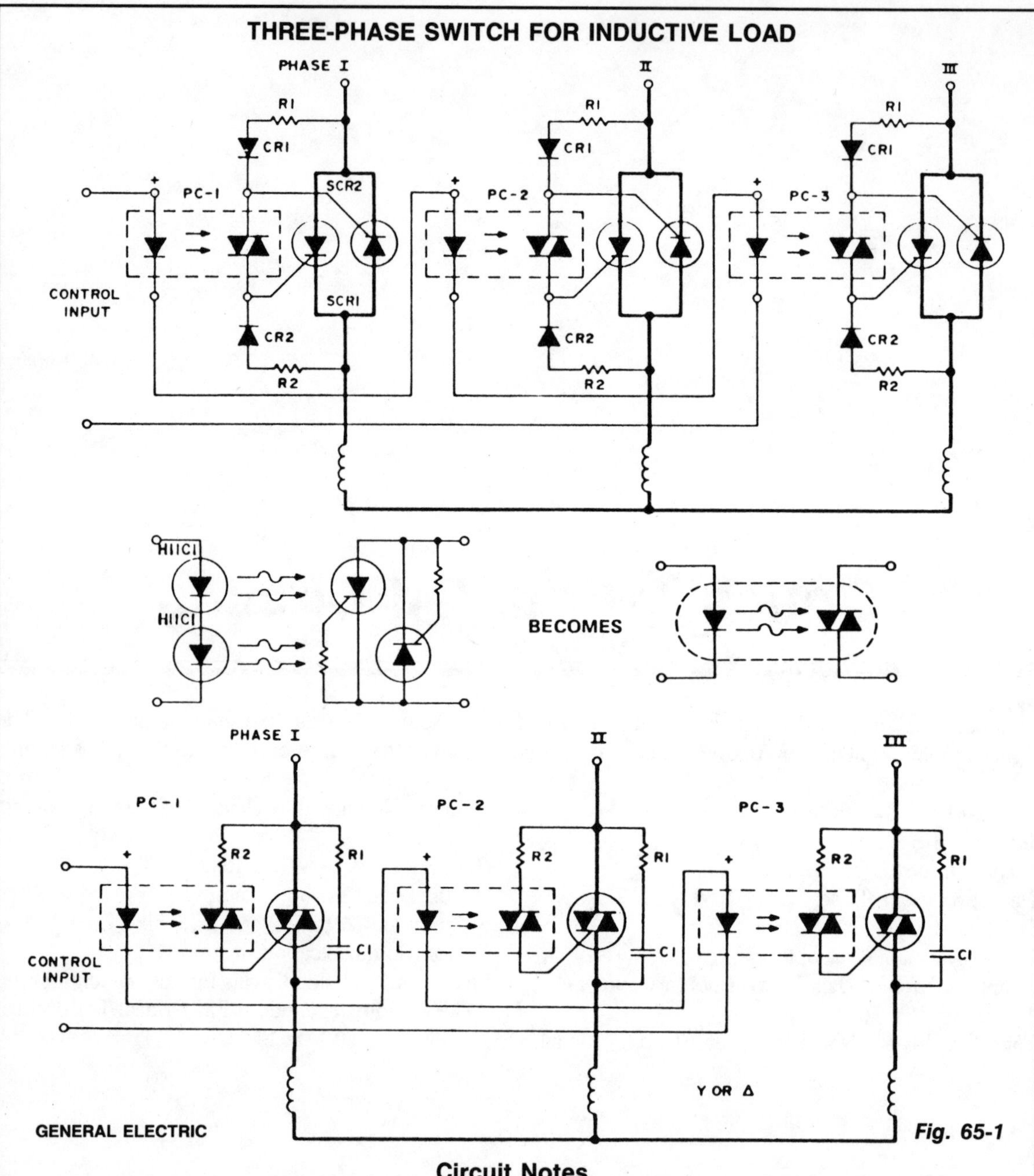

Fig. 65-1

Circuit Notes

The following are three-phase switches for low voltage. Higher currents can be obtained by using inverse parallel SCRs which would be triggered as shown. To simplify the following schematics and facilitate easy understanding of the principles involved, the following schematic substitution is used (Note the triac driver is of limited use at 3 ϕ voltage levels).

INTEGRATED SOLID STATE RELAY

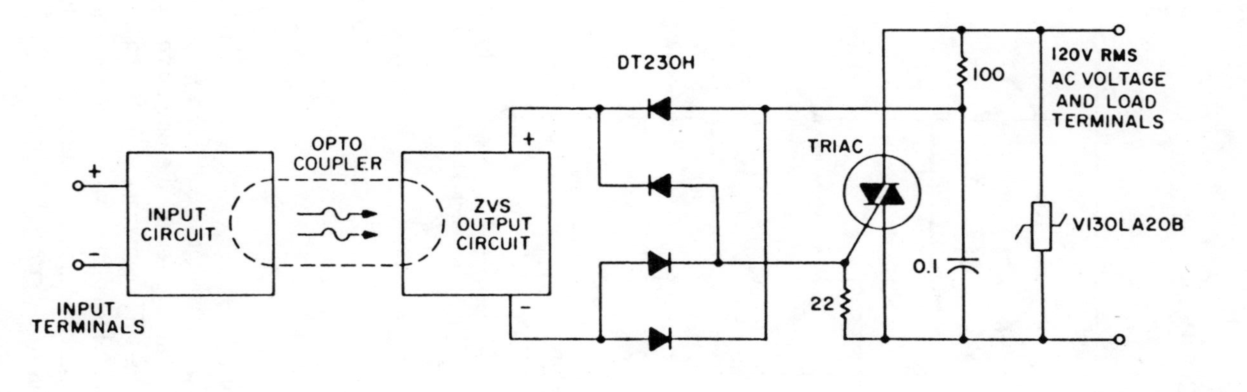

GENERAL ELECTRIC

Fig. 65-2

Circuit Notes

A complete zero-voltage switch solid-state relay contains an input circuit, an output circuit, and the power thyristor. The circuit illustrates a triac power thyristor with snubber circuit and GE-MOV^R II Varistor transient over-voltage protection. The 22 ohm resistor shunts di/dt currents, passing through the bridge diode capacitances, from the triac gate, while the 100 ohm resistor limits surge and gate currents to safe levels. Although the circuits illustrated are for 120-V rms operation, relays that operate on 220 V require higher voltage ratings on the MOV, rectifier diodes, triac, and pilot SCR. The voltage divider that senses zero crossing must also be selected to minimize power dissipation in the transistor optisolator circuit for 220-V operation.

STABLE OPTOCOUPLER

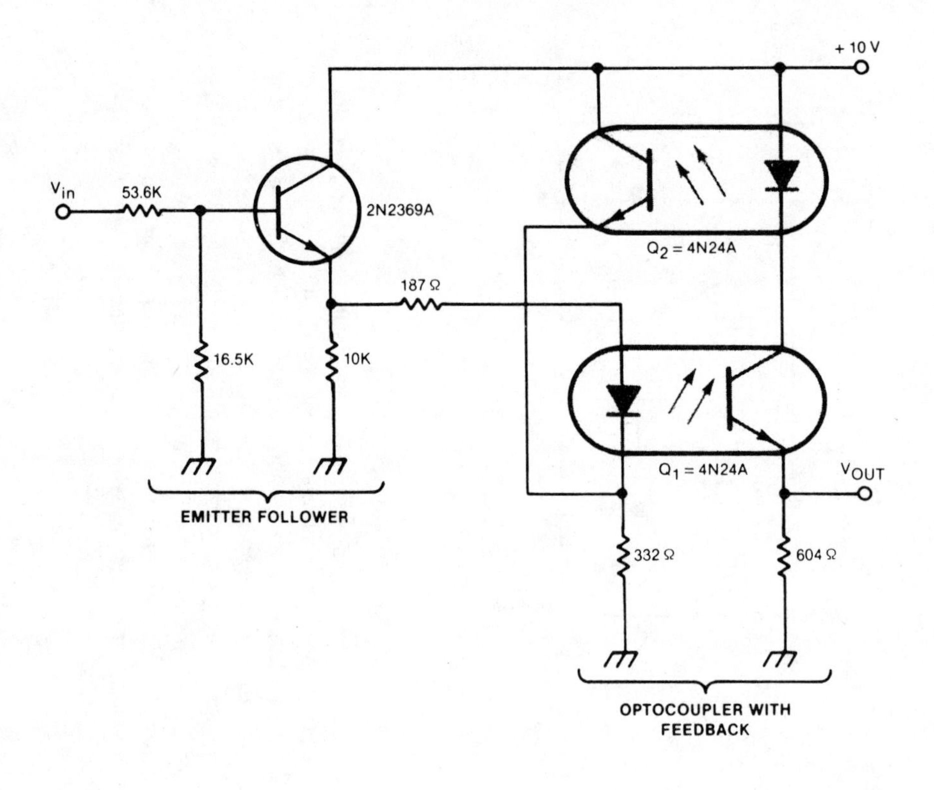

NASA

Fig. 65-3

Circuit Notes

A circuit stabilizes the current-transfer ratio (CTR) of an optically coupled isolator used as a linear transducer. The optocoupler produces a voltage output that is proportional to—but electrically isolated from—the voltage input. However, the output voltage is directly affected by changes in the CTR, and the CTR can change substantially with temperature and current. To a lesser extent the CTR changes with time over the life of the optocoupler. The circuit employs a feedback circuit containing a second optocoupler. The feedback signal tends to oppose changes in the overall CTR.

MICROPROCESSOR TRIAC ARRAY DRIVER

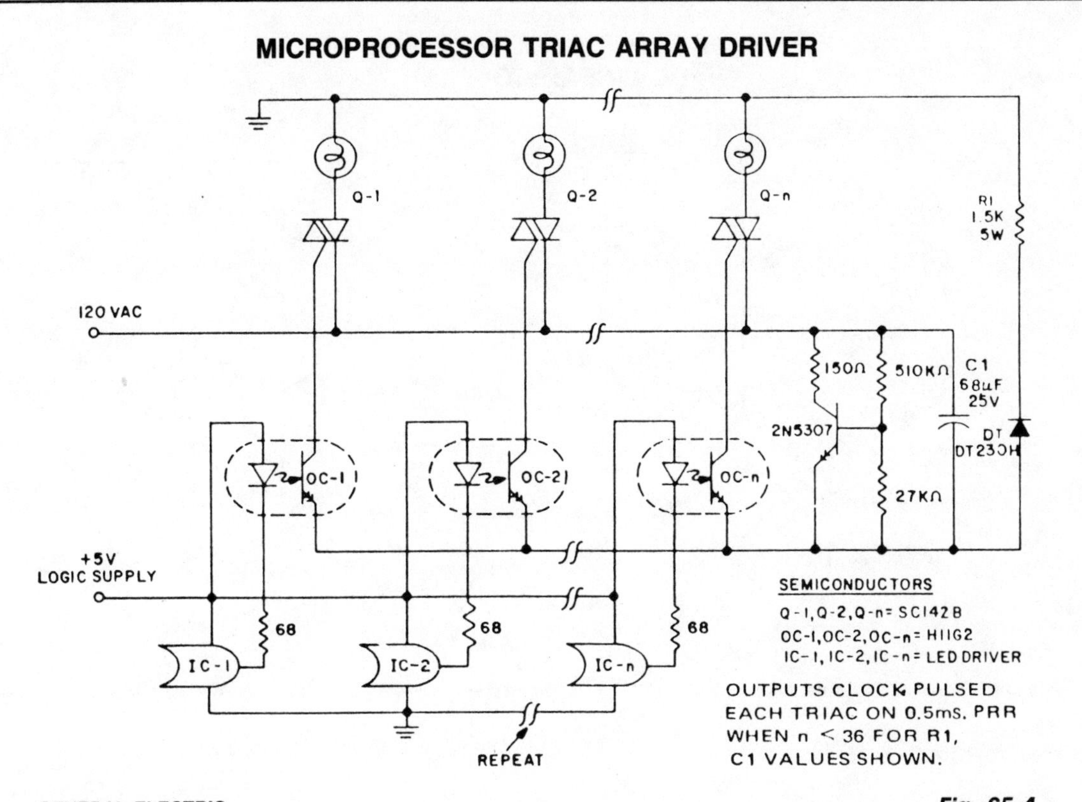

GENERAL ELECTRIC

Fig. 65-4

Circuit Notes

In microprocessor control of multiple loads, the minimum cost per load is critical. A typical application example is a large display involving driving arrays of incandescent lamps. This circuit provides minimal component cost per stage and optocoupler triggering of triac power switches from logic outputs. The minimal component cost is attained by using more complex software in the logic. A darlington output optocoupler provides gate current pulses to the triac, with cost advantages gained from eliminating the current limiting resistor and from the low cost coupler. The trigger current source is a dipped tantalum capacitor, charged from the line via a series resistor with coarse voltage regulation being provided by the darlington signal transistor. The resistor and capacitor are shared by all the darlington-triac pairs and are small in size and cost due to the low duty cycle of pulsing. Coupler IRED current pulses are supplied for the duration of one logic clock pulse (2-10 μsec), at 0.4 to 1 msec intervals, from a LED driver I.C. The pulse timing is derived from the clock waveform when the logic system requires triac conduction. A current limiting resistor is not used, which prevents Miller effect slowdown of the H11G2 switching speed to the extent the triac is supplied insufficient current to trigger. Optodarlington power dissipation is controlled by the low duty cycle and the capacitor supply characteristics.

DC LINEAR COUPLER

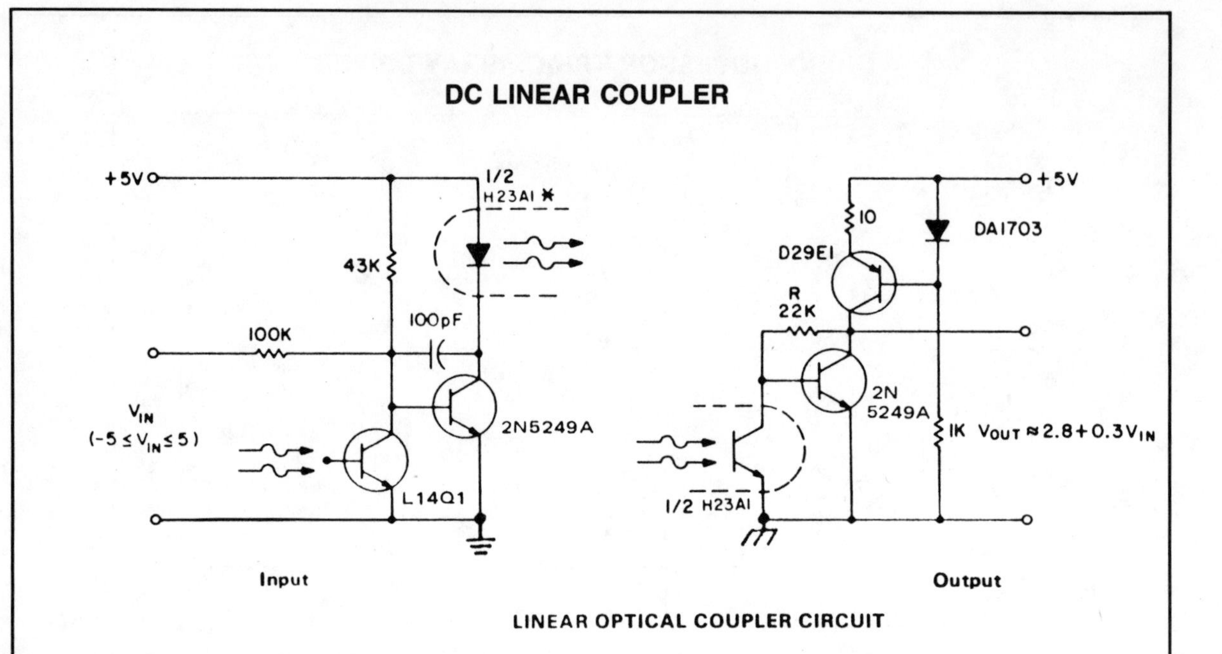

LINEAR OPTICAL COUPLER CIRCUIT

*Closely positioned to illuminate L14Q1 and H23A1 Detector, such that $V_{OUT} \cong 2.8V$ at $V_{IN} = 0$.

GENERAL ELECTRIC

Fig. 65-5

Circuit Notes

The accuracy of direct linear coupling of analog current signals via an optocoupler is determined by the coupler linearity and its temperature coefficient. Use of an additional coupler for feedback can provide linearity only if the two couplers are perfectly matched and identically biased. These are not practical constraints in most equipment designs and indicate the need for a different design approach. One of the most successful solutions to this problem can be illustrated by using a H23 emitter-detector pair and a L14H4. The H23 detector and L14H4 are placed so both are illuminated by the H23 IRED emitter. Ideally, the circuit is mechanically designed such that the H23 emitter may be positioned to provide $V_{OUT} = 2.8$ V when $V_{IN} = 0$, thereby insuring collector current matching in the detectors. Then all three devices are locked in position relative to each other. Otherwise, R may be adjusted to provide the proper null level, although temperature tracking should prove worse when R is adjusted. Note that the input bias is dependent on power supply voltage, although the output is relatively independent of supply variations. Testing indicated linearity was better than could be resolved, due to alignment motion caused by using plastic tape to lock positions. The concept of feedback control of IRED power output is useful for both information transmission and sensing circuitry.

LINEAR AC ANALOG COUPLER

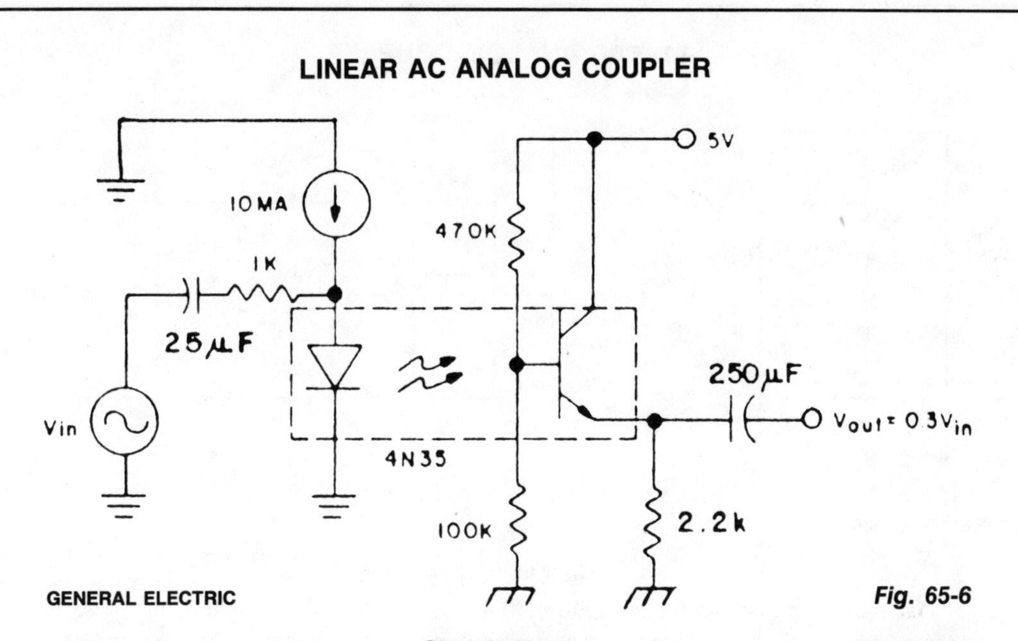

GENERAL ELECTRIC

Fig. 65-6

Circuit Notes

With the coupler biased in the linear region by the 10 mA dc bias on the IRED and the voltage divider on the phototransistor base, photodiode current flows out of the base into the voltage divider, producing an ac voltage proportional to the ac current in the IRED. The transistor is biased as an emitter follower and requires less than 10% of the photodiode current to produce the low impedance ac output across the emitter resistor. Note that the H11AV1 may be substituted for the 4N35 to provide VDE line voltage rated isolation of less than 0.5 pF.

SIMPLE AC RELAY USING TWO PHOTON COUPLERS

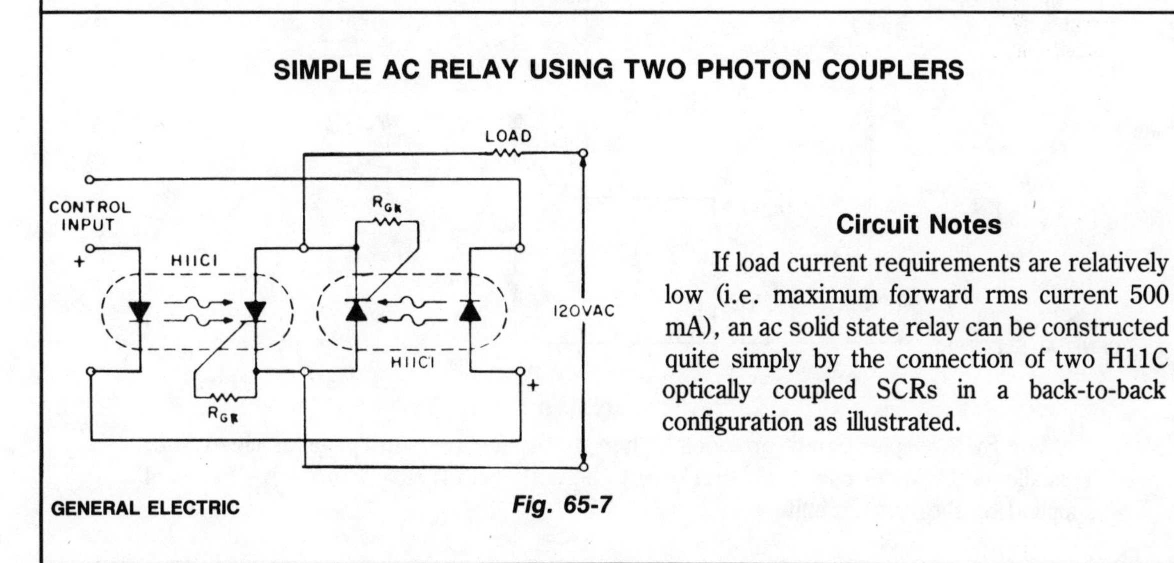

GENERAL ELECTRIC

Fig. 65-7

Circuit Notes

If load current requirements are relatively low (i.e. maximum forward rms current 500 mA), an ac solid state relay can be constructed quite simply by the connection of two H11C optically coupled SCRs in a back-to-back configuration as illustrated.

LINEAR ANALOG COUPLER

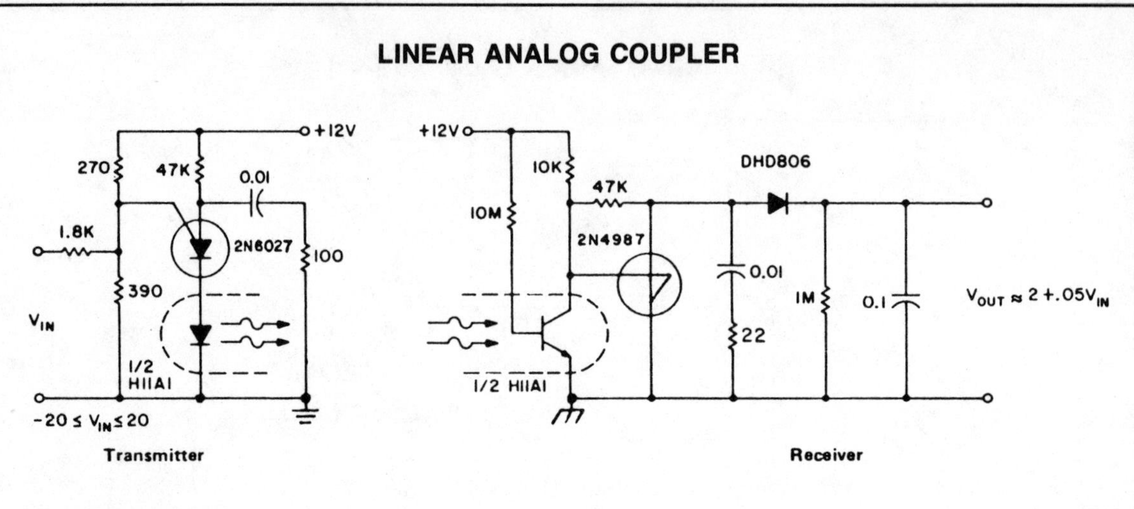

GENERAL ELECTRIC

Fig. 65-8

Circuit Notes

The minimum parts count version of this system provides isolated, linear signal transfer useful at shorter distances or with an optocoupler for linear information transfer. Although the output is low level and cannot be loaded significantly without harming accuracy, a single I.C. operational or instrumentation amplifier can supply both the linear gain and buffering for use with a variety of loads.

HIGH SENSITIVITY, NORMALLY OPEN, TWO TERMINAL, ZERO VOLTAGE SWITCHING, HALF-WAVE CONTACT CIRCUIT

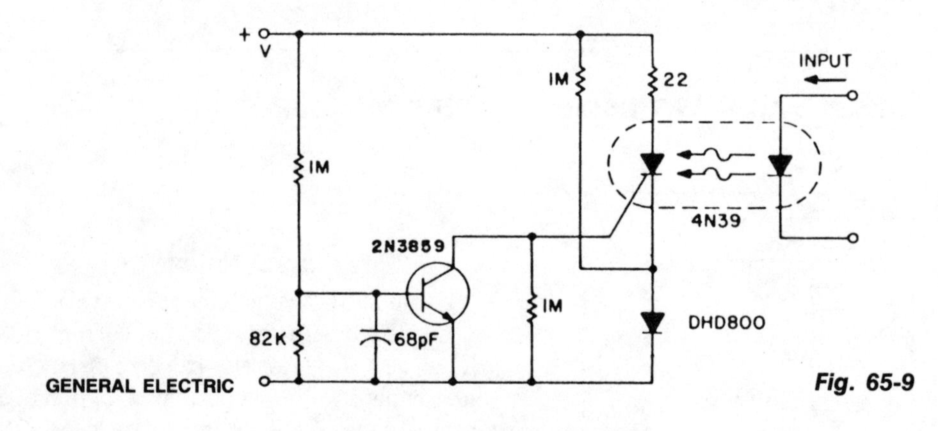

GENERAL ELECTRIC

Fig. 65-9

Circuit Notes

The SCR coupler circuit provides higher sensitivity to input signals as illustrated. This allows the lower cost 4N39 (H11C3) to be used with the > 7 mA drive currents supplied by the input circuit.

HIGH SPEED PAPER TAPE READER

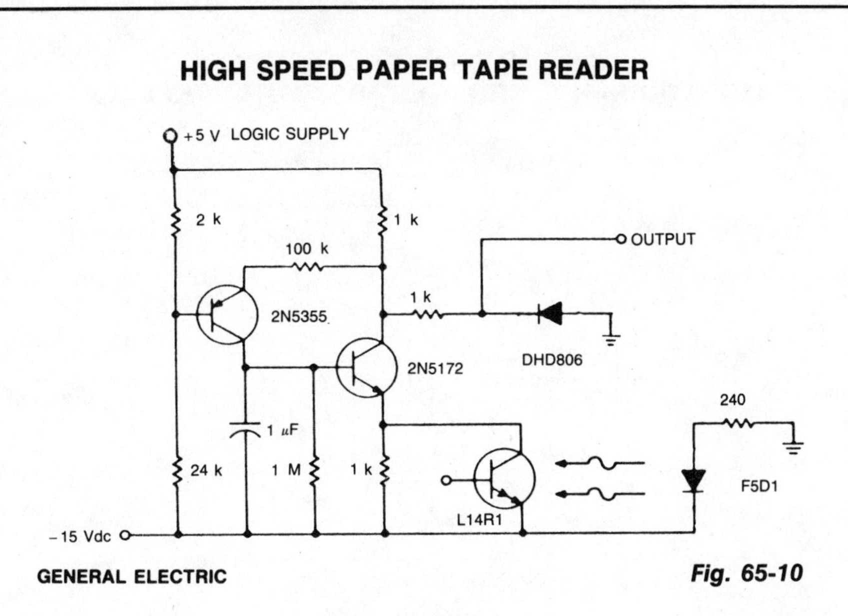

GENERAL ELECTRIC

Fig. 65-10

Circuit Notes

When computer peripheral equipment is interfaced, it is convenient to work with logic signal levels. With a nominal 4 V at the output dropping to −0.6 V on illumination, this circuit reflects the requirements of a high-speed, paper tape optical reader system. The circuit operates at rates of up to 1000 bits per second. It will also operate at tape translucency such that 50% of the incident light is transmitted to the sensor, and provide a fixed threshold signal to the logic circuit, all at low cost. Several circuit tricks are required. Photodarlington speed is enhanced by cascode constant voltage biasing. The output threshold and tape translucency requirements are provided for by sensing the output voltage and operating to 2000 bits per second at ambient light levels equal to signal levels.

DIGITAL TRANSMISSION ISOLATOR

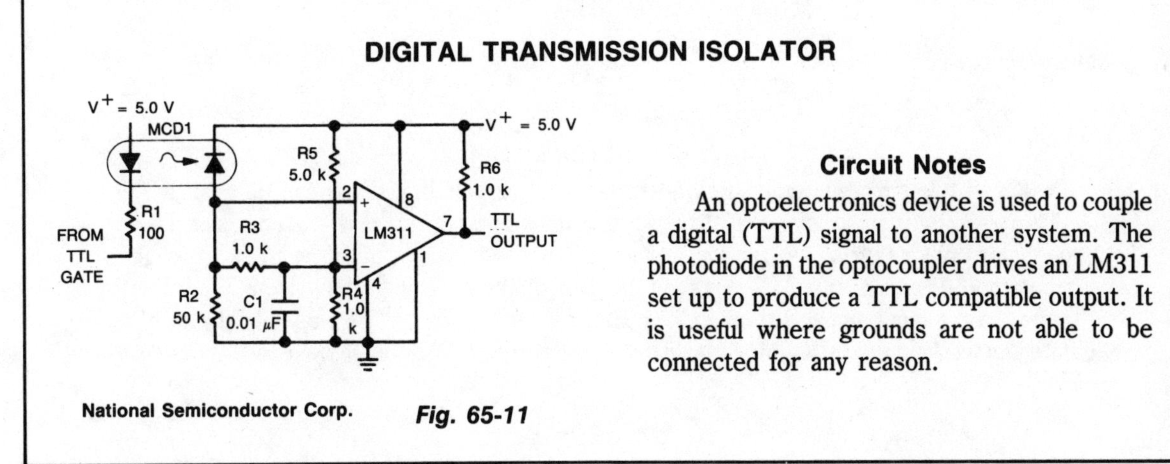

National Semiconductor Corp.

Fig. 65-11

Circuit Notes

An optoelectronics device is used to couple a digital (TTL) signal to another system. The photodiode in the optocoupler drives an LM311 set up to produce a TTL compatible output. It is useful where grounds are not able to be connected for any reason.

ISOLATION AND ZERO VOLTAGE SWITCHING LOGIC

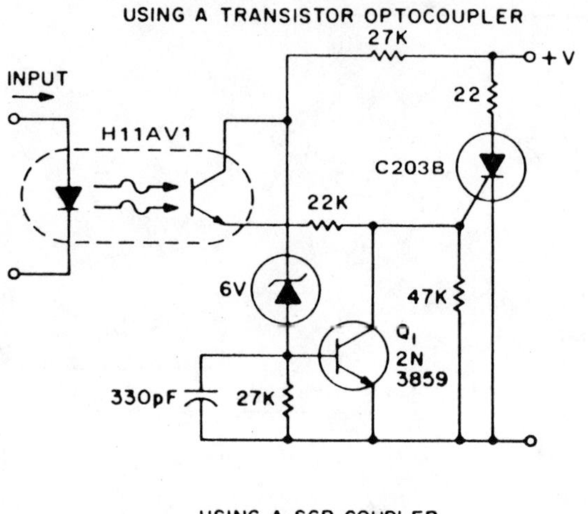

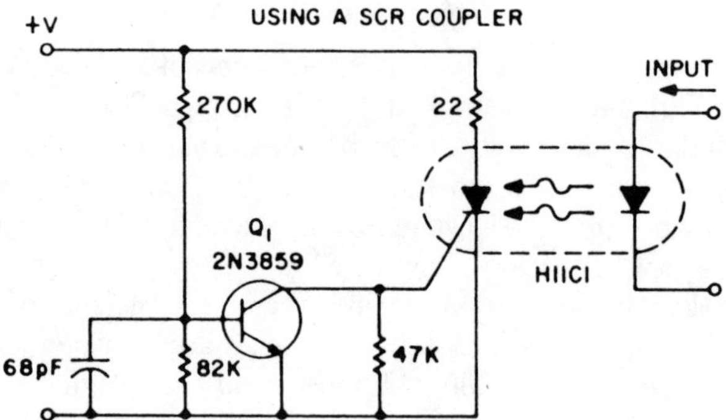

NORMALLY OPEN, TWO TERMINAL, ZERO VOLTAGE SWITCHING HALF WAVE CONTACT CIRCUITS

GENERAL ELECTRIC *Fig. 65-12*

Circuit Notes

These two simple circuits provide zero voltage switching. They can be used with full wave bridges or in antiparallel to provide full wave control and are normally used to trigger power thyristors. If an input signal is present during the time the ac voltage is between 0 to 7 V, the SCR will turn on. But, if the ac voltage has risen above this range and the input signal is then applied, the transistor, Q1, will be biased to the "on" state and will hold the SCR and, consequently, the relay "off" until the next zero crossing.

OPTICAL COMMUNICATION SYSTEM

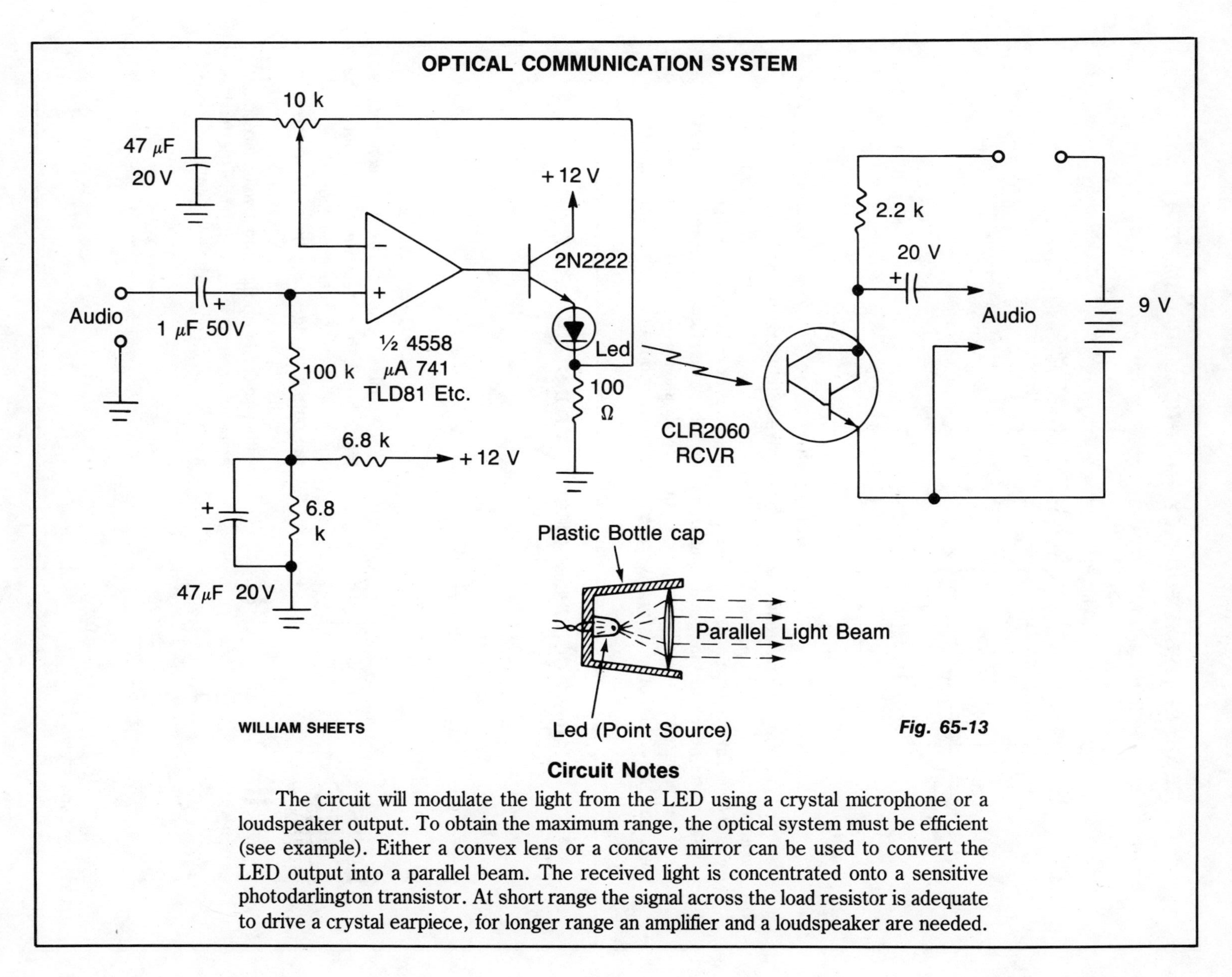

WILLIAM SHEETS

Fig. 65-13

Circuit Notes

The circuit will modulate the light from the LED using a crystal microphone or a loudspeaker output. To obtain the maximum range, the optical system must be efficient (see example). Either a convex lens or a concave mirror can be used to convert the LED output into a parallel beam. The received light is concentrated onto a sensitive photodarlington transistor. At short range the signal across the load resistor is adequate to drive a crystal earpiece, for longer range an amplifier and a loudspeaker are needed.

50 kHz CENTER FREQUENCY FM OPTICAL TRANSMITTER

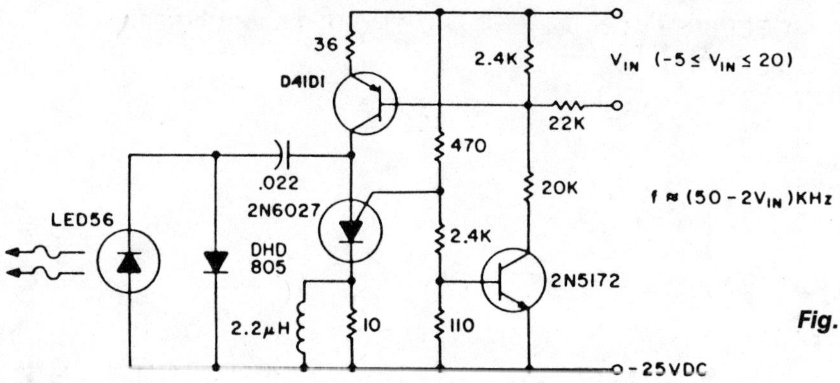

V_{IN} $(-5 \le V_{IN} \le 20)$

$f \approx (50 - 2V_{IN})$ KHz

Fig. 65-14

GENERAL ELECTRIC

Circuit Notes

The pulse repetition rate is relatively insensitive to temperature, and power supply voltage and is a linear function of V_{IN}, the modulating voltage. Useful information transfer was obtained in free air ranges of 12 feet ($\approx$ 4m). Lenses or reflectors at the light emitter and detector increases range and minimizes stray light noise effects. Greater range can also be obtained by using a higher power output IRED such as the F5D1 in combination with the L14P2 phototransistor. Average power consumption of the transmitter circuit is less than 3 watts.

LINEAR OPTOCOUPLER CIRCUIT FOR INSTRUMENTATION

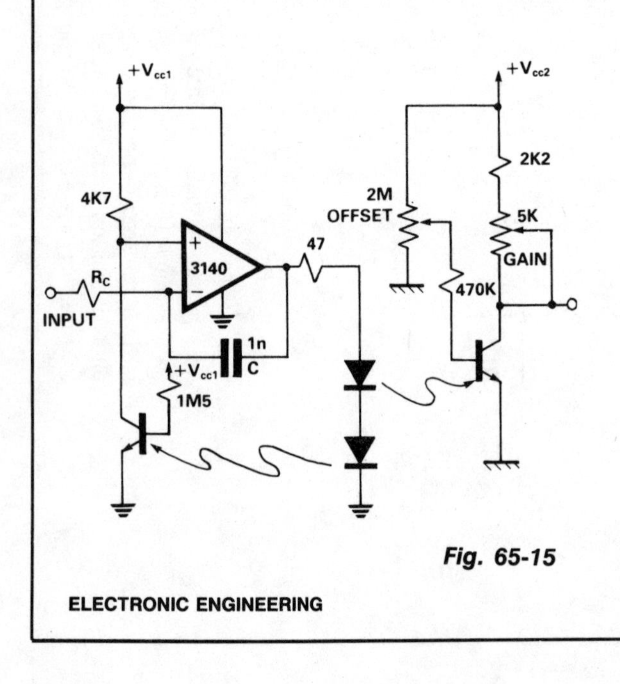

Fig. 65-15

ELECTRONIC ENGINEERING

Circuit Notes

A dual optocoupler is used in a configuration which has the same current throughout as the LEDs. Assuming similar optocoupler features the output voltage must be equal to the non-inverting input voltage. Since the op amp is within a closed loop the output voltage becomes equal to the input voltage. R_c and C perform as a compensation network to prevent oscillations.

RECEIVER FOR 50 kHz FM OPTICAL TRANSMITTER

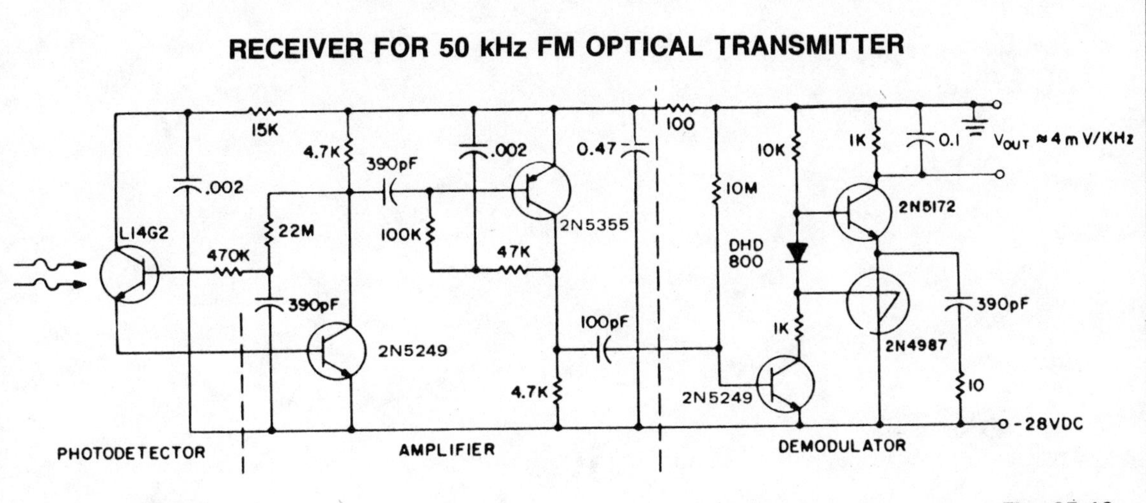

GENERAL ELECTRIC *Fig. 65-16*

Circuit Notes

For maximum range, the receiver must be designed in the same manner as a radio receiver front end, since the received signals will be similar in both frequency component and in amplitude of the photodiode current. The major constraint on the receiver performance is signal to noise ratio, followed by e.m. shielding, stability, bias points, parts layout, etc. These become significant details in the final design. This receiver circuit consists of a L14G2 detector, two stages of gain, and a FM demodulator which is the tachometer circuit, modified to operate up to 100 kHz. Better sensitivity can be obtained using more stages of stabilized gain with AGC, lower cost and sensitivity may be obtained by using an H23A1 emitter-detector pair and/or by eliminating amplifier stages. For some applications, additional filtering of the output voltage may be desired.

66

Oscillators

The sources of the following circuits are contained in the Sources section beginning on page 694. The figure number contained in the box of each circuit correlates to the source entry in the Sources section.

RF-GENIE

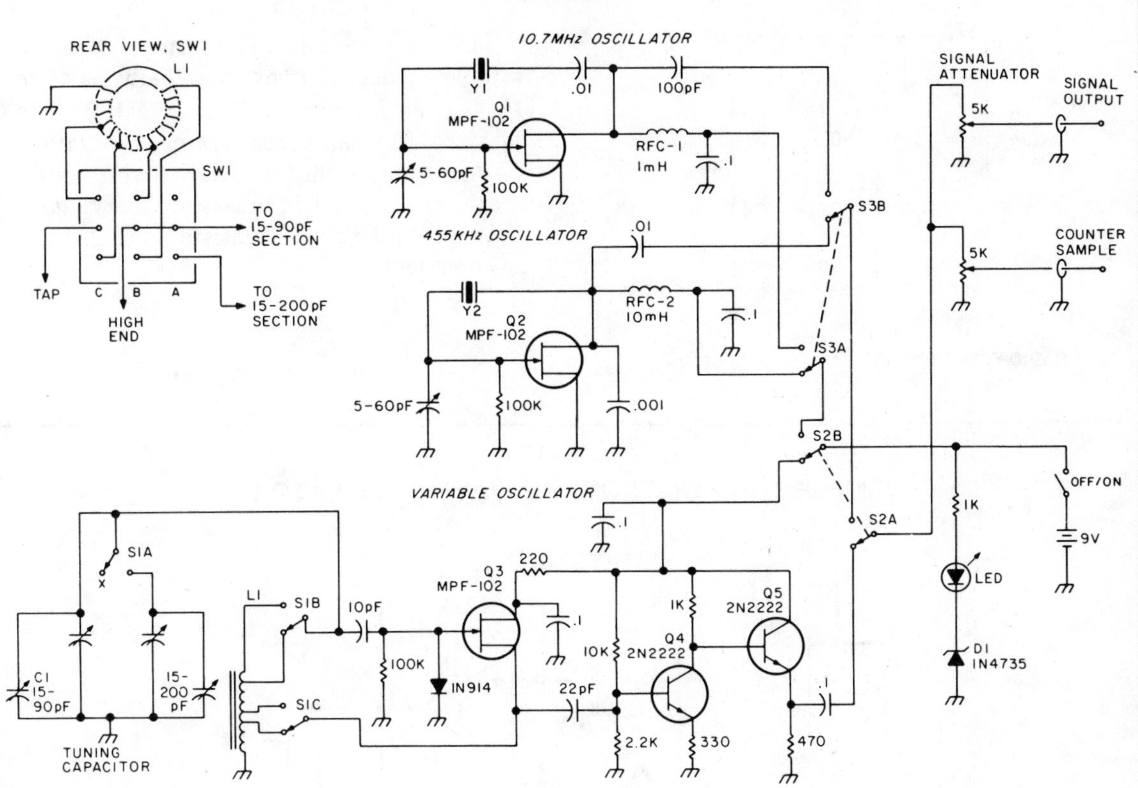

73 MAGAZINE *Fig. 66-1*

Circuit Notes

A variable oscillator covers 3.2 to 22 MHz in two bands—providing coverage of 80 through 15 meters plus most crystal-filter frequencies. Optional 455 kHz and 10.7 MHz crystal oscillators can be switched on-line for precise *if* alignment. Generator output is on the order of 4 volts p-p into a 500 ohm load. A simple voltage-divider attenuator controls the generator's output level, and a second output provides sufficient drive for an external frequency counter.

EMITTER-COUPLED BIG LOOP OSCILLATOR

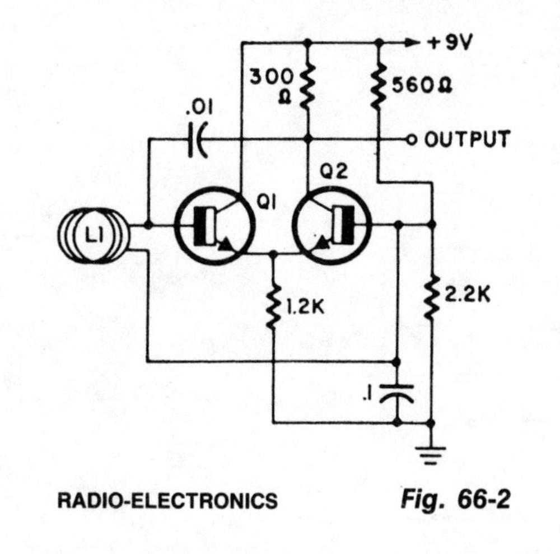

RADIO-ELECTRONICS *Fig. 66-2*

Circuit Notes

L1 is a loop of 10 to 20 turns of insulated wire with a diameter anywhere between 4″ to 4′. Oscillator frequency (7 to 30 MHz) shifts substantially when a person comes near or into the loop. This oscillator together with a resonant detector might make a very good anti-personnel alarm. Transistors are 2N2926 or equivalent.

SIMPLE TRIANGLE SQUARE WAVE OSCILLATOR

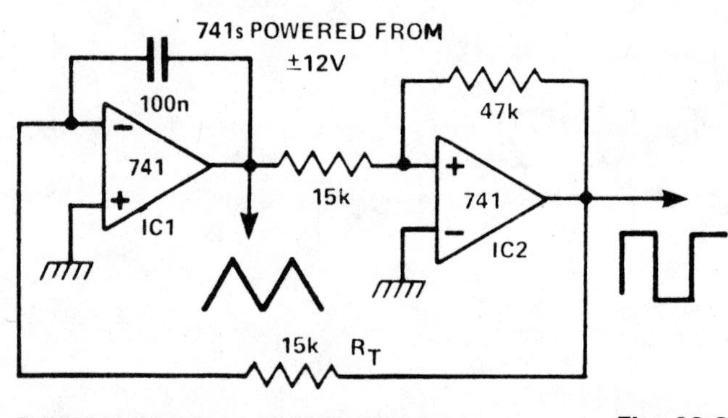

ELECTRONICS TODAY INTERNATIONAL *Fig. 66-3*

Circuit Notes

This circuit generates simultaneously, a triangle and a square waveform. It is self starting and has no latch up problems. IC1 is an integrator with a slew rate determined by CT and RT and IC2 is a Schmitt trigger. The output of IC1 ramps up and down between the hysteresis levels of the Schmitt, the output of which drives the integrator. By making RT variable, it is possible to alter the operating frequency over a 100 to 1 range. Three resistors, one capacitor, and a dual op amp is all that is needed to make a versatile triangle and square wave oscillator with a possible frequency range of 0.1 Hz to 100 kHz.

OSCILLATOR ADJUSTABLE OVER 10:1 RANGE

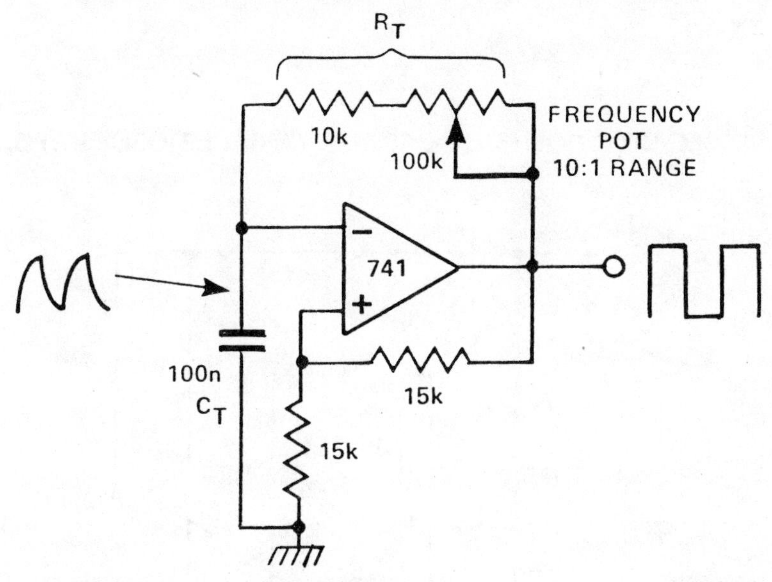

ELECTRONICS TODAY INTERNATIONAL

Fig. 66-4

Circuit Notes

In this circuit, there are two feedback paths around an op amp. One is positive dc feedback which forms a Schmitt trigger. The other is a CR timing network. Imagine that the output voltage is +10 V. The voltage at the noninverting terminal is +15 V. The voltage at the inverting terminal is a rising voltage with a time constant of $C_T R_T$. When this voltage exceeds +5 V, the op amp's output will go low and the Schmitt trigger action will make it snap into its negative state. Now the output is −10 V and the voltage at the inverting terminal falls with the time constant as before. By changing this time constant with a variable resistor, a variable frequency oscillation may be produced.

ONE SECOND, 1 kHz OSCILLATOR

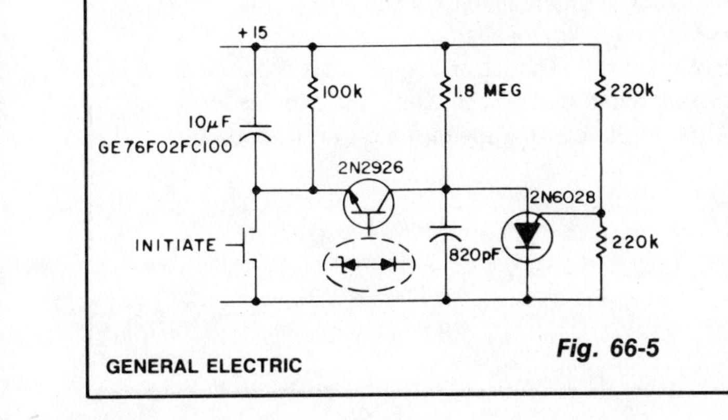

GENERAL ELECTRIC

Fig. 66-5

Circuit Notes

This circuit operates as an oscillator and a timer. The 2N6028 is normally on due to excess holding current through the 100 k resistor. When the switch is momentarily closed, the 10 μF capacitor is charged to a full 15 volts and 2N2926 starts oscillating (1.8 M and 820 pF). The circuit latches when 2N2926 zener breaks down again.

SINGLE CONTROL FOUR-DECADE VARIABLE OSCILLATOR

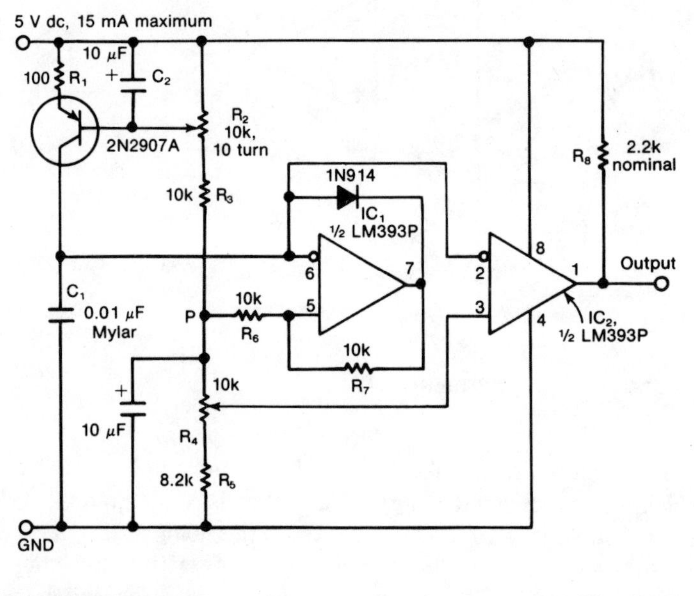

ELECTRONIC DESIGN

Fig. 66-6

Circuit Notes

The circuit consists of a variable current source that charges a capacitor, which is rapidly discharged by a Schmitt-trigger comparator. The sawtooth waveform thus produced is fed to another comparator, one with a variable switching level. The output from the second comparator is a pulse train with an independently adjustable frequency and duty cycle. The variable-frequency ramp generator consists of capacitor C1, which is charged by a variable and nonlinear current source. The latter comprises a 2N2907A pnp transistor, plus resistor R1 and the potentiometer R2. Capacitor C2 eliminates any ripple or noise at the base of the transistor that might cause frequency jitter at the output.

TUNABLE FREQUENCY OSCILLATORS

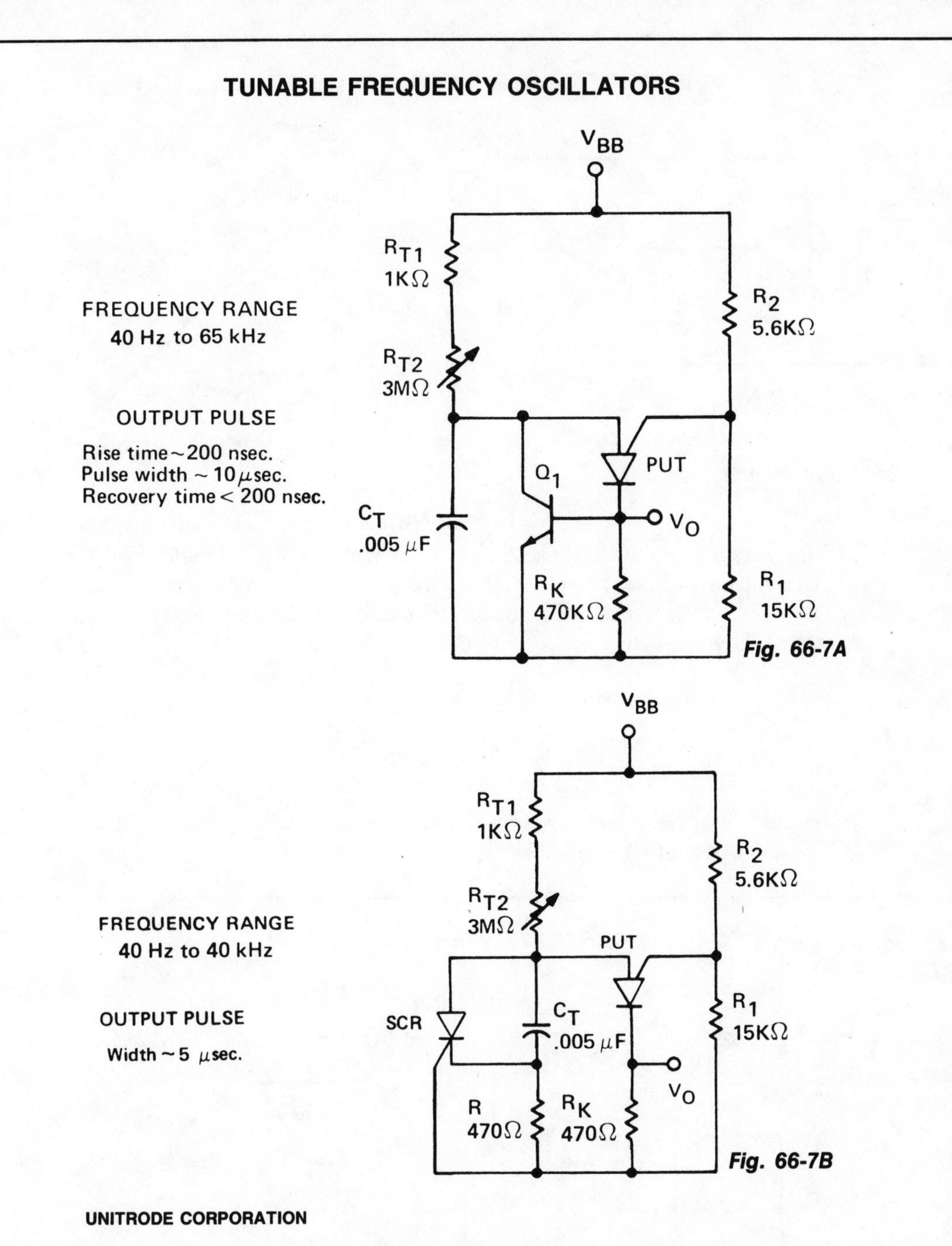

Circuit Notes

The variable oscillator circuit includes active elements for discharging the timing capacitor C_T shown in Fig. 66-7A. A second method is given as in Fig. 66-7B.

RESISTANCE CONTROLLED DIGITAL OSCILLATOR

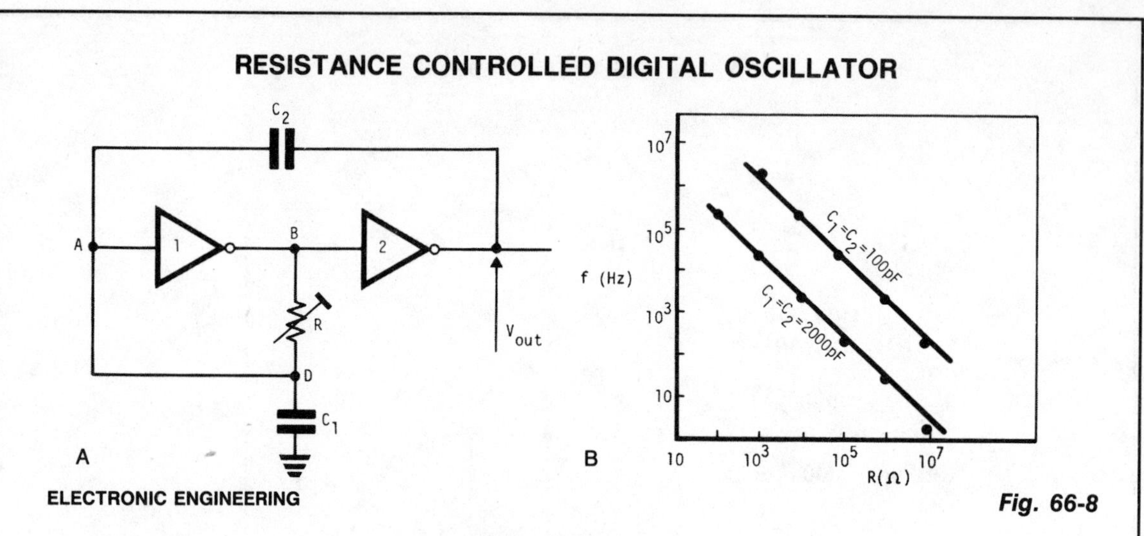

ELECTRONIC ENGINEERING

Fig. 66-8

Circuit Notes

This very simple, low cost oscillator, is built with two CMOS buffer inverters, two capacitors and a variable resistance. The circuit can work with voltages ranging from 4 V up to 18 V. If C1 = C2, the frequency of oscillation, (ignoring the output and input impedance) is given by:

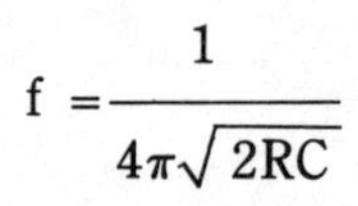

$$f = \frac{1}{4\pi\sqrt{2RC}}$$

The graph in Fig. B shows how the output frequency varies with resistance when C1 = C2 = 100 pF and C1 = C2 = 2000 pF.

CASSETTE BIAS OSCILLATOR

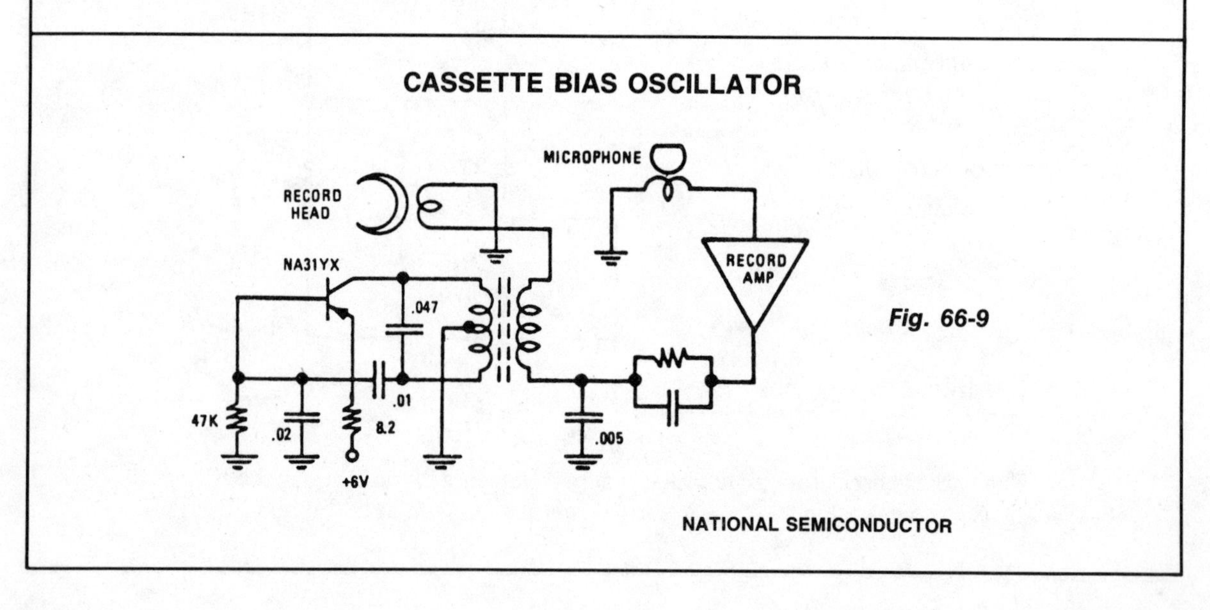

Fig. 66-9

NATIONAL SEMICONDUCTOR

1 kHz OSCILLATOR

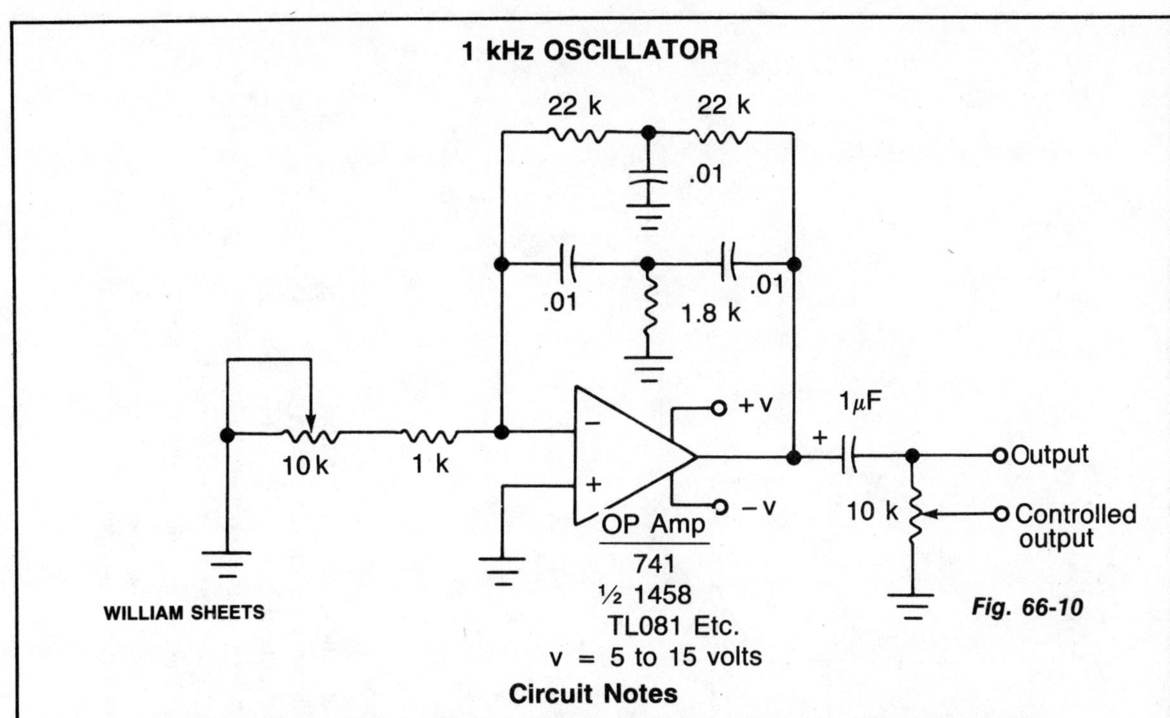

WILLIAM SHEETS

Fig. 66-10

Circuit Notes

If fine output control is desired, add the 10 K potentiometer. When the oscillator is connected to a dc circuit then connect a dc blocking capacitor in series with the potentiometer's wiper arm.

INEXPENSIVE OSCILLATOR IS TEMPERATURE STABLE

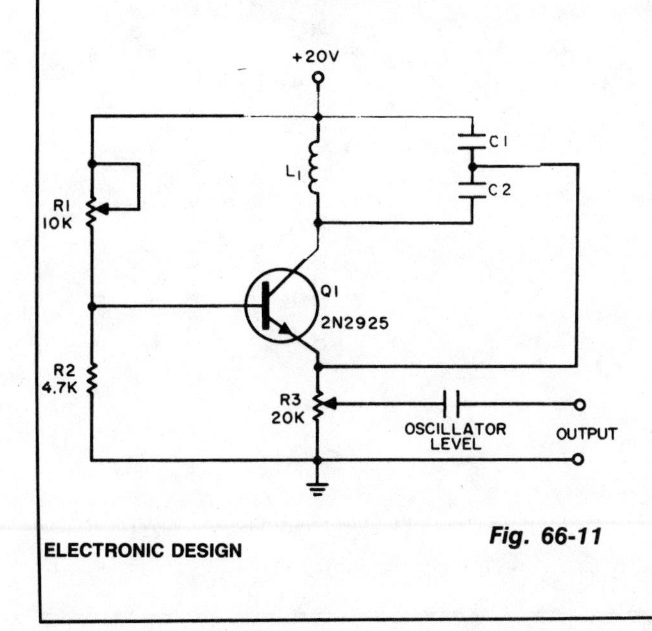

ELECTRONIC DESIGN

Fig. 66-11

Circuit Notes

The Colpitts sinusoidal oscillator provides stable output amplitude and frequency from 0°F to +150°F. In addition, output amplitude is large and harmonic distortion is low. Oscillation is sustained by feedback from the collector tank circuit to the emitter. The oscillator's frequency is determined by:

$$f = \frac{1}{2\pi\sqrt{\frac{L1C1C2}{C1 + C2}}}$$

Potentiometer R3 is an output level control. Control R1 may be used to adjust base bias for maximum-amplitude output. The circuit was operated at 50 kHz with L1 = 10mH, C1 = 3500 pF, and C2 = 1500 pF.

CODE PRACTICE OSCILLATOR

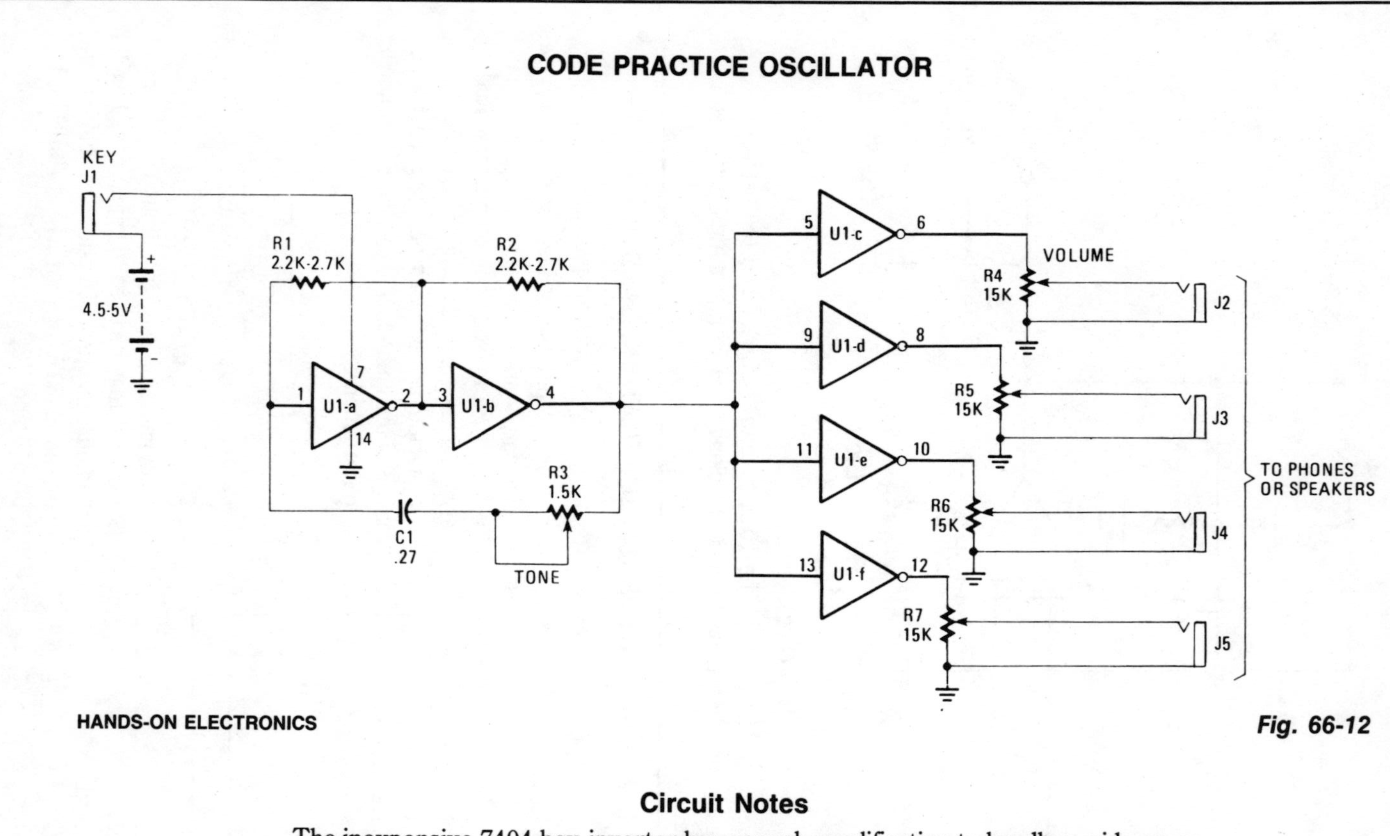

HANDS-ON ELECTRONICS

Fig. 66-12

Circuit Notes

The inexpensive 7404 hex-inverter has enough amplification to handle a wide range of transducers. Closing the key completes the battery circuit and applies four to five volts to the 7404. Bias for the first two inverter amps (U1a and U1b) comes from the two resistors, R1 and R2, connected between their inputs and outputs. The capacitor and rheostat (R3/C1) close the feedback loop from the input to the properly-phased output. The signal leaving U1b drives the remaining four inverter amplifiers, U1c through U1f; they, in turn, drive the phones or speakers. The volume control potentiometers, R4-R7, may have any value from 1500 ohms to 10,000 ohms. The smaller values work best when speakers, or low impedance phones, are used.

WIDE RANGE VARIABLE OSCILLATOR

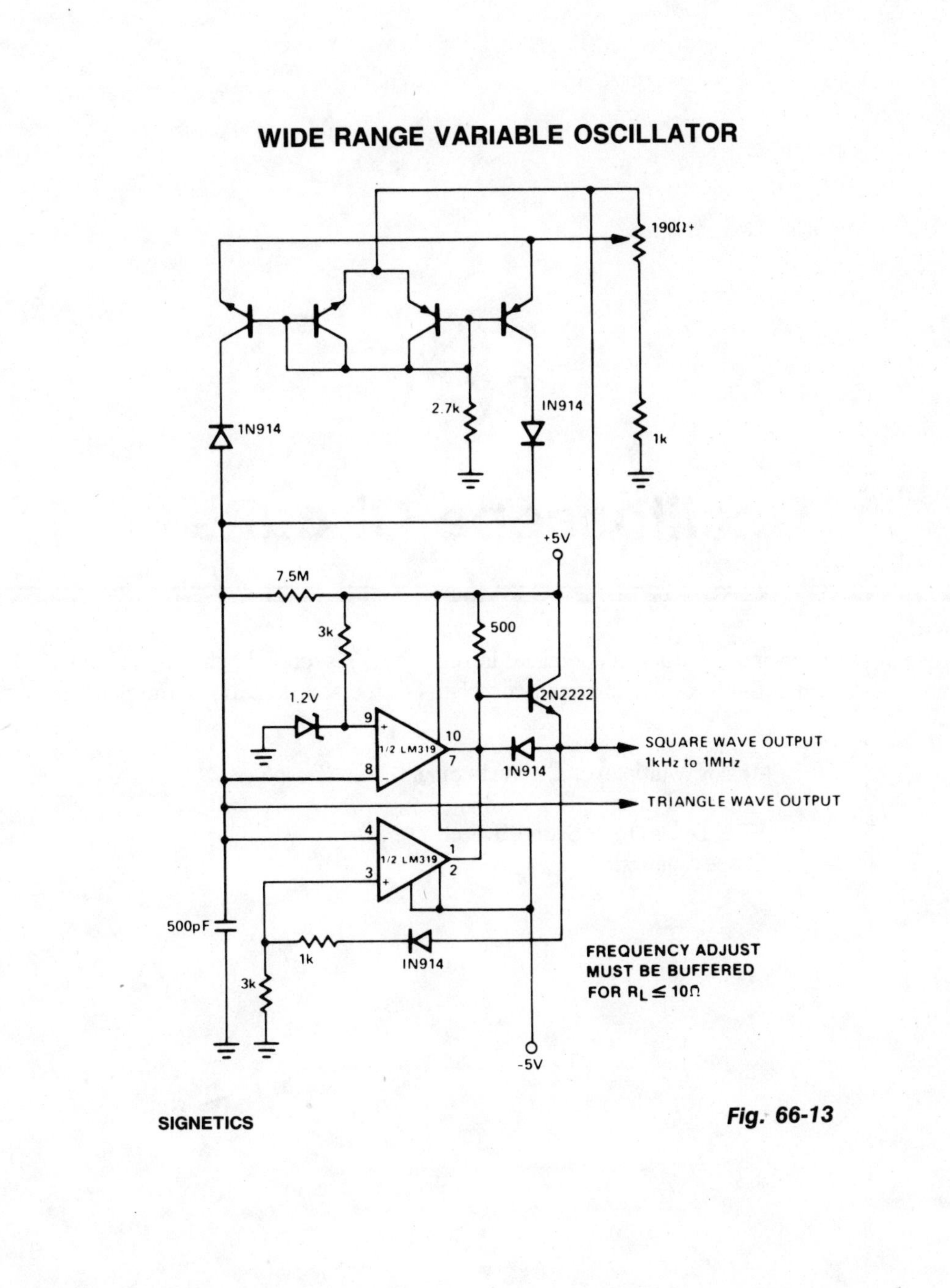

SIGNETICS

Fig. 66-13

67

Oscilloscope Circuits

The sources of the following circuits are contained in the Sources section beginning on page 694. The figure number contained in the box of each circuit correlates to the source entry in the Sources section.

ANALOG MULTIPLEXER CONVERTS SINGLE-TRACE SCOPE TO FOUR-TRACE

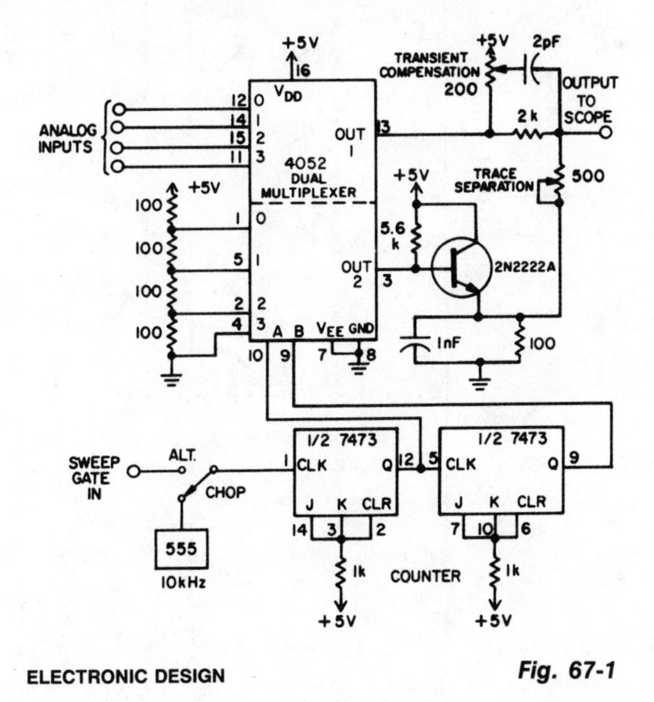

ELECTRONIC DESIGN *Fig. 67-1*

Circuit Notes

This adapter circuit, based on a dual four-channel analog multiplexer handles digital signals to at least 1 MHz, and analog signals at least through the audio range. The dual multiplexer's upper half selects one input for display. The lower half generates a staircase to offset the baselines of each channel, keeping them separate on the screen. The emitter-follower buffers the staircase, which is then summed with the selected signal. A two-bit binary counter addresses the CMOS 4052 multiplexer.

FET DUAL-TRACE SCOPE SWITCH

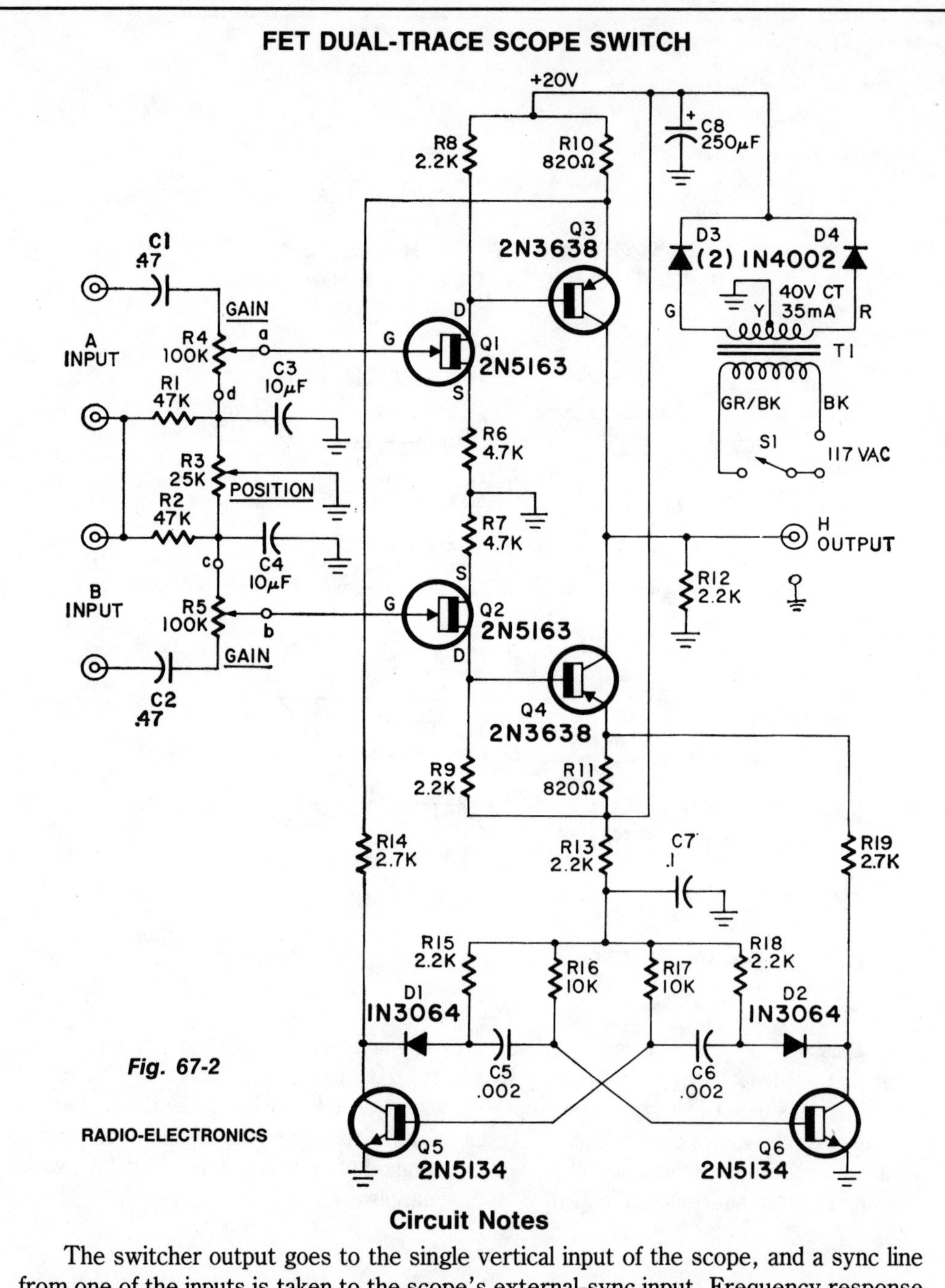

Fig. 67-2

RADIO-ELECTRONICS

Circuit Notes

The switcher output goes to the single vertical input of the scope, and a sync line from one of the inputs is taken to the scope's external-sync input. Frequency response of the input amplifiers is 300 kHz over the range of the gain controls. With the gain controls wide open so no attenuation of the signal takes place, the frequency response is up to 1 MHz.

SCOPE CALIBRATOR

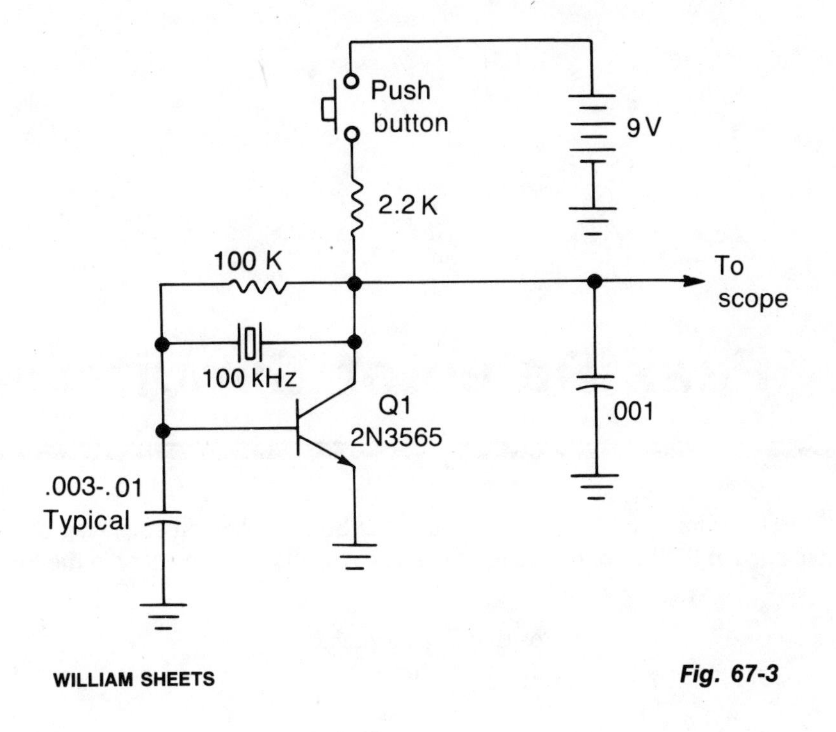

WILLIAM SHEETS

Fig. 67-3

Circuit Notes

The calibrator operates on exactly 100 kHz providing a reference for calibrating the variable time base oscillator of general purpose scopes. For example, if the scope is set so that one cycle of the signal fills exactly 10 graticule divisions then each division represents 1 MHz, or 1 microsecond. If the scope is adjusted for 10 cycles on 10 graticule divisions. (1 cycle per division) then each division represents 100 kHz or 10 microseconds.

68

Peak Detector Circuits

The sources of the following circuits are contained in the Sources section beginning on page 694. The figure number contained in the box of each circuit correlates to the source entry in the Sources section.

POSITIVE PEAK DETECTOR

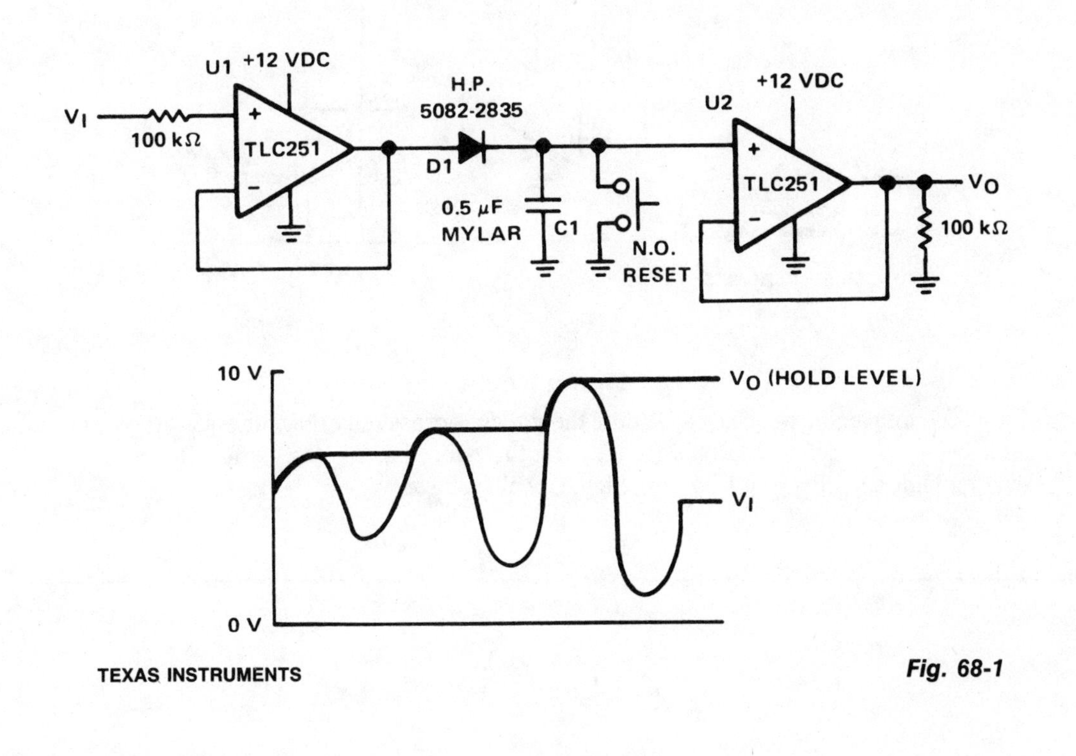

TEXAS INSTRUMENTS

Fig. 68-1

Circuit Notes

The purpose of the circuit is to hold the peak of the input voltage on capacitor C1, and read the value, V_O, at the output of U2. Op amps U1 and U2 are connected as voltage followers. When a signal is applied to V_I, C1 will charge to this same voltage through diode D1. This positive peak voltage on C1 will maintain V_O at this level until the capacitor is reset (shorted). Of course, higher positive peaks will raise this level while lower power peaks will be ignored. C1 can be reset manually with a switch, or electronically with an FET that is normally off. The capacitor specified for C1 should have low leakage and low dielectric absorption. Diode D1 should also have low leakage. Peak values of negative polarity signals may be detected by reversing D1.

PEAK DETECTOR

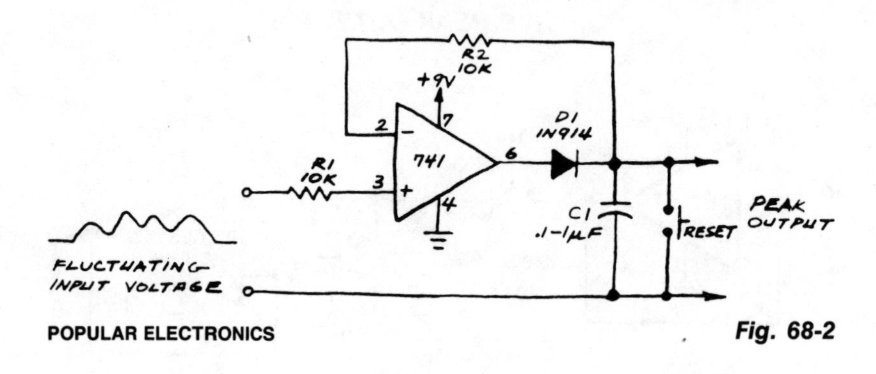

POPULAR ELECTRONICS

Fig. 68-2

Circuit Notes

The comparator will charge C1 until the voltage across the capacitor equals the input voltage. If subsequent input voltage exceeds that stored in C1, the comparator voltage will go high and charge C1 to new higher peak voltage.

69

Phase Sequence Circuits

The sources of the following circuits are contained in the Sources section beginning on page 694. The figure number contained in the box of each circuit correlates to the source entry in the Sources section.

RC CIRCUIT DETECTS PHASE SEQUENCE REVERSAL

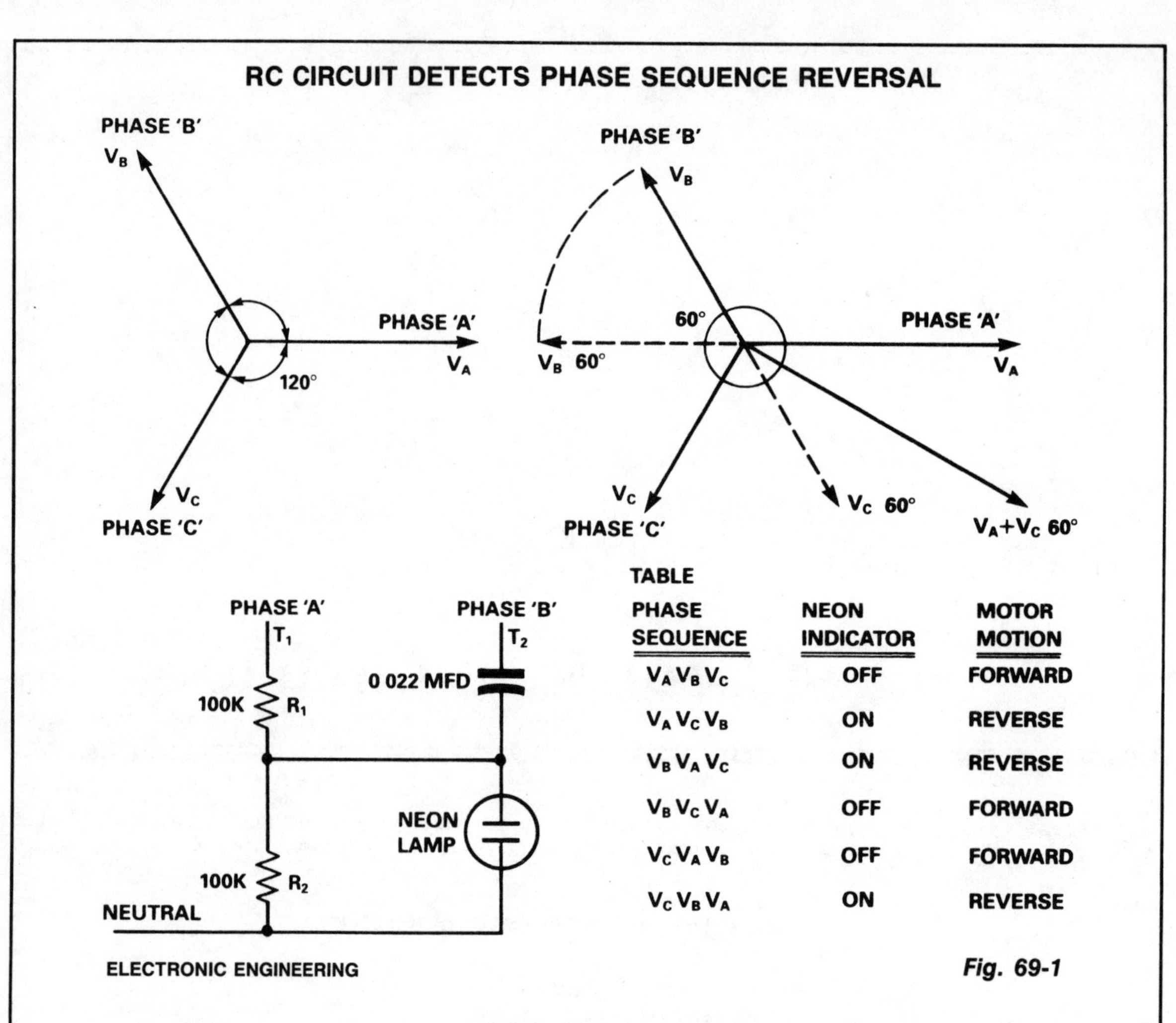

TABLE

PHASE SEQUENCE	NEON INDICATOR	MOTOR MOTION
$V_A V_B V_C$	OFF	FORWARD
$V_A V_C V_B$	ON	REVERSE
$V_B V_A V_C$	ON	REVERSE
$V_B V_C V_A$	OFF	FORWARD
$V_C V_A V_B$	OFF	FORWARD
$V_C V_B V_A$	ON	REVERSE

ELECTRONIC ENGINEERING

Fig. 69-1

Circuit Notes

Assume the correct phase sequence to be V_A-V_B-V_C. The circuit terminals are connected such that T1 gets connected to phase A and T2 to phase B. The capacitor advances the voltage developed across R2 due to phase "B" by ~ 60°, while the voltages developed across it by phase "A" is in phase with V_A as shown in Fig. 69-1. The net voltage developed across R2 ~ zero, the neon lamp is not energized, thereby signaling correct phase sequence. If terminal T2 gets connected to phase C, a large voltage, $K(V_A + V_C\ 60°)$, gets developed across R2, energizing the neon indicator to signal reverse phase sequence.

The motor terminals can be connected to the three phases in six different combinations. A three-phase motor will run in the forward direction for three such combinations, while for the other three it will operate in the reverse direction. As shown in the table, the circuit detects all three reverse combinations. This circuit can be wired into any existing motor starter where the operator can see whether the phase sequence has been altered, before starting the machine.

PHASE INDICATOR

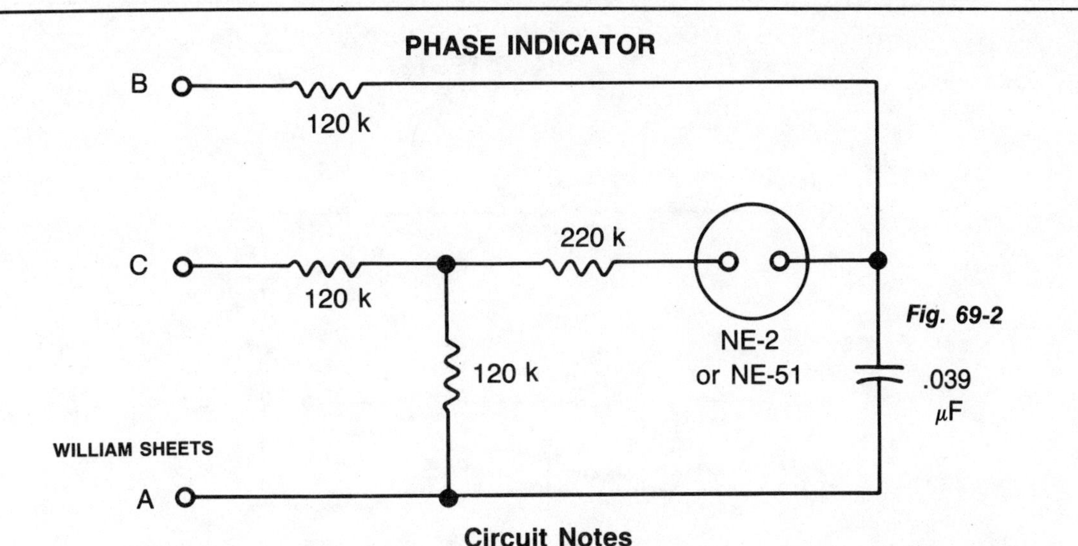

Fig. 69-2

Circuit Notes

The circuit provides a simple means of determining the succession of phases of a 3-phase 120 V source used in synchro work. Terminals A, B, and C are connected to the three terminals of the source to be checked. If the neon bulb lights, interchange any two leads; the light then extinguishes and A, B, and C indicate the correct sequence. If power on any one line is lost, the neon bulb will light. This feature may be useful for monitoring purposes.

PHASE SEQUENCE DETECTOR

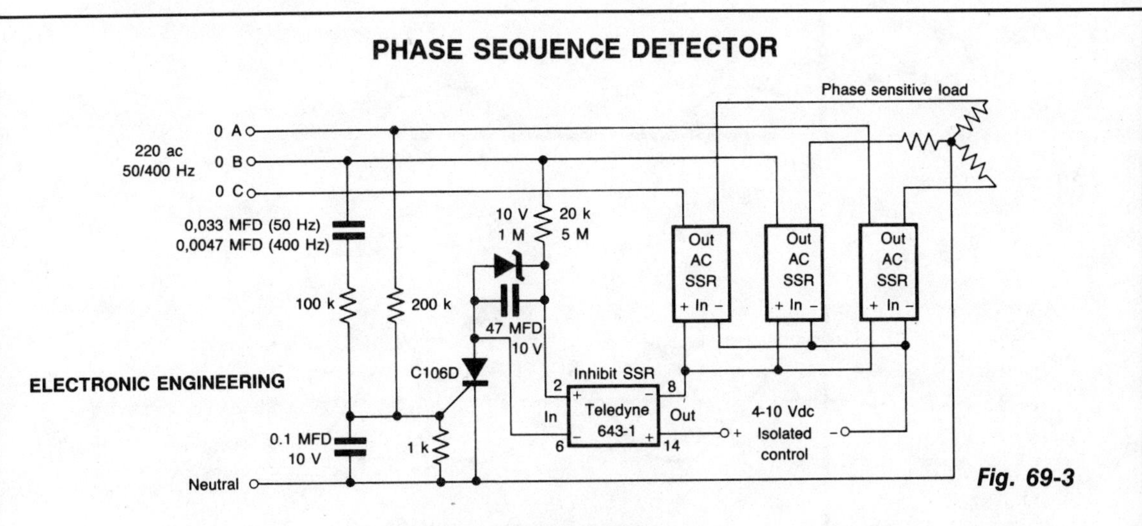

Fig. 69-3

Circuit Notes

This circuit prevents damage to the load due to incorrect phasing. The three power SSR's are only permitted to turn-on for a phase sequence of phase A leading phase B. If phase A lags phase B the input currents will cancel, causing the SCR and the inhibit SSR to remain off until the sequence is reversed. The inhibit SSR is included to maintain isolation at the input.

THREE PHASE TESTER

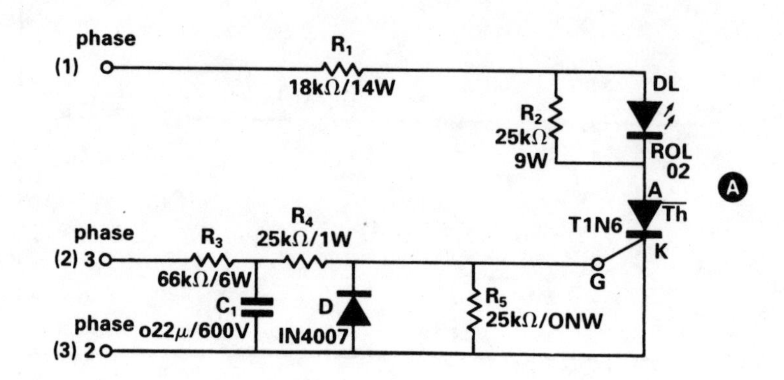

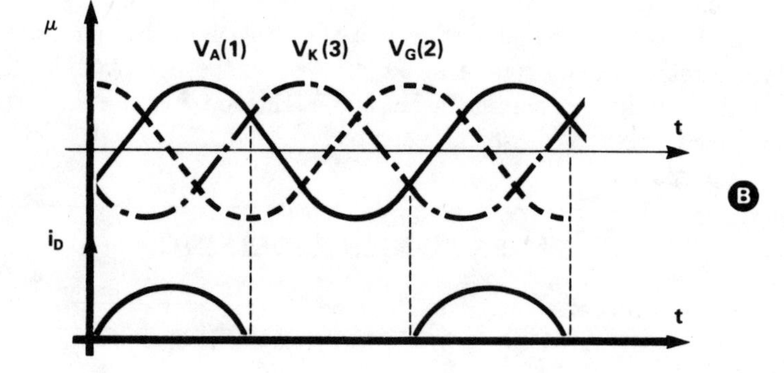

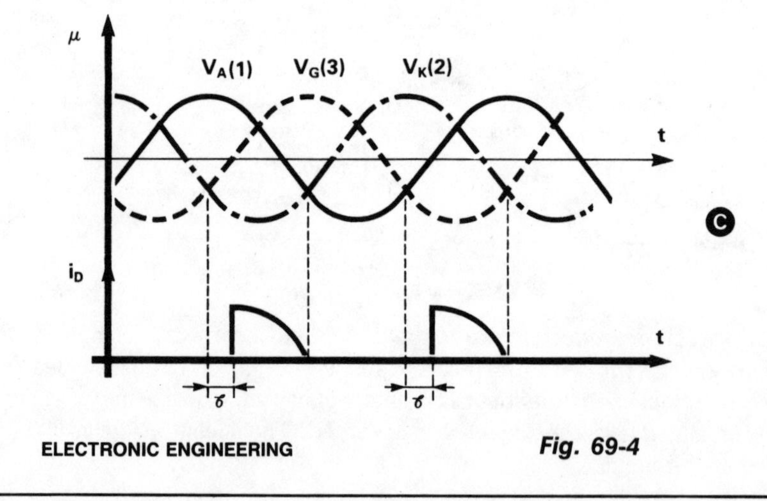

ELECTRONIC ENGINEERING *Fig. 69-4*

THREE PHASE TESTER, Continued.

Circuit Notes

This simple three-phase tester, uses only a small current thyristor as a main element for testing the right or wrong succession of the three phases, and there is no need for a supplementary power supply.

The basic circuit is shown in Fig. 69-4A. When connecting to the thyristor anode, grid, and cathode the three phases of the supply network in the sequence phase 1, phase 3, phase 2, are considered as correct, the mean value of the current through the thyristor is relatively high (since it is turned on during an entire half-period of one phase). The result is that the LED will emit a normal light.

The wave shapes for the three voltages and the current through the LED for this situation are shown in Fig. 69-4B.

If the three phases are not correctly connected—phase 1 to the anode, phase 2 to the grid, and phase 3 to the cathode, for instance—the thyristor will be turned on for a very short time and the LED will produce a very poor light. The wave shapes for this case are shown in Fig. 69-4C. The delay time is given by the R3-R1-R4 group.

When any of the three phases is missing, there is no current through the thyristor and the LED will emit no light.

PHASE-SEQUENCE DETECTOR II

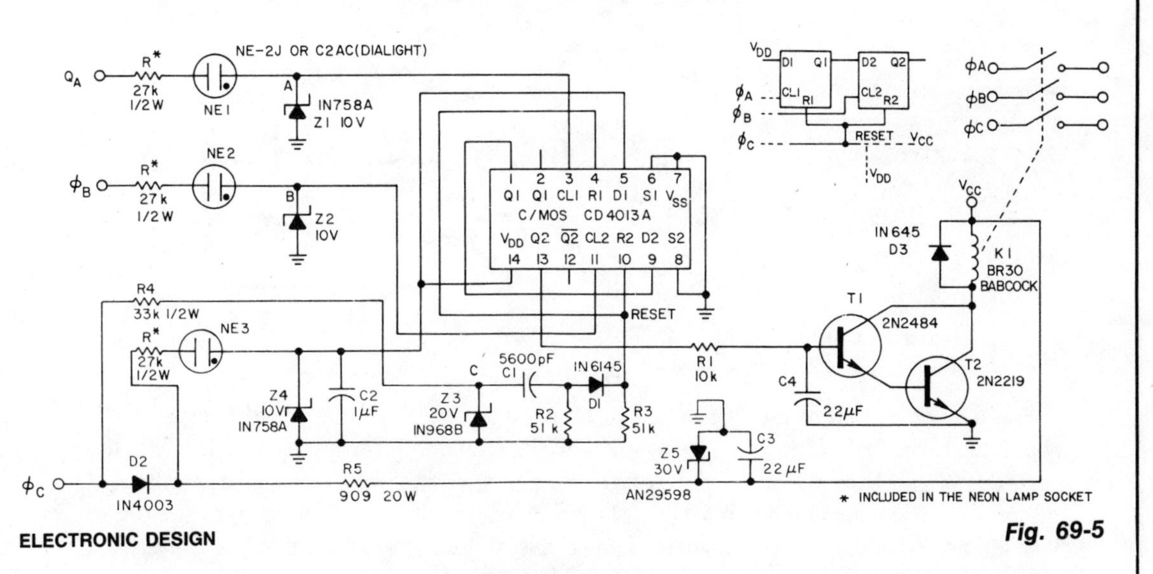

ELECTRONIC DESIGN

Fig. 69-5

Circuit Notes

This circuit derives its supply voltage, V_{cc} and V_{dd} from ϕ_C. This factor, together with the neon lamps and zener diodes in the phase inputs, establishes 50% threshold that detects low voltage or absence of one or more phases. Relay K1 energizes for correct phase volts.

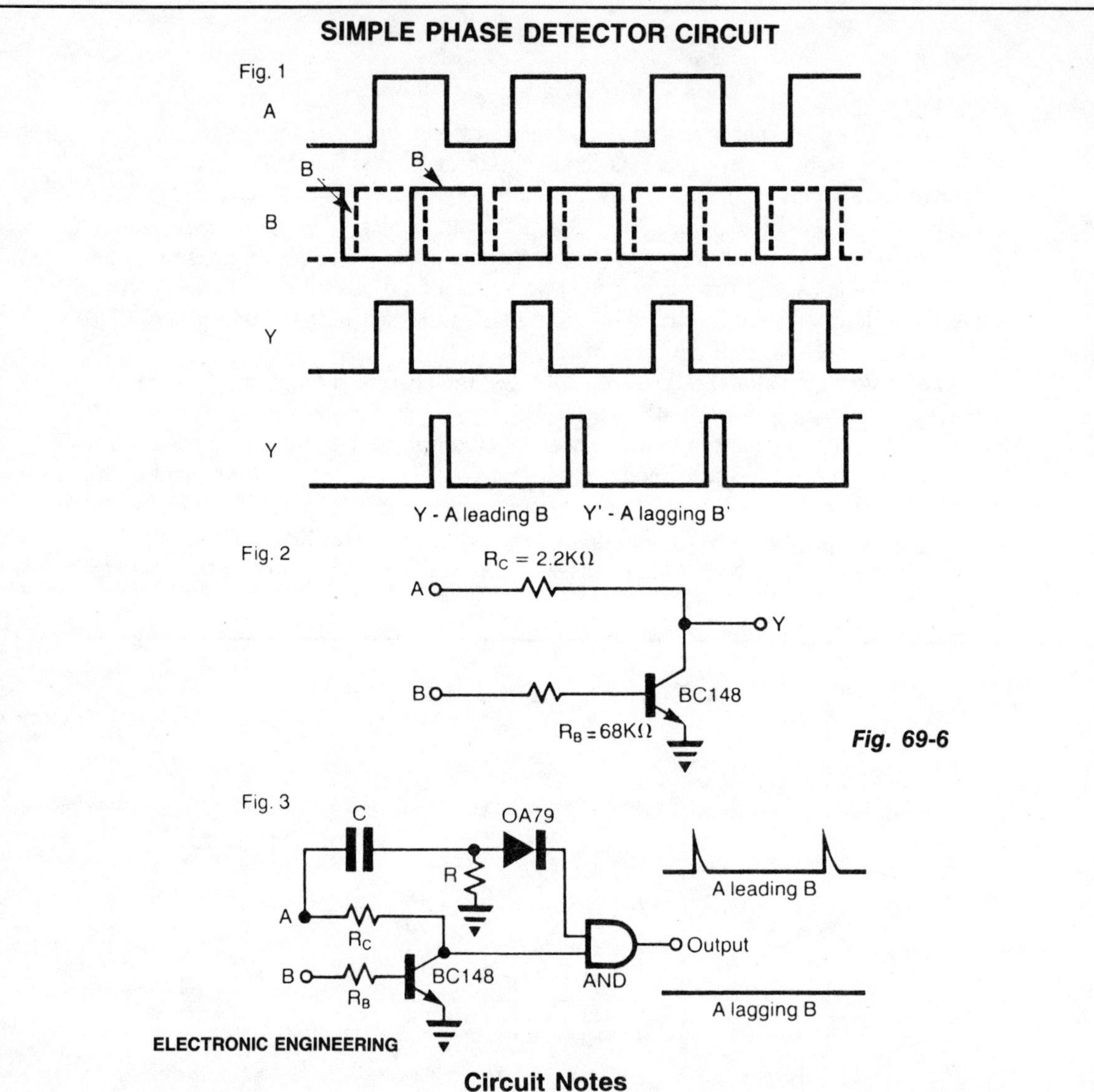

Fig. 69-6

Circuit Notes

The operation of the circuit is like an enabled inverter, that is, the output Y = B provided A is high. If A is low, output is low (independent of the state of B). When the signals A and B or B1 are connected to the inputs A and B of this gate the output Y is a pulse train signal (shown a Y or Y1) which has a pulse duration equal to the phase difference between the two signals. The circuit is directly suitable for phase difference measurement from zero to 180°. This performance is similar to the circuits like the Exclusive OR gate used for this purpose. With this method leading and lagging positions of the signals can also be found using an AND gate. Phase difference measured along with the leading and lagging information gives complete information about the phases of the two signals between zero and 360°.

70

Photography-Related Circuits

The sources of the following circuits are contained in the Sources section beginning on page 694. The figure number contained in the box of each circuit correlates to the source entry in the Sources section.

Auto-Advance Projector
Shutter-Speed Tester
Enlarger Timer
Contrast Meter
Electronic Flash Trigger
Sound Trigger for Flash Unit

AUTO-ADVANCE PROJECTOR

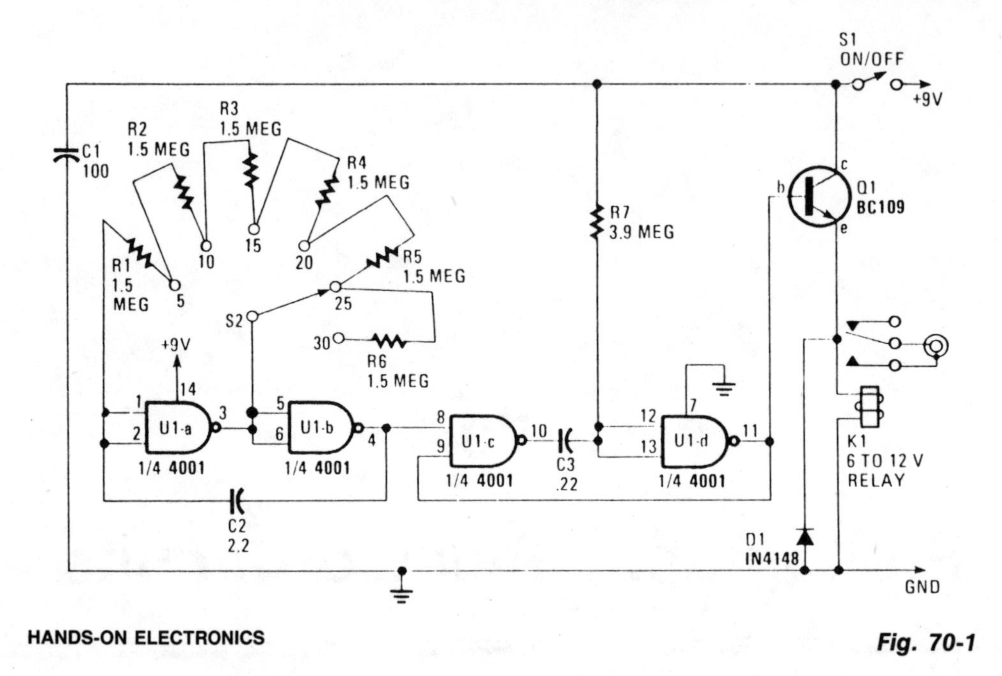

HANDS-ON ELECTRONICS

Fig. 70-1

Circuit Notes

The circuit is built around a 4001 quad two-input NOR gate, it provides switch selectable auto-advance times of 5, 10, 15, 20, 25 or 30 seconds through the remote-control socket of your projector. U1a and U1b form an astable multivibrator, with its operating frequency dependent on the number of timing resistors switched into the circuit via S2. The frequency is about one cycle for every five seconds with a single timing resistor, one every ten seconds with two resistors, etc., providing six switched time intervals. The output of the astable at pin 4 of U1b is fed to the input of a monostable multivibrator, consisting of the second pair of gates, U1c and U1d. R7 and C3 are the timing components; they set the length of the (positive) output pulse of the monostable at a little more than half a second. The monostable is triggered by each positive-going input it receives from the astable. The output from the monostable therefore, consists of a series of short pulses, the interval between the pulses being controlled using S2. The output of the monostable (at pin 11) controls a relay by way of Q1, which is configured as an emitter-follower buffer stage. The projector is controlled via the normally-open contacts of relay K1. When the output of the monostable goes positive, the relay contacts close, triggering the slide-change mechanism of the projector. The monostable assures that the power to the relay is applied only briefly by the timer, so that multiple operation of the projector is avoided.

SHUTTER-SPEED TESTER

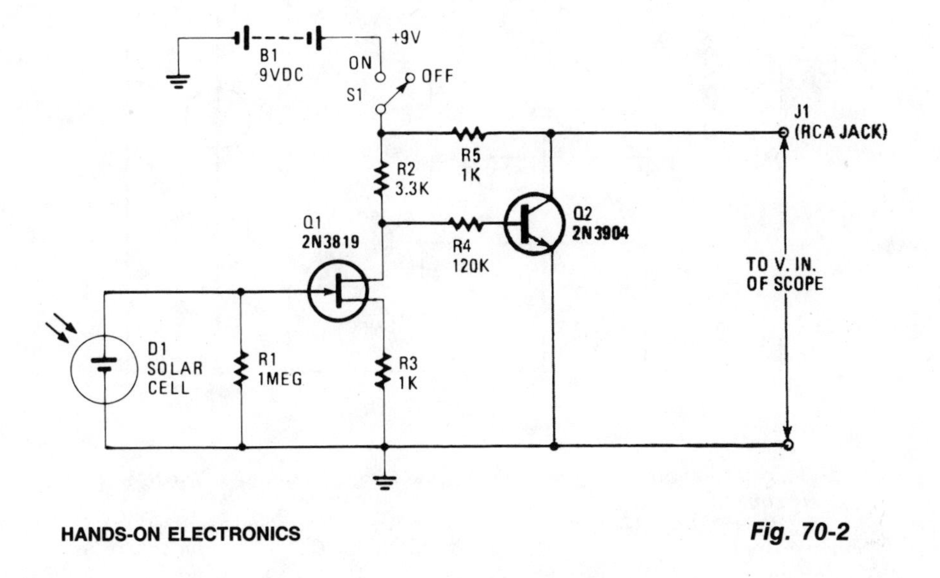

HANDS-ON ELECTRONICS *Fig. 70-2*

Circuit Notes

The solar cell is connected across the input of the FET (field-effect transistor), Q1, so that it will produce positive dc voltage to the gate when activated by light shining through the open shutter, decreasing the negative gate-source bias already established by the source resistor, and causes an increase in drain current. The drain voltage goes more negative which causes a decrease in Q2's base current. Q2's collector current decreases, and its collector voltage becomes more positive. There is an amplified positive-going voltage output at the collector, and it's applied directly to the oscilloscope's vertical input, producing a waveform that is displaced vertically whenever light strikes the cell.

ENLARGER TIMER

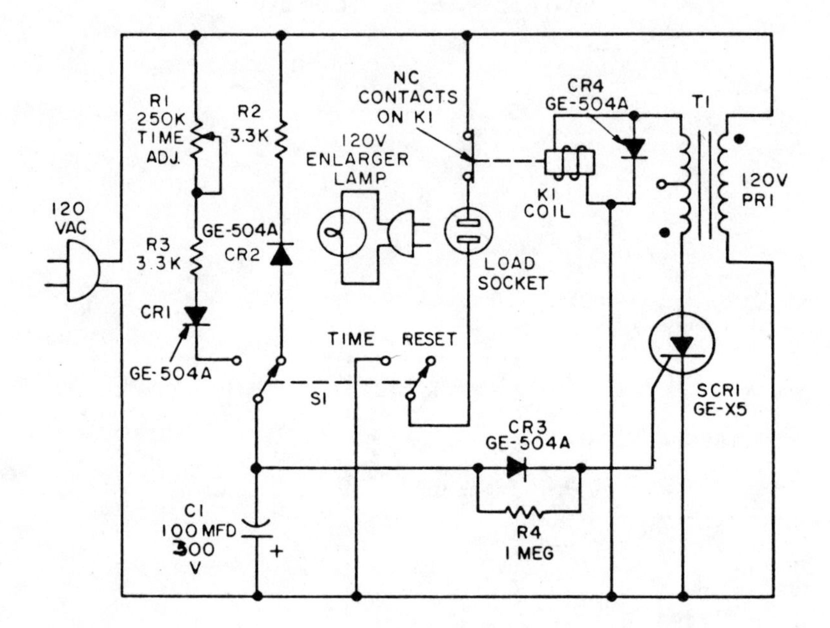

Parts List

C1 — 100-mfd, 300-volt electrolytic capacitor
CR1 thru CR4 — GE-504A rectifier diode
K1 —12 -volt a-c relay (Potter & Brumfield No. MR5A, or equivalent)
R1 — 250K-ohm, 2-watt potentiometer
R2, R3 — 3.3K-ohm, 1/2-watt resistor
R4 — 1-megohm, 1/2-watt resistor
S1 — DPDT toggle switch
SCR1 — GE-X5 silicon controlled rectifier
T1 — Filament transformer: primary, 120-volts a-c; secondary, 12.6-volts center tapped (Triad F25X, or equivalent)
Line cord, vectorboard, minibox etc.

GENERAL ELECTRIC

Fig. 70-3

Circuit Notes

This precision, solid state, time delay circuit has delayed *off* and delayed *on* switching functions that are interchangeably available by simply interchanging the relay contacts.

CONTRAST METER

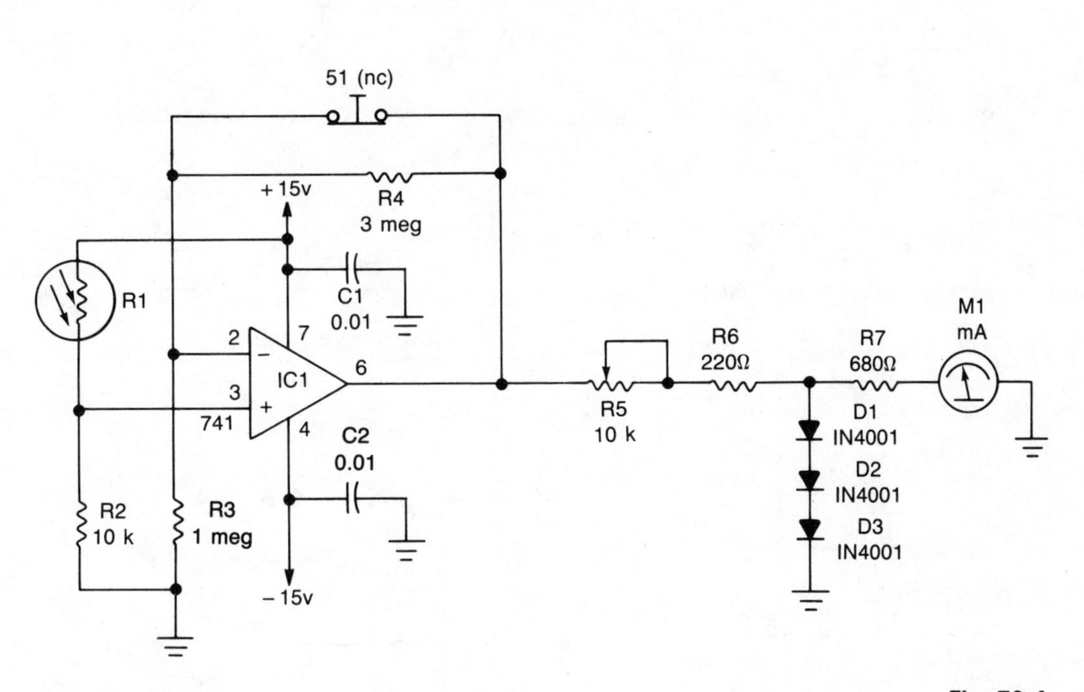

RADIO ELECTRONICS

Fig. 70-4

Circuit Notes

One leg of the photocell (R1) is tied to the +15 volt supply and the other end is connected to ground through resistor R2, forming a voltage-divider network. The non-inverting input of the 741 op amp, IC1, is tied to the junction formed by R1 and R2, while its inverting input is grounded through resistor R3. When switch S1 is pressed, another divider network is formed, reducing the voltage applied to the inverting input of the op amp. When light hits the photocell its resistance begins to decrease causing a greater voltage drop across R2 and a higher voltage to be presented to the non-inverting input of IC1. This causes IC1 to output a voltage proportional to the two inputs. The circuit gives a meter reading that depends on the intensity of light hitting photocell R1; therefore, R1 should be mounted in a bottle cap so that the light must pass through a 3/16 inch hole. Potentiometer R5 is used to adjust the circuit for the negative you're working with.

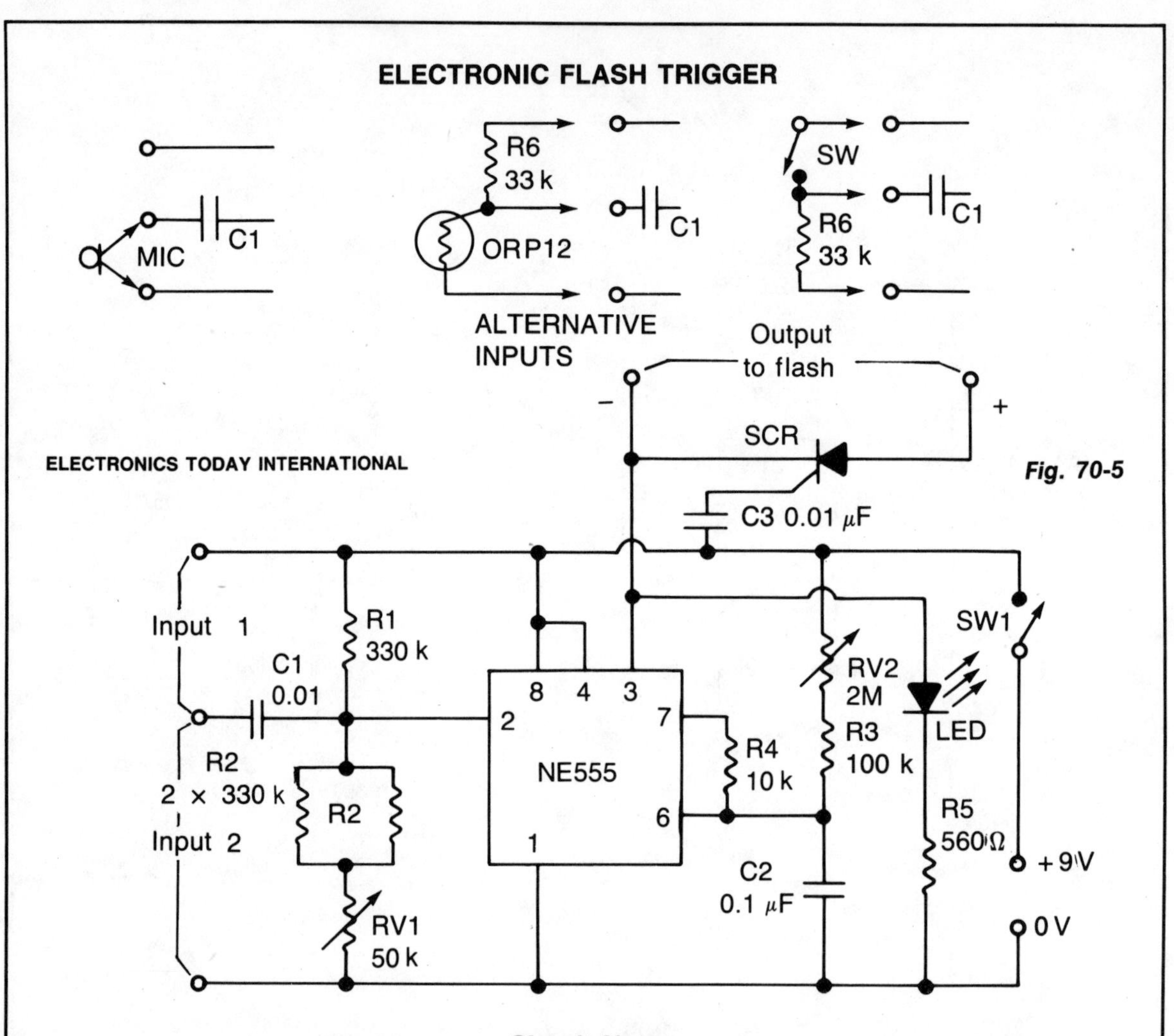

Circuit Notes

A negative pulse at the input is fed via capacitor C1 to the input pin (2) of the IC. Pin 2 is held slightly above its triggering voltage of 1/3 V_{cc} by the voltage divider comprising R1, R2 and RV1. The negative pulse triggers the IC and the output (pin 3) goes high for a time period controlled by RV2, R3 and C2. When the output goes low again at the end of the time interval, capacitor C3 charges through the gate cathode circuit of the SCR switching it on and firing the flash. Capacitor C1 isolates the input from the voltage divider so that the unit isn't sensitive to the dc level at the input. RV1 acts as a sensitivity control by allowing the voltage to be adjusted to a suitable level so that the input signal will trigger the IC. Resistor R4 limits the discharge current from C2 at the end of the timing cycle protecting the IC. The LED and its protective resistor R5 act as an indicator to show that the unit has triggered, simplifying the setting up process and minimizing the number of times the strobe has to be fired. This means that the strobe needn't be fired until a photo is to be taken.

SOUND TRIGGER FOR FLASH UNIT

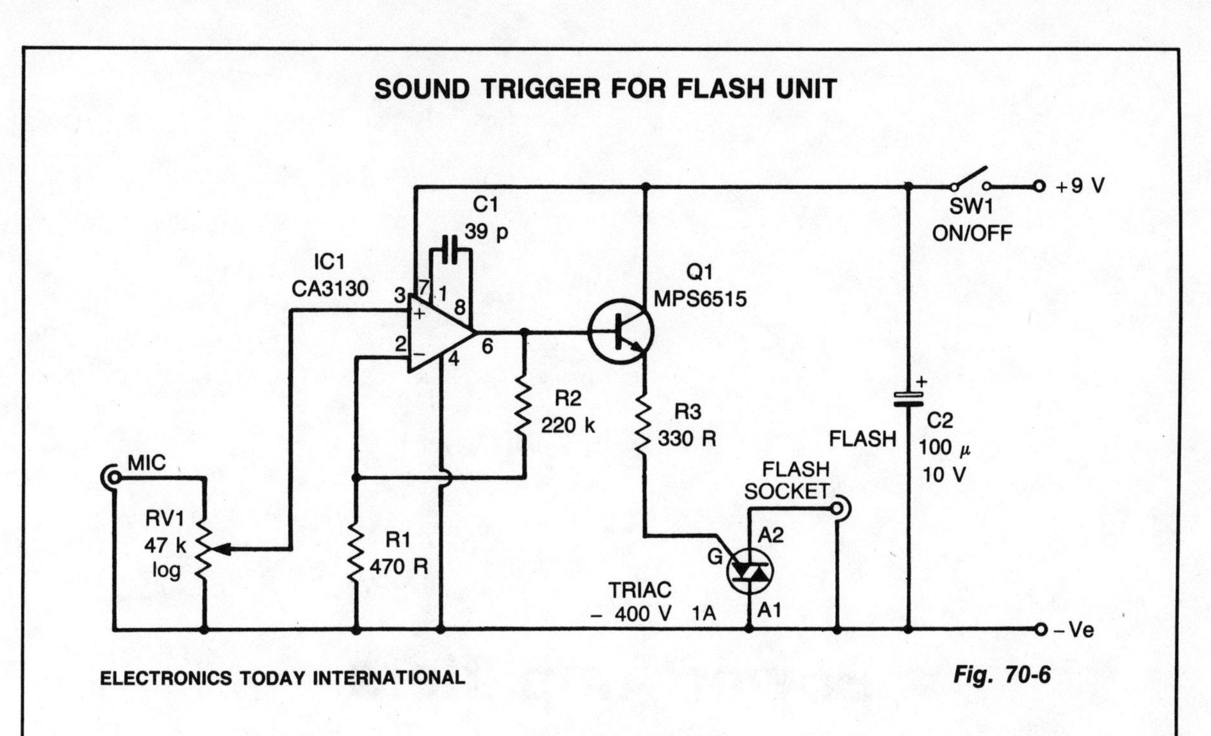

ELECTRONICS TODAY INTERNATIONAL

Fig. 70-6

Circuit Notes

The circuit is based on operational amplifier IC1 used in the noninverting amplifier mode. R1 and 2 set the gain at about 500. RV1 (sensitivity) biases the noninverting input to the negative supply. Q1 provides the relatively high trigger current required by the triac. When a signal is received by the microphone, the signals are amplified (by IC1). The triac is triggered and a low resistance appears across its A1 and A2 terminals which are connected via the flashlead to the strobe. The circuit operates almost instantly, giving very little delay between the commencement of the sound and the flashgun being triggered.

71

Power Amplifiers

The sources of the following circuits are contained in the Sources section beginning on page 694. The figure number contained in the box of each circuit correlates to the source entry in the Sources section.

2 TO 6 W AUDIO AMPLIFIER WITH PREAMPLIFIER

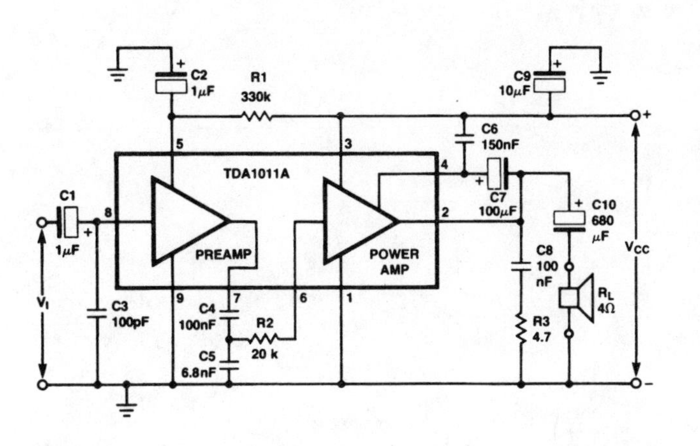

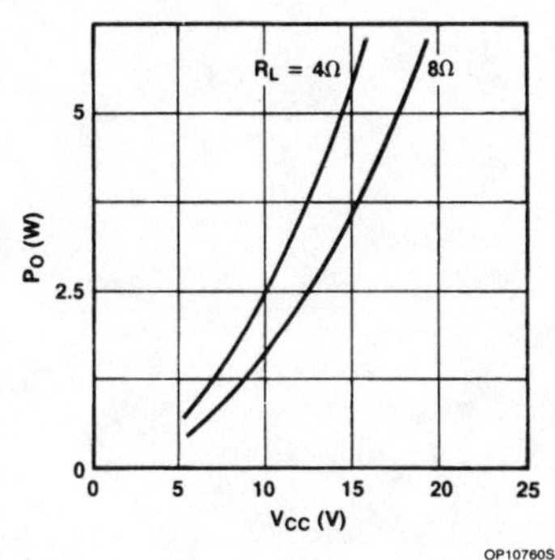

NOTES:
d_{TOT} = 10%; typical values. The available output power is 5% higher when measured at Pin 2 (due to series resistance of C1). *

Output Power Across R_L as a Function of Supply Voltage with Bootstrap

SIGNETICS

Fig. 71-1

Circuit Notes

The monolithic integrated audio amplifier circuit is especially designed for portable radio and recorder applications and delivers up to 4 W in a 4 ohm load impedance.

AUDIO POWER AMPLIFIER

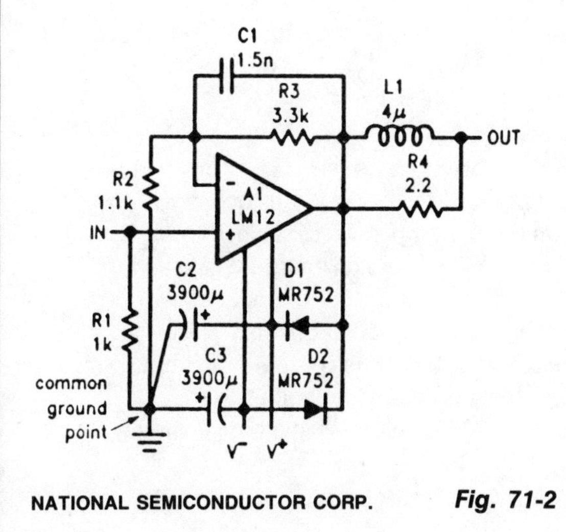

NATIONAL SEMICONDUCTOR CORP.

Fig. 71-2

Circuit Notes

Output-clamp diodes are mandatory because loudspeakers are inductive loads. Output LR isolation is also used because audio amplifiers are usually expected to handle up to 2 μF load capacitance. Large, supply-bypass capacitors located close to the IC are used so that the rectified load current in the supply leads does not get back into the amplifier, increasing high-frequency distortion. Single-point grounding for all internal leads plus the signal source and load is recommended to avoid ground loops that can increase distortion.

25 WATT AMPLIFIER

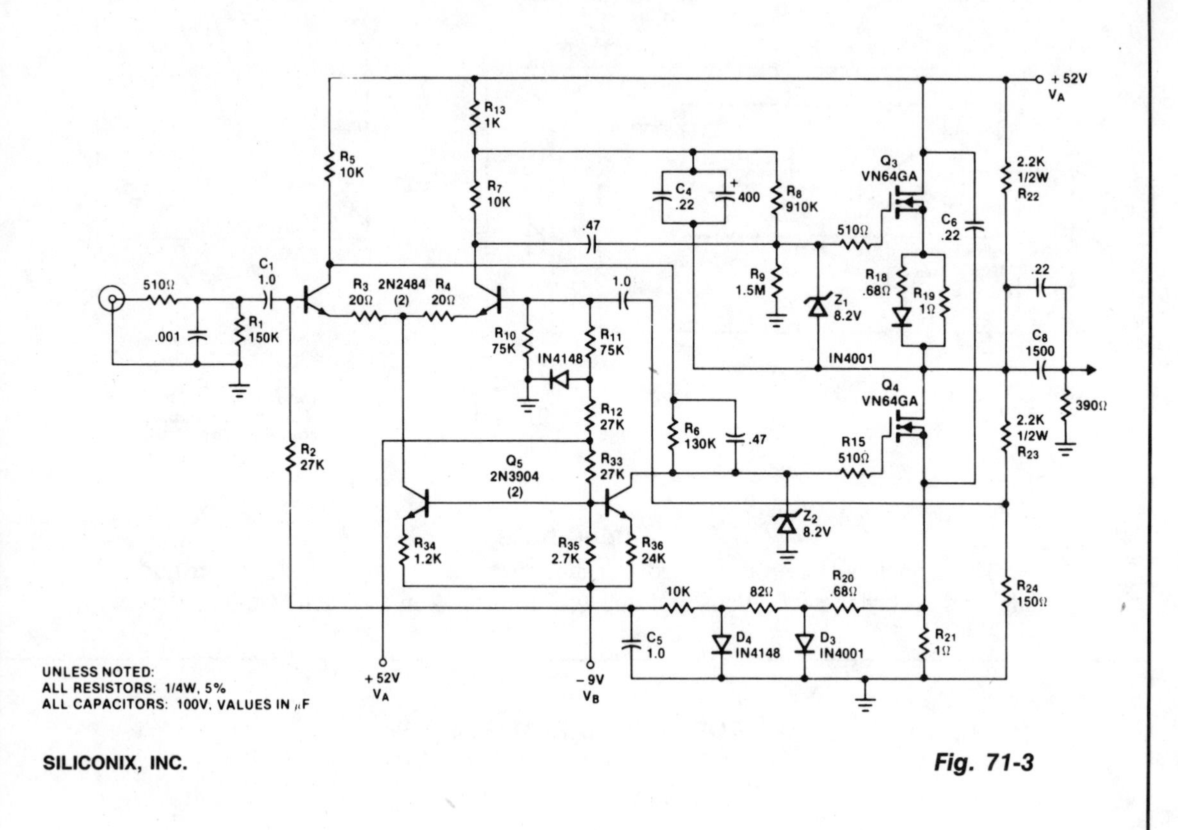

UNLESS NOTED:
ALL RESISTORS: 1/4W, 5%
ALL CAPACITORS: 100V, VALUES IN μF

SILICONIX, INC.

Fig. 71-3

Circuit Notes

Transistors are used for current sources. Base drive for these transistors is derived from the main power supply V_A, so that their collector current is proportional to the rail voltage. This feature holds the voltage on the diff-amp collectors close to $V_A/2$. The sensitivity of I_Q to V_A is about 3.4 mA/volt when V_B is held constant; the sensitivity of I_Q to V_B is -15 mA/volt when V_A is held constant. In a practical amplifier with a non-regulated supply, variations in power output will cause fluctuations in V_A, but will not affect V_B; therefore, having I_Q increase slightly with power output will tend to compensate for the 3.4 mA/volt I_QV_A sensitivity. In the case of line voltage variations, since V_A is about five times V_B, the sensitivities tend to cancel, leaving a net sensitivity of about 2 mA/volt.

BULL HORN

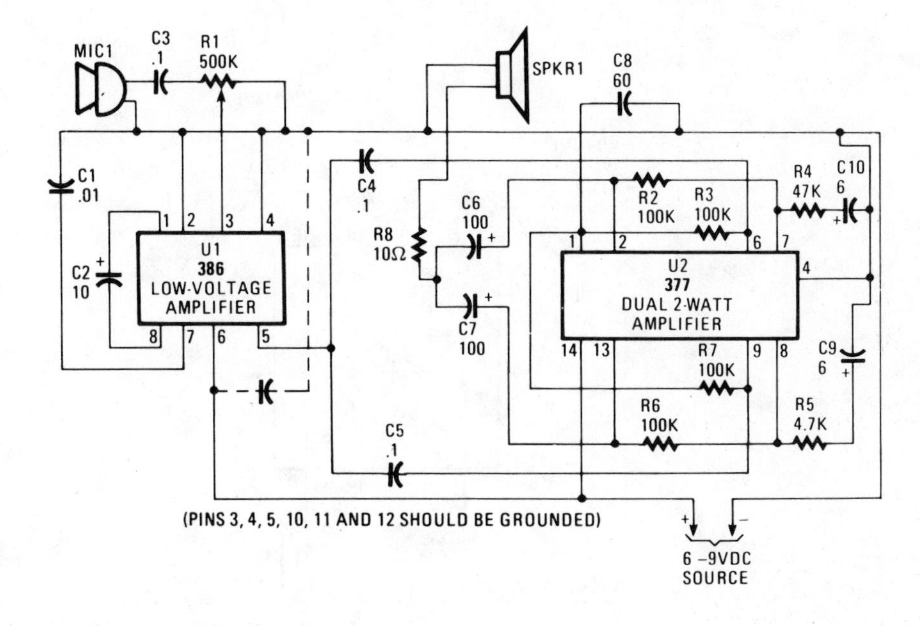

HANDS-ON ELECTRONICS *Fig. 71-4*

Circuit Notes

The input audio signal is fed to pin 3 of U1, an LM386 low-voltage amplifier, via C3 and R1. Potentiometer R1 sets the drive or volume level. U1, which serves as a driver stage, can be set for a gain of from 20 to 200. The output of U1 at pin 5 is fed to U2—a 377 dual two-watt amplifier connected in parallel to produce about four watts of output power—at pins 6 and 9 via C4 and C5. Frequency stability is determined by R2, R4, and C10 on one side, and the corresponding components R6, R5, and C9 on the other side. The outputs of the two amplifiers (at pins 2 and 13) are capacitively coupled to SPKR1 through C6 and C7.

LOW POWER AUDIO AMPLIFIER

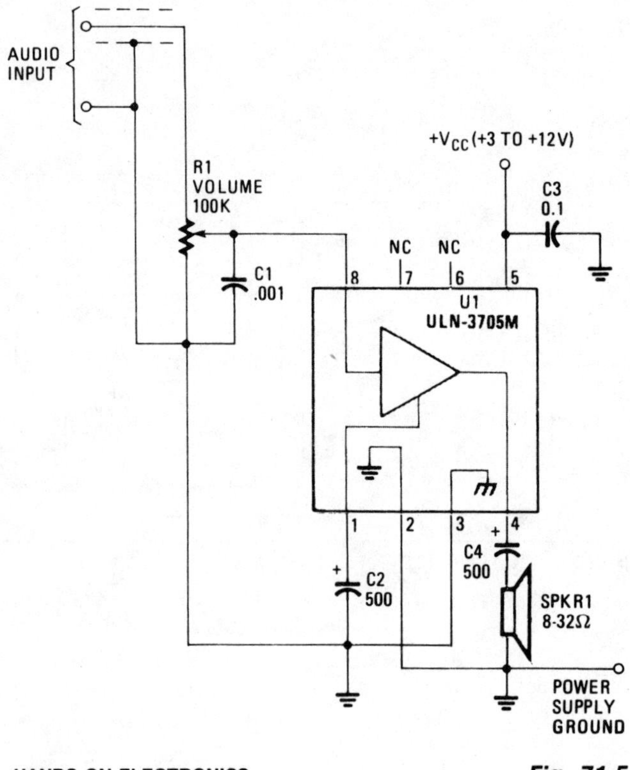

HANDS-ON ELECTRONICS

Fig. 71-5

Circuit Notes

The amplifier operates from supplies ranging up to 12 volts, and operates (with reduced volume) from supply voltages as low as 1.8 volts without having distortion rise to unacceptable levels. (Its power requirements make it suitable for solar-cell application.) Components external to the integrated circuit, U1, consist of four capacitors and a potentiometer for volume control. Capacitor C3 is for decoupling, low-frequency roll-off, and power-supply ripple rejection. Capacitor C4 is an electrolytic type that couples the audio output to an 8 to 32 ohm speaker that is efficient.

AUDIO BOOSTER

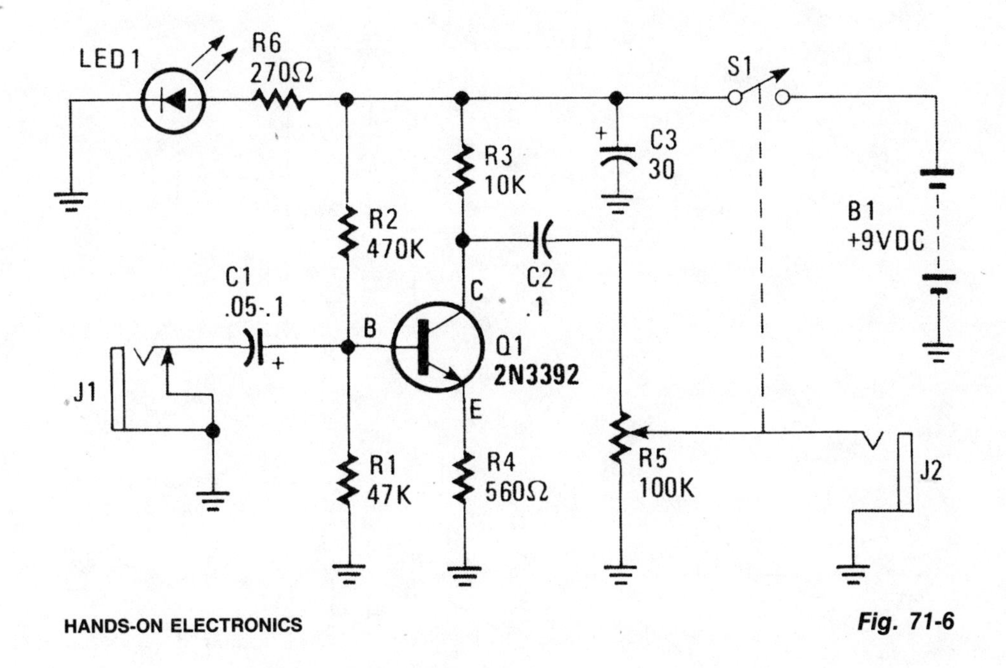

HANDS-ON ELECTRONICS

Fig. 71-6

Circuit Notes

The amplifier's gain is nominally 20 dB. Its frequency response is determined primarily by the value of just a few components—primarily C1 and R1. The values of the schematic diagram provide a response of ±3.0 dB from about 120 Hz to better than 20,000 Hz. Actually, the frequency response is ruler flat from about 170 Hz to well over 20,000 Hz; it's the low end that deviates from a flat frequency response. The low end's roll-off is primarily a function of capacitor C1 (since R1's resistive value is fixed). If C1's value is changed to 0.1 μF, the low end's corner frequency—the frequency at which the low-end roll-off starts—is reduced to about 70 Hz. If you need an even deeper low-end roll-off, change C1 to a 1.0 μF capacitor; if it's an electrolytic type, make certain that it's installed into the circuit with the correct polarity, with the positive terminal connected to Q1's base terminal.

WALKMAN AMPLIFIER

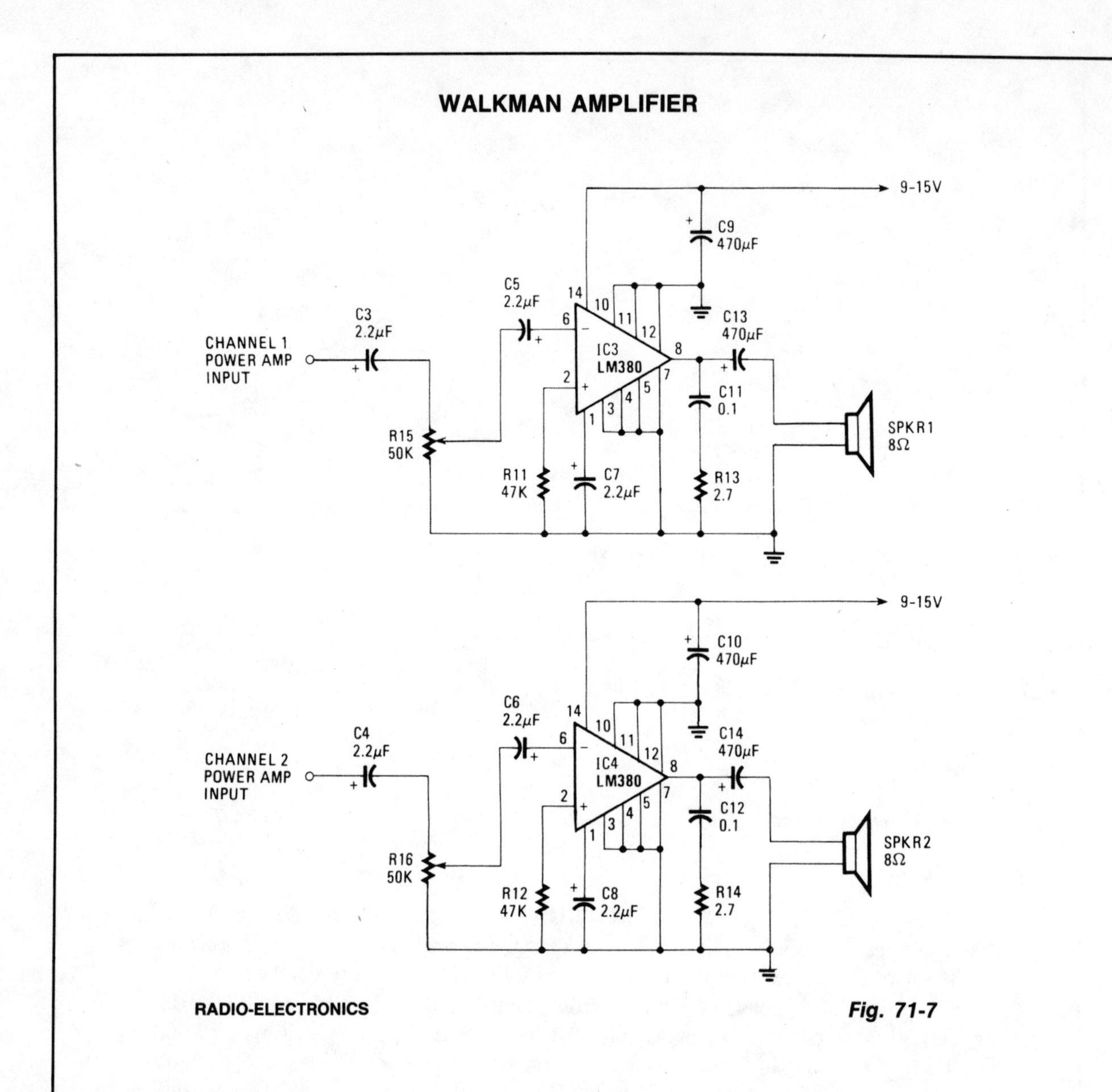

RADIO-ELECTRONICS

Fig. 71-7

Circuit Notes

The gain of the low-cost IC is internally fixed so that it is not less than 34 dB (50 times). A unique input stage allows input signals to be referenced to ground. The output is automatically self centering to one half the supply voltage. The output is also short-circuit proof with internal thermal limiting. With a maximum supply of 15 volts and an 8 ohm load, the output is around 1.5 watts per channel. The input stage is usable with signals from 50 mV to 500 mV rms. If the amplifier is to be used with a source other

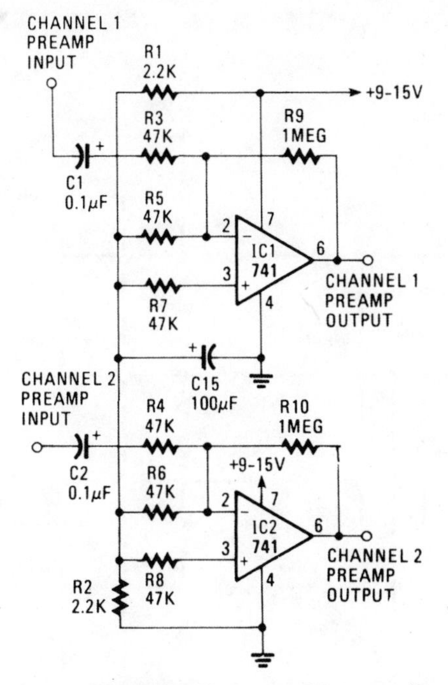

–THE PREAMP. If you wish to amplify low-level signals, such as the output of a turntable, the signal will first have to be fed to the preamp shown here.

than a personal stereo, such as a phonograph or an electric guitar, some type of preamplifier is required. A suitable circuit is shown. In that circuit, two 741 op amps have been configured as input amplifiers. Their input stages referenced to a common point—half the supply voltage. That voltage is derived from a voltage divider made up of R1 and R2, two 2.2 k resistors. The gain of each of the 741's has been fixed at 21 by the input resistors (R9, R10). Input capacitors, C1 and C2, are used to filter out any dc component from the input signal.

REAR SPEAKER AMBIENCE (4-CHANNEL) AMPLIFIER

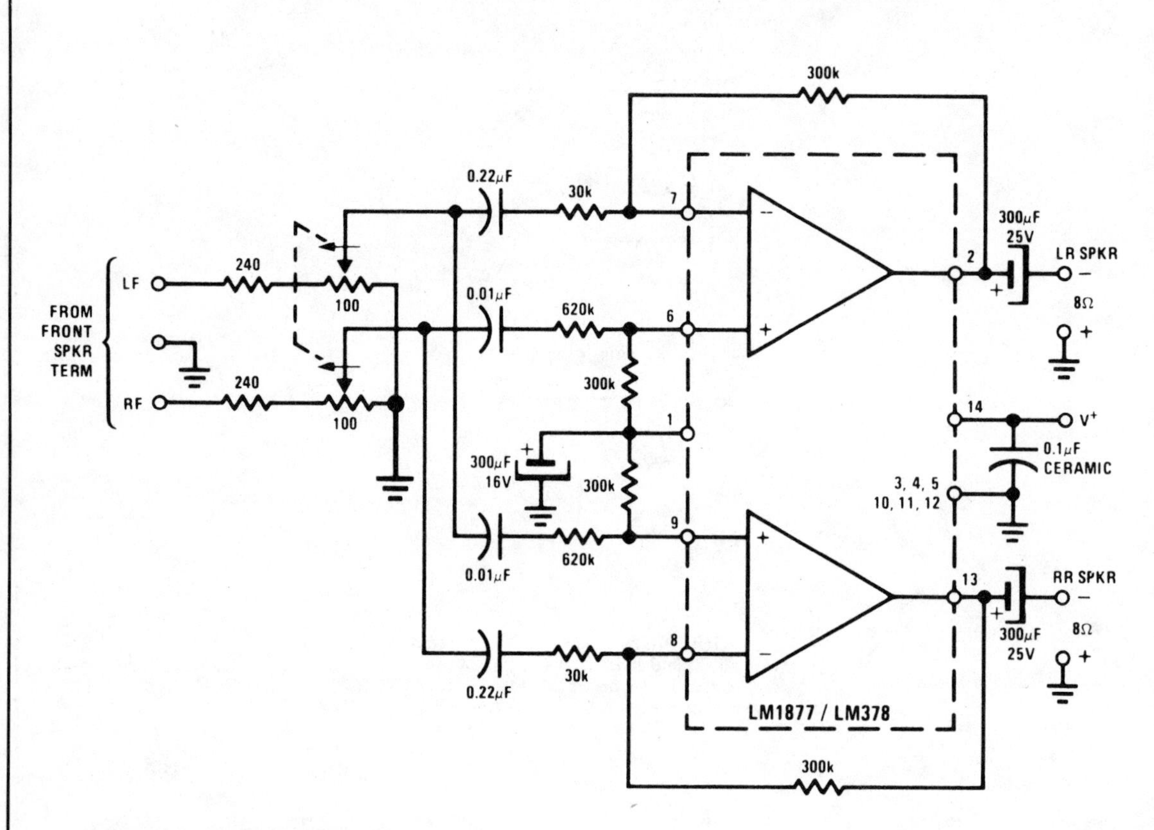

NATIONAL SEMICONDUCTOR CORP.

Fig. 71-8

Circuit Notes

Rear channel "ambience" can be added to an existing stereo system to extract a difference signal (R - L or L - R) which, when combined with some direct signal (R or L), adds fullness, or "concert hall realism" to the reproduction of recorded music. Very little power is required at the rear channels, hence an LM1877 suffices for most "ambience" applications. The inputs are merely connected to the existing speaker output terminals of a stereo set, and two more speakers are connected to the ambience circuit outputs. The rear speakers should be connected in the opposite phase to those of the front speakers, as indicated by the +/− signs.

90 W AUDIO POWER AMPLIFIER WITH SAFE AREA PROTECTION

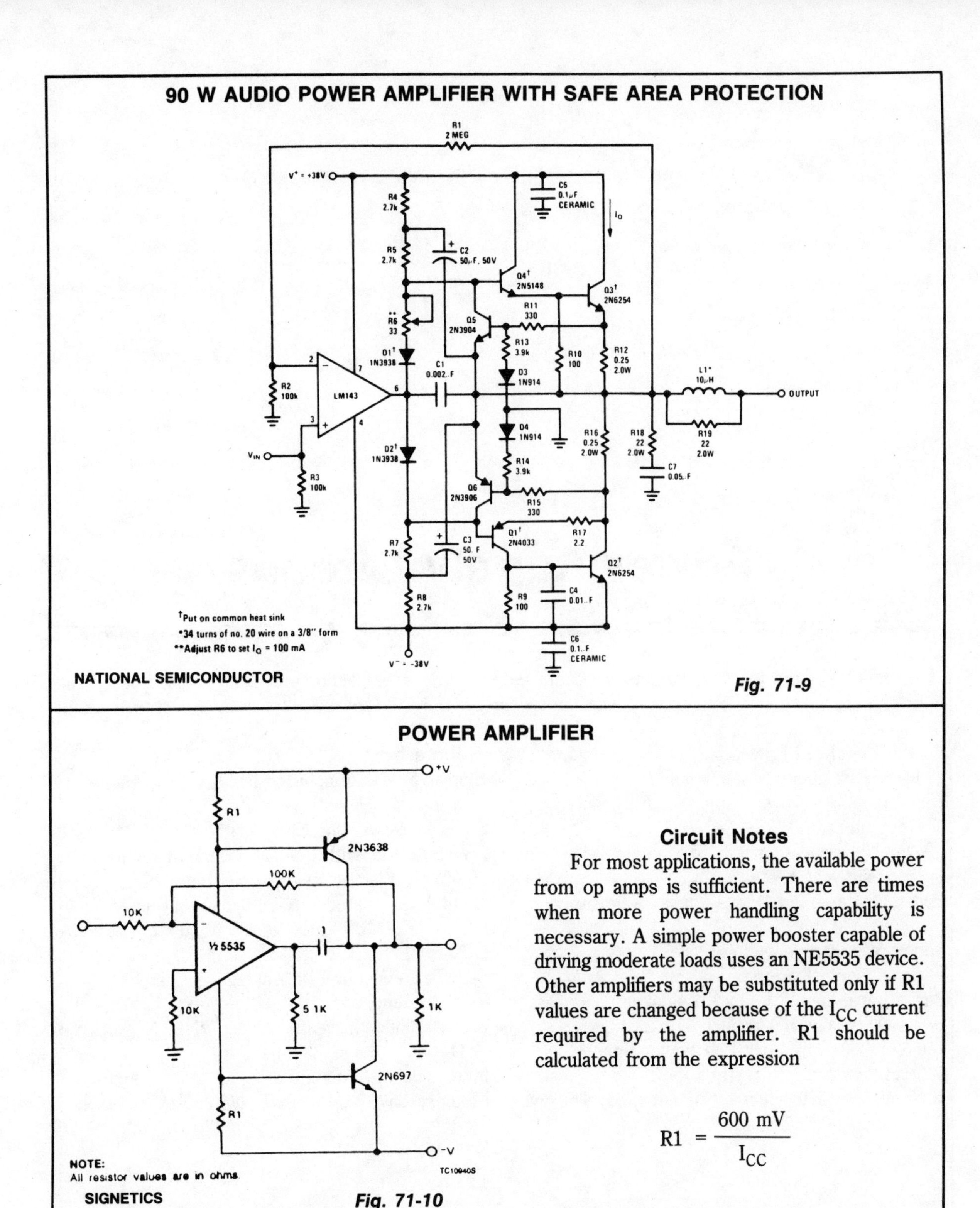

NATIONAL SEMICONDUCTOR

Fig. 71-9

POWER AMPLIFIER

SIGNETICS

Fig. 71-10

Circuit Notes

For most applications, the available power from op amps is sufficient. There are times when more power handling capability is necessary. A simple power booster capable of driving moderate loads uses an NE5535 device. Other amplifiers may be substituted only if R1 values are changed because of the I_{CC} current required by the amplifier. R1 should be calculated from the expression

$$R1 = \frac{600 \text{ mV}}{I_{CC}}$$

72

Power Supply Circuits

The sources of the following circuits are contained in the Sources section beginning on page 694. The figure number contained in the box of each circuit correlates to the source entry in the Sources section.

8-AMP REGULATED POWER SUPPLY FOR OPERATING MOBILE EQUIPMENT

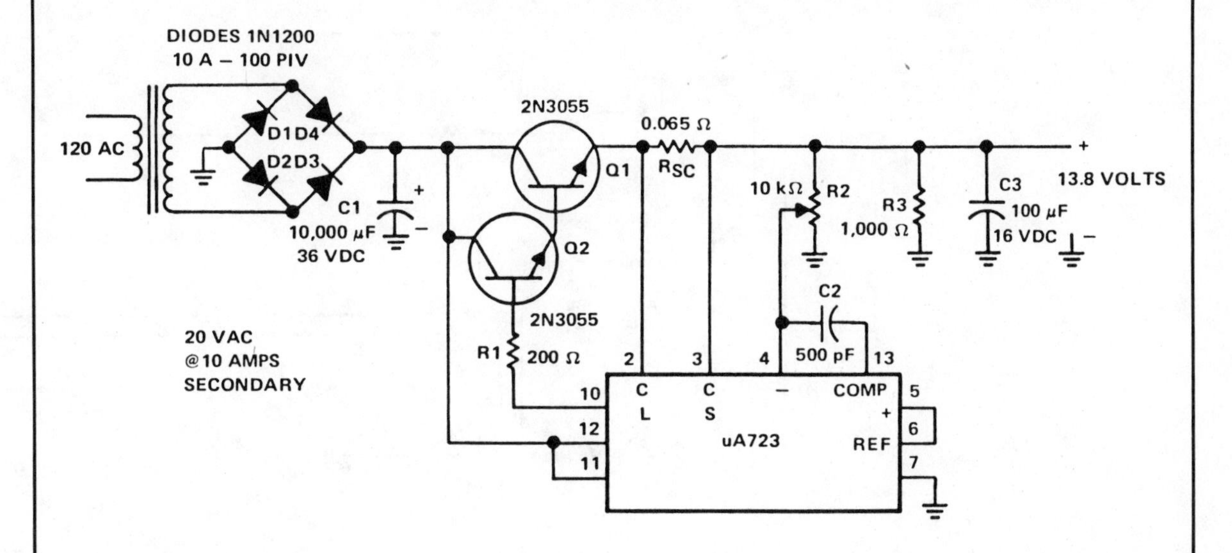

TEXAS INSTRUMENTS

Fig. 72-1

Circuit Notes

This supply is powered by a transformer operating from 120 Vac on the primary and providing approximately 20 Vac on the primary, and providing approximately 20 Vac on the secondary. Four 10-A diodes with a 100 PIV rating are used in a full-wave bridge rectifier. A 10,000 µF/36 Vdc capacitor completes the filtering, providing 28 Vdc. The dc voltage is fed to the collectors of the Darlington connected 2N3055's. Base drive for the pass transistors is from pin 10 of the µA723 through a 200 ohm current limiting resistor, R1. The reference terminal (pin 6) is tied directly to the non-inverting input of the error amplifier (pin 5), providing 7.15 V for comparison.

The inverting input to the error amplifier (pin 4) is fed from the center arm of a 10 k ohm potentiometer connected across the output of the supply. This control is set for the desired output voltage of 13.8 V. Compensation of the error amplifier is accomplished with a 500 pF capacitor connected from pin 13 to pin 4. If the power supply should exceed 8 A or develop a short circuit, the µA723 regulator will bias the transistors to cutoff and the output voltage will drop to near zero until the short circuit condition is corrected.

UNINTERRUPTIBLE POWER SUPPLY FOR PERSONAL COMPUTERS

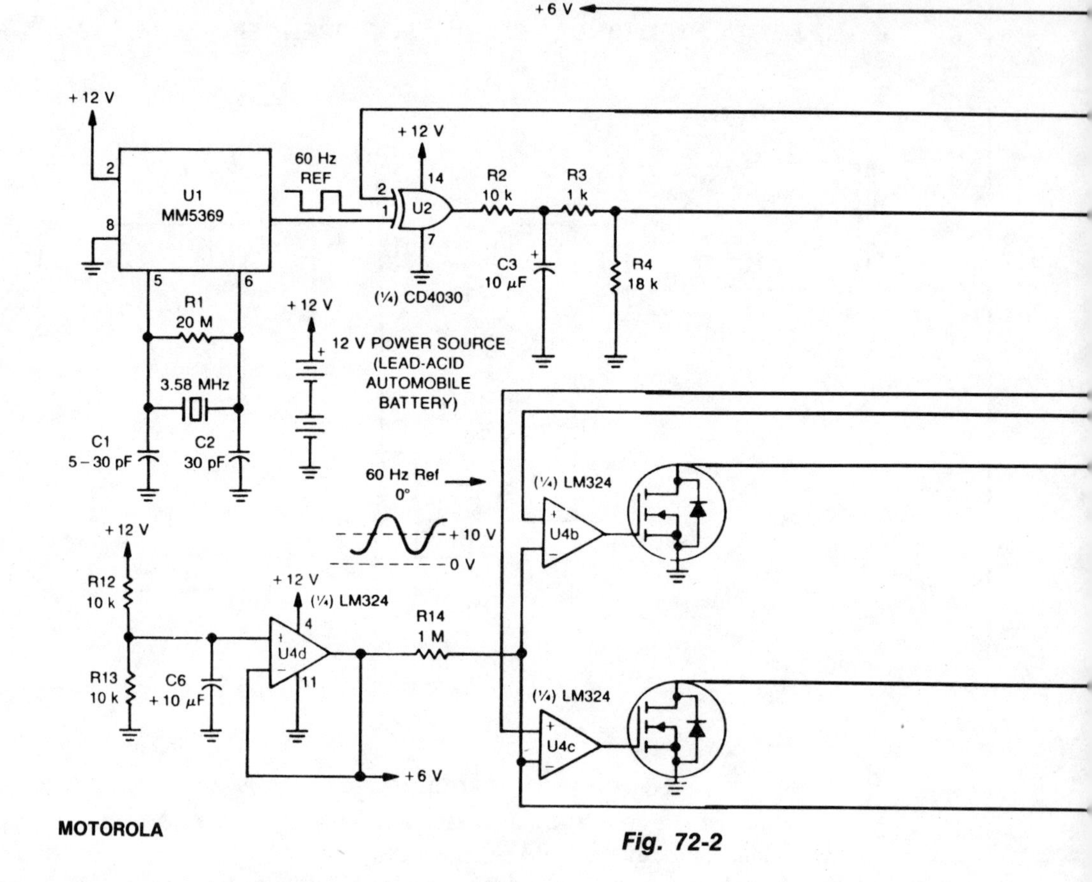

Fig. 72-2

Circuit Notes

The UPS is basically an ac inverter that is powered by a 12-V, lead-acid automobile battery. During power outages, it can supply several minutes of power for an average personal computer. It incorporates a crystal-controlled 60 Hz time base, so that a computer with a real time clock can maintain its accuracy. It isolates the ac line from the computer, so it can be used to operate sensitive electronic equipment on noisy power sources.

Two MTM60N06 Power FETs (Q1 and Q2) alternately switch current through a center-tapped 120-V to 12-V filament transformer (T1) with its primary and secondary reversed. The 120-V output is compared with a 60 Hz reference in a closed-loop configuration that maintains a constant output at optimum efficiency.

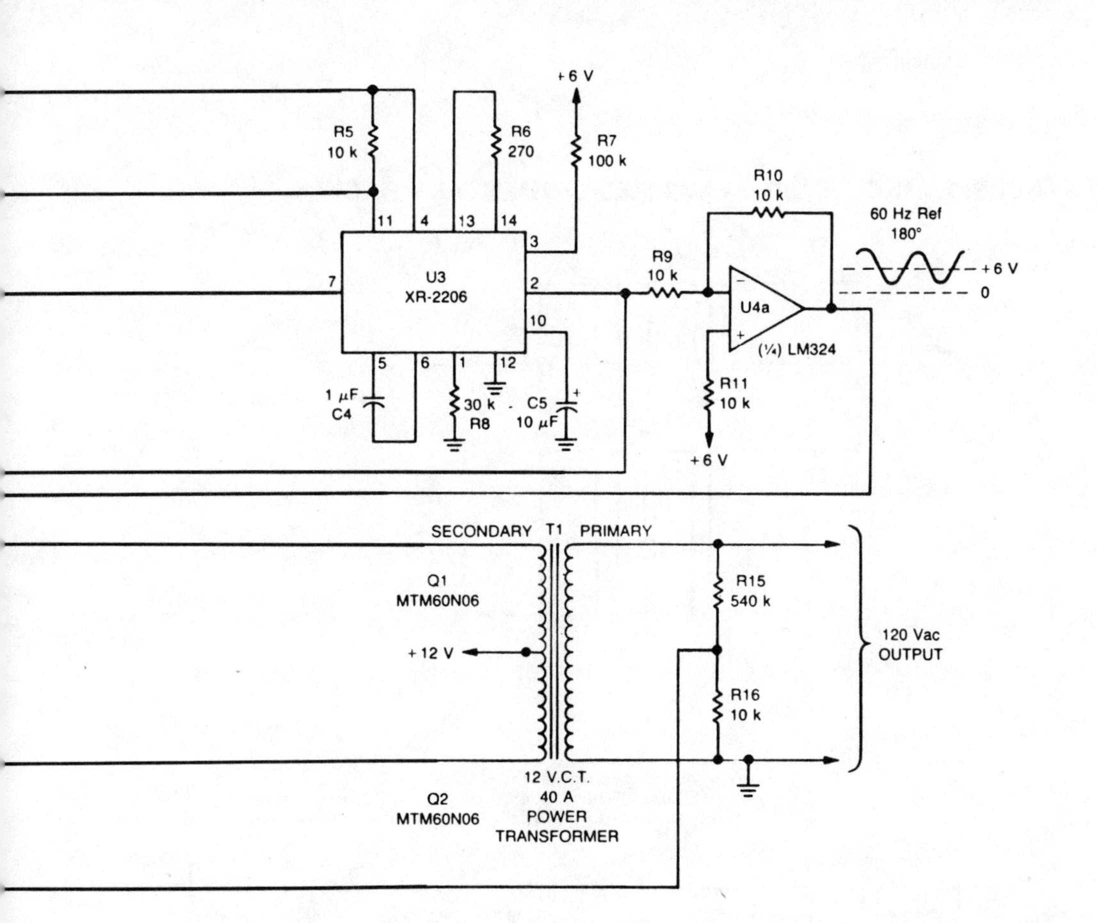

A 60 Hz reference frequency is derived from a crystal oscillator and divider circuit, U1. An inexpensive 3.58 MHz color burst crystal provides the time base that can be accurately adjusted by C1. The 60 Hz output from U1 is applied to the exclusive-OR gate, U2, and then to the XR-2206 function generator (U3) that converts the square wave into a sine wave. U2 and U3 form a phase-locked loop that synchronizes the sine wave output of U3 with the 60 Hz square wave reference of U1. The sine wave is then inverted by op amp U4a, so that two signals 180 out of phase can be applied to U4b and U4c that drive Q1 and Q2. Due to the closed-loop configuration of the drive circuits, Q1 and Q2 conduct only during the upper half of the sine wave. Therefore, one TMOS device conducts during the first half of the sine wave and the other conducts during the second half.

5 V SUPPLY INCLUDING STABILIZED MOMENTARY BACKUP

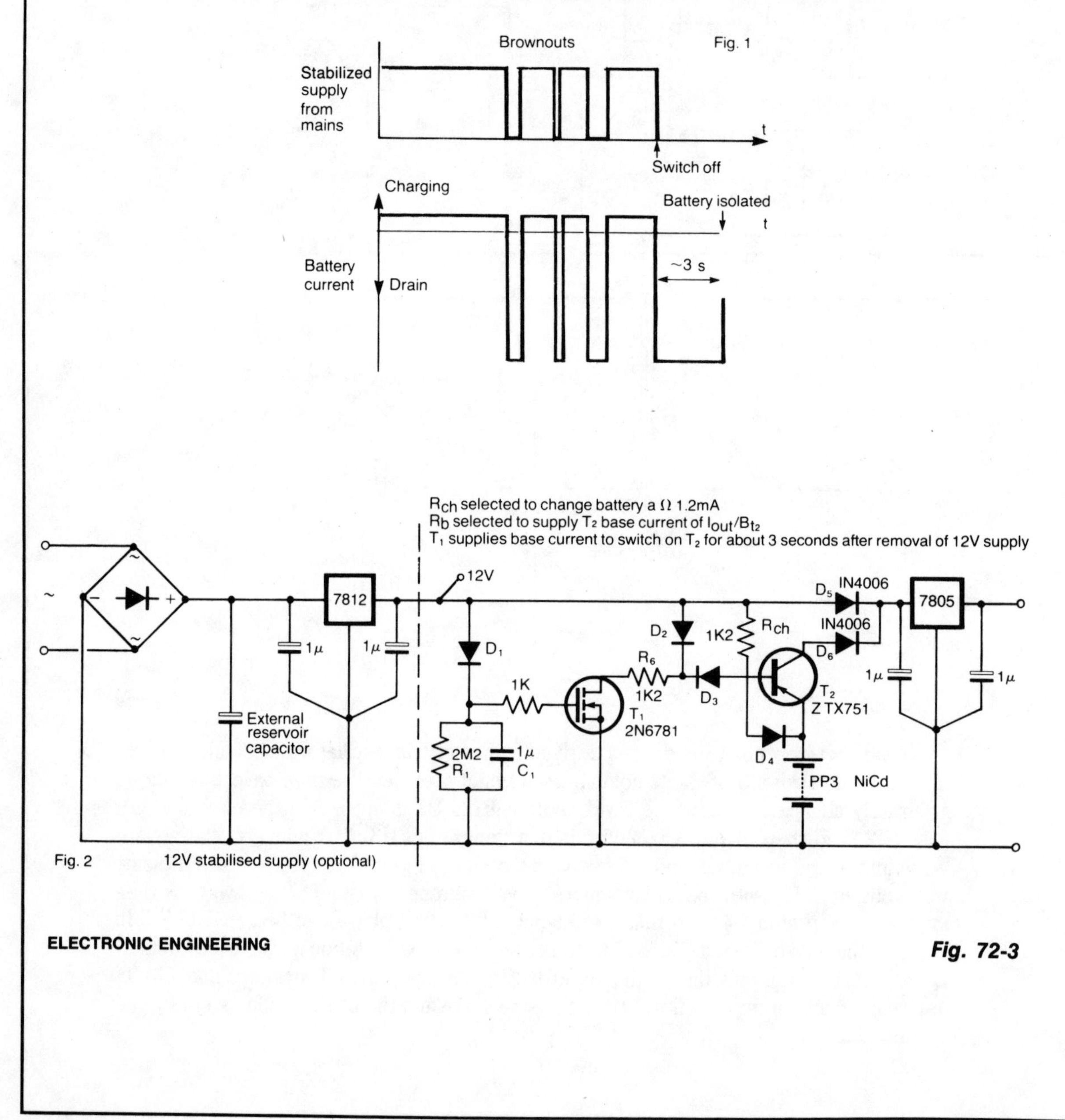

ELECTRONIC ENGINEERING

Fig. 72-3

5 V SUPPLY INCLUDING STABILIZED MOMENTARY BACKUP , Continued.

Circuit Notes

This circuit protects microprocessor systems from "brownouts" without the expense of an uninterruptible power supply. Designed around a small 9-V nickel cadmium battery the circuit continues to provide a constant 5-V output during brownouts of up to a few seconds. Load currents of up to 500 mA may be drawn using the components shown. With this mains-derived supply present, D5 is forward biased so that the stabilized supply powers the 5-V regulator and hence the circuitry to be protected. FET T_1 is held on by D1, its drain current being provided from the dc supply via R_b and D2. Diode D3 is reverse-biased so that T2 is off, and the battery is isolated from D6. R_{CH} and D4 serve to trickle charge the battery with approximately 1.2 mA.

When the 12-V supply is removed, R1 and C1 initially keep T1 switched on. D3 is now forward biased, so that T1 drain current is drawn via R_b, D3 and T2 from the battery. This switches T2 on, allowing the load circuitry to draw current from the battery via D6 and the 5-V regulator. After a few seconds C1 has discharged (via R1) such that V_{gs} falls below the threshold value for the FET, and T1 switches off. There is then no path for T2 base current, so that it also switches off, isolating the battery.

POWER-SWITCHING CIRCUIT

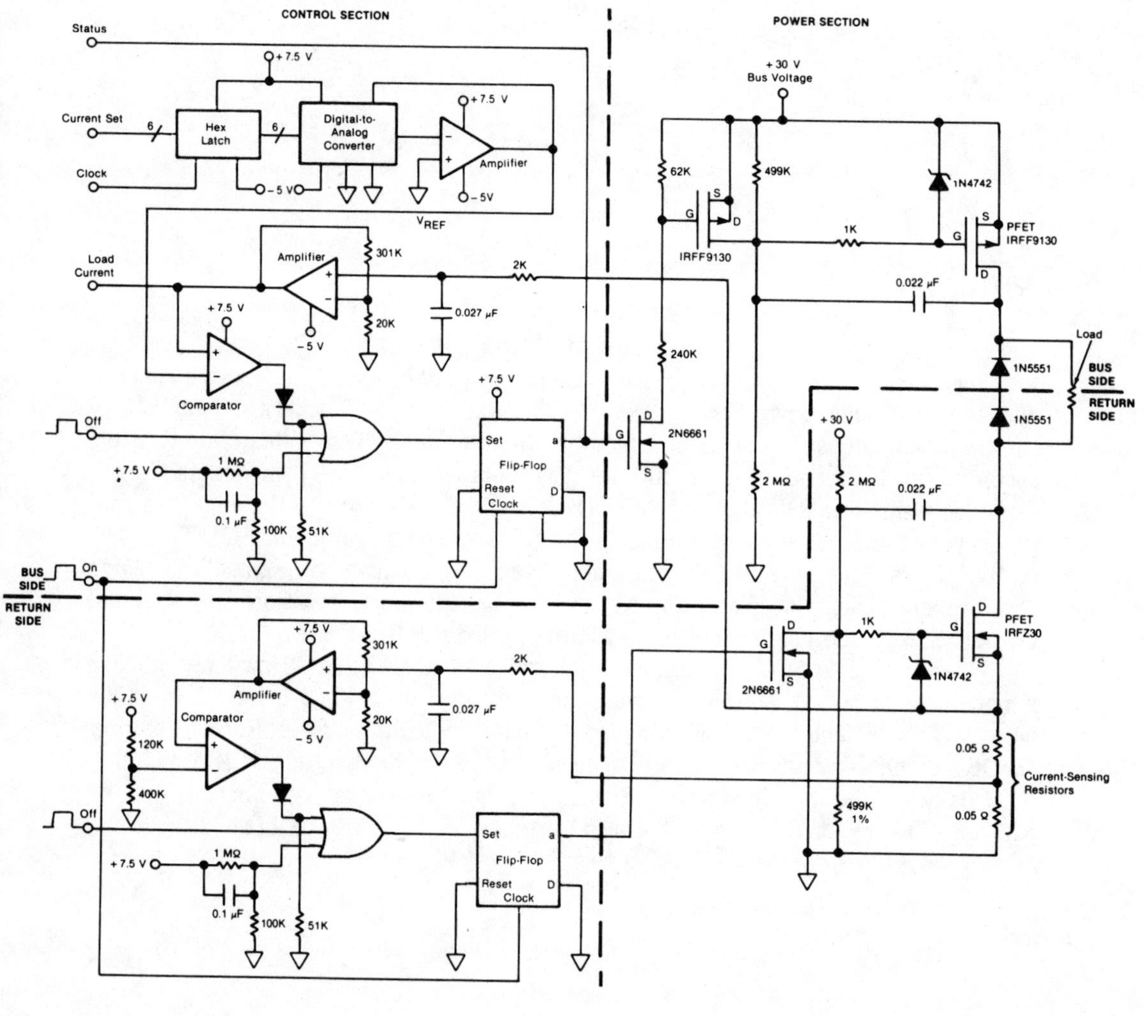

NASA

Fig. 72-4

Circuit Notes

This circuit provides on/off switching, soft starting, current monitoring, current tripping, and protection against overcurrent for a 30 Vdc power supply at normal load currents up to 2 A. The switch is turned on by an "on" command pulse; it is turned off by an "off" command pulse. An overcurrent trip can also be set on the bus side by a 6-digit binary signal, which is converted to an analog voltage and compared with the amplified voltage developed across a load-current-sensing resistor. Resistor/capacitor combinations (0.027 μF, 2 kΩ) at the inputs of the current-sensing amplifiers act as low-pass filters: this introduces a few hundred μs of delay in the response to overcurrent, thereby providing some immunity to noise. The 0.022 μF capacitors connected to the drain terminals of the PFETs provide a Miller effect, which reduces the rate of change of the drain voltage and therefore the rate of rise of current at turn-on. The soft-turn-on time depends upon the load impedance and is typically 100 to 200 ms.

RADIATION-HARDENED, 125 A LINEAR REGULATOR

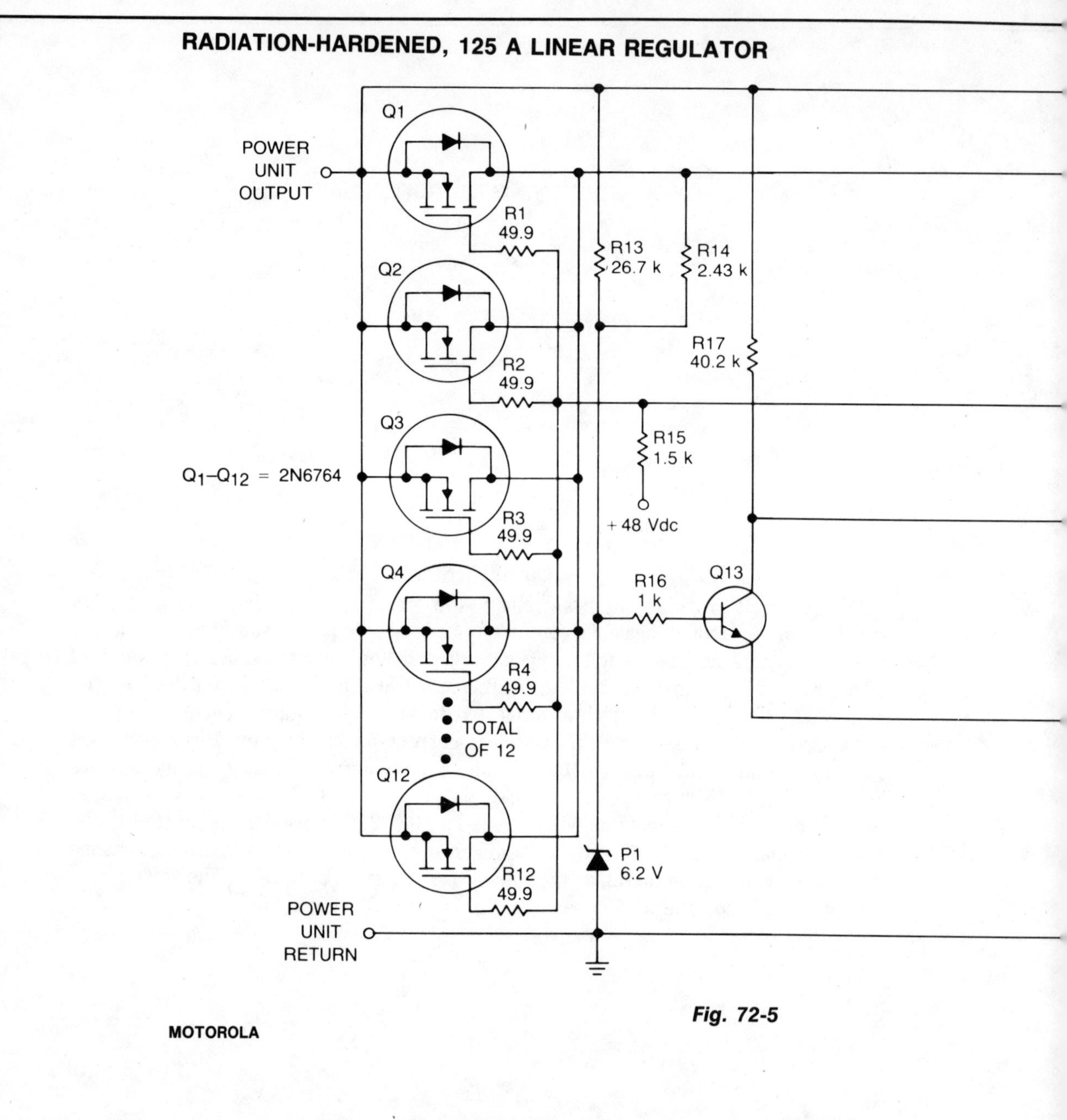

MOTOROLA

Fig. 72-5

Circuit Notes

Intended for extreme temperature, radiation-hardened environments, this linear supply is capable of supplying 28 Vdc at 125 A from an ac-driven power unit.

In operation, power supply output voltage is sensed by the voltage divider consisting of R24 to R28 and fed to one input of a discrete differential amplifier composed of Q13

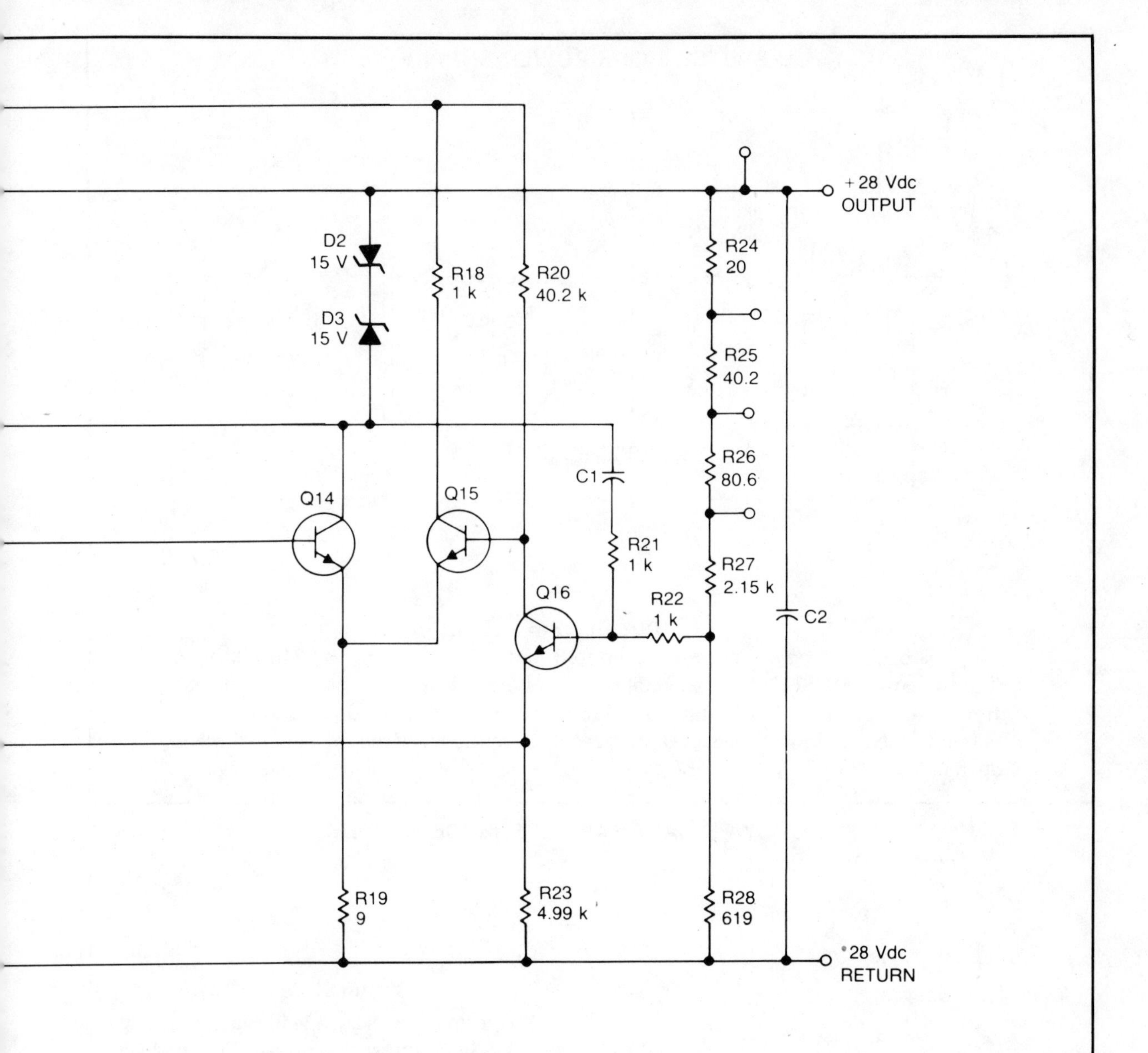

through Q16. The other input of the amplifier is connected to a radiation-hardened zener diode, D1. Local feedback using R21 and C1 produces gain to phase shift that are independent of individual component parameters, which provides stable operation into the required loads.

SWITCH MODE POWER SUPPLY

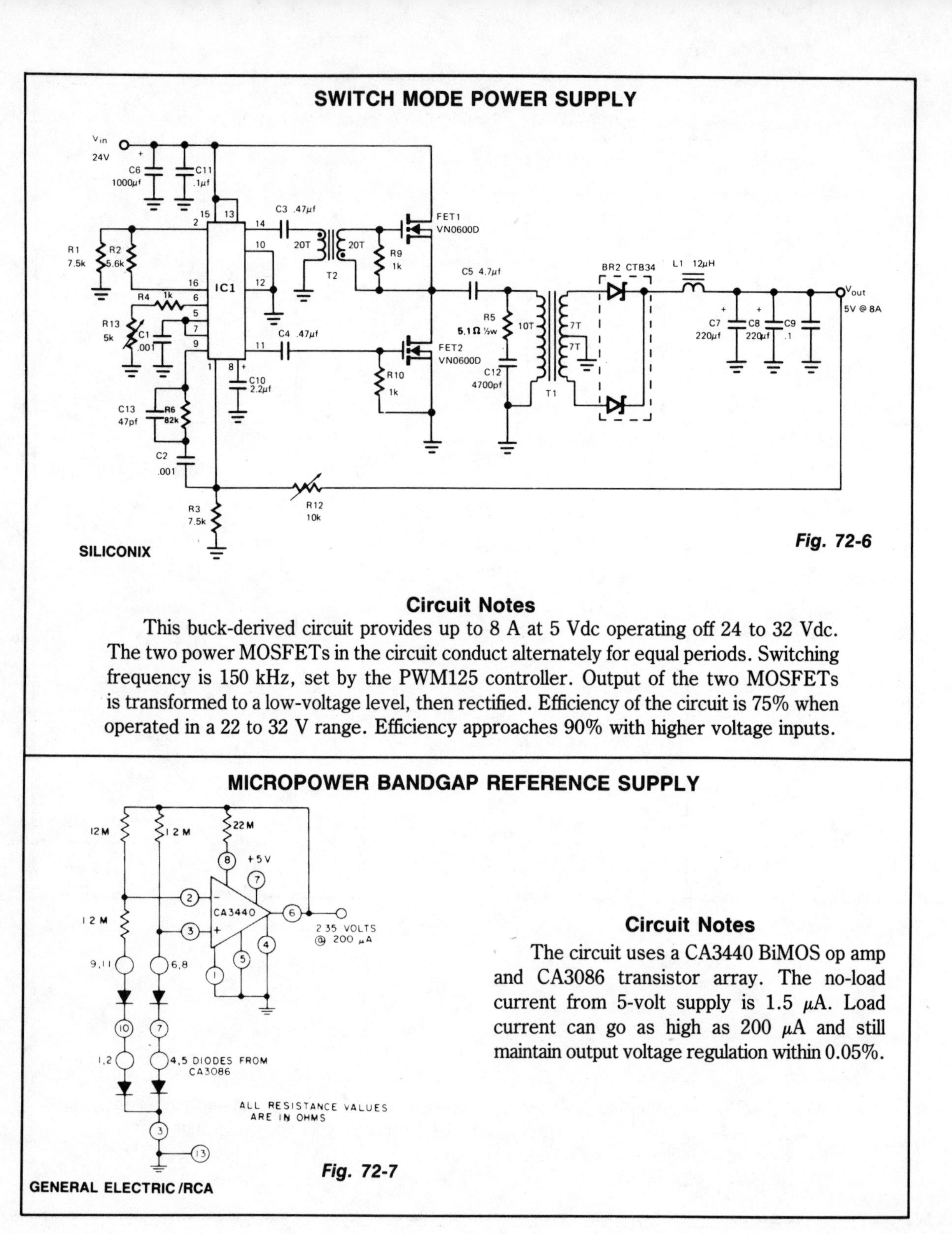

Fig. 72-6

Circuit Notes

This buck-derived circuit provides up to 8 A at 5 Vdc operating off 24 to 32 Vdc. The two power MOSFETs in the circuit conduct alternately for equal periods. Switching frequency is 150 kHz, set by the PWM125 controller. Output of the two MOSFETs is transformed to a low-voltage level, then rectified. Efficiency of the circuit is 75% when operated in a 22 to 32 V range. Efficiency approaches 90% with higher voltage inputs.

MICROPOWER BANDGAP REFERENCE SUPPLY

Fig. 72-7

Circuit Notes

The circuit uses a CA3440 BiMOS op amp and CA3086 transistor array. The no-load current from 5-volt supply is 1.5 μA. Load current can go as high as 200 μA and still maintain output voltage regulation within 0.05%.

VARIABLE CURRENT SOURCE, 100 mA TO 2 AMP

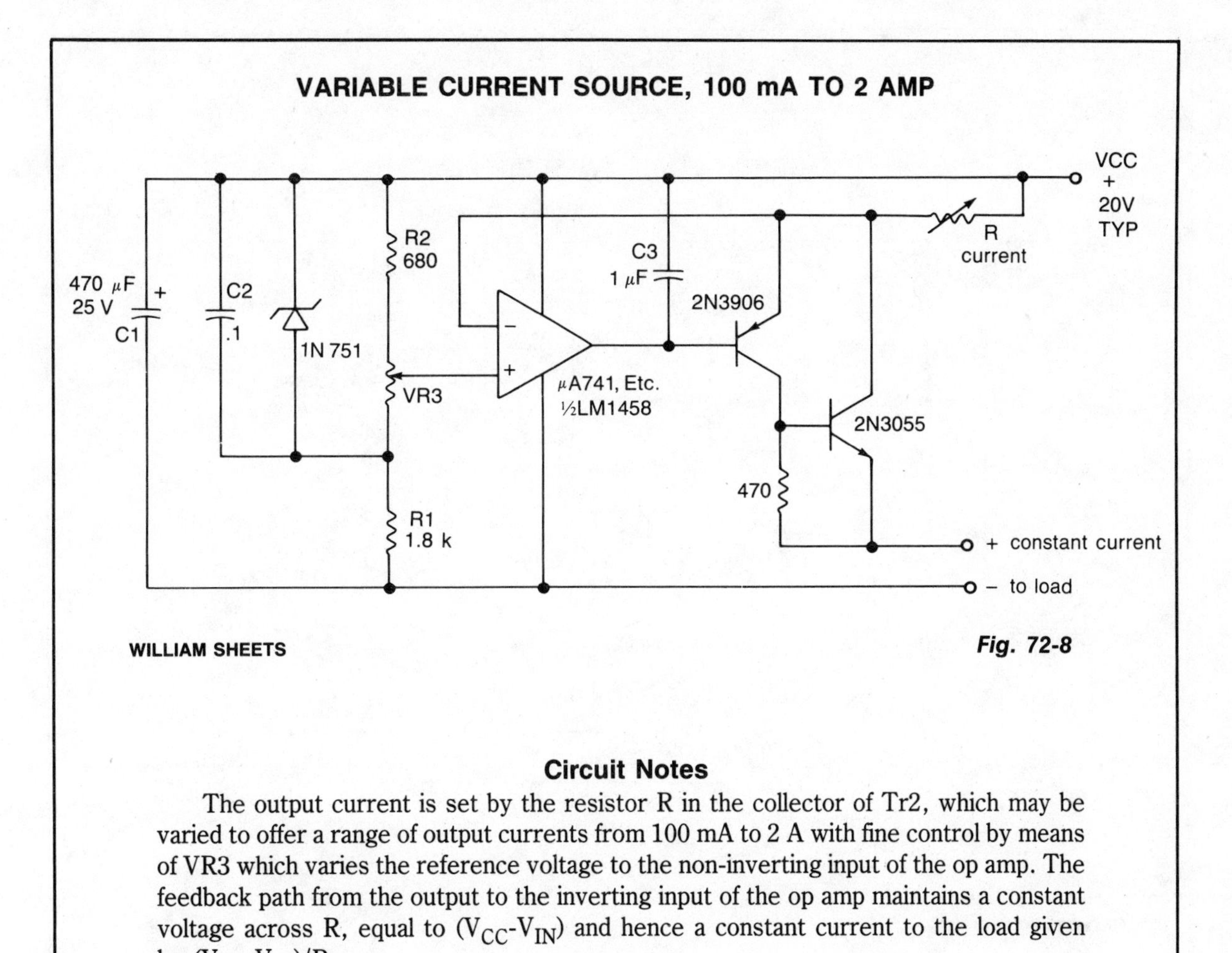

WILLIAM SHEETS

Fig. 72-8

Circuit Notes

The output current is set by the resistor R in the collector of Tr2, which may be varied to offer a range of output currents from 100 mA to 2 A with fine control by means of VR3 which varies the reference voltage to the non-inverting input of the op amp. The feedback path from the output to the inverting input of the op amp maintains a constant voltage across R, equal to (V_{CC}-V_{IN}) and hence a constant current to the load given by (V_{CC}-V_{IN})/R.

BASIC SINGLE-SUPPLY VOLTAGE REGULATOR

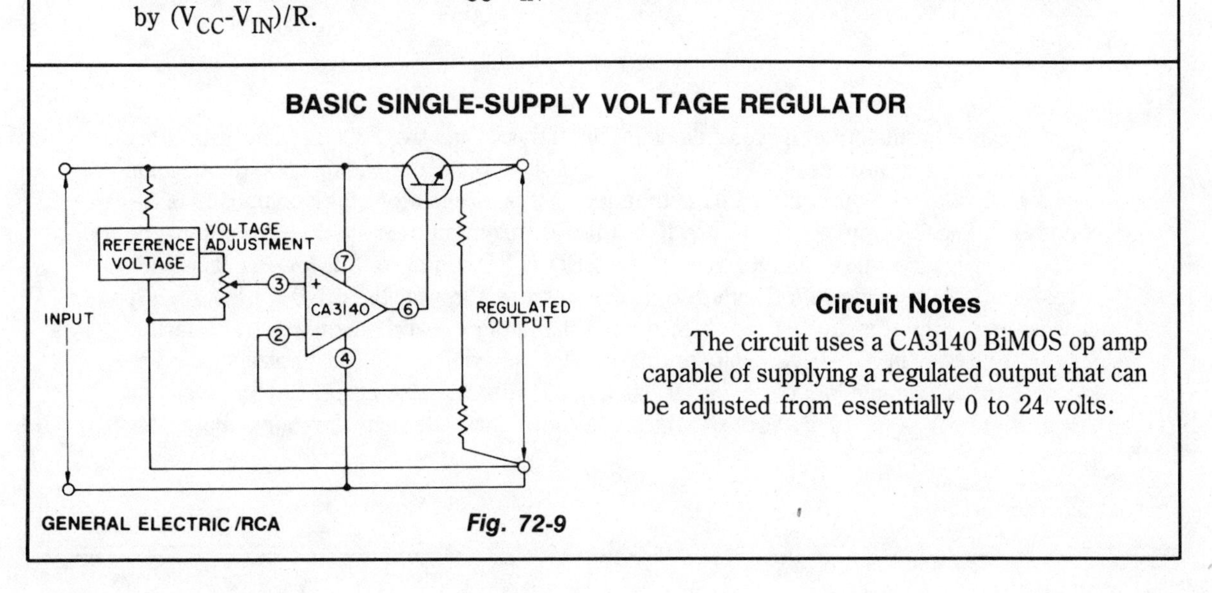

GENERAL ELECTRIC /RCA

Fig. 72-9

Circuit Notes

The circuit uses a CA3140 BiMOS op amp capable of supplying a regulated output that can be adjusted from essentially 0 to 24 volts.

BENCH TOP POWER SUPPLY

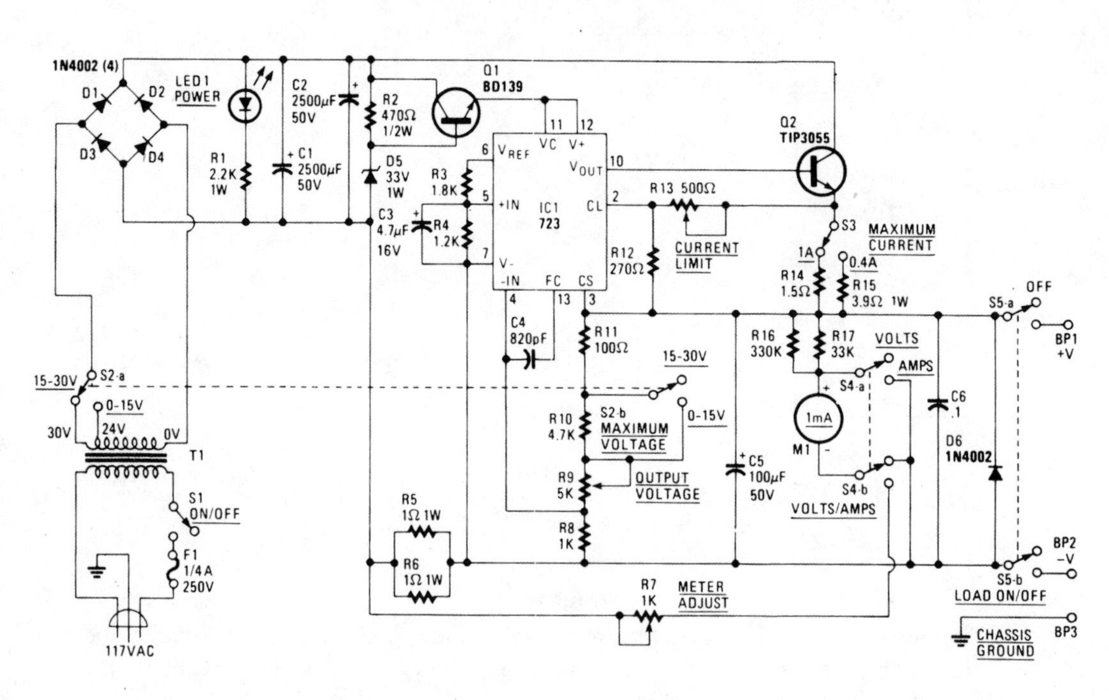

RADIO-ELECTRONICS *Fig. 72-10*

Circuit Notes

A tapped transformer drives a diode bridge (D1-D4) and two 2500 μF filter capacitors (C1 and C2), that provide a no-load voltage of 37 or 47 volts, depending upon the position of switch S2a. The unregulated dc is then fed to a pre-regulator stage composed of Q1 and D5. Those components protect IC1 (the 723) from an over-voltage condition; the 723 can't handle more than 40 volts. The LED (LED1) and its 2.2 k current-limiting resistor (R1) provide on/off indication. The current through the LED varies slightly according to the transformer tap selected, but that's of no real consequence. The series-pass transistor in IC1 drives voltage-follower Q2, providing current amplification. The transistor can handle lots of power. It has a maximum collector current of 15 amps and a maximum V_{CE} of 70 V, both of which are more than adequate for our supply.

400-VOLT, 60-WATT PUSH-PULL POWER SUPPLY

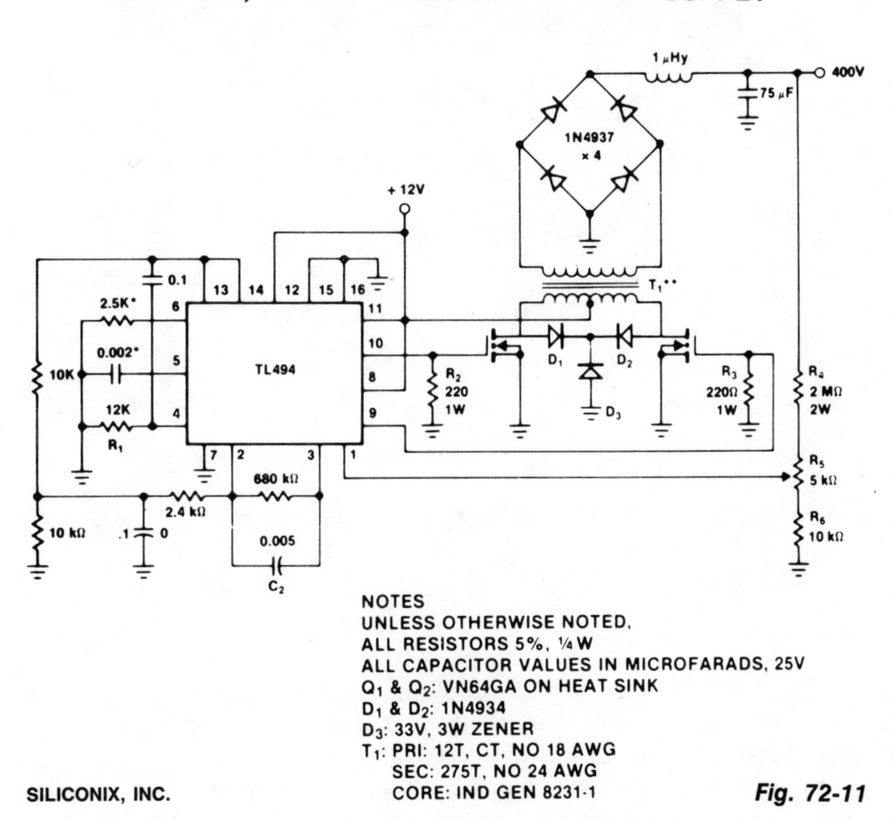

SILICONIX, INC.

Fig. 72-11

Circuit Notes

The design delivers a regulated 400-V, 60-W output. The TL494 switching regulator governs the operating frequency and regulates output voltage. R1 and C1 determine switching frequency, which is approximately 0.5RC—100 kHz for the values shown. The TL494 directly drives the FET's gates with a voltage-controlled, pulse-width-modulated signal. After full-wave rectification, the output waveform is filtered by a choke-input arrangement. The 1 μH, 75 μF filter accomplishes the job nicely at 100 kHz. A feedback scheme using R4, R5 and R6 provides for output-voltage regulation adjustment, with loop compensation handled by C2. Diodes D1 and D2 provide isolation and steering for the 33-V zener transient clamp, D3. Output regulation is typically 1.25% from no-load to the full 60-W design rating. Regulation is essentially determined by the TL494. Output noise and ripple consists mainly of positive and negative 0.8-V spikes occurring when the output stage switches.

500 kHz SWITCHING INVERTER FOR 12 V SYSTEMS

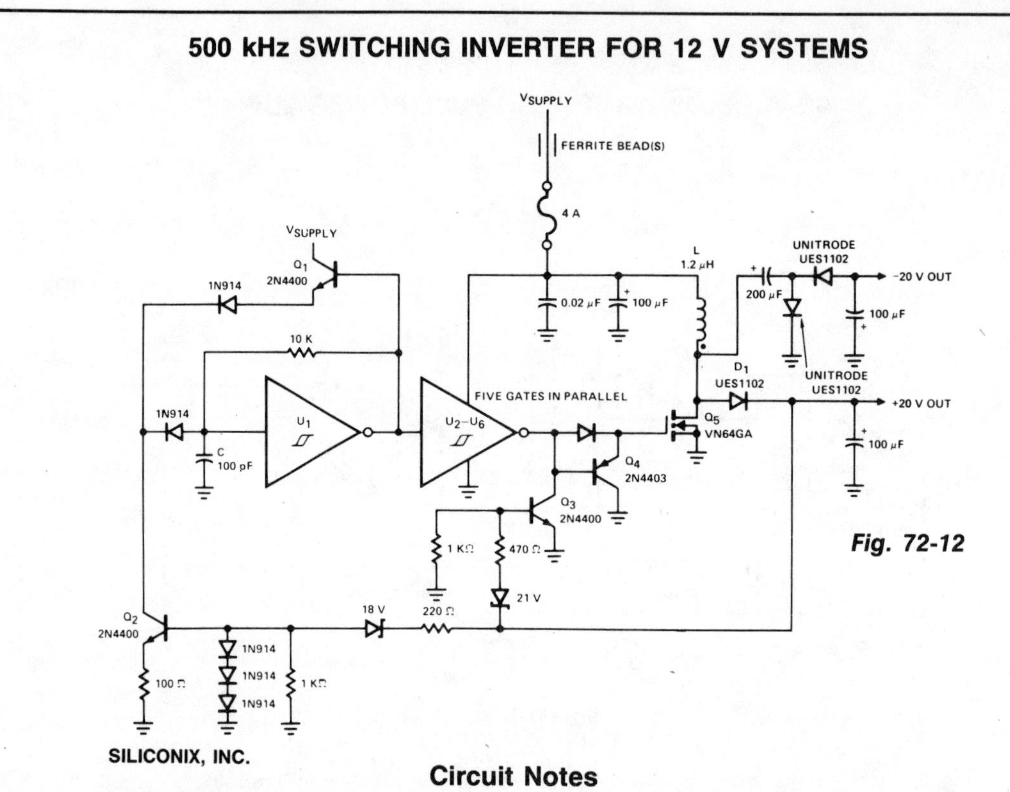

SILICONIX, INC.

Fig. 72-12

Circuit Notes

This PWM control circuit provides the control pulse to the DMOS Power Switch in the flyback circuit. The output of the PWM is a pulse whose width is proportional to the input control voltage and whose repetition rate is determined by an external clock signal. To provide the control input to the PWM and to prevent the output voltage from soaring or sagging as the load changes the error amplifier and reference voltage complete the design. They act as the feedback loop in this control circuit much like that of a servo control system.

10 AMP-REGULATOR WITH CURRENT AND THERMAL PROTECTION

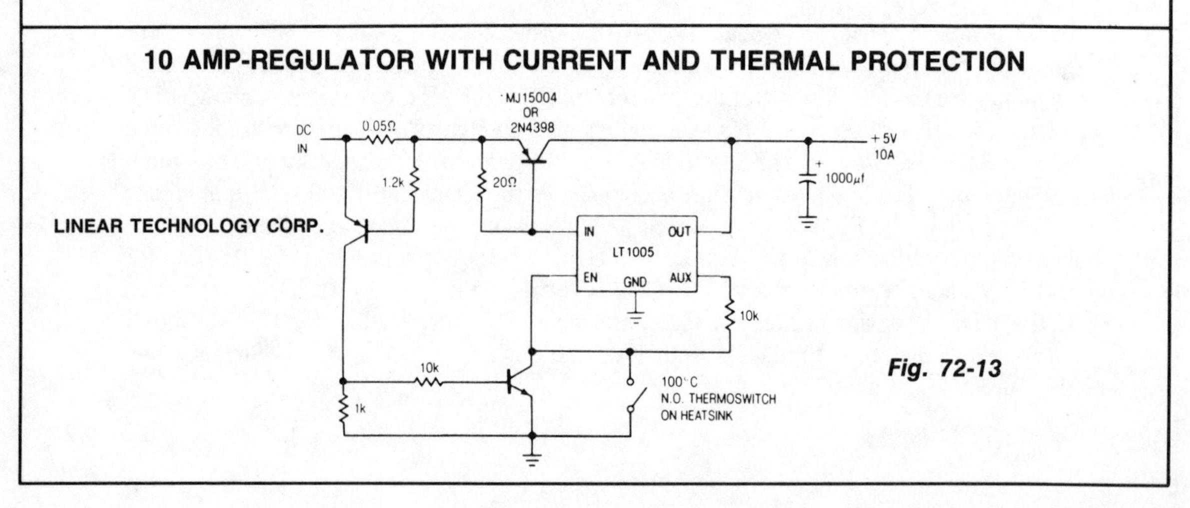

LINEAR TECHNOLOGY CORP.

Fig. 72-13

BIPOLAR POWER SUPPLY FOR BATTERY INSTRUMENTS

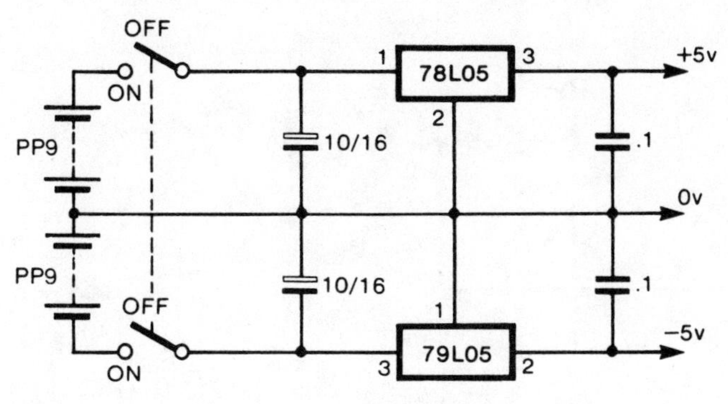

Fig. 1

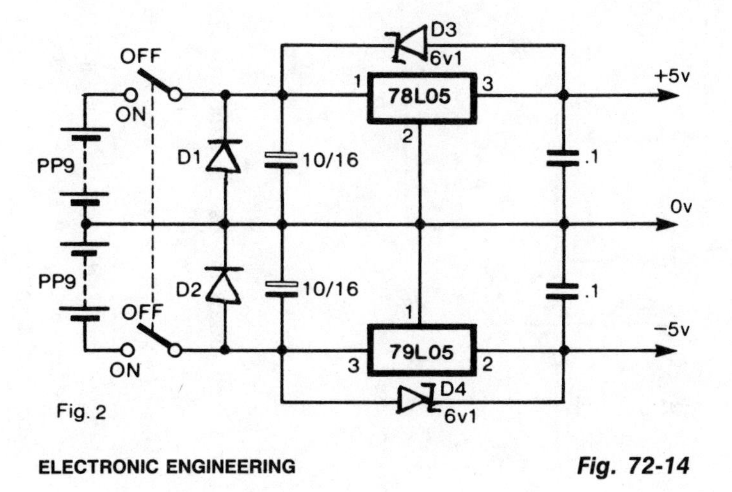

Fig. 2

ELECTRONIC ENGINEERING

Fig. 72-14

Circuit Notes

To generate regulated ±5-V supplies from a pair of dry batteries, the circuit of Fig. 1 is commonly used. In order to give protection from inadvertent reverse connection of a battery, a diode in series with each battery would produce an unacceptable voltage drop. The more effective approach is to fit diodes D1 and D2 as shown in Fig. 2, in parallel with each battery.

When the supply is switched off, there is the risk of a reverse bias being applied across the regulators, if there is significant inductance or capacitance in the load circuit. Diodes across the regulators prevent damage. When the power supply is switched on, the two switches do not act in unison. There is a probability that one or the other regulators will be latched hard off by the other. To prevent this, D3 and D4 are Zener diodes so that ±5-V rails are pulled up by the batteries until the regulators establish the correct levels.

POWER SUPPLY FOR 25-WATT ARC LAMP

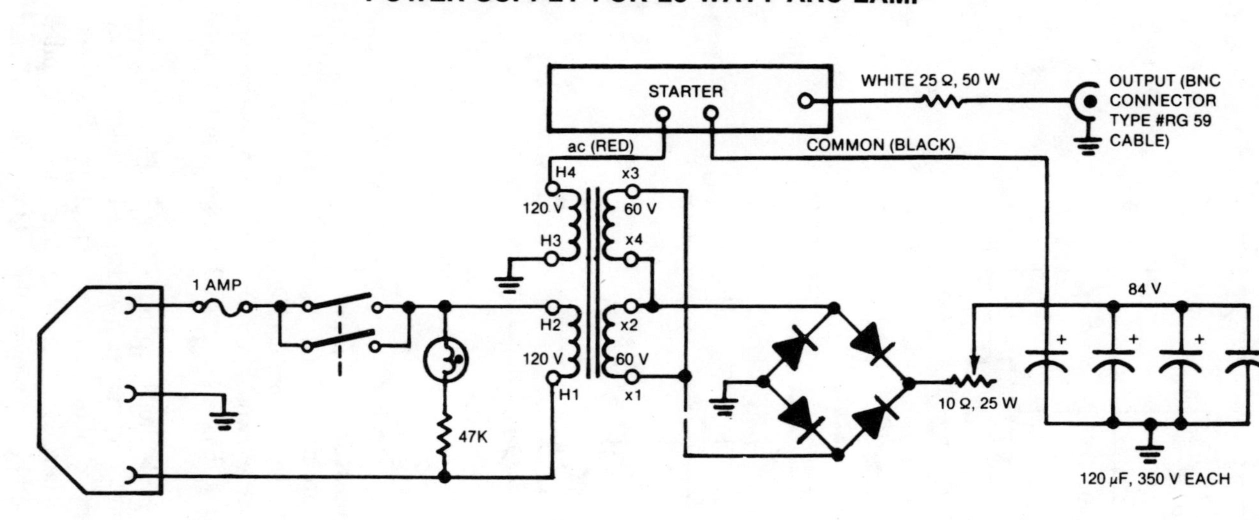

NASA

Fig. 72-15

Circuit Notes

A dual-voltage circuitry both strikes and maintains the arc. The lamps require a starting voltage in excess of 1,000 volts. Once stabilized, the voltage drop across the lamp is near 20 volts. Power supply consists of two main sections. The first section, the low-voltage power supply section, is an 84-volt direct-current supply. This supply powers the stabilized arc. Current is limited by the 10 ohm adjustable and 25 ohm fixed resistance. The second section, the high-voltage starter circuit, is a Cockroft-Walton voltage multiplier. With no load, the output voltage is 2,036 volts. However, when the arc is established, the heavy current drain maintains a forward bias on all of the diodes, and the circuit becomes a straight path with a voltage drop of 7.2 volts. The small value of the capacitors used in the multiplier guarantees that the diodes will be forward-biased once the arc is established.

STAND-BY POWER FOR NON-VOLATILE CMOS RAMs

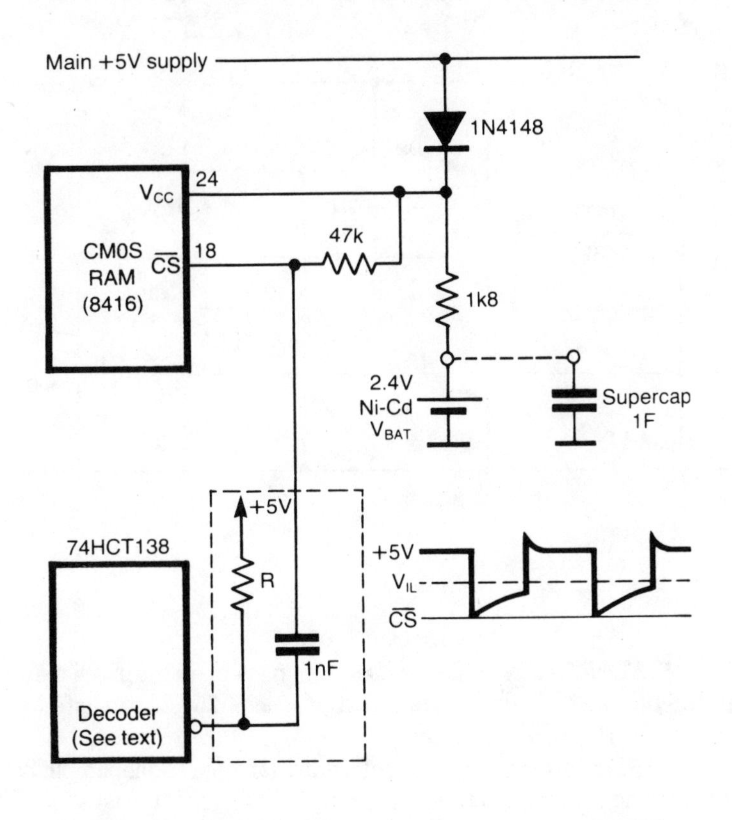

ELECTRONIC ENGINEERING

Fig. 72-16

Circuit Notes

To prevent loss of data when a CMOS RAM is switched from normal operation ($V_{CC} = 5$ volts) to stand-by mode ($V_{CC} = V_{BAT}$) it must be ensured that the CS pin goes near the V_{CC} rail at all times. Ac coupling to the chip select is made through capacitor C, breaking the dc current path between V_{CC} (and hence V_{BAT}) and the decoder output. So, whatever the impedance state of the decoder in power down, the battery will provide current only for the RAM, low enough to keep the voltage at CS near to V_{CC}.

HV REGULATOR WITH FOLDBACK CURRENT LIMITING

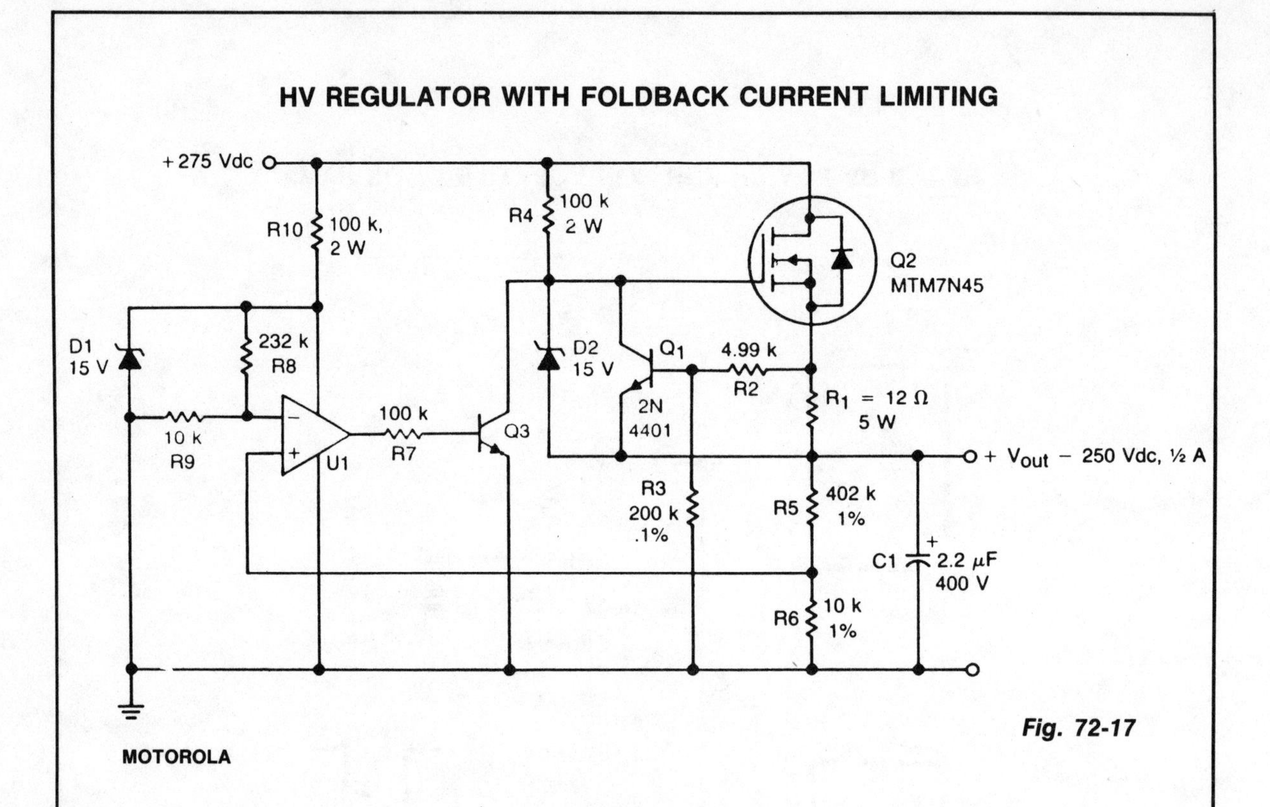

Fig. 72-17

Circuit Notes

A TMOS MTM7N45 (Q2) is used as a series pass element in a linear high voltage supply that accepts +275-V unregulated and produces 250 V regulated with foldback current limiting.

A 15-V zener, D1, provides the dc reference for operational amplifier U1, whose other input is obtained from a fraction of the output voltage. U1 drives Q3, which drives the gate of Q2. Foldback current limiting is achieved by R1, R2, R3, R4, Q1, and D2. The formula to establish the current "knee" for limiting is:

$$I_{KNEE} = \frac{V_{OUT}(R2/R2 + R3) + 0.5\text{ V}}{R1}$$

Short circuit current is:

$$I_{SC} = \frac{0.5\text{ V}}{R1}$$

90 V rms VOLTAGE REGULATOR USING A PUT

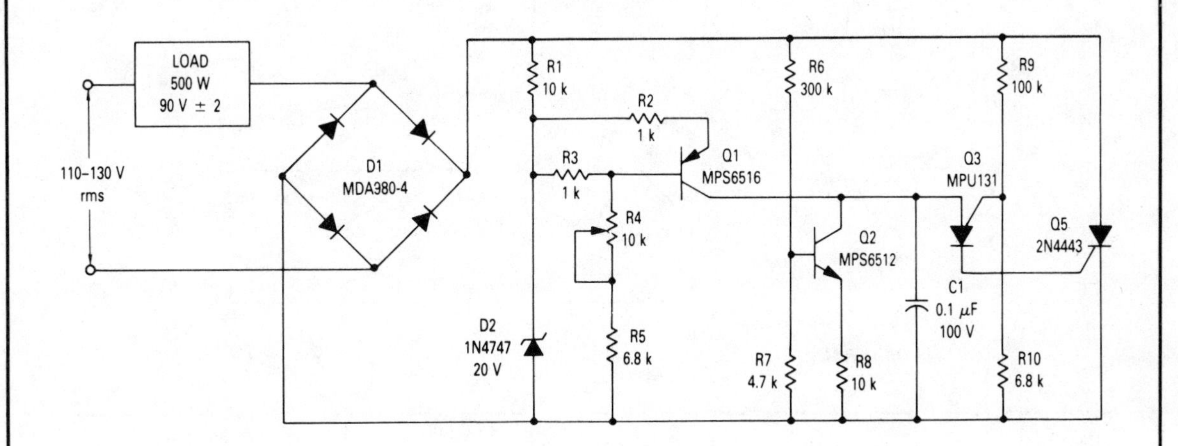

MOTOROLA

Fig. 72-18

Circuit Notes

The circuit is an open loop rms voltage regulator that will provide 500 watts of power at 90 V rms with good regulation for an input voltage range of 110-130 V rms. With the input voltage applied, capacitor C1 charges until the firing point of Q3 is reached causing it to fire. This turns Q5 on which allows current to flow through the load. As the input voltage increases, the voltage across R10 increases which increases the firing point of Q3. This delays the firing of Q3 because C1 now has to charge to a higher voltage before the peak-point voltage is reached. Thus the output voltage is held fairly constant by delaying the firing of Q5 as the input voltage increases. For a decrease in the input voltage, the reverse occurs.

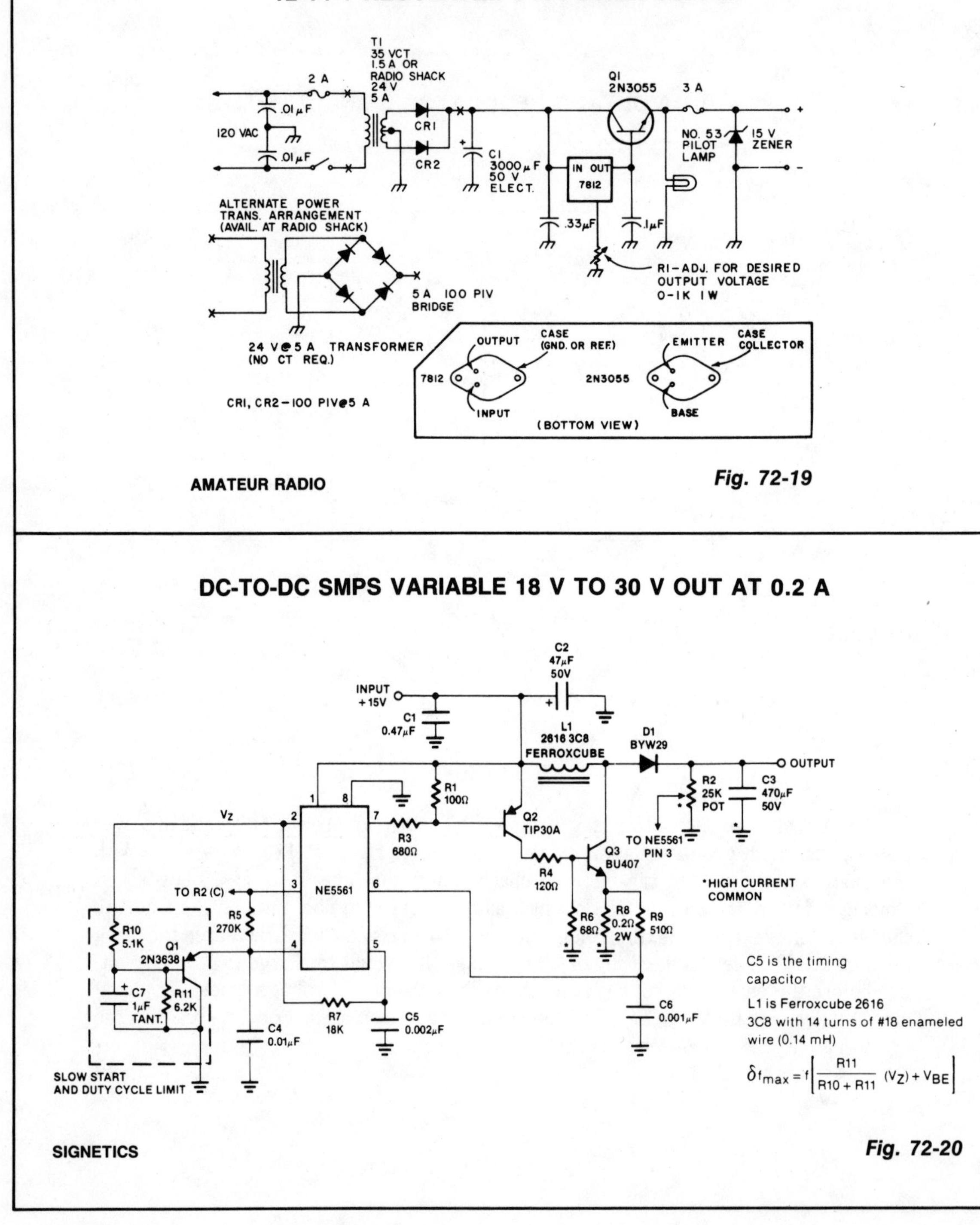
12-14 V REGULATED 3 A POWER SUPPLY
T1
35 VCT
1.5 A OR
RADIO SHACK
24 V
5 A
2 A
.01 μF
120 VAC
.01 μF
CR1
CR2
C1
3000 μF
50 V
ELECT.
Q1
2N3055
3 A
IN OUT
7812
NO. 53
PILOT
LAMP
15 V
ZENER
.33 μF
.1 μF
R1 – ADJ. FOR DESIRED
OUTPUT VOLTAGE
0 – 1K 1W
ALTERNATE POWER
TRANS. ARRANGEMENT
(AVAIL. AT RADIO SHACK)
5 A 100 PIV
BRIDGE
24 V @ 5 A TRANSFORMER
(NO CT REQ.)
CR1, CR2 – 100 PIV @ 5 A
OUTPUT
CASE
(GND. OR REF.)
7812
INPUT
EMITTER
CASE
COLLECTOR
2N3055
BASE
(BOTTOM VIEW)
AMATEUR RADIO
Fig. 72-19
DC-TO-DC SMPS VARIABLE 18 V TO 30 V OUT AT 0.2 A
INPUT
+15V
C1
0.47μF
C2
47μF
50V
L1
2616 3C8
FERROXCUBE
D1
BYW29
OUTPUT
R2
25K
POT
C3
470μF
50V
R1
100Ω
Q2
TIP30A
R3
680Ω
VZ
NE5561
TO NE5561
PIN 3
Q3
BU407
R4
120Ω
*HIGH CURRENT
COMMON
TO R2 (C)
R5
270K
R6
68Ω
R8
0.2Ω
2W
R9
510Ω
R10
5.1K
Q1
2N3638
C7
1μF
TANT.
R11
6.2K
C4
0.01μF
R7
18K
C5
0.002μF
C6
0.001μF
SLOW START
AND DUTY CYCLE LIMIT
C5 is the timing
capacitor
L1 is Ferroxcube 2616
3C8 with 14 turns of #18 enameled
wire (0.14 mH)
$\delta f_{max} = f\left[\frac{R11}{R10 + R11}(V_Z) + V_{BE}\right]$
SIGNETICS
Fig. 72-20

OFF-LINE FLYBACK REGULATOR

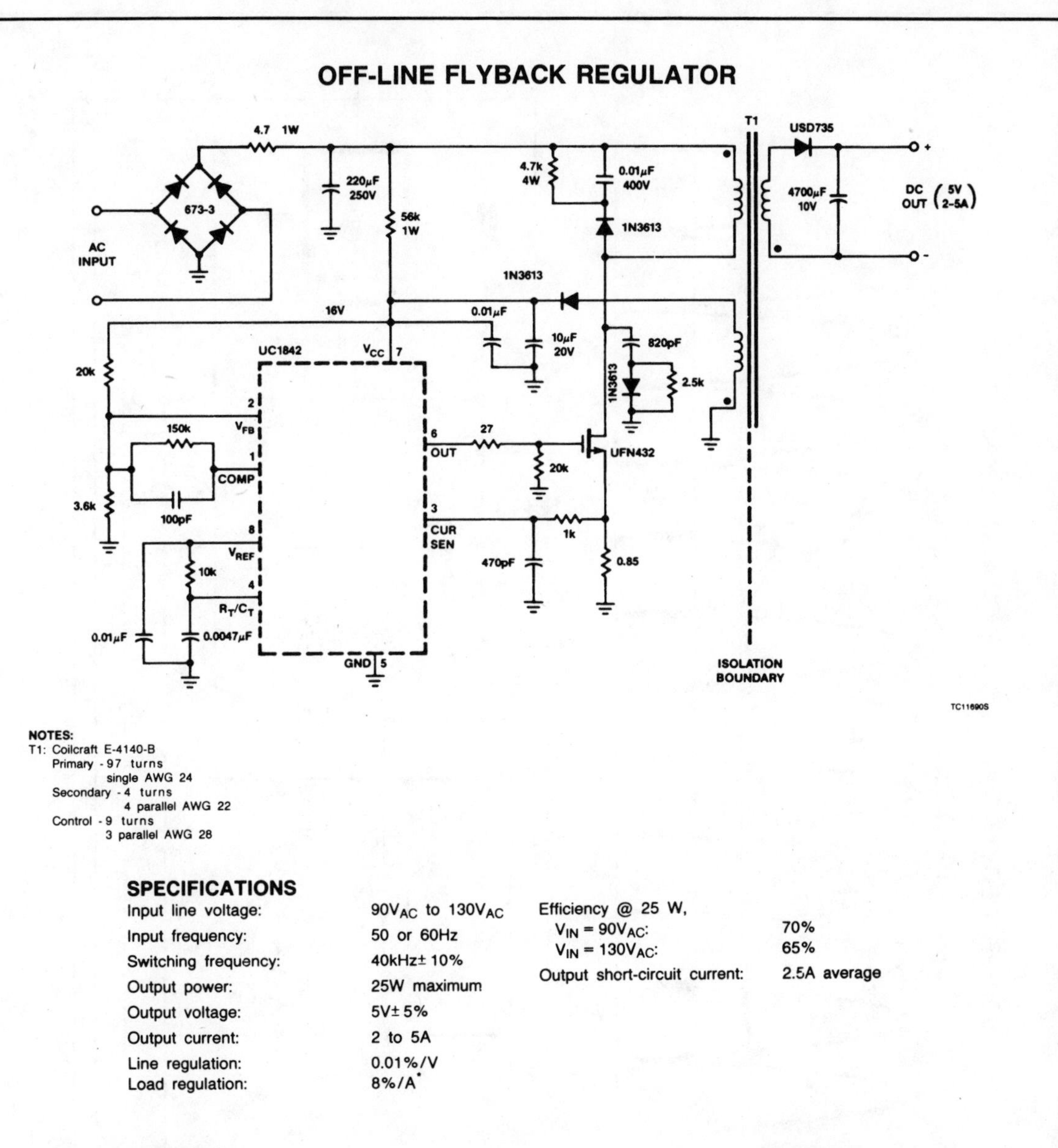

NOTES:
T1: Coilcraft E-4140-B
Primary - 97 turns
single AWG 24
Secondary - 4 turns
4 parallel AWG 22
Control - 9 turns
3 parallel AWG 28

SPECIFICATIONS

Input line voltage:	$90V_{AC}$ to $130V_{AC}$
Input frequency:	50 or 60Hz
Switching frequency:	40kHz± 10%
Output power:	25W maximum
Output voltage:	5V± 5%
Output current:	2 to 5A
Line regulation:	0.01%/V
Load regulation:	8%/A*

Efficiency @ 25 W,	
$V_{IN} = 90V_{AC}$:	70%
$V_{IN} = 130V_{AC}$:	65%
Output short-circuit current:	2.5A average

SIGNETICS

Fig. 72-21

Circuit Notes

This circuit uses a low-cost feedback scheme in which the dc voltage developed from the primary-side control winding is sensed by the UC1842 error amplifier. Load regulation is therefore dependent on the coupling between secondary and control windings, and on transformer leakage inductance. For applications requiring better load regulation, a UC1901 Isolated Feedback Generator can be used to directly sense the output voltage.

SCR PREREGULATOR FITS ANY POWER SUPPLY

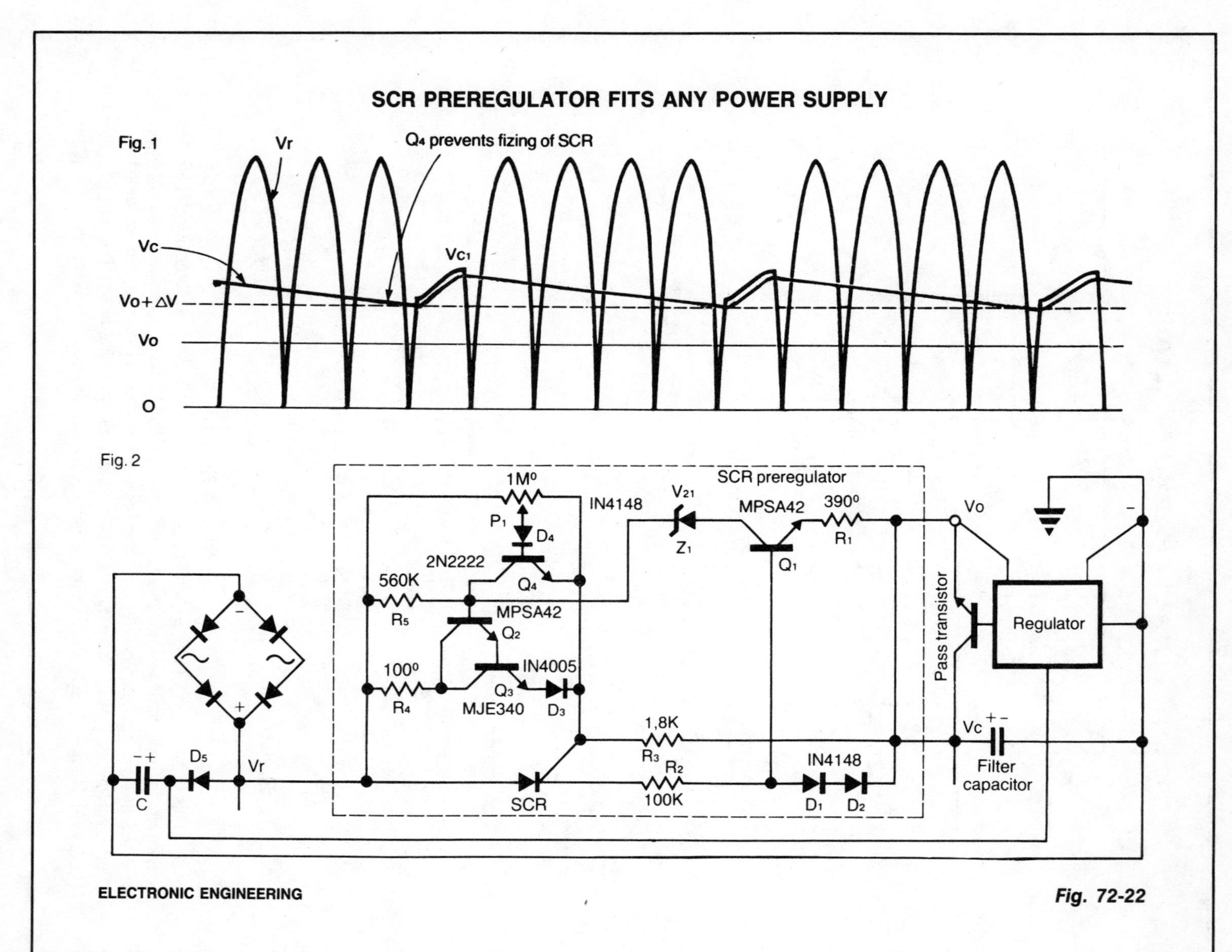

ELECTRONIC ENGINEERING

Fig. 72-22

Circuit Notes

This SCR pre-regulator keeps the filter capacitor V_c, in a variable output power supply, a few volts above the output voltage V_o. The benefits include: less heat dissipated by the pass transistor and therefore small heatsink, cooler operation and higher efficiency, especially at low output voltages.

Q1, R1, R2, D1 and D2 form a constant current source for zener Z1, so that the contribution to the output current is always a few mA (2 - 3 mA).

The Darlington pair Q2, Q3 keeps the SCR off. The voltage V_c decreases until $V_c = V_o = V$ at which point the Darlington pair fires the SCR, charging the filter capacitor to a higher voltage V_c^1 in less than half the period of the input voltage. The component values, shown are for a 0 - 250-V, 3-A power supply.

VOLTAGE REGULATOR

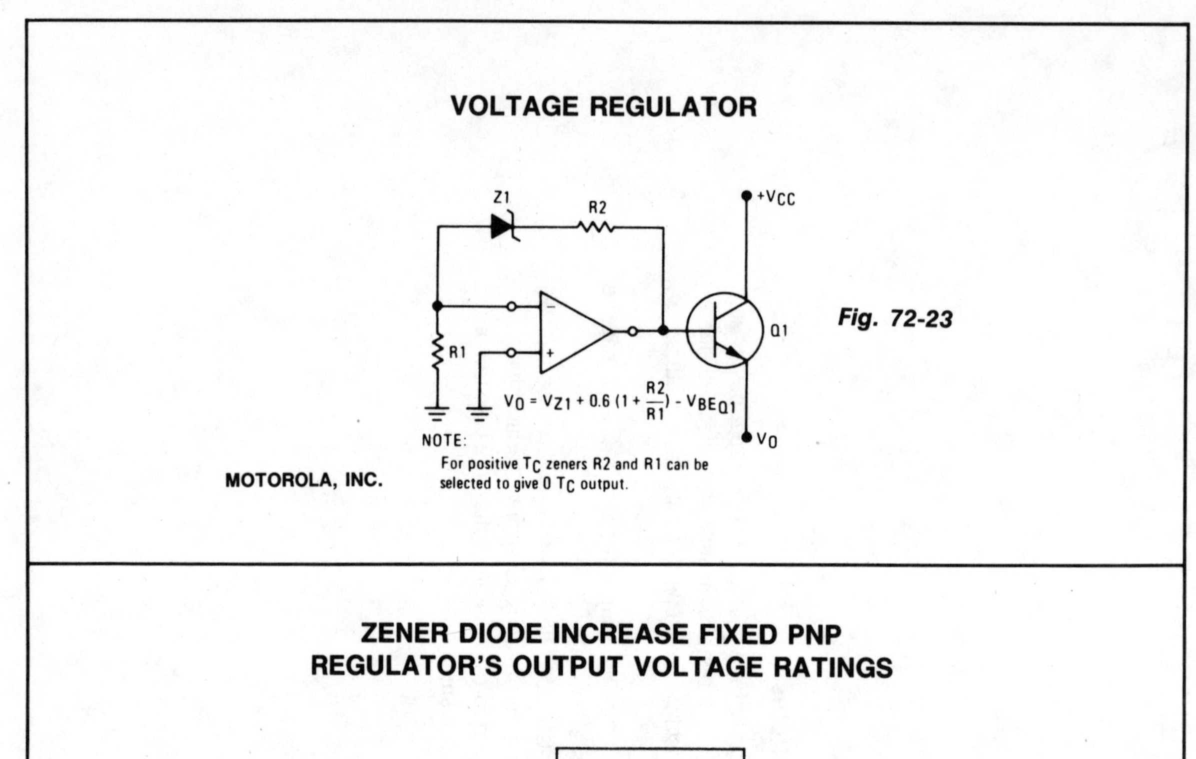

MOTOROLA, INC.

Fig. 72-23

ZENER DIODE INCREASE FIXED PNP REGULATOR'S OUTPUT VOLTAGE RATINGS

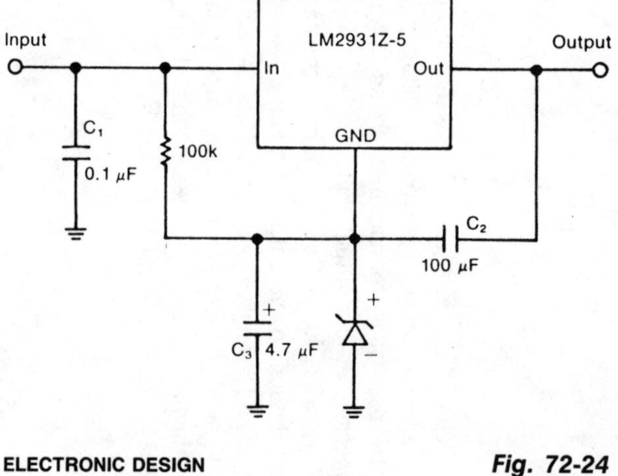

ELECTRONIC DESIGN

Fig. 72-24

Circuit Notes

A zener diode in the ground lead of a fixed pnp regulator varies the voltage output of that device without a significant sacrifice in regulation. The technique also allows the regulator to operate with output voltages beyond its rated limit.

INCREASING THE POWER RATING OF ZENER DIODES

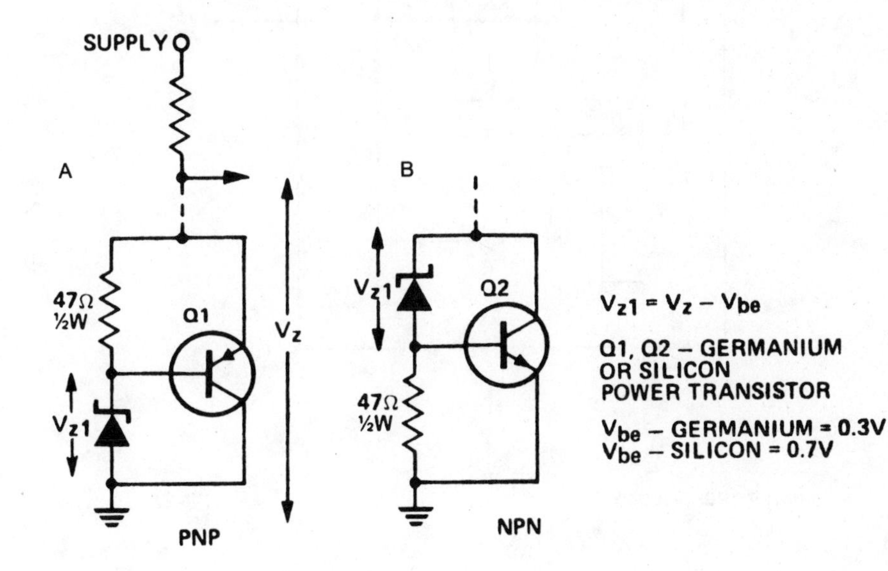

ELECTRONICS TODAY INTERNATIONAL

Fig. 72-25

Circuit Notes

A power transistor can be used to provide a high powered zener voltage from a low wattage zener. A 400 mW zener can be used where a 10 watt zener is required or a 1 W zener can be used where a 50 to 80 watt zener is required by using appropriate transistors for Q1 and Q2 in the circuits shown. Where low rating is required, Q1 would be an ASZ 15 (germanium) or an AY9140 (silicon). Q2 could be a 2N2955 (silicon). For higher powers, Q1 should be an ASZ18 (germanium) or a 2N2955 (silicon) and Q2 a 2N3055 (silicon) or an AY8149 (silicon). A heatsink on the transistor is required. The circuit in *A* has the advantage that power transistors can be bolted directly on to a chassis which may serve as a heatsink.

MEMORY SAVE ON POWER-DOWN

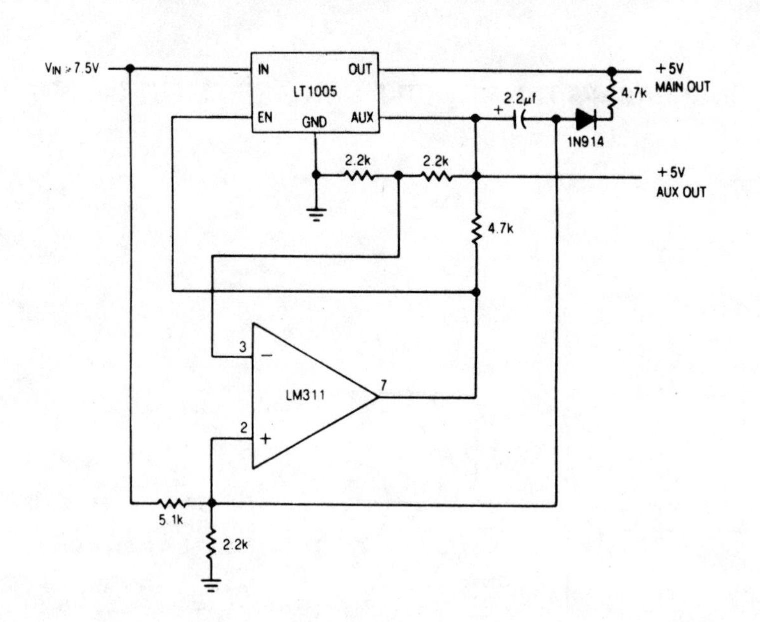

LINEAR TECHNOLOGY

Fig. 72-26

Circuit Notes

The auxiliary output powers the memory, while the main output powers the system and is connected to the memory store pin. When power goes down, the main output goes low, commanding the memory to store. The auxiliary output then drops out.

73
Power Supply Circuits (High Voltage)

The sources of the following circuits are contained in the Sources section beginning on page 694. The figure number contained in the box of each circuit correlates to the source entry in the Sources section.

LOW COST ULTRA HIGH VOLTAGE GENERATOR

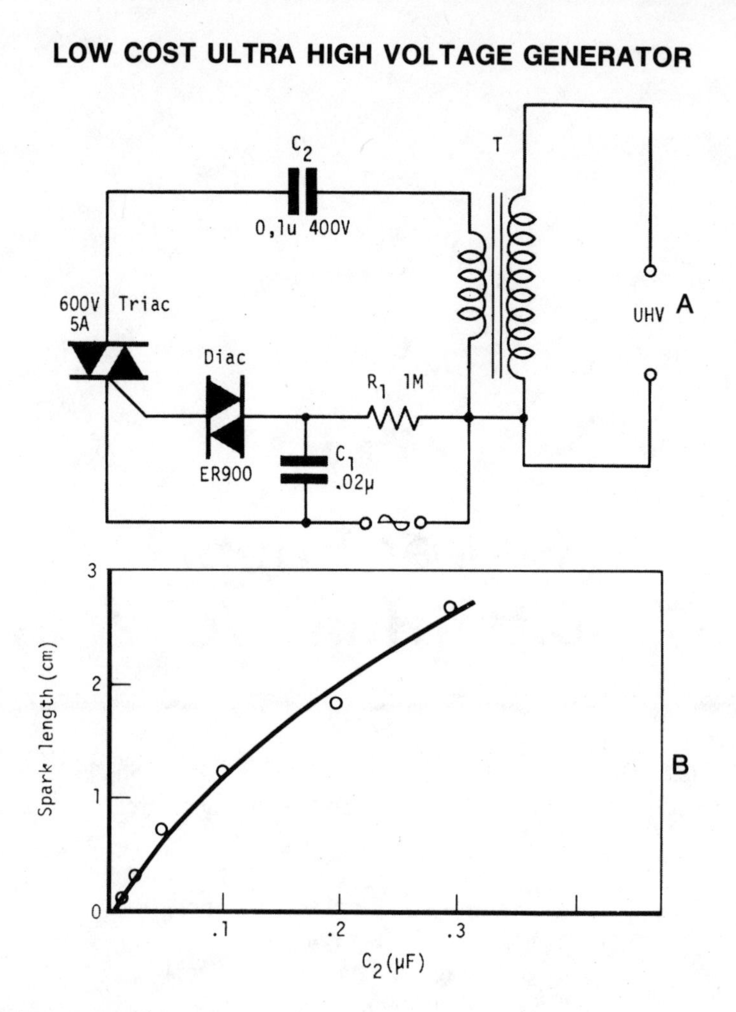

ELECTRONIC ENGINEERING

Fig. 73-1

Circuit Notes

By repetitively charging and discharging a capacitor through the primary of an induction coil with a high voltage, an ultra high emf is induced in the secondary. Switching is performed by the triac, triggered by the disc at times set by C1 and R1. With a 12 V car ignition coil for example, the length of sparkgap obtained is 12 mm of air for C2 = 0.1 μF. If the dielectric strength of air is assumed to be 3 kV/mm, this spark-gap length corresponds to 36 kV. From the curve shown in Fig. B, care must be taken in keeping the value of C2 below 1 μF as the coil is liable to be seriously damaged at this value of C2. Power consumption is only about one watt.

SIMPLE HIGH-VOLTAGE SUPPLY

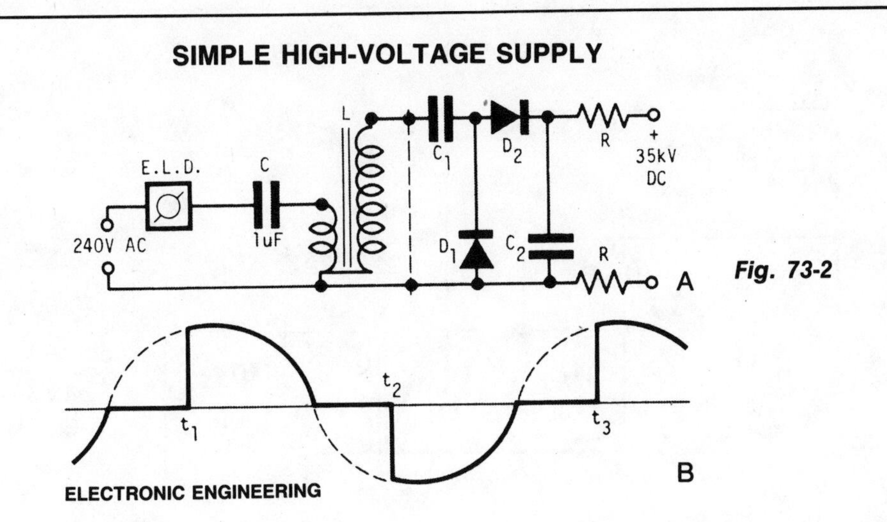

Fig. 73-2

Circuit Notes

A light dimmer, a 1 μF capacitor and a 12 V car ignition coil form the simple line powered HV generator. The current in the dimmer is shown in Fig. B. At times t_1, t_2, . . . , set by the dimmer switch, the inner triac of the dimmer switches on, and a very high and very fast current pulse charges the capacitor through the primary of the induction coil. Then at a rate of 120 times per second for a 60 Hz line, a very high voltage pulse appears at the secondary of the coil. To obtain an HV dc output, use a voltage doubler. D1 and D2 are selenium rectifiers (TV 18 Siemens or ITT) used for the supply of television sets. High value output shock protection resistors, R, are recommended when suitable.

HIGH VOLTAGE GEIGER COUNTER SUPPLY

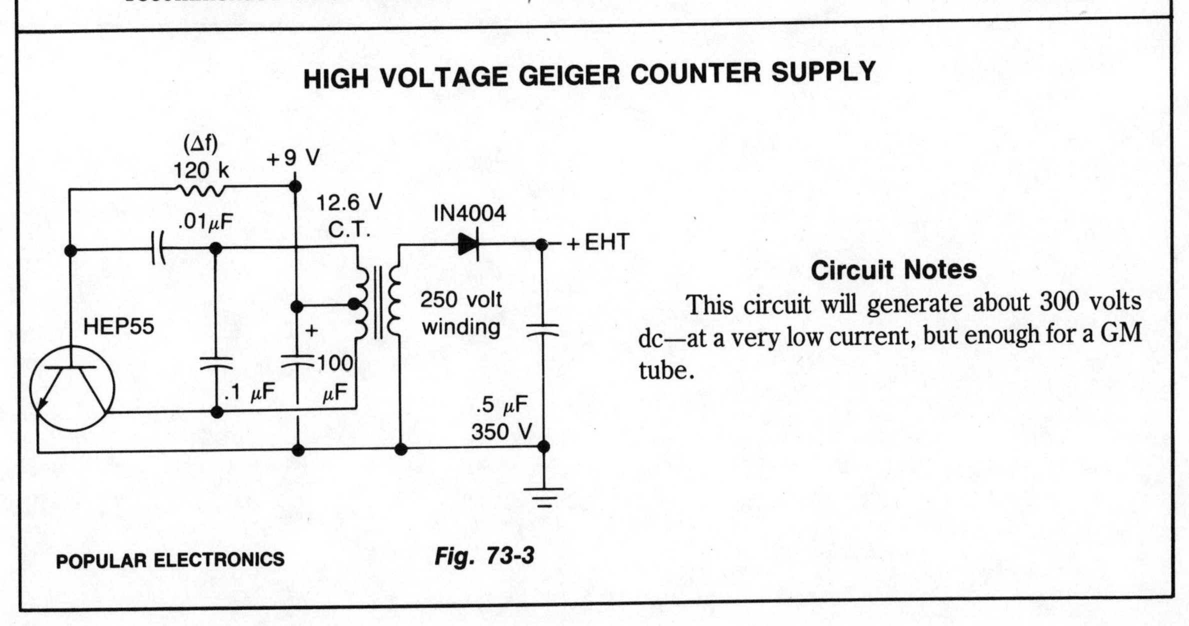

Fig. 73-3

Circuit Notes

This circuit will generate about 300 volts dc—at a very low current, but enough for a GM tube.

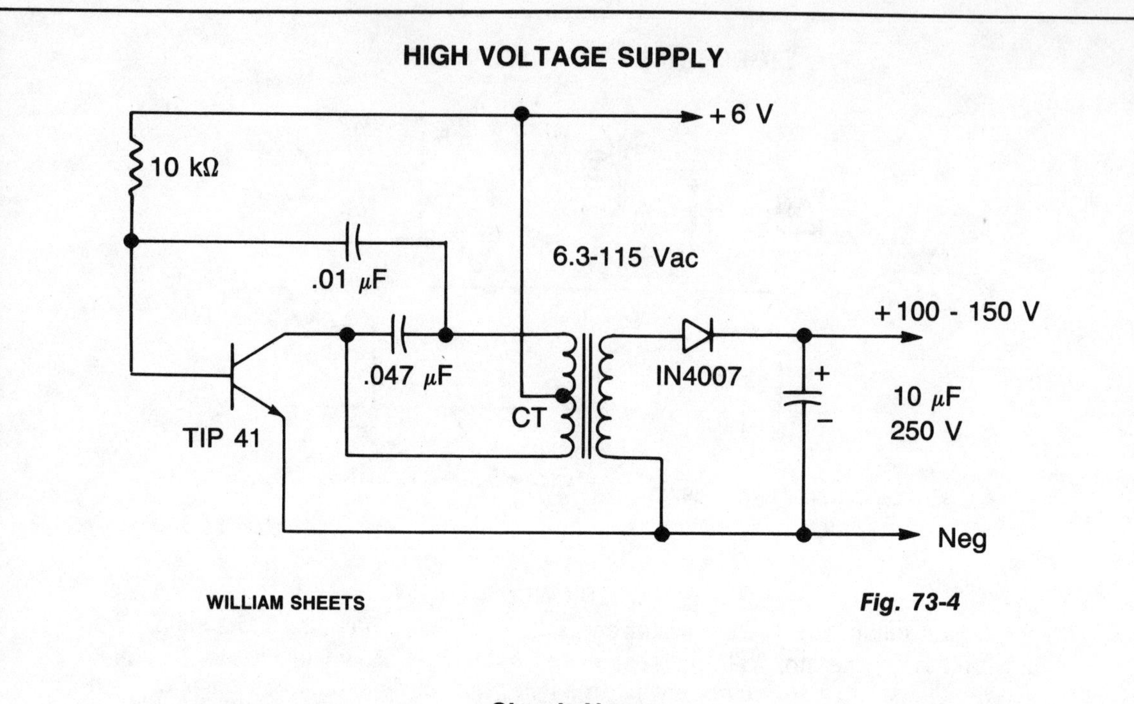

Fig. 73-4

Circuit Notes

A 6 V battery can provide 100-150 Vdc center-tapped at a high internal impedance (not dangerous though it can inflict an unpleasant jolt). A 6.3 V transformer is connected "in reverse" with a transistor used in a Hartley oscillator configuration. The frequency of operation may be controlled by varying the value of the 10 K ohm resistor. The 10 μF capacitor must have a working voltage of at least 250 Vdc.

74

Power Supply Monitors

The sources of the following circuits are contained in the Sources section beginning on page 694. The figure number contained in the box of each circuit correlates to the source entry in the Sources section.

POWER SUPPLY MONITOR

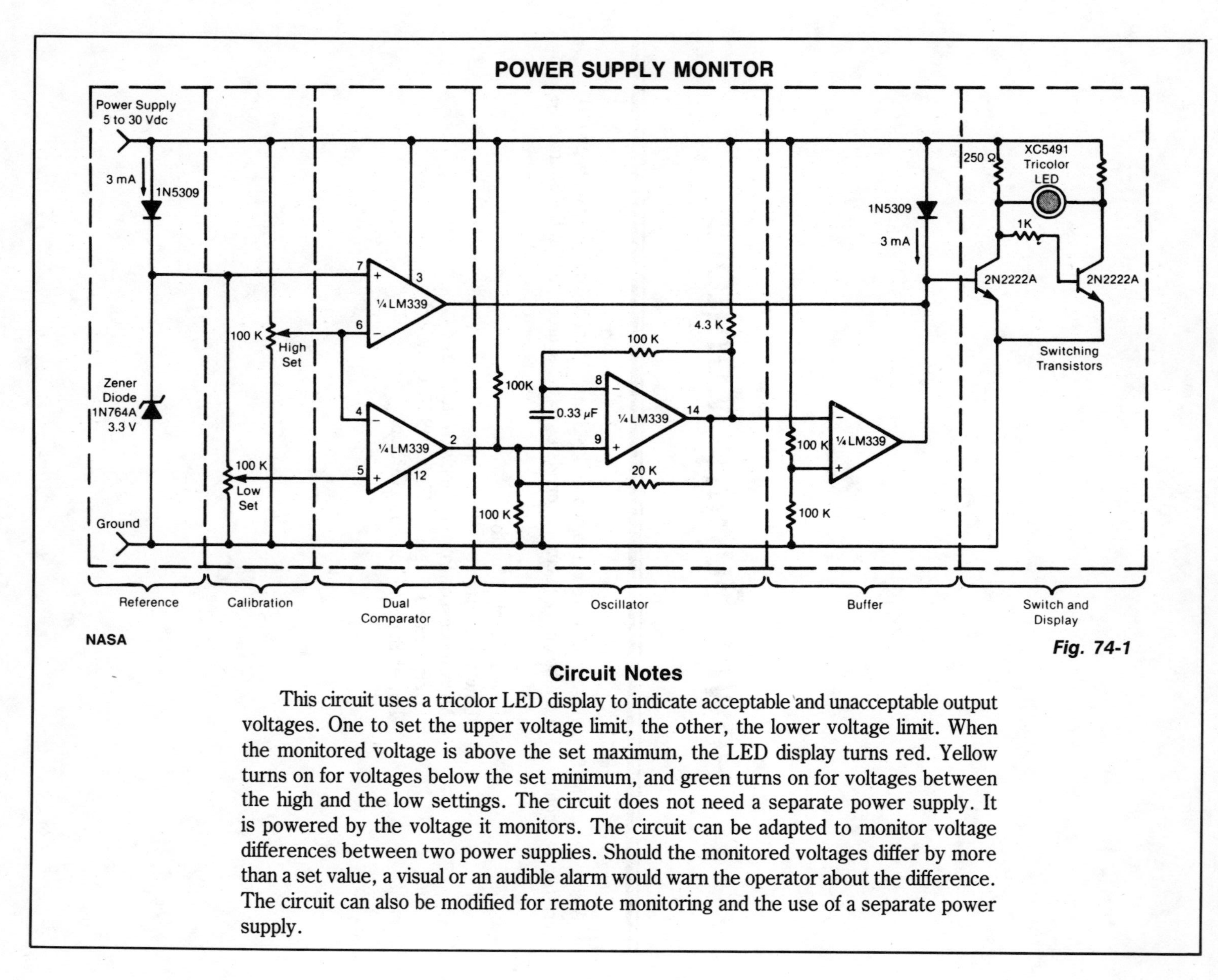

Fig. 74-1

Circuit Notes

This circuit uses a tricolor LED display to indicate acceptable and unacceptable output voltages. One to set the upper voltage limit, the other, the lower voltage limit. When the monitored voltage is above the set maximum, the LED display turns red. Yellow turns on for voltages below the set minimum, and green turns on for voltages between the high and the low settings. The circuit does not need a separate power supply. It is powered by the voltage it monitors. The circuit can be adapted to monitor voltage differences between two power supplies. Should the monitored voltages differ by more than a set value, a visual or an audible alarm would warn the operator about the difference. The circuit can also be modified for remote monitoring and the use of a separate power supply.

LOW-VOLTS ALARM

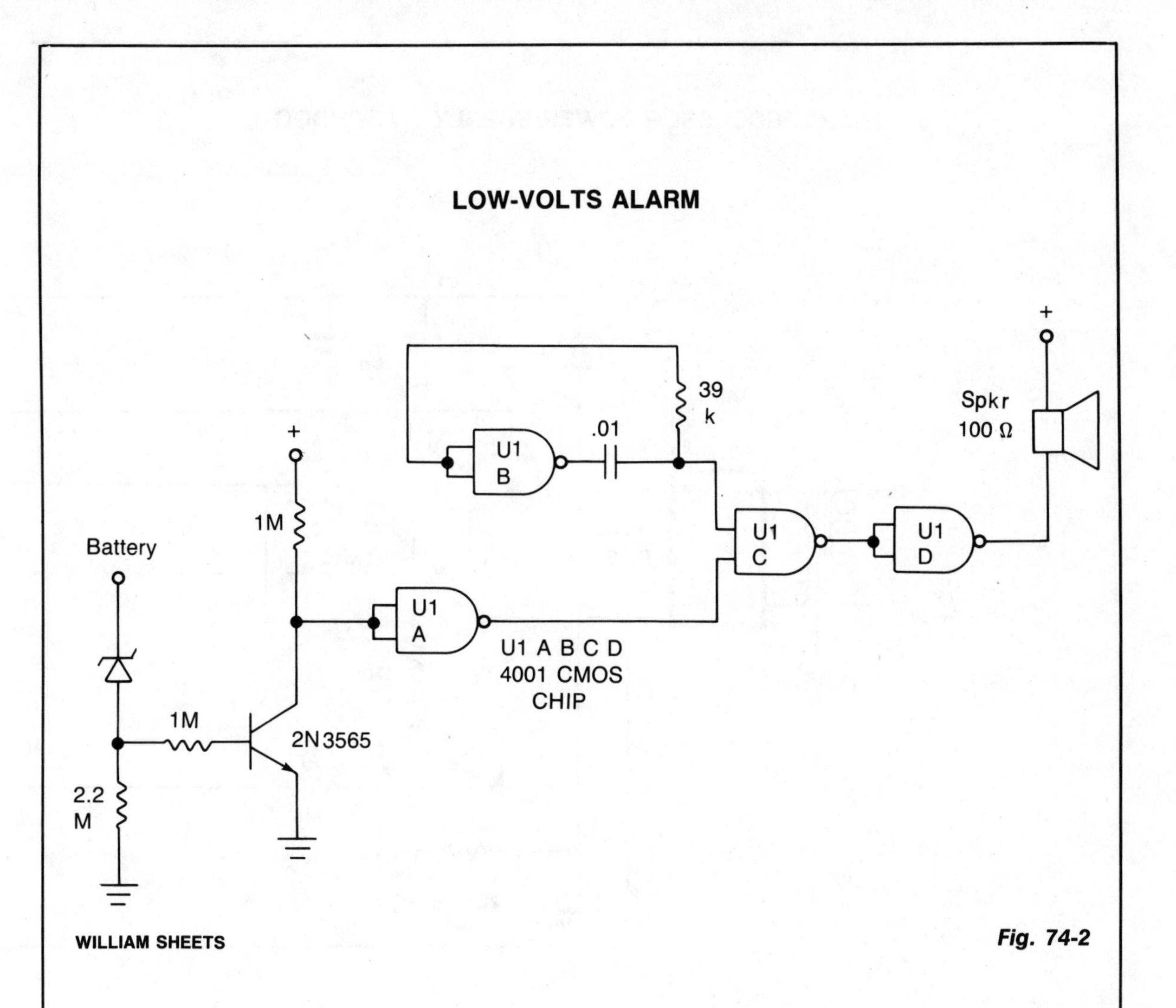

WILLIAM SHEETS

Fig. 74-2

Circuit Notes

This inexpensive dc supply-voltage monitor sounds a warning when the voltage falls below a preset value. It is ideal for monitoring rechargeable batteries since it draws only a few microamperes when not sounding. The voltage at which the alarm sounds is determined by the zener diode. When the voltage falls below the zener voltage, the alarm sounds. The alarm tone is determined by the RC time constant of the 39 k resistor and 0.01 mf capacitor.

MICROPROCESSOR POWER SUPPLY WATCHDOG

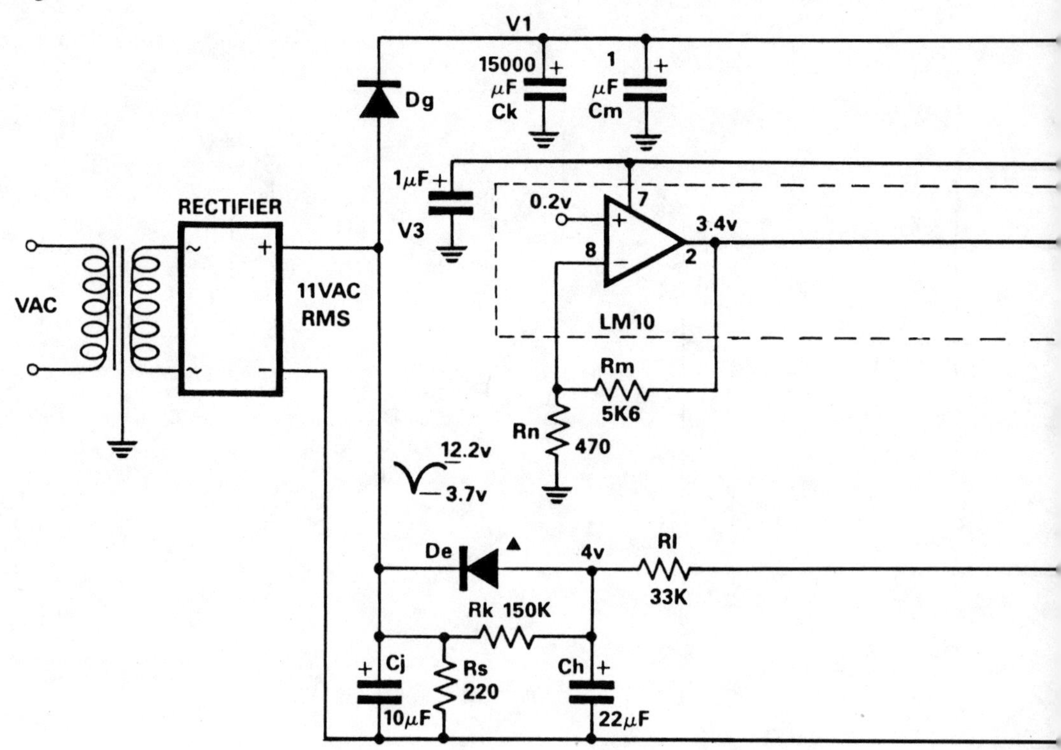

ELECTRONIC ENGINEERING

Fig. 74-3

Circuit Notes

The circuit monitors the input to the microprocessor 5 V regulated supply for voltage drops and initiates a reset sequence before supply regulation is lost. In operation, the resistor capacitor combination Rs and Cj form a short time constant smoothing network for the output of the fullwave bridge rectifier. An approximately triangular, voltage waveform appears across C and Rs and it is the minimum excursion of this that initiates

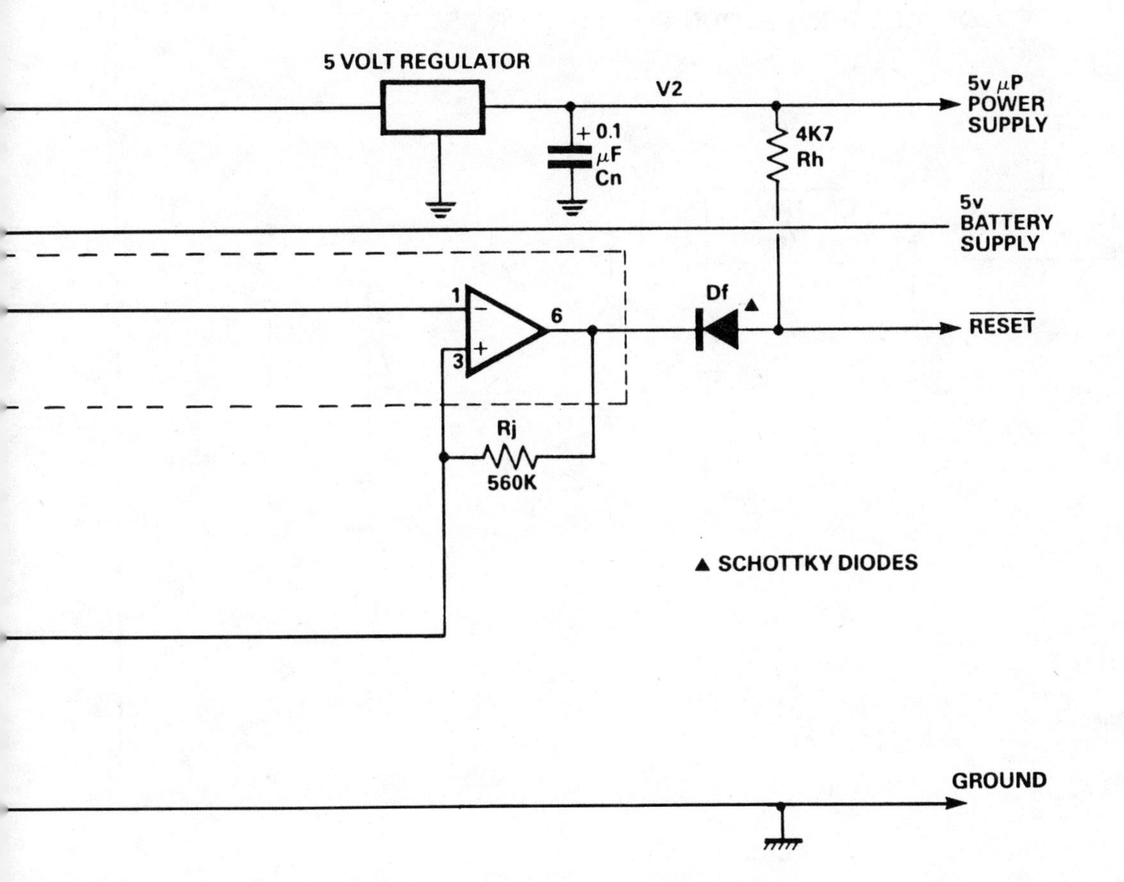

the reset. Diode Dg prevents charge sharing between capacitors Cj and Ck. Resistors Rn and Rm form a feedback network around the voltage reference section of the LM10C, setting a threshold voltage of 3.4 volts. The threshold voltage is set at 90% of the minimum voltage of the triangular waveform. When the triangular wave trough, at the comparator's non-inverting input, dips below the threshold, the comparator output is driven low. This presents a reset to the microprocessor. Capacitor Ch is charged slowly through resistor Rk and discharged rapidly through diode De.

OVERVOLTAGE PROTECTION CIRCUIT (SCR CROWBAR)

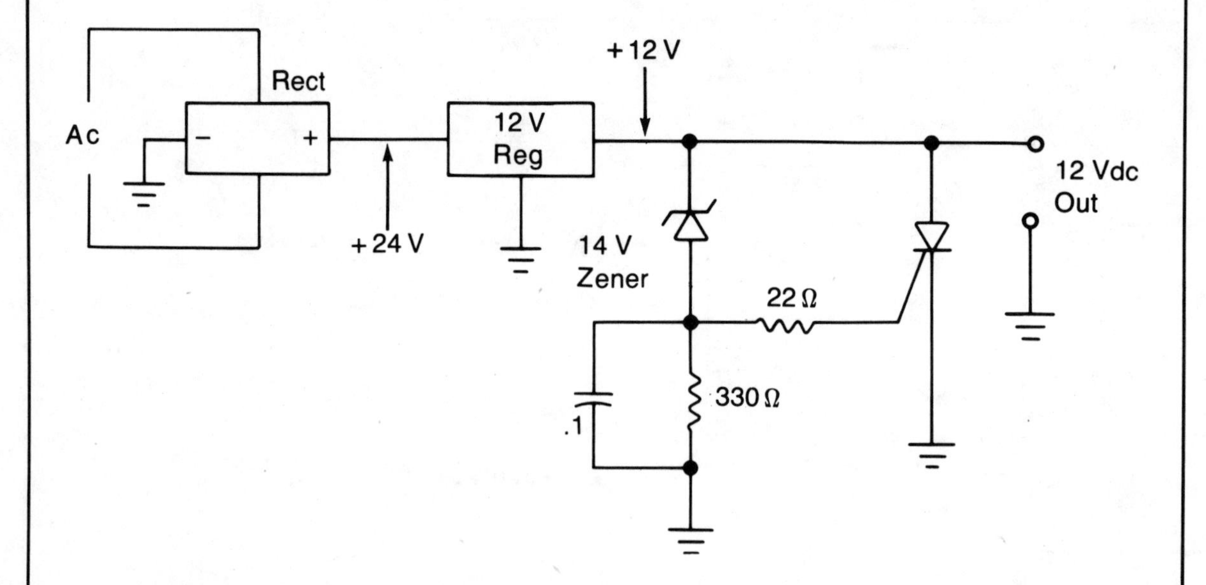

WILLIAM SHEETS

Fig. 74-4

Circuit Notes

The silicon controlled rectifier (SCR) is rated to handle at least the current of the power supply. It is connected in parallel across the 12 V dc output lines, but remains inert until a voltage appears at the gate terminal. This triggering voltage is supplied by the zener diode. At potentials less than 14 V the zener will not conduct current. But, at potentials greater than 14 Vdc the zener conducts and creates a voltage drop across the 330 ohm resistor that will fire the SCR. When the SCR turns on, the output lines of the power supply are shorted to ground. This will blow the primary fuse or burn out the transformer if there is no primary fuse.

POWER SUPPLY PROTECTION CIRCUIT

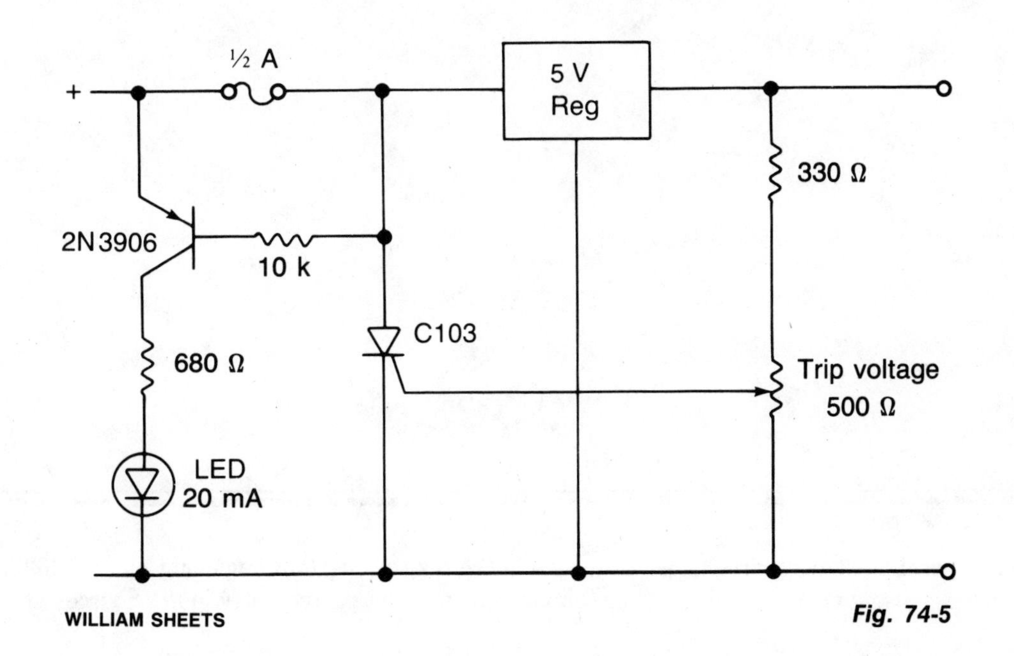

WILLIAM SHEETS

Fig. 74-5

Circuit Notes

When using a regulated supply to reduce a supply voltage there is always the danger of component failure in the supply and consequent damage to the equipment. A fuse will protect when excess current is drawn, but might be too slow to cope with overvoltage conditions. The values shown are for a 12 V supply being dropped to 5 V. The trip voltage is set to 5.7 V to protect the equipment in the event of a regulator fault. The 330 ohm resistor and the 500 ohm potentiometer form a potential divider which samples the output voltage as set by adjustment of the potentiometer. The SCR is selected to carry at least twice the fuse rating. The full supply voltage is connected to the input of the regulator. The 2N2906 is held bias off by the 10 k resistor and the SCR so that the LED is held off. If the output voltage rises above a set trip value then the SCR will conduct, the fuse will blow, and the 2N3906 will be supplied with base current via the 10 k resistor, and the LED will light up.

75

Probes

The sources of the following circuits are contained in the Sources section beginning on page 694. The figure number contained in the box of each circuit correlates to the source entry in the Sources section.

MICROVOLT PROBE

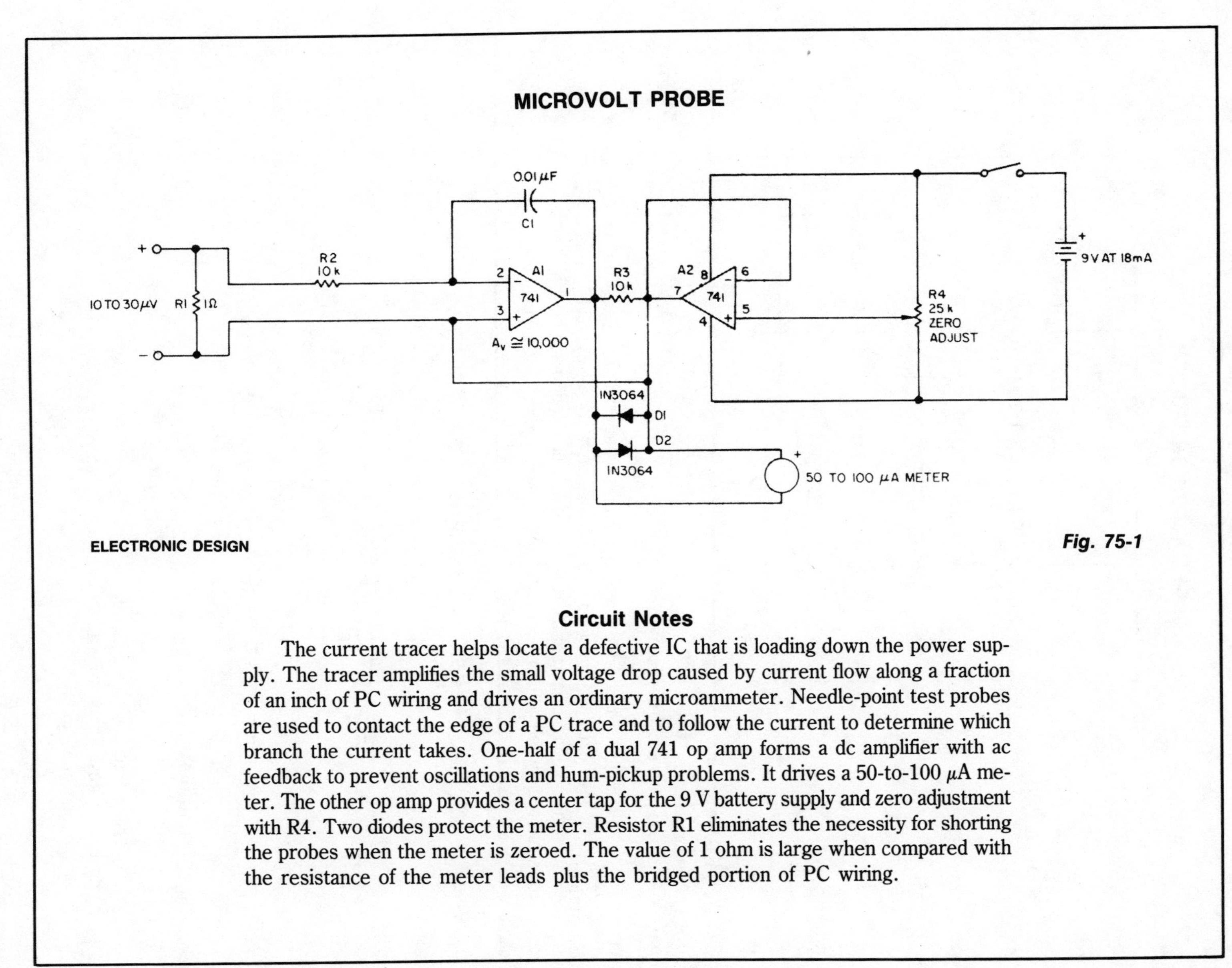

ELECTRONIC DESIGN

Fig. 75-1

Circuit Notes

The current tracer helps locate a defective IC that is loading down the power supply. The tracer amplifies the small voltage drop caused by current flow along a fraction of an inch of PC wiring and drives an ordinary microammeter. Needle-point test probes are used to contact the edge of a PC trace and to follow the current to determine which branch the current takes. One-half of a dual 741 op amp forms a dc amplifier with ac feedback to prevent oscillations and hum-pickup problems. It drives a 50-to-100 μA meter. The other op amp provides a center tap for the 9 V battery supply and zero adjustment with R4. Two diodes protect the meter. Resistor R1 eliminates the necessity for shorting the probes when the meter is zeroed. The value of 1 ohm is large when compared with the resistance of the meter leads plus the bridged portion of PC wiring.

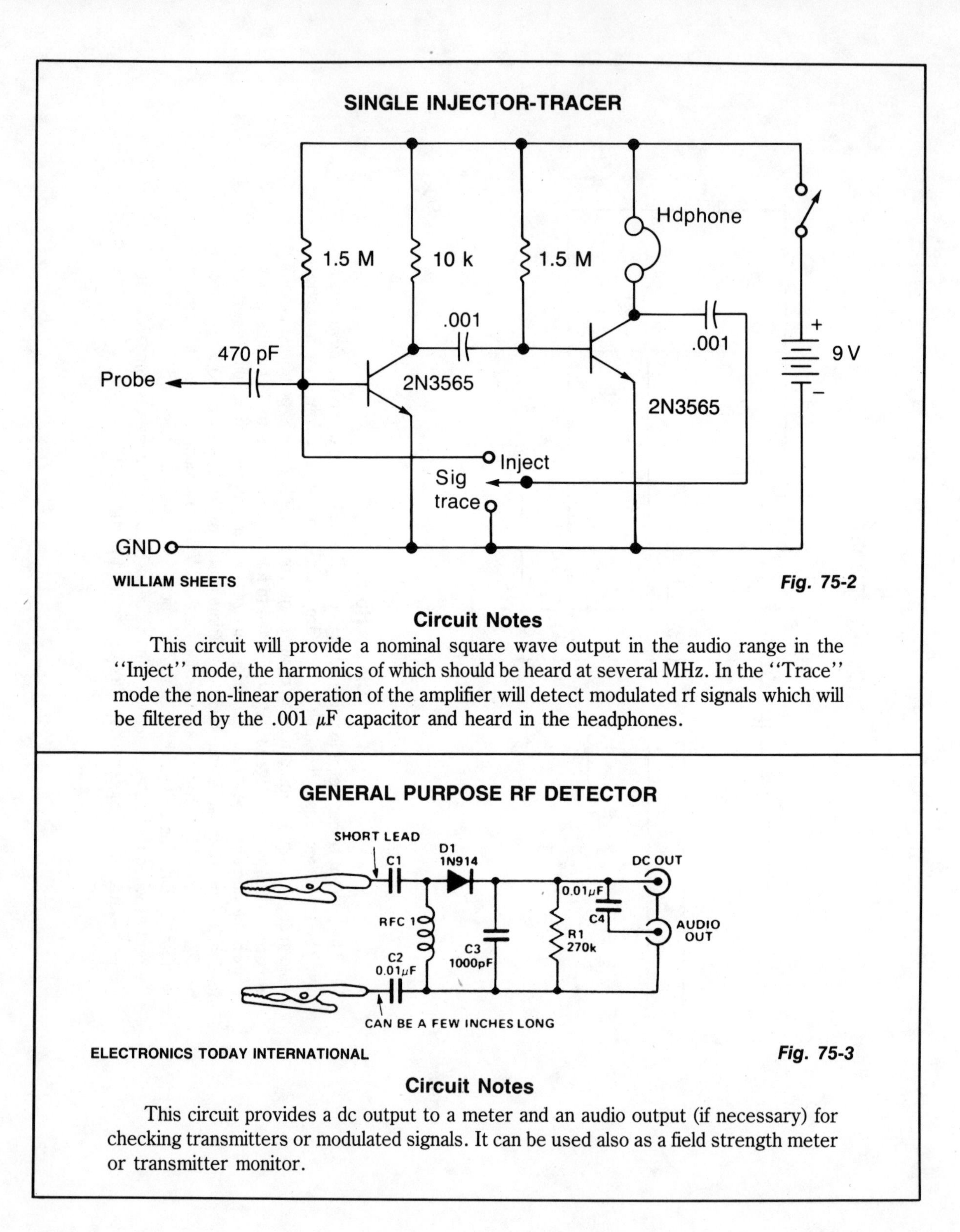

SINGLE INJECTOR-TRACER

WILLIAM SHEETS

Fig. 75-2

Circuit Notes

This circuit will provide a nominal square wave output in the audio range in the "Inject" mode, the harmonics of which should be heard at several MHz. In the "Trace" mode the non-linear operation of the amplifier will detect modulated rf signals which will be filtered by the .001 μF capacitor and heard in the headphones.

GENERAL PURPOSE RF DETECTOR

ELECTRONICS TODAY INTERNATIONAL

Fig. 75-3

Circuit Notes

This circuit provides a dc output to a meter and an audio output (if necessary) for checking transmitters or modulated signals. It can be used also as a field strength meter or transmitter monitor.

CLAMP-ON-CURRENT PROBE COMPENSATOR

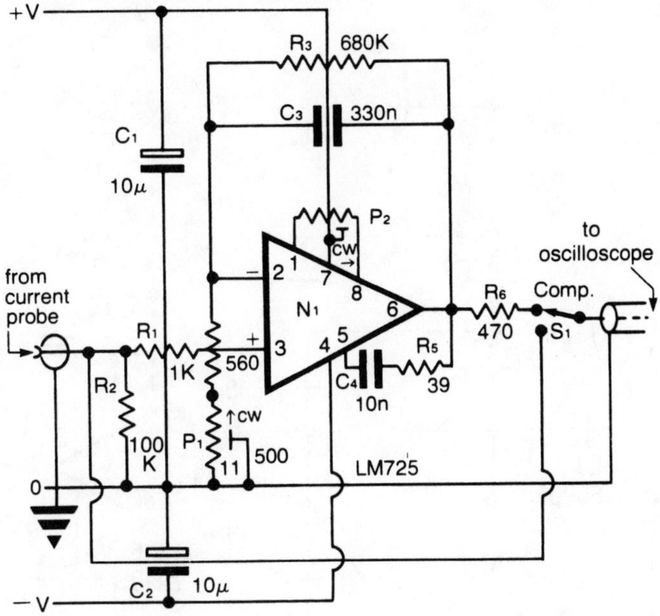

Table

Tek P6021, on its own	with Tek amp. 134	with compensator
120Hz to 60MHz @ 10mA/mV 450Hz to 60MHz @ 2mA/mV	12Hz to 38MHz (switched 1mA to 1A/div for 50mV/div output	1Hz to 100kHz @ 2mA/mV

ELECTRONIC ENGINEERING

Fig. 75-4

Circuit Notes

A clamp-on ''current probe'' such as the Tektronix P6021 is a useful means of displaying current waveforms on an oscilloscope. Unfortunately, the low-frequency response is somewhat limited, as shown in the Table.

The more sensitive range on the P6021 is 2 mA/mV, but it has a roll-off of 6 dB per octave below 450 Hz. The compensator counteracts the low-frequency attenuation, and this is achieved by means of C3 and R4 + P1 in the feedback around op amp N1. The latter is a low-noise type, such as the LM725 shown, and even so it is necessary at some point to limit the increasing gain with decreasing frequency; otherwise amplifier noise and drift will overcome the signal. The values shown for C3R3 give a lower limit below 1 Hz. A test square wave of ± 10 mA is fed to the current probe so that P1 can be adjusted for minimum droop or overshoot in the output waveform. At high frequencies, the response begins to fall off at 100 kHz.

650 MHz AMPLIFYING PRESCALER PROBE

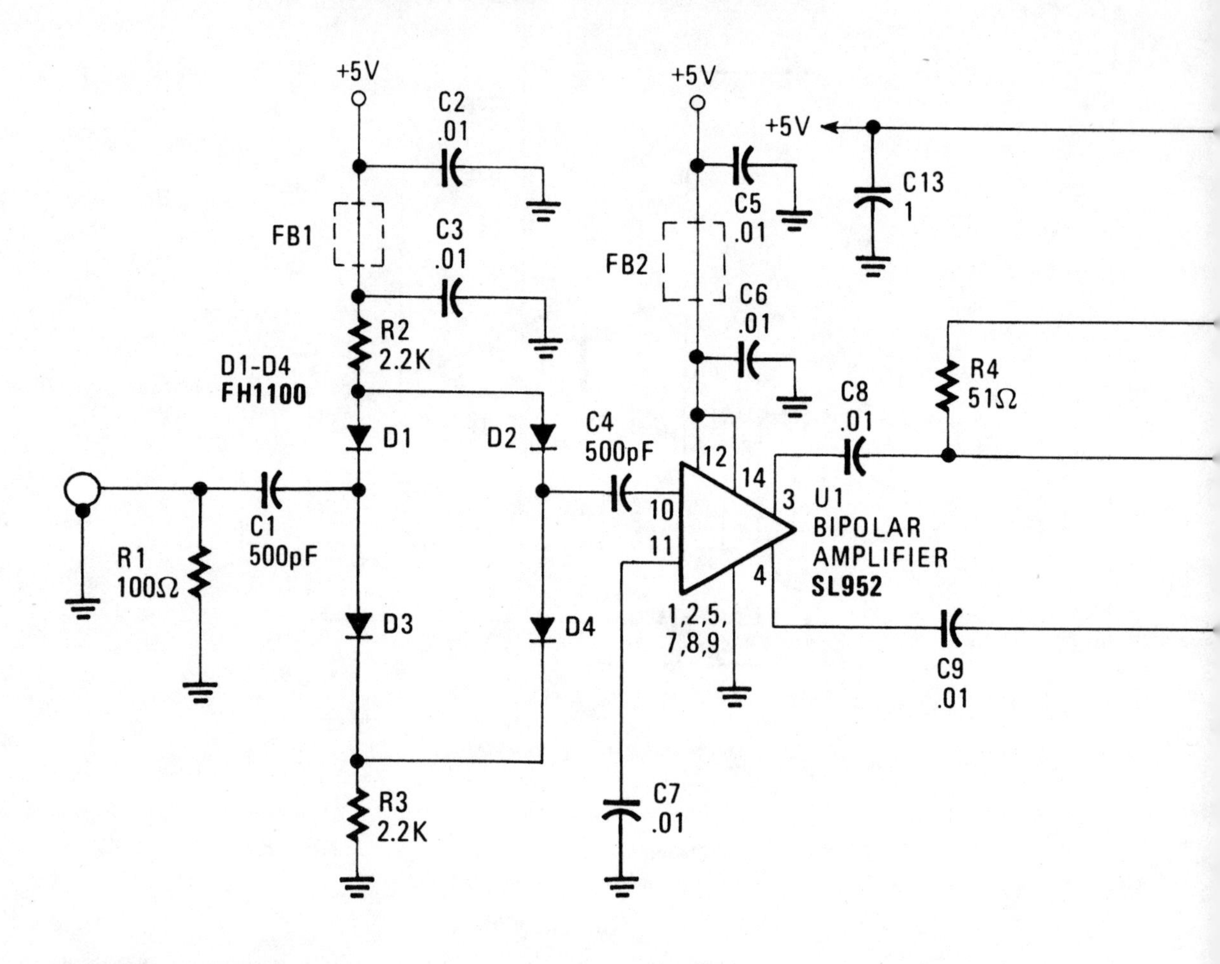

HANDS-ON ELECTRONICS

Fig. 75-5

Circuit Notes

The 650 MHz Prescaler Probe's input is terminated by resistor R1 and is fed through C1 to the diode limiter composed of D1 through D4. Those diodes are forward-biased by the +5 volt supply for small-input signals and, in turn, feed the signal to U1. However, for larger input signals, diodes D1 through D4 will start to turn off, passing less of the signal, and, thus, attenuating it. But even in a full-off state, the FH1100-type diodes will always pass a small part of the input to U1 because of capacitive leakage within the diodes. Integrated circuit U1, a Plessey SL952 bipolar amplifier, capable of 1 GHz operation, provides 20 to 30 dB of gain. The input signal is supplied to pin 10, U1 with

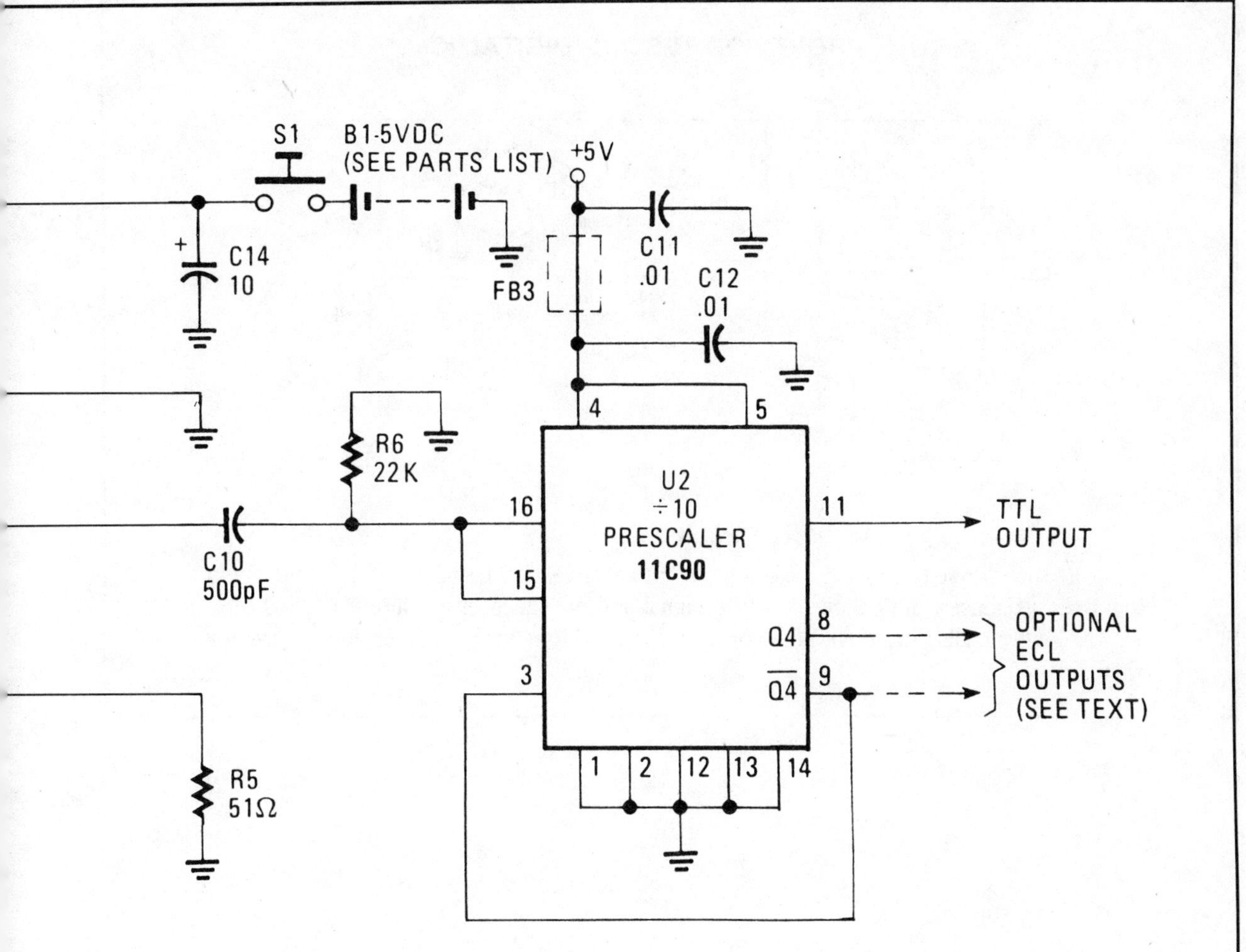

the other input (pin 11) is bypassed to ground. The output signal is taken at pin 3 and pin 4, with pin 3 loaded by R4 and pin 4 by R5.

Integrated circuit 11C90, U2, is a high-speed prescaler capable of 650 MHz operation configured for a divide-by-10 format. A reference voltage internally generated appears at pin 15 and is tied to pin 16, the clock input. This centers the capacitive-coupled input voltage from U1 around the switching threshold-voltage level. An ECL-to-TTL converter in U1 provides level conversion to drive TTL input counters by typing pin 13 low. Therefore, no external ECL to TTL converter is required at the pin 11 output. On the other hand, ECL outputs are available at U2, pin 8 (Q4) and at pin 9 (Q4), if desired. In that circuit configuration, pin 13 is left open, and U2 will use less power.

TONE PROBE FOR TESTING DIGITAL ICs

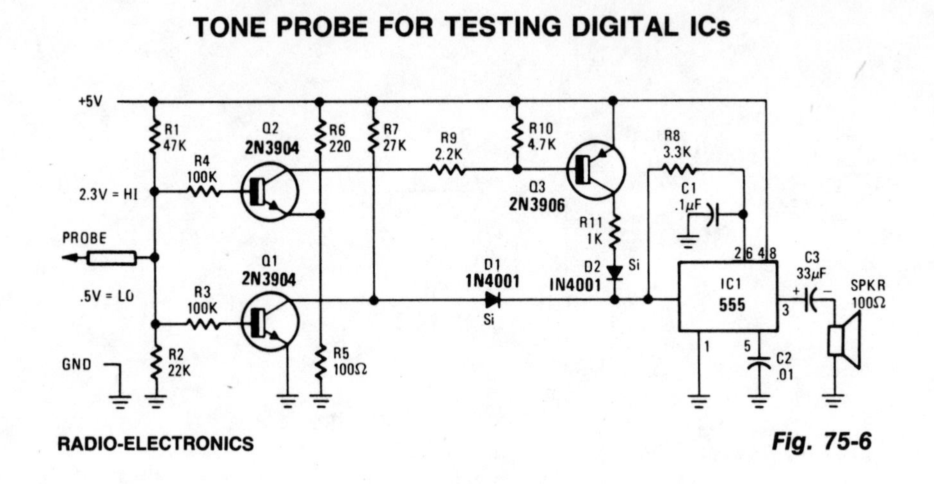

Fig. 75-6

Circuit Notes

The tone probe uses sound to tell the status of the signal being probed. The probe's input circuit senses the condition of the signal and produces either a low-pitched tone for low-level signals (less than 0.8 V) or a high-pitched tone for high-level signals (greater than 2 V).

76

Proximity Sensors

The sources of the following circuits are contained in the Sources section beginning on page 694. The figure number contained in the box of each circuit correlates to the source entry in the Sources section.

PROXIMITY ALARM

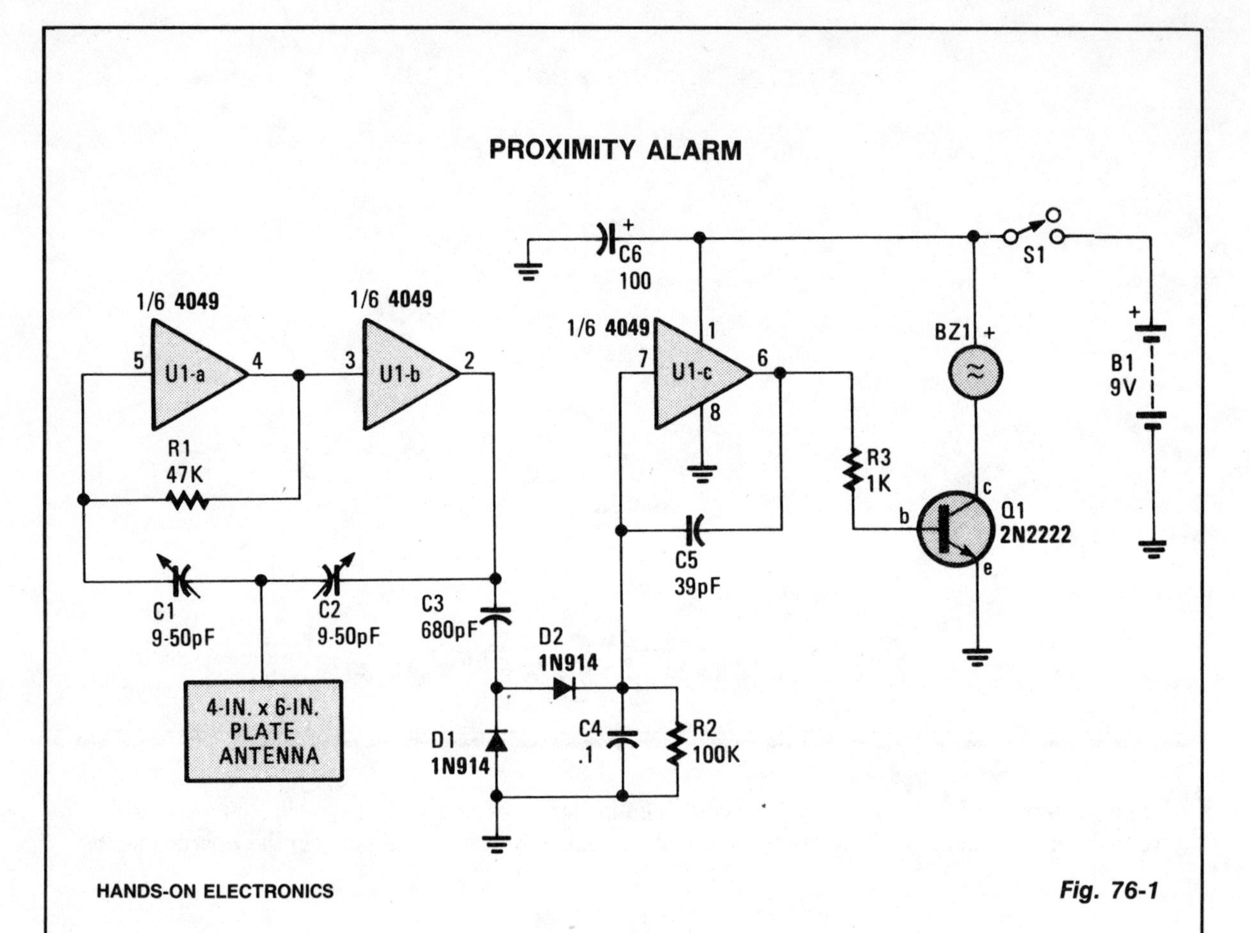

HANDS-ON ELECTRONICS

Fig. 76-1

Circuit Notes

Inverters U1a and U1b are connected in a simple RC oscillator circuit. The frequency is determined by the values of R1, C1, C2, and the internal characteristics of the integrated circuit. As long as the circuit is oscillating, a positive dc voltage is developed at the output of the voltage-coupler circuit: C3, D1, D2 and C4. The dc voltage is applied to the input of U1c—the third inverter amplifier—keeping its output in a low state, which keeps Q1 turned off so that no sound is produced by BZ1. With C1 and C2 adjusted to the most sensitive point, the pickup plate will detect a hand 3 to 5-inches away and sound an alert. Set C1 and C2 to approximately one-half of their maximum value and apply power to the circuit. The circuit should oscillate and no sound should be heard. Using a non-metallic screwdriver, carefully adjust C1 and C2, one at a time, to a lower value until the circuit just ceases oscillation: Buzzer BZ1 should sound off. Back off either C1 or C2 just a smidgen until the oscillator starts up again—that is the most sensitive setting of the circuit.

FIELD DISTURBANCE SENSOR/ALARM

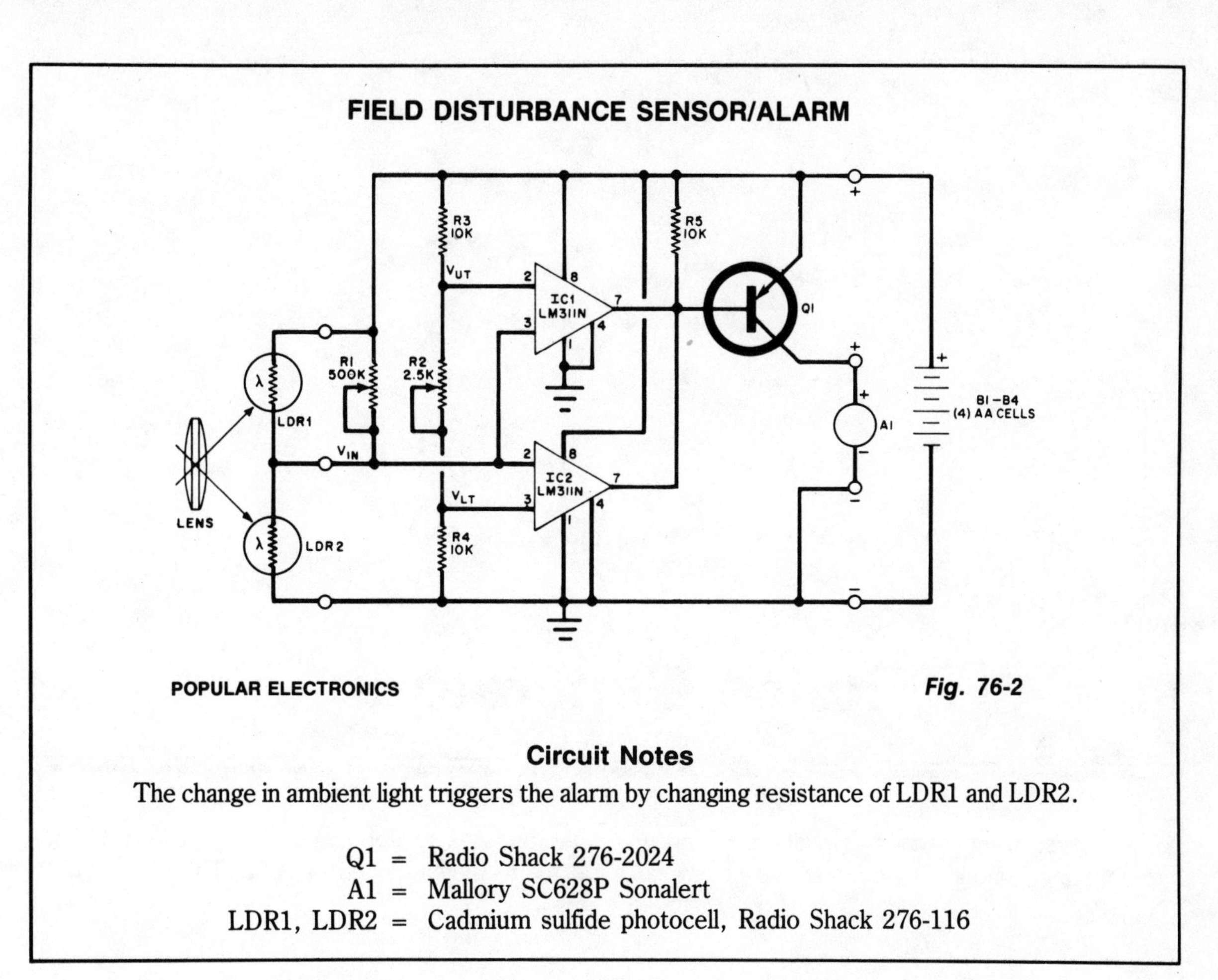

POPULAR ELECTRONICS *Fig. 76-2*

Circuit Notes

The change in ambient light triggers the alarm by changing resistance of LDR1 and LDR2.

Q1 = Radio Shack 276-2024
A1 = Mallory SC628P Sonalert
LDR1, LDR2 = Cadmium sulfide photocell, Radio Shack 276-116

77

Pulse Generators

The sources of the following circuits are contained in the Sources section beginning on page 694. The figure number contained in the box of each circuit correlates to the source entry in the Sources section.

DELAYED PULSE GENERATOR

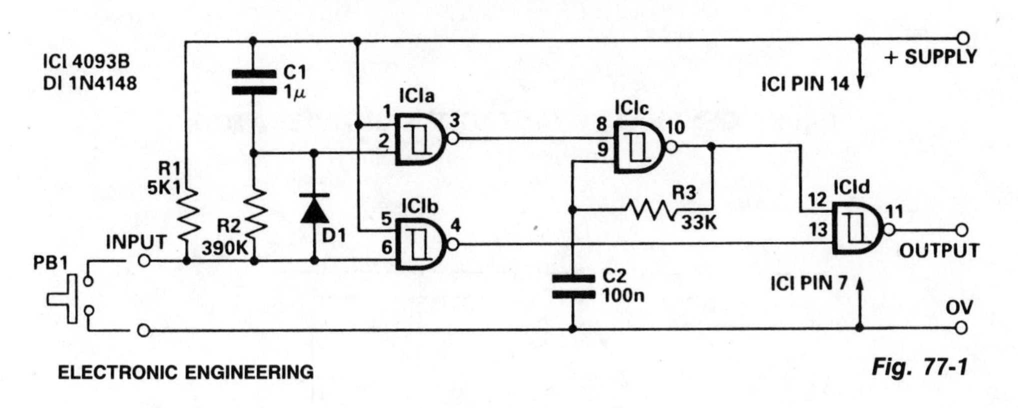

ELECTRONIC ENGINEERING

Fig. 77-1

Circuit Notes

The circuit offers independent control of initial delay and pulse rate. IC1c is connected as a pulse generator whose operation is inhibited by the normally low O/P of the IC1a. When the circuit input goes low i.e., by pressing PB1, IC1b O/P goes high and the circuit O/P goes low thus replicating the input. When the input is kept low capacitor C1 charges via R2 to a point where IC1a O/P goes low. This allows the pulse generator IC1c to start and "rapid fire" pulses appear at the circuit O/P. When the circuit input returns to the high state C1 is rapidly discharged via D1 and R1. The value of R2 and C1 control the initial delay while R3 and C2 control the pulse rate. The values given will give a delay of around 0.5 seconds and a pulse rate of 200/300 Hz depending on supply voltage. PB1 may be replaced by an open collector TTL gate or a common emitter transistor stage if required.

PULSE GENERATOR (ASTABLE MULTIVIBRATOR)

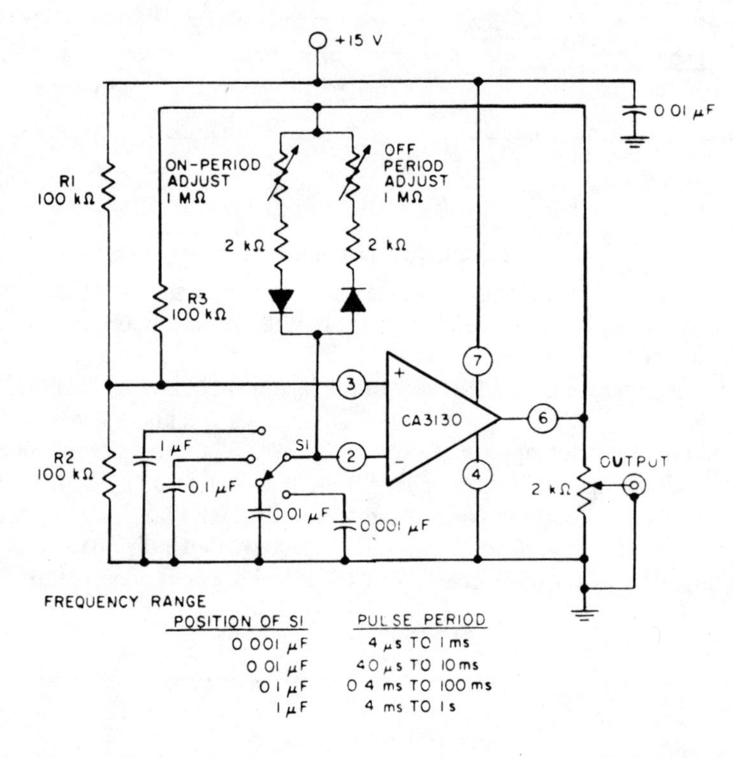

POSITION OF S1	PULSE PERIOD
0 001 μF	4 μs TO 1 ms
0 01 μF	40 μs TO 10 ms
0 1 μF	0 4 ms TO 100 ms
1 μF	4 ms TO 1 s

RCA *Fig. 77-2*

Circuit Notes

Resistors R1 and R2 bias the CA3130 to the mid-point of the supply-voltage, and R3 is the feedback resistor. The pulse repetition rate is selected by positioning S1 to the desired position and the rate remains essentially constant when the resistors which determine on-period and off-period are adjusted.

NON-INTEGER PROGRAMMABLE PULSE DIVIDER

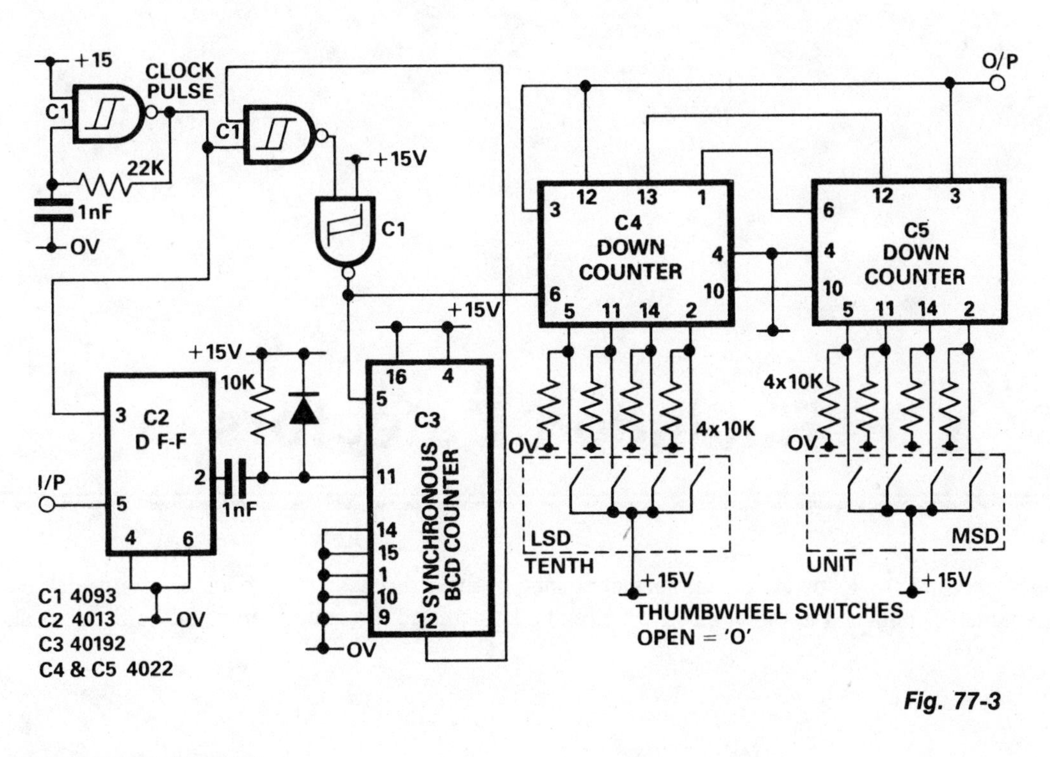

Fig. 77-3

ELECTRONIC ENGINEERING

Circuit Notes

In applications where the period of the input pulses is uneven and the divider is required to cover a wide range of frequencies, the non-integer programmable pulse divider shown can be used. The purpose of the D-type flip-flop (IC2) is to synchronize the input signal with the clock pulse. When the clock pulse changes from low to high and the input is high, IC2 output goes high. Subsequently, IC3 resets to zero and starts counting up. The number of pulses at the output of IC3 is ten time the input pulse. IC4 and IC5 are cascaded to form a two decade programmable down counter.

78

Radiation Detectors

The sources of the following circuits are contained in the Sources section beginning on page 694. The figure number contained in the box of each circuit correlates to the source entry in the Sources section.

MICROPOWER RADIOACTIVE RADIATION DETECTOR

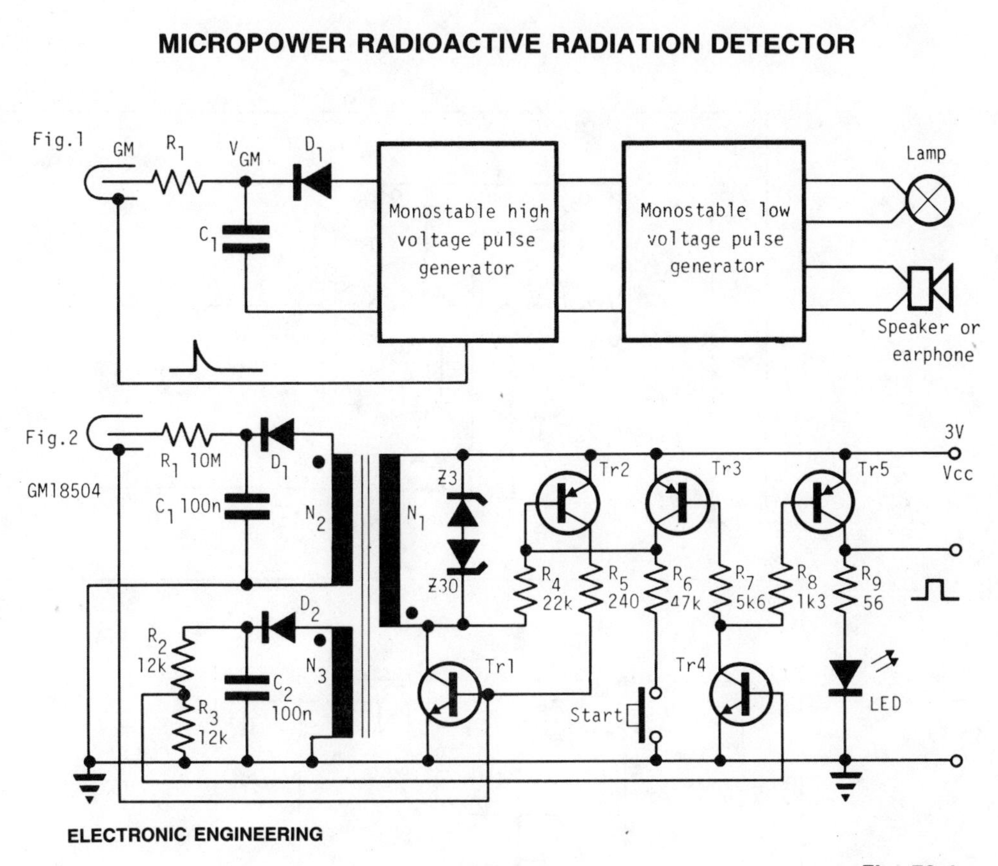

ELECTRONIC ENGINEERING

Fig. 78-1

Circuit Notes

In the absence of radiation, no current is drawn. At normal background radiation levels the power consumption is extremely low. The instrument may be left on for several months without changing batteries. In this way the detector is always ready to indicate an increase in radiation. An LED is used as an indicator lamp. With background radiation it draws less than 50 μA. A ferrite pot core is used for the transformer with N1 = 30, N2 = 550, and N3 = 7. Using two 1.5V batteries with 0.5 Ah total capacity, the detector can work at background radiation levels for 0.5 Ah $\div$ 50μA = 10,000 hours, which is more than a year.

POCKET-SIZED GEIGER COUNTER

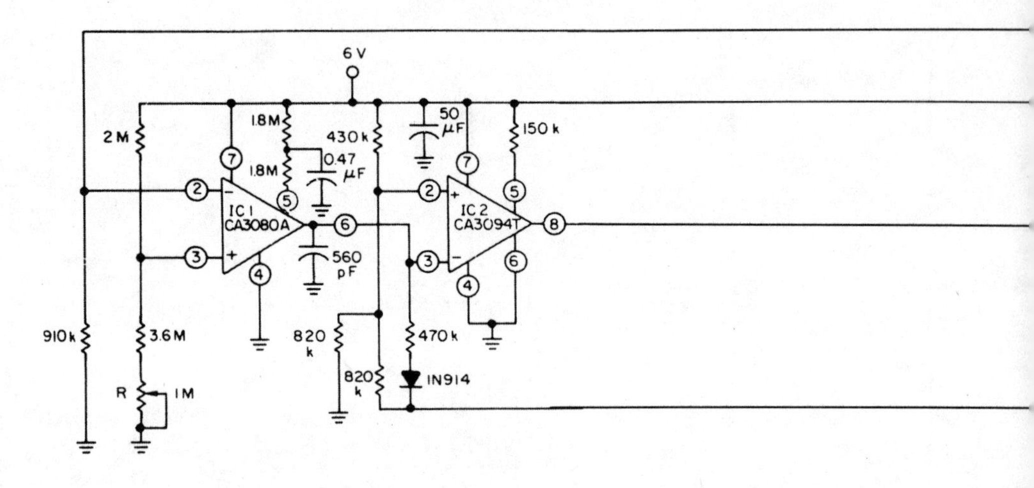

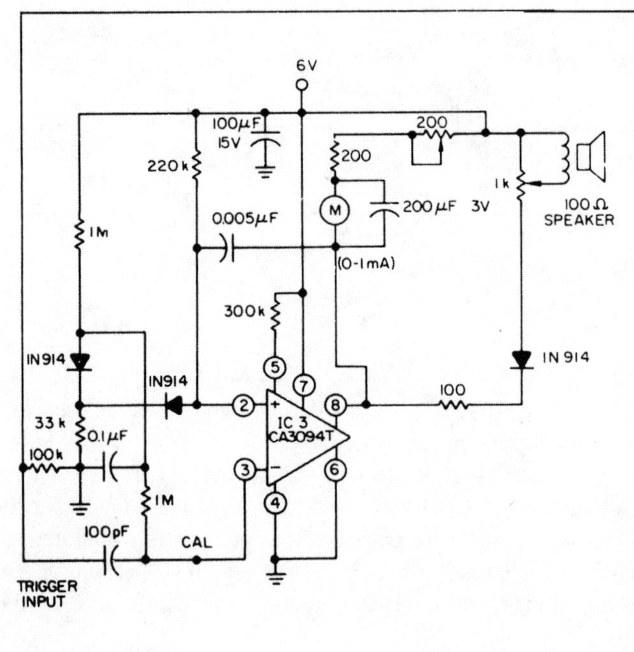

ELECTRONIC DESIGN

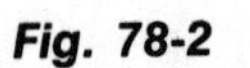

Fig. 78-2

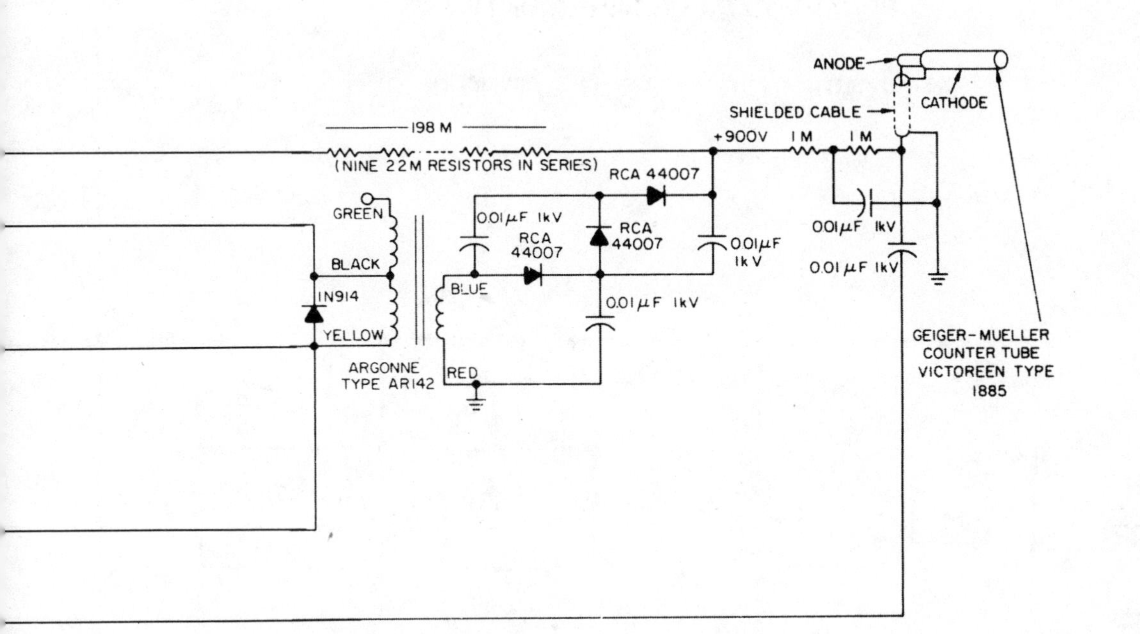

Circuit Notes

A single 6.75 V mercury battery powers the counter, which features a 1 mA count-rate meter as well as an aural output. A regulated 900 V supply provides stable operation of the counter tube. A multivibrator, built around a differential power amplifier IC2, drives the step-up transformer. Comparator IC1 varies the multivibrator duty cycle to provide a constant 900 V. The entire regulated supply draws less than 2 mA. A one-shot multivibrator, built with IC3, provides output pulses that have constant width and amplitude. Thus the average current through the meter is directly proportional to the pulse-rate output from the counter tube. And the constant-width pulses also drive the speaker. Full-scale meter deflection (1 mA) represents 5000 counts/min, or 83.3 pulses/s. A convenient calibration checkpoint can be provided on the meter scale for 3600 ppm (60 pulses/s.)

PHOTOMULTIPLIER OUTPUT-GATING CIRCUIT

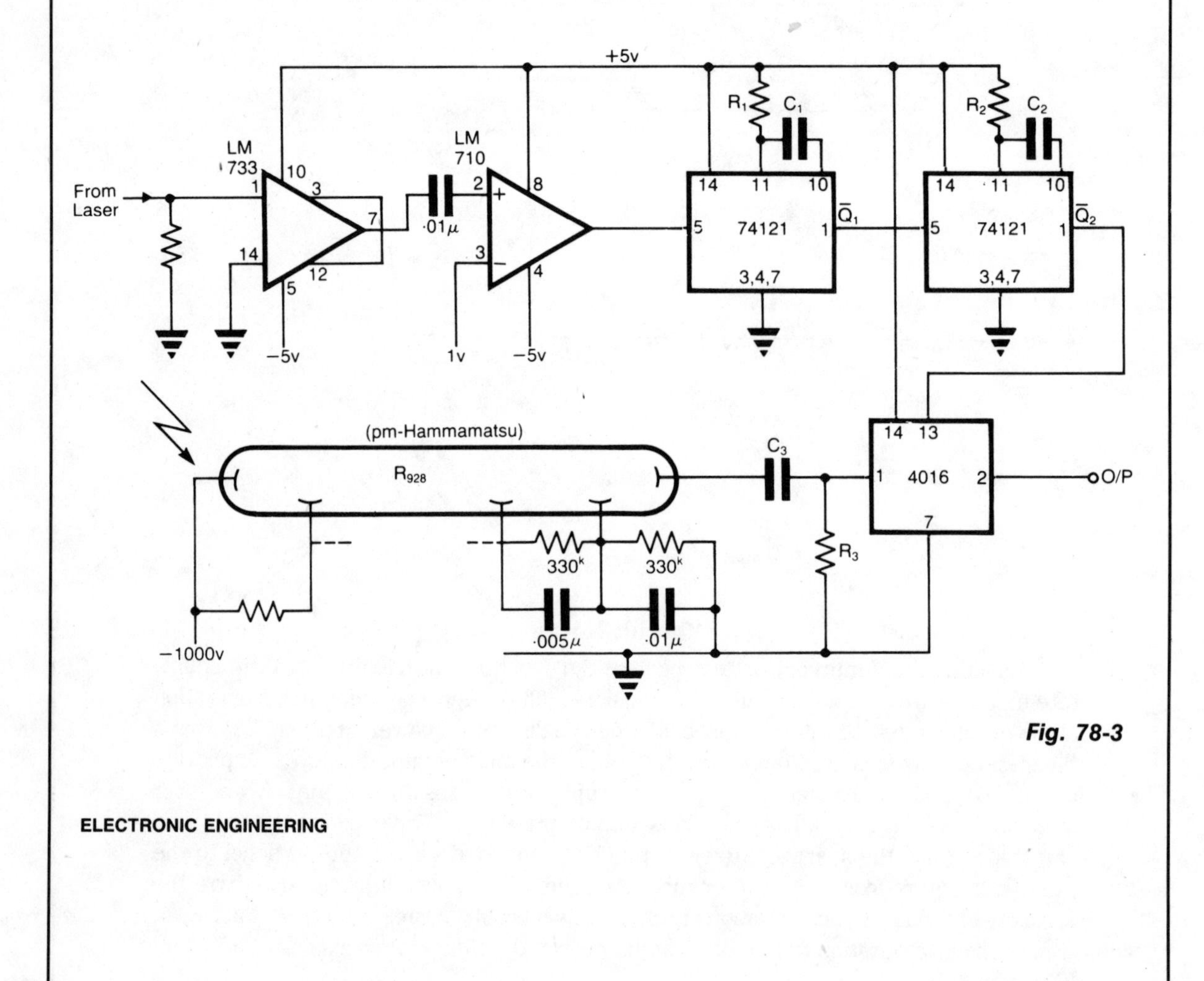

Fig. 78-3

ELECTRONIC ENGINEERING

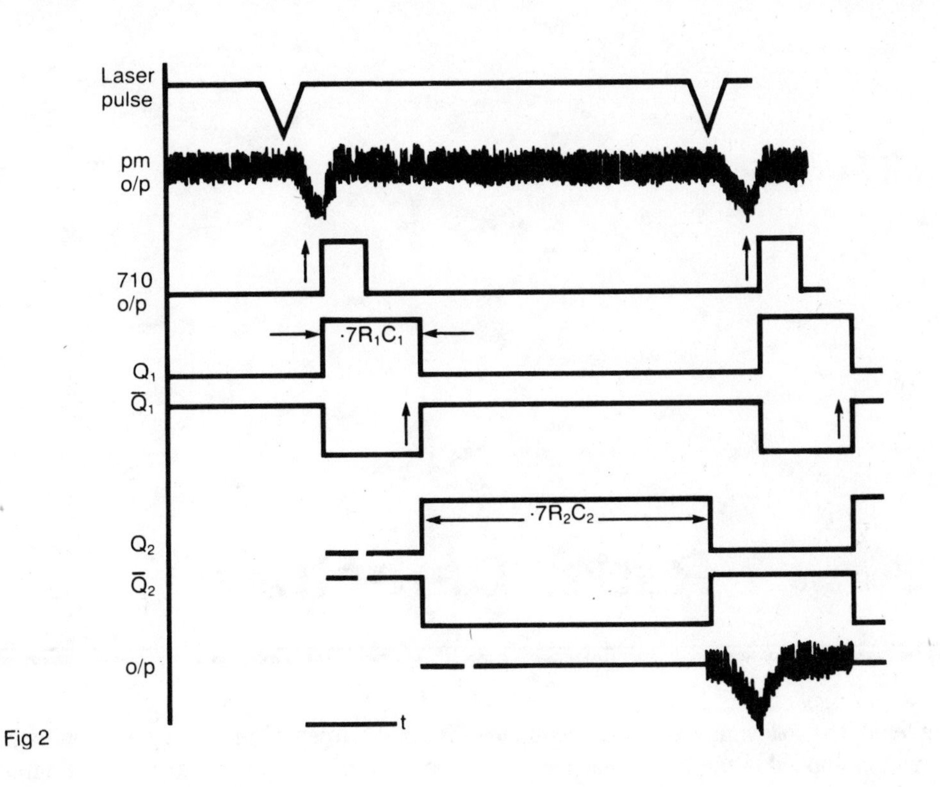

Fig 2

Circuit Notes

The application involves observing the light pulse emerging from a thick specimen after transillumination by a laser pulse. Pulses derived from the laser source are amplified using a Video Amplifier LM733. The reference level is set to 1 V in the comparator LM 710, to provide the necessary trigger pulses for the monostable multivibrator 74121. The laser pulses have a repetition frequency of 500 Hz and suitable values are as below:

R1 = 33 k ohm, C1 = 22 pF
R2 = 33 k ohm, C2 = 68 nF

The pulse width for each monostable is approximately given by tw = 0.7 RC. R3 and C3 is a high pass filter. The method therefore permits the use of low cost components having moderate response times for extracting the pulse of interest.

79
Radar Detectors

The sources of the following circuits are contained in the Sources section beginning on page 694. The figure number contained in the box of each circuit correlates to the source entry in the Sources section.

ONE-CHIP RADAR DETECTION CIRCUIT

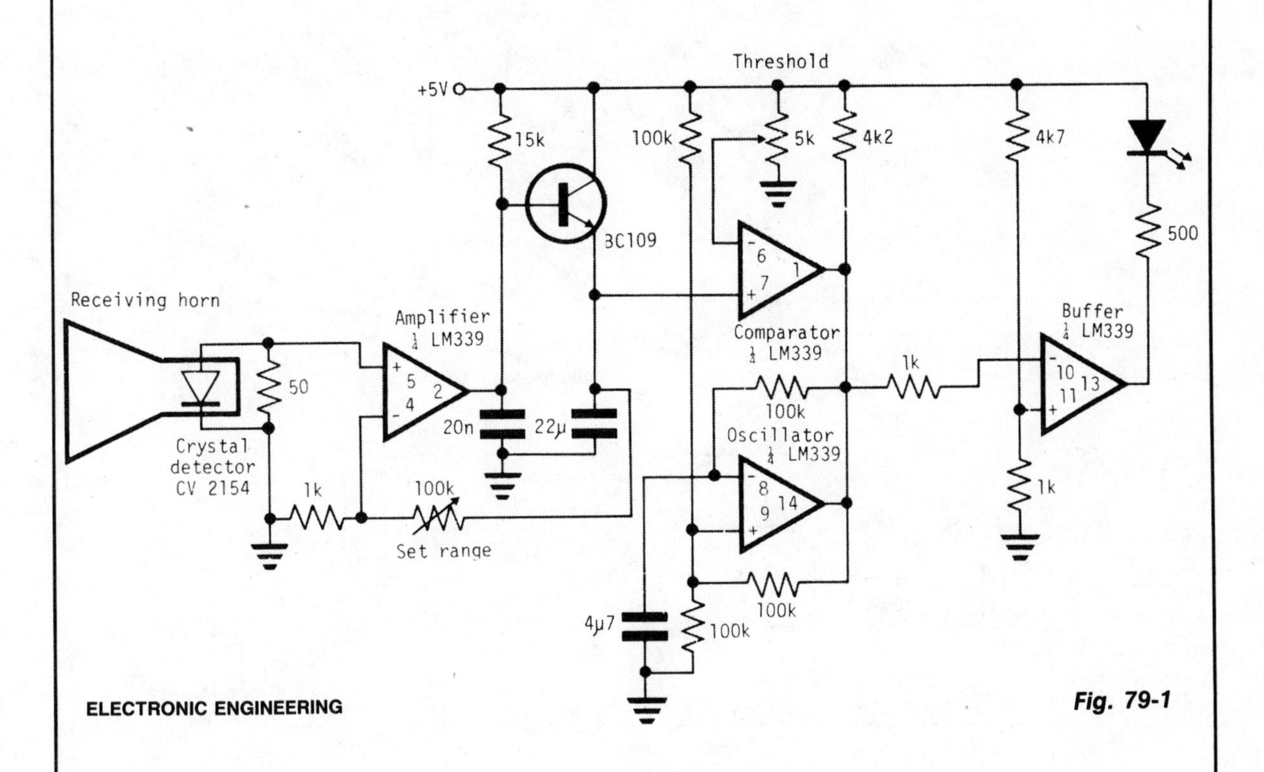

ELECTRONIC ENGINEERING

Fig. 79-1

Circuit Notes

A simple X-band radar detector is capable of indicating changes in rf radiation strength at levels down to 2 mW/cm^2. Radiation falling on the detector diode, produces a voltage at the input of an amplifier whose gain may be adjusted to vary the range at which the warning is given. The amplifier output drives a voltage comparator with a variable threshold set to a level that avoids false alarms. The comparator output is connected in the wired-OR configuration with the open collector output of an oscillator running at a frequency of 2 Hz. In the absence of a signal, the comparator output level is low, inhibiting the oscillator output stage and holding the buffer so the lamp is off. When a signal appears, the comparator output goes high, removing the lock from the oscillator which free-runs, switching the lamp on and off at 2 Hz.

RADAR SIGNAL DETECTOR

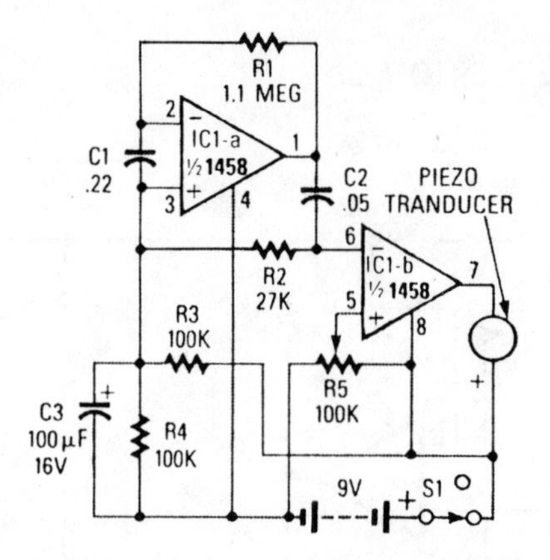

FIG. 1—THE ECONOMY RADAR DETECTOR needs only one IC and a few discrete components.

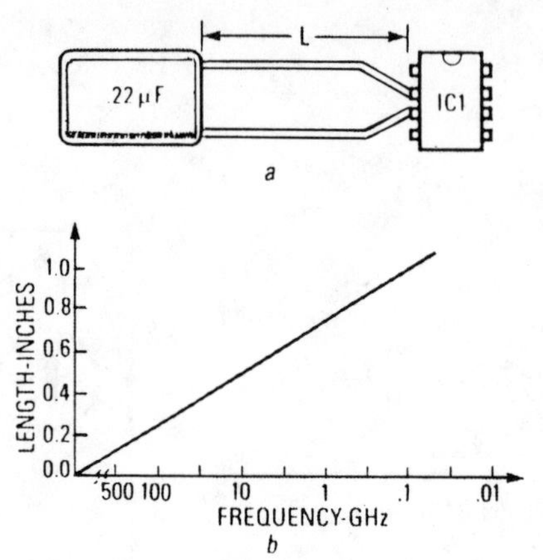

FIG. 3—VARY THE LEAD LENGTHS OF C1 to tune the input circuit.

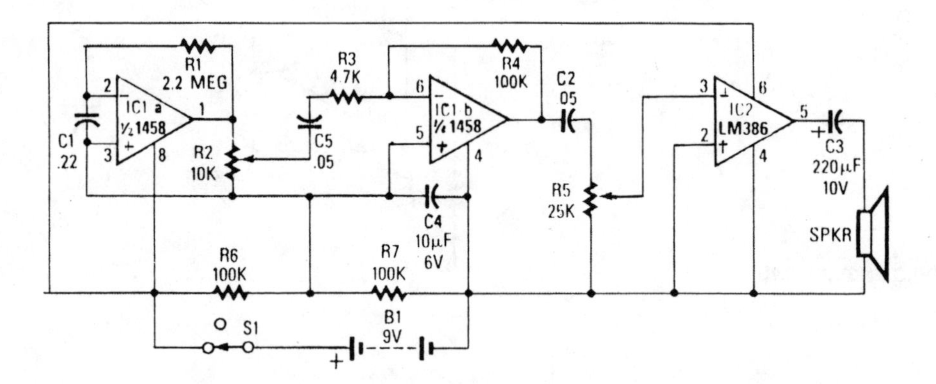

FIG. 2—DELUXE RADAR DETECTOR adds a buffer amplifier and an audio power amp to drive a speaker.

RADIO-ELECTRONICS *Fig. 79-2*

Circuit Notes

The circuit can be tuned to respond to signals between 50 MHz and 500 GHz. The economy model is shown in Fig. 1, and the deluxe model is shown in Fig. 2. The first op amp in each circuit functions as a current-to-voltage converter. In the economy model IC1b buffers the output to drive the piezo buzzer. The deluxe model functions in a similar manner except that IC1b is configured as a ×20 buffer amplifier to drive the LM386. In both circuits C1 functions as a "transmission line" that intercepts the incident radar signal. The response may be optimized by trimming C1's lead length for the desired frequency. Typically the capacitor's leads should be 0.5-0.6 inches long.

80
Ramp Generators

The sources of the following circuits are contained in the Sources section beginning on page 694. The figure number contained in the box of each circuit correlates to the source entry in the Sources section.

RAMP GENERATOR

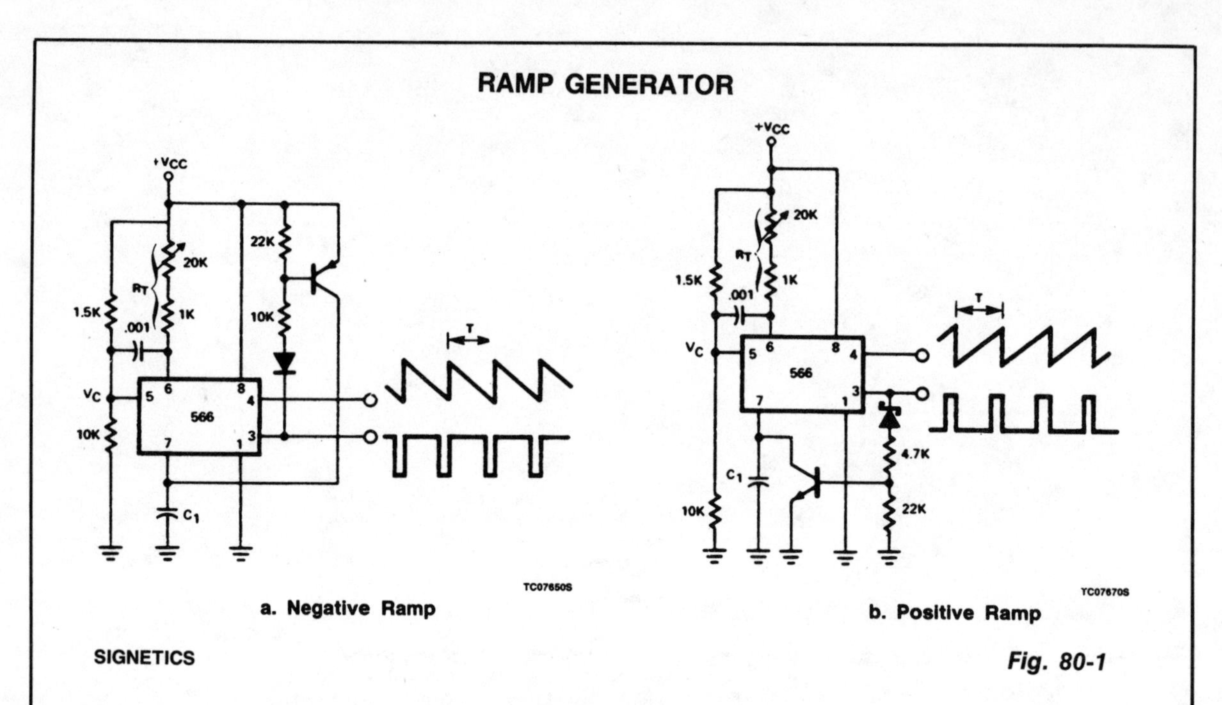

SIGNETICS

Fig. 80-1

Circuit Notes

The 566 can be wired as a positive or negative ramp generator. In the positive ramp generator, the external transistor driven by the Pin 3 output rapidly discharges C1 at the end of the charging period so that charging can resume instantaneously. The pnp transistor of the negative ramp generator likewise rapidly charges the timing capacitor C1 at the end of the discharge period. Because the circuits are reset so quickly, the temperature stability of the ramp generator is excellent. The period

$$T \text{ is } \frac{1}{2f_o}$$

where f_o is the 566 free-running frequency in normal operation. Therefore,

$$T = \frac{1}{2f_o} = \frac{R_T C_1 V_{CC}}{5(V_{CC} - V_C)} \tag{1}$$

where V_C is the bias voltage at Pin 5 and R_T is the total resistance between Pin 6 and V_{CC}. Note that a short pulse is available at Pin 3. (Placing collector resistance in series with the external transistor collector will lengthen the pulse.)

VOLTAGE CONTROLLED RAMP GENERATOR

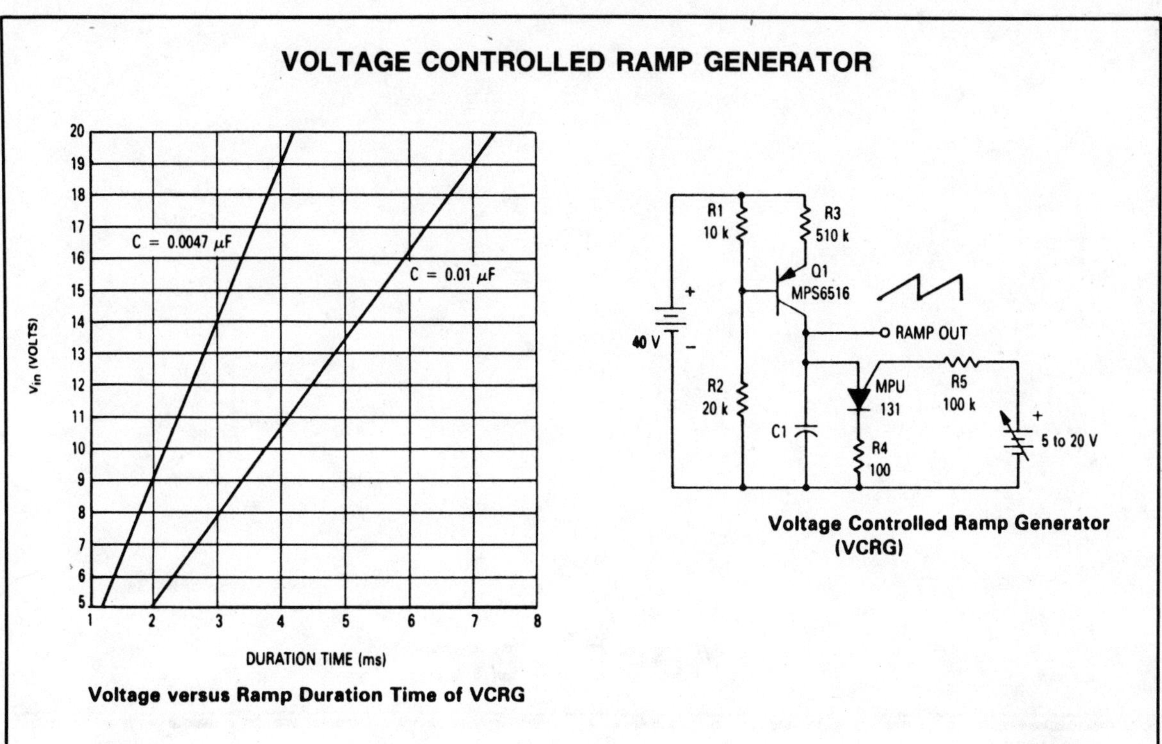

Voltage versus Ramp Duration Time of VCRG

Voltage Controlled Ramp Generator (VCRG)

MOTOROLA ***Fig. 80-2***

Circuit Notes

The current source formed by Q1 in conjunction with capacitor C1 set the duration time of the ramp. As the positive dc voltage at the gate is changed, the peak point firing voltage of the PUT is changed, which changes the duration time, i.e., increasing the supply voltage increases the peak point firing voltage causing the duration time to increase.

81
Receivers

The sources of the following circuits are contained in the Sources section beginning on page 694. The figure number contained in the box of each circuit correlates to the source entry in the Sources section.

CAR RADIO WITH CAPACITIVE DIODE TUNING AND ELECTRONIC MW/LW SWITCHING

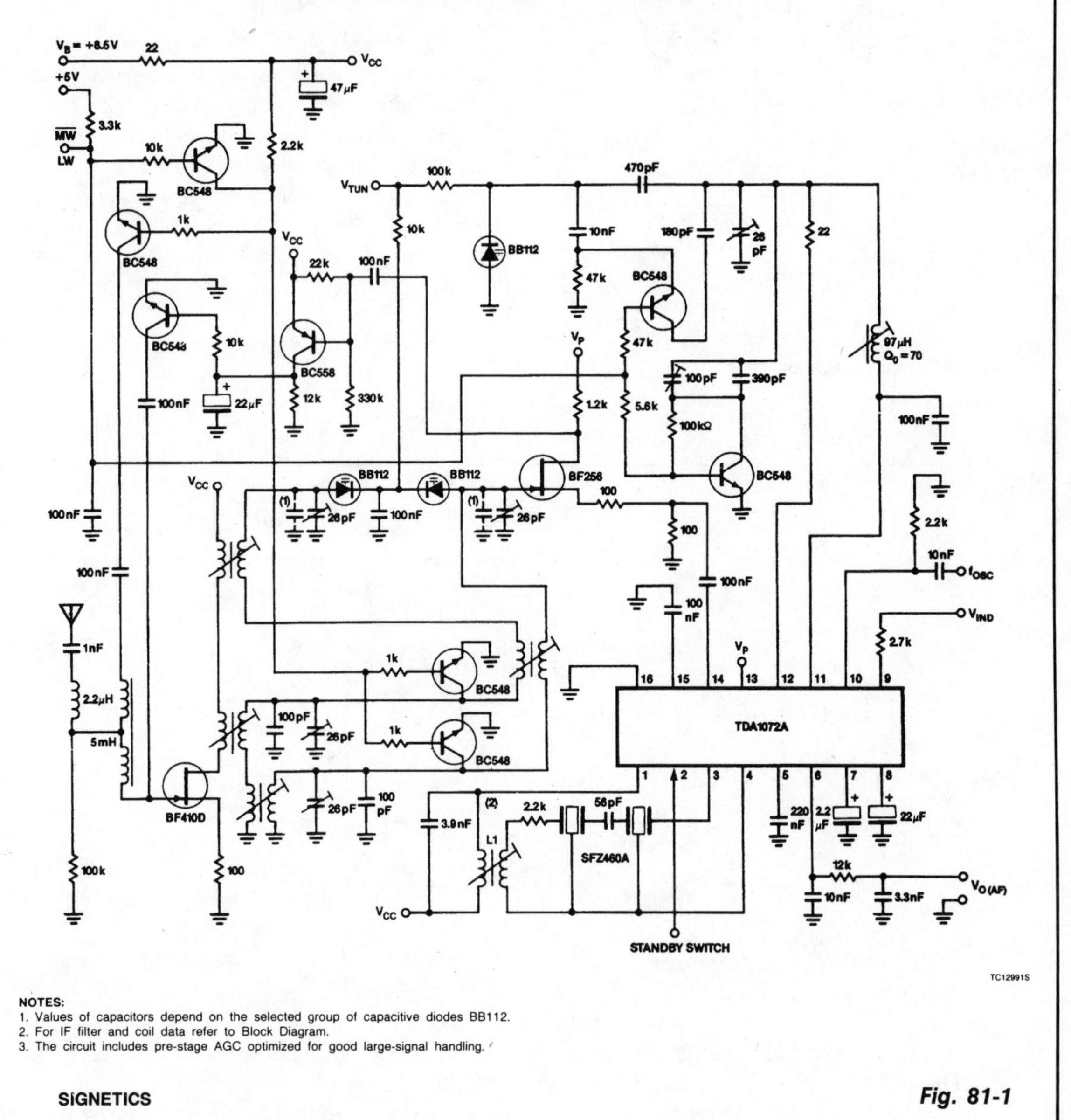

NOTES:

1. Values of capacitors depend on the selected group of capacitive diodes BB112.
2. For IF filter and coil data refer to Block Diagram.
3. The circuit includes pre-stage AGC optimized for good large-signal handling.

SIGNETICS

Fig. 81-1

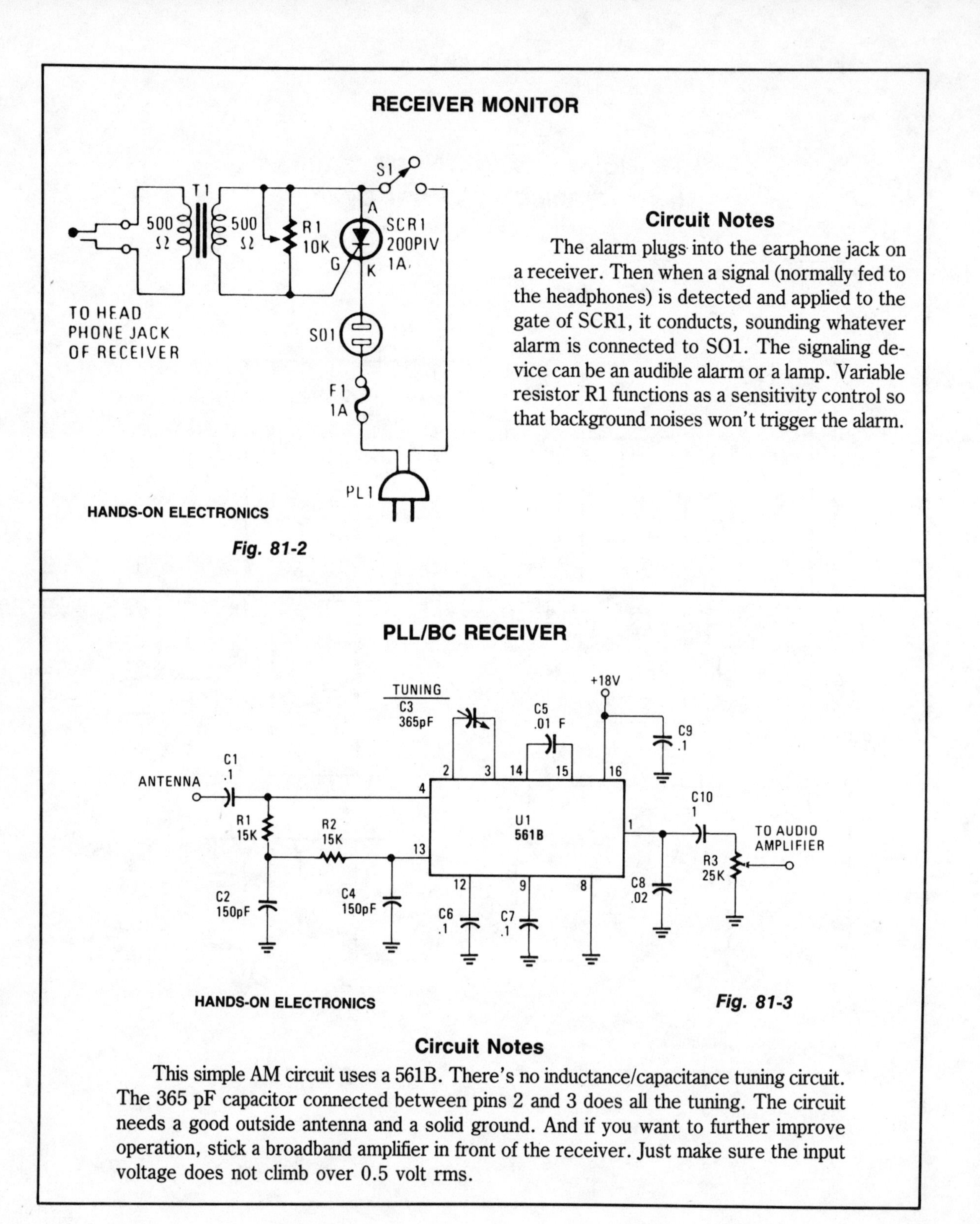

RECEIVER MONITOR

Fig. 81-2

Circuit Notes

The alarm plugs into the earphone jack on a receiver. Then when a signal (normally fed to the headphones) is detected and applied to the gate of SCR1, it conducts, sounding whatever alarm is connected to SO1. The signaling device can be an audible alarm or a lamp. Variable resistor R1 functions as a sensitivity control so that background noises won't trigger the alarm.

PLL/BC RECEIVER

Fig. 81-3

Circuit Notes

This simple AM circuit uses a 561B. There's no inductance/capacitance tuning circuit. The 365 pF capacitor connected between pins 2 and 3 does all the tuning. The circuit needs a good outside antenna and a solid ground. And if you want to further improve operation, stick a broadband amplifier in front of the receiver. Just make sure the input voltage does not climb over 0.5 volt rms.

82
Rectifier Circuits

The sources of the following circuits are contained in the Sources section beginning on page 694. The figure number contained in the box of each circuit correlates to the source entry in the Sources section.

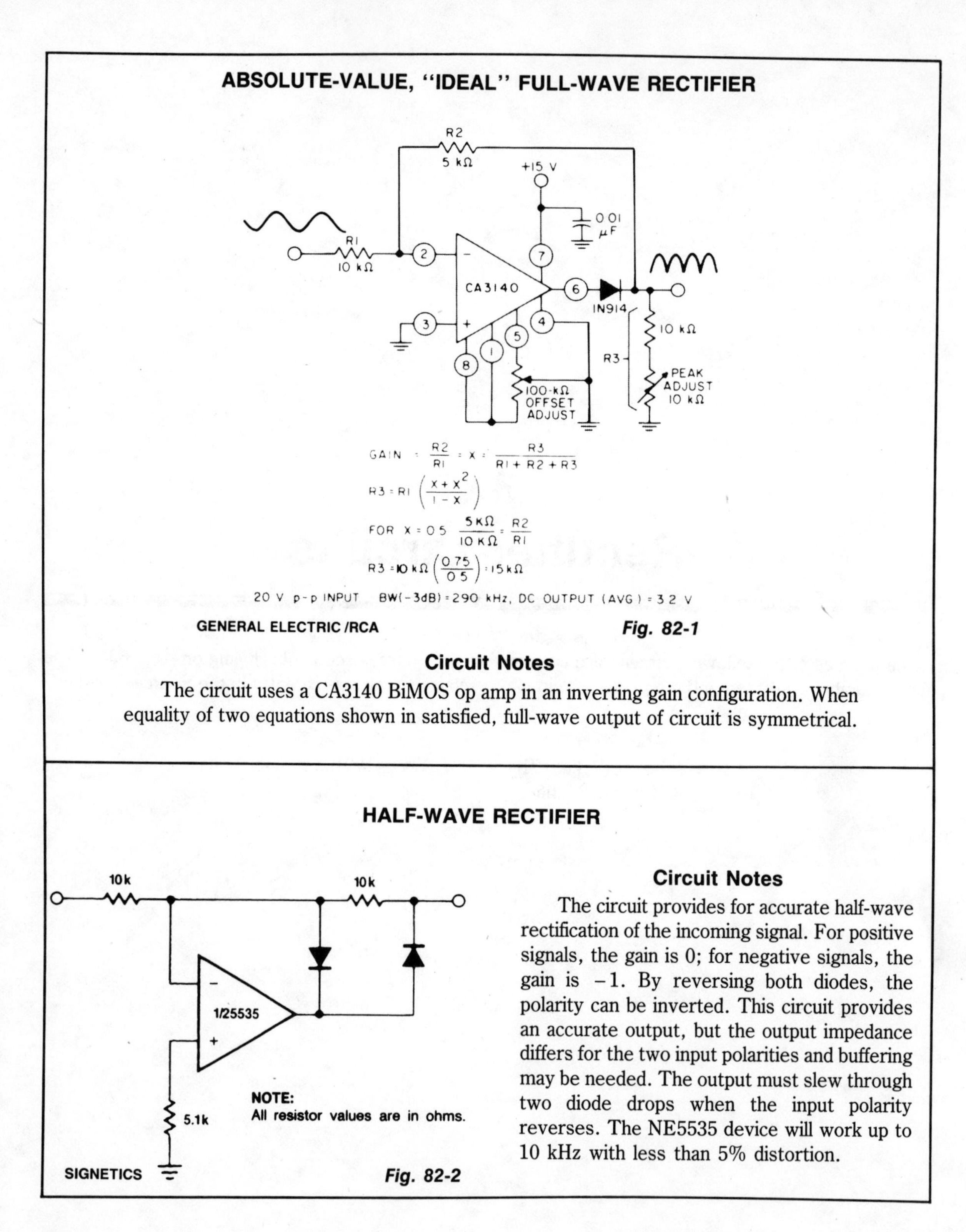

ABSOLUTE-VALUE, "IDEAL" FULL-WAVE RECTIFIER

GENERAL ELECTRIC/RCA

Fig. 82-1

Circuit Notes

The circuit uses a CA3140 BiMOS op amp in an inverting gain configuration. When equality of two equations shown in satisfied, full-wave output of circuit is symmetrical.

HALF-WAVE RECTIFIER

Circuit Notes

The circuit provides for accurate half-wave rectification of the incoming signal. For positive signals, the gain is 0; for negative signals, the gain is -1. By reversing both diodes, the polarity can be inverted. This circuit provides an accurate output, but the output impedance differs for the two input polarities and buffering may be needed. The output must slew through two diode drops when the input polarity reverses. The NE5535 device will work up to 10 kHz with less than 5% distortion.

SIGNETICS

Fig. 82-2

83
Relay Circuits

The sources of the following circuits are contained in the Sources section beginning on page 694. The figure number contained in the box of each circuit correlates to the source entry in the Sources section.

RELAY DRIVER PROVIDES DELAY AND CONTROLS CLOSURE TIME

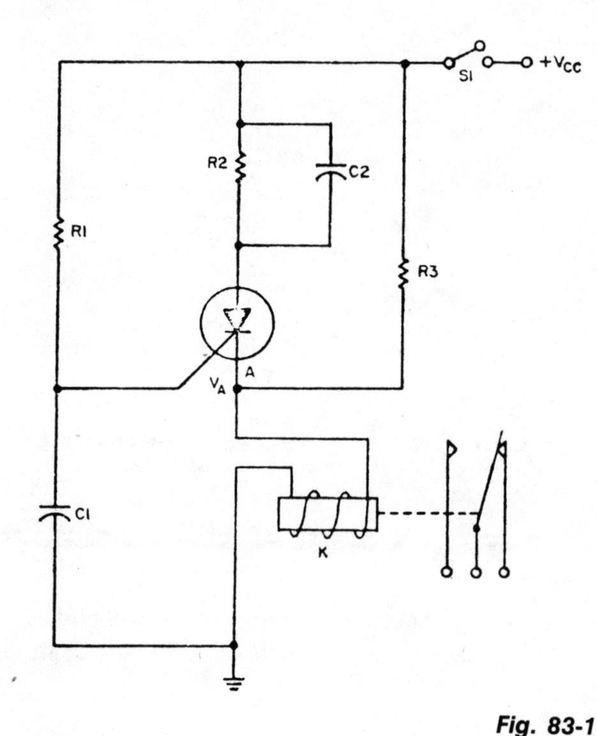

Fig. 83-1

ELECTRONIC DESIGN

Circuit Notes

The relay operates a certain time, t_d, after power is applied to it, and then it operates for a length of time, t_c. The SCR fires when the voltage on C1 reaches V_A. This operates the relay, which stays activated until the current charging C2 drops below the dropout current. To keep the relay in its activated position indefinitely ($t_c = \infty$), eliminate C2 and choose R2 just large enough to keep the relay coil current within its related limits. Typical component values for t_d = 30 seconds and t_c = 2 seconds are: R1 = 1.5 megohms, R2 = 10 k ohms, R3 = 3 k ohms, C1 = 47 μF, and C2 = 100 μF. The SCR is a 2N1877 and the relay is a Potter Brumfield PW-5374. A value of 12 Vdc is assumed for V_{CC}.

TRIAC RELAY-CONTACT PROTECTION

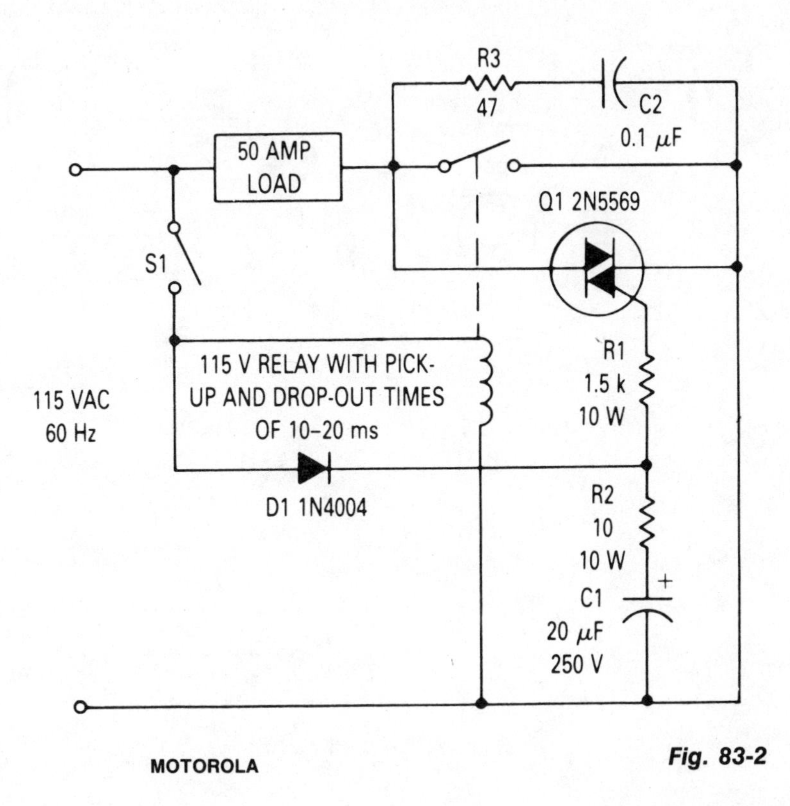

MOTOROLA

Fig. 83-2

Circuit Notes

This circuit can be used to prevent relay contact arcing for loads up to 50 amperes. There is some delay between the time a relay coil is energized and the time the contacts close. There is also a delay between the time the coil is de-energized and the time the contacts open. For the relay used in this circuit both times are about 15 ms. The TRIAC across the relay contacts will turn on as soon as sufficient gate current is present to fire it. This occurs after switch S1 is closed but before the relay contacts close. When the contacts close, the load current passes through them, rather than through the TRIAC, even though the TRIAC is receiving gate current. If S1 should be closed during the negative half cycle of the ac line, the TRIAC will not turn on immediately but will wait until the voltage begins to go positive, at which time diode D1 conducts providing gate current through R1. The maximum time that could elapse before the TRIAC turns on is 8-⅓ ms for the 60 Hz supply. This is adequate to ensure that the TRIAC will be on before the relay contact closes.

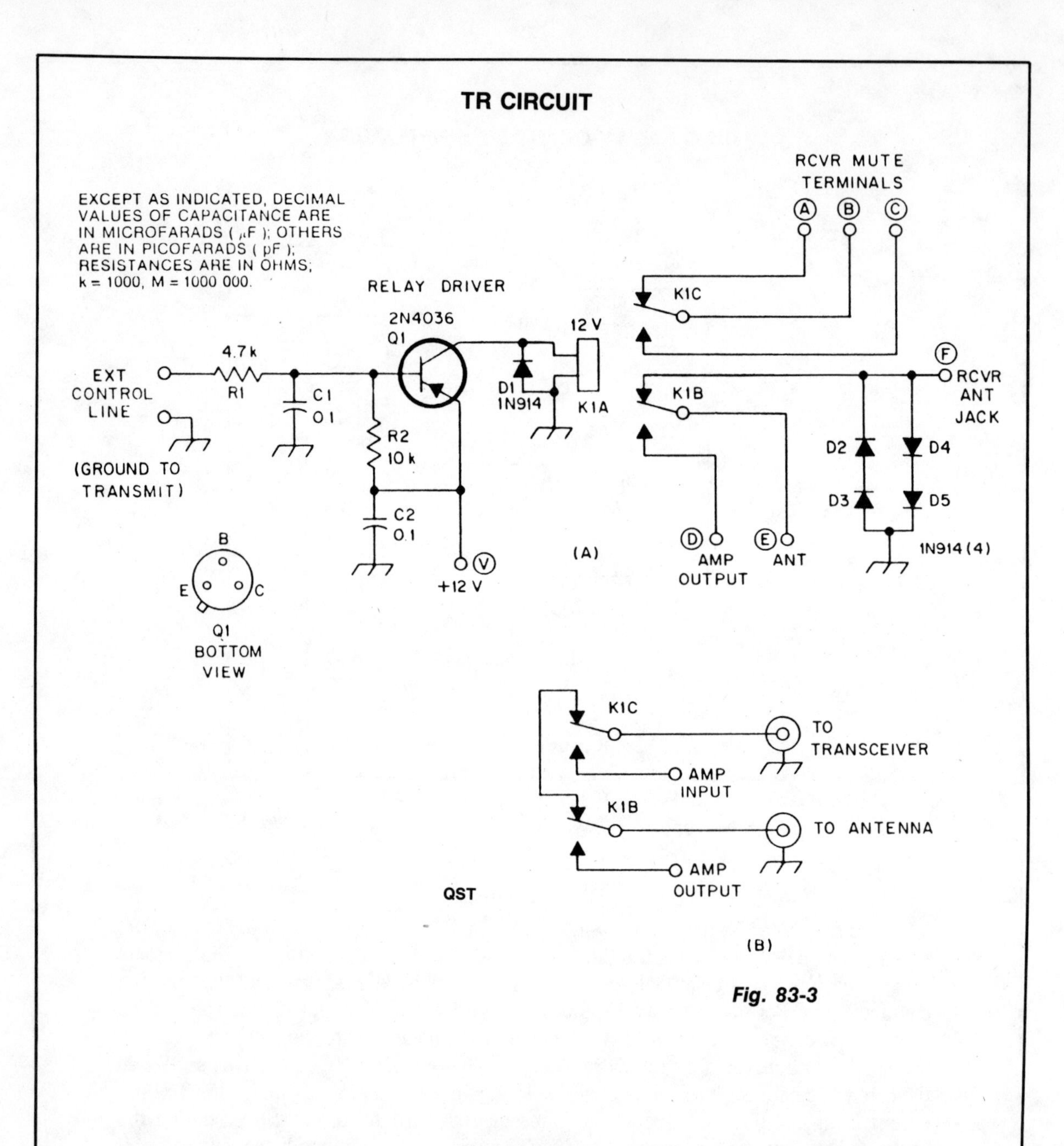

Fig. 83-3

Circuit Notes

C1 and C2 are disc ceramic. R1 and R2 are ¼ or ½ W carbon composition resistors. K1 is a 12 V DPDT DIP relay. Illustration A shows how to connect the relay contacts for use with a separate transmitter-receiver combination. The circuit at B is for amplifier use with a transceiver.

84
Resistance/Continuity Meters

The sources of the following circuits are contained in the Sources section beginning on page 694. The figure number contained in the box of each circuit correlates to the source entry in the Sources section.

SINGLE CHIP CHECKS RESISTANCE

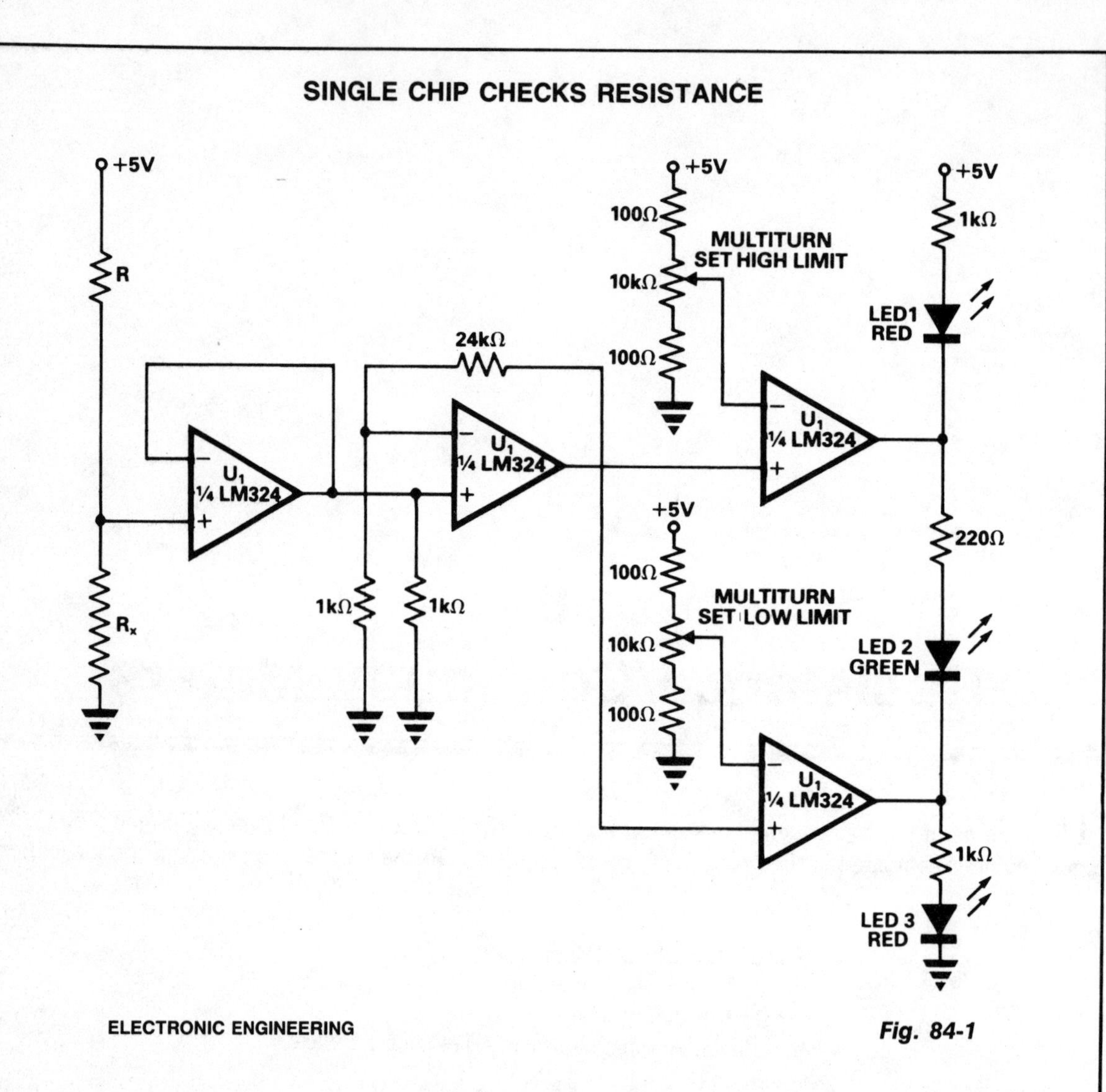

ELECTRONIC ENGINEERING

Fig. 84-1

Circuit Notes

A simple tester can be used for routine checks for resistance on production lines of relays, coils, or similar components where frequent changes in resistance to be tested are not required. The tester is built around a single quad op amp chip, the LM324. R, which is chosen to be around 80 times the resistance to be checked, and the 5 V supply form the current source. The first op amp buffers the voltage generated across the resistance under test, R_X. The second op amp amplifies this voltage. The third and fourth op amps compare the amplified voltage with high and low limits. The high and low limits are set on multiturn presets with high and low limit resistors connected in place of R_X. LED 1 (red) lights when the resistance is high. LED 2 (green) shows that the resistance is within limits. LED 3 (red) indicates that the resistance is low.

SIMPLE CONTINUITY TESTER FOR PCBs

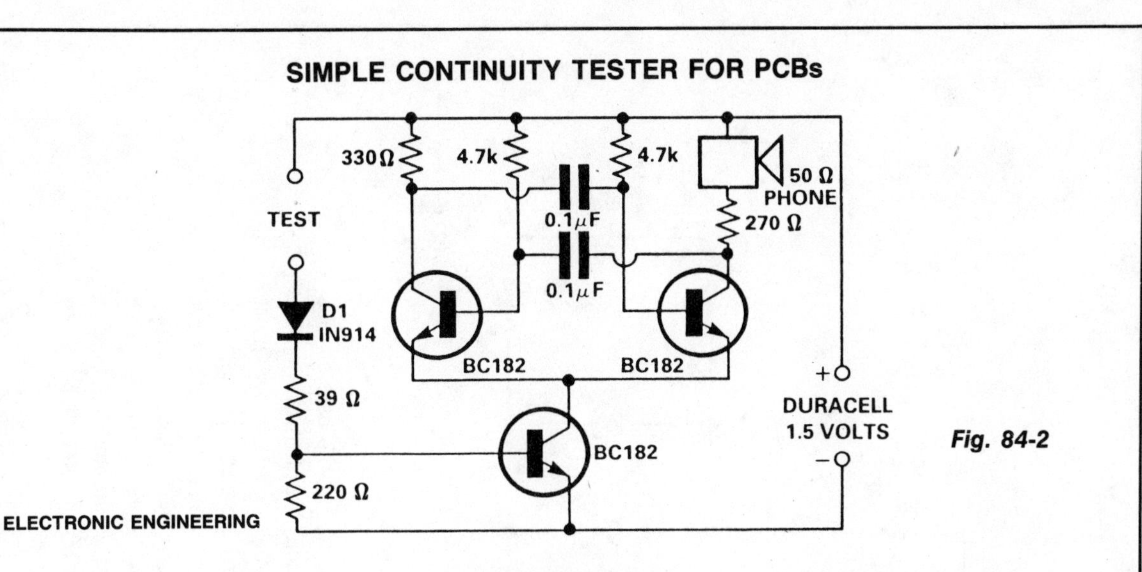

Fig. 84-2

Circuit Notes

This tester is for tracing wiring on Printed Circuit Boards. Resistors below 50 ohms act as a short circuit; above 100 ohms as an open circuit. The circuit is a simple multivibrator switched on by transistor T3. The components in the base of T3 are D1, R1, R2, and the test resistance. With a 1.5 volt supply, there is insufficient voltage to turn on a semiconductor connected to the test terminals.

SIMPLE CONTINUITY TESTER

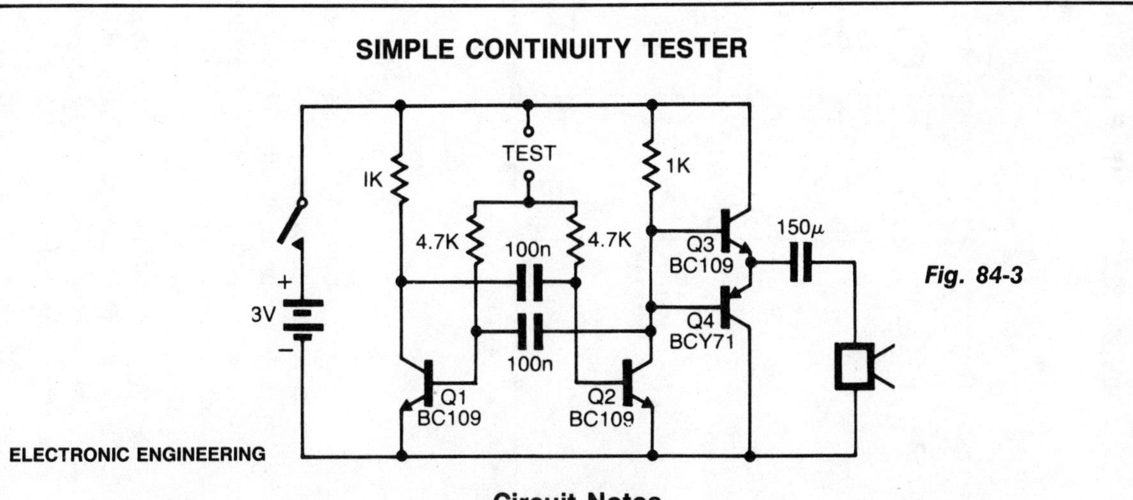

Fig. 84-3

Circuit Notes

The pitch of the tone is dependent upon the resistance under test. The tester will respond to resistance of hundreds of kilohms, yet it is possible to distinguish differences of just a few tens of ohms in low-resistance circuits. Q1 and Q2 form a multivibrator, the frequency of which is influenced by the resistance between the test points. The output stage Q3 and Q4 will drive a small loudspeaker or a telephone earpiece.

ADJUSTABLE, AUDIBLE CONTINUITY TESTER FOR DELICATE CIRCUITS

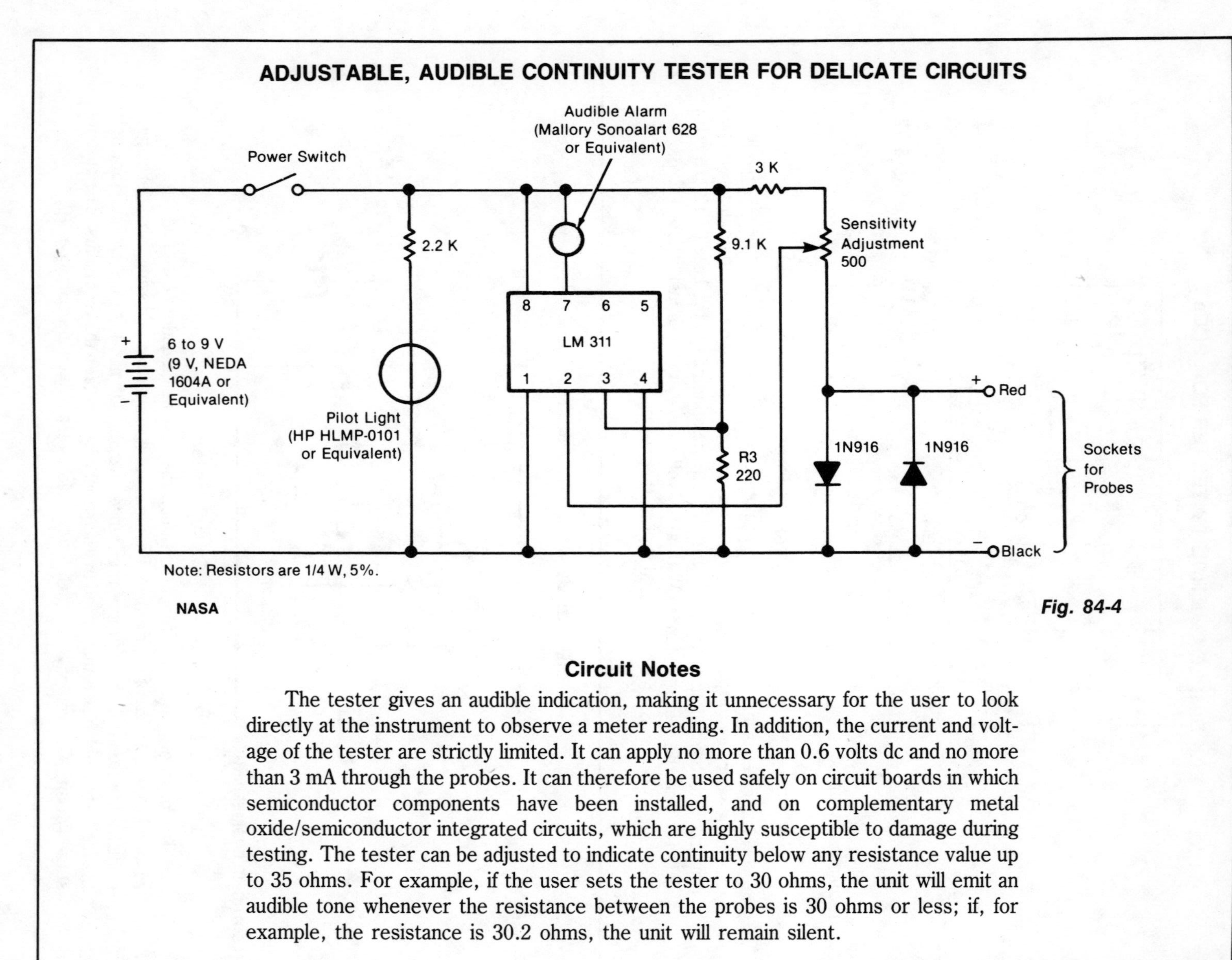

Fig. 84-4

Circuit Notes

The tester gives an audible indication, making it unnecessary for the user to look directly at the instrument to observe a meter reading. In addition, the current and voltage of the tester are strictly limited. It can apply no more than 0.6 volts dc and no more than 3 mA through the probes. It can therefore be used safely on circuit boards in which semiconductor components have been installed, and on complementary metal oxide/semiconductor integrated circuits, which are highly susceptible to damage during testing. The tester can be adjusted to indicate continuity below any resistance value up to 35 ohms. For example, if the user sets the tester to 30 ohms, the unit will emit an audible tone whenever the resistance between the probes is 30 ohms or less; if, for example, the resistance is 30.2 ohms, the unit will remain silent.

85
RF Amplifiers

The sources of the following circuits are contained in the Sources section beginning on page 694. The figure number contained in the box of each circuit correlates to the source entry in the Sources section.

LOW-DISTORTION 1.6 TO 30 MHz SSB DRIVER

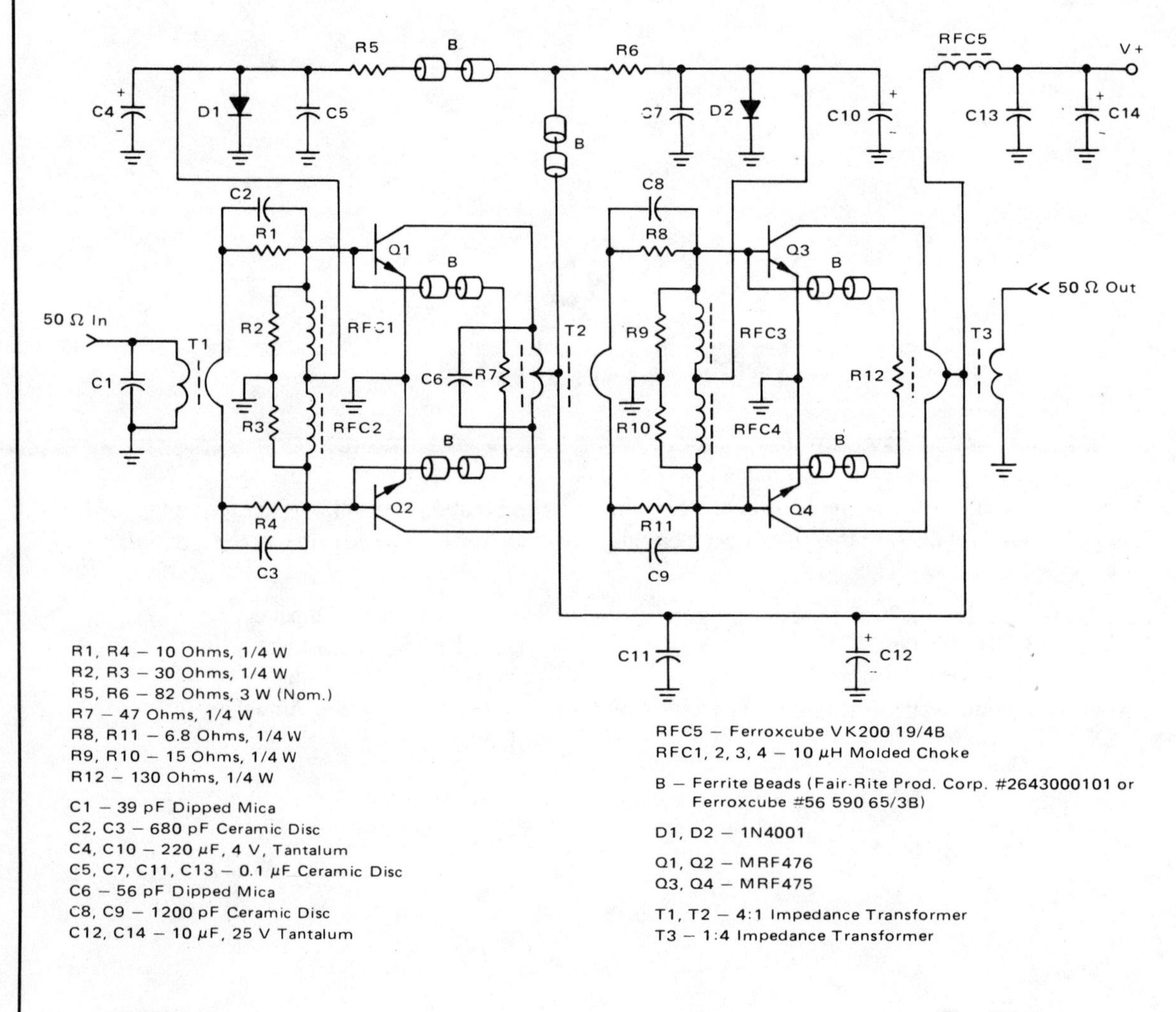

MOTOROLA

Fig. 85-1

Circuit Notes

The amplifier provides a total power gain of about 25 dB, and the construction technique allows the use of inexpensive components throughout. The MRF476 is specified as a 3 watt device and the MRF475 has an output power of 12 watts. Both are extremely tolerant to overdrive and load mismatches, even under CW conditions. Typical IMD numbers are better than −35 dB, and the power gains are 18 dB and 12 dB, respectively, at 30 MHz. The bias currents of each stage are individually adjustable with R5 and R6. Capacitors C4 and C10 function as audio-frequency bypasses to further reduce the source impedance at the frequencies of modulation. Gain leveling across the band is achieved with simple RC networks in series with the bases, in conjunction with negative feedback. The amplitude of the out-of-phase voltages at the bases is inversely proportional to the frequency as a result of the series inductance in the feedback loop and the increasing input impedance of the transistor at low frequencies. Conversely, the negative feedback lowers the effective input impedance presented to the source (not the input impedance of the device itself) and with proper voltage slope would equalize it. With this technique, it is possible to maintain an input VSWR of 1.5:1 or less than 1.6 to 30 MHz.

1 WATT, 2.3 GHz AMPLIFIER

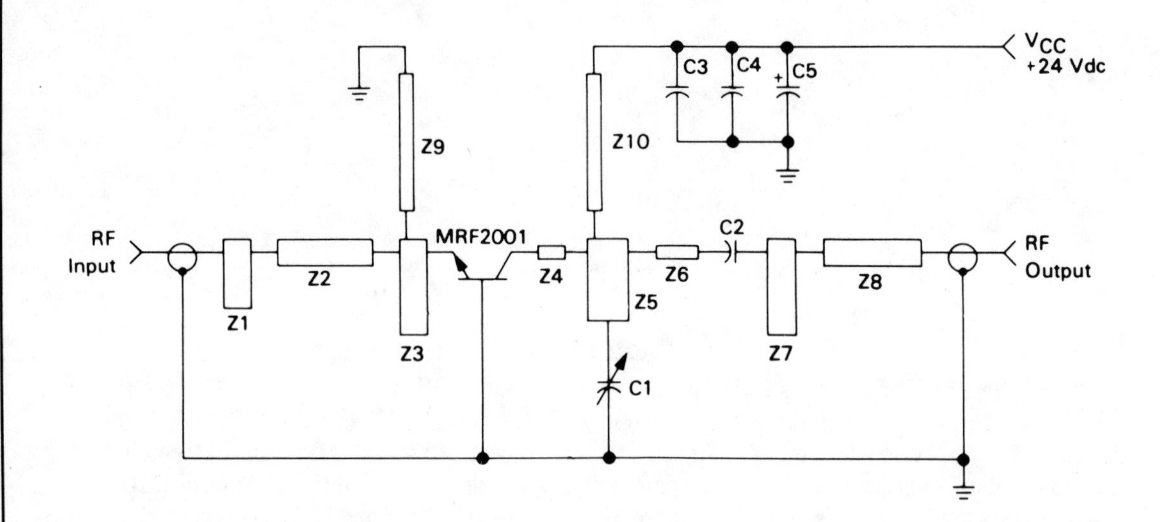

C1 — 0.4-2.5 pF Johanson 7285*
C2, C3 — 68 pF, 50 mil ATC**
C4 — 0.1 μF, 50 V
C5 — 4.7 μF, 50 V Tantalum

Z1-Z10 — Microstrip; see Photomaster, Figure 3

Board Material — 0.0625" 3M Glass Teflon,***
$\epsilon_r = 2.5 \pm 0.05$
*Johanson Manufacturing Corp., 400 Rockaway Valley Road, Boonton, NJ 07005
**American Technical Ceramics, One Norden Lane, Huntington Station, NY 11746
***Registered Trademark of Du Pont

MOTOROLA

Fig. 85-2

Circuit Notes

Simplicity and repeatability are featured in this 1 watt S-band amplifier design. The design uses an MRF2001 transistor as a common base, Class C amplifier. The amplifier delivers 1 watt output with 8 dB minimum gain at 24 V, and is tunable from 2.25 to 2.35 GHz. Applications include microwave communications equipment and other systems requiring medium power, narrow band amplification. The amplifier circuitry consists almost entirely of distributed microstrip elements. A total of six additional components, including the MRF2001, are required to build a working amplifier. The input and output impedances of the transistor are matched to 50 ohms by double section low pass networks. The networks are designed to provide about 3% 1 dB power bandwidth while maintaining a collector efficiency of approximately 30%. There is one tuning adjustment in the amplifier—C1 in the output network. Ceramic chip capacitors, C2 and C3, are used for dc blocking and power supply decoupling. Additional low frequency decoupling is provided by capacitors C4 and C5.

5-W RF POWER AMPLIFIER

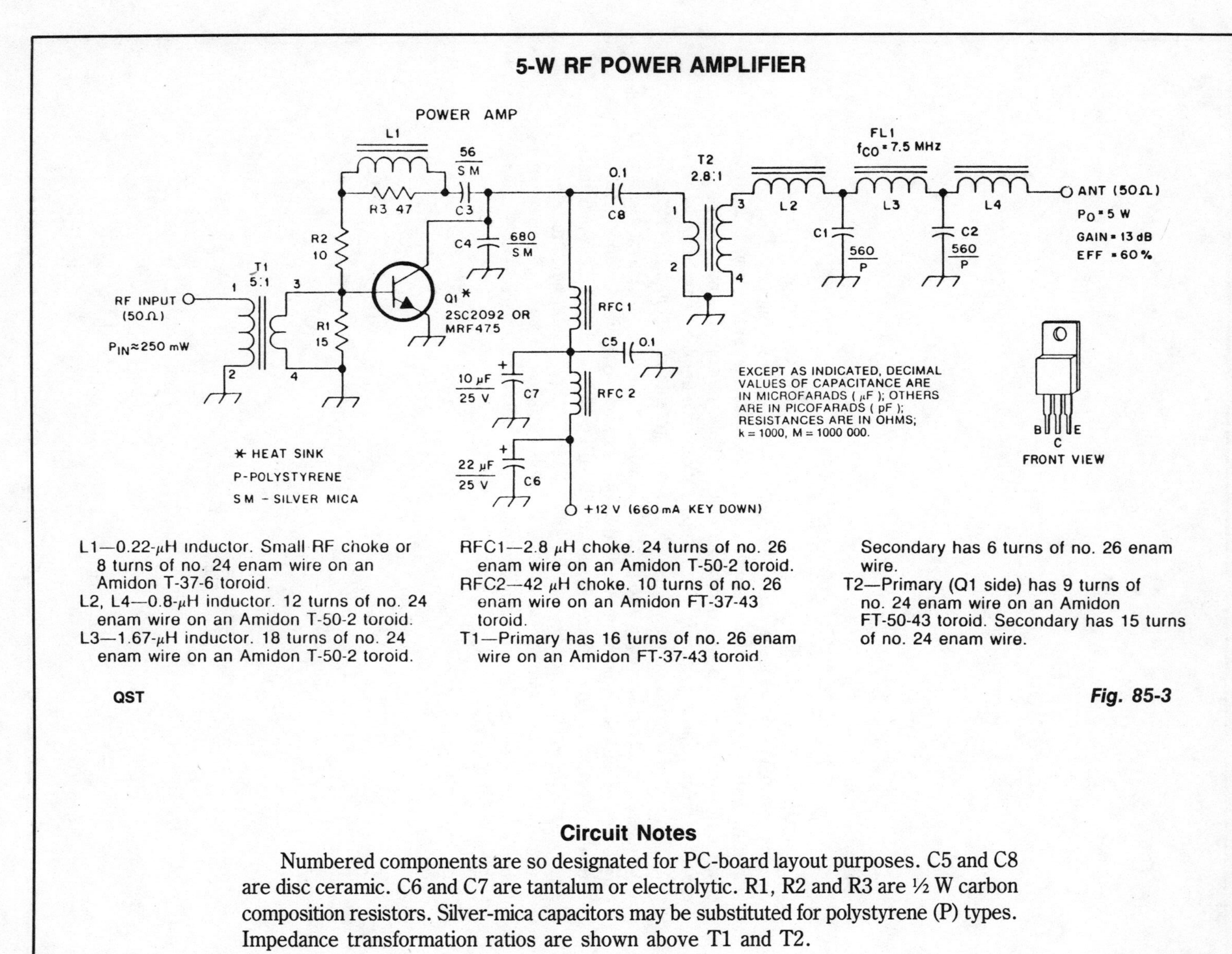

L1—0.22-μH inductor. Small RF choke or 8 turns of no. 24 enam wire on an Amidon T-37-6 toroid.
L2, L4—0.8-μH inductor. 12 turns of no. 24 enam wire on an Amidon T-50-2 toroid.
L3—1.67-μH inductor. 18 turns of no. 24 enam wire on an Amidon T-50-2 toroid.
RFC1—2.8 μH choke. 24 turns of no. 26 enam wire on an Amidon T-50-2 toroid.
RFC2—42 μH choke. 10 turns of no. 26 enam wire on an Amidon FT-37-43 toroid.
T1—Primary has 16 turns of no. 26 enam wire on an Amidon FT-37-43 toroid. Secondary has 6 turns of no. 26 enam wire.
T2—Primary (Q1 side) has 9 turns of no. 24 enam wire on an Amidon FT-50-43 toroid. Secondary has 15 turns of no. 24 enam wire.

QST

Fig. 85-3

Circuit Notes

Numbered components are so designated for PC-board layout purposes. C5 and C8 are disc ceramic. C6 and C7 are tantalum or electrolytic. R1, R2 and R3 are ½ W carbon composition resistors. Silver-mica capacitors may be substituted for polystyrene (P) types. Impedance transformation ratios are shown above T1 and T2.

6-METER PREAMPLIFIER PROVIDES 20 dB GAIN AND LOW NF

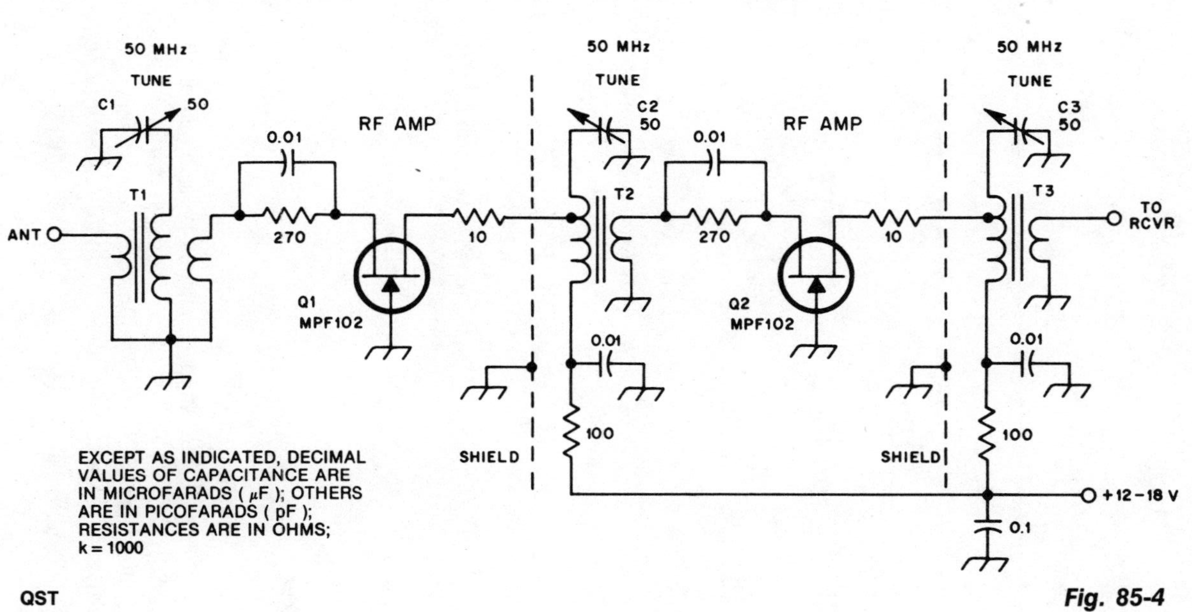

QST

Fig. 85-4

Circuit Notes

C1, C2, and C3 are miniature ceramic or plastic trimmers. T1 (main winding) is 0.34 μH. Use of 11 turns of no. 24 enameled wire on a T37-10 toroid core. The antenna winding has one turn, and Q1 the source winding has three turns. T2 primary consists of 11 turns of no. 24 enameled wire on a T37-10 toroid. Tap Q1 drain is three turns from C2 the end of the winding. The secondary has three turns. T3 is the same as T2, except its secondary has one turn.

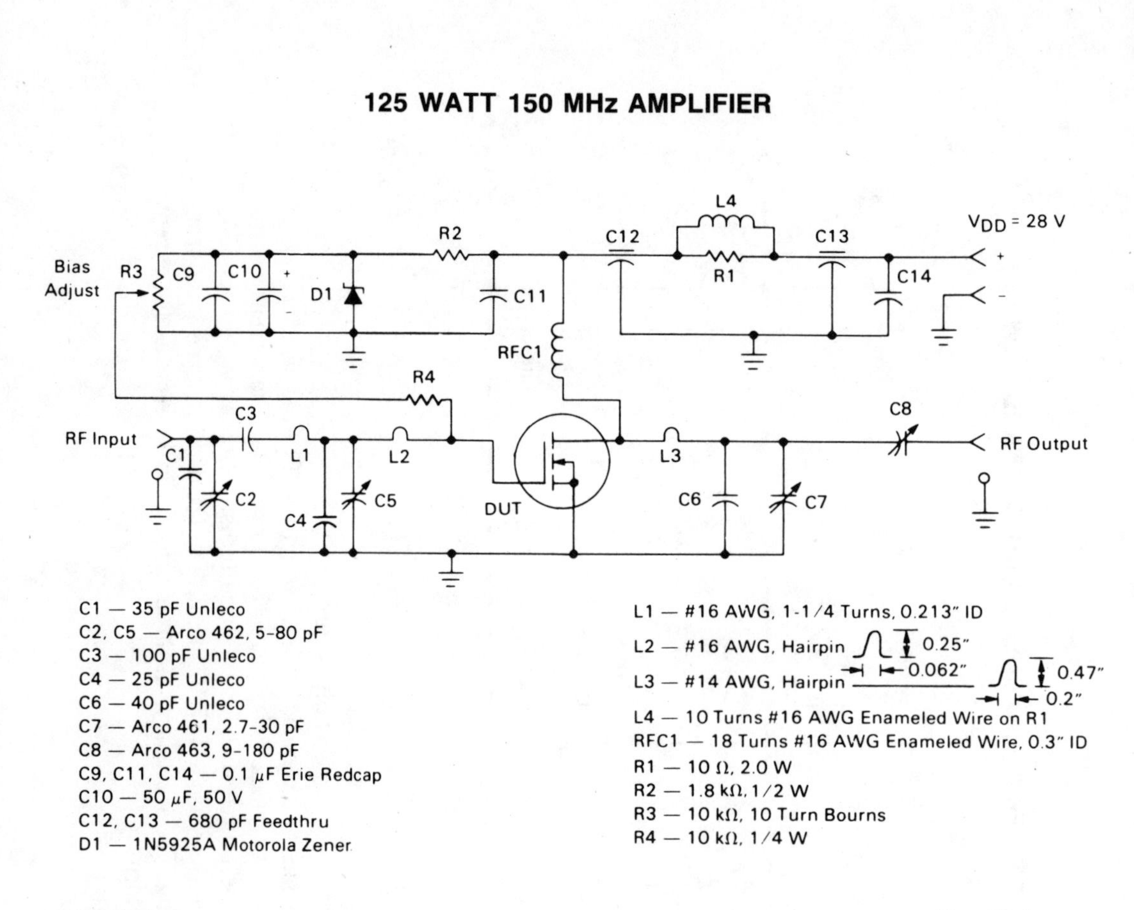

MOTOROLA *Fig. 85-5*

Circuit Notes

This amplifier operates from a 28 Vdc supply. It has a typical gain of 12 dB, and can survive operation into a 30:1 VSWR load at any phase angle with no damage. The amplifier has an AGC range in excess of 20 dB. This means that with input power held constant at the level that provides 125 watts output, the output power may be reduced to less than 1.0 watt continuously by driving the dc gate voltage negative from its I_{DQ} value.

6-METER KILOWATT AMPLIFIER

Circuit Notes

The amplifier uses a grounded grid circuit with either the Eimac 3CX1000A7 or 8877, ceramic/metal triodes intended for linear service in the HF and VHF ranges. The amp provides the legal power output of 1500 watts PEP and CW service with no effort and requires a driver delivering between 50 and 80 watts at 50 MHz. With a plate voltage of 3000 volts at 0.8 amps the amplifier performs at 60 percent efficiency. The grid is grounded by means of the grid ring of the 3CX1000A7 socket providing a low-inductance path to ground. The amplifier is completely stable.

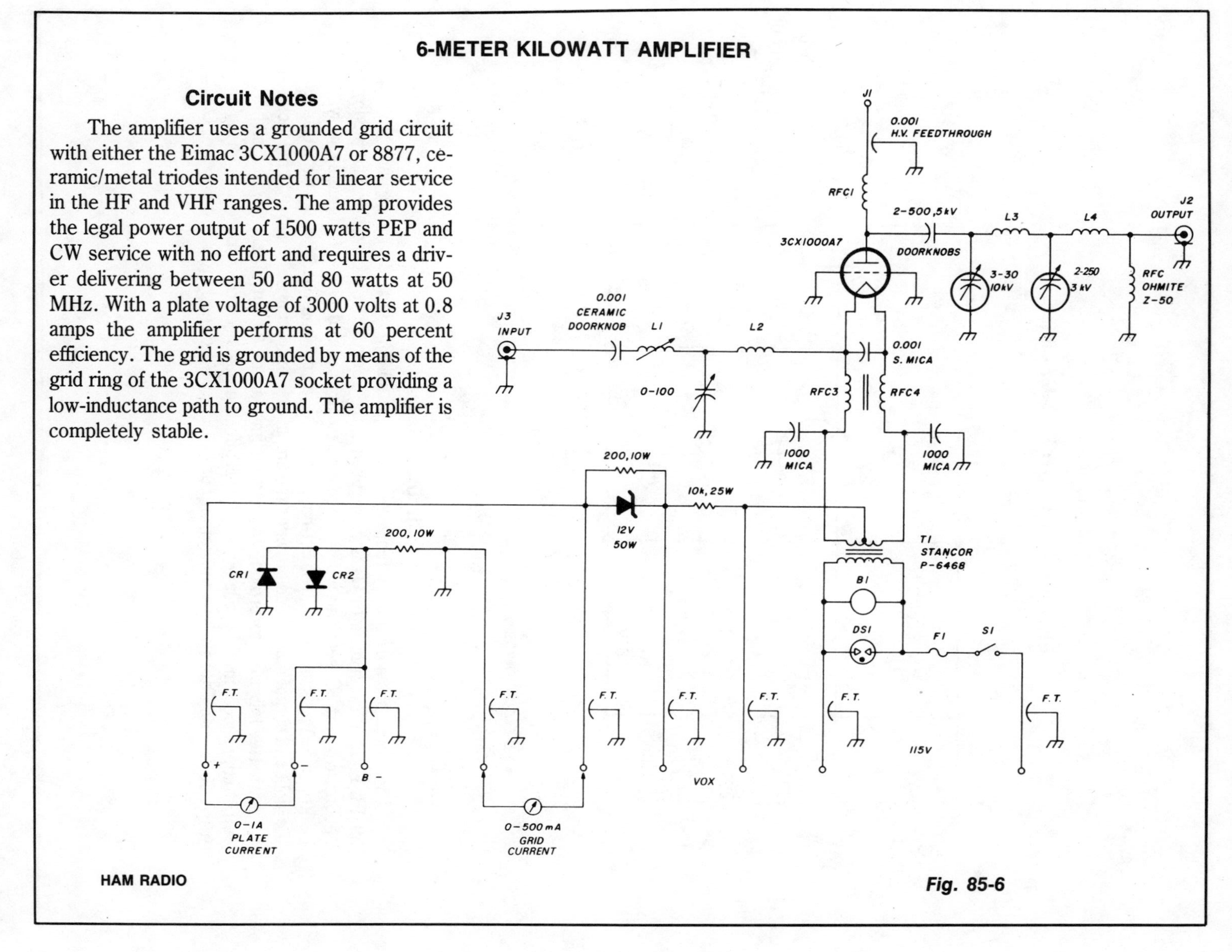

HAM RADIO

Fig. 85-6

BROADCAST-BAND RF AMPLIFIER

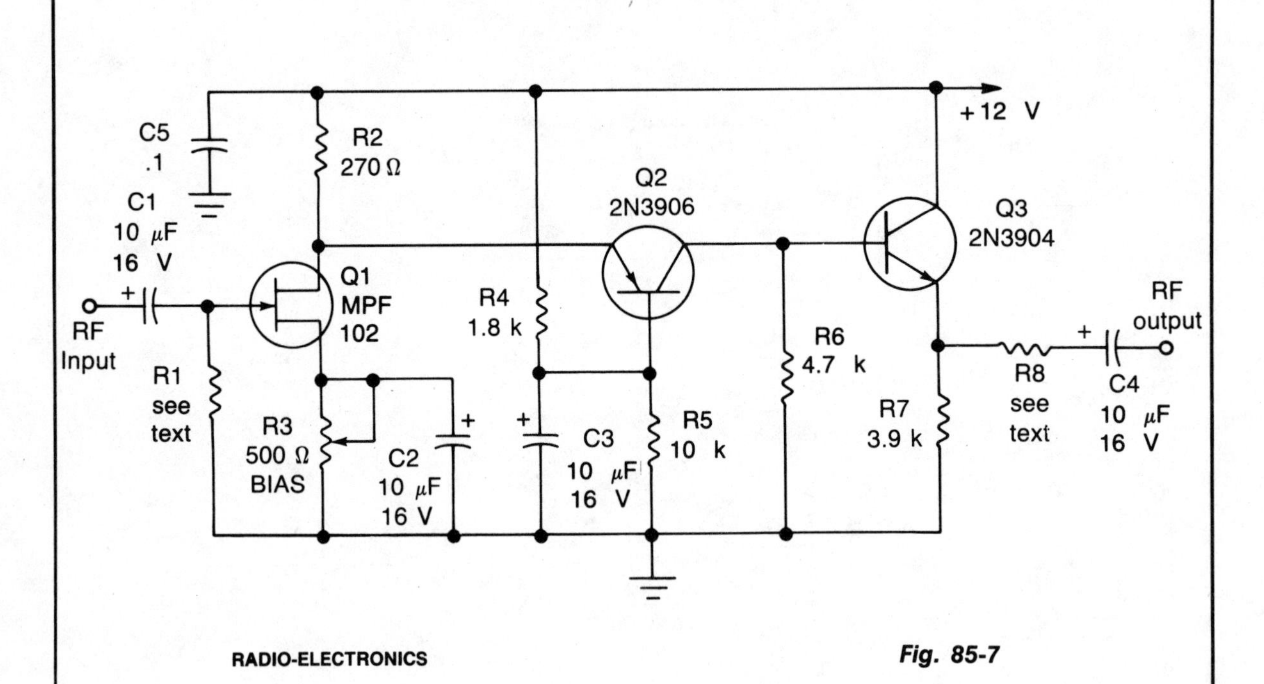

RADIO-ELECTRONICS

Fig. 85-7

Circuit Notes

The circuit has a frequency response ranging from 100 Hz to 3 MHz; gain is about 30 dB. Field-effect transistor Q1 is configured in the common-source self-biased mode; optional resistor R1 sets the input impedance to any desired value. Commonly, it will be 50 ohms. The signal is then direct-coupled to Q2, a common-base circuit that isolates the input and output stages and provides the amplifier's exceptional stability. Q3 functions as an emitter follower, to provide low output impedance (about 50 ohms). For higher output impedance, include resistor R8. It will affect impedance according to this formula: R8 $\sim R_{out} - 50$. Otherwise, connect output capacitor C4 directly to the emitter of Q3.

IMPROVED RF ISOLATION AMPLIFIER

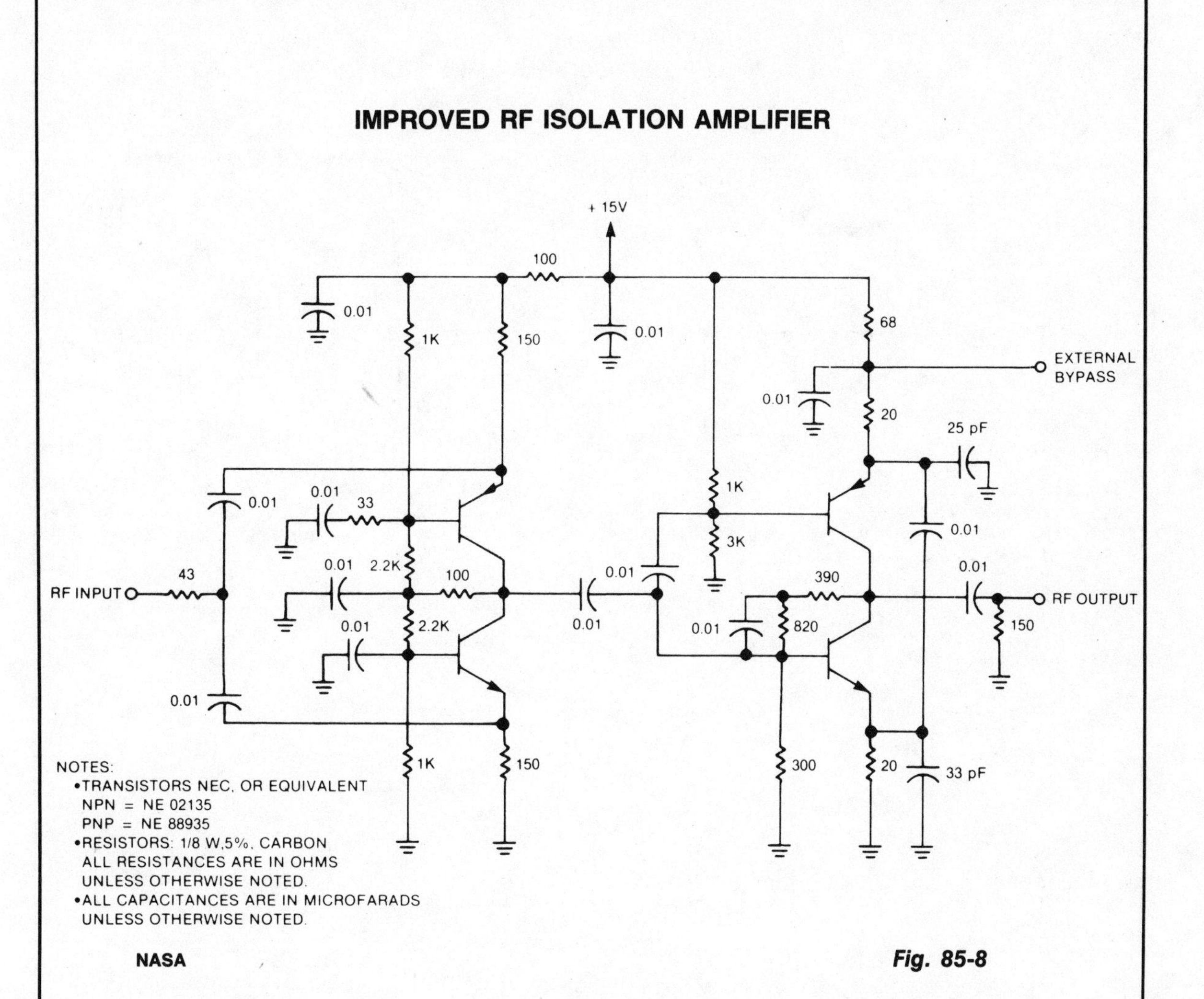

NASA *Fig. 85-8*

Circuit Notes

This wideband RF isolation amplifier has a frequency response of 0.5 to 400 MHz ± 0.5 dB. This two stage amplifier can be used in applications requiring high reverse isolation, such as receiver intermediate-frequency (IF) strips and frequency distribution systems. Both stages use complementary-symmetry transistor arrangements. The input stage is a common-base connection for the complementary circuit. The output stage, which supplies the positive gain, is a common-emitter circuit using emitter degeneration and collector-base feedback for impedance control.

A 10 WATT 225-400 MHz AMPLIFIER

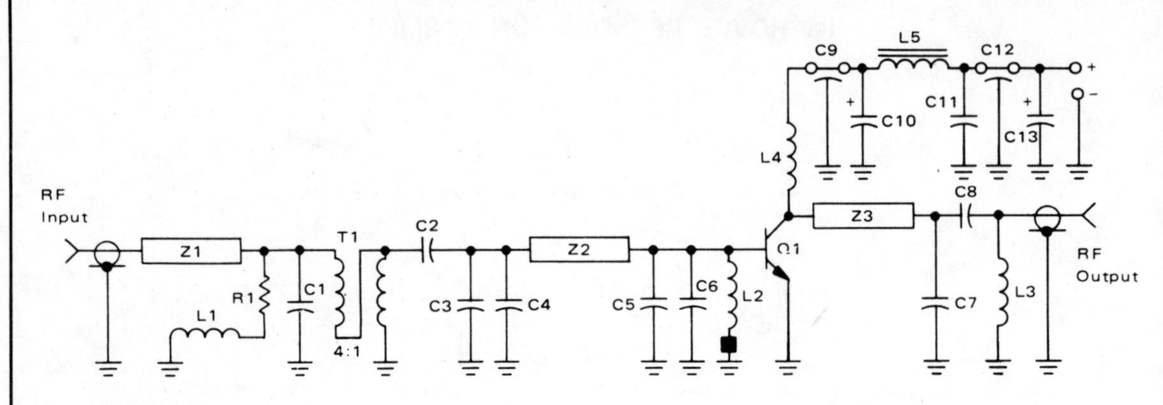

C1 – 8.2 pF Chip*
C2 – 270 pF Chip*
C3 – 36 pF Chip*
C4, C7 – 15 pF Chip*
C5, C6 – 50 pF Chip*
C8 – 82 pF Chip*
C9, C12 – 680 pF Feedthru
C10, C13 – 1.0 μF 50 V Tantalum
C11 – 0.1 μF Erie Redcap

L1, L3 – 3 Turns #22 AWG 1/8" (3.175 mm) ID
L2 – 0.15 μH Molded Choke
L4 – 0.15 μH Molded Choke with Ferroxcube Bead (Ferroxcube 56 590-65/4B on Ground End of Choke)
L5 – Ferroxcube VK200-19/4B

*100 mil A.C.I. Chip Capacitors

R1 – 36 Ω 1/4 Watt

T1 – 25 Ω Subminiature Coax (Type UT34-25) – 1.75 inches (44.45 mm) long

Z1 – Microstrip Line
720 mils L X 162 mils W
18.29 mm L X 4.115 mm W

Z2 – Microstrip Line
680 mils L X 162 mils W
17.27 mm L X 4.115 mm W

Z3 – Microstrip Line
2200 mils L X 50 mils W
55.88 mm L X 1.27 mm W

Board – 0.0625" (1.588 mm) Glass Teflon, $\epsilon_r = 2.56$

Q1 – MRF331

MOTOROLA

Fig. 85-9

SCHEMATIC REPRESENTATION

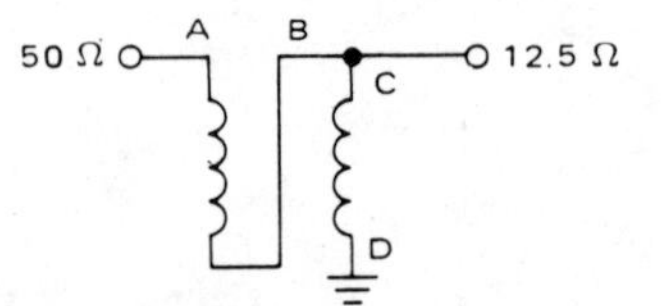

ASSEMBLY AND PICTORIAL

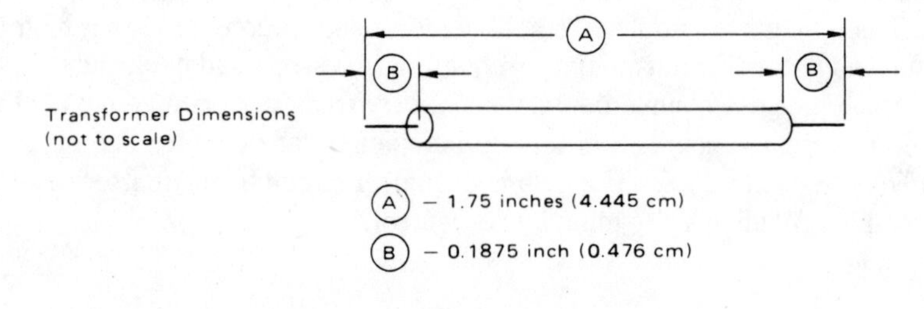

A 10 WATT 225-400 MHz AMPLIFIER, Continued.

Transformer Connections

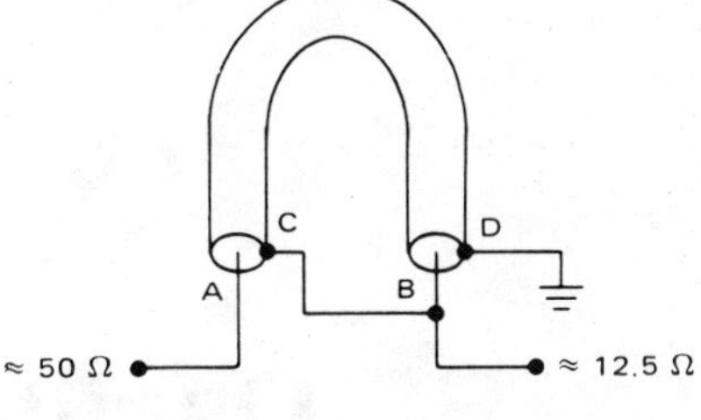

Circuit Notes

This broadband amplifier covers the 225-400 MHz military communications band producing 10 watt RF output power and operating from a 28 volt supply. The amplifier can be used as a driver for higher power devices such as 2N6439 and MRF327. The circuit is designed to be driven by a 50 ohm source and operate into a nominal 50 ohm load. The input matching network consists of a section composed of C3, C4, Z2, C5 and C6. C2 is a dc blocking capacitor, and T1 is a 4:1 impedance ratio coaxial transformer. Z1 is a 50 ohm transmission line. A compensation network consisting of R1, C1, and L1 is used to improve the input VSWR and flatten the gain response of the amplifier. L2 and a small ferrite bead make up the base bias choke. The output network is made up of a microstrip L-section consisting of Z3 and C7, and a high pass section consisting of C8 and L3. C8 also serves as a dc blocking capacitor. Collector decoupling is accomplished through the use of L4, L5, C9, C10, C11, C12, and C13.

86

RF Oscillators

The sources of the following circuits are contained in the Sources section beginning on page 694. The figure number contained in the box of each circuit correlates to the source entry in the Sources section.

5 MHz VFO

5 MHz VFO

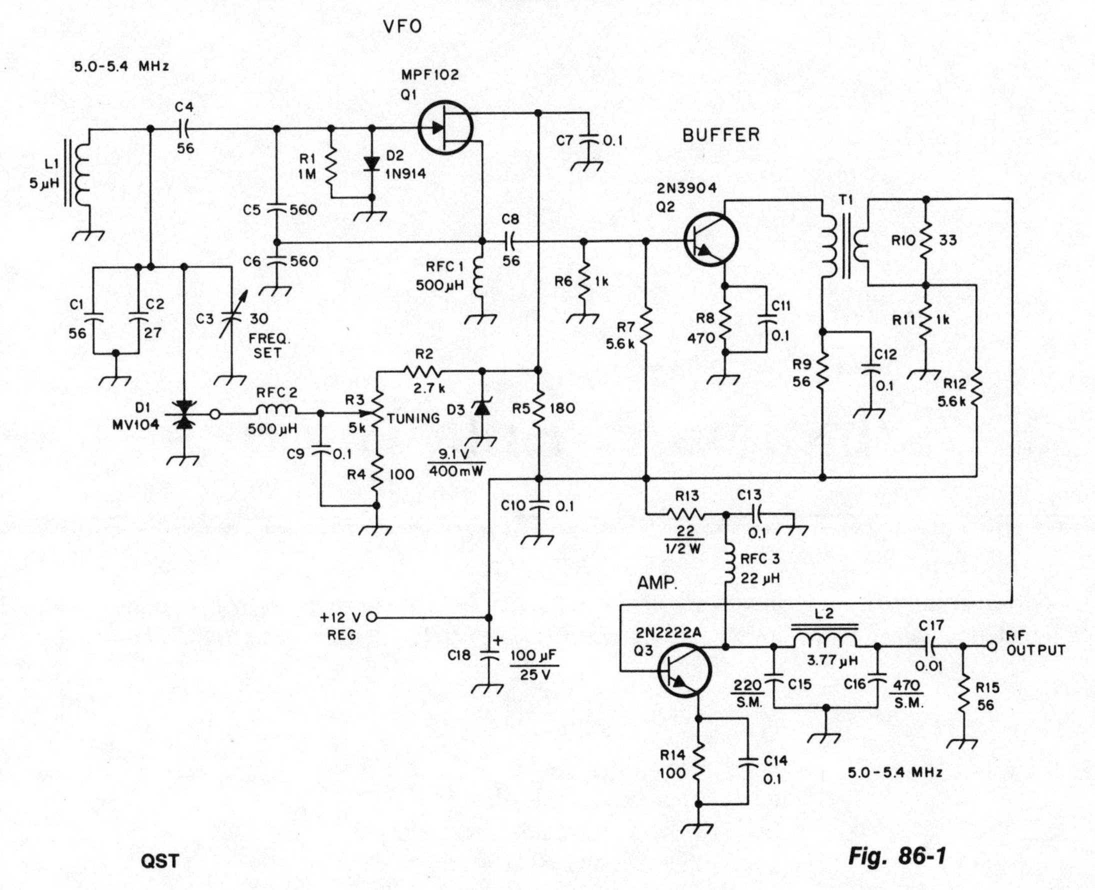

QST

Fig. 86-1

Circuit Notes

A JFET (Q1) serves as the oscillator. D2 helps to stabilize the transistor by limiting positive sinewave peaks and stabilizing the bias. Output from Q1 is supplied to a class A buffer, Q2. It operates as a broadband amplifier by means of T1, which is untuned. Output amplifier Q3 is also a class A stage. A low-pass, single-section filter is used at the output of Q3 to remove some of the harmonic currents generated within the system. The filter output impedance is 50 ohms. The injection level to the mixer is 600 mV p-p.

87

Sample-and-Hold Circuits

The sources of the following circuits are contained in the Sources section beginning on page 694. The figure number contained in the box of each circuit correlates to the source entry in the Sources section.

SAMPLE-AND-HOLD CIRCUIT

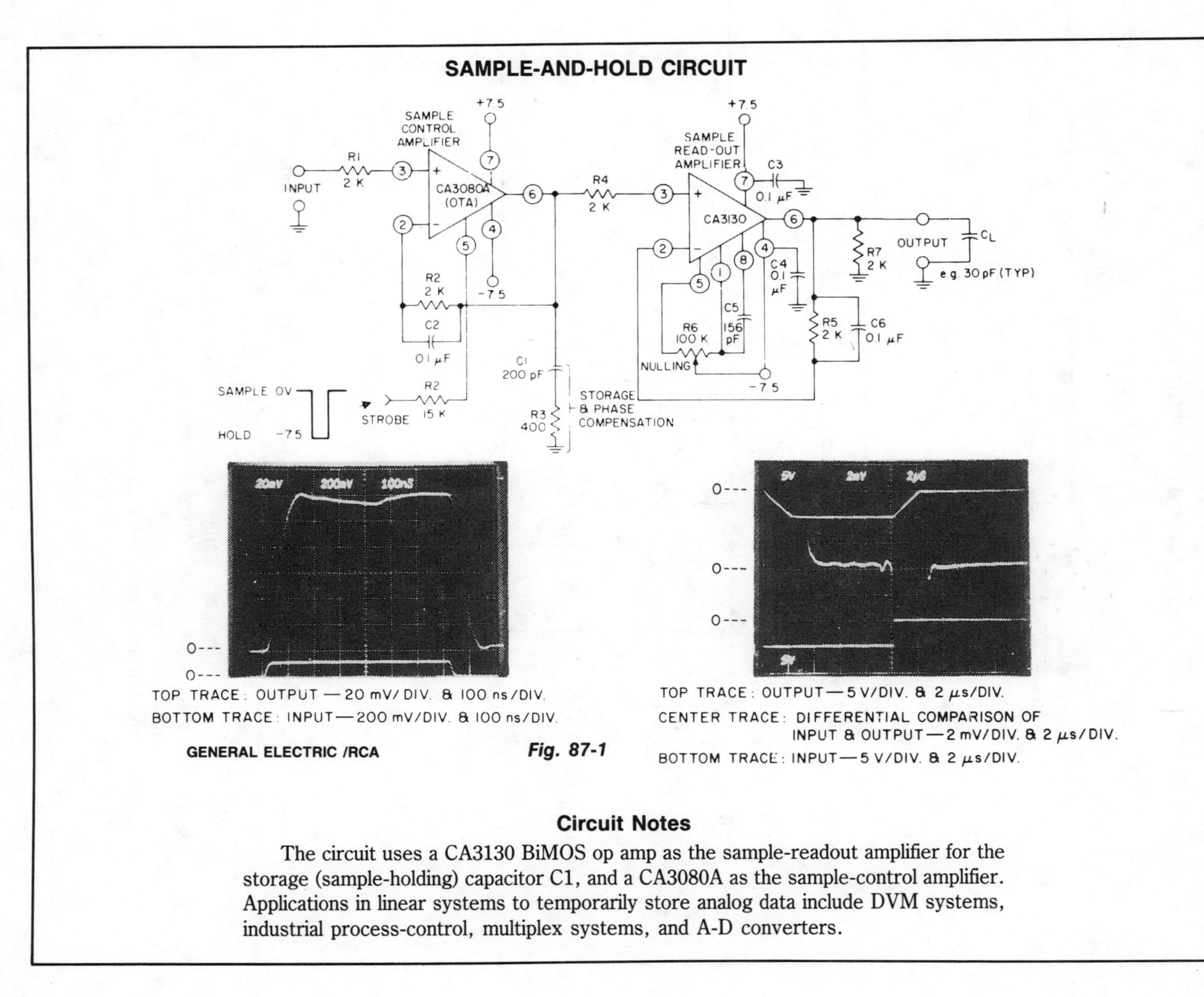

TOP TRACE: OUTPUT — 20 mV/DIV. & 100 ns/DIV.
BOTTOM TRACE: INPUT — 200 mV/DIV. & 100 ns/DIV.

TOP TRACE: OUTPUT — 5 V/DIV. & 2 μs/DIV.
CENTER TRACE: DIFFERENTIAL COMPARISON OF INPUT & OUTPUT — 2 mV/DIV. & 2 μs/DIV.
BOTTOM TRACE: INPUT — 5 V/DIV. & 2 μs/DIV.

GENERAL ELECTRIC /RCA

Fig. 87-1

Circuit Notes

The circuit uses a CA3130 BiMOS op amp as the sample-readout amplifier for the storage (sample-holding) capacitor C1, and a CA3080A as the sample-control amplifier. Applications in linear systems to temporarily store analog data include DVM systems, industrial process-control, multiplex systems, and A-D converters.

SAMPLE AND HOLD CIRCUIT II

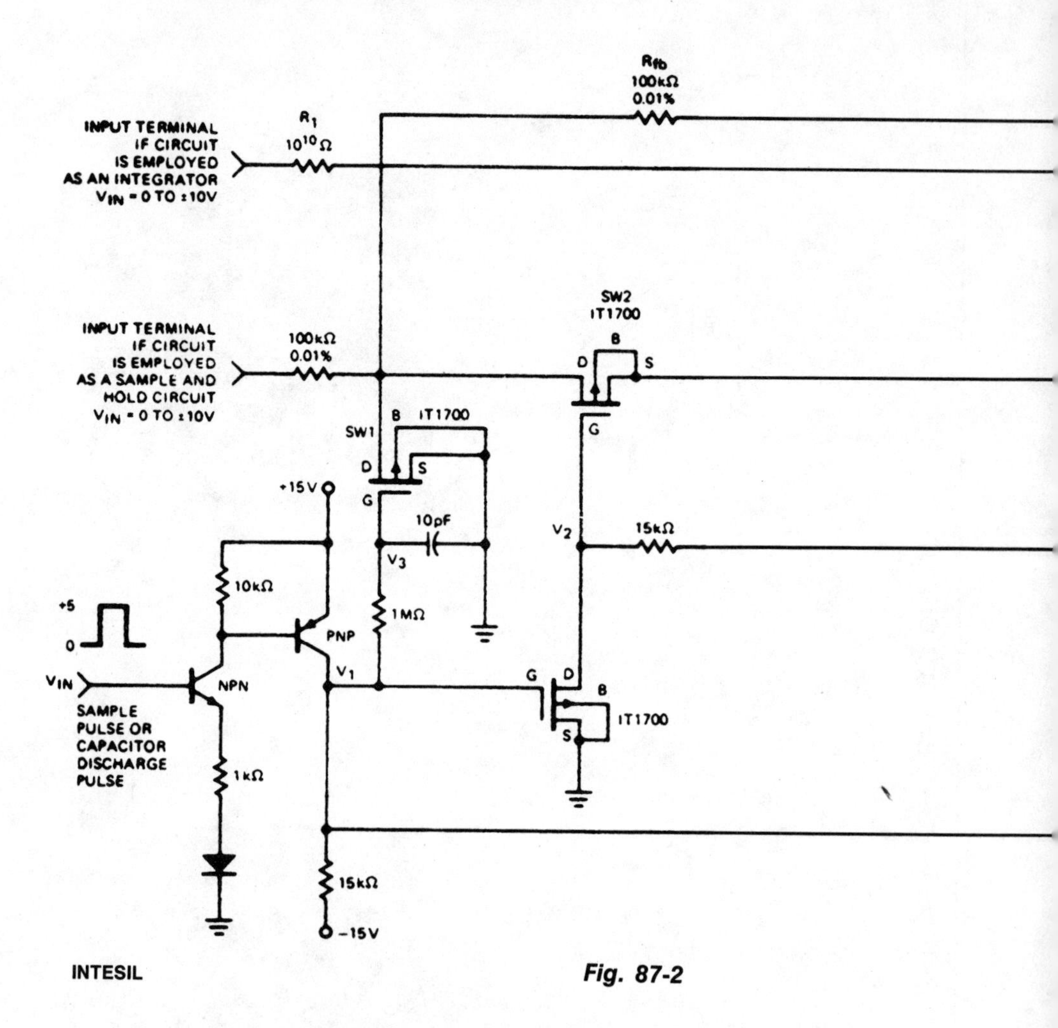

INTESIL

Fig. 87-2

Circuit Notes

This circuit rapidly charges capacitor C_{STO} to a voltage equal to an input signal. The input signal is then electrically disconnected from the capacitor with the charge still remaining on C_{STO}. Since C_{STO} is in the negative feedback loop of the operational amplifier, the output voltage of the amplifier is equal to the voltage across the capacitor. Ideally, the voltage across C_{STO} should remain constant causing the output of the amplifier to remain constant as well. However, the voltage across C_{STO} will decay at a rate proportional to the current being injected or taken out of the current summing node of the amplifier. This current can come from four sources: leakage resistance of

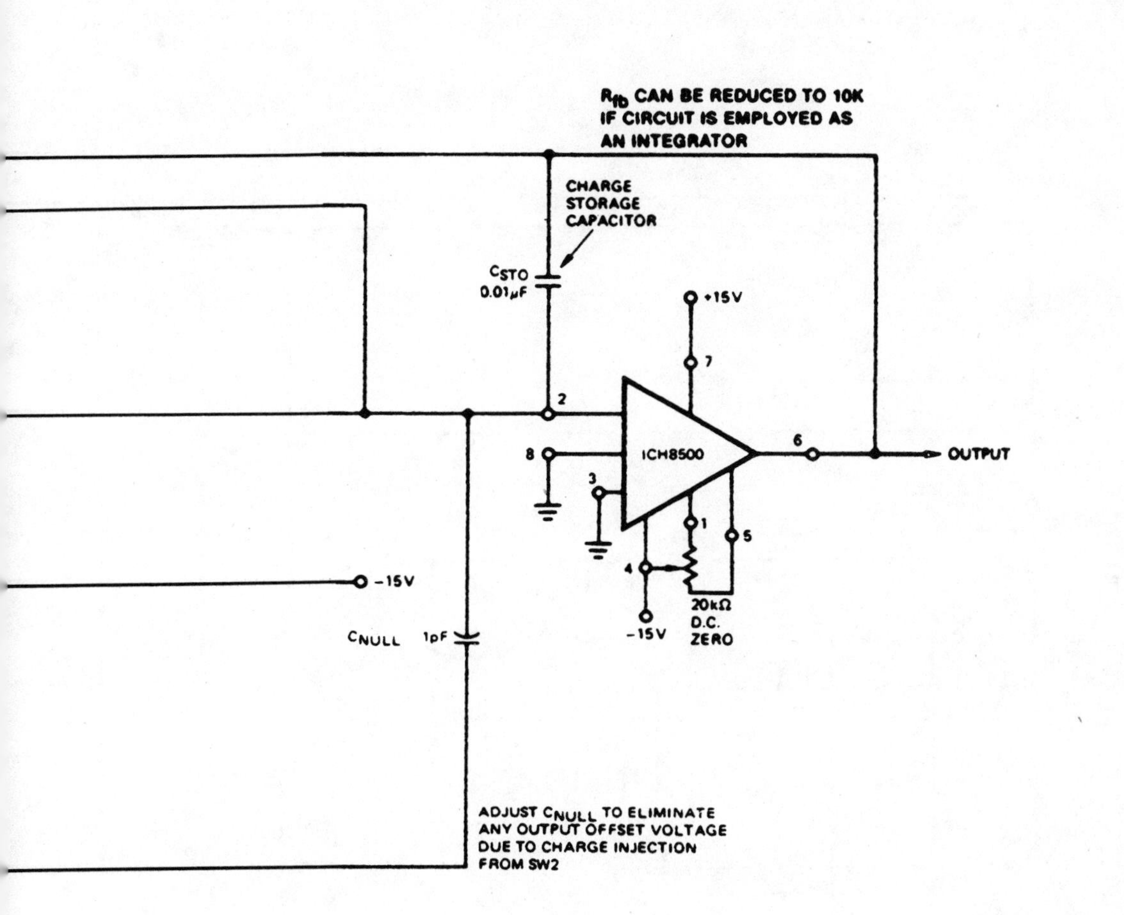

C_{STO}, leakage current due to the solid state switch SW2, currents due to high resistance paths on the circuit fixture, and most important, bias current of the operational amplifier. If the ICH8500A operational amplifier is employed, this bias current is almost non-existent (less than 0.01pA). Note that the voltages on the source, drain and gate of switch SW2 are zero or near zero when the circuit is in the hold mode. Careful construction will eliminate stray resistance paths and capacitor resistance can be eliminated if a quality capacitor is selected. The net result is a low drift sample and hold circuit.

The circuit can double as an integrator. In this application the input voltage is applied to the integrator input terminal. The time constant of the circuit is the product of R1 and C_{STO}.

FAST, PRECISION SAMPLE-HOLD

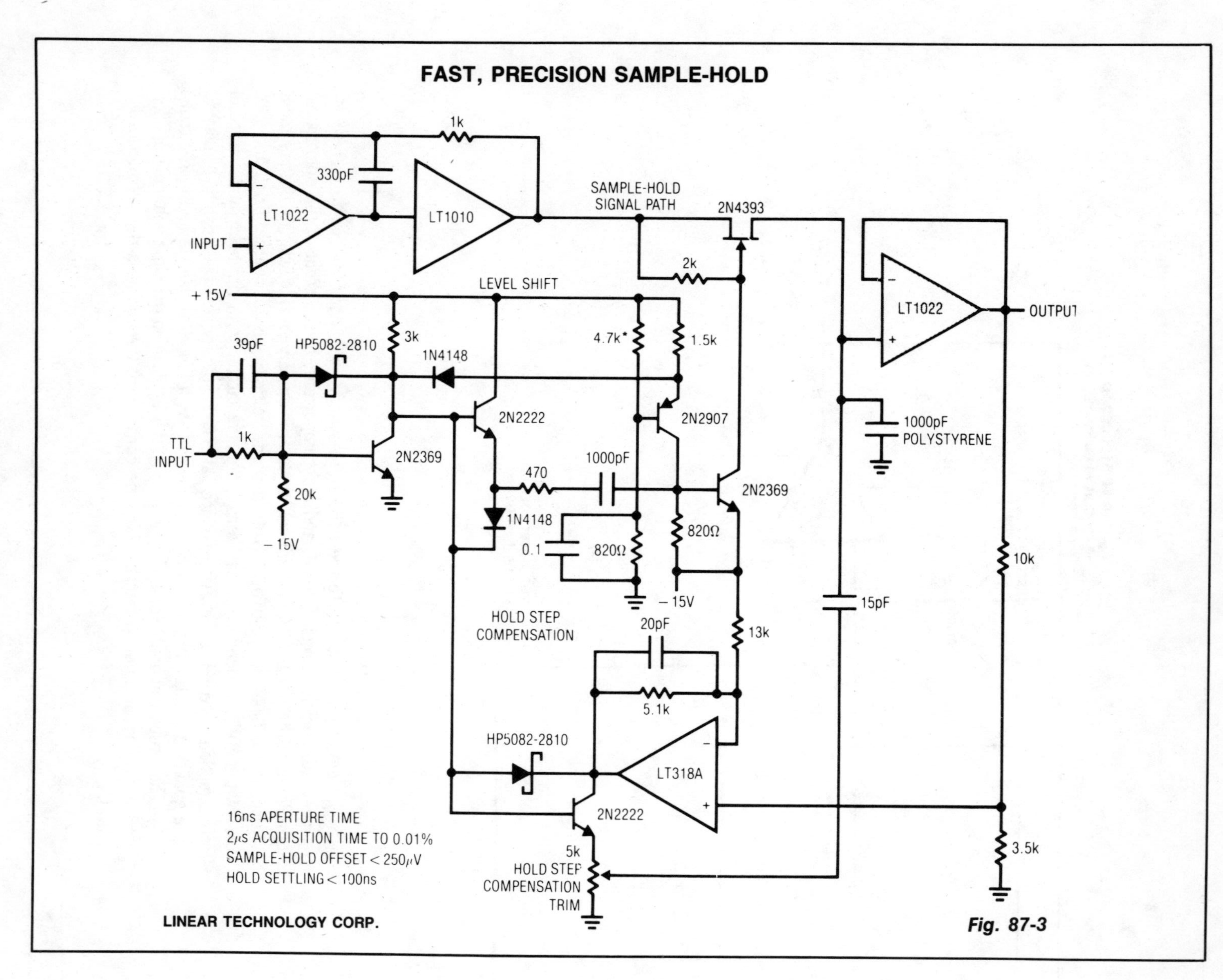

Fig. 87-3

HIGH PERFORMANCE SAMPLE AND HOLD

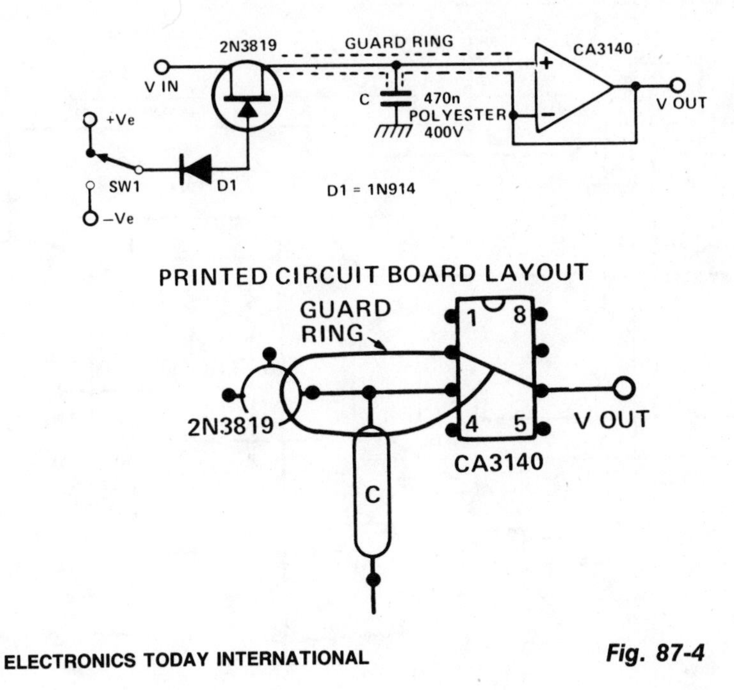

ELECTRONICS TODAY INTERNATIONAL

Fig. 87-4

Circuit Notes

When switch SW1 is positive, the FET is turned on, and has a resistance of about 400 ohm. The input voltage charges up the capacitor through the FET. When SW1 is negative, the FET is turned off (pinched off). To get a long storage time, the op amp must have a very low input bias current. For the CA3140, this current is about 10 pico amps. The rate at which the capacitor will be discharged by this current is based on the equation, C (dv/dt) = i where dv/dt is the rate of change of voltage on the capacitor. Therefore:

$$\frac{dv}{dt} = \frac{i}{C} = \frac{10^{-11}}{0.47 \times 10^{-6}} = 22\mu V/s$$

INFINITE SAMPLE AND HOLD AMPLIFIER

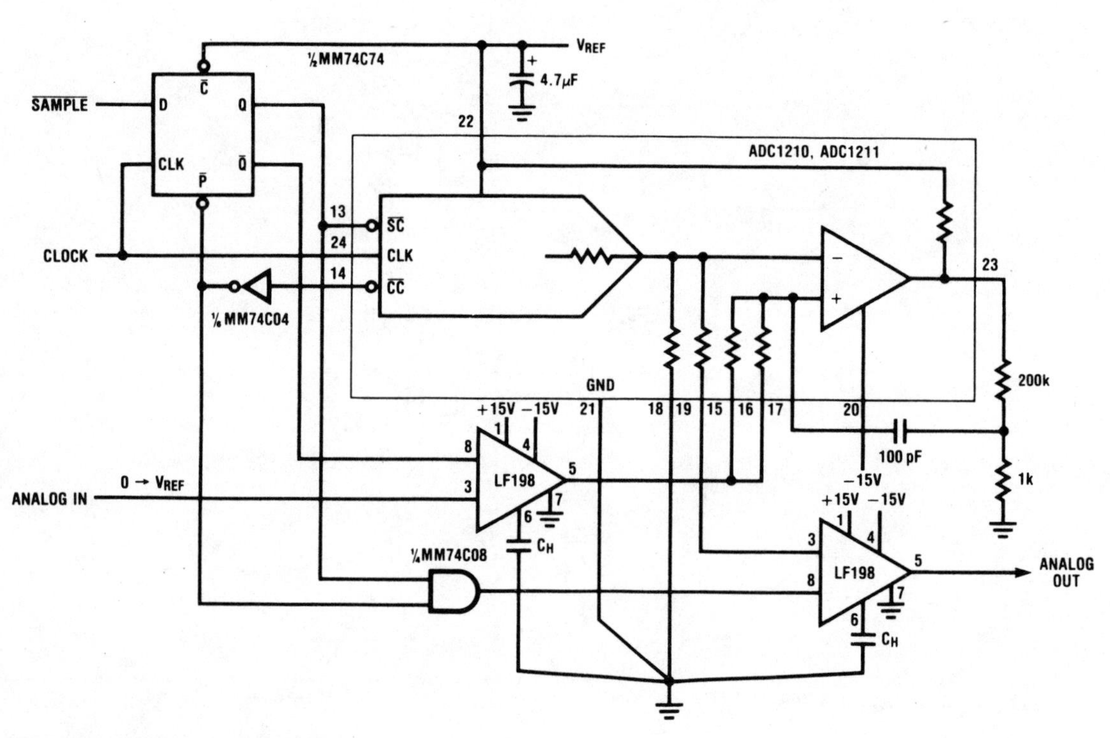

NATIONAL SEMICONDUCTOR CORP.

Fig. 87-5

Circuit Notes

During normal "hold" mode, the replicated analog voltage is buffered straight through the S/H amplifier to the output. Upon issuance of a SAMPLE signal, the S/H amplifier is placed in the hold mode, holding the voltage until the new analog voltage is valid. The same SAMPLE signal triggers an update to the input sample-and-hold amplifier. The most current analog voltage is captured and held for conversion. The previously determined voltage is held stable at the output during the conversion cycle while the SAR/D-to-A converter continuously adjusts to replicate the new input voltage. At the end of the conversion, the output sample-and-hold amplifier is once again placed in the track mode. The new analog voltage is then regenerated.

CHARGE COMPENSATED SAMPLE AND HOLD

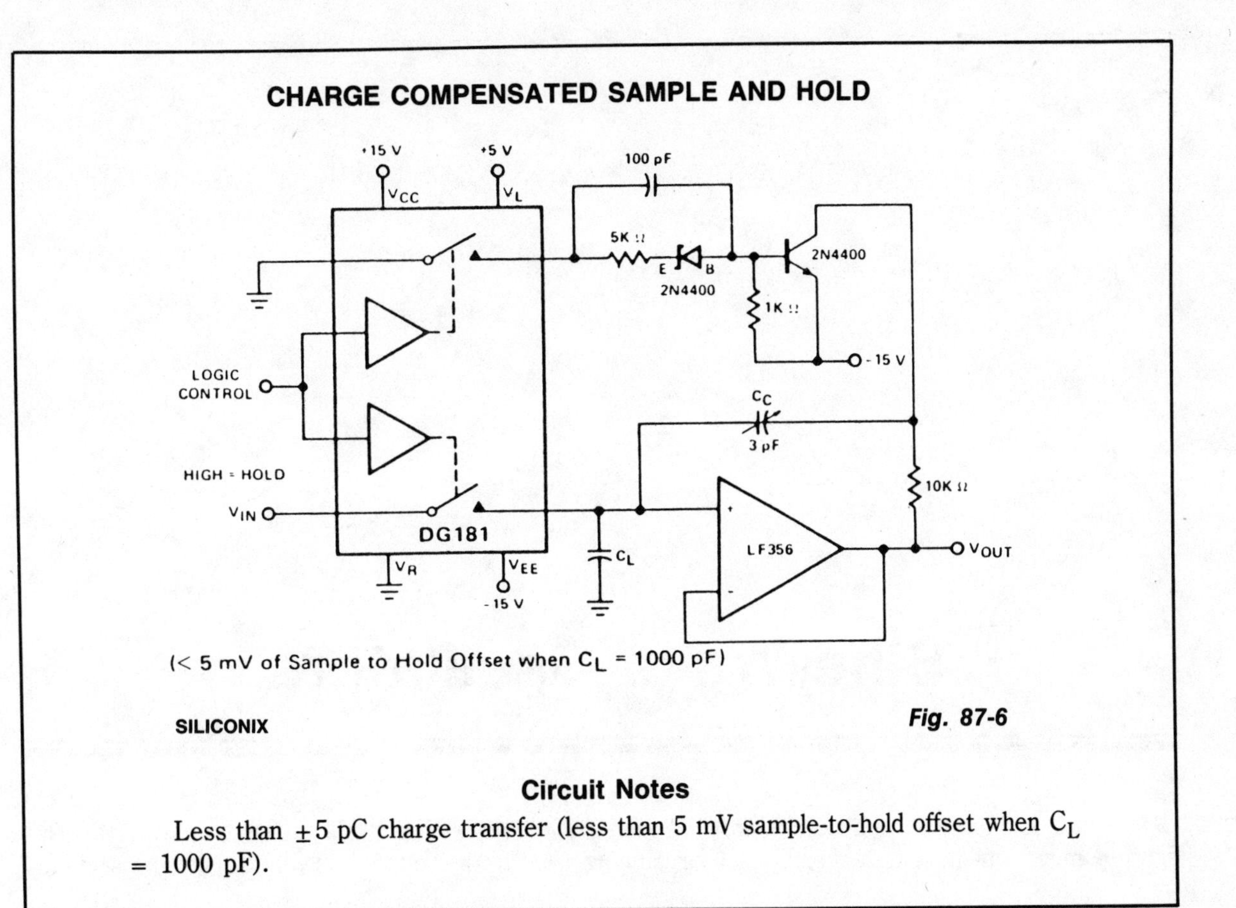

(< 5 mV of Sample to Hold Offset when C_L = 1000 pF)

SILICONIX *Fig. 87-6*

Circuit Notes

Less than ±5 pC charge transfer (less than 5 mV sample-to-hold offset when C_L = 1000 pF).

88

Sine-Wave Oscillators

The sources of the following circuits are contained in the Sources section beginning on page 694. The figure number contained in the box of each circuit correlates to the source entry in the Sources section.

SINE-WAVE SHAPER

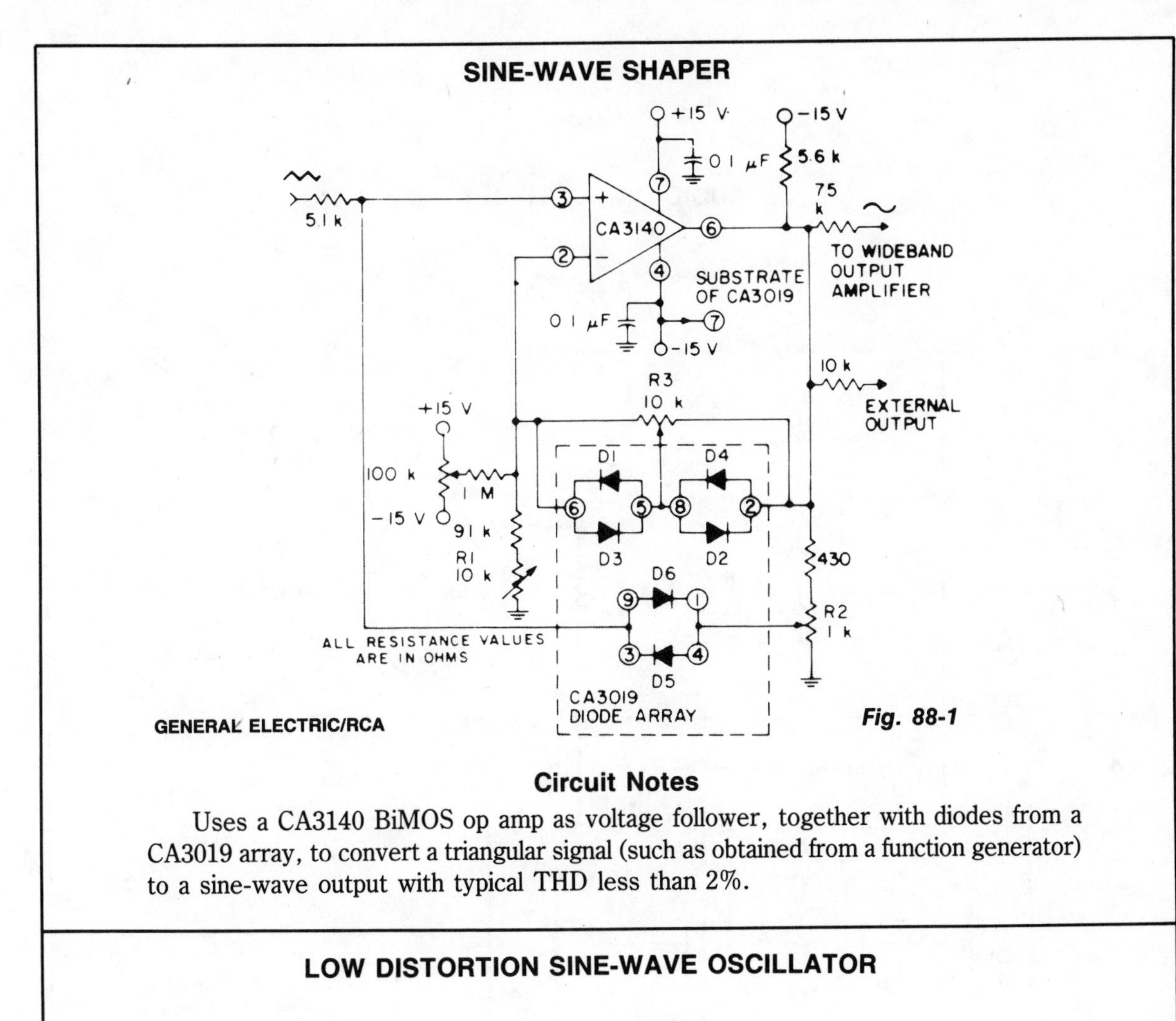

GENERAL ELECTRIC/RCA

Fig. 88-1

Circuit Notes

Uses a CA3140 BiMOS op amp as voltage follower, together with diodes from a CA3019 array, to convert a triangular signal (such as obtained from a function generator) to a sine-wave output with typical THD less than 2%.

LOW DISTORTION SINE-WAVE OSCILLATOR

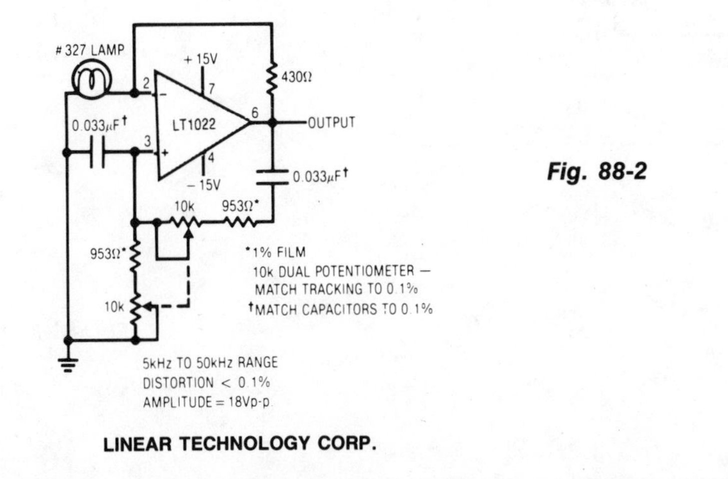

Fig. 88-2

LINEAR TECHNOLOGY CORP.

AUDIO OSCILLATOR

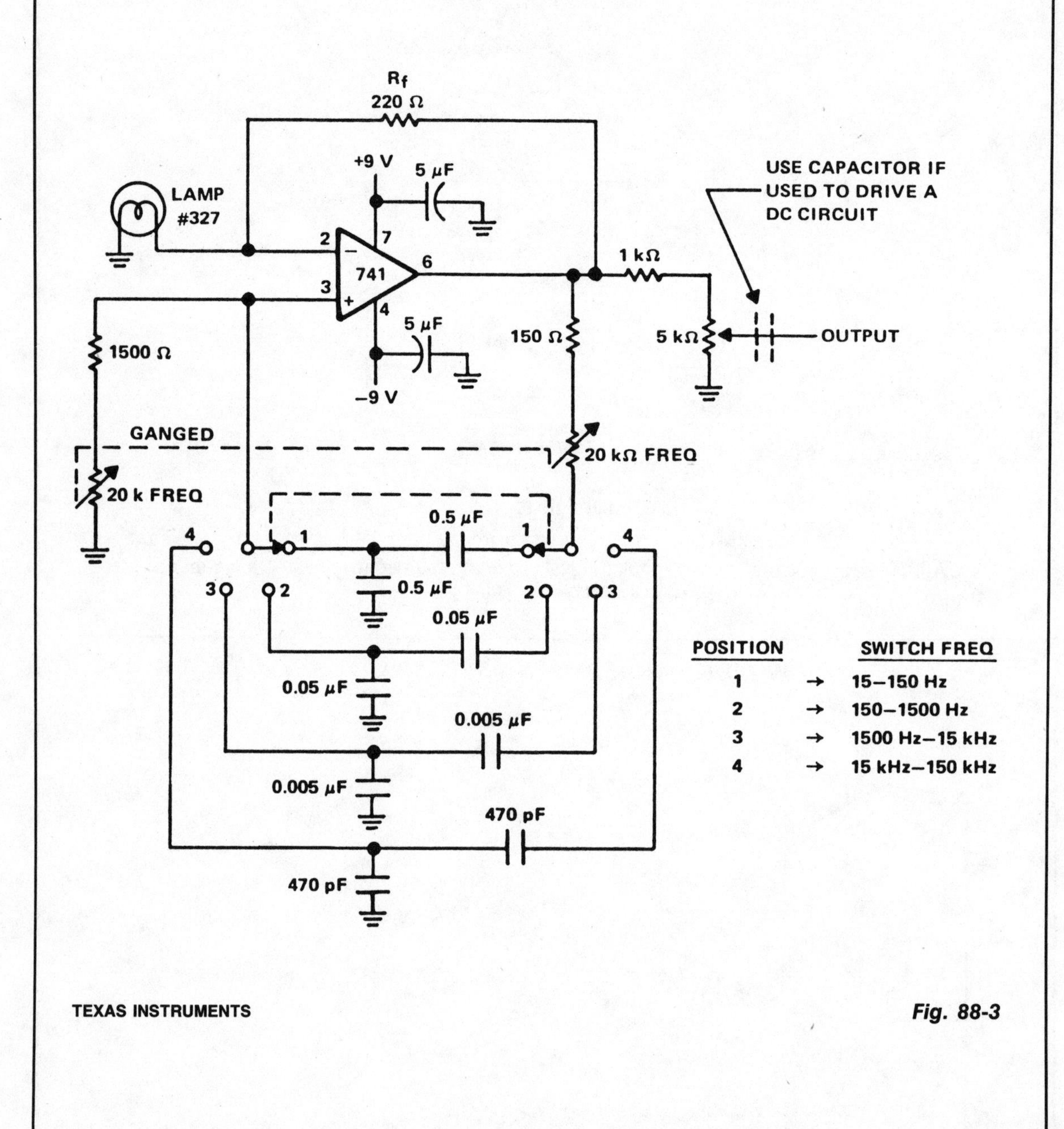

POSITION		SWITCH FREQ
1	→	15–150 Hz
2	→	150–1500 Hz
3	→	1500 Hz–15 kHz
4	→	15 kHz–150 kHz

TEXAS INSTRUMENTS

Fig. 88-3

Circuit Notes

A Wien bridge oscillator produces sine waves with very low distortion level. The Wien bridge oscillator produces zero phase shift at only one frequency ($f = ½ \pi RC$) which will be the oscillation frequency. Stable oscillation can occur only if the loop gain remains at unity at the oscillation frequency. The circuit achieves this control by using the positive temperature coefficient of a small lamp to regulate gain (R_f/R_{LAMP}) as the oscillator attempts to vary its output. The oscillator shown here has four frequency bands covering about 15 Hz to 150 kHz. The frequency is continuously variable within each frequency range with ganged 20 k ohm potentiometers. The oscillator draws only about 4.0 mA from the 9-V batteries. Its output is from 4 to 5 V with a 10 k ohm load and the R_f (feedback resistor) is set at about 5% below the point of clipping. As shown, the center arm of the 5 k ohm output potentiometer is the output terminal. To couple the oscillator to a dc type circuit, a capacitor should be inserted in series with the output lead.

SIMPLE AUDIO SINE-WAVE GENERATOR

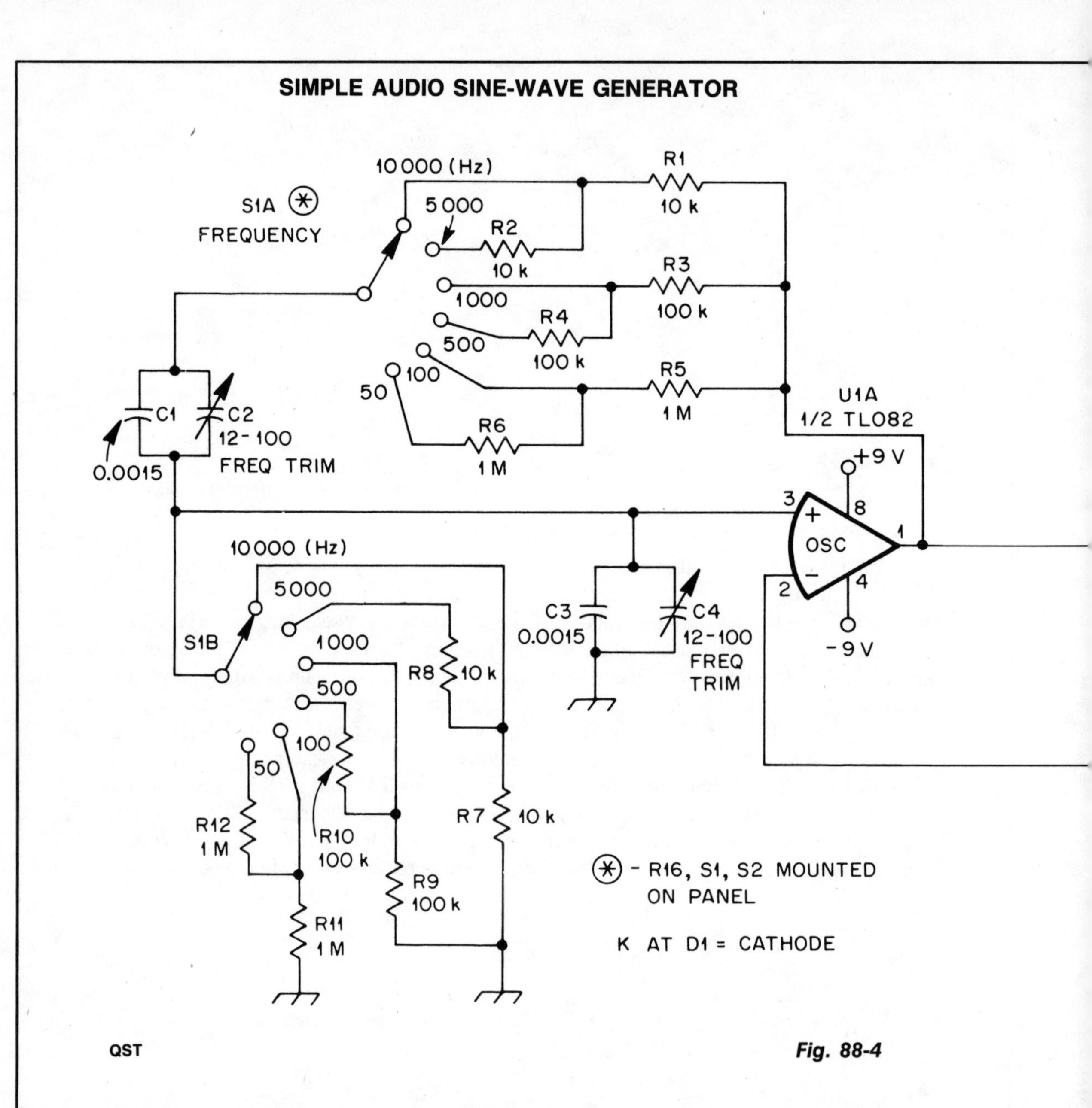

QST

Fig. 88-4

Circuit Notes

U1A, an op amp, oscillates at the frequency at which the phase shift in the Wien bridge network is exactly zero degrees. Changing bridge component values changes the oscillator frequency. In this circuit, we need change only the two resistors to do this. S1A chooses a value among R1 through R6, and S1B similarly selects a value from R7 through R12. U1A must provide enough gain to overcome losses in the bridge, but not so much gain that oscillation builds up to the point of overload and distortion. U2 and

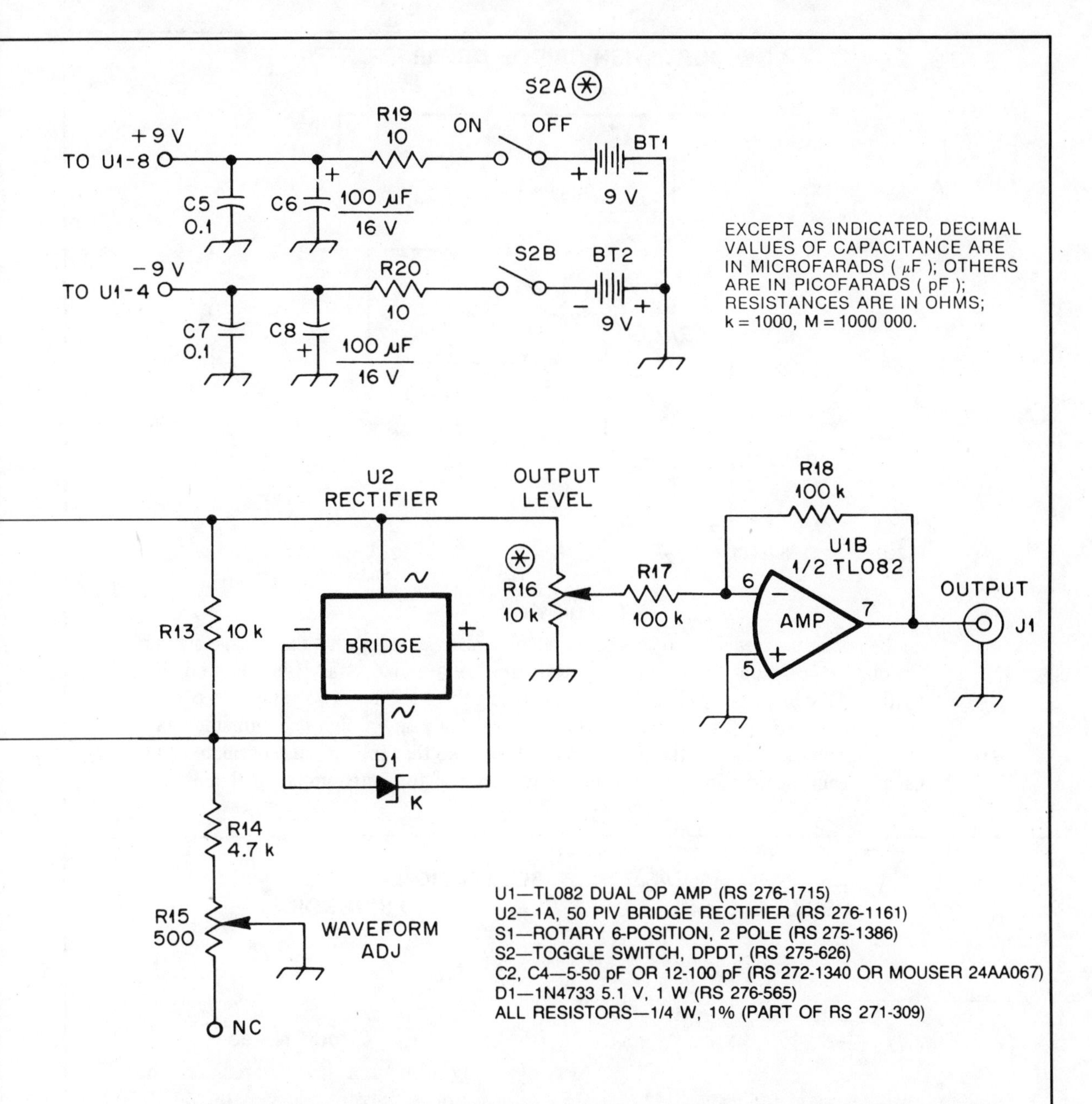

C1 automatically regulate circuit gain to maintain oscillation. U2 places D1 across R13 with the proper polarity on both positive and negative alterations of the signal at pin 1 of U1. As the voltage at pin 1 of U1 approaches its peak value, D1 enters its Zener breakdown region, effectively shunting R13 with a resistive load. This increases the amount of negative feedback around U1, reducing its gain. R15, WAVEFORM ADJ, allows you to optimize circuit operation for lowest distortion. U1B provides isolation between oscillator and load. With the values shown for R17 and R18, U1B operates at unity gain.

LOW COST WIEN BRIDGE OSCILLATOR

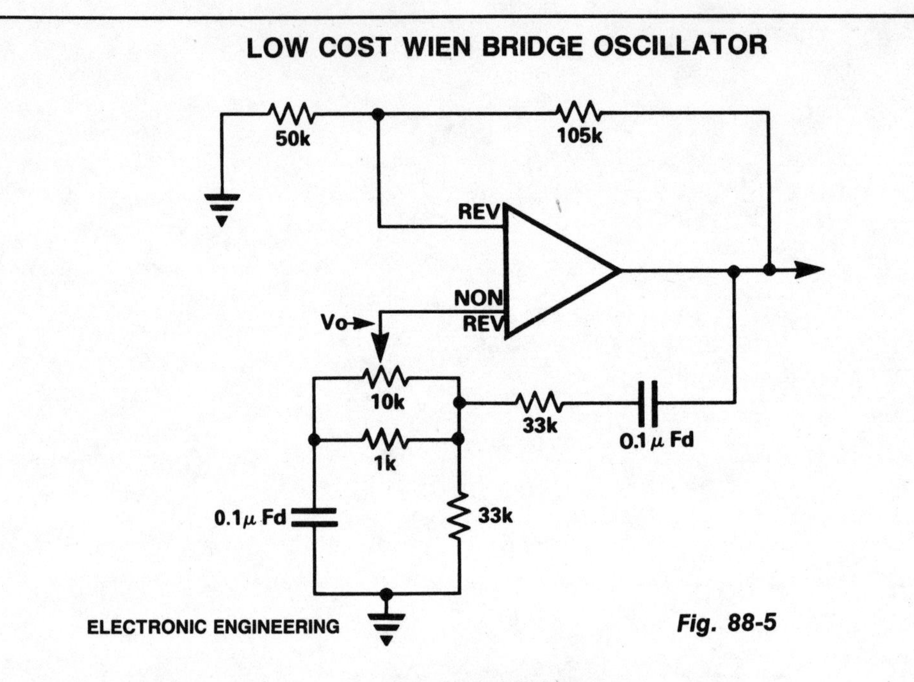

ELECTRONIC ENGINEERING

Fig. 88-5

Circuit Notes

In the circuit the frequency trimming component is arranged so that the voltage across it is in quadrature with the voltage V_0 from the bridge so that as it is adjusted the attenuation of the bridge only changes a little, avoiding the need for a two gang component. The range of variation of frequency is very limited. By using a high gain amplifier and metal film feedback resistors the loop gain can be set so that the unit just oscillates and the use of an automatic gain setting component, a thermistor for example, is eliminated.

MODIFIED UJT RELAXATION OSCILLATOR PRODUCES CLEAN AUDIO SINUSOIDS

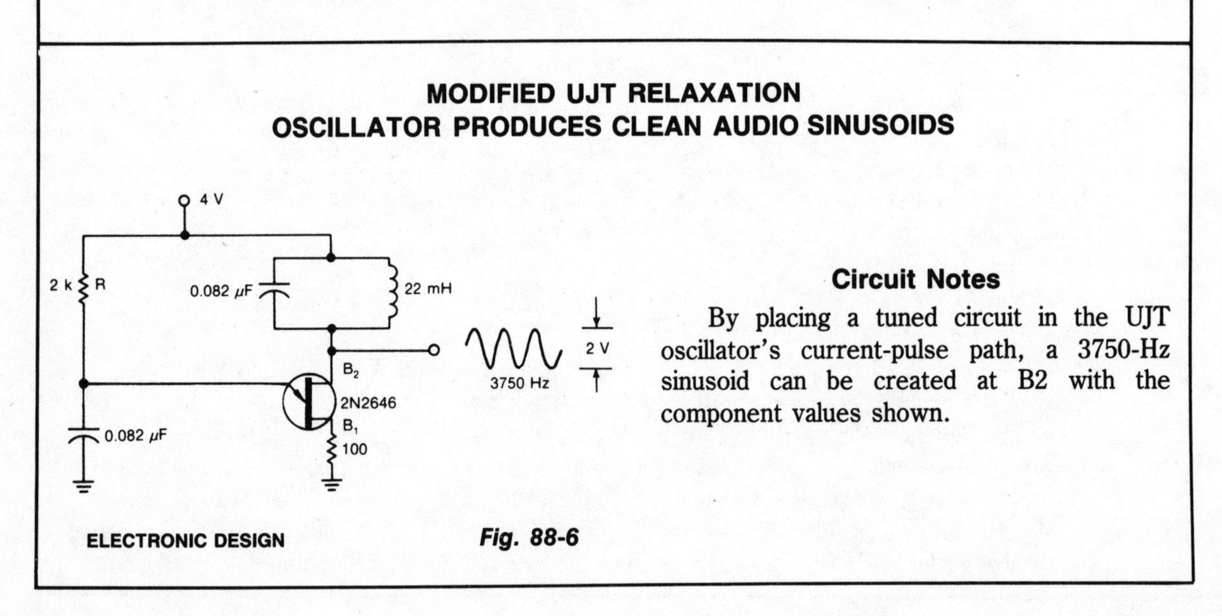

Circuit Notes

By placing a tuned circuit in the UJT oscillator's current-pulse path, a 3750-Hz sinusoid can be created at B2 with the component values shown.

ELECTRONIC DESIGN

Fig. 88-6

A 555 USED AS AN RC AUDIO OSCILLATOR

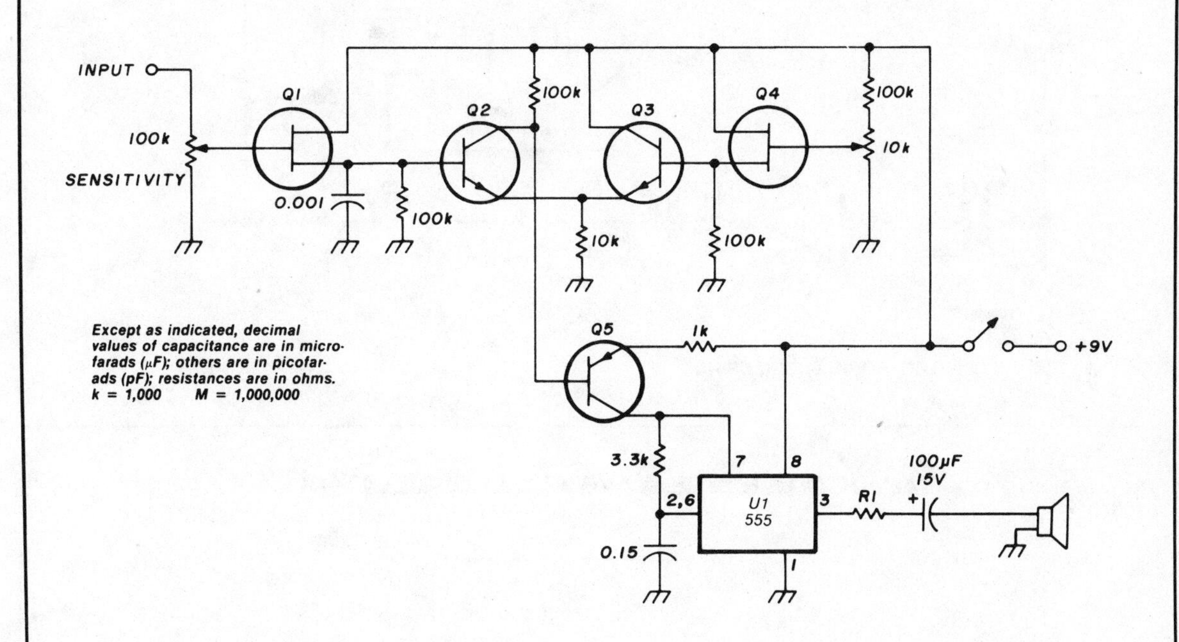

HAM RADIO

Fig. 88-7

Circuit Notes

Transistor Q5 and the 1000 ohm resistor form the variable element needed for controlling the frequency of VCO by limiting the charging current flowing into the 0.15 µF timing capacitor according to the forward bias being applied to Q5. As the voltage on pins 2 and 6 of U1 reach ⅔ V_{CC} (about 6 volts with a 9-volt supply) the timer will fire and pin 3 will be pulled low. Pin 7, an open collector output, goes low and begins to discharge the timing capacitor—through the 3.3 kilohm resistor. The discharge time provided by this resistor assures a reasonable, although asymmetrical, waveform for the aural signal generated by U1. At ⅓ V_{CC} the internal flip-flop resets, the output on pin 3 goes high, the open collector output on pin 7 floats, and the timing cycle begins again.

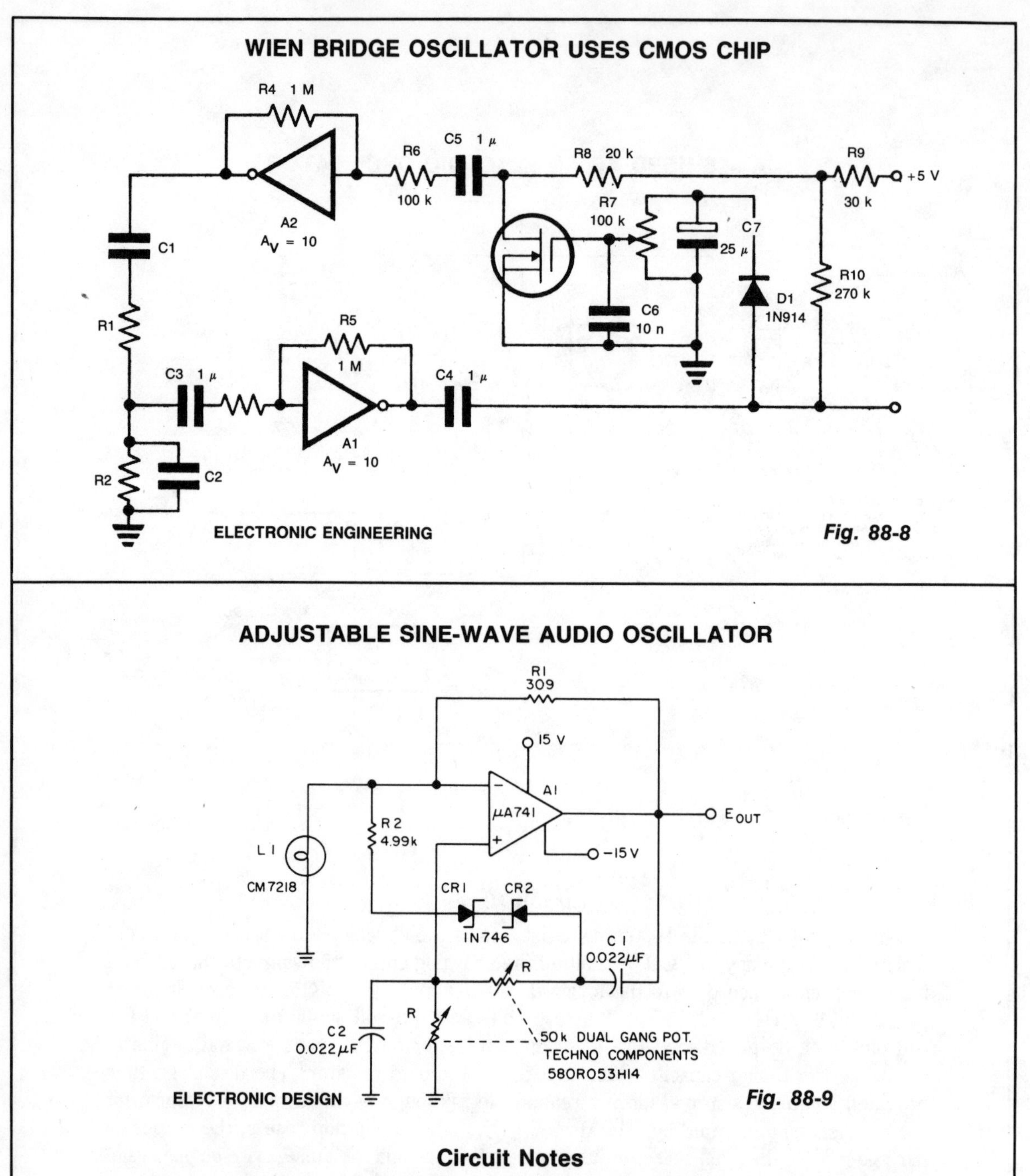

Fig. 88-8

Fig. 88-9

Circuit Notes

Waveform purity at low frequencies for a Wien bridge oscillator is enhanced by diode limiting. Lamp L1 stabilizes the loop gain at higher frequencies while the limiting action of R2, CR1, and CR2 prevents clipping at low frequencies and increases the frequency adjustment range from about 3:1 to greater than 10:1.

ONE-IC AUDIO GENERATOR

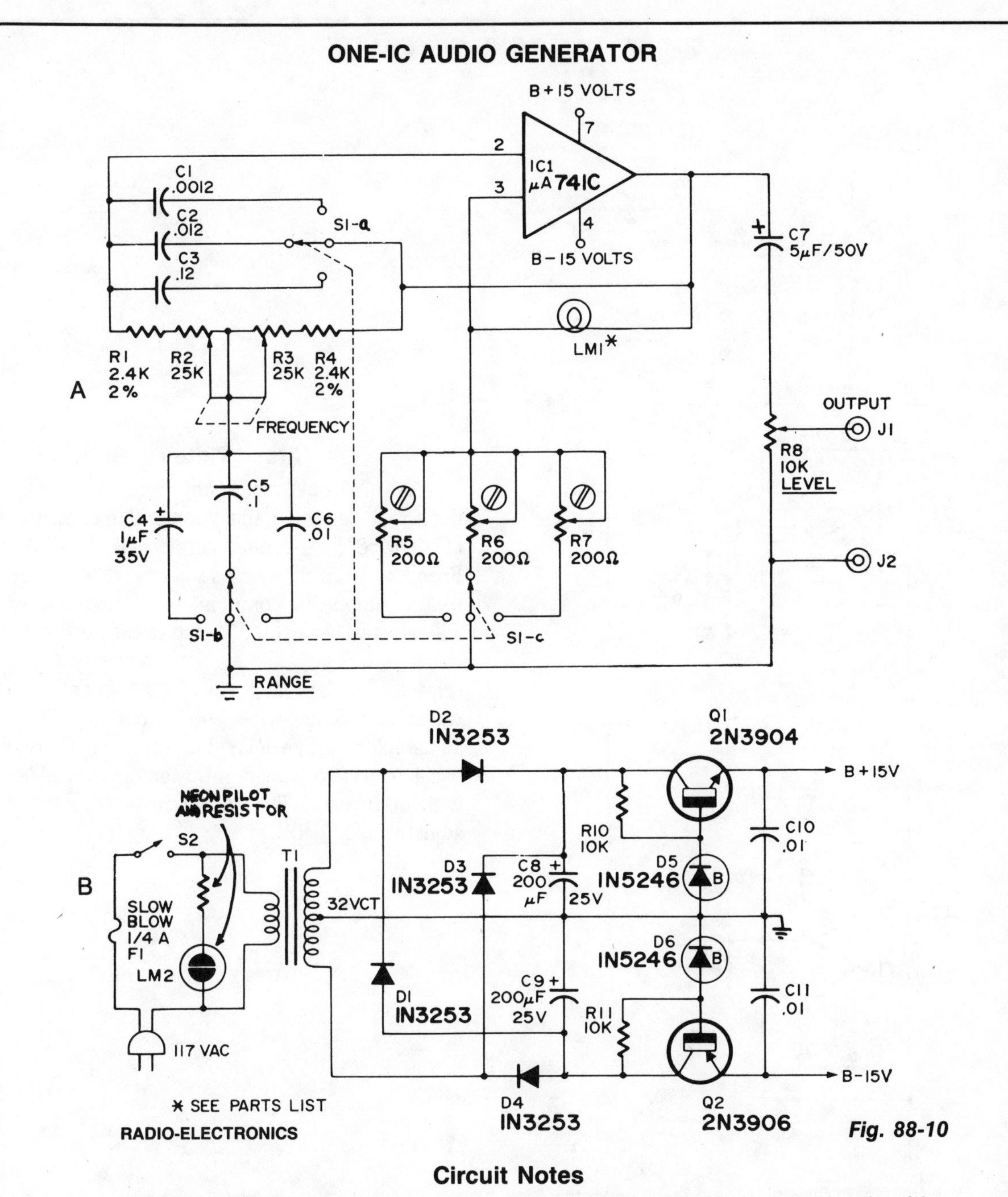

Fig. 88-10

Circuit Notes

This high-quality low-cost generator covers 20 Hz to 20 kHz in three bands with less than 1% distortion. LM1—10 V, 14 mA (344, 1869, 914) or 10 V, 10 mA (913, 367).

A = oscillator
B = power supply

SIMPLE TWO-TONE GENERATOR

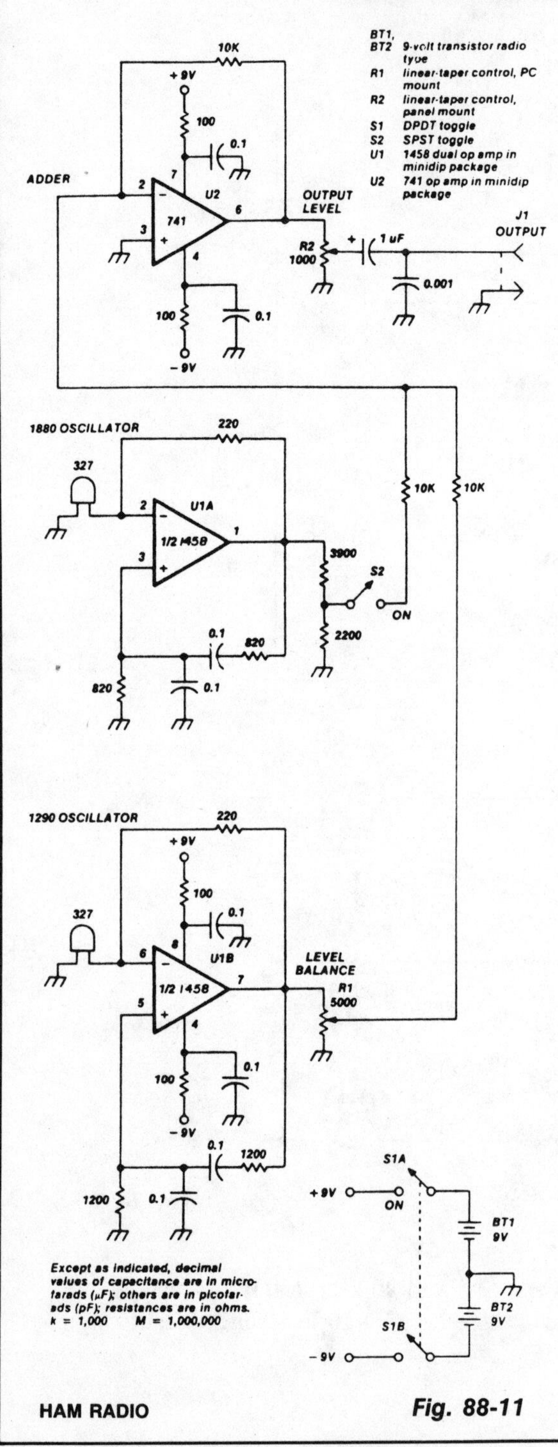

HAM RADIO

Fig. 88-11

Circuit Notes

Two 741 operational amplifiers are used for the active element in this Wien bridge oscillator. (The 1458 is the dual version of the 741.) Frequencies of the two oscillators were chosen to fit standard component values. Other frequencies between 500 and 2000 Hz can be employed. They should not be harmonically related. The output level of U1A is set by a resistive divider, while the output of U1B is adjustable through R1. The output of the two oscillators is combined in U2, an op-amp adder with unity gain. The output from U2 can be adjusted using R2.

89

Sirens, Warblers and Wailers

The sources of the following circuits are contained in the Sources section beginning on page 694. The figure number contained in the box of each circuit correlates to the source entry in the Sources section.

WARBLE GENERATOR

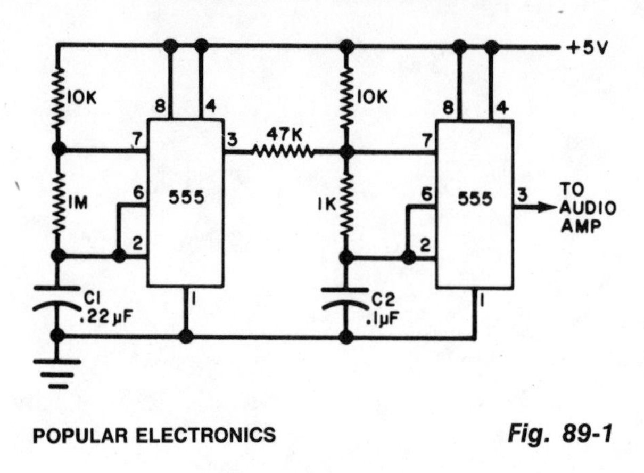

POPULAR ELECTRONICS

Fig. 89-1

Circuit Notes

The circuit uses a pair of 555 timers or a single dual timer. Capacitor C1 controls the speed of the warble, while C2 determines the pitch. The values shown should produce quite a distinctive signal.

WAILING ALARM

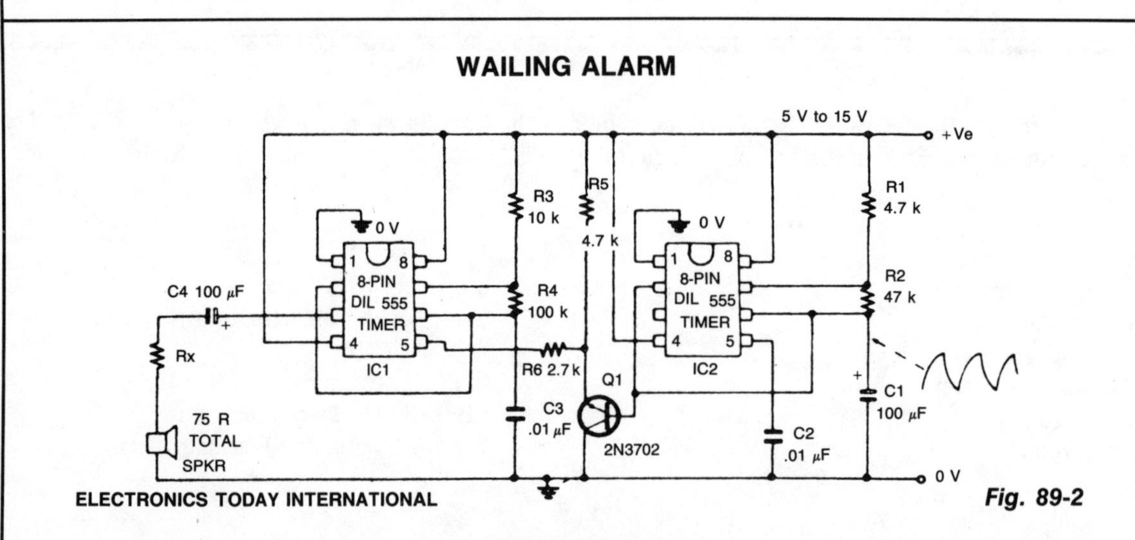

ELECTRONICS TODAY INTERNATIONAL

Fig. 89-2

Circuit Notes

This circuit simulates the sound of an American police siren. IC2 is wired as a low frequency astable that has a cycling period of about 6 seconds. The slowly varying ramp waveform on C1 is fed to pnp emitter follower Q1, and is then used to frequency modulate alarm generator IC1 via R6. IC1 has a natural center frequency of about 800Hz. Circuit action is such that the alarm output signal starts at a low frequency, rises for 3 seconds to a high frequency, then falls over 3 seconds to a low frequency again, and so on ad infinitum.

WARBLE-TONE ALARM

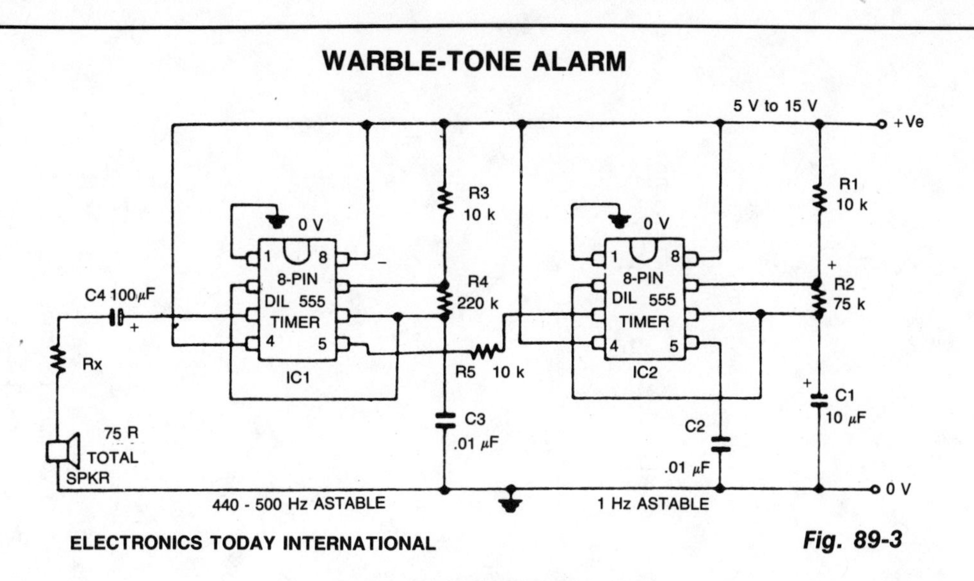

ELECTRONICS TODAY INTERNATIONAL

Fig. 89-3

Circuit Notes

The circuit generates a warble-tone alarm signal that simulates the sound of a British police siren. IC1 is wired as an alarm generator and IC2 is wired as a 1 Hz astable multivibrator. The output of IC2 is used to frequency modulate IC1 via R5. The action is such that the output frequency of IC1 alternates symmetrically between 500 Hz and 440 Hz, taking one sound to complete each alternating cycle.

WARBLING TONE GENERATOR

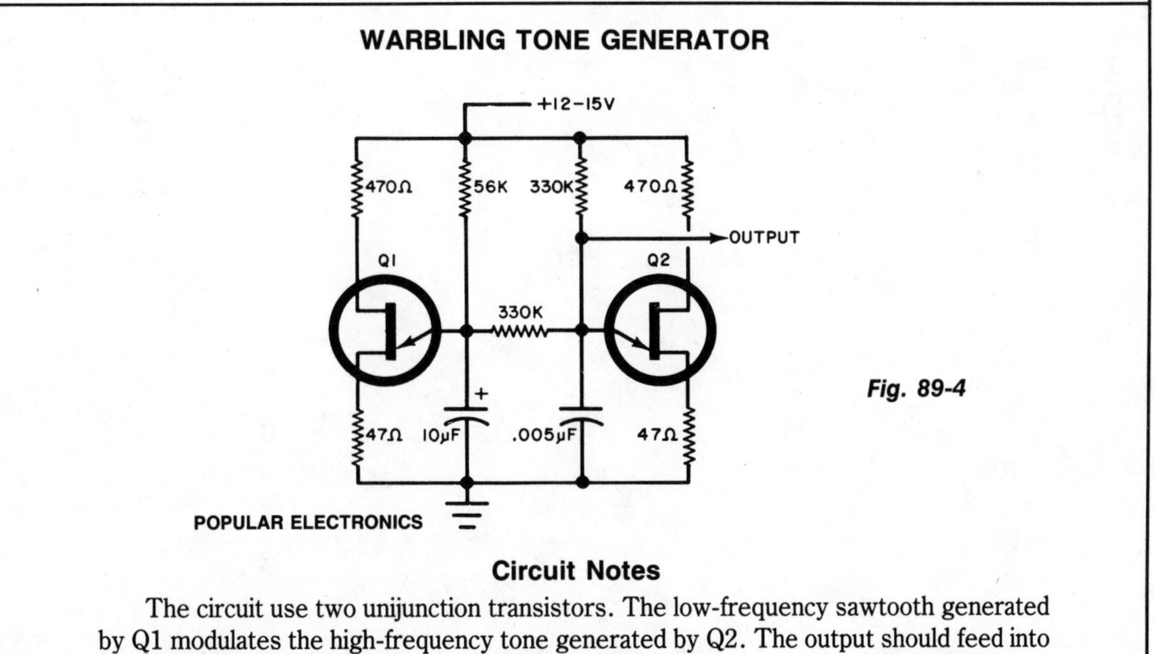

POPULAR ELECTRONICS

Fig. 89-4

Circuit Notes

The circuit use two unijunction transistors. The low-frequency sawtooth generated by Q1 modulates the high-frequency tone generated by Q2. The output should feed into a high-impedance amplifier. Q1 = Q2 = 2N4871.

MULTIFUNCTION SIREN SYSTEM

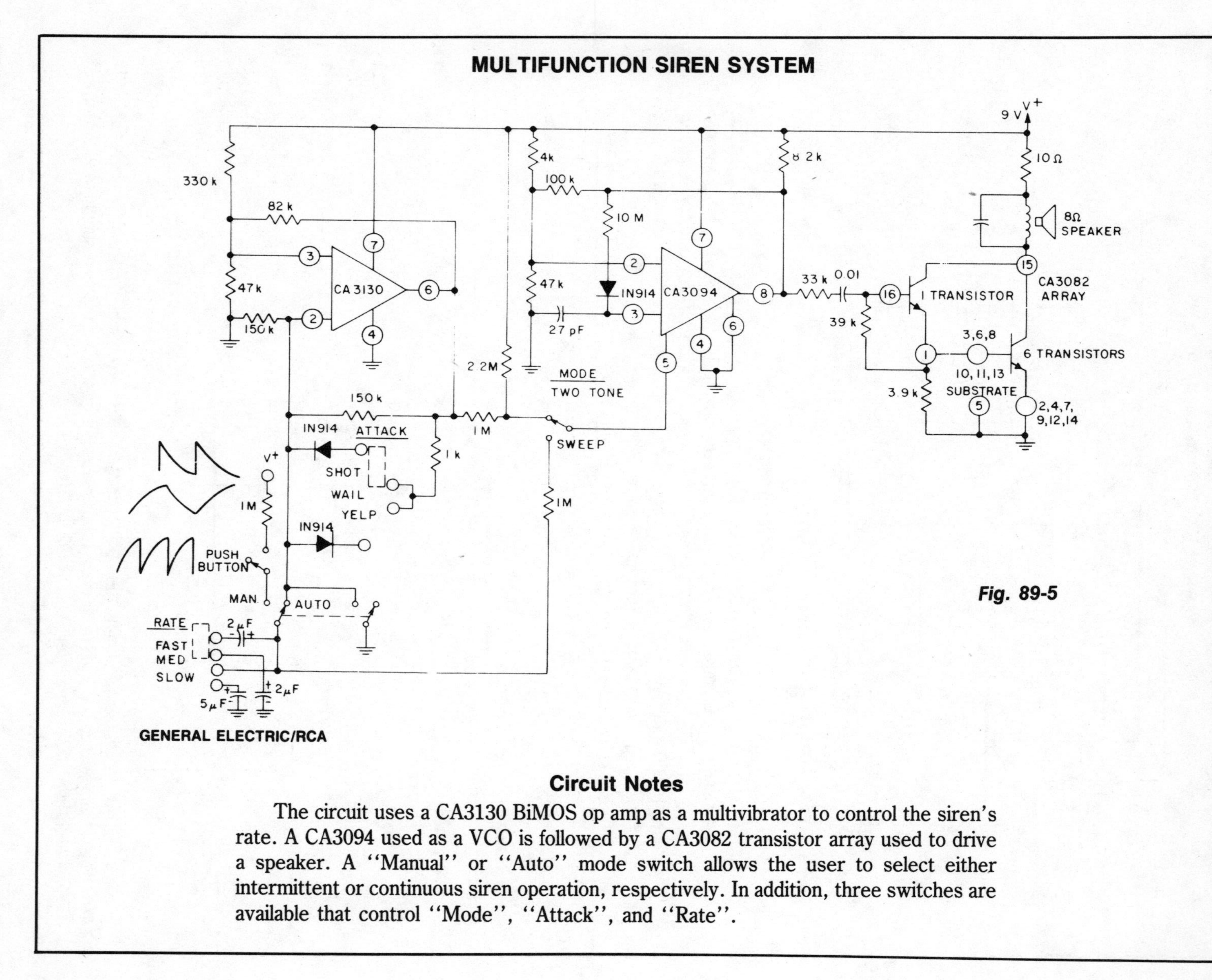

GENERAL ELECTRIC/RCA

Fig. 89-5

Circuit Notes

The circuit uses a CA3130 BiMOS op amp as a multivibrator to control the siren's rate. A CA3094 used as a VCO is followed by a CA3082 transistor array used to drive a speaker. A ''Manual'' or ''Auto'' mode switch allows the user to select either intermittent or continuous siren operation, respectively. In addition, three switches are available that control ''Mode'', ''Attack'', and ''Rate''.

7400 SIREN

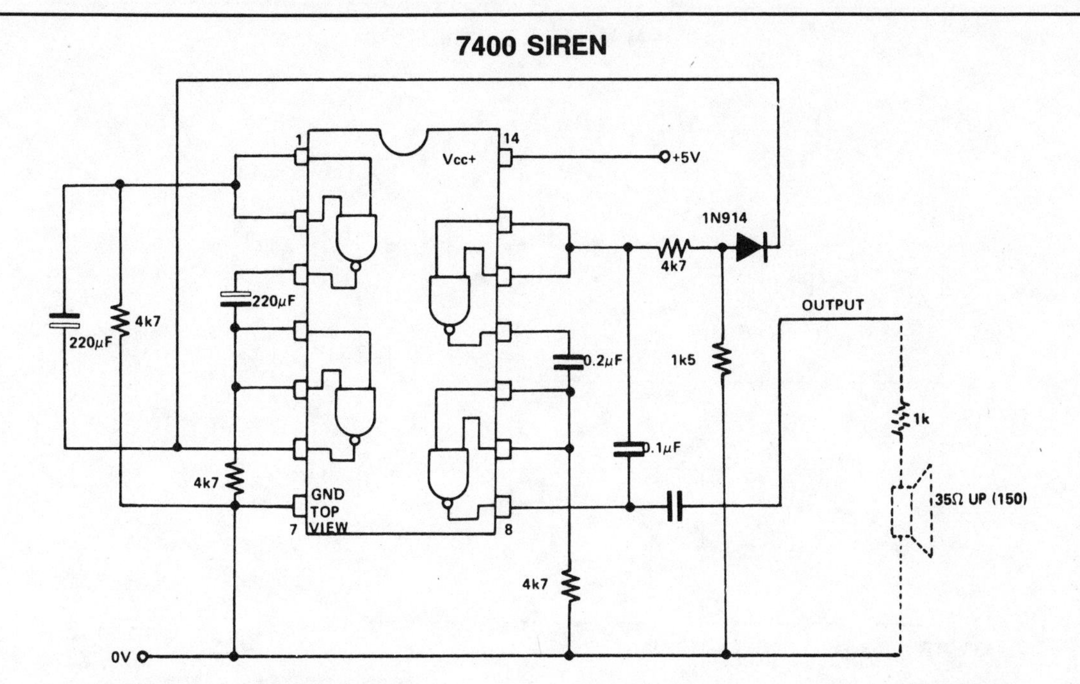

ELECTRONICS TODAY INTERNATIONAL *Fig. 89-6*

Circuit Notes

Two NAND gates are used for the oscillator, and two as the control. If the two-tone speed needs to be altered, the 220 µF capacitors can be changed (larger for slower operation). If the frequency of the oscillator is to be changed, the 0.2 and 0.1 µF capacitors can be varied and the value of R1 can be increased. To change frequency range between the two notes, alter the 1.5 k (1,500) resistor.

TOY SIREN

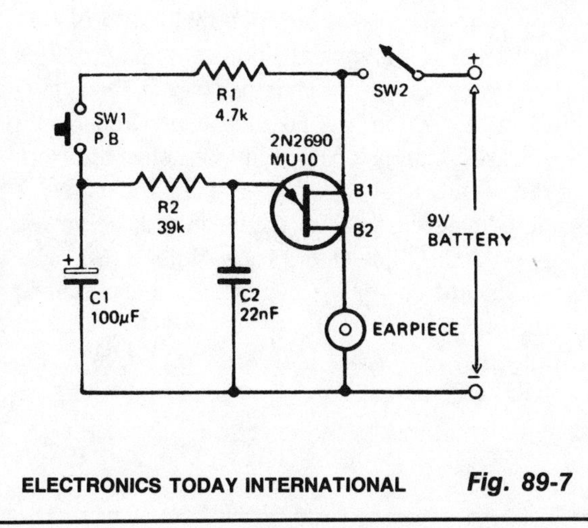

ELECTRONICS TODAY INTERNATIONAL *Fig. 89-7*

Circuit Notes

This circuit can be built small enough to fit inside a toy. The circuit consists of a relaxation oscillator utilizing one unijunction transistor (2N2646, MU10, TIS43). R2 and C2 determine the frequency of the tone. Pushing the button, SW1 charges up the capacitor and the potential at the junction of R2 and C2 rises, causing an upswing in the frequency of oscillation. On releasing the pushbutton the charge on C2 will drop slowly with a proportional reduction in the frequency of oscillation. Manual operation of the button at intervals of approximately 2 seconds will produce a siren sound.

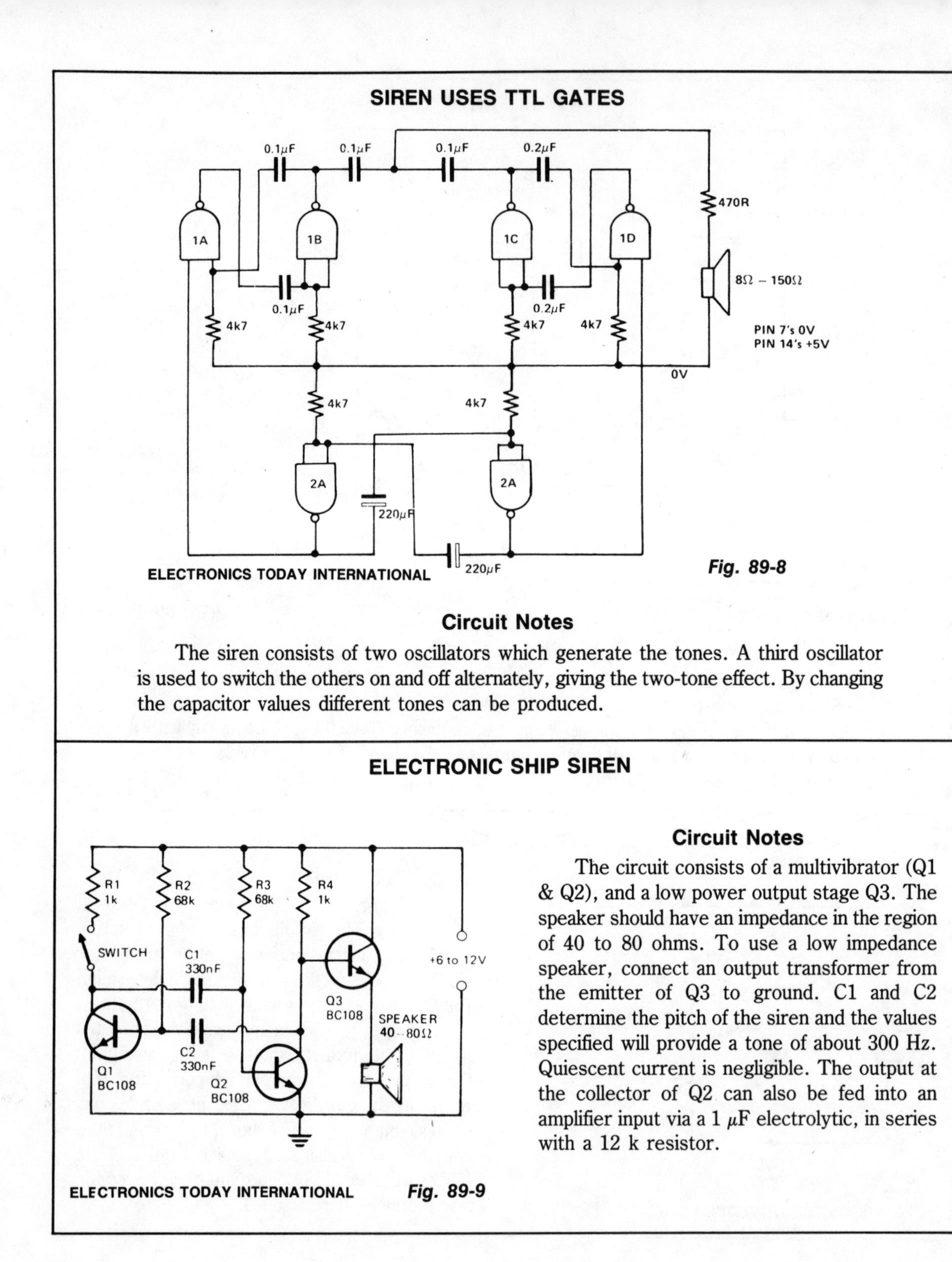

SIREN USES TTL GATES

Fig. 89-8

Circuit Notes

The siren consists of two oscillators which generate the tones. A third oscillator is used to switch the others on and off alternately, giving the two-tone effect. By changing the capacitor values different tones can be produced.

ELECTRONIC SHIP SIREN

Fig. 89-9

Circuit Notes

The circuit consists of a multivibrator (Q1 & Q2), and a low power output stage Q3. The speaker should have an impedance in the region of 40 to 80 ohms. To use a low impedance speaker, connect an output transformer from the emitter of Q3 to ground. C1 and C2 determine the pitch of the siren and the values specified will provide a tone of about 300 Hz. Quiescent current is negligible. The output at the collector of Q2 can also be fed into an amplifier input via a 1 μF electrolytic, in series with a 12 k resistor.

SIREN ALARM SIMULATES STAR TREK RED ALERT

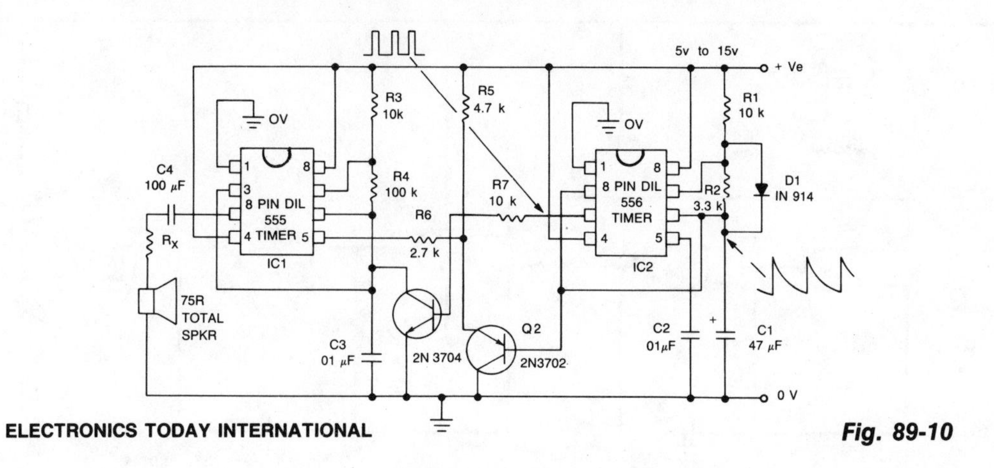

ELECTRONICS TODAY INTERNATIONAL *Fig. 89-10*

Circuit Notes

The signal starts at a low frequency, rises for about 1.15 seconds to a high frequency, ceases for about 0.35 seconds, then starts rising again from a low frequency, and so on ad infinitum.

YELP OSCILLATOR

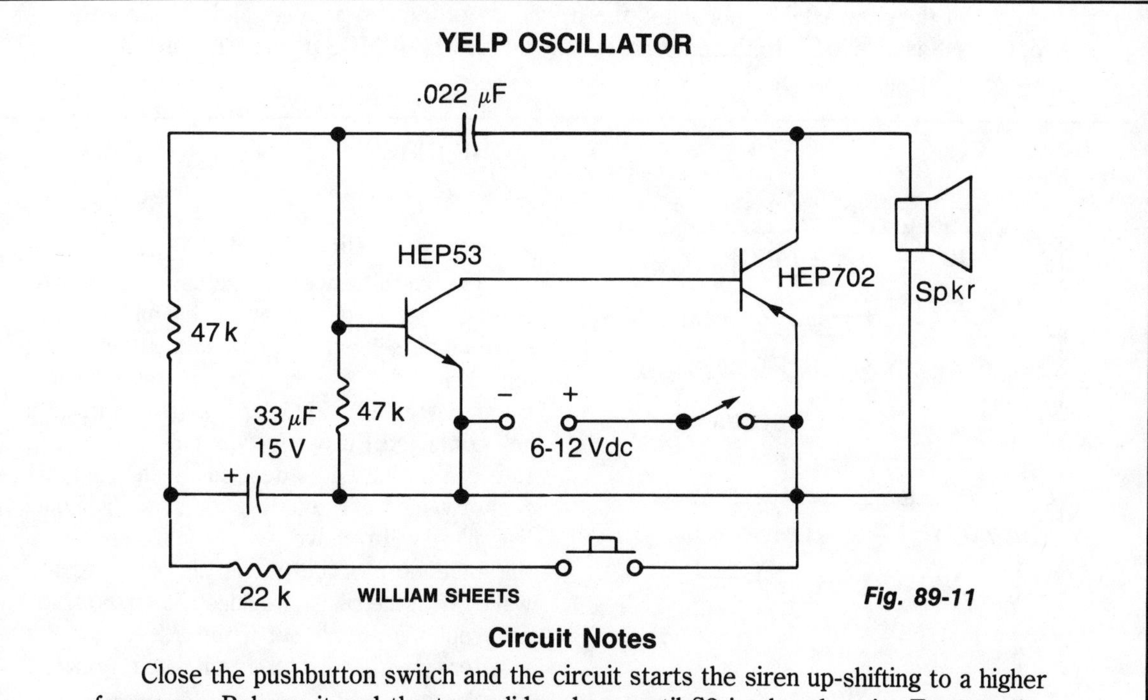

WILLIAM SHEETS *Fig. 89-11*

Circuit Notes

Close the pushbutton switch and the circuit starts the siren up-shifting to a higher frequency. Release it and the tone slides down until S2 is closed again. Tone quality is adjusted by changing the 0.022 µF capacitor.

HIGH POWER SIREN

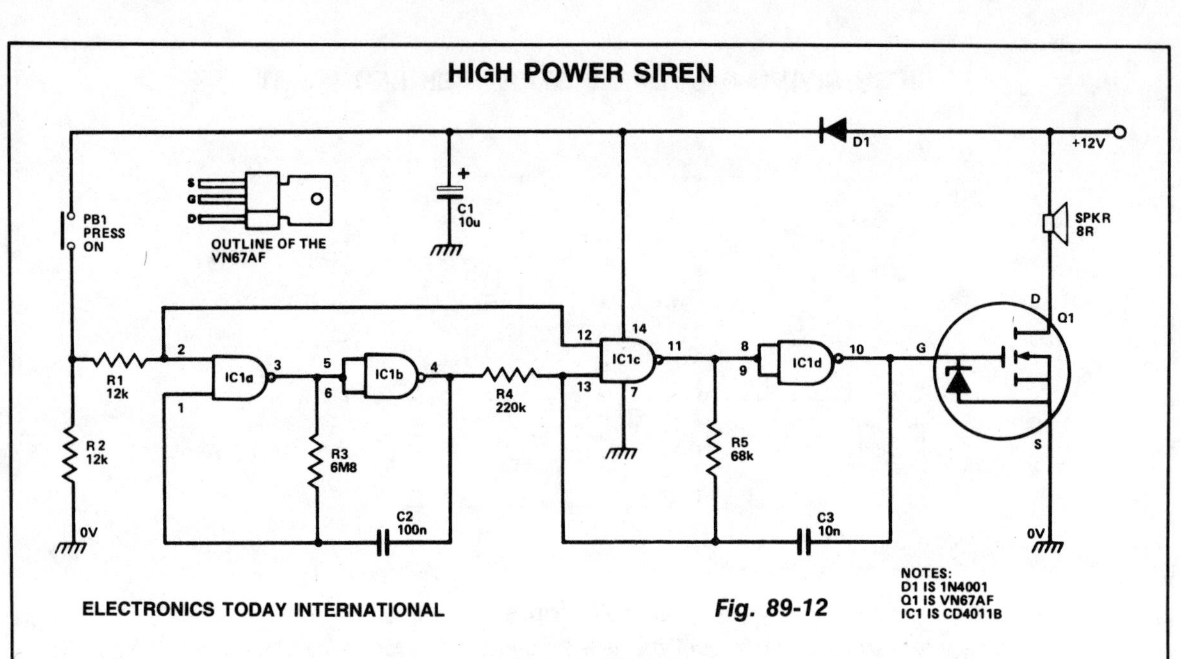

ELECTRONICS TODAY INTERNATIONAL

Fig. 89-12

Circuit Notes

IC1a and IC1b are wired as a slow astable multivibrator and IC1c-IC1d are wired as a fast astable. Both are "gated" types, which can be turned on and off via PB1. The output of the slow astable modulates the frequency of the fast astable, and the output of the fast astable is fed to the external speaker via the Q1 VMOS power FET amplifier stage.

"HEE-HAW" TWO-TONE SIREN

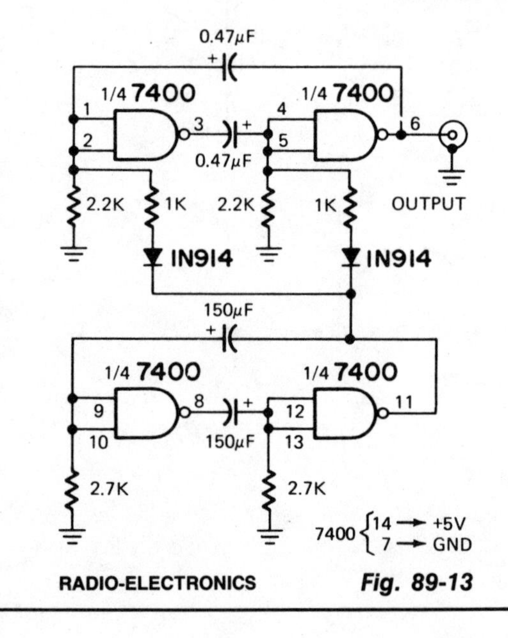

RADIO-ELECTRONICS

Fig. 89-13

Circuit Notes

The circuit uses two gates of a 7400 IC cross-connected to form an astable multivibrator driven by the 1-pulse per second output of the digital clock IC. The hee-haw circuit has a low frequency astable modulator added to make a self-contained European-type siren. Tone and rate can be varied as desired by changing capacitor values. If the tone is too harsh, a simple R-C filter will remove the harmonic content—the multivibrator output is almost a square wave. With the resistor values shown, no start-up problems occur; but if the 2.2 k or 2.7 k resistors are changed too much, latch-up can be a problem.

VARYING-FREQUENCY WARNING ALARM

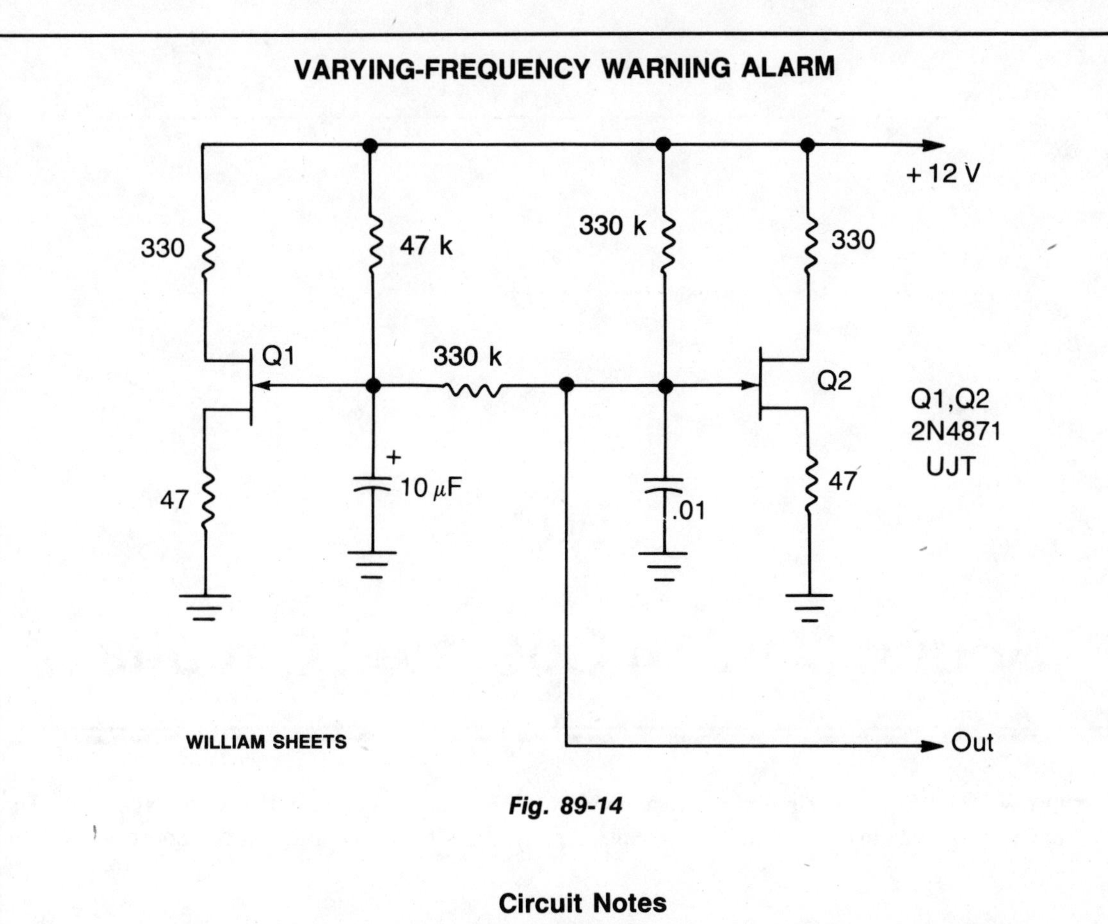

Fig. 89-14

Circuit Notes

The output frequency changes continuously. Low frequency oscillator (Q1) modulates high frequency oscillator Q2 and its associated timing capacitor.

90

Sound (Audio) Operated Circuits

The sources of the following circuits are contained in the Sources section beginning on page 694. The figure number contained in the box of each circuit correlates to the source entry in the Sources section.

SOUND-ACTIVATED SWITCH

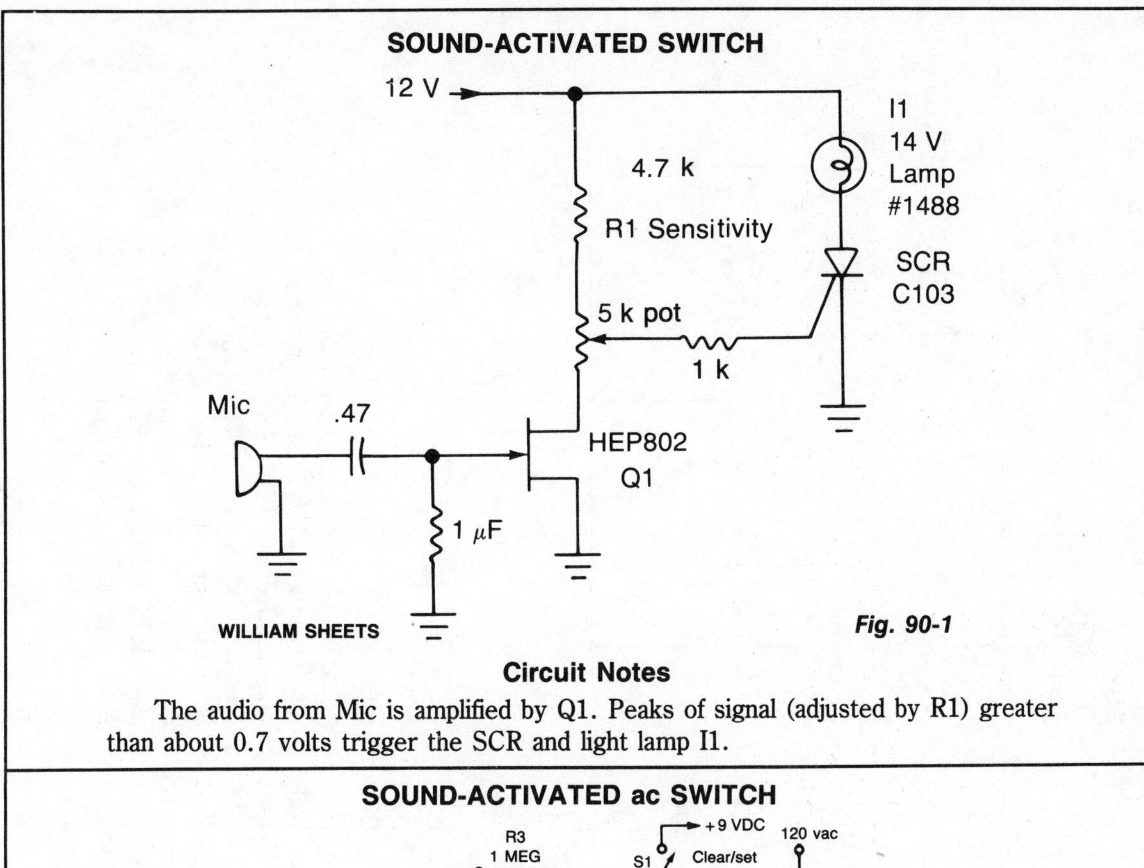

Fig. 90-1

Circuit Notes

The audio from Mic is amplified by Q1. Peaks of signal (adjusted by R1) greater than about 0.7 volts trigger the SCR and light lamp I1.

SOUND-ACTIVATED ac SWITCH

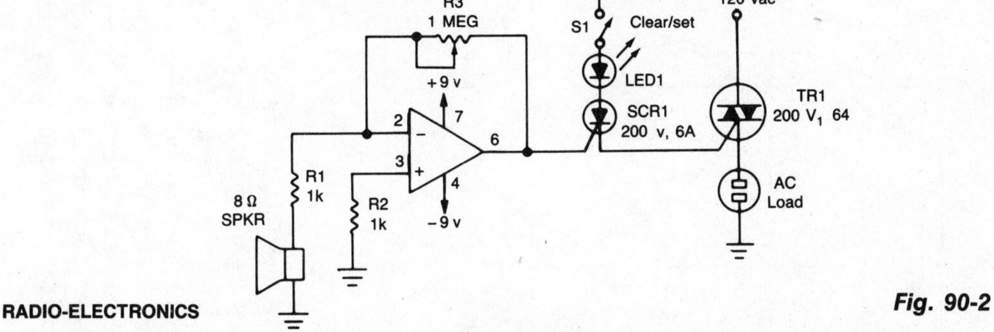

Fig. 90-2

Circuit Notes

The circuit uses a 741 op amp operating as an inverting amplifier to amplify the voltage produced by an 8-ohm speaker used to detect any sounds. The feedback resistor R3, a 1-megohm potentiometer used to vary the gain of the amplifier determines the sensitivity of the circuit. When S1 is closed in the (SET) position and a sound is applied to the speaker, SCR1 is turned on. It will remain in conduction until the anode voltage is removed by opening S1, putting it in its RESET position. (Once an SCR is turned on, the gate or trigger has no control over the circuit.) As long as the SCR conducts, the Triac, TR1, will remain on and supply voltage to the load.

VOX BOX

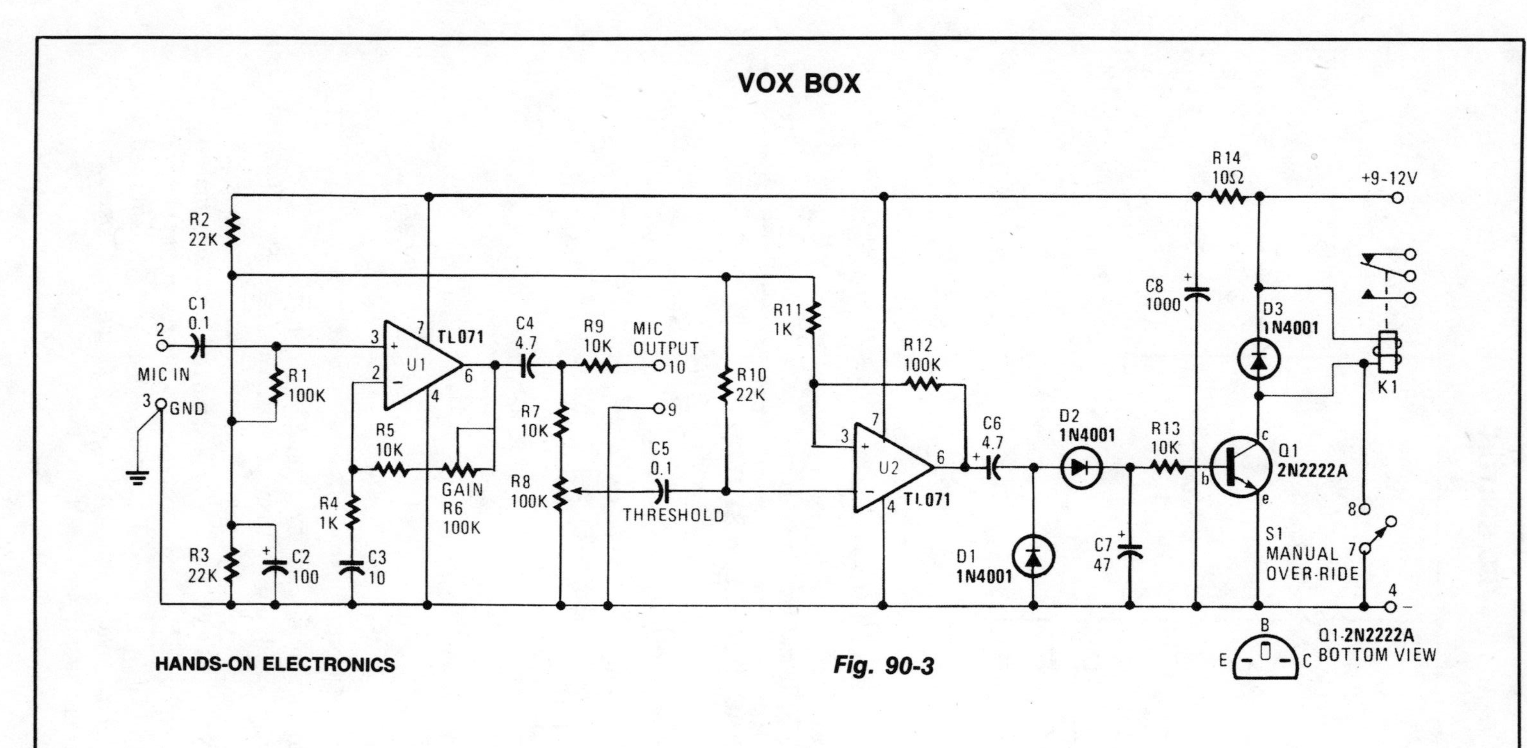

HANDS-ON ELECTRONICS

Fig. 90-3

Circuit Notes

The electronic circuit in the VOX Box consists of three parts: a microphone preamplifier, a Schmitt trigger, and a relay driver. Input signals (MIC INPUT terminals) to the microphone preamplifier (U1) are amplified and fed to a THRESHOLD control (R8). When the preselected threshold voltage level is exceeded, the output of the Schmitt trigger (U2) immediately goes high. The signal from U2 is rectified and the voltage developed across C7 turns on the relay energizer transistor (Q1). That transistor action passes pull-down current through the coil of relay K1. The changing of the relay SPDT contacts can be used to either make or break an external ac or dc circuit.

COLOR ORGAN

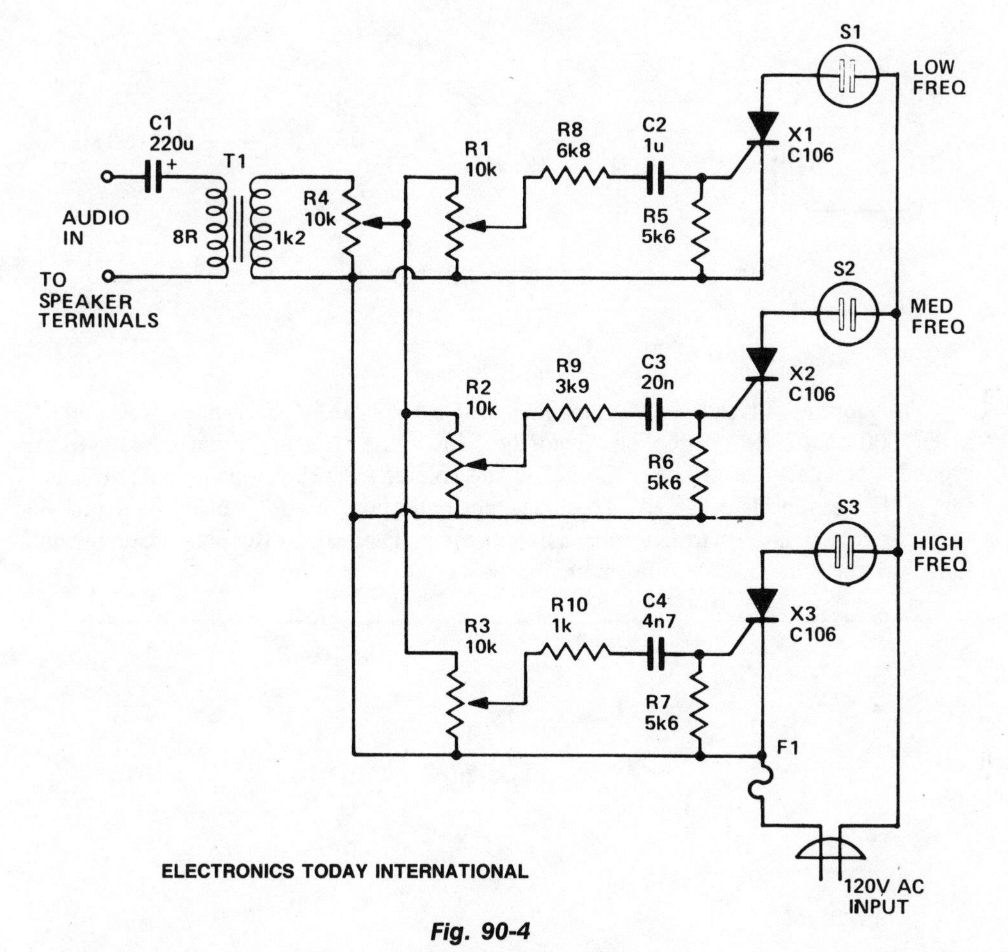

ELECTRONICS TODAY INTERNATIONAL

Fig. 90-4

Circuit Notes

Three lights are controlled by the three channels. One light will pulse in response to the bass, another illuminates with mid-range sounds, and the last lights for high notes. Four level controls allow adjustment of overall light level and each channel individually. Up to 200 watts per channel can be handled.

BASIC COLOR ORGAN

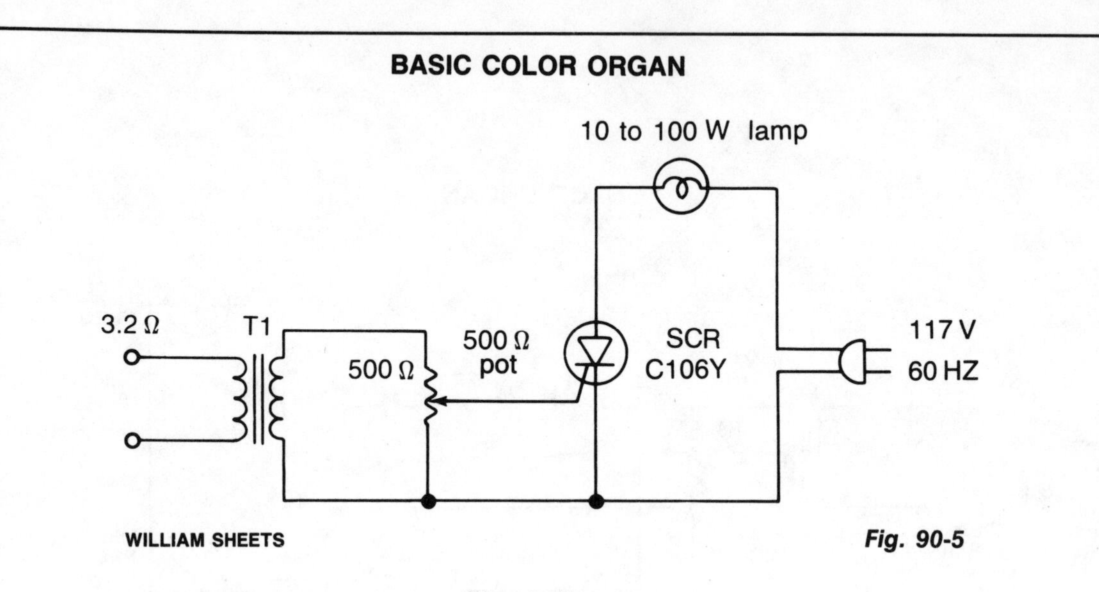

Fig. 90-5

Circuit Notes

Transformer T1 can be any matching transistor type in the range of 500/500 to 2500/2500 ohms. No connections from the SCR or its components are connected to ground. For safety's sake, keep the 117-V line voltage from the amplifier connections—that is the reason for using T1. To adjust, set potentiometer R1 ''off'' and adjust the amplifier volume control for a normal listening level. Then adjust the potentiometer until the lamp starts to throb in step with the beat.

91

Sound Effect Circuits

The sources of the following circuits are contained in the Sources section beginning on page 694. The figure number contained in the box of each circuit correlates to the source entry in the Sources section.

- Sound Effects Generator
- Electronic Bongos
- Train Chuffer
- Bird Chirp
- Steam Locomotive Whistle
- WAA-WAA Circuit
- Unusual Fuzz
- Autodrum
- Twang-Twang Circuit
- Steam Train/Prop Plane
- Funk Box

SOUND EFFECTS GENERATOR

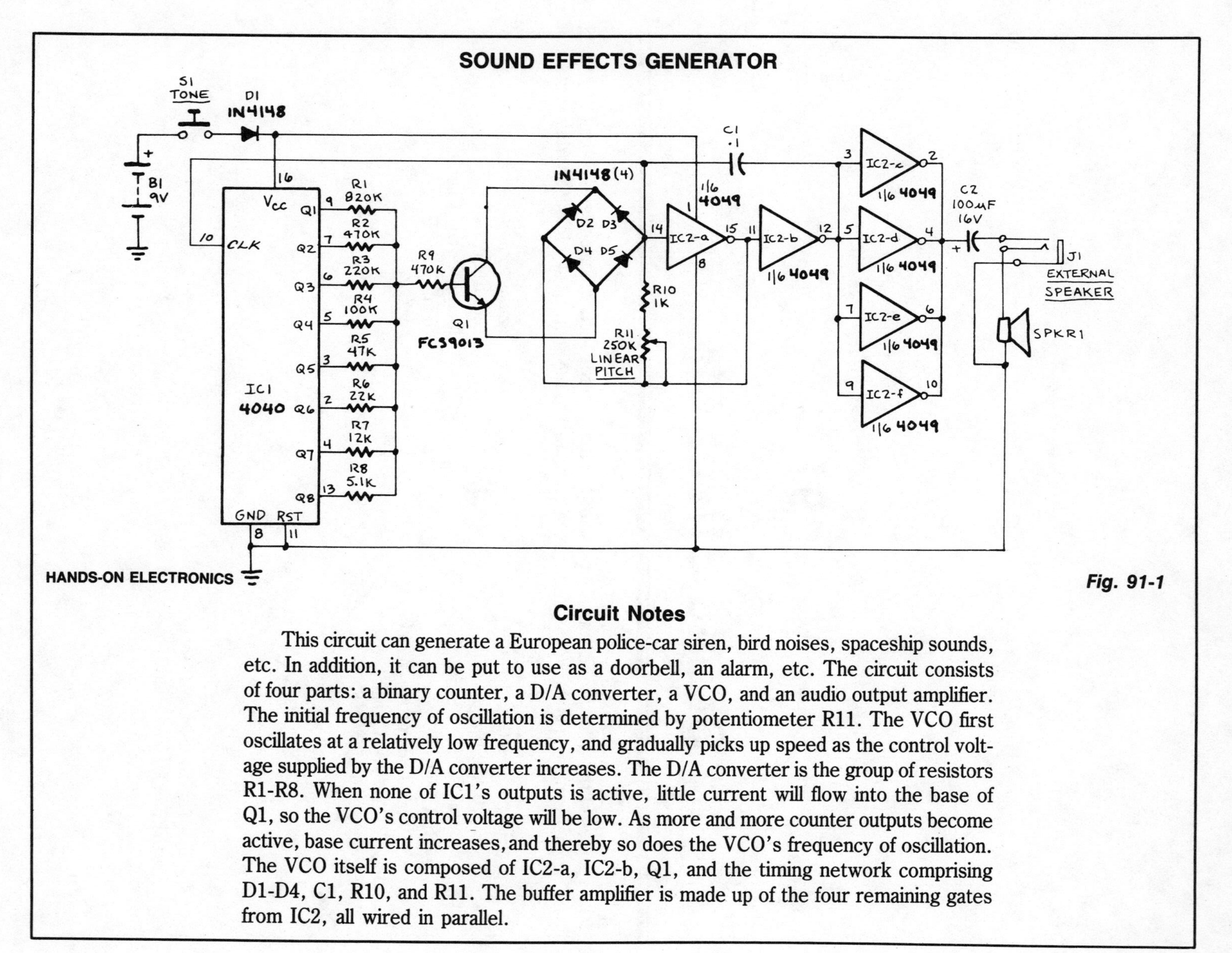

HANDS-ON ELECTRONICS

Fig. 91-1

Circuit Notes

This circuit can generate a European police-car siren, bird noises, spaceship sounds, etc. In addition, it can be put to use as a doorbell, an alarm, etc. The circuit consists of four parts: a binary counter, a D/A converter, a VCO, and an audio output amplifier. The initial frequency of oscillation is determined by potentiometer R11. The VCO first oscillates at a relatively low frequency, and gradually picks up speed as the control voltage supplied by the D/A converter increases. The D/A converter is the group of resistors R1-R8. When none of IC1's outputs is active, little current will flow into the base of Q1, so the VCO's control voltage will be low. As more and more counter outputs become active, base current increases, and thereby so does the VCO's frequency of oscillation. The VCO itself is composed of IC2-a, IC2-b, Q1, and the timing network comprising D1-D4, C1, R10, and R11. The buffer amplifier is made up of the four remaining gates from IC2, all wired in parallel.

ELECTRONIC BONGOS

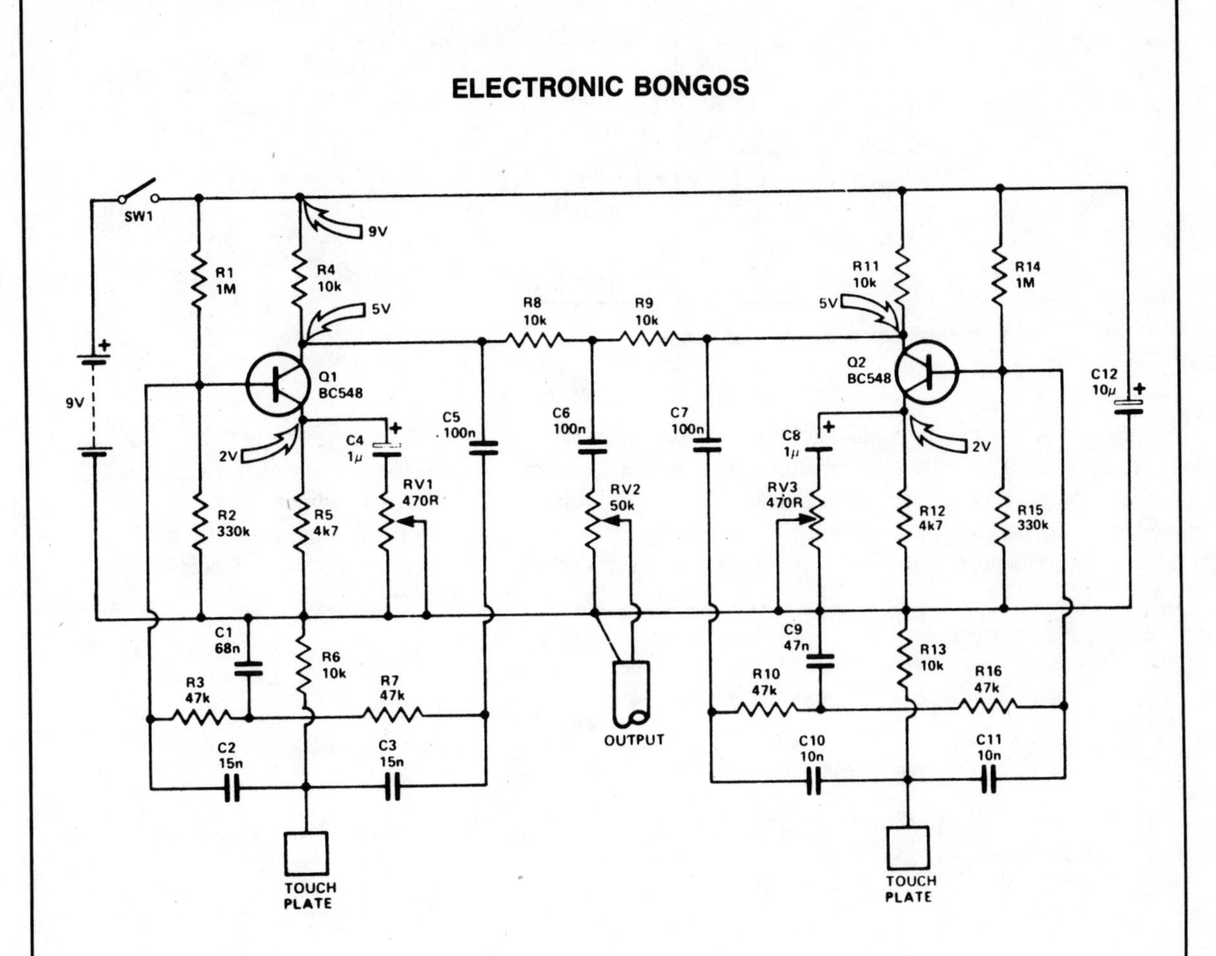

ELECTRONICS TODAY INTERNATIONAL

Circuit Notes

This circuit consists of twin-T sine-wave oscillators. Each oscillator has a filter in the feedback loop. If the loop gain is greater than unity, the circuit will oscillate. Gain is adjusted to be just less than unity. Touching the touch plate starts the oscillator, but the moment your finger is removed from the touch plate the oscillations will die away. The rate of decay is a function of circuit gain and controlled by RV1 (and RV3).

TRAIN CHUFFER

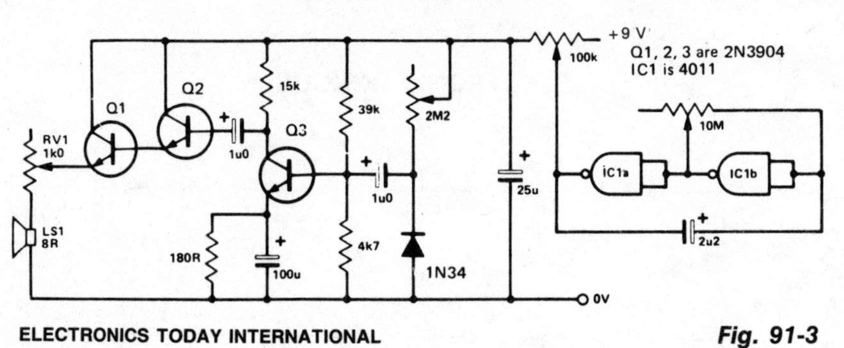

ELECTRONICS TODAY INTERNATIONAL *Fig. 91-3*

Circuit Notes

The circuit consists of a white noise generator which only switches on with the high part of the square wave output from the clock circuit. The frequency of the clock is adjusted with the 10 M pot and the output voltage of the clock is adjusted by the 100 k pot (rate and volume of chuff respectively). The 2M2 pot controls the amount of noise produced and the 1 k pot on the speaker controls the pitch of the average noise.

BIRD CHIRP

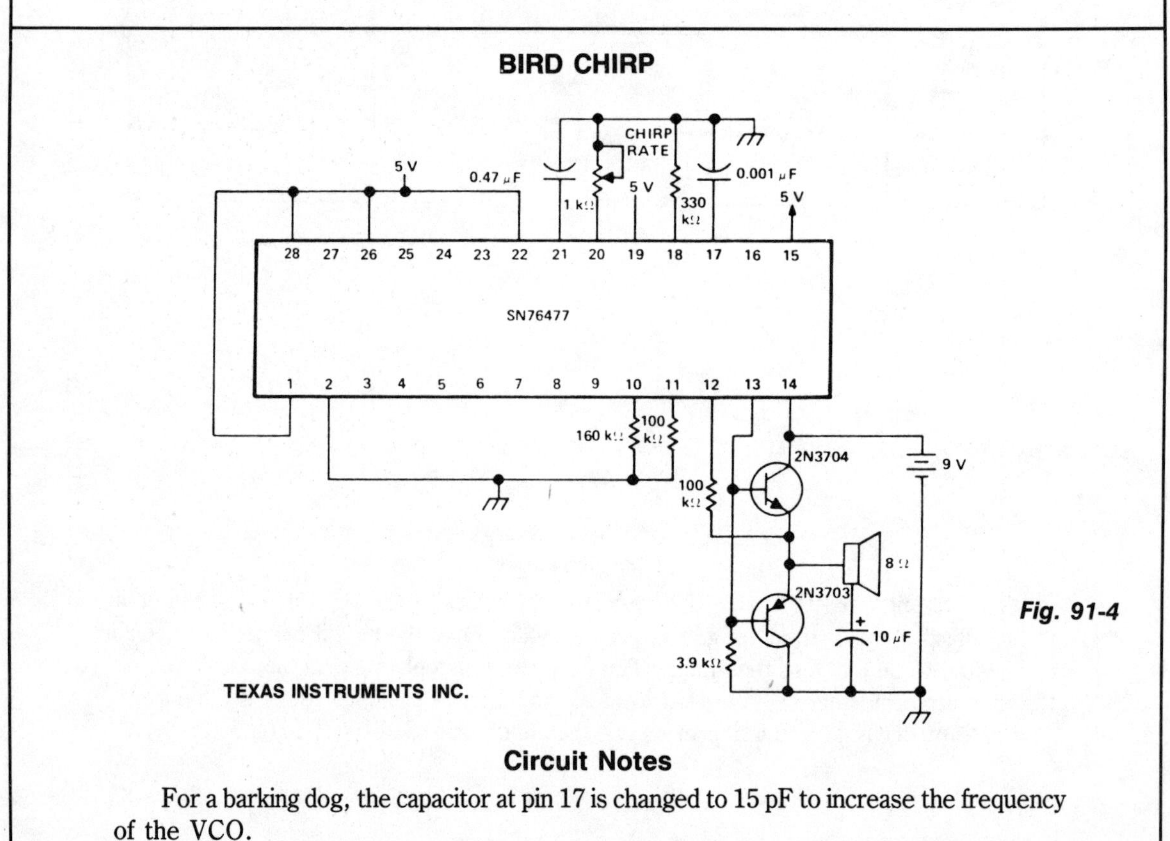

TEXAS INSTRUMENTS INC. *Fig. 91-4*

Circuit Notes

For a barking dog, the capacitor at pin 17 is changed to 15 pF to increase the frequency of the VCO.

STEAM LOCOMOTIVE WHISTLE

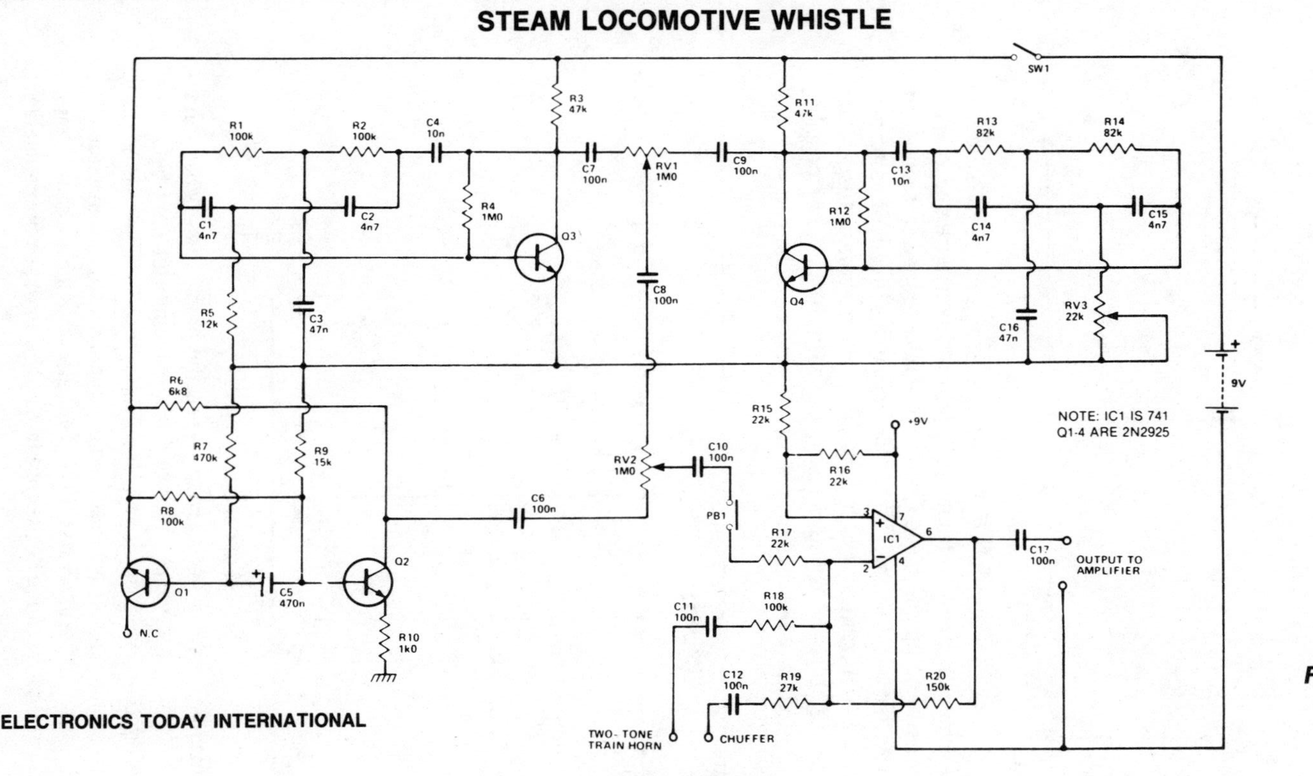

ELECTRONICS TODAY INTERNATIONAL

Fig. 91-5

Circuit Notes

The waveform of a steam whistle is a complex combination of white noise and an audio frequency oscillation. The noise generator is a transistor (Q1) biased into zener mode. The audio frequency oscillation is a straightforward mixture of two similar (but not identical) sine waves, which after their addition produce a more complex waveshape. The sine wave generators are twin-t oscillators. Preset RV1 mixes the two sine waves so that an appropriate waveform is obtained. RV2 mixes this waveform with the white noise. Adjustment of all three presets will result in the required sound. Integrated circuit IC1 is an operational amplifier used as a simple mixer/amplifier which combines the steam whistle, chuffer, (generated elsewhere) and two-tone horn sounds into one, suitable for amplification by an external amplifier.

WAA-WAA CIRCUIT

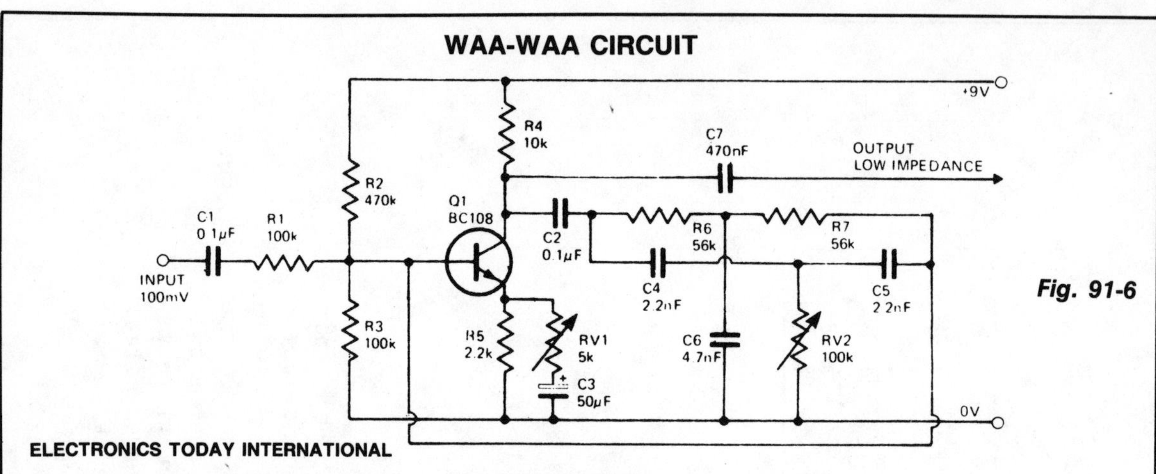

ELECTRONICS TODAY INTERNATIONAL

Fig. 91-6

Circuit Notes

The waa-waa effect is achieved as certain frequencies are amplified more than others. A phase shift RC oscillator makes up the basic circuit. Negative feedback is obtained by feeding part of the signal back to the base. When adjusting initially, RV1 is turned to minimum. RV2 is adjusted to a point at which an audible whistle appears indicating oscillation. RV1 is then adjusted till the oscillation just disappears. It should be possible to set RV2 to any value without any oscillation, this should also be achieved with the minimum possible value of RV1.

UNUSUAL FUZZ

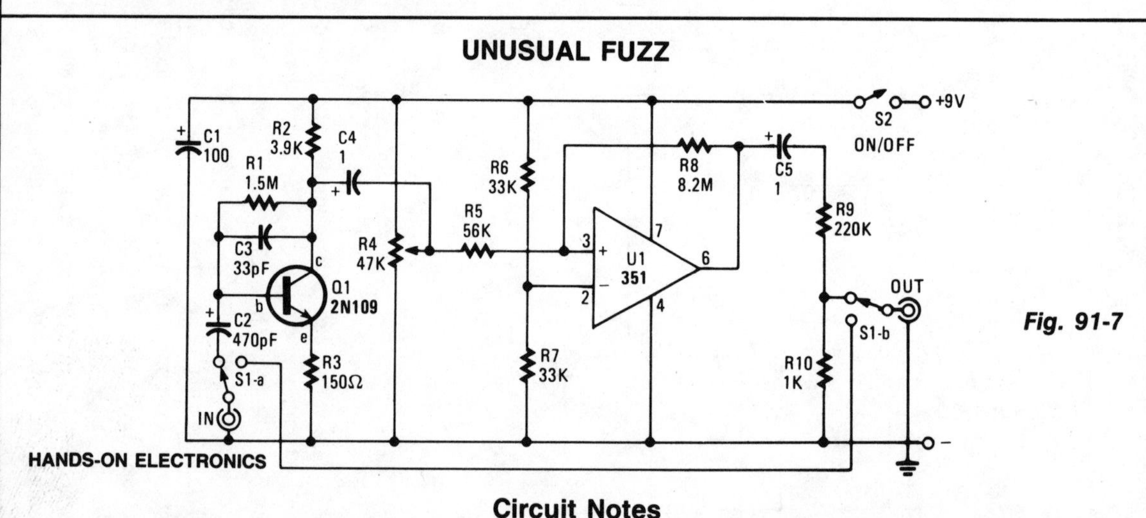

HANDS-ON ELECTRONICS

Fig. 91-7

Circuit Notes

It seems that guitar fuzz boxes have been around since the beginning of rock, and have seen little improvement over the years. This one is somewhat different because rather than simply distorting the sound, it also pulses in step with the peaks of the waveform from the pickup because of the Schmitt trigger op amp circuit. Capacitor C2 requires some explanation. It should normally be a 1- or 2-μF electrolytic capacitor. However, we show the value as 470 pF because it's recommended as an experimental value giving far out effects.

AUTODRUM

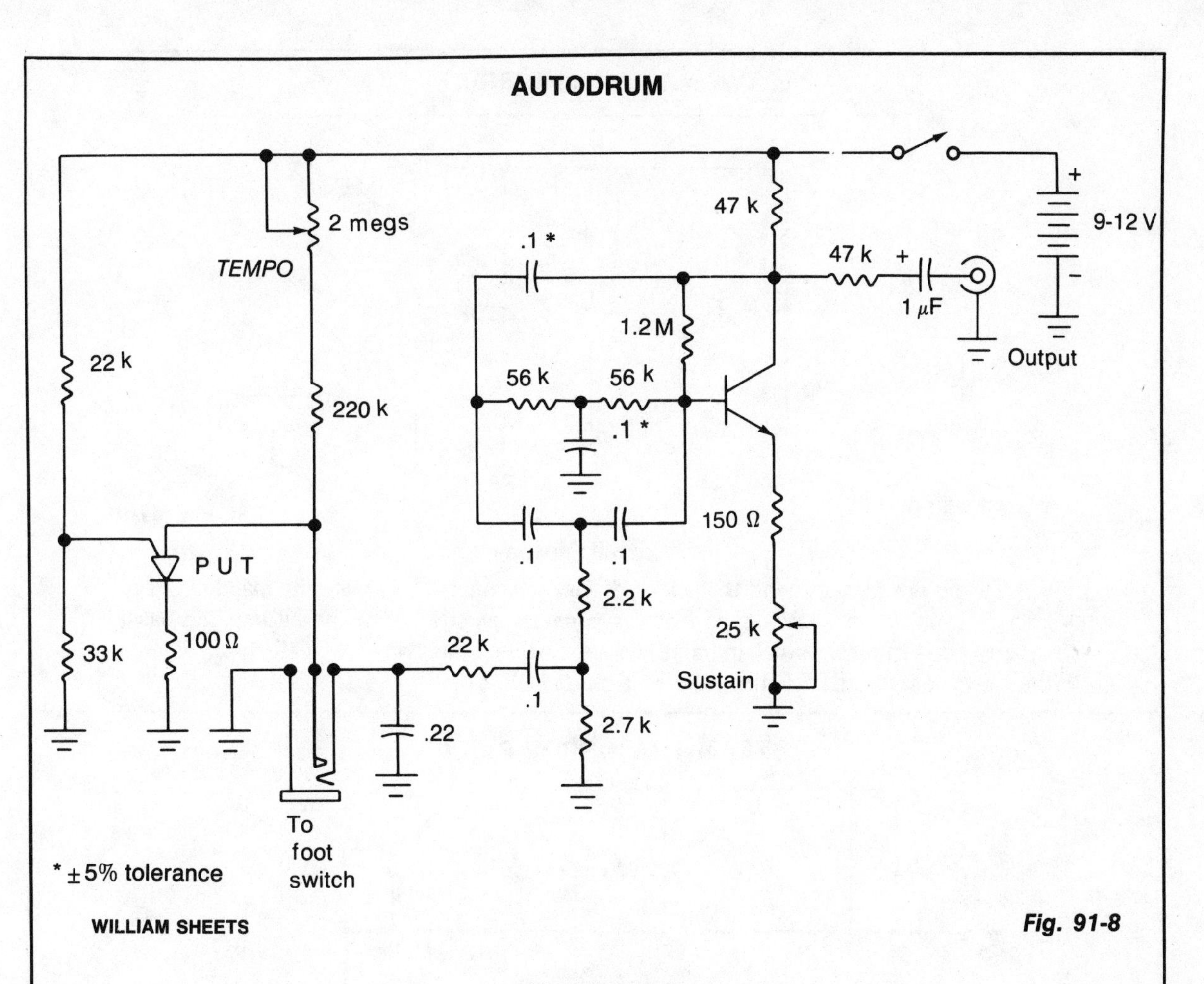

WILLIAM SHEETS

Fig. 91-8

Circuit Notes

This unit generates a drum-like damped oscillation that sounds best when fed into a higher power amplifier. The beat rate may be determined by operating a foot pedal in much the same manner as for a real drum, or by means of an internal oscillator, the speed of which may be preset.

TWANG-TWANG CIRCUIT

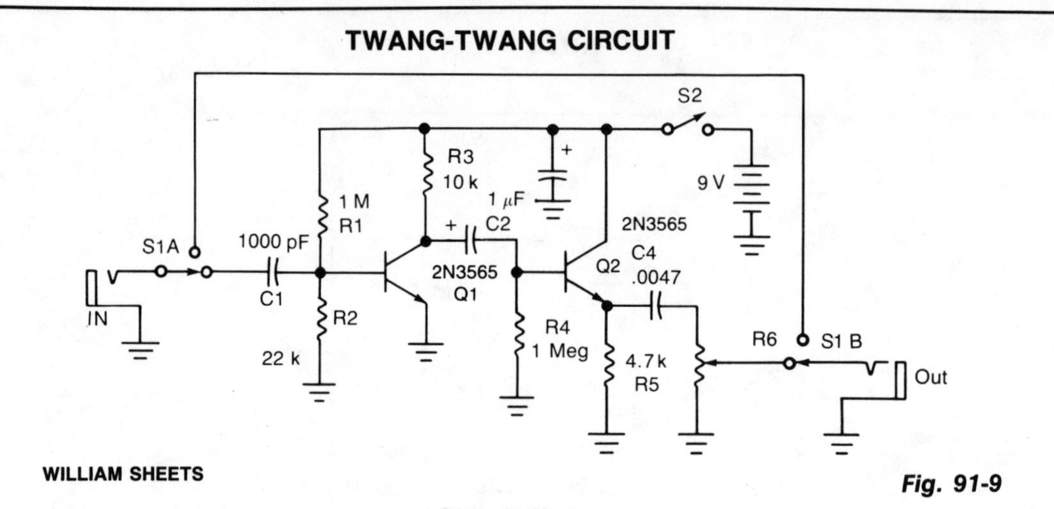

WILLIAM SHEETS

Fig. 91-9

Circuit Notes

Twang is a guitar sound that more or less approximates a banjo or mandolin. The circuit produces unusual sounds from an ordinary electric guitar by cutting the bass, severely distorting the midband and highs, and then amplifying the distortion. S1 cuts the effect in and out, S2 turns the unit on and off.

STEAM TRAIN/PROP PLANE

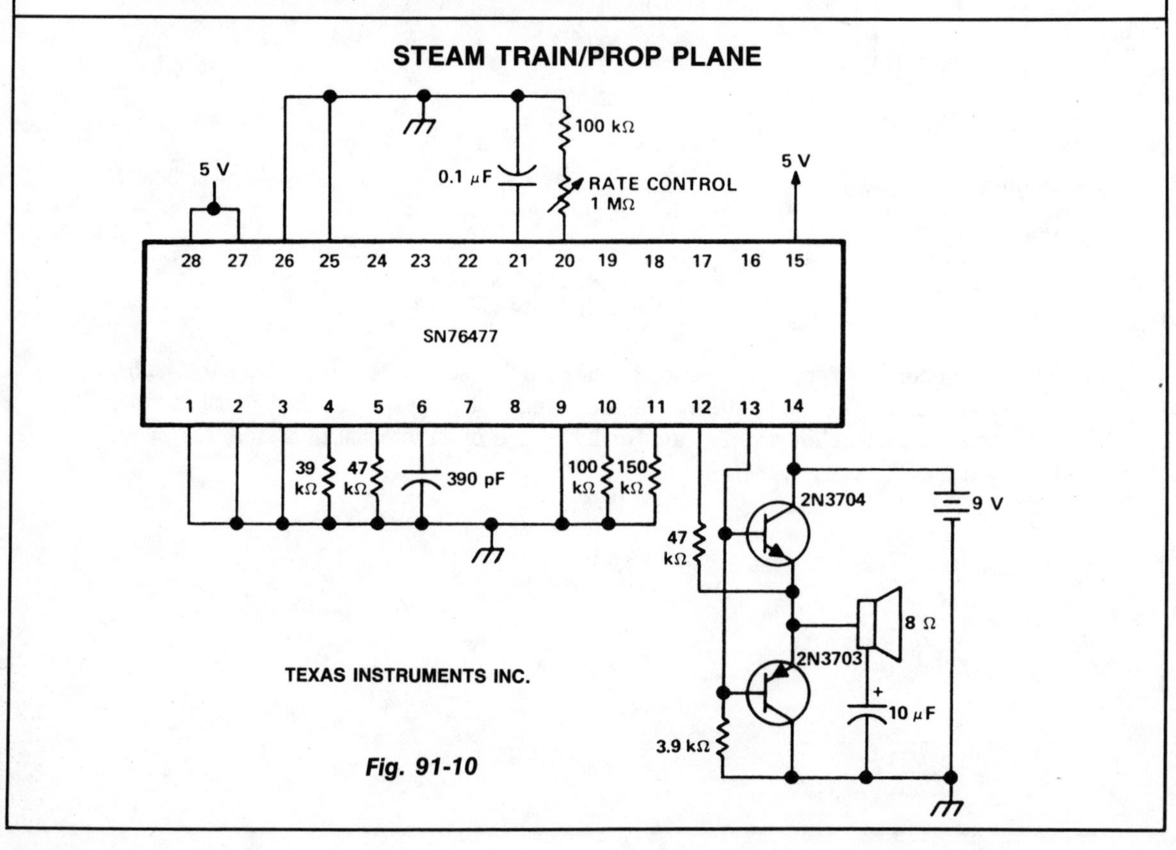

TEXAS INSTRUMENTS INC.

Fig. 91-10

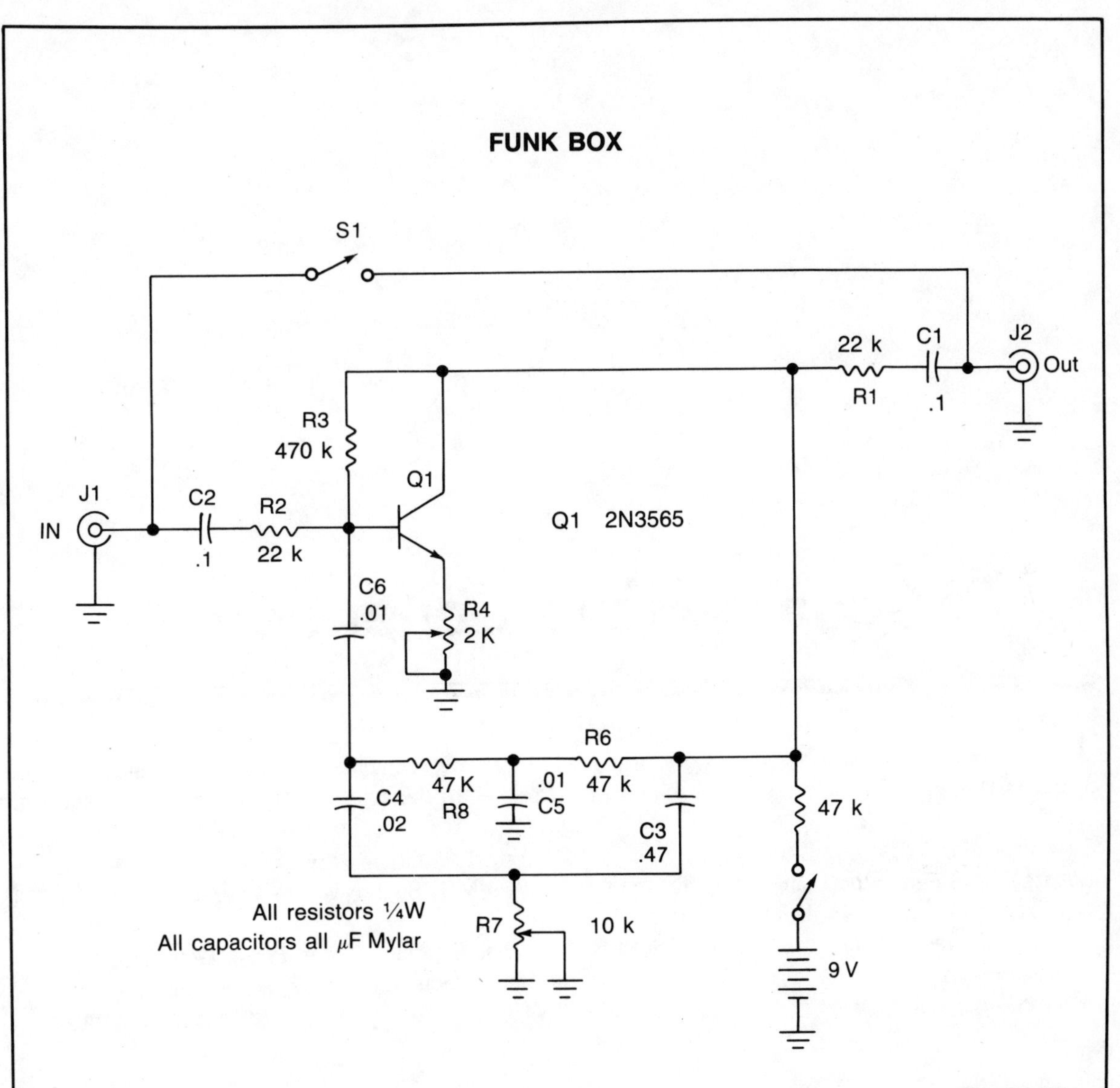

Fig. 91-11

Circuit Notes

Adjusting potentiometer R7 adds extra twang from way down low to way up high. To set the unit, adjust potentiometer R4 until you hear a whistle (oscillation); then back off R4 until the oscillation just ceases. The effect can be varied from bass to treble by R7.

92
Square-Wave Generators

The sources of the following circuits are contained in the Sources section beginning on page 694. The figure number contained in the box of each circuit correlates to the source entry in the Sources section.

LOW FREQUENCY TTL OSCILLATOR

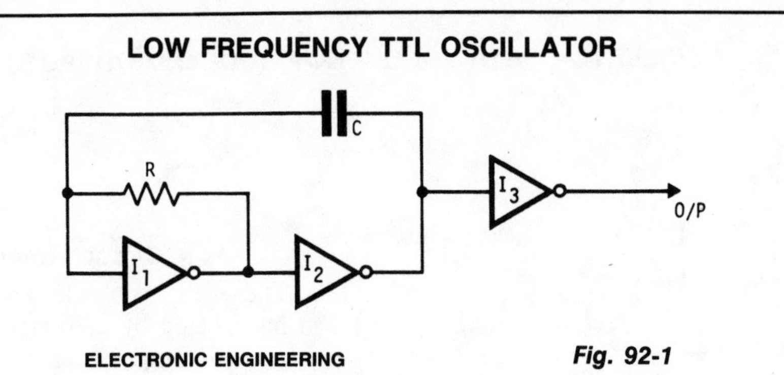

ELECTRONIC ENGINEERING *Fig. 92-1*

Circuit Notes

This oscillator uses standard inverters, one resistor and one capacitor, and has no minimum operating frequency. R and C must be chosen such that currents into the gates are below recommended operating limits and that leakage current into the gates and into C are small in comparison with the current in R also the output should be buffered (I3) to prevent variations in load affecting frequency. This circuit may also be used to square up slowly changing logic levels by use of multi input gates (NANDS, NORS Etc).

SQUARE-WAVE GENERATOR USING A 555 TIMER

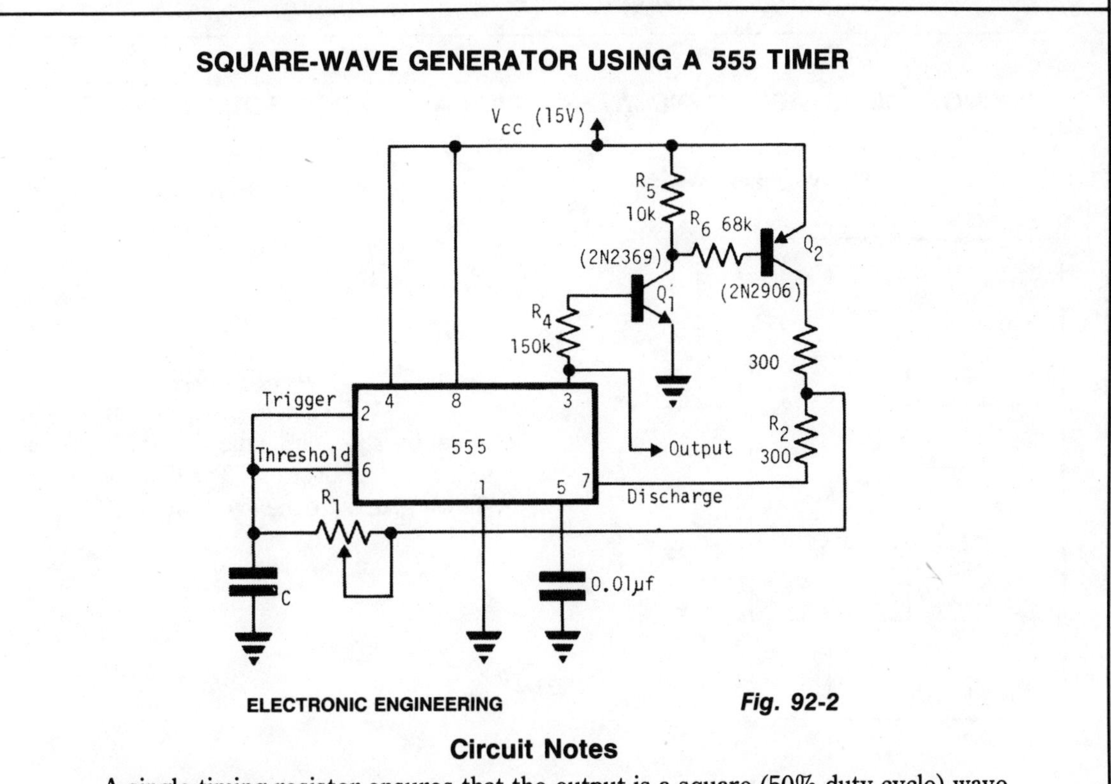

ELECTRONIC ENGINEERING *Fig. 92-2*

Circuit Notes

A single timing resistor ensures that the output is a square (50% duty cycle) wave at all frequency settings.

OSCILLATOR WITH FREQUENCY DOUBLED OUTPUT

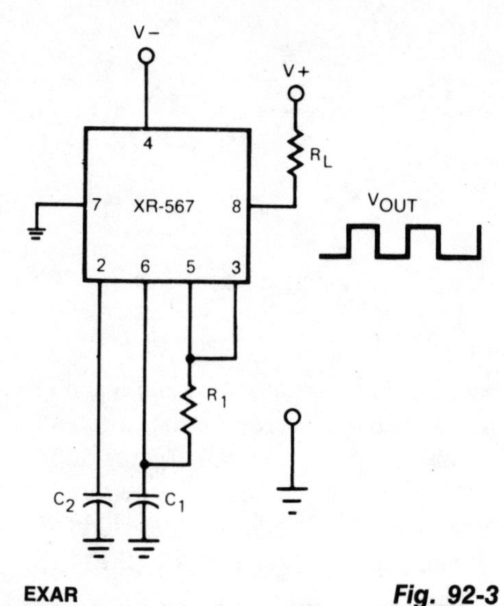

EXAR

Fig. 92-3

Circuit Notes

The current-controlled oscillator frequency can be doubled by applying a portion of the square-wave output at pin 5 back to the input at pin 3, as shown. In this manner, the quadrature detector functions as a frequency doubler and produces an output of 2 f_o at pin 8.

CMOS 555 ASTABLE GENERATES TRUE RAIL-TO-RAIL SQUARE WAVES

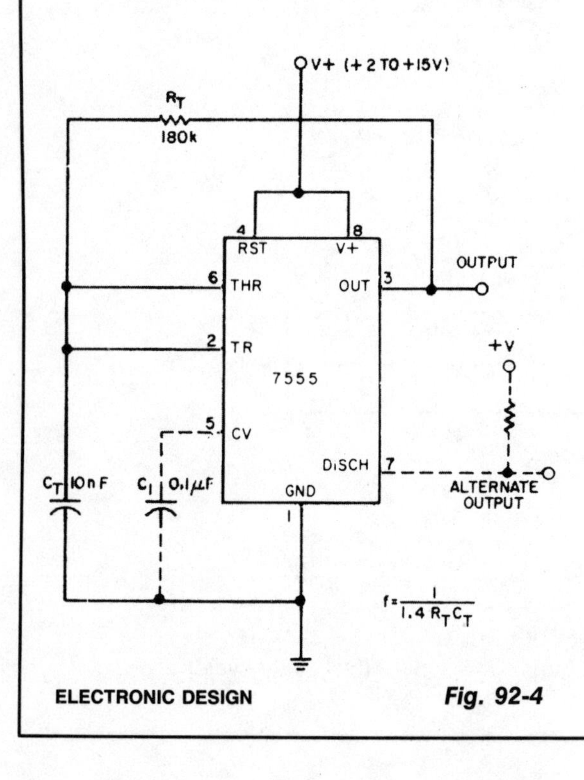

ELECTRONIC DESIGN

Fig. 92-4

Circuit Notes

A CMOS timer generates true square waves because, unlike the bipolar 555, its output swings from rail to rail. The component values shown give a frequency of about 400 Hz.

SQUARE-WAVE OSCILLATOR

ASTABLE MULTIVIBRATOR

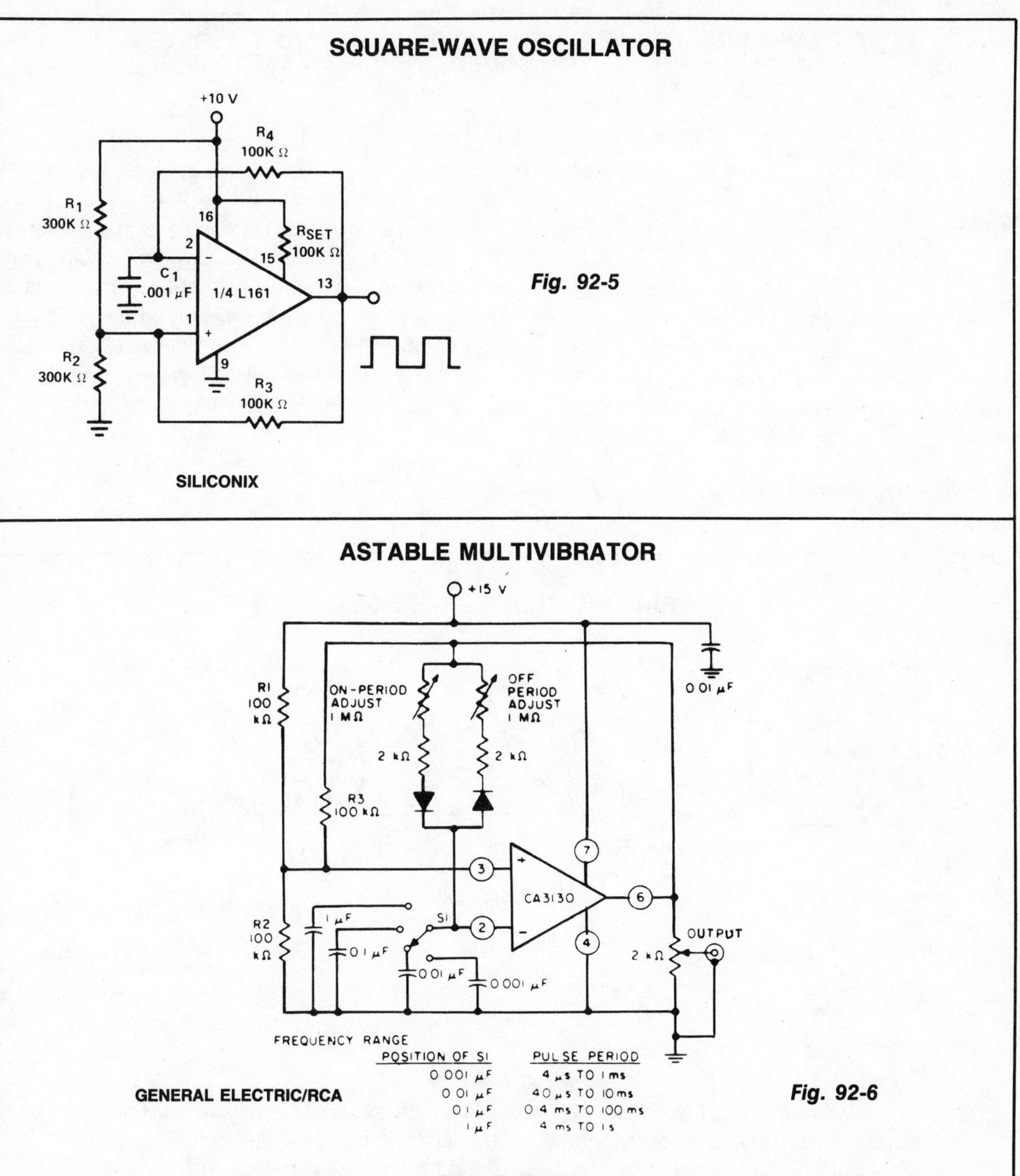

SILICONIX

Fig. 92-5

GENERAL ELECTRIC/RCA

Fig. 92-6

Circuit Notes

The circuit with independent control of "ON" and "OFF" periods uses the CA3130 BiMOS op amp for filters, oscillators, and long-duration timers. With input current at 50 pA, oscillators can utilize large-resistor/small-capacitor combinations without loading effects.

TWO-MHz SQUARE-WAVE GENERATOR USES TWO TTL GATES

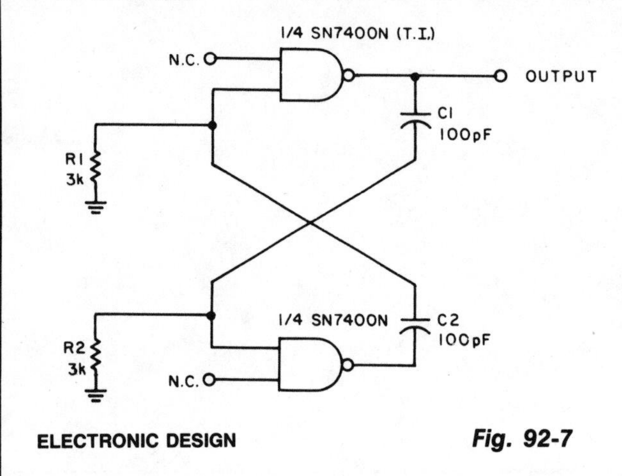

ELECTRONIC DESIGN

Fig. 92-7

Circuit Notes

With the values shown the circuit generates a 2-MHz symmetrical square wave. Changing capacitors C1 and C2 to 0.01 μF results in a frequency of 500 Hz. For the particular integrated circuits and power supply voltages (5.0 V), the reliable operating range of R1 = R2 is 2 k ohm to 4 k ohm.

PHASE TRACKING THREE-PHASE GENERATOR

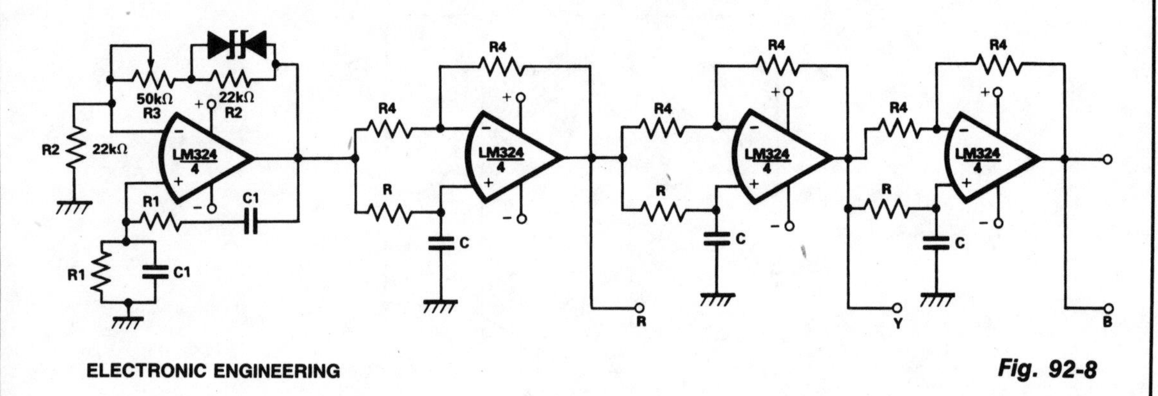

ELECTRONIC ENGINEERING

Fig. 92-8

Circuit Notes

Using a single chip LM324 can, with active R-C networks, reduce the size of a 3-phase waveform generator, and prove useful in compact and stable 3-phase inverters. One quarter of an LM324 is used as a Wien bridge oscillator generating a pure sinusoidal waveform while the remaining parts of the LM324 are used as three 120° fixed phase shifters. Initially potentiometer R3 should be varied to adjust the loop gain of the oscillator in order to start the oscillator.

LINE FREQUENCY SQUARE-WAVE GENERATOR

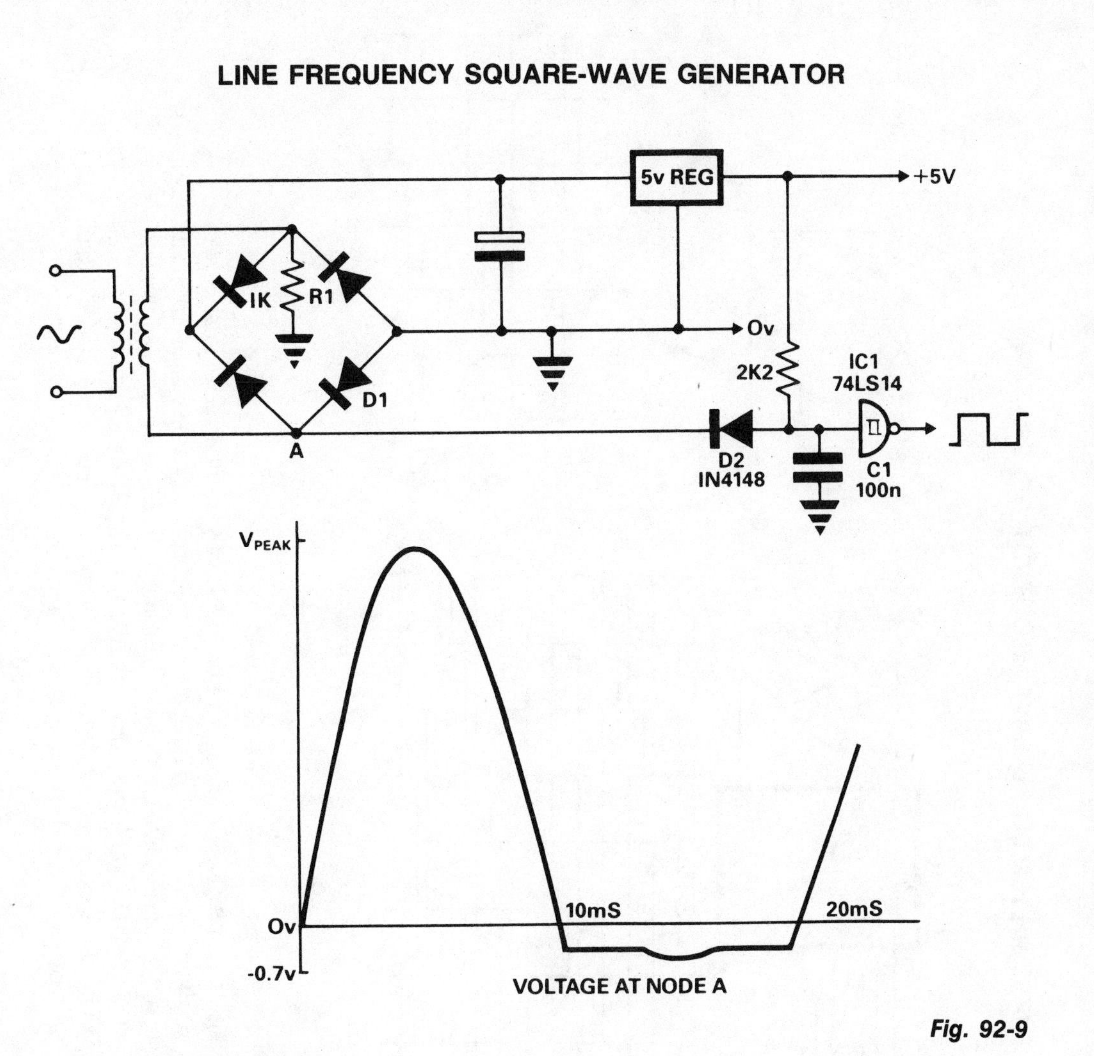

Fig. 92-9

ELECTRONIC ENGINEERING

Circuit Notes

With only three components and a buffer, a line frequency square wave having a 1:1 duty cycle may be derived from the power supply. During the alternate half-cycle, however, A is effectively clamped to −0.7 V by D1 in the bridge which offsets the forward voltage across D2 giving an input to IC1 of approximately 0 V. When A rises above +5 V, D2 is reverse biased and remains at +5 V. R1 is needed to load the transformer secondary maintaining a distortion-free waveform at A during the time the diode bridge is not conducting. C1 although not essential may be required to remove transients.

THREE PHASE SQUARE-WAVE OUTPUT GENERATOR

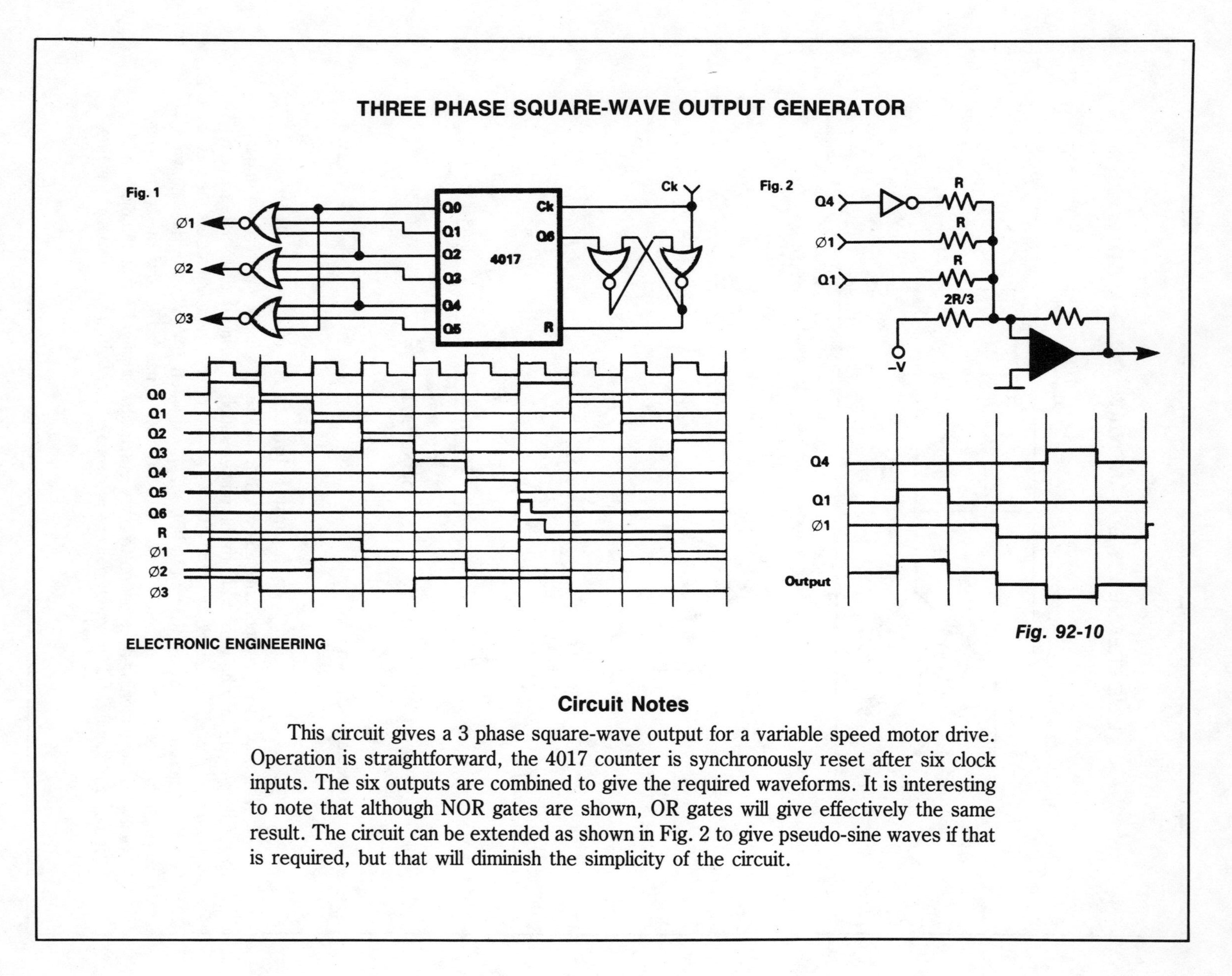

ELECTRONIC ENGINEERING

Fig. 92-10

Circuit Notes

This circuit gives a 3 phase square-wave output for a variable speed motor drive. Operation is straightforward, the 4017 counter is synchronously reset after six clock inputs. The six outputs are combined to give the required waveforms. It is interesting to note that although NOR gates are shown, OR gates will give effectively the same result. The circuit can be extended as shown in Fig. 2 to give pseudo-sine waves if that is required, but that will diminish the simplicity of the circuit.

93
Staircase Generator Circuits

The sources of the following circuits are contained in the Sources section beginning on page 694. The figure number contained in the box of each circuit correlates to the source entry in the Sources section.

STAIRCASE GENERATOR

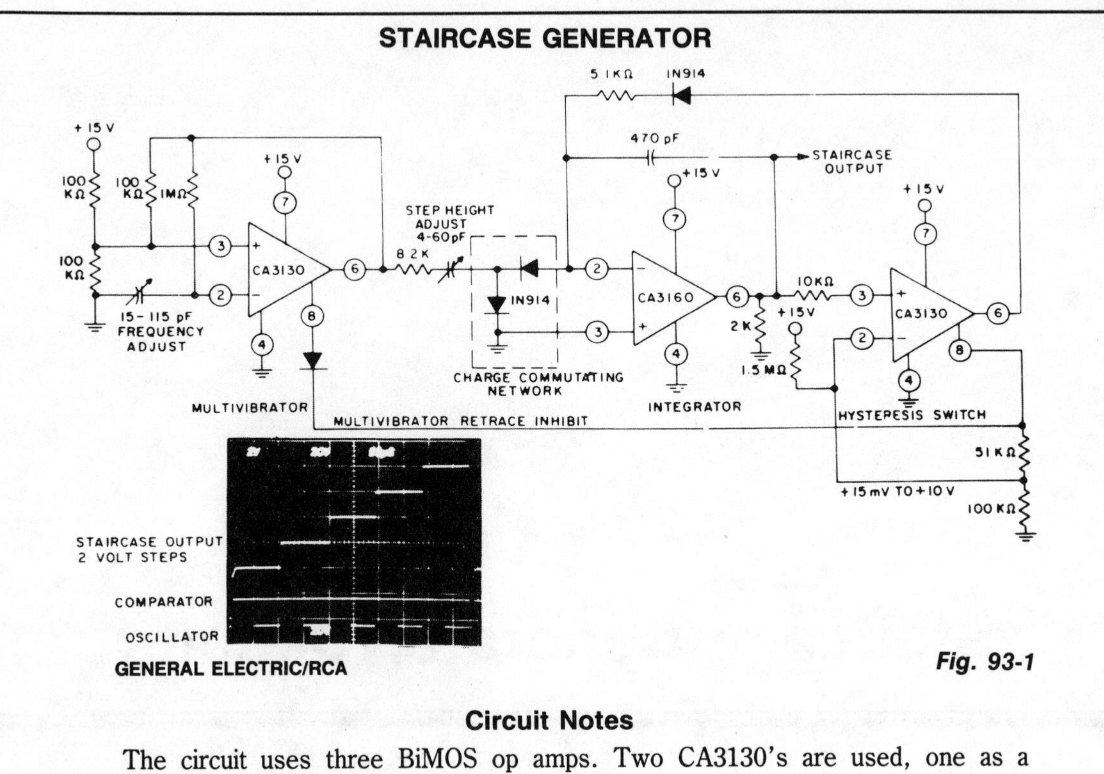

GENERAL ELECTRIC/RCA

Fig. 93-1

Circuit Notes

The circuit uses three BiMOS op amps. Two CA3130's are used, one as a multivibrator and the other as a hysteresis switch. The third amplifier, a CA3160, is used as a linear staircase generator.

STAIRCASE GENERATOR II

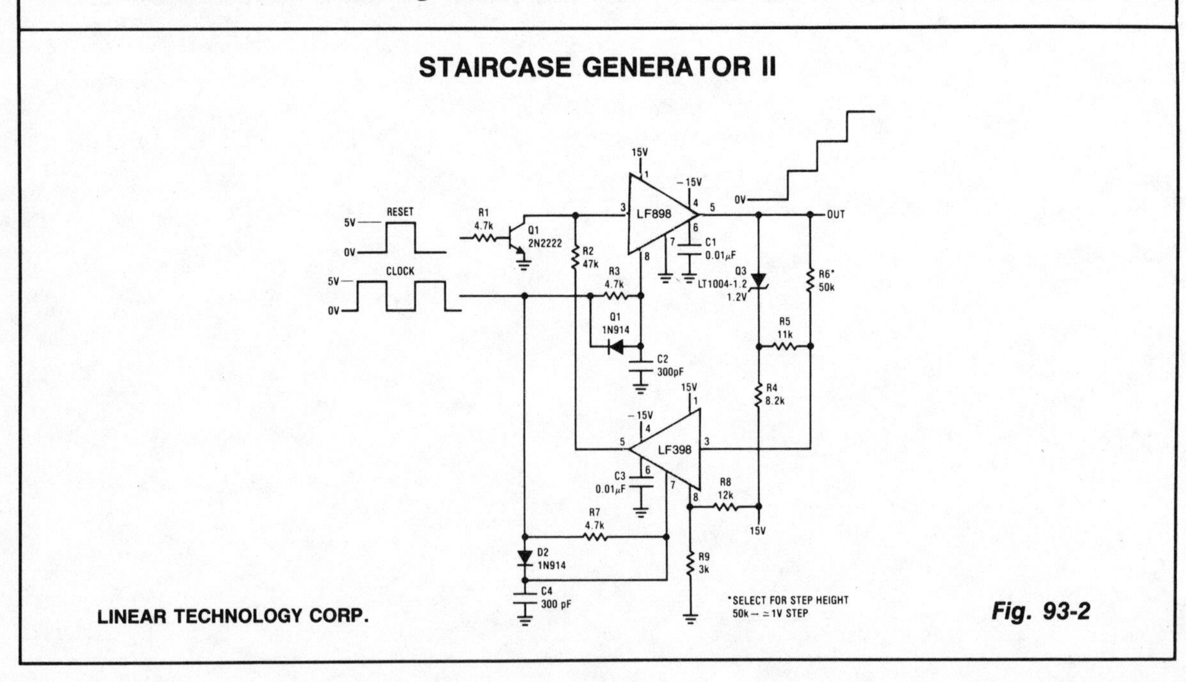

LINEAR TECHNOLOGY CORP.

Fig. 93-2

94
Stereo Balance Circuits

The sources of the following circuits are contained in the Sources section beginning on page 694. The figure number contained in the box of each circuit correlates to the source entry in the Sources section.

STEREO BALANCE TESTER

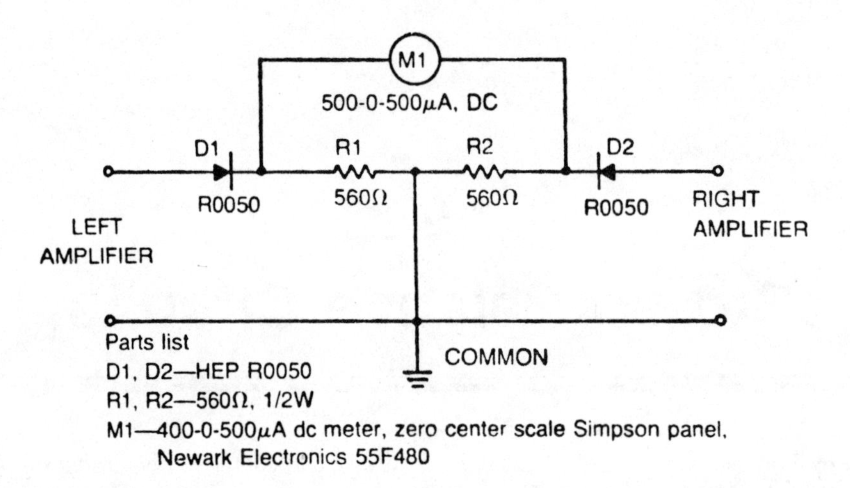

TAB BOOKS INC. *Fig. 94-1*

Circuit Notes

The meter will show volume and tone control balance between left and right stereo amplifiers. For maximum convenience the meter is a zero-center type. Resistors are five percent or better and the diodes a matched pair. Optimum stereo level and phase balance occurs for matched speakers when the meter indicates zero. If the meter indicates either side of zero, the levels are not matched or the wires are incorrectly phased. Check phasing by making certain the meter leads are connected to the amplifier hot terminals and the common leads go to ground.

STEREO BALANCE METER

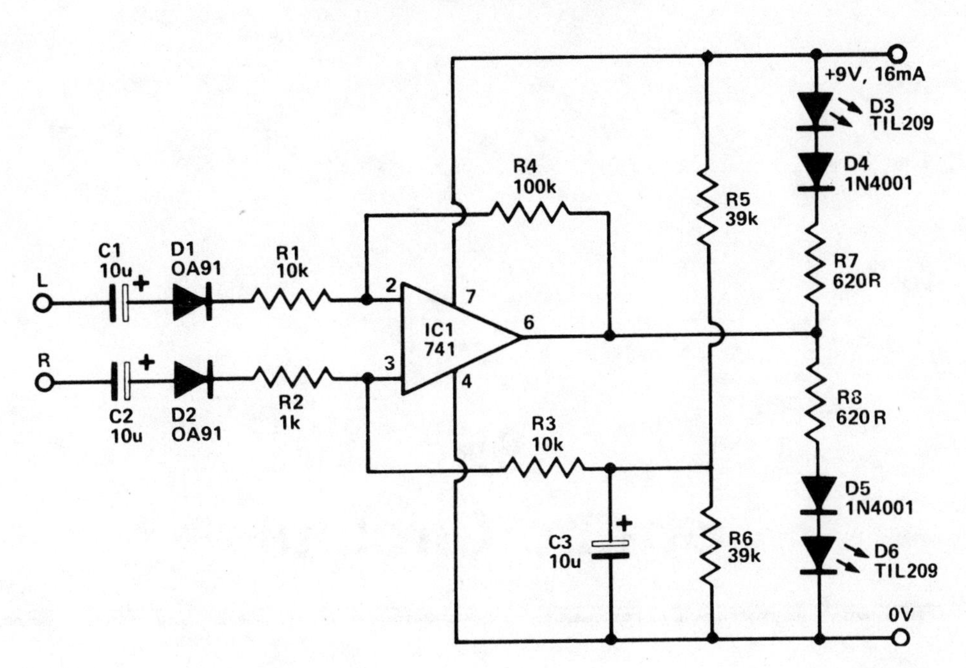

ELECTRONICS TODAY INTERNATIONAL

Fig. 94-2

Circuit Notes

To use the indicator, switch the amplifier to mono mode and adjust the balance control until both LEDs are equally illuminated. The amplifier is now in perfect stereo mode balance.

95 Strobe Circuits

The sources of the following circuits are contained in the Sources section beginning on page 694. The figure number contained in the box of each circuit correlates to the source entry in the Sources section.

SIMPLE STROBE

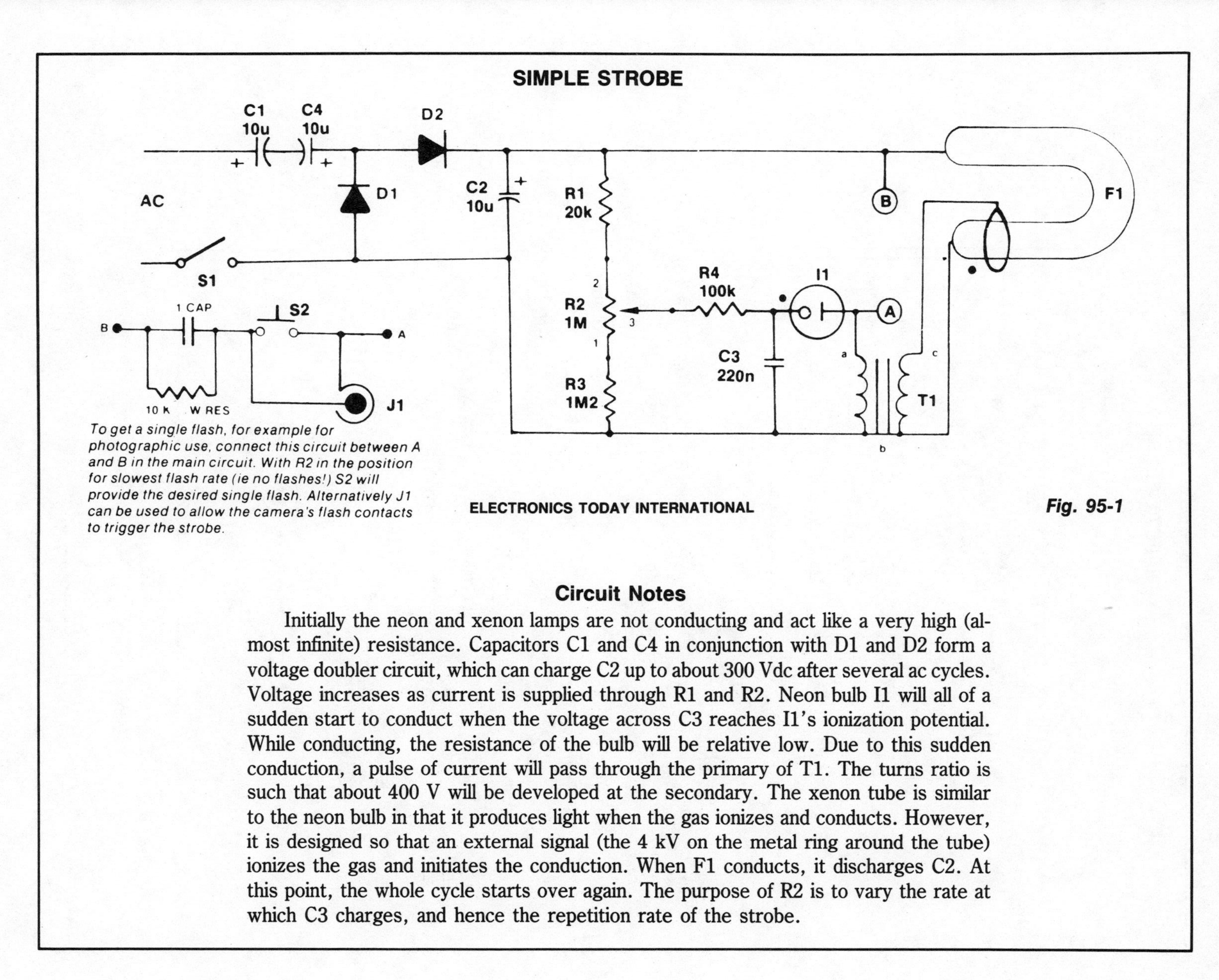

To get a single flash, for example for photographic use, connect this circuit between A and B in the main circuit. With R2 in the position for slowest flash rate (ie no flashes!) S2 will provide the desired single flash. Alternatively J1 can be used to allow the camera's flash contacts to trigger the strobe.

ELECTRONICS TODAY INTERNATIONAL

Fig. 95-1

Circuit Notes

Initially the neon and xenon lamps are not conducting and act like a very high (almost infinite) resistance. Capacitors C1 and C4 in conjunction with D1 and D2 form a voltage doubler circuit, which can charge C2 up to about 300 Vdc after several ac cycles. Voltage increases as current is supplied through R1 and R2. Neon bulb I1 will all of a sudden start to conduct when the voltage across C3 reaches I1's ionization potential. While conducting, the resistance of the bulb will be relative low. Due to this sudden conduction, a pulse of current will pass through the primary of T1. The turns ratio is such that about 400 V will be developed at the secondary. The xenon tube is similar to the neon bulb in that it produces light when the gas ionizes and conducts. However, it is designed so that an external signal (the 4 kV on the metal ring around the tube) ionizes the gas and initiates the conduction. When F1 conducts, it discharges C2. At this point, the whole cycle starts over again. The purpose of R2 is to vary the rate at which C3 charges, and hence the repetition rate of the strobe.

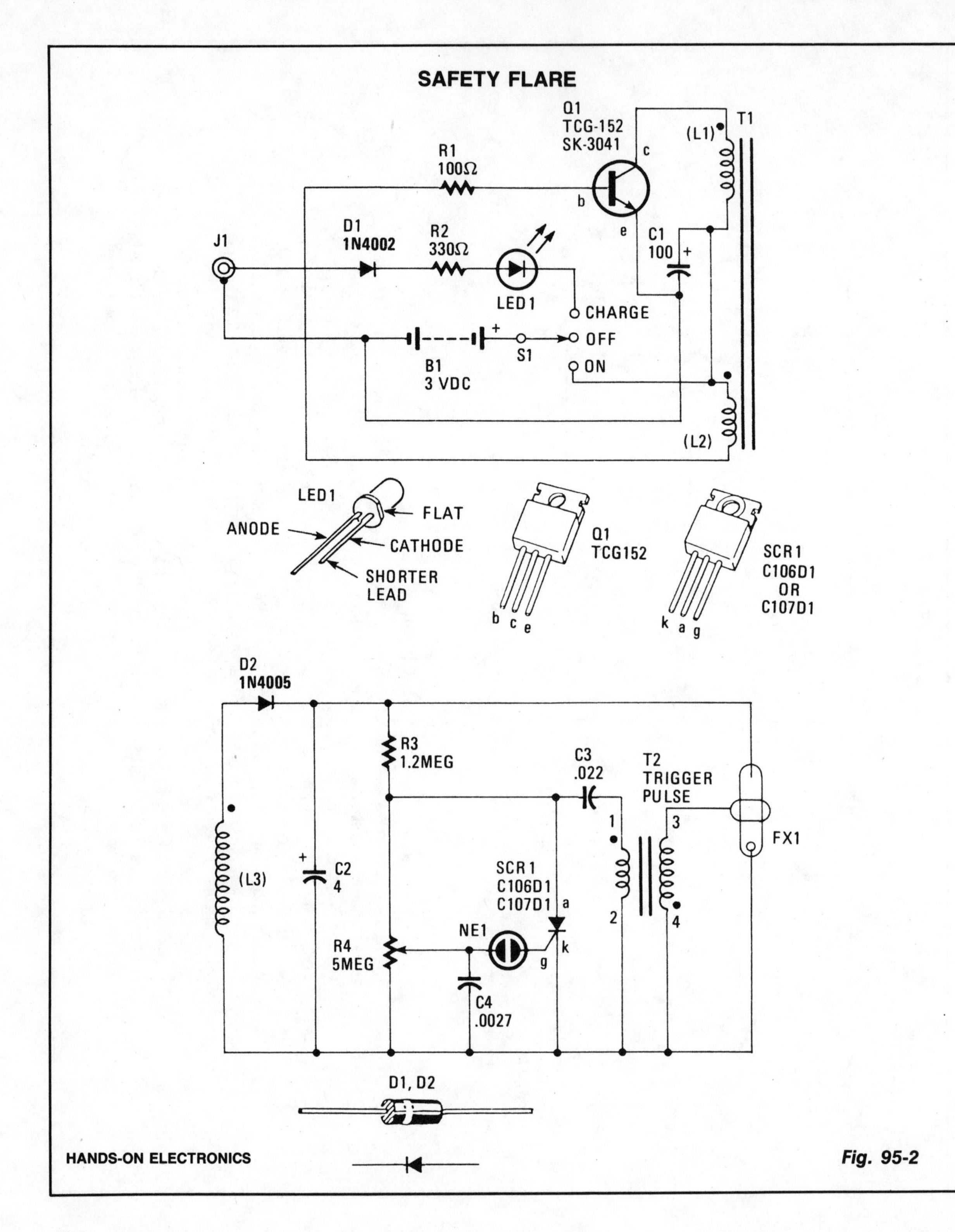

HANDS-ON ELECTRONICS

Fig. 95-2

Circuit Notes

When S1 is on, power is applied to an oscillator composed of Q1, R1, C1, L1, and L2. Coil L1 is the primary winding of T1, and L2 is the feedback winding. When Q1 turns on, its collector current saturates T1's ferrite core. That, in turn, removes the base drive to Q1 through L2. Transistor Q1 then turns off. As the field around L1 and L2 decays, Q1 will eventually turn on again, and the cycle repeats over, and over. Transformer T1 is a step-up, ferrite-core, potted-type unit whose secondary-winding (L3) output is rectified by D2 and filtered by C2. That capacitor charges up to around 250 to 300 volts, which is applied to the resistor divider composed of R3 and R4, along with the flash tube FX1. Capacitors C3 and C4 will charge up to around 200 and 100 volts, through R3 and R4, respectively. Flash rate is adjustable via R4. When the charge on C4 gets to around 100 volts, neon lamp NE1 fires discharging C4 into the gate circuit of silicon control rectifier SCR1. The SCR1 turns on discharging C3 into the primary winding of trigger-pulse transformer T2. Transformer T2 is another step-up, pulse-type unit providing an output of around 4 kW across transformer T2's secondary winding. The xenon gas inside FX1 is ionized and a bright flash is emitted. Finally, C3 quickly discharges through L4, and the cycle repeats over, and over.

DISCO-STROBE LIGHT

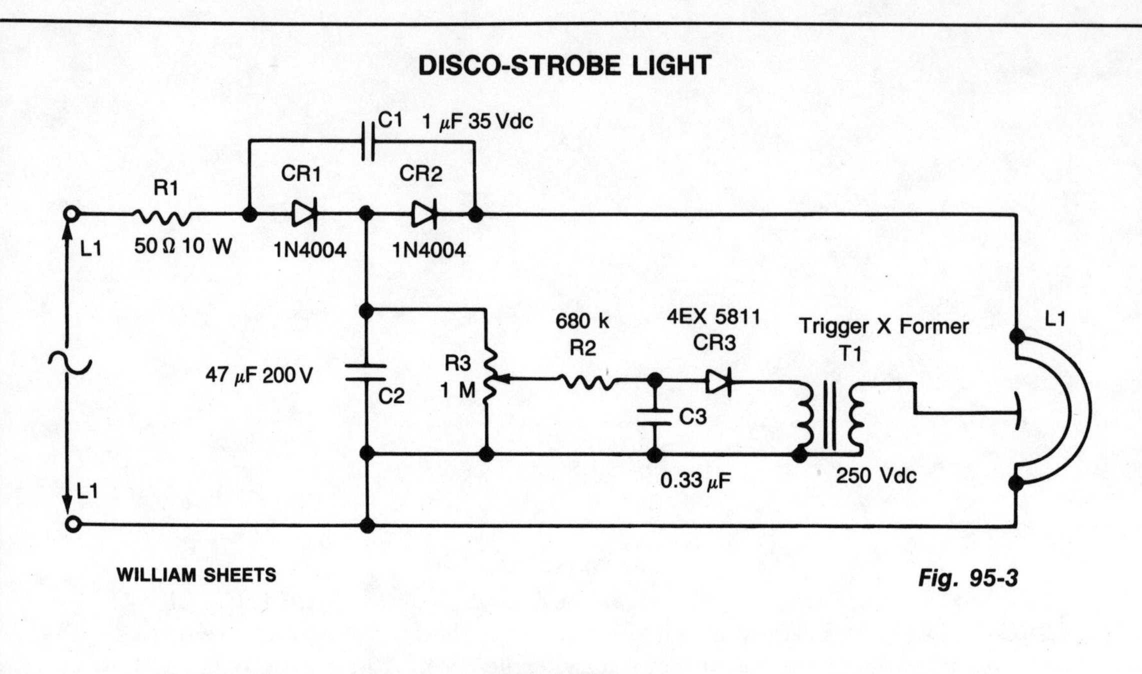

WILLIAM SHEETS

Fig. 95-3

Circuit Notes

This circuit uses a voltage doubler CR1 and CR2 to obtain about 280 V dc across C1. C2 and R3 form a voltage divider to obtain a dc voltage to change C3 thru R2. When CR3 fires, a high voltage is generated in T1, firing L1.

96
Switch Circuits

The sources of the following circuits are contained in the Sources section beginning on page 694. The figure number contained in the box of each circuit correlates to the source entry in the Sources section.

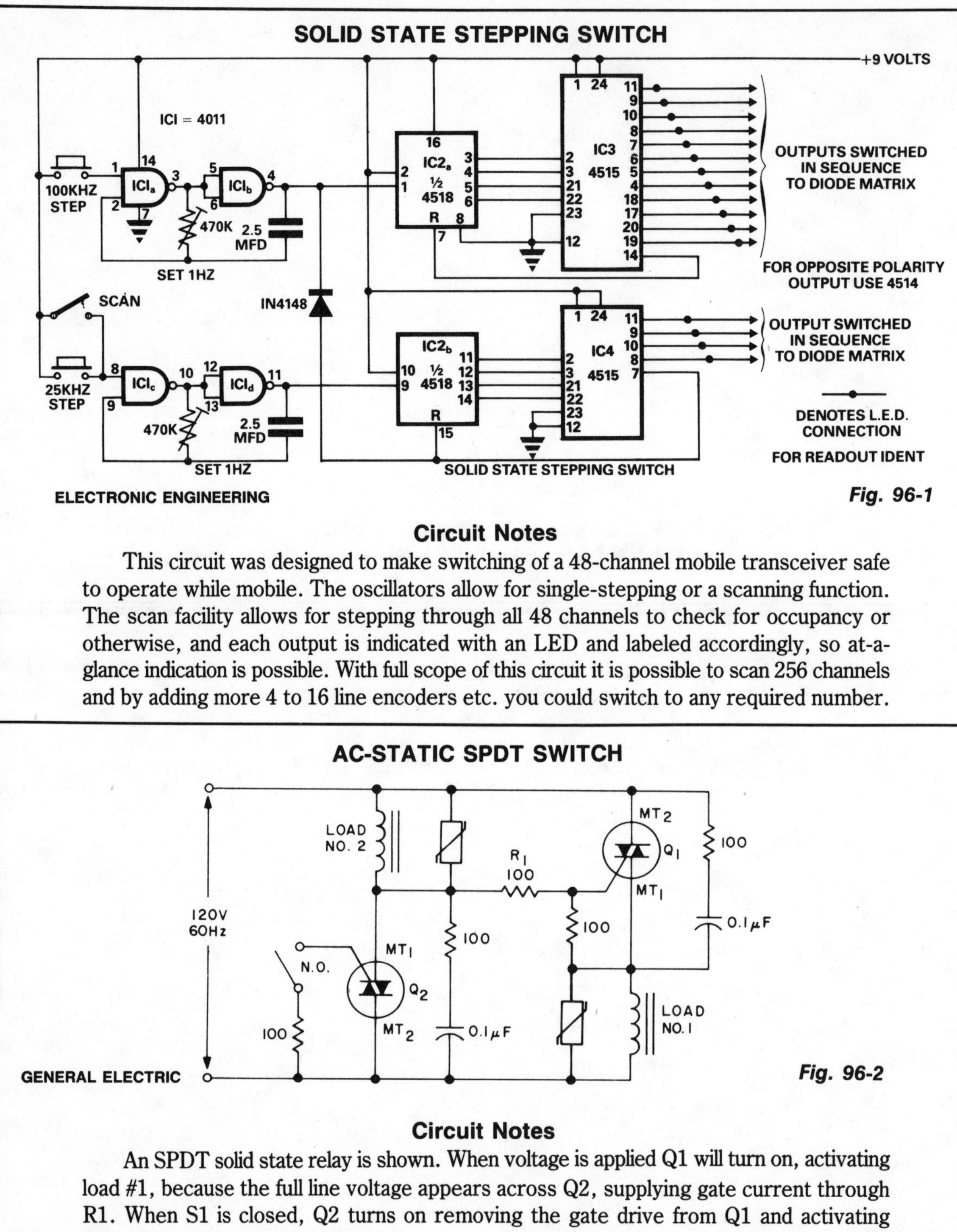

Fig. 96-1

Circuit Notes

This circuit was designed to make switching of a 48-channel mobile transceiver safe to operate while mobile. The oscillators allow for single-stepping or a scanning function. The scan facility allows for stepping through all 48 channels to check for occupancy or otherwise, and each output is indicated with an LED and labeled accordingly, so at-a-glance indication is possible. With full scope of this circuit it is possible to scan 256 channels and by adding more 4 to 16 line encoders etc. you could switch to any required number.

Fig. 96-2

Circuit Notes

An SPDT solid state relay is shown. When voltage is applied Q1 will turn on, activating load #1, because the full line voltage appears across Q2, supplying gate current through R1. When S1 is closed, Q2 turns on removing the gate drive from Q1 and activating load #2.

97
Tape Recorder Circuits

The sources of the following circuits are contained in the Sources section beginning on page 694. The figure number contained in the box of each circuit correlates to the source entry in the Sources section.

TAPE RECORDER INTERFACE

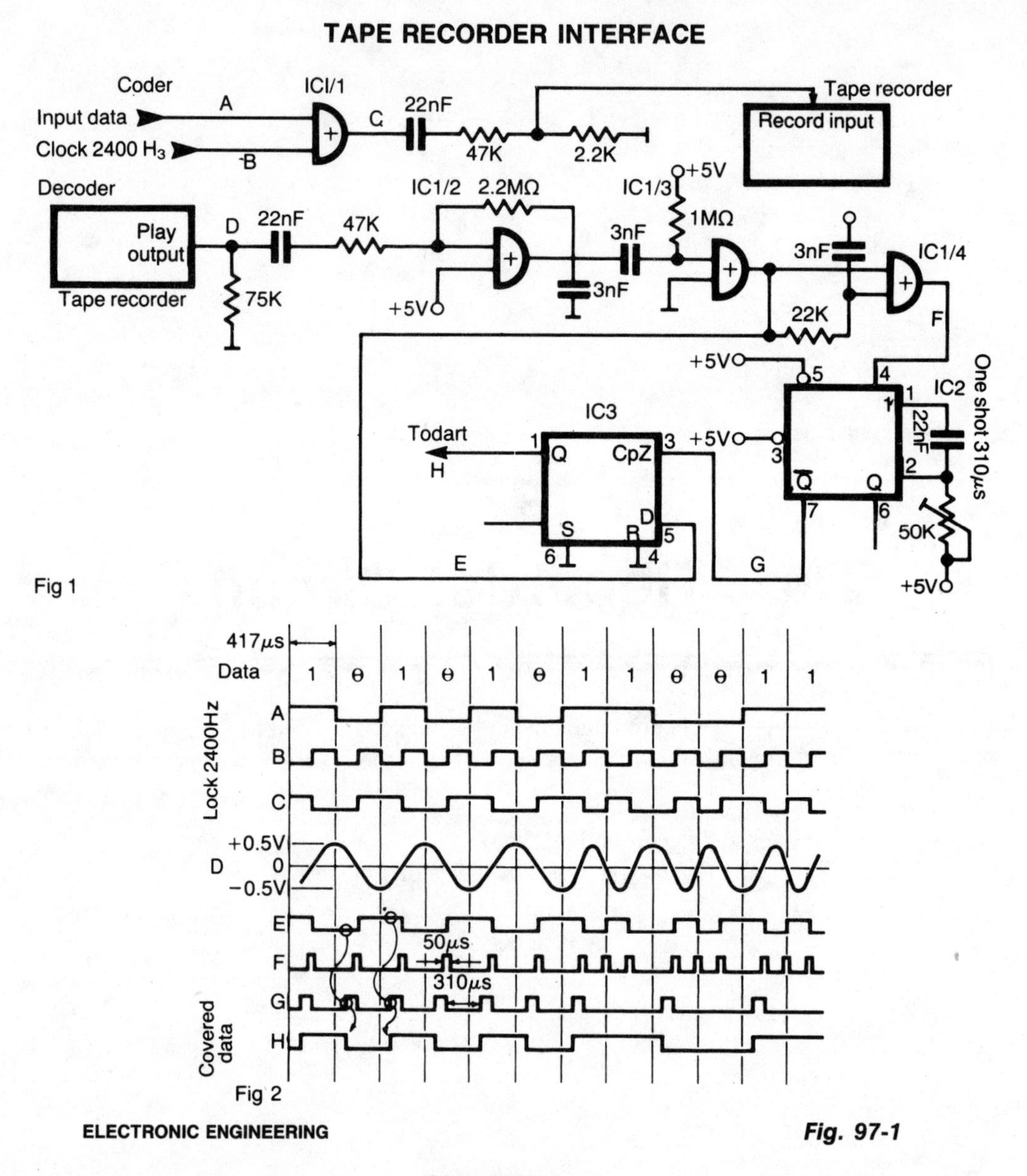

ELECTRONIC ENGINEERING *Fig. 97-1*

Circuit Notes

The interface allows data to be saved on an ordinary tape recorder at a speed of 2400 bit/s.

The serial stream of data Fig. 1 (A) is coded with a clock of 2400 Hz (B), by means of XOR gate IC 1/1. Logical "high" and "low" appear as shown in Fig. 2 (C). These impulses are lowered in amplitude and feed into the record input of a low cost tape recorder.

TAPE RECORDER INTERFACE, Continued.

During the playback, pulses (D) are amplified with CMOS gate IC 1/2 connected as a linear amplifier, and providing a TTL level signal shown in (E). On both positive and negative transitions IC 1/4 forms short pulses as shown in (F) (approx. 50 μs) that triggers one shot IC2. A monostable one shot pulse width is adjusted to be ¾ of bit length (310 μs). A change from "high" to "low" in a coded stream generates a "low" pulse width of one bit cell. The same is for change from "low" to "high" that generates a "high" pulse of the same width. During this pulse one shot latches the state of line E in D type flip-flop IC3 (G). When a stream consists of multiple "ones" or "zeros," the one shot is retriggered before it comes to the end of the quasistable state and the state of the flip-flop remains unchanged. The original data stream is available at the output of the flip-flop (H). Z80 the DUART that receives these pulses is programmed so that the receiver clock is 16 times the data rate (38.4 kHz).

TAPE RECORDER POSITION INDICATOR/CONTROLLER

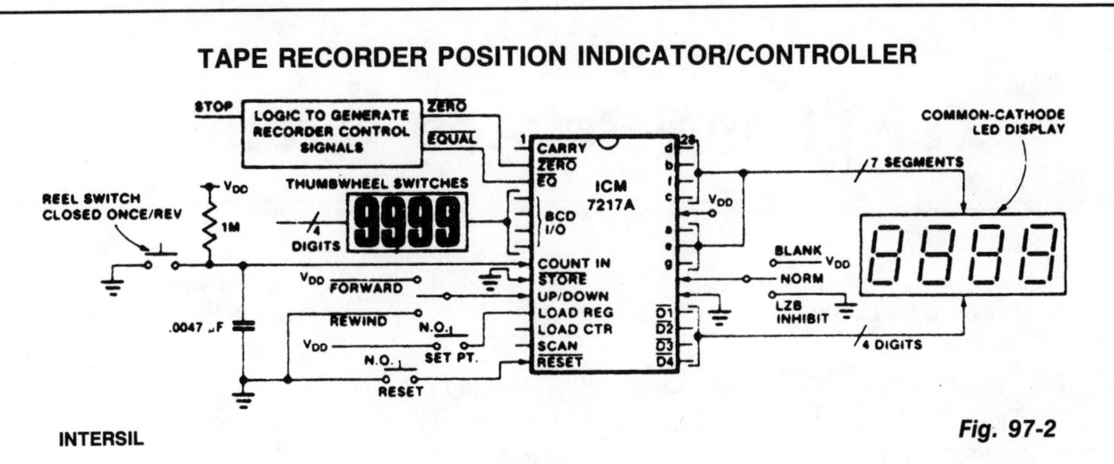

INTERSIL

Fig. 97-2

Circuit Notes

This circuit is representative of the many applications of up/down counting in monitoring dimensional position. In the tape recorder application, the LOAD REGISTER, EQUAL, and ZERO outputs are used to control the recorder. To make the recorder stop at a particular point on the tape, the register can be set with the stop at a particular point on the tape, the register can be set with the stop point and the EQUAL output used to stop the recorder either on fast forward, play or rewind.

To make the recorder stop before the tape comes free of the reel on rewind, a leader should be used. Resetting the counter at the starting point of the tape, a few feet from the end of the leader, allows the ZERO output to be used to stop the recorder on rewind, leaving the leader on the reel. The 1 M ohm resistor and .0047 μF capacitor on the COUNT INPUT provide a time constant of about 5 ms to debounce the reel switch. The Schmitt trigger on the COUNT INPUT of the ICM7217 squares up the signal before applying it to the counter. This technique may be used to debounce switchclosure inputs in other applications.

98
Telephone-Related Circuits

The sources of the following circuits are contained in the Sources section beginning on page 694. The figure number contained in the box of each circuit correlates to the source entry in the Sources section.

Speech Activity Detector for Telephone Lines
Scramble Phone
Musical Telephone Ringer
Dual Tone Decoding
Automatic Telephone Recording Device
Telephone Ringing Detector, Frequency and Volume Controlled
Music on Hold
Circuit Monitors Blinking Phone Lights
Phone Light
High Isolation Telephone Ringer
Remote Telephone Monitor
Plug-In Remote Telephone Ringer
Telephone Hold Button
Telephone Blinker
Telephone "In Use" Indicator
Tone Ringer
Tone Ringer II
Speakerphone
Speech Network
Programmable Multi-Tone Telephone Ringer

SPEECH ACTIVITY DETECTOR FOR TELEPHONE LINES

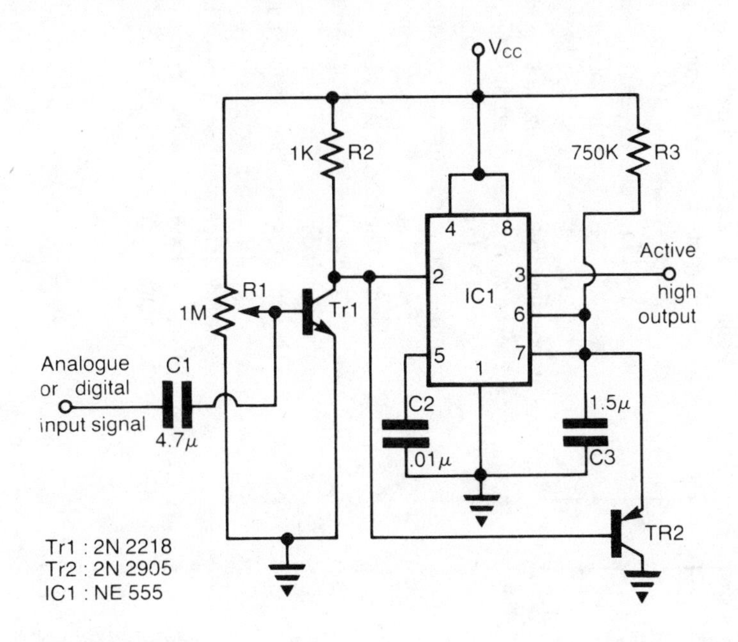

ELECTRONIC ENGINEERING

Fig. 98-1

Circuit Notes

The circuit can be used in telephone lines for speech activity detection purposes. This detection is very useful in the case of half-duplex conversation between two stations, in the case of simultaneous transmission of voice and data over the same pair of cables by the method of interspersion data on voice traffic, and also in echo suppressor devices. The circuit consists of a class-A amplifier in order to amplify the weak analog signals (in the range 25-400 mW of an analog telephone line).

The IC1 is connected as a retriggerable monostable multivibrator with the Tr2 discharging the timing capacitor C3, if the pulse train reaches the trigger input 2 of IC1 with period less than the time:

$$T_{high} = 1.1 \text{ (R3 C3)}$$

The output 3 of IC1 is active ON when an analog or digital signal is presented at the output and it drops to low level, T_{high} seconds after the input signal has ceased to exist.

SCRAMBLE PHONE

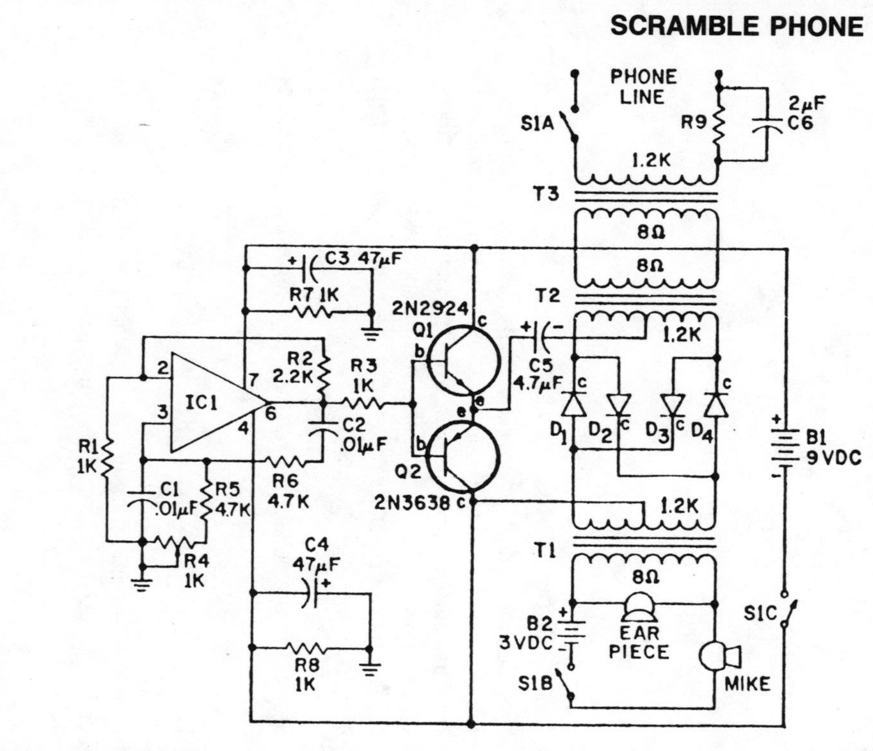

B1—9-volt battery, Eveready 216 or equiv.
B2—3-volt battery, two AA penlight cells in series
C1, C2—0.01 uF polystyrene capacitor, 100 VDC or better
C3, C4—47 uF electrolytic capacitor, 25 VDC or better
C5—4.7 uF electrolytic capacitor, 25 VDC or better
C6—2 uF paper or mylar capacitor, 50 VDC or better
D1 to D4—Diode, IN914, HEP-156
IC1—Integrated circuit, Signetics N5741K or equiv.
Q1—NPN transistor, 2N2924, HEP-724
Q2—PNP transistor, 2N3638, HEP-716
R1, R3, R7, R8—1000-ohm, ½-watt resistor
R2—2,200-ohm ½-watt resistor
R4—1000-ohm potentiometer
R5, R6—4,700-ohm, ½-watt resistor
R9—Limit line current to 25mA (see text)
S1A, S1B, S1C—Phone hook switch (see text)
T1 to T3—Small transistor audio transformer; 8-ohm primary, 1,200-ohm center taped secondary.
Misc.—Surplus telephone (see Lafayette, Radio Shack, EDI, BA catalogs), battery holders, hardware, knob, wire, solder, etc.

TAB BOOKS, INC.

Fig. 98-2

Circuit Notes

IC-1 and the associated circuitry form a stable audio tone generator that feeds a buffer amplifier, Q1 and Q2. The tone output is taken from the emitters of the transistor pair to supply a carrier voltage for a balanced modulator made up of four diodes—D1 through D4—and T1 and T2. If the two transformers and the four diodes are perfectly matched (which is almost impossible to achieve and not necessary in any case) no carrier will appear at the input or output of T1 or T2. In a practical circuit, a small amount of unbalance will occur and produce a low-level carrier tone at the input and output of the balanced modulator. A telephone carbon mike and earpiece are connected to the low impedance winding of T1, with a three volt battery supplying the necessary mike current. Trim potentiometer R4 is used to make a fine frequency adjustment of the oscillator so that two scrambler units may be synchronized to the same carrier frequency. Rg limits line current to 25 mA.

MUSICAL TELEPHONE RINGER

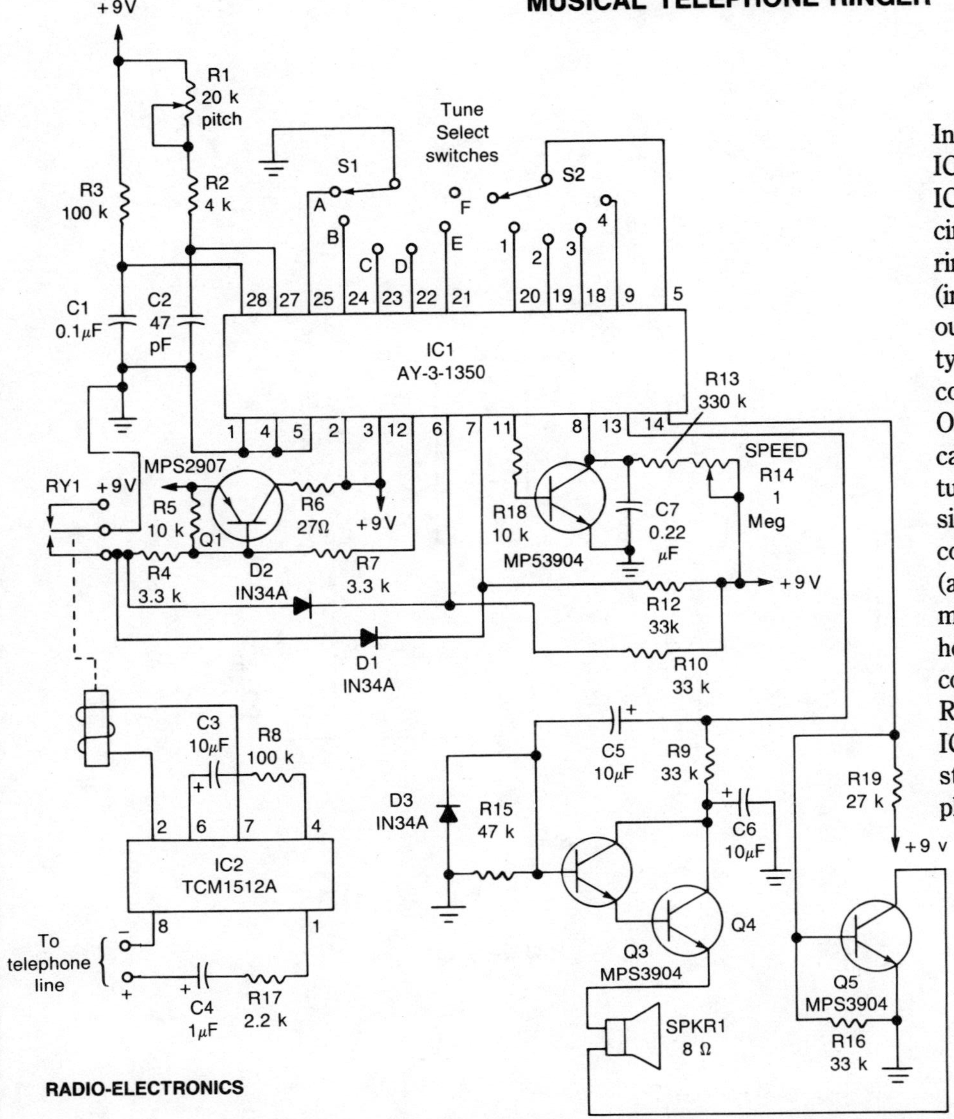

Circuit Notes

The heart of the circuit is IC1, General Instrument's AY-3-1350 melody-synthesizer IC. IC2 is a TCM1512 telephone ring detector IC that is powered by the telephone line. The circuit's operation begins when IC2 senses a ring pulse on the telephone line. The detector (internally) rectifies the ring signal and then outputs a voltage to relay RY1 (an SPST reed-type relay with 5 volt contacts), causing its contacts to close. That pulls pin 12 (the ON/OFF control) of IC1 low (logic "0"), causing it to output a signal—the selected tune—to transistor amplifier Q2. The amplified signal is then fed to the speaker. The melody continues to play either until the tune is finished (at which time IC1 returns to the standby mode), or until someone takes the phone off the hook. Taking the phone off the hook discontinues the ring pulses to IC2, which opens RY1. When the relay contacts open, pin 12 of IC1 goes high, returning the circuit to the standby mode to wait for the next incoming phone call.

RADIO-ELECTRONICS

Fig. 98-3

DUAL TONE DECODING

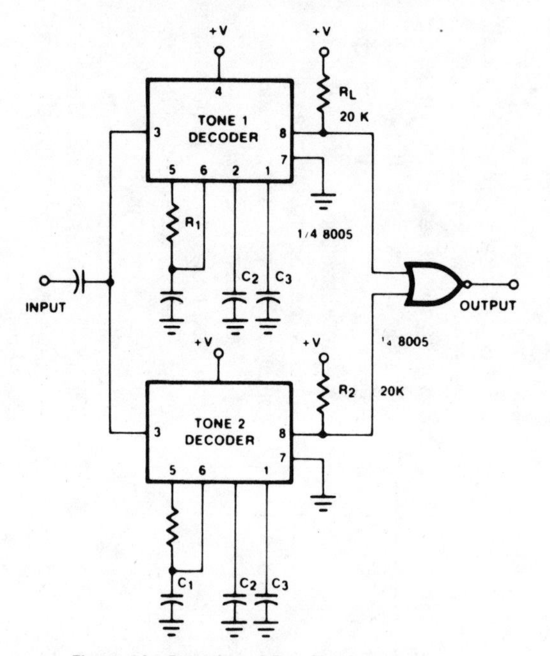

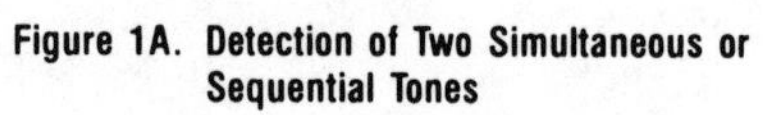

Figure 1A. Detection of Two Simultaneous or Sequential Tones

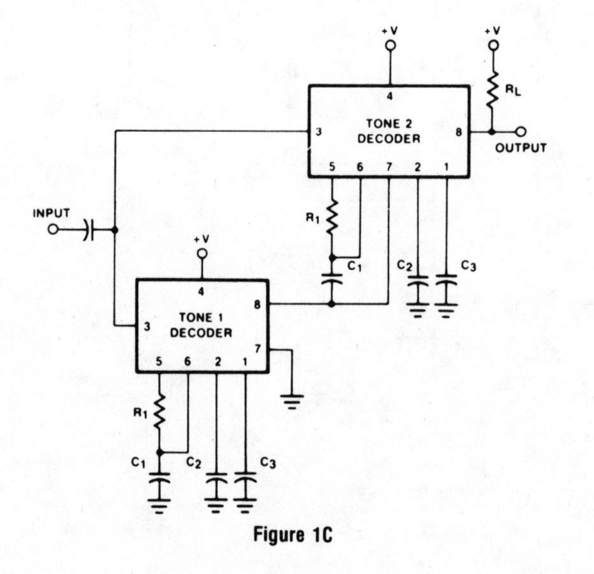

Figure 1C

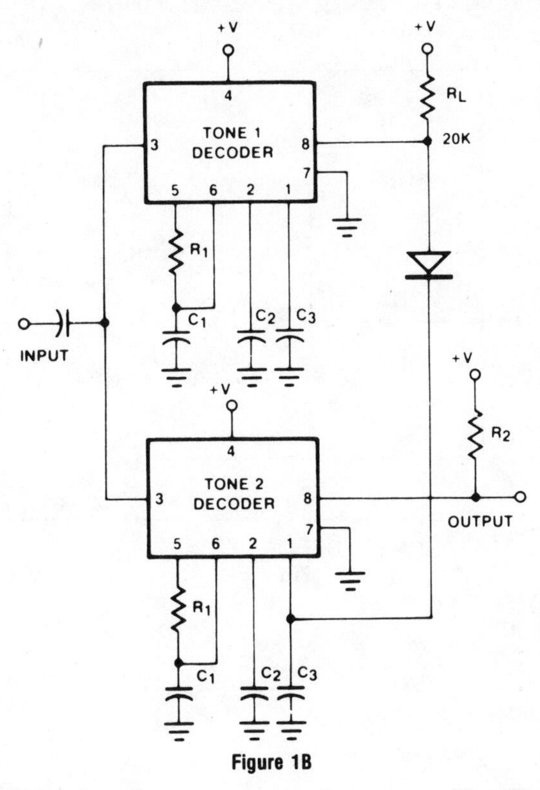

Figure 1B

EXAR

Fig. 98-4

Circuit Notes

Two integrated tone decoders, XR-567 units, can be connected (as shown in Fig. 1A) to permit decoding of simultaneous or sequential tones. Both units must be on before an output is given. R1C1 and R′1C′1 are chosen, respectively, for Tones 1 and 2. If sequential tones (1 followed by 2) are to be decoded, then C3 is made very large to delay turn-off of Unit 1 until Unit 2 has turned on and the NOR gate is activated. Note that the wrong sequence (2 followed by 1) will not provide an output since Unit 2 will turn off before Unit 1 comes on. Figure 1B shows a circuit variation which eliminates the NOR gate. The output is taken from Unit 2, but the Unit 2 output stage is biased off by R2 and CR1 until activated by Tone 1. A further variation is given in Fig. 1C. Here, Unit 2 is turned on by the Unit 1 output when Tone 1 appears, reducing the standby power to half. Thus, when Unit 2 is on, Tone 1 is or was present. If Tone 2 is now present, Unit 2 comes on also and an output is given. Since a transient output pulse may appear at Unit 1 turn-on, even if Tone 2 is not present, the load must be slow in response to avoid a false output due to Tone 1 alone. The XR-267 Dual Tone Decoder can replace two integrated tone decoders in this application.

AUTOMATIC TELEPHONE RECORDING DEVICE

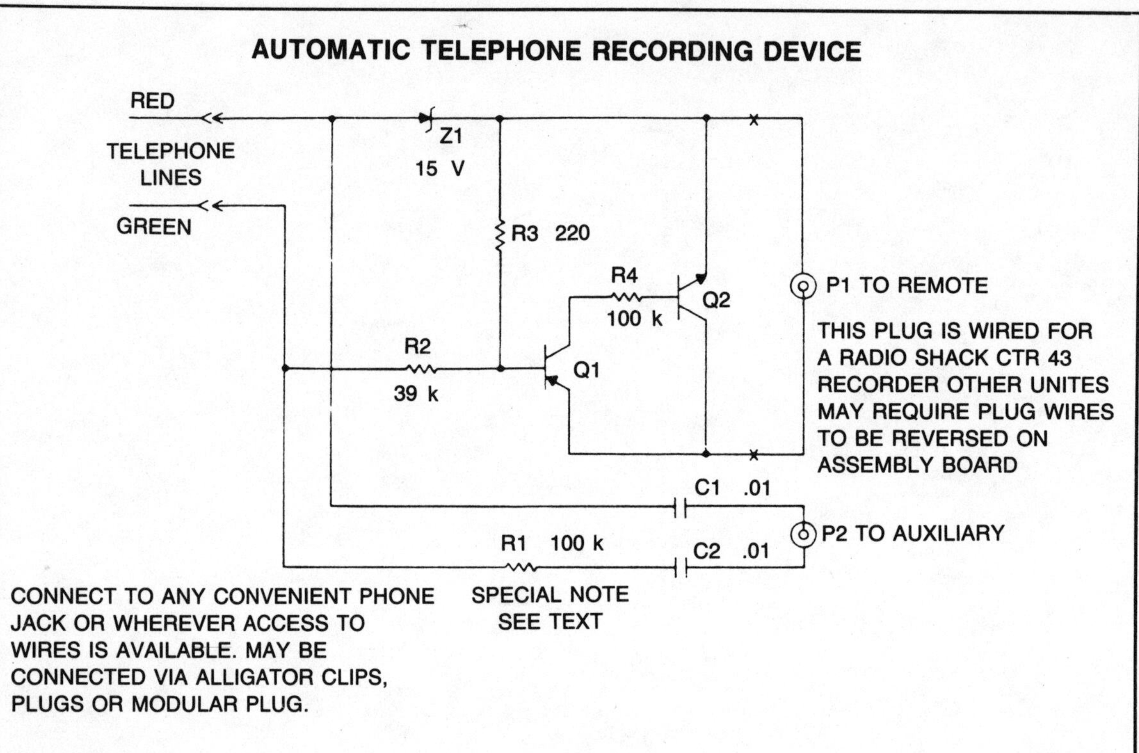

TAB BOOKS, INC.

Fig. 98-5

Circuit Notes

The device is a dc switch that is normally on via the forward biasing of Q1 via R3. Q1 now clamps Q2 into a forward state by biasing its complement well into a saturated state via R4. The dc switch is turned off via a negative voltage above that of the zener (D1). This voltage is usually about 48 and is the on-hook value of the phone line. This negative voltage overrides the effect of R3 and keeps the circuit ''off.'' When the phone is off the hook, the 48 volts drops to 10 volts, that is below the zener voltage of D1 and R3 now turns the circuit on. The audio signal is via attenuator resistor R1 and dc isolating capacitors C1, C2. The device is a high impedance switch that isolates the recording controlled device from the phone line via some relatively simple electronic circuitry. It requires no battery and obtains power for operating via the remote jack that in most recorders is a source of 6 volts. When clamped to ground it initiates recorder operation. The unit interfaces with most portable cassette recorders providing they contain a remote control jack.

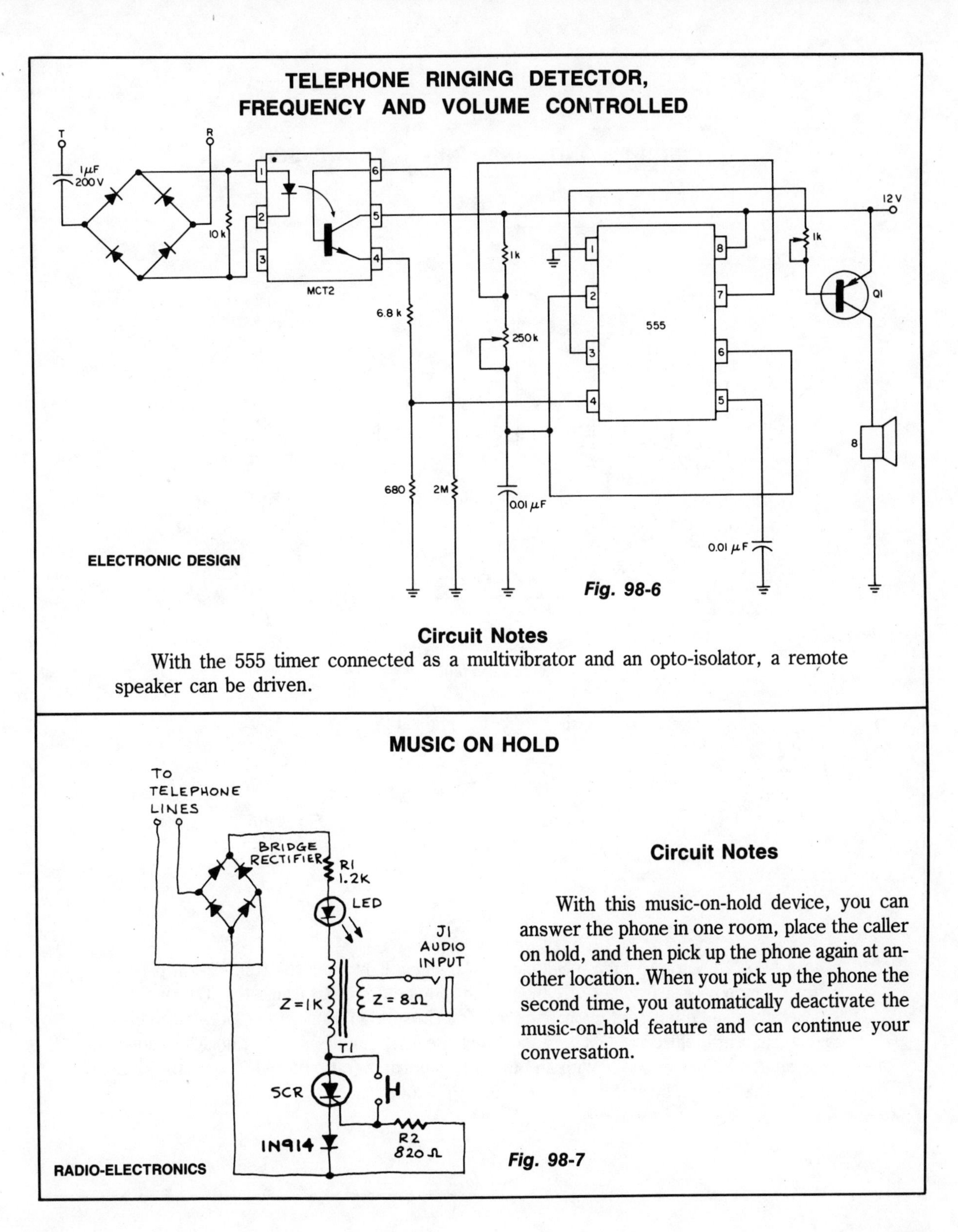

TELEPHONE RINGING DETECTOR, FREQUENCY AND VOLUME CONTROLLED

Fig. 98-6

Circuit Notes

With the 555 timer connected as a multivibrator and an opto-isolator, a remote speaker can be driven.

MUSIC ON HOLD

Circuit Notes

With this music-on-hold device, you can answer the phone in one room, place the caller on hold, and then pick up the phone again at another location. When you pick up the phone the second time, you automatically deactivate the music-on-hold feature and can continue your conversation.

Fig. 98-7

CIRCUIT MONITORS BLINKING PHONE LIGHTS

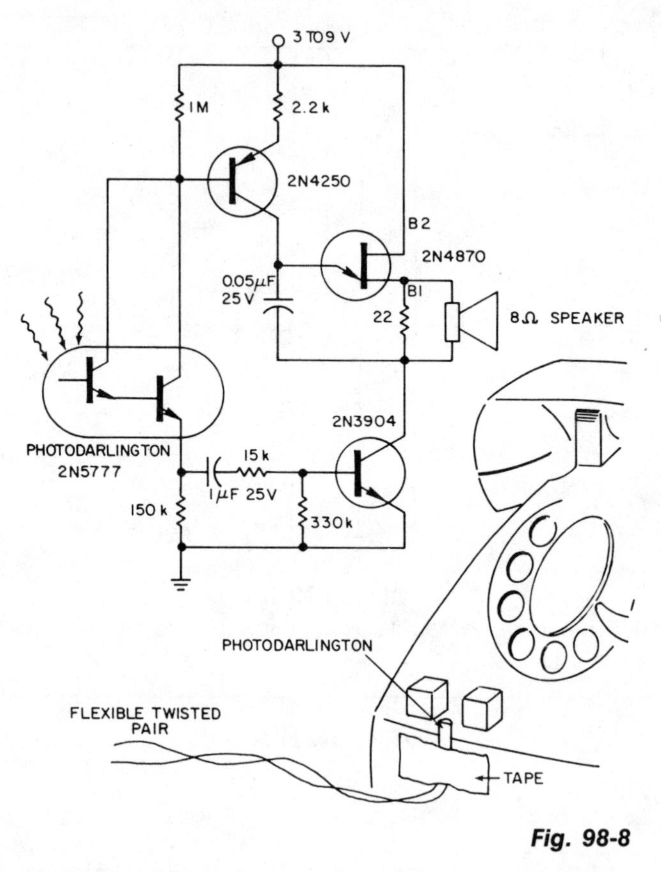

Fig. 98-8

ELECTRONIC DESIGN

Circuit Notes

A 2N5777 photo-Darlington cell picks up blinking light from the transparent plastic buttons. The power is switched ON and OFF by a hi-beta 2N3904 transistor. The circuit's 9 V battery can be left continuously connected. Less than a micro-ampere is drawn—even with normal, office ambient light and the phone lights not flashing. For noisy locations, the tone can be made louder with an output transformer (ratio of 250:8) or a 100 ohm speaker that replaces the 22 ohm resistor in the output.

PHONE LIGHT

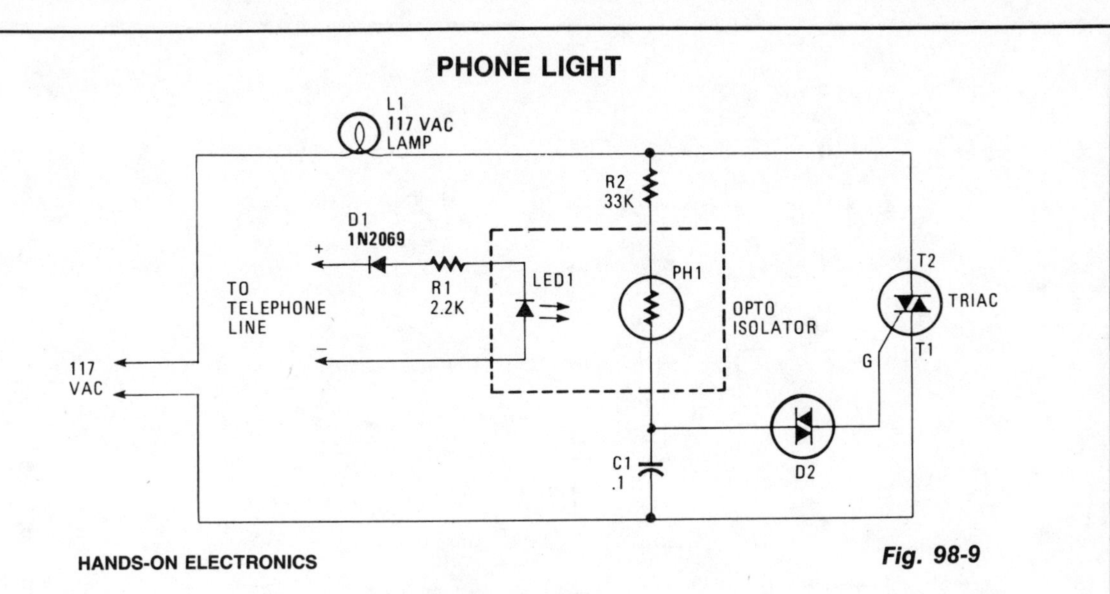

HANDS-ON ELECTRONICS

Fig. 98-9

Circuit Notes

When the phone does ring the triac is triggered into conduction by a signal applied to its gate (G) through a bilateral switch (diac), D2. The triac acts as a switch, conducting only when a signal is present at the gate.

HIGH ISOLATION TELEPHONE RINGER

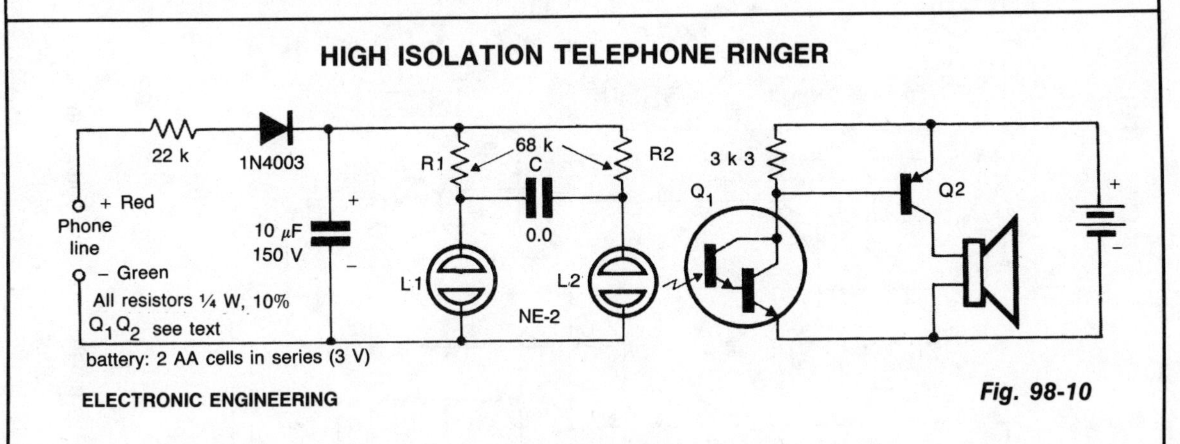

ELECTRONIC ENGINEERING

Fig. 98-10

Circuit Notes

The diode rectifies the ringing signal to supply the operating power to the audio relaxation oscillator made up of L1, L2, R1, R2, and C. Moreover, L2 together with Q1 acts as an opto-isolator, totally isolating the telephone line from the rest of the circuit. The oscillator audio frequency is optically coupled to the photo-Darlington which drives Q2 and thus the speaker. The 10 μF capacitor is not large enough to smooth the ringing ripple completely. This results in frequency modulation of the audio oscillator giving it an attention-getting warble.

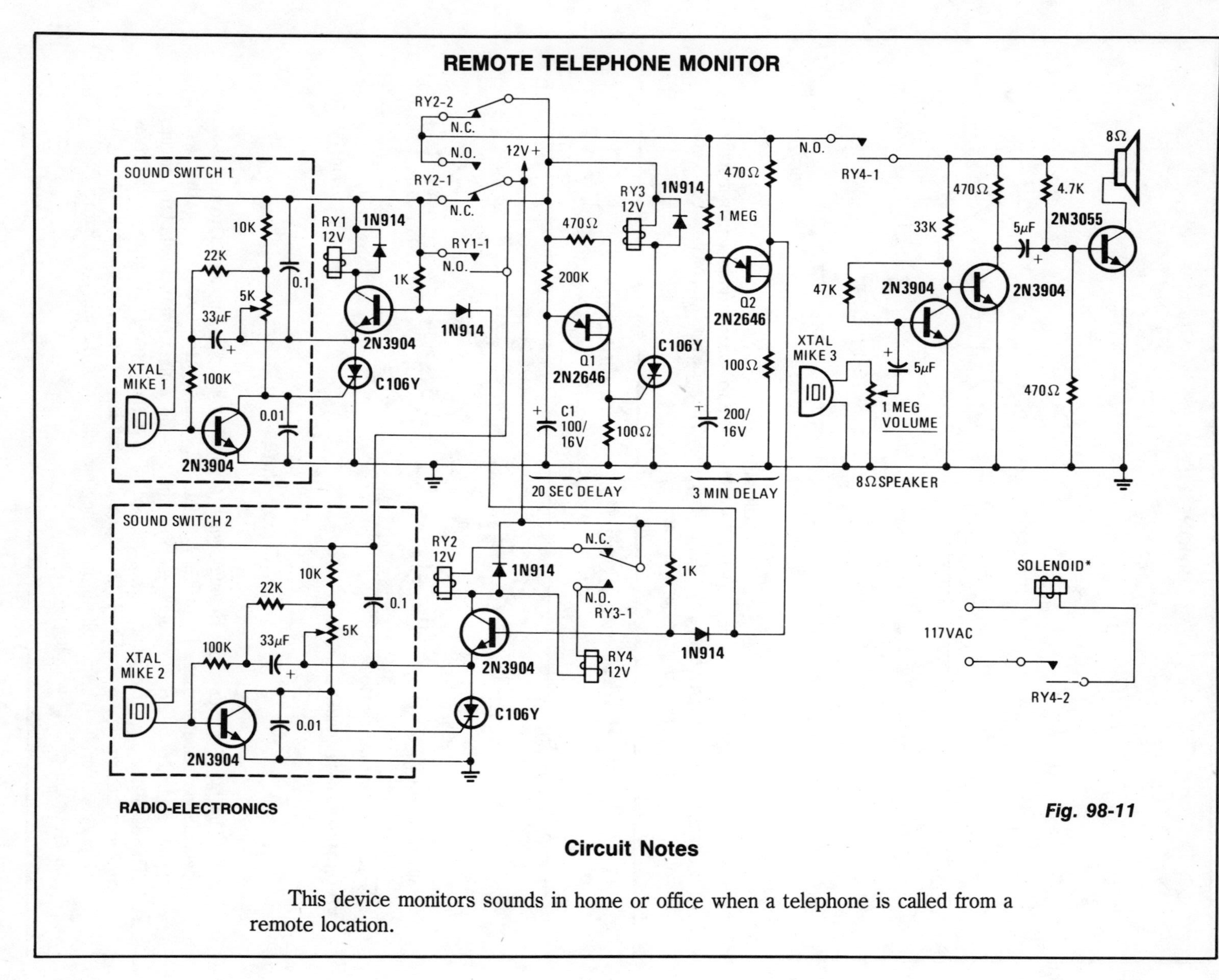

Fig. 98-11

Circuit Notes

This device monitors sounds in home or office when a telephone is called from a remote location.

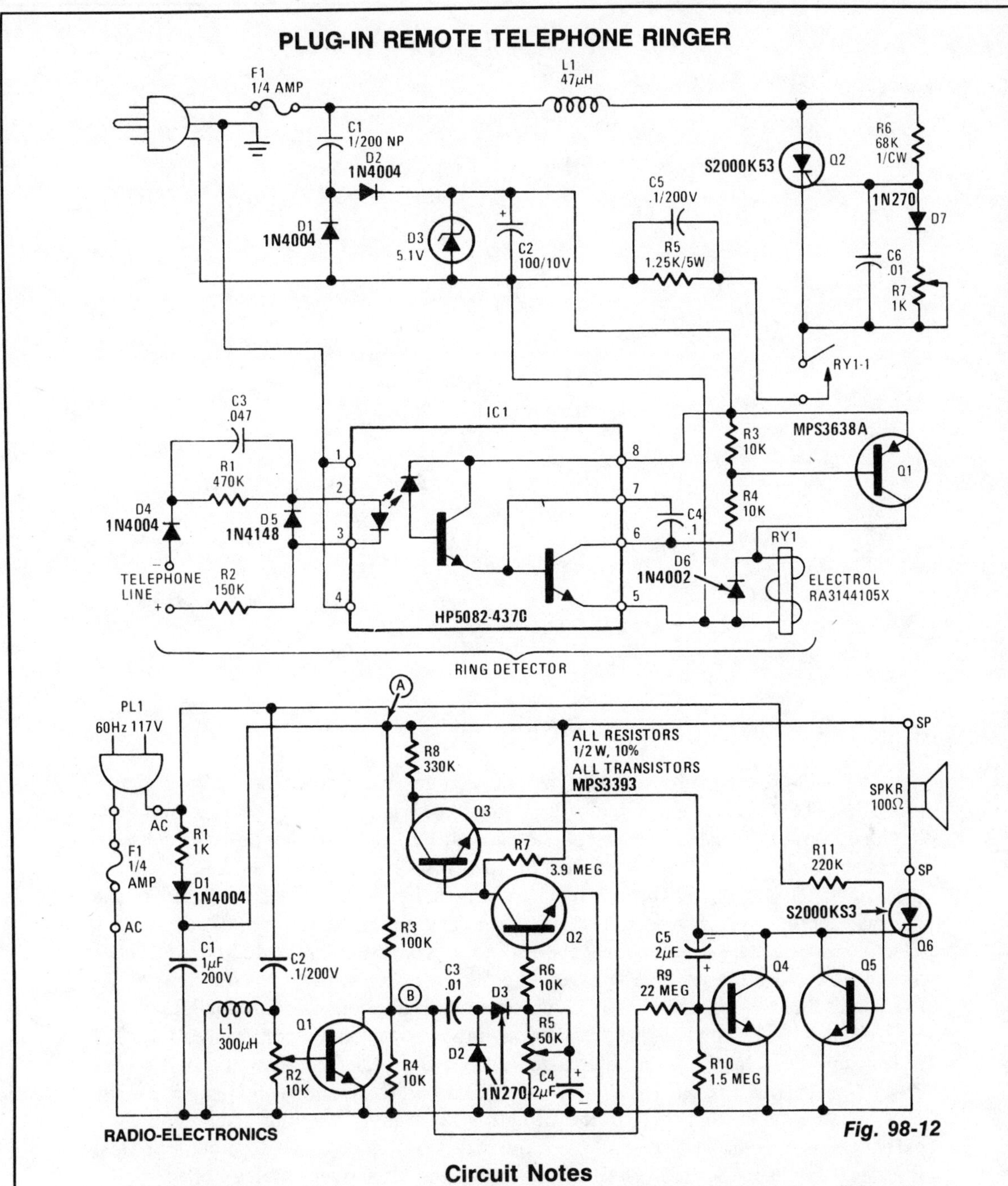

Fig. 98-12

Circuit Notes

This device consists of a ring detector connected to the telephone line. When the telephone rings, the ring detector impresses high-frequency pulses on the ac power line. A receiver placed anywhere on the same power line detects these pulses and emits an audible tone in synchronization with the telephone signal.

TELEPHONE HOLD BUTTON

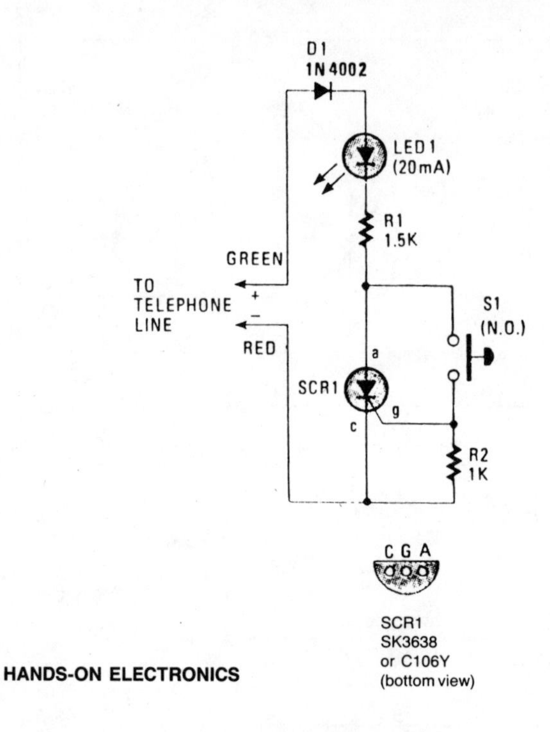

HANDS-ON ELECTRONICS

Fig. 98-13

Circuit Notes

The on-hook (no load) voltage across the red-green wires will be 48 V or slightly less when all telephones are on-hook (disconnected). When any telephone goes off-hook the load current flowing in the telephone causes the voltage to fall below 5 volts dc. Although the telephone hold is connected across the red-green wires, silicon control rectifier SCR1 is open; so there is no current path across the telephone line. To hold the call, depress normally-open switch S1 and hang up the telephone (still depressing S1). When the phone goes on-hook the red-green voltage jumps to 48 volts dc. Since switch S1 is closed, a positive voltage is applied to SCR1's gate, which causes SCR1 to conduct, thereby completing the circuit across the telephone line through D1, LED1, R1, and SCR1. The current that flows through those components also causes the LED to light up—indicating that the telephone line is being held. The effective load across the red-green wires is the 1500 ohm value of R1, which is sufficient to seize the line while limiting the current through the LED to a safe value. When the telephone, or an extension, is once again placed off-hook the red-green voltage falls to 5 volts or less. But diode D1 has a normal voltage drop—called the breakover voltage—of 0.7 volts, and the LED has a forward drop of 2.0 volts. Excluding the voltage drop across R1 there is a maximum of 2.3 volts available for SCR1, which is too low to maintain conduction; so SCR1 automatically opens the hold circuit when any telephone goes off-hook.

TELEPHONE BLINKER

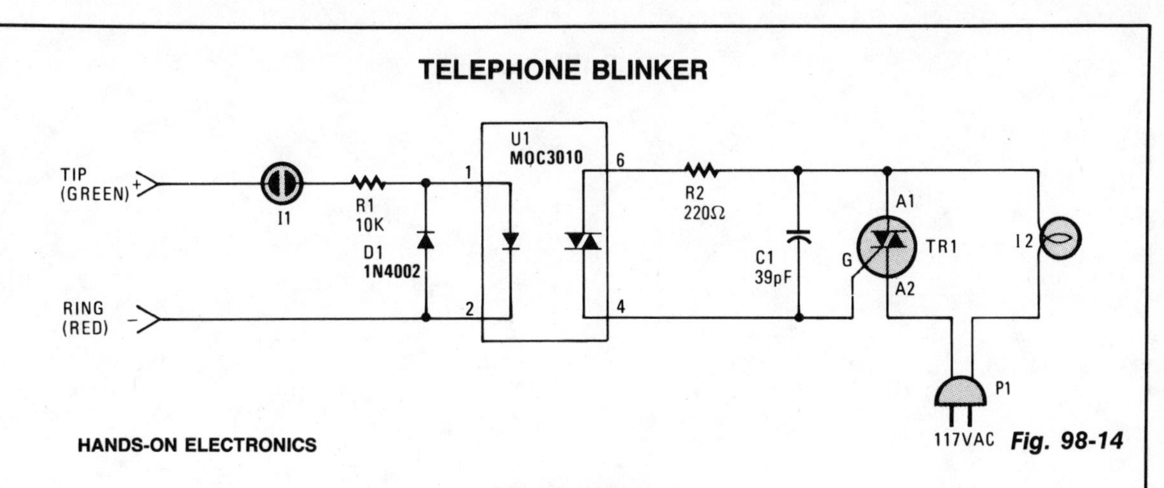

HANDS-ON ELECTRONICS

Fig. 98-14

Circuit Notes

A small neon lamp is triggered into conduction by the telephone's ringing voltage, passes just enough current to activate the LED in optocoupler U1, which in turn triggers the 6-A Triac that controls I2—a 117-Vac lamp or bell. (Capacitor C1 is necessary only when the circuit is used to drive a bell.) The lamp will flash off-and-on at the ringing rate, which is normally around 20 Hz. If a 117 Vac bell is used, connect it in place of the lamp.

TELEPHONE "IN USE" INDICATOR

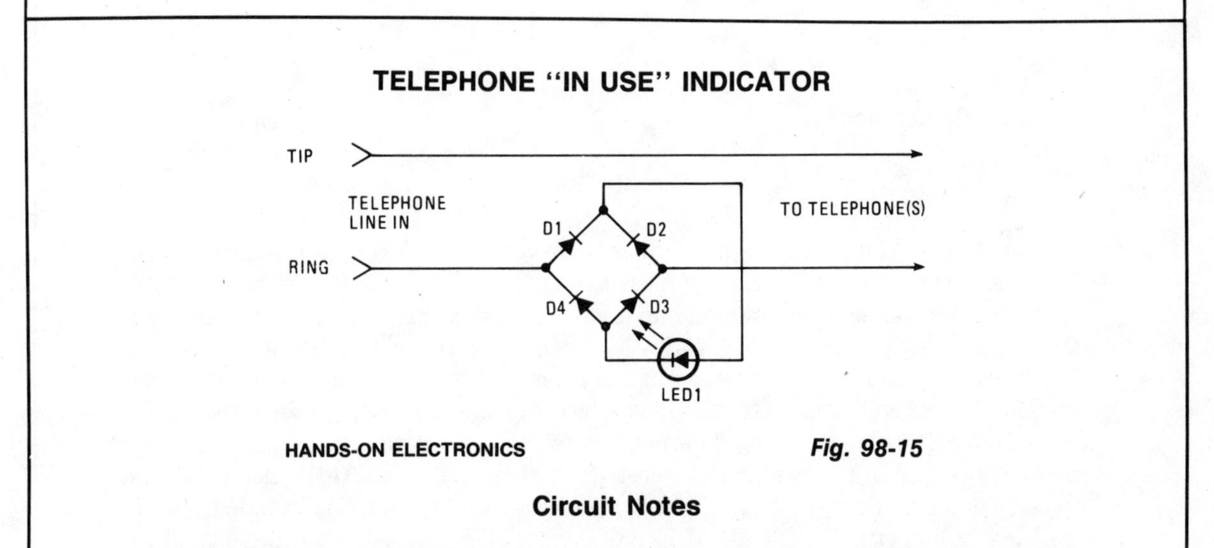

HANDS-ON ELECTRONICS

Fig. 98-15

Circuit Notes

This circuit functions as a line-current sensor and can be connected in series with either of the phone lines. For the circuit to indicate an "in use" status for all phones on a single line, it must be connected in series with the phone line before, or ahead of all phones on the line. Since the power for the circuit is supplied by the phone company, a circuit could be added to each phone as an off-hook indicator.

TONE RINGER

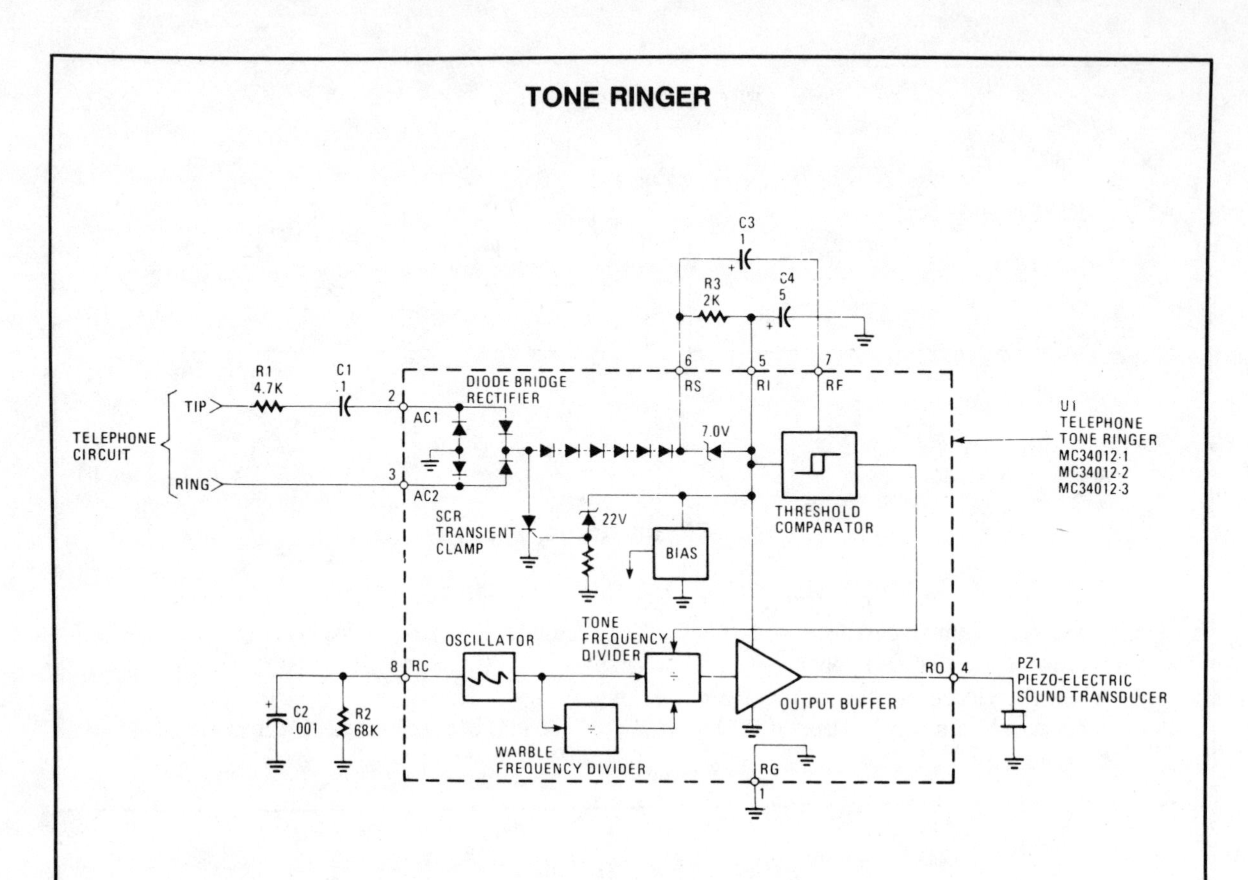

HANDS-ON ELECTRONICS *Fig. 98-16*

Circuit Notes

The MC34012 tone-ringer chip derives its power by rectifying the ac ringing signal. That signal is normally at 20 Hz and measures between 70 and 130 volts rms. It uses that power for the tone generator and to drive the piezoelectric transducer. The sound that is produced is a warble that varies between two frequencies, $f_o/4$ (f_o – 4) $f_o/5$. The clock, or fundamental, frequency, f_o, is generated by a relaxation oscillator. That oscillator has R2 and C2 as its frequency setting components providing a selectable range of 1 kHz to 10 kHz. Selecting different values for R2 and/or C2 changes the clock frequency, which in turn varies the warble frequencies. The MC34012 chip comes in three different warble rates at which the warble frequencies ($f_o/4$, $f_o/5$) are varied. These warble rates are $f_o/320$, $f_o/640$, or $f_o/160$ and the different chips are designated as MC34012-1, -2, and -3, respectively. For example: with a 4.40 kHz oscillator frequency, the MC34012-1 produces 800 Hz and 1000 Hz tones with a 12.5 Hz warble rate. The MC34012-2 generates 1600 Hz and 2000 Hz tones with a similar 12.5 Hz warble frequency from an 8.0 kHz oscillator frequency. MC34012-3 will produce 400 Hz and 500 Hz tones with a 12.5 warble rate from a 2.0 kHz oscillator frequency.

TONE RINGER II

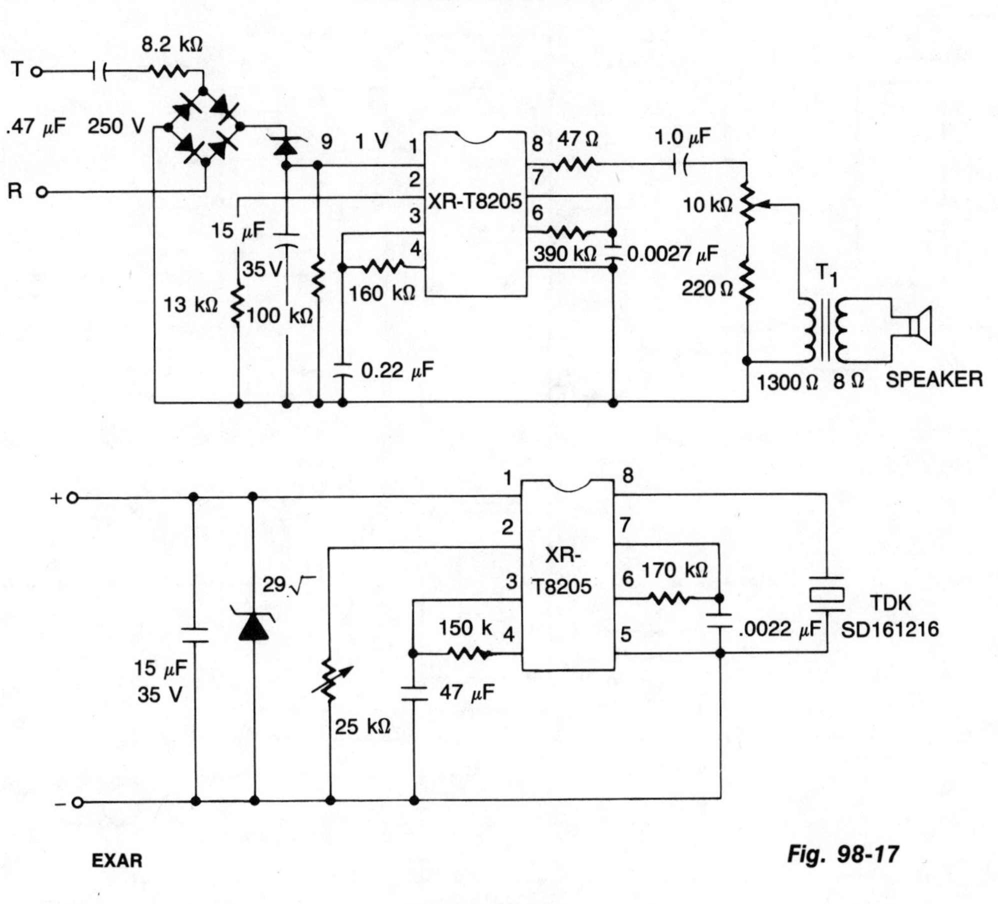

EXAR

Fig. 98-17

Circuit Notes

The XR-T8205 Tone Ringer is primarily intended as a replacement for the mechanical telephone bell. The device can be powered directly from telephone ac ringing voltage or from a separate dc supply. An adjustable trigger level is provided with an external resistor. The circuit is designed for nominal 15 volt operation.

SPEAKERPHONE

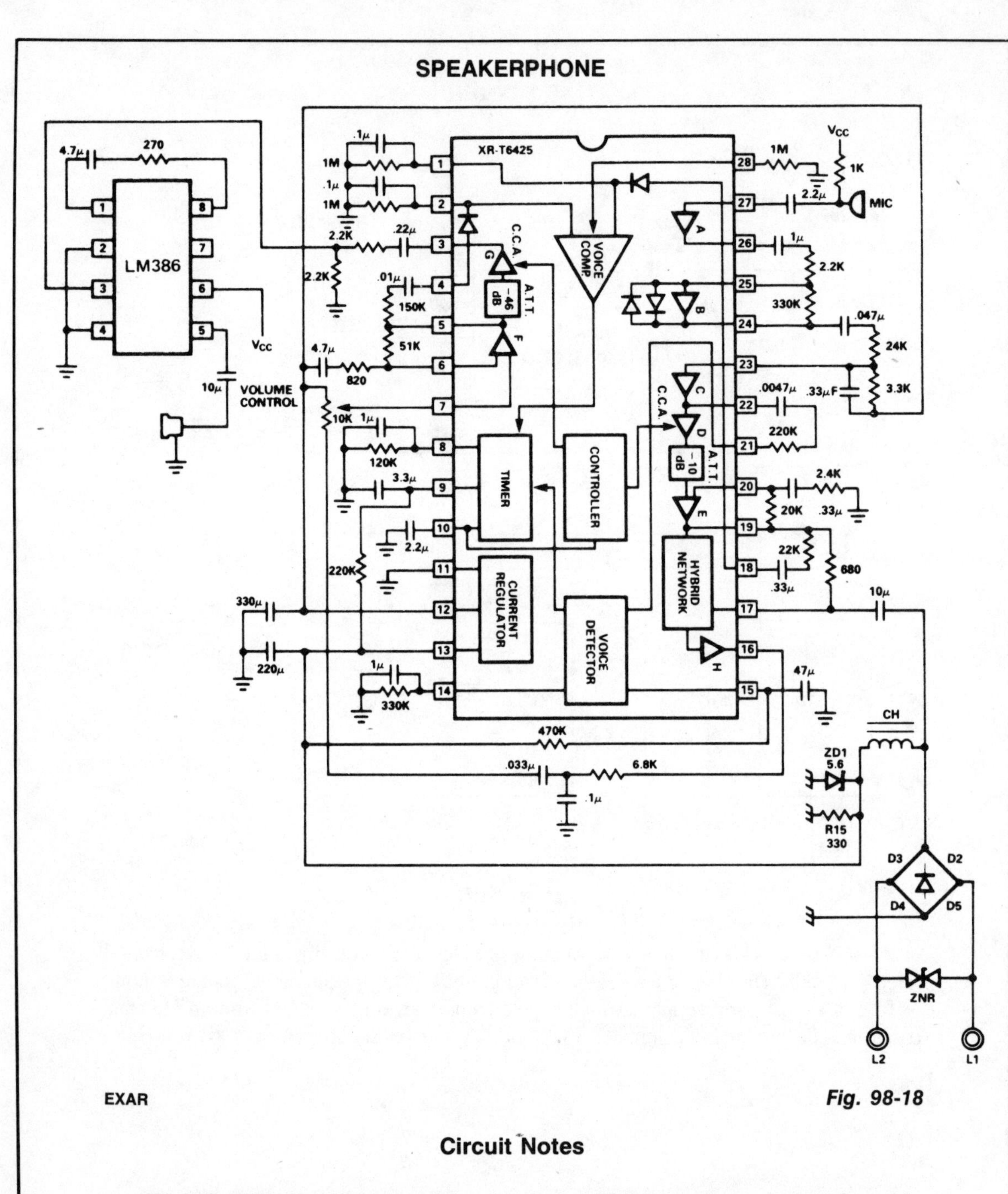

EXAR

Fig. 98-18

Circuit Notes

The XR-T6425 Speakerphone IC makes it possible to carry on conversation without using the handset, while the user is talking into a microphone and listening from a loudspeaker. It is ideal for hands-free conference calls. The XR-T6425 contains most of the circuits to eliminate singing and excessive background noise.

SPEECH NETWORK

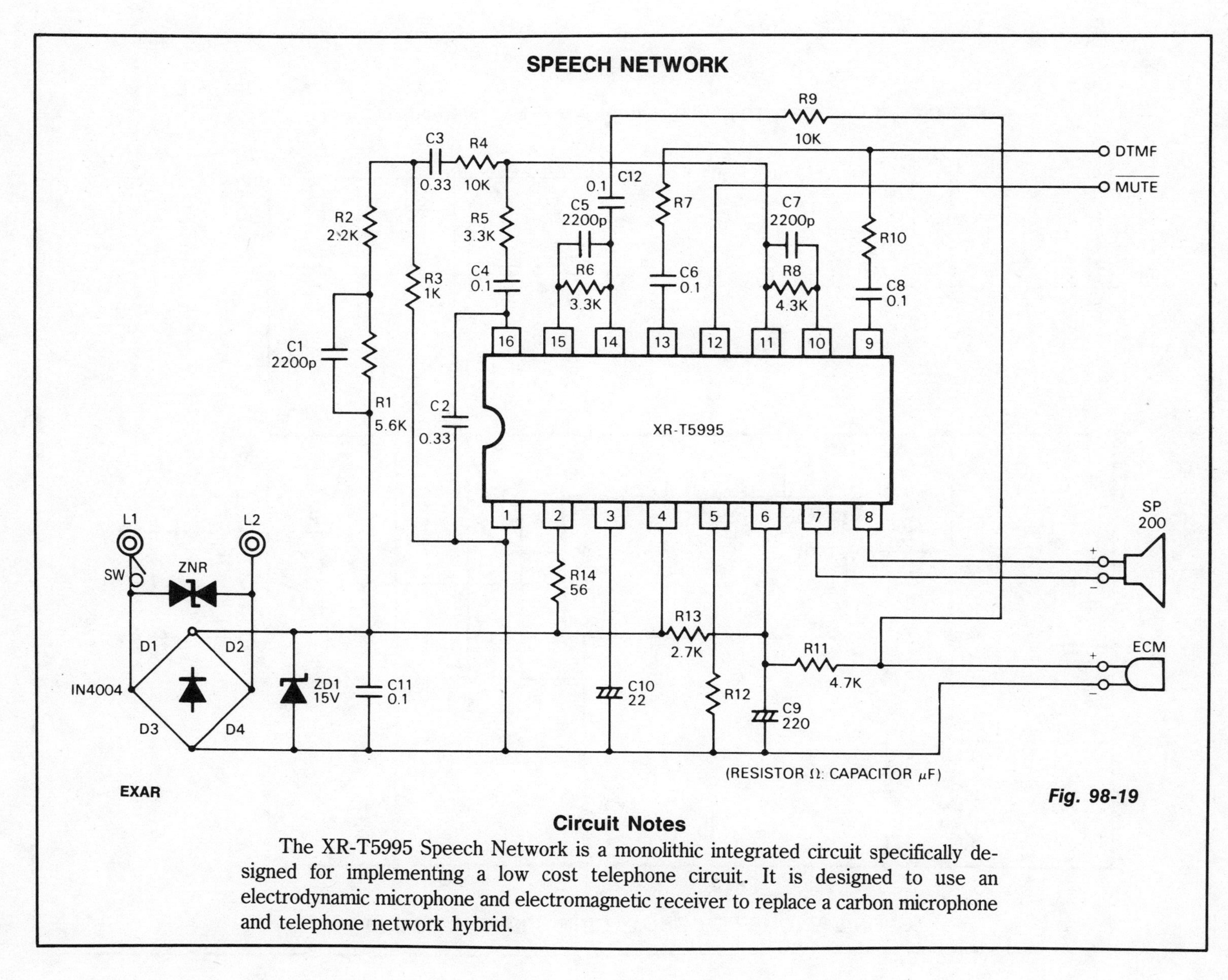

Fig. 98-19

Circuit Notes

The XR-T5995 Speech Network is a monolithic integrated circuit specifically designed for implementing a low cost telephone circuit. It is designed to use an electrodynamic microphone and electromagnetic receiver to replace a carbon microphone and telephone network hybrid.

PROGRAMMABLE MULTI-TONE TELEPHONE RINGER

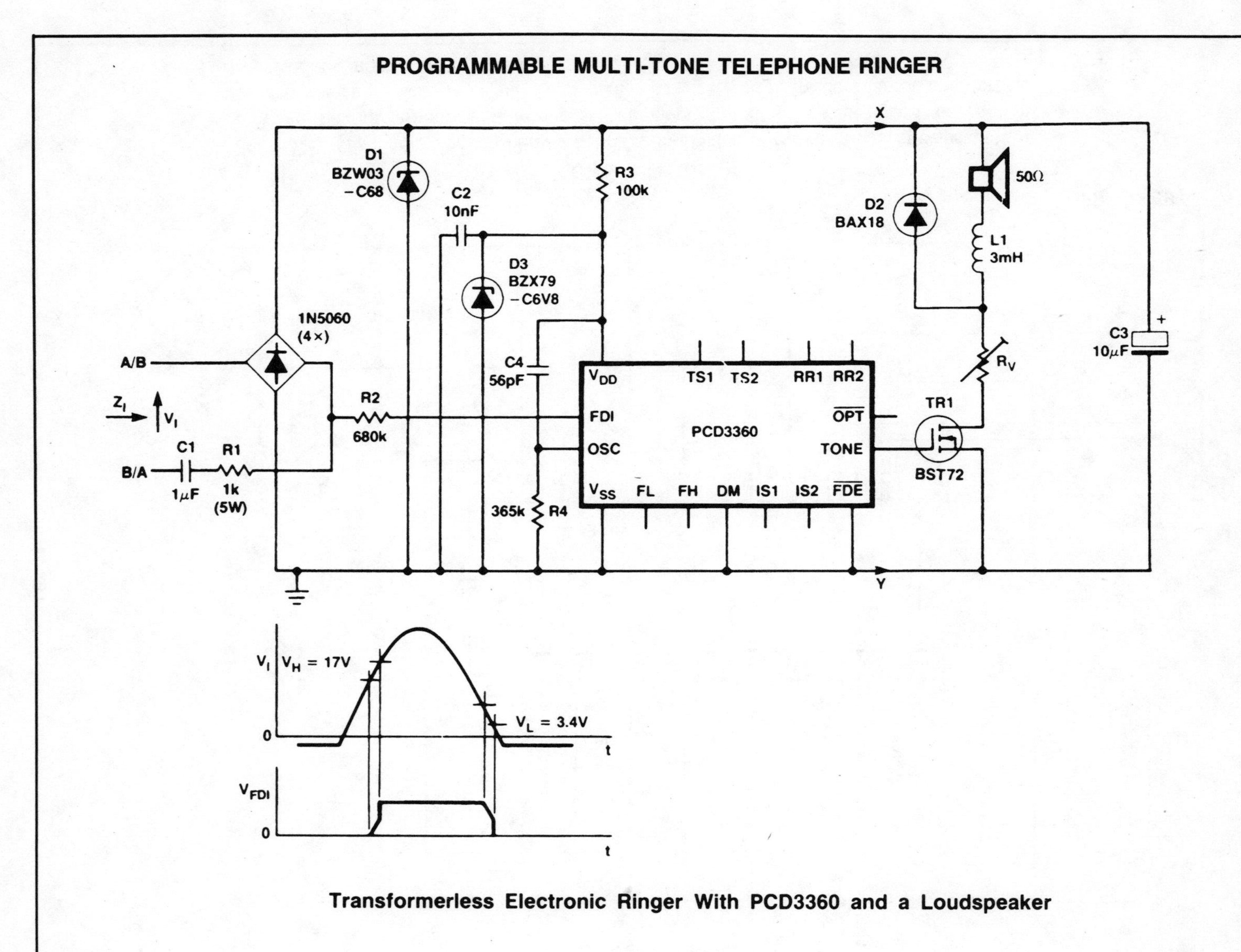

Transformerless Electronic Ringer With PCD3360 and a Loudspeaker

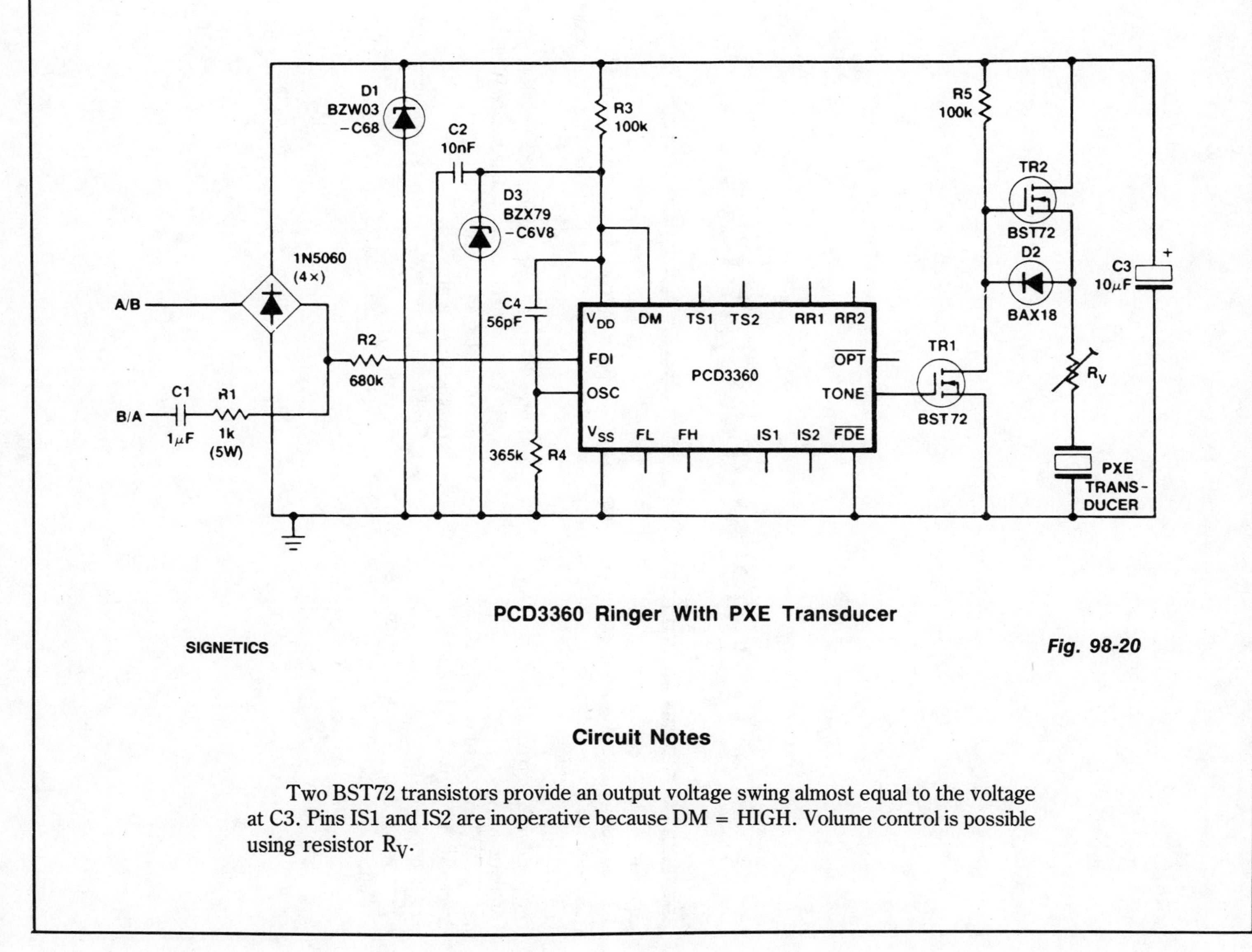

PCD3360 Ringer With PXE Transducer

SIGNETICS

Fig. 98-20

Circuit Notes

Two BST72 transistors provide an output voltage swing almost equal to the voltage at C3. Pins IS1 and IS2 are inoperative because DM = HIGH. Volume control is possible using resistor R_V.

99
Temperature Controls

The sources of the following circuits are contained in the Sources section beginning on page 694. The figure number contained in the box of each circuit correlates to the source entry in the Sources section.

Temperature-Controlling Circuit
Temperature Control
Low-Cost Temperature Controller
Precision, Linearized Platinum RTD Signal Conditioner
Low Power Zero Voltage Switch Temperature Controller
Heater Element Temperature Controller
Dual-Time Chip Controls Temperature While Monitoring Liquid Level
Temperature Alarm
Adjustable Threshold Temperature Alarm

TEMPERATURE-CONTROLLING CIRCUIT

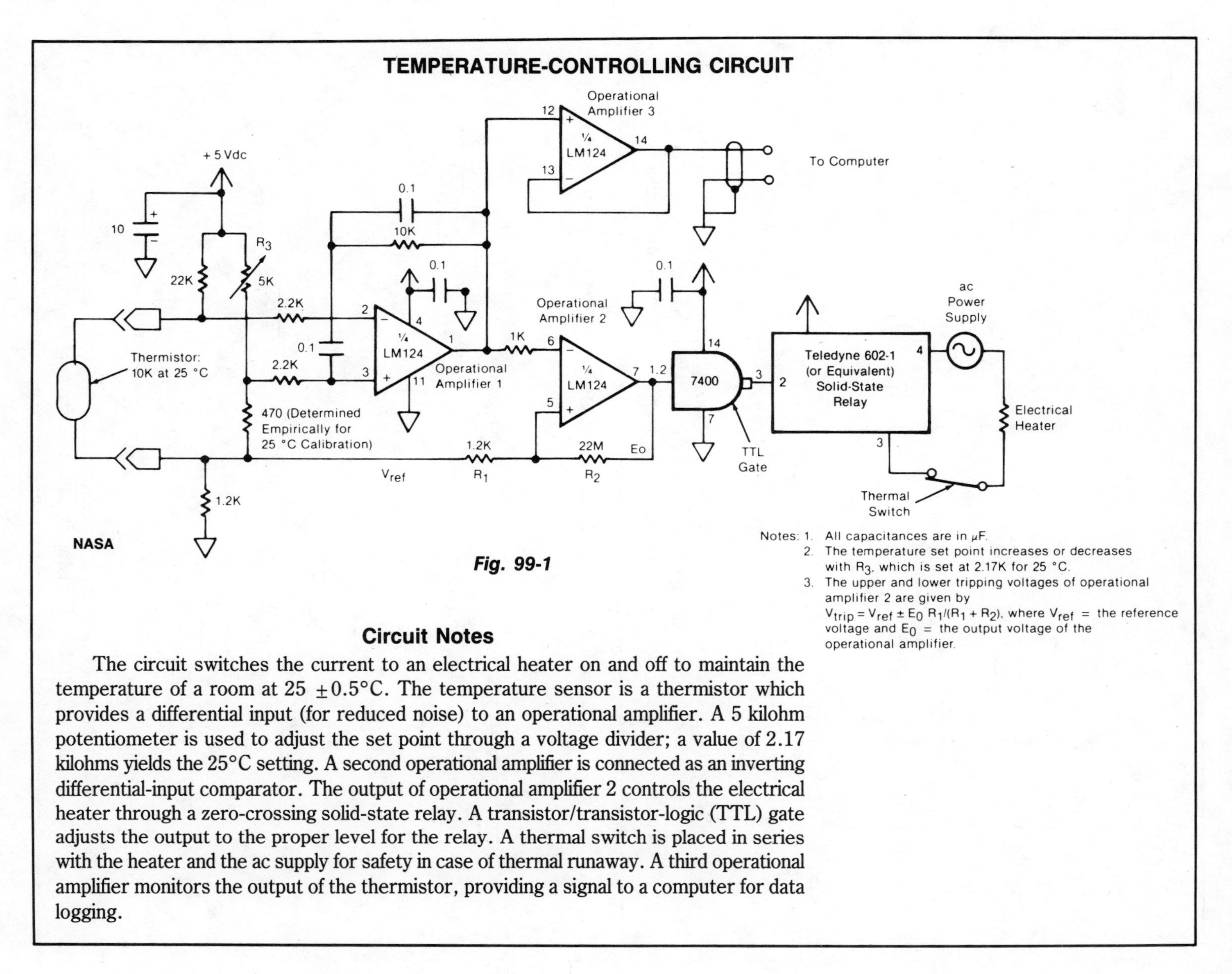

Notes: 1. All capacitances are in μF.
2. The temperature set point increases or decreases with R_3, which is set at 2.17K for 25 °C.
3. The upper and lower tripping voltages of operational amplifier 2 are given by $V_{trip} = V_{ref} \pm E_0 R_1/(R_1 + R_2)$, where V_{ref} = the reference voltage and E_0 = the output voltage of the operational amplifier.

Fig. 99-1

Circuit Notes

The circuit switches the current to an electrical heater on and off to maintain the temperature of a room at 25 ±0.5°C. The temperature sensor is a thermistor which provides a differential input (for reduced noise) to an operational amplifier. A 5 kilohm potentiometer is used to adjust the set point through a voltage divider; a value of 2.17 kilohms yields the 25°C setting. A second operational amplifier is connected as an inverting differential-input comparator. The output of operational amplifier 2 controls the electrical heater through a zero-crossing solid-state relay. A transistor/transistor-logic (TTL) gate adjusts the output to the proper level for the relay. A thermal switch is placed in series with the heater and the ac supply for safety in case of thermal runaway. A third operational amplifier monitors the output of the thermistor, providing a signal to a computer for data logging.

TEMPERATURE CONTROL

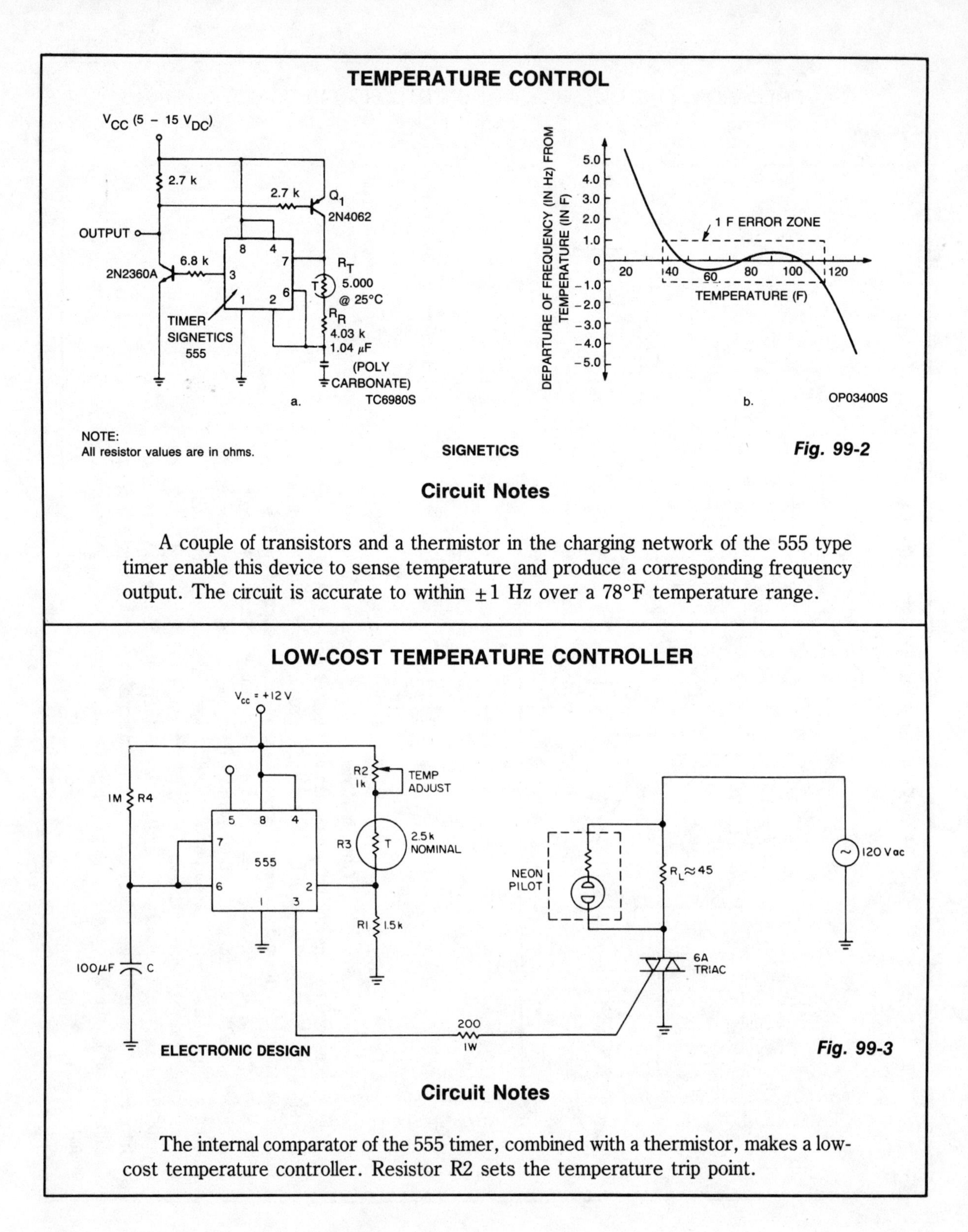

Fig. 99-2

Circuit Notes

A couple of transistors and a thermistor in the charging network of the 555 type timer enable this device to sense temperature and produce a corresponding frequency output. The circuit is accurate to within ±1 Hz over a 78°F temperature range.

LOW-COST TEMPERATURE CONTROLLER

Fig. 99-3

Circuit Notes

The internal comparator of the 555 timer, combined with a thermistor, makes a low-cost temperature controller. Resistor R2 sets the temperature trip point.

PRECISION, LINEARIZED PLATINUM RTD SIGNAL CONDITIONER

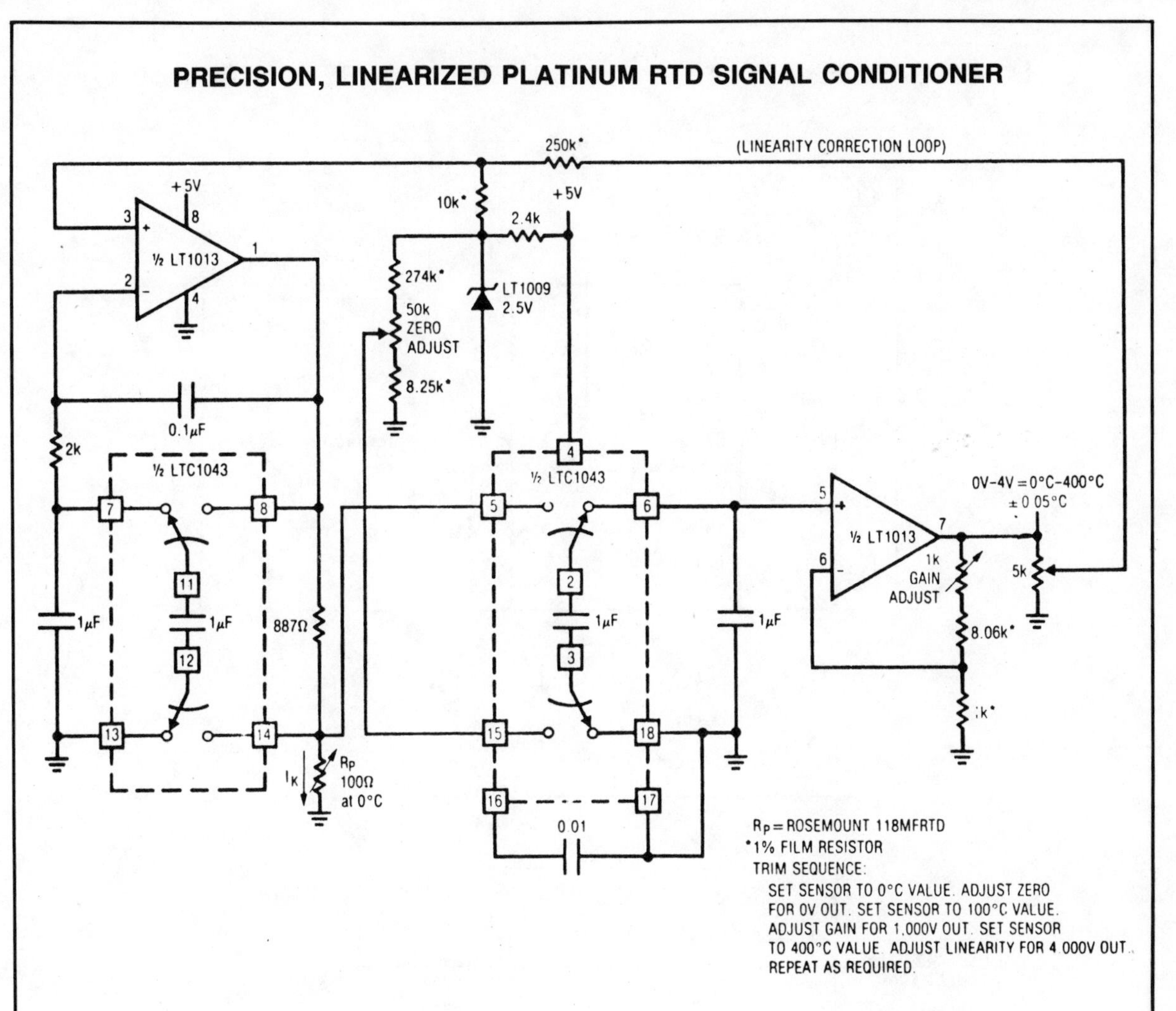

LINEAR TECHNOLOGY CORPORATION ***Fig. 99-4***

Circuit Notes

The circuit provides complete, linearized signal conditioning for a platinum RTD. This LTC1043 based circuit is considerably simpler than instrumentation or multi-amplifier based designs, and will operate form a single 5 V supply. A1 serves as a voltage-controlled ground referred current source by differentially sensing the voltage across the 998 phm feedback resistor. The LTC1043 section which does this presents a single-ended signal to A1's negative input, closing a loop. The 2 k 0.1 μF combination sets amplifier roll-off well below the LTC1043's switching frequency and the configuration is stable. Because A1's loop forces a fixed voltage across the 887 ohm resistor, the current through R_P is constant. A1's operating point is primarily fixed by the 2.5 V LT1009 voltage reference.

LOW POWER ZERO VOLTAGE SWITCH TEMPERATURE CONTROLLER

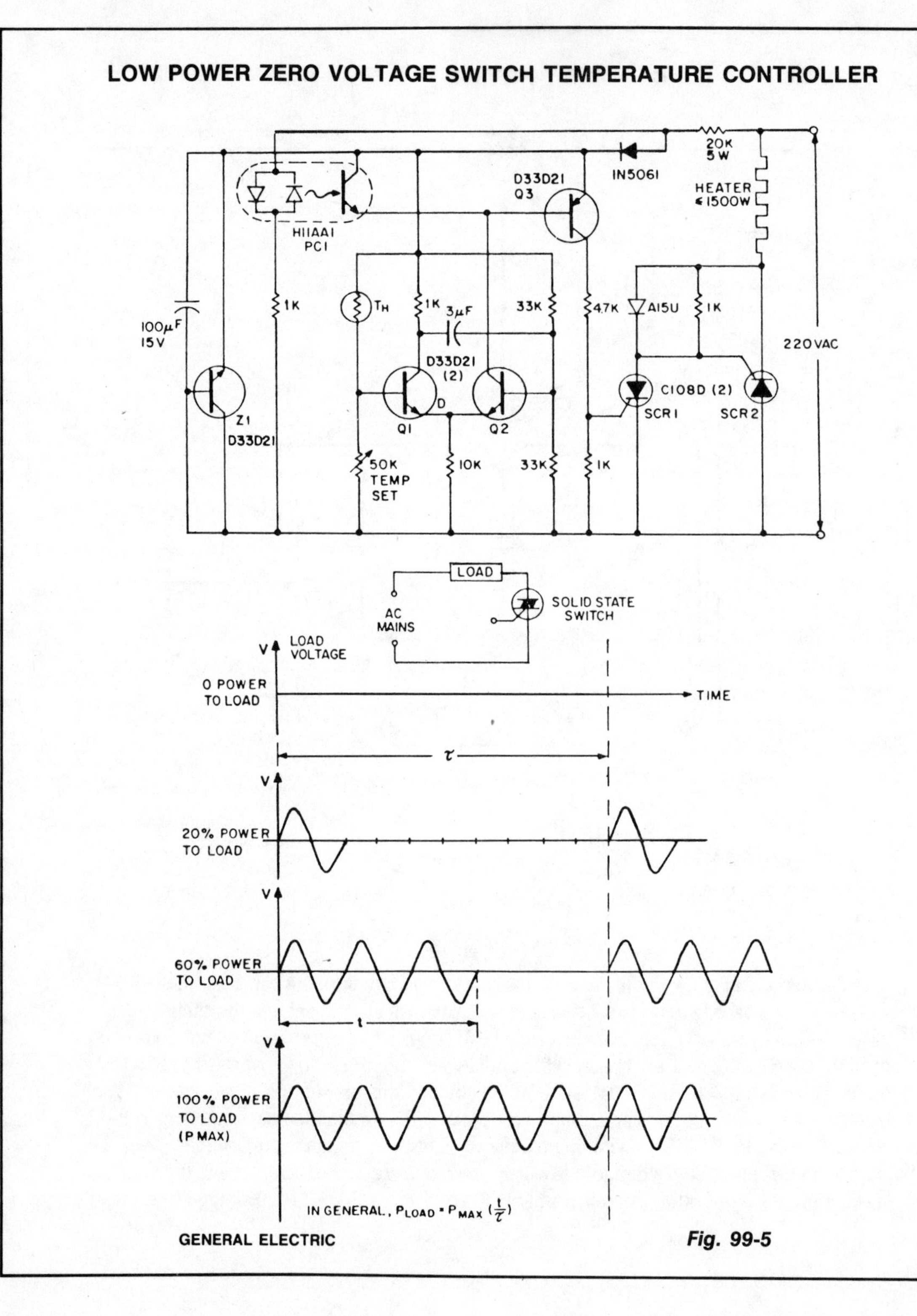

GENERAL ELECTRIC *Fig. 99-5*

Circuit Notes

The "zero voltage switching" technique is widely used to modulate heating and similar types of ac loads where the time constant associated with the load (tens of seconds to minutes) is sufficiently long to allow smooth proportional modulation by time ratio control, using one complete cycle of the ac input voltage as the minimum switching movement. Despite its attractions, the traditional triac-based ZVS is virtually unusable for the control of very low power loads, especially from 220 volt ac inputs due to the triac's reluctance to latch-on into the near-zero instantaneous currents that flow through it and the load near the ac voltage zero crossover points. The circuit side-steps the latching problem by employing a pair of very sensitive low current reverse blocking thyristors (C106) connected in antiparallel; these are triggered by a simple thermistor modulated differential amplifier (Q1, Q2), with zero voltage logic furnished by an H11AA1 ac input optocoupler. With the NTC thermistor TH calling for heat, transistor Q1 is cut off and Q2 is on, which would normally provide continuous base drive to Q3, with consequent triggering of either SCR, or of SCR 2 via SCR1, depending on phasing of the ac input.

Note that when the ac input voltage is positive with respect to SCR 2, SCR 1 is reverse biased and, in the presence of "gate" current from Q3, behaves as a remote base transistor, whose output provides via blocking diode CR1, positive gate trigger current for SCR 2. When the ac input polarity is reversed (SCR 1's anode positive), SCR 1 behaves as a direct fired conventional thyristor. "Trigger" current to SCR 1, however, is not continuous, even when TH is calling for heat and Q2 is delivering base current to Q3. In this situation, Q3 is inhibited from conduction by the clamping action of PC1, an H11AA photocoupler, except during those brief instants when the ac input voltage is near zero and the coupler input diodes are deprived of current.

Triggering of either SCR can occur only at ac voltage crossing points, and RFI-less operation results. The proportional control feature is injected via the positive feedback action of capacitor CM, which converts the differential amplifier Q1, Q2 into a simple multivibrator, whose duty cycle varies from one to 99 percent according to the resistance of TH. Zener diode Z1 is operational, being preferred when maximum immunity from ac voltage induced temperature drift is desired.

HEATER ELEMENT TEMPERATURE CONTROLLER

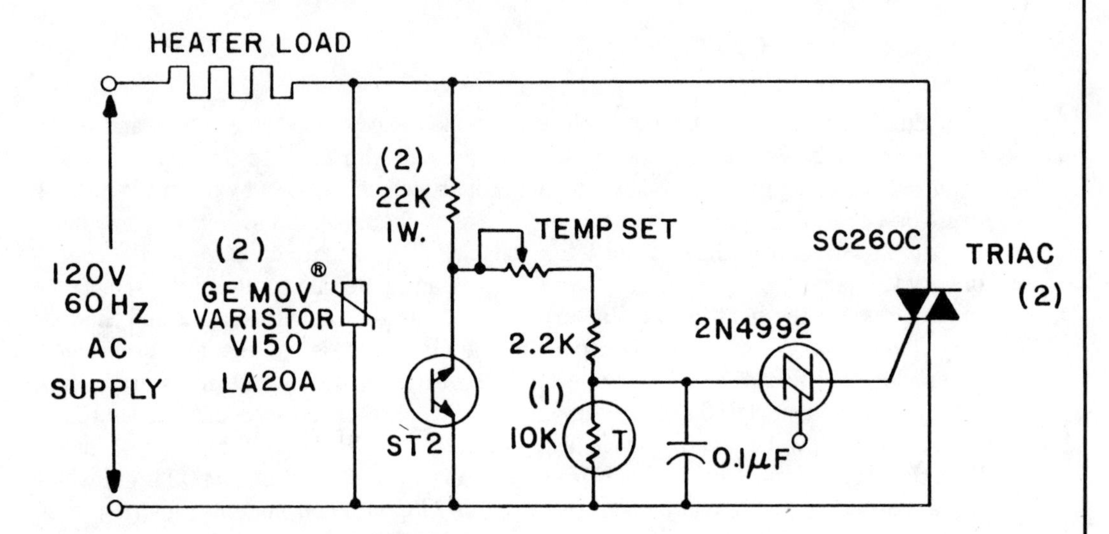

NOTES: 1. Thermistor National Lead type 1D101, or equivalent.
2. Component values for 220V operation:
Resistor – 47K, 2W
GE-MOV® Varistor – V275LA20A
Triac – SC260E

GENERAL ELECTRIC *Fig. 99-6*

Circuit Notes

The circuit can control up to 6 kW of heating, with moderate gain, using a 25-amp triac (SC260D). Feedback is provided by the negative temperature co-efficient (NTC) thermistor, which is mounted adjacent to the environment being temperature controlled. The temperature set potentiometer is initially adjusted to the desired heating level. As the thermistor becomes heated by the load, its resistance drops, phasing back the conduction angle of the triac, so the load voltage is reduced. The ST2 diac is used as a back-to-back zener diode. Its negative resistance region in its E-I characteristic provides a degree of line voltage stabilization. As the input line voltage increases, the diac triggers earlier in the cycle and, hence, the average charging voltage to the 0.1 μF capacitor, decreases.

DUAL-TIMER CHIP CONTROLS TEMPERATURE WHILE MONITORING LIQUID LEVEL

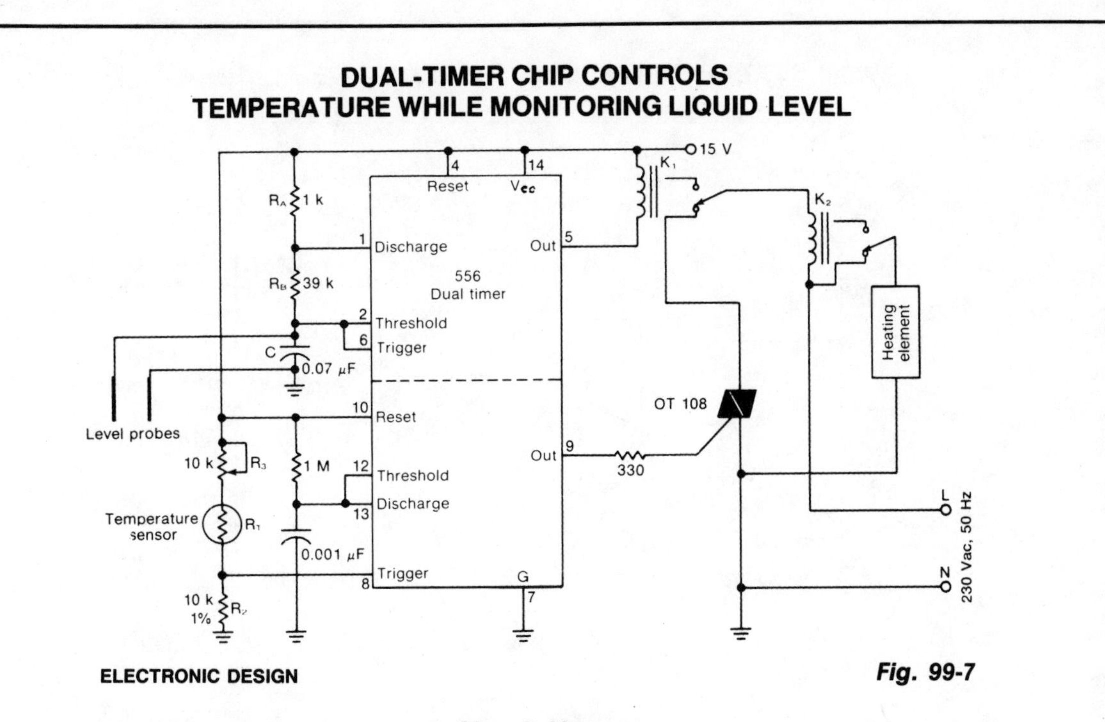

ELECTRONIC DESIGN

Fig. 99-7

Circuit Notes

One-half of a 556 dual timer monitors the temperature of a liquid bath, controlling a heating element that maintains temperature within ±2°C over a 32° - 200°C range. The other half monitors the liquid level, disconnecting the heater when the level drops below a preset point.

TEMPERATURE ALARM

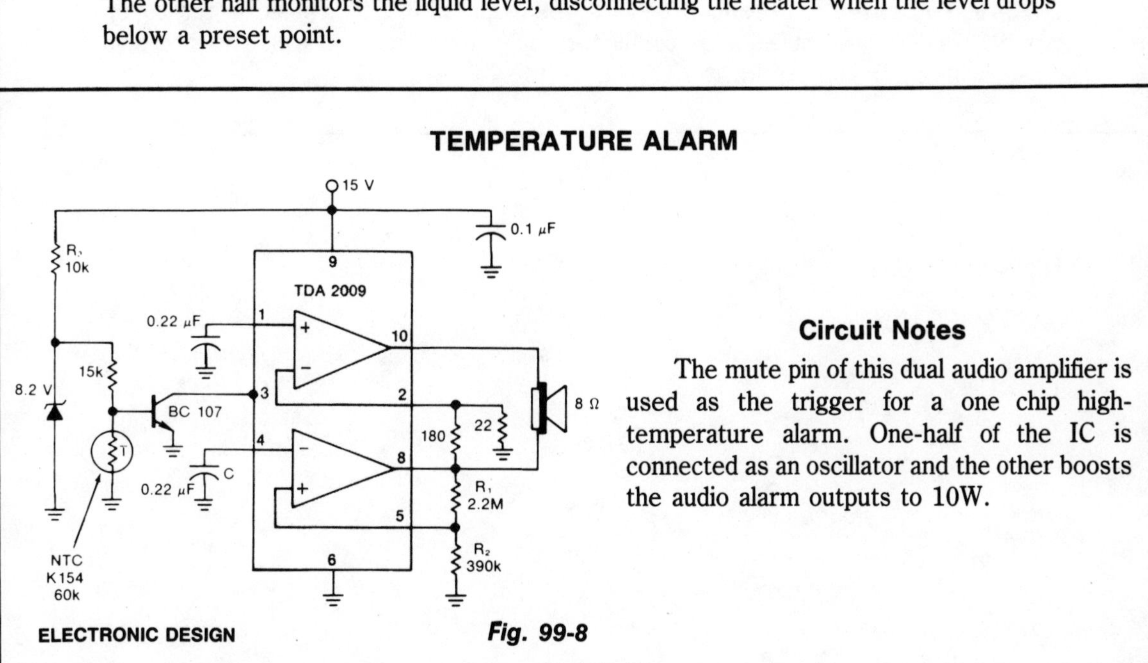

ELECTRONIC DESIGN

Fig. 99-8

Circuit Notes

The mute pin of this dual audio amplifier is used as the trigger for a one chip high-temperature alarm. One-half of the IC is connected as an oscillator and the other boosts the audio alarm outputs to 10W.

ADJUSTABLE THRESHOLD TEMPERATURE ALARM

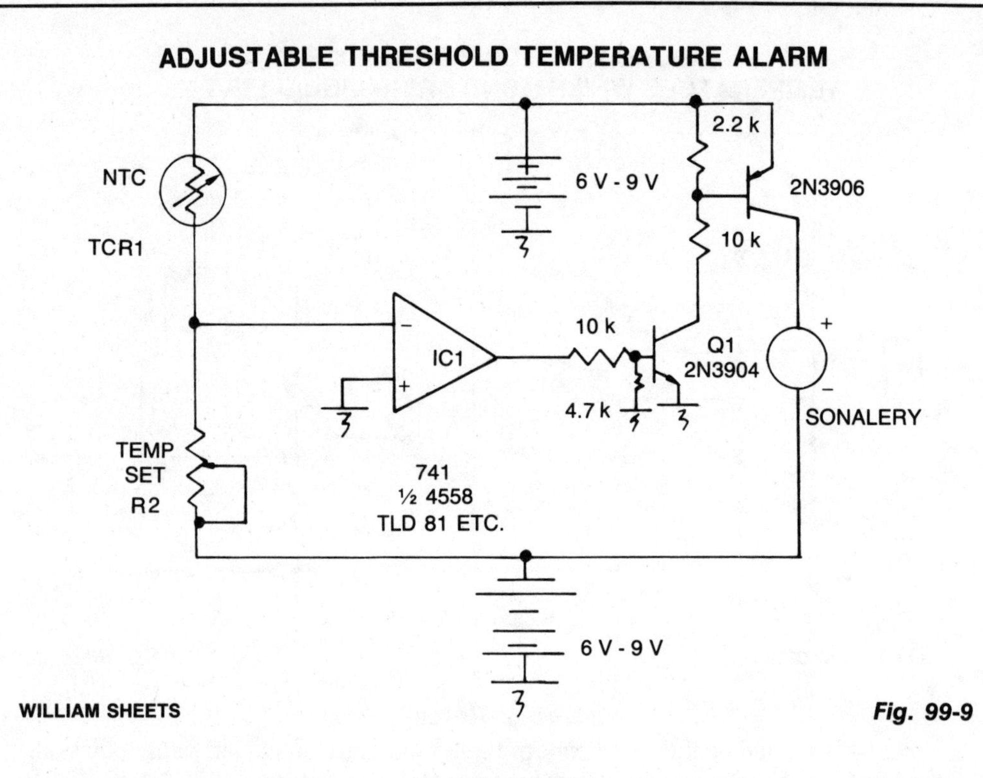

WILLIAM SHEETS

Fig. 99-9

Circuit Notes

When R1 increases as temp decreases, the output of IC1 goes positive, turning on Q1. Q1 conducts and causes Q2 to conduct, turning on the audible alarm. The threshold is set with potentiometer R2.

100

Temperature Sensors

The sources of the following circuits are contained in the Sources section beginning on page 694. The figure number contained in the box of each circuit correlates to the source entry in the Sources section.

DUAL OUTPUT, OVER-UNDER TEMPERATURE MONITOR

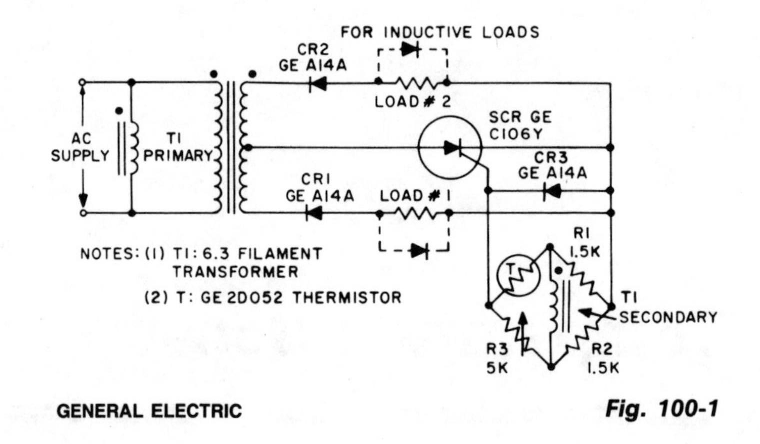

GENERAL ELECTRIC

Fig. 100-1

Circuit Notes

This circuit is ideal for use as an over-under temperature monitor, where its dual output feature can be used to drive HIGH and LOW temperature indicator lamps, relays, etc. T1 is a 6.3 volt filament transformer whose secondary winding is connected inside a four arm bridge. When the bridge is balanced, ac output is zero, and C5 (or C7) receives no gate signal. If the bridge is unbalanced by raising or lowering the thermistor's ambient temperature, and ac voltage will appear across the SCR's gate cathode terminals. Depending in which sense the bridge is unbalanced, the positive gate voltage will be in phase with, or 180° out of phase with the ac supply. If the positive gate voltage is in phase, the SCR will deliver load current through diode CR1 to load (1), diode CR2 blocking current to load (2). Conversely, if positive gate voltage is 180° out of phase, diode CR2 will conduct and deliver power to load (2), CR1 being reverse biased under these conditions. With the component values shown, the circuit will respond to changes in temperature of approximately 1-2°C. Substitution of other variable-resistance sensors, such as cadmium sulfide light dependent resistors (LDR) or strain gauge elements, for the thermistor shown is permissible.

TEMPERATURE SENSOR AND DVM INTERFACE

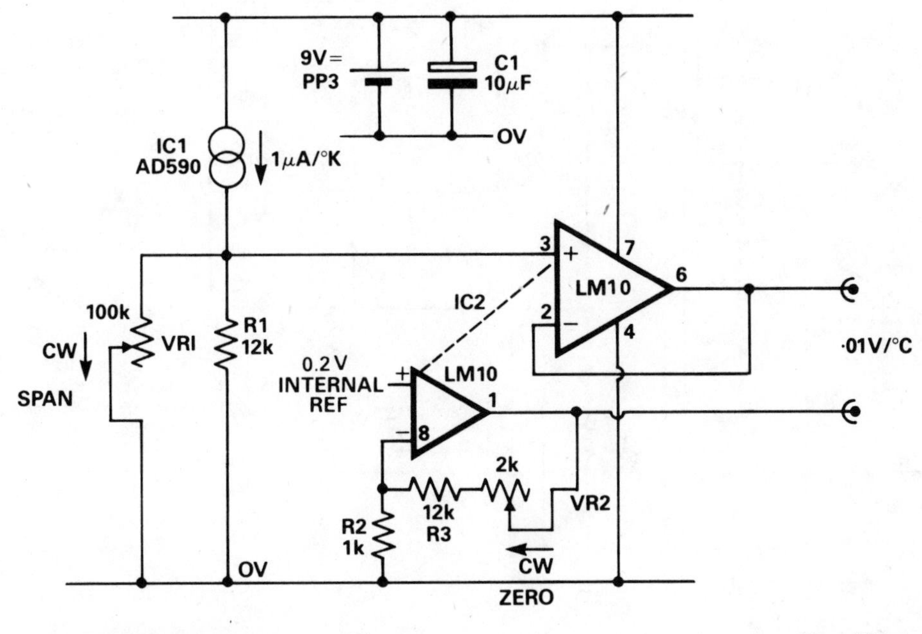

ELECTRONIC ENGINEERING

Fig. 100-2

Circuit Notes

The DVM gives a direct indication of the temperature of the sensor in degrees Centigrade. The temperature sensor IC1 gives a nominal 1 μA per degree Kelvin which is converted to 10 mV per degree Kelvin by R1 and VR1. IC2 is a micropower, low input drift op amp with internal voltage reference and amplifier. The main op amp in IC1 is connected as a voltage follower to buffer the sensor voltage at R1.

The second amplifier in IC1 is used to amplify the .2 V internal reference up to 2.73 V in order to offset the 273 degrees below 0°C. The output voltage of the unit is the differential output of the two op amps and is thus equal to 0.01 V per °C.

CURVATURE CORRECTED PLATINUM RTD THERMOMETER

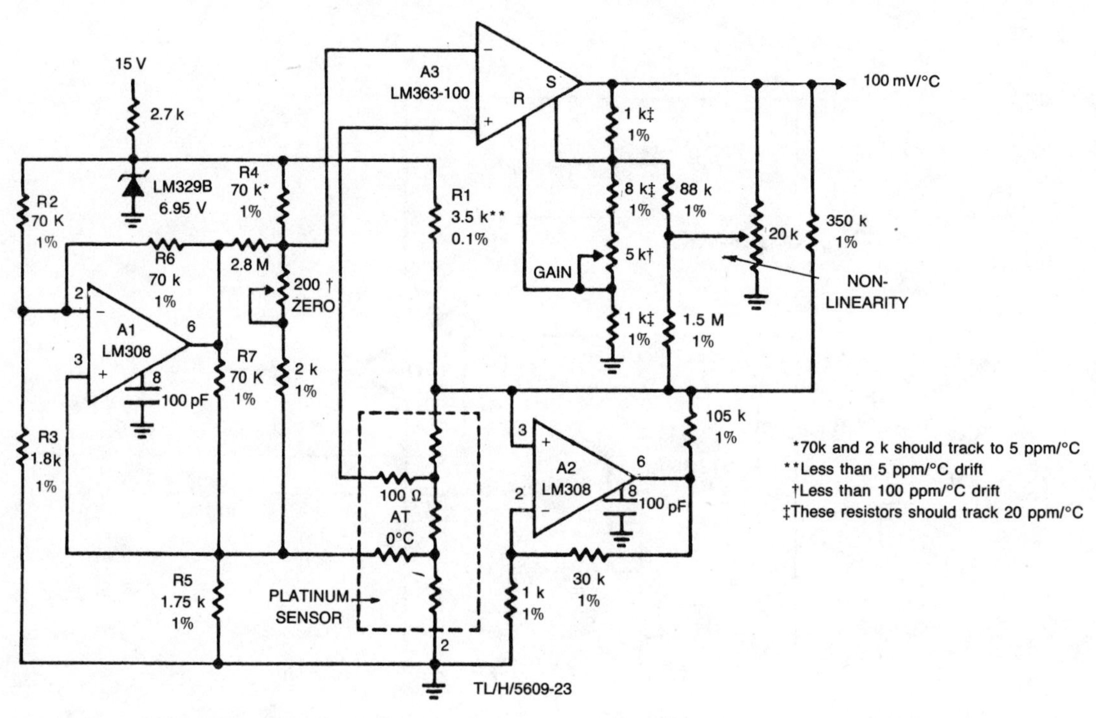

NATIONAL SEMICONDUCTOR CORP. *Fig. 100-3*

Circuit Notes

This thermometer is capable of 0.01°C accuracy over −50°C to +150°C. A unique trim arrangement eliminates cumbersome trim interactions so that zero gain, and nonlinearity correction can be trimmed in one even trip. Extra op amps provide full Kelvin sensing on the sensor without adding drift and offset terms found in other designs. A1 is configured as a Howland current pump, biasing the sensor with a fixed current. Resistors R2, R3, R4 and R5 form a bridge driven into balance by A1. In balance, both inputs of A2 are at the same voltage. Since R6 = R7, A1 draws equal currents from both legs of the bridge. Any loading of the R4/R5 leg by the sensor would unbalance the bridge; therefore, both bridge taps are given to the sensor open circuit voltage and no current is drawn.

THERMOCOUPLE AMPLIFIER WITH COLD JUNCTION COMPENSATION

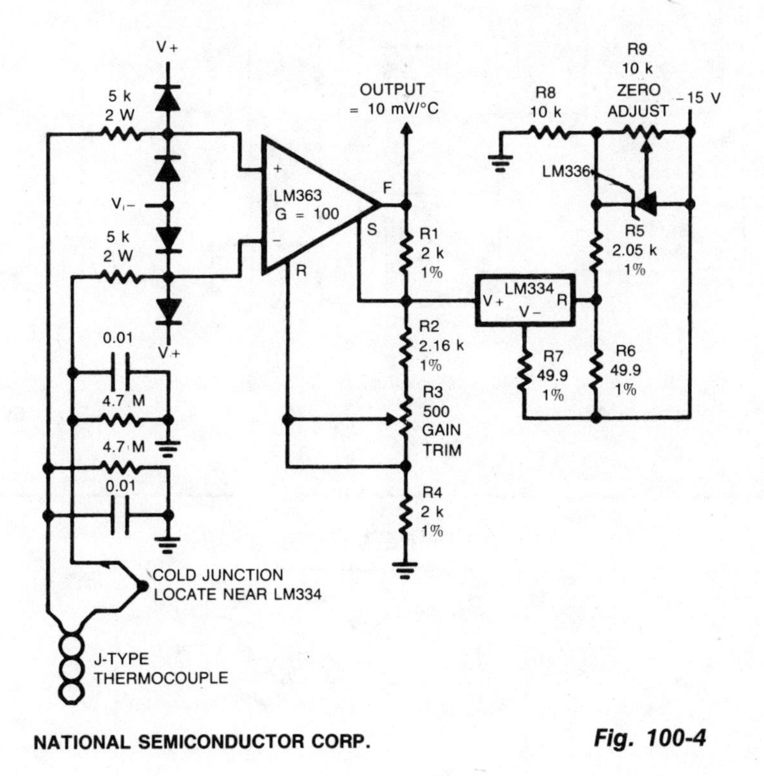

NATIONAL SEMICONDUCTOR CORP. ***Fig. 100-4***

Circuit Notes

Input protection circuitry allows thermocouple to short to 120 Vac without damaging the amplifier.

Calibration:

1. Apply a 50 mV signal in place of the thermocouple. Trim R3 for $V_{OUT} = 12.25$ V.
2. Reconnect the thermocouple. Trim R9 for correct output.

5-V POWERED, LINEARIZED PLATINUM RTD SIGNAL CONDITIONER

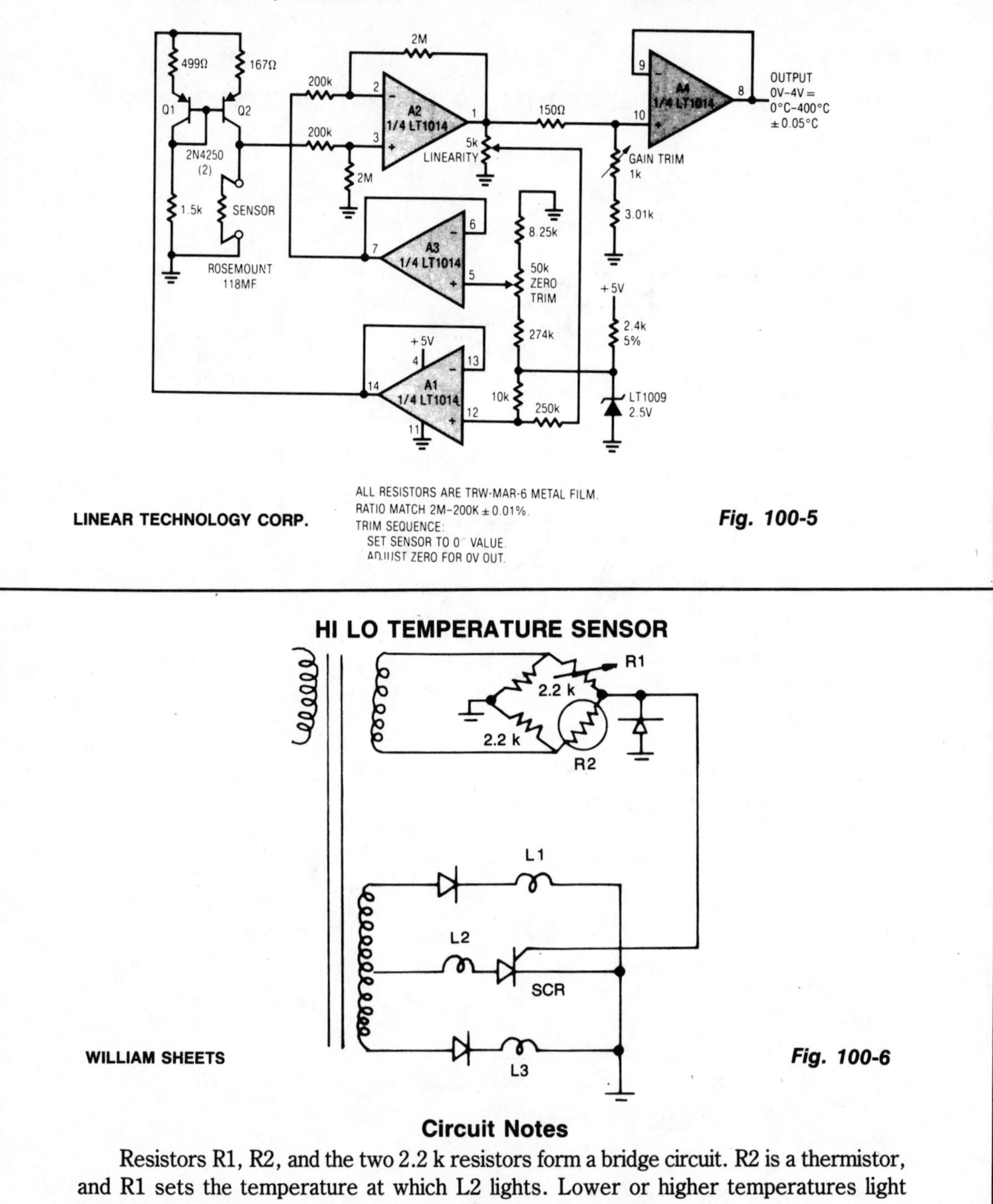

LINEAR TECHNOLOGY CORP.

Fig. 100-5

HI LO TEMPERATURE SENSOR

WILLIAM SHEETS

Fig. 100-6

Circuit Notes

Resistors R1, R2, and the two 2.2 k resistors form a bridge circuit. R2 is a thermistor, and R1 sets the temperature at which L2 lights. Lower or higher temperatures light L1 or L3 to indicate an over- or under-temperature condition.

101

Temperature-to-Frequency Converters

The sources of the following circuits are contained in the Sources section beginning on page 694. The figure number contained in the box of each circuit correlates to the source entry in the Sources section.

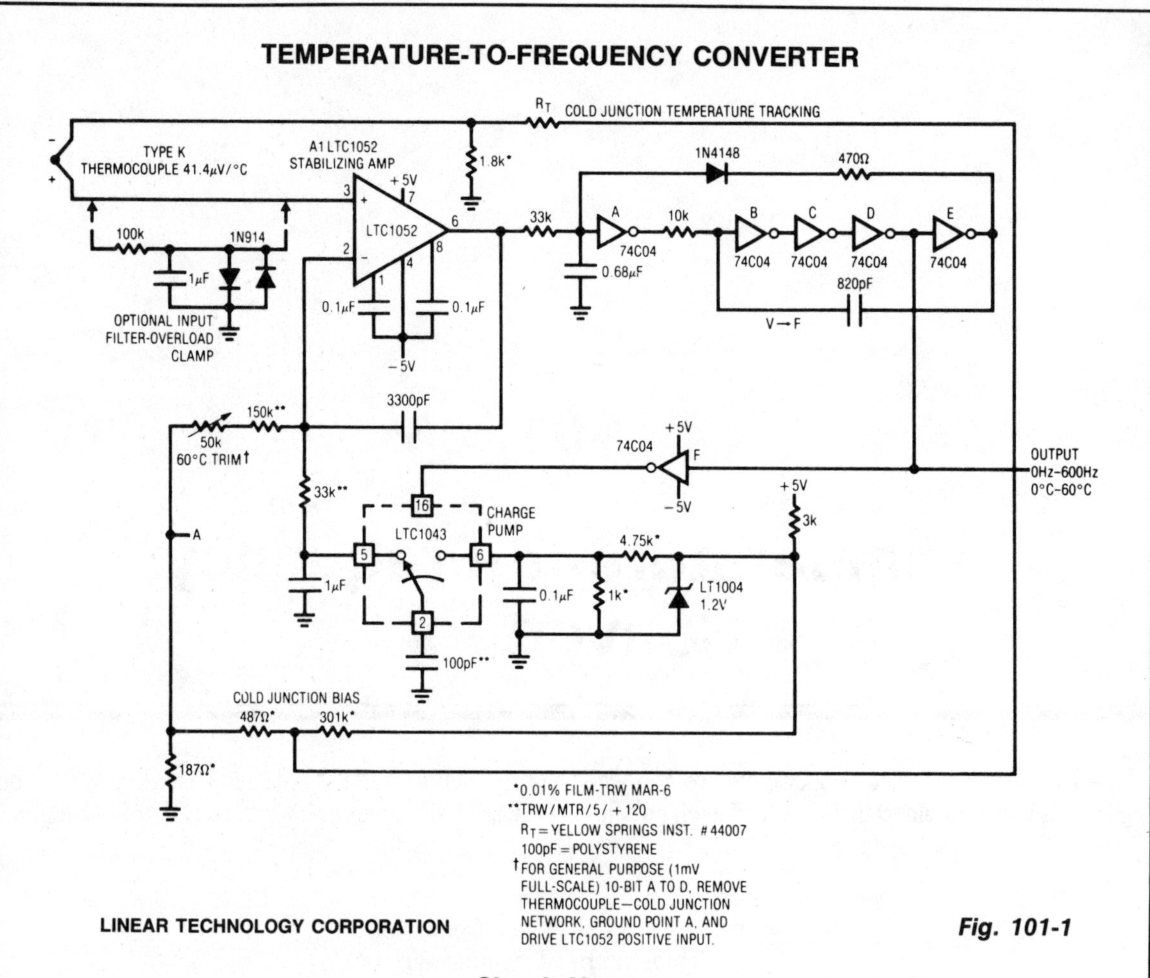

Fig. 101-1

Circuit Notes

A1's positive input is biased by the thermocouple. A1's output drives a crude V → F converter, comprised of the 74C04 inverters and associated components. Each V → F output pulse causes a fixed quantity of charge to be dispensed into the 1 μF capacitor from the 100 pF capacitor via the LT1043 switch. The larger capacitor integrates the packets of charge, producing a dc voltage at A1's negative input. A1's output forces the V→ F converter to run at whatever frequency is required to balance the amplifier's inputs. This feedback action eliminates drift and nonlinearities in the V → F converter as an error item and the output frequency is solely a function of the dc conditions at A1's inputs. The 3300 pF capacitor forms a dominant response pole at A1, stabilizing the loop.

DIGITAL TEMPERATURE MEASURING CIRCUIT

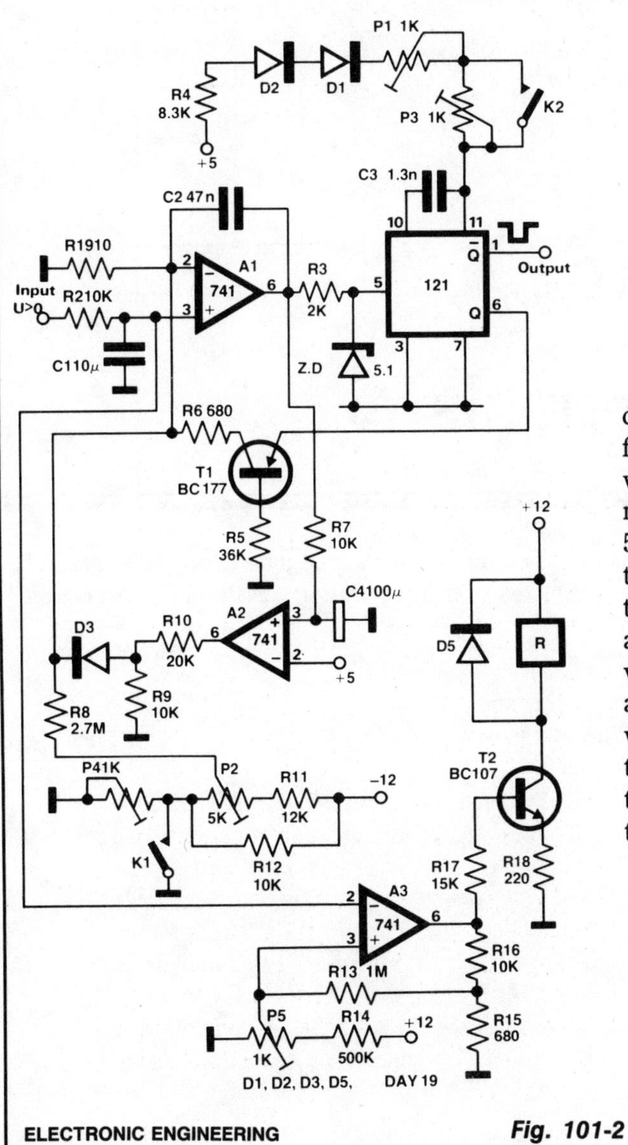

ELECTRONIC ENGINEERING

Fig. 101-2

Circuit Notes

The output voltage of a thermocouple is converted into frequency measured by a digital frequency meter. The measuring set connected with Ni-NiCr thermocouple permits you to measure the temperatures within the range of 5°C - 800°C with ±1°C error. The output thermocouple signal is proportional to the temperature difference between the hot junction and the thermostat kept at 0°C, it drives the voltage-to-frequency converter changing the analogue input signal into the output frequency with the conversion ratio adjusted in such a way, that the frequency is equal to the measured temperature in Celsius degrees, e.g., for 350°C the frequency value is 350 Hz.

102

Theremins

The sources of the following circuits are contained in the Sources section beginning on page 694. The figure number contained in the box of each circuit correlates to the source entry in the Sources section.

ELECTRONIC THEREMIN

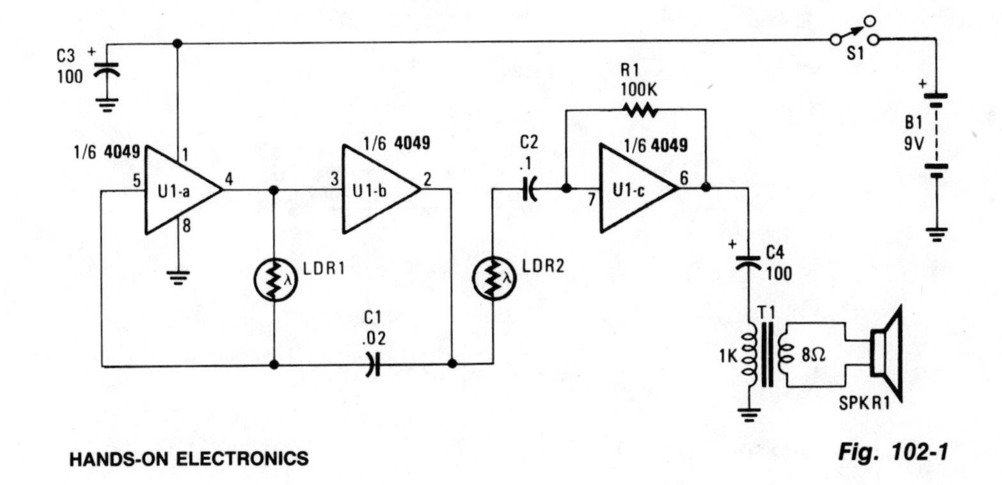

HANDS-ON ELECTRONICS

Fig. 102-1

Circuit Notes

This circuit has the CMOS IC doing double-duty performance. The first two inverters operate as a digital audio oscillator; the third operates as a low-gain linear audio amplifier. As the intensity of the light falling on photoresistor LDR1 increases the oscillator's frequency increases; similarly, the illumination falling on photoresistor LDR2 determines the volume level from the loudspeaker: The more illumination the more volume. If you flop and wave your hands between the two photocells and a light source, a special kind of electronic music will be produced.

DIGITAL THEREMIN

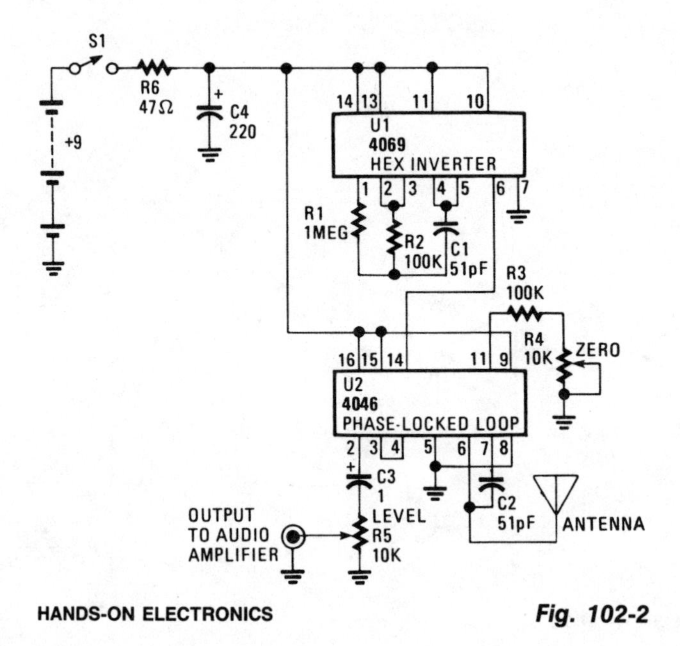

HANDS-ON ELECTRONICS

Fig. 102-2

Circuit Notes

The CD4069 or 74C04 hex inverter—is used as a fixed-frequency oscillator centered around 100 kHz. U2 contains the variable frequency oscillator and balanced modulator. The CD4046 is a phase-locked loop and R3, R4, and C2 determine the center frequency of the on-chip oscillator. The antenna forms a parallel capacitance with C2, which allows the frequency to be shifted several kilohertz by bringing a hand near the antenna. R4, the ZERO control, allows the variable oscillator to be set to the same frequency as the fixed oscillator. When the difference frequency is below 15 Hz, it is below the lower frequency limit of the ear. By setting both oscillators to the same frequency, the Theremin remains silent until the performer brings his or her hand near the antenna. The oscillators are mixed by an exclusive OR gate inside the 4046. That gate acts as a digital balanced modulator, which produces the sum and difference frequencies. The output of the gate is then ac coupled by C3 to LEVEL control R5 and an output jack for connection to an audio amplifier or stereo receiver.

103
Thermometer Circuits

The sources of the following circuits are contained in the Sources section beginning on page 694. The figure number contained in the box of each circuit correlates to the source entry in the Sources section.

DIGITAL THERMOCOUPLE THERMOMETER

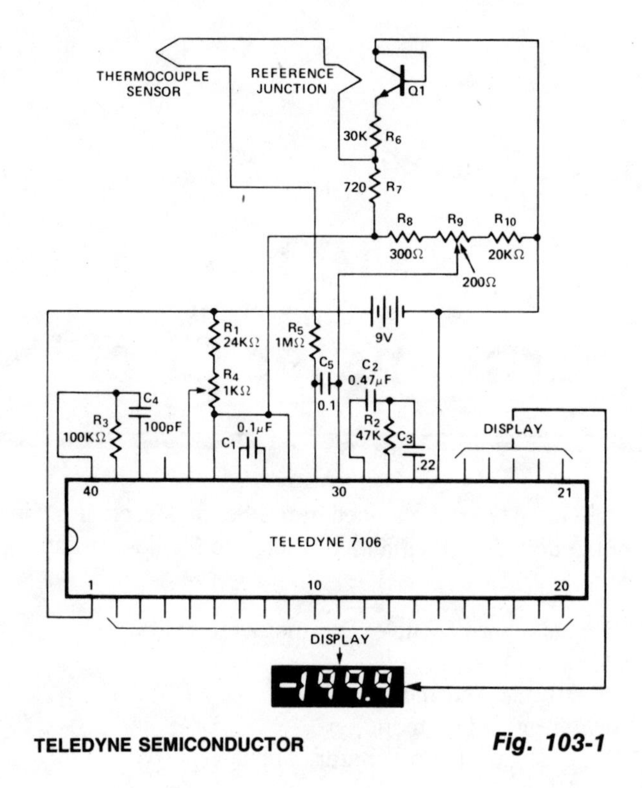

TELEDYNE SEMICONDUCTOR *Fig. 103-1*

Circuit Notes

This digital thermocouple thermometer uses one active component and 15 passive components. With this circuit, both type J and type K thermocouples may be used. The type J will measure over the temperature range of 10 to 530°C with a conformity of ±2°C. The type K will measure over a temperature range of 0°C to 1000°C with a conformity of ±3°C.

REMOTE THERMOMETER

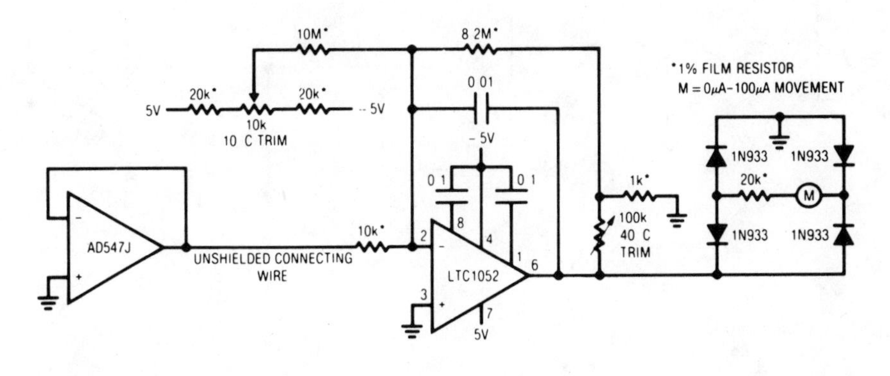

LINEAR TECHNOLOGY

Fig. 103-2

Circuit Notes

The low output impedance of a closed loop op amp gives ideal line-noise immunity, while the op amp's offset voltage drift provides a temperature sensor. Using the op amp in this way requires no external components and has the additional advantages of a hermetic package and unit-to-unit mechanical uniformity if replacement is ever required. The op amp's offset drift is amplified to drive the meter by the LTC1052. The diode bridge connection allows either positive or negative op amp temperature sensor offsets to interface directly with the circuit. In this case, the circuit is arranged for a +10°C to +40°C output, although other ranges are easily accommodated. To calibrate this circuit, subject the op amp sensor to a +10°C environment and adjust the 10°C trim for an appropriate meter indication. Next, place the op amp sensor in a +40°C environment and trim the 40°C adjustment for the proper reading. Repeat this procedure until both points are fixed. Once calibrated, this circuit will typically provide accuracy within ±2°C, even in high noise environments.

ELECTRONIC THERMOMETER

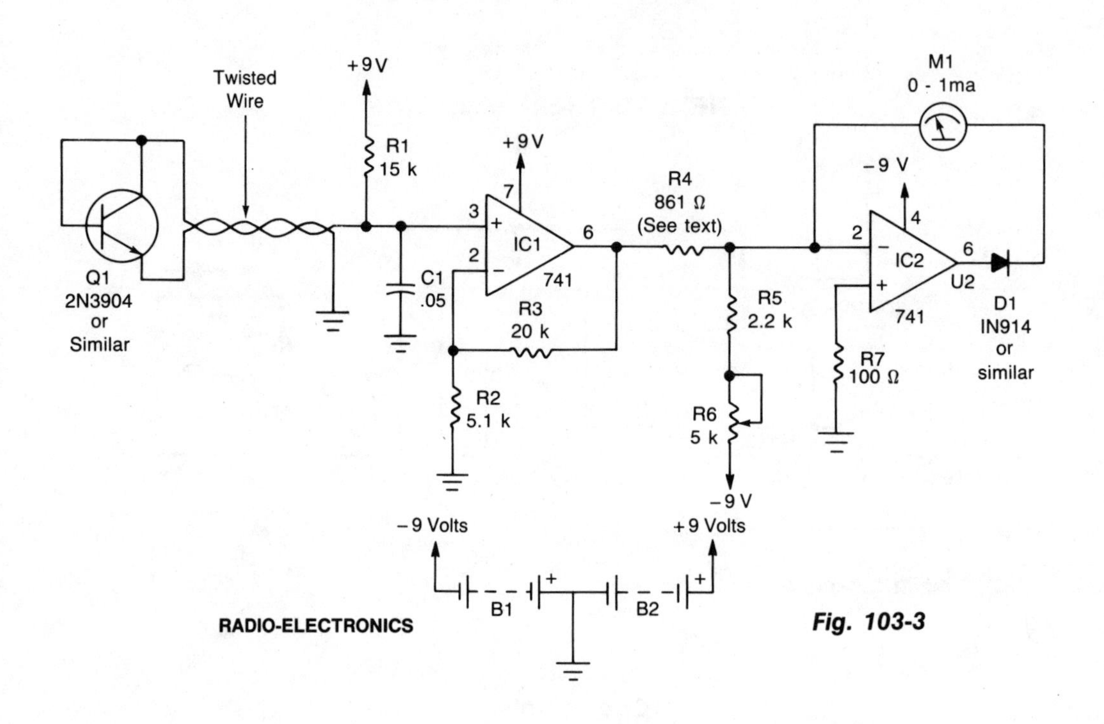

Fig. 103-3

Circuit Notes

An inexpensive electronic thermometer is capable of measuring temperatures over a range of from −30°F to +120°F. A diode-connected 2N3904 transistor used as the temperature sensor forms a voltage divider with R1. As temperature increases, the voltage drop across the transistor changes by approximately −1.166 millivolts-per°F. As a result, the current at pin 3 of IC1, a 741 op amp with a gain of 5, decreases as the temperature measured by the sensor increases.

A second 741 op amp, IC2 is configured as an inverting amplifier. Resistors R5 and R6 calibrate the circuit. Calibration is also straightforward. When properly done, a temperature of −30°F will result in a meter reading of 0 milliamps, while a temperature of 120°F will result in a meter reading of 1 milliamp. Divide the scale between those points into equal segments and mark the divisions with the appropriate corresponding temperatures. The calibration is completed by placing the sensor in an environment with a known temperature, such as an ice-point bath. Place the sensor in the bath and adjust R6 until you get the correct meter reading.

DIFFERENTIAL THERMOMETER

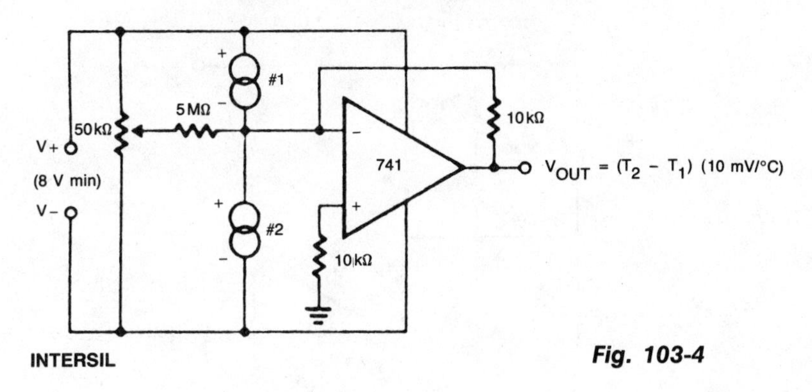

INTERSIL

Fig. 103-4

Circuit Notes

The 50 k ohm pot trims offsets in the devices whether internal or external, so it can be used to set the size of the difference interval. This also makes it useful for liquid-level detection (where there will be a measurable temperature difference).

BASIC DIGITAL THERMOMETER, KELVIN SCALE WITH ZERO ADJUST

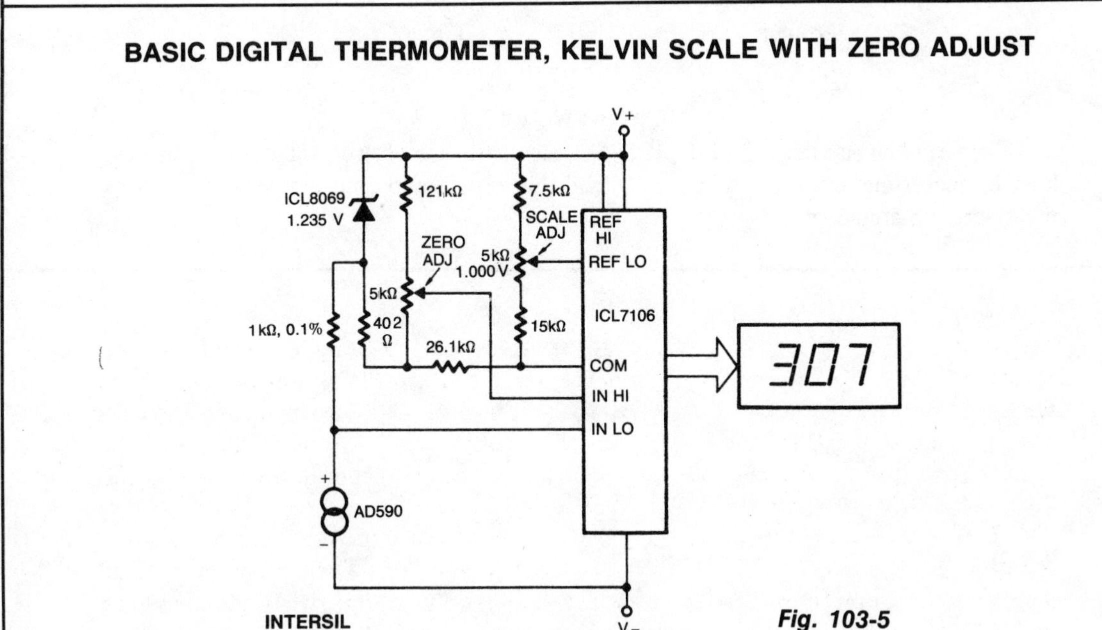

INTERSIL

Fig. 103-5

Circuit Notes

This circuit allows "zero adjustment" as well as slope adjustment. The ICL8069 brings the input within the common-mode range, while the 5 k ohm pots trim any offset at 218° K (−55°C), and set the scale factor.

CENTIGRADE THERMOMETER (0°C-100°C)

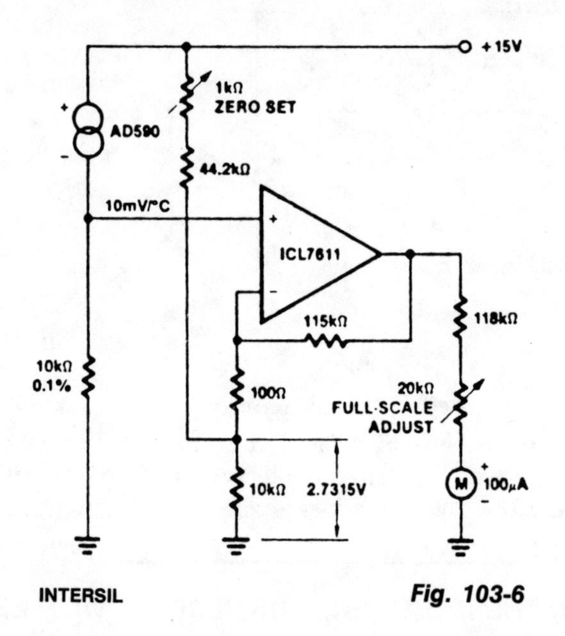

INTERSIL

Fig. 103-6

Circuit Notes

The ultra-low bias current of the ICL7611 allows the use of large-value gain-resistors, keeping meter-current error under ½%, and therefore saving the expense of an extra meter-driving amplifier.

104

Tilt Meters

The sources of the following circuits are contained in the Sources section beginning on page 694. The figure number contained in the box of each circuit correlates to the source entry in the Sources section.

TILTMETER INDICATES SENSE OF SLOPE

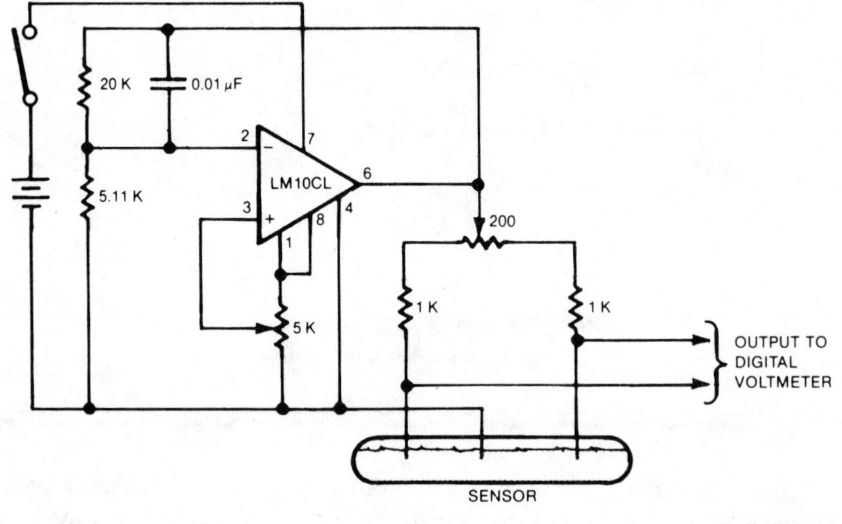

NASA

Fig. 104-1

Circuit Notes

Electrodes are immersed in an electrolyte that remains level while the sensor follows the tilt of the body on which it is placed, more of one outer electrode and less of the other are immersed and their resistances fall or rise, respectively. The resistance change causes a change in the output voltage of the bridge circuit. The sensor forms the two lower legs of the bridge, and two 1000 ohm metal film resistors and a 200 ohm ceremet balance potentiometer form the two upper legs. In preparation for use, the bridge is balanced by adjusting the balance potentiometer so that the bridge output voltage is zero when the sensor is level. The bridge input voltage (dc excitation) is adjusted to provide about 10 millivolts output per degree of slope, the polarity indicating the sense of the slope. This scaling factor allows the multimeter to read directly in degrees if the user makes a mental shift of the meter decimal point. The scaling-factor calibration is done at several angles to determine the curve of output voltage versus angle.

DIFFERENTIAL CAPACITANCE MEASUREMENT CIRCUIT

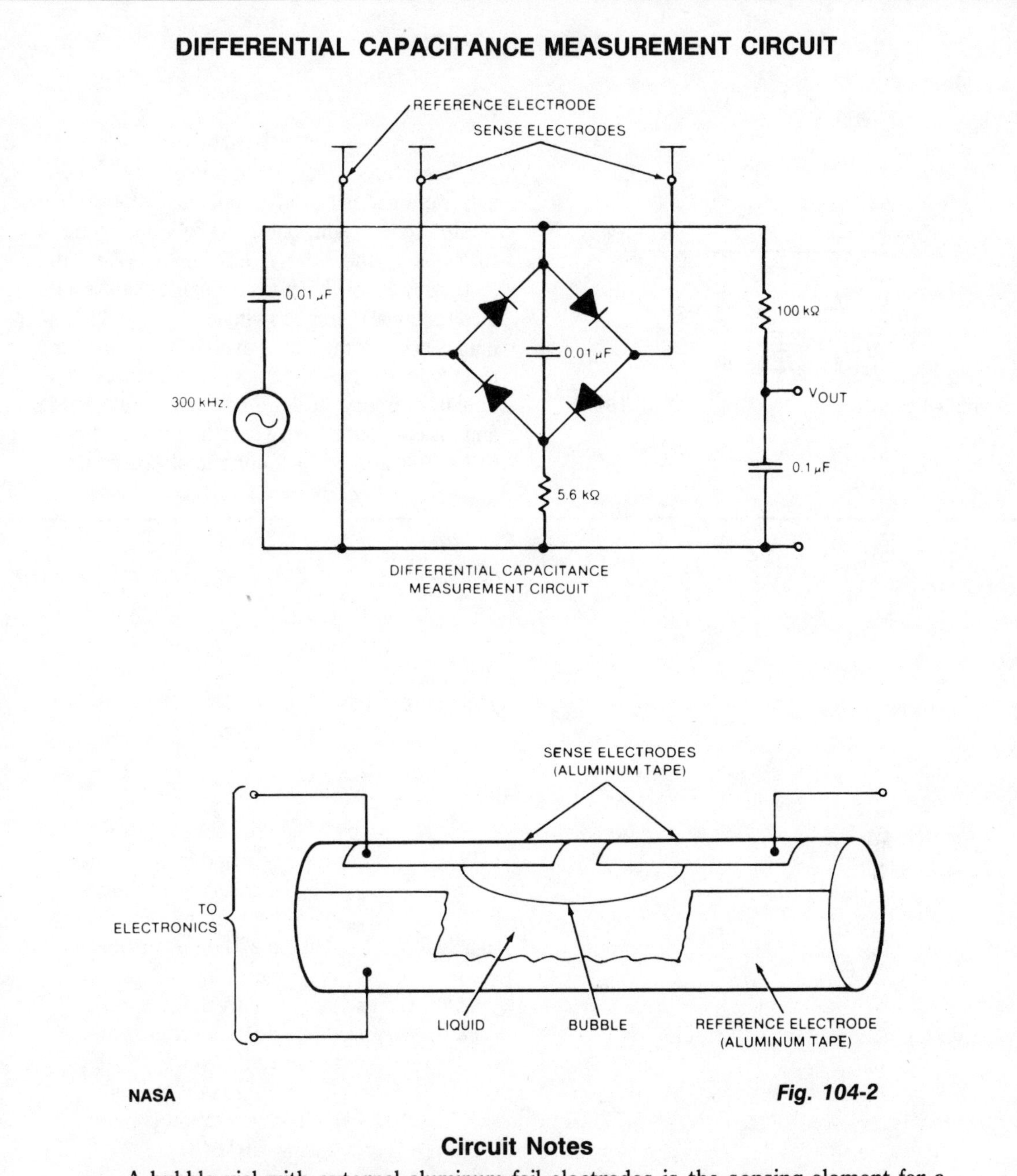

NASA ***Fig. 104-2***

Circuit Notes

A bubble vial with external aluminum-foil electrodes is the sensing element for a simple indicating tiltmeter. To measure bubble displacement, a bridge circuit detects the difference in capacitance between the two sensing electrodes and the reference electrode. Using this circuit, a tiltmeter level vial with 2 mm deflection for 5 arc-seconds of tilt easily resolves 0.05 arc-second. The four diodes are CA3039, or equivalent.

ULTRA-SIMPLE LEVEL

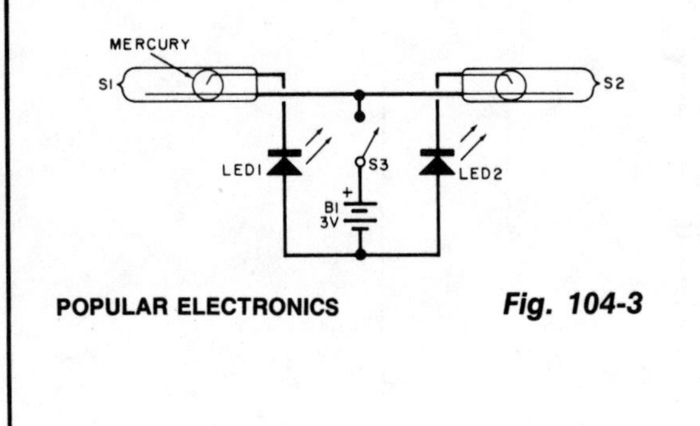

POPULAR ELECTRONICS *Fig. 104-3*

Circuit Notes

This electronic level uses two LED indicators instead of an air bubble. If the surface is tilted to the right, one LED lights; if it's tilted to the left, the other LED lights. When the surface is level, both LEDs light. It uses two unidirectional mercury switches, S1 and S2. The unidirectional mercury switch has one long electrode and one short, angled electrode. The pool of mercury "rides" on the long electrode and makes contact between the two electrodes if the unit is held in a horizontal position.

105

Time-Delay Circuits

The sources of the following circuits are contained in the Sources section beginning on page 694. The figure number contained in the box of each circuit correlates to the source entry in the Sources section.

HOUR TIME-DELAY SAMPLING CIRCUIT

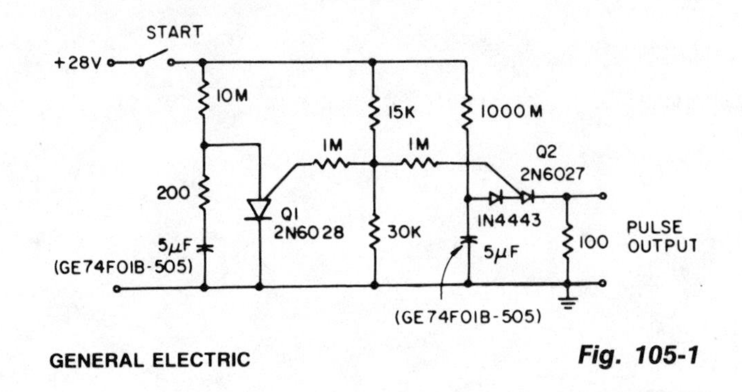

GENERAL ELECTRIC

Fig. 105-1

Circuit Notes

The circuit lowers the effective peak current of the output PUT, Q2. By allowing the capacitor to charge with high gate voltage and periodically lowering gate voltage, when Q1 fires, the timing resistor can be a value which supplies a much lower current than I_P. The triggering requirement here is that minimum charge to trigger flow through the timing resistor during the period of the Q1 oscillator. This is not capacitor size dependent, only capacitor leakage and stability dependent.

TIME DELAY WITH CONSTANT CURRENT CHARGING

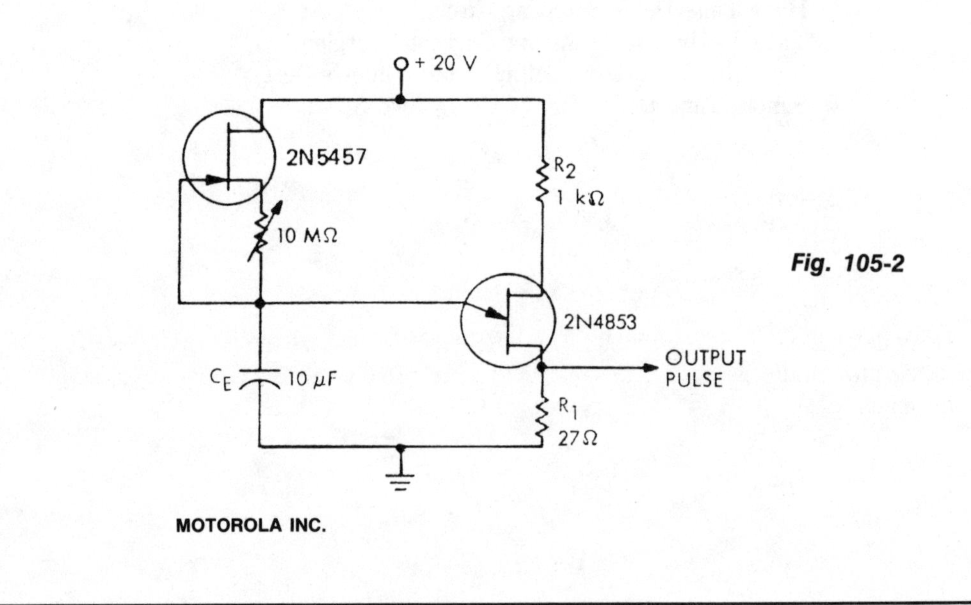

MOTOROLA INC.

Fig. 105-2

LOW-COST INTEGRATOR MULTIPLIES 555 TIMER'S DELAY

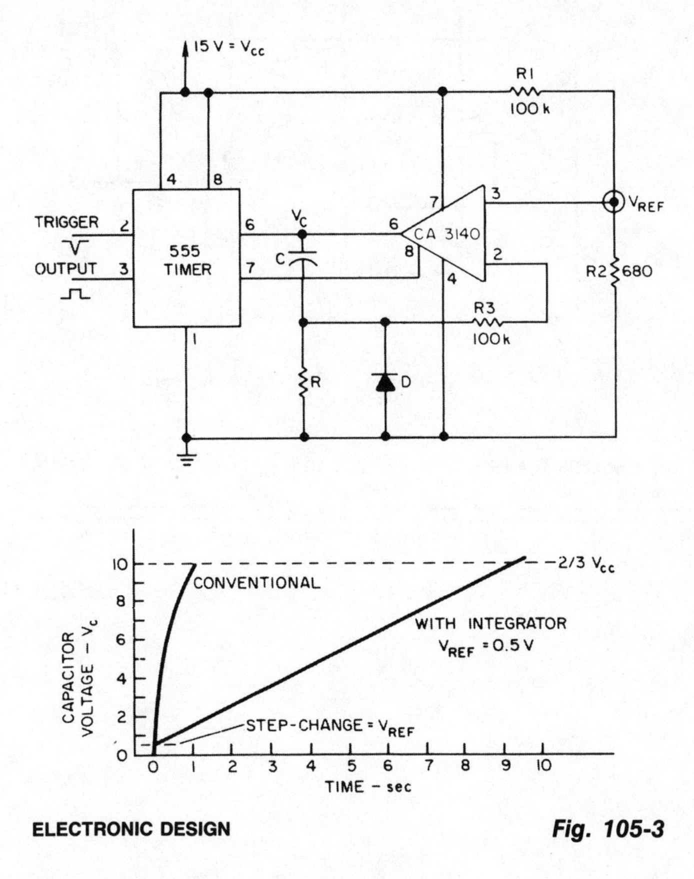

ELECTRONIC DESIGN *Fig. 105-3*

Circuit Notes

Long delay times can be derived from a 555 timer with reasonably sized capacitors if an integrator circuit is used. The capacitor's charging time with an integrator circuit can be much longer than with a conventional 555-timer configuration.

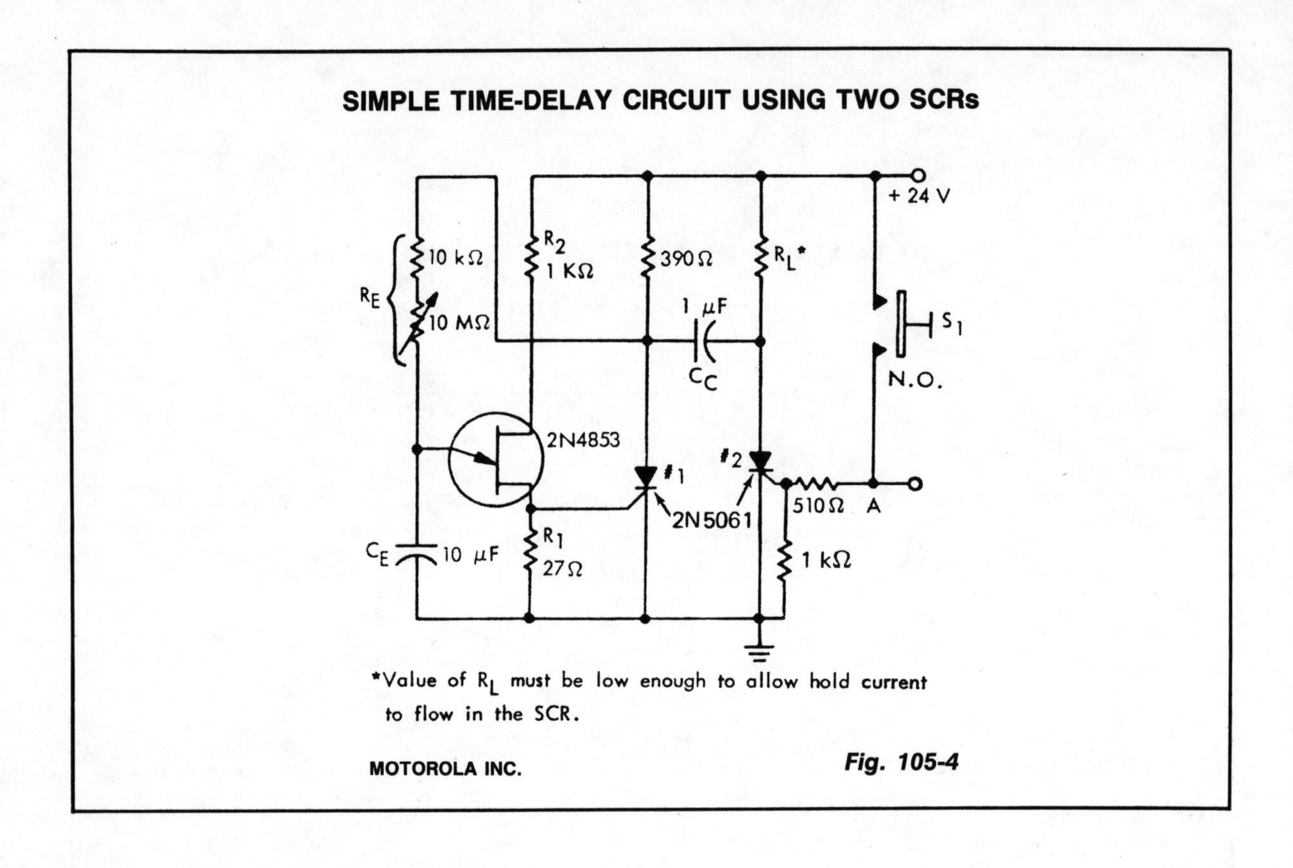
SIMPLE TIME-DELAY CIRCUIT USING TWO SCRs
+ 24 V
10 kΩ
R2
1 KΩ
390 Ω
RL*
RE
10 MΩ
1 μF
S1
CC
N.O.
2N4853
#2
#1
510 Ω
A
2N5061
CE
10 μF
R1
27 Ω
1 kΩ
*Value of RL must be low enough to allow hold current to flow in the SCR.
MOTOROLA INC.
Fig. 105-4

106

Timers

The sources of the following circuits are contained in the Sources section beginning on page 694. The figure number contained in the box of each circuit correlates to the source entry in the Sources section.

LONG-TERM ELECTRONIC TIMER

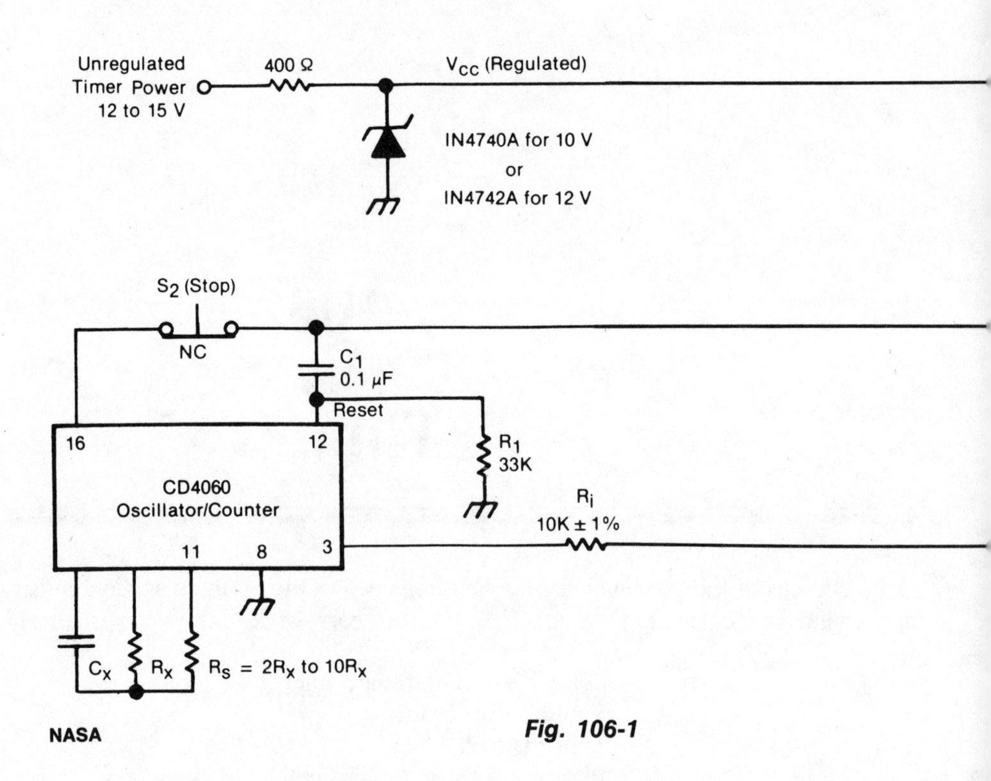

NASA

Fig. 106-1

Circuit Notes

The timer includes an oscillator and a counter in an integrated circuit. The timing interval equals the oscillator period multiplied by the number of cycles to be counted. The oscillator frequency depends upon resistor RS and capacitor CX. The number of oscillator cycles to be counted before the counter output changes state is determined by the selection of the counter output terminal, shown here as pin 3. The interval can be set anywhere in the range from fractions of a second to months; it is given by $T = 0.55\ R_S C_X 2^n$, where n is an integer determined by the counter-output selection. Operation is initiated by the closure of momentary switch S1 (or by a command signal having a similar effect). This grounds one side of relay K1, thereby activating the relay

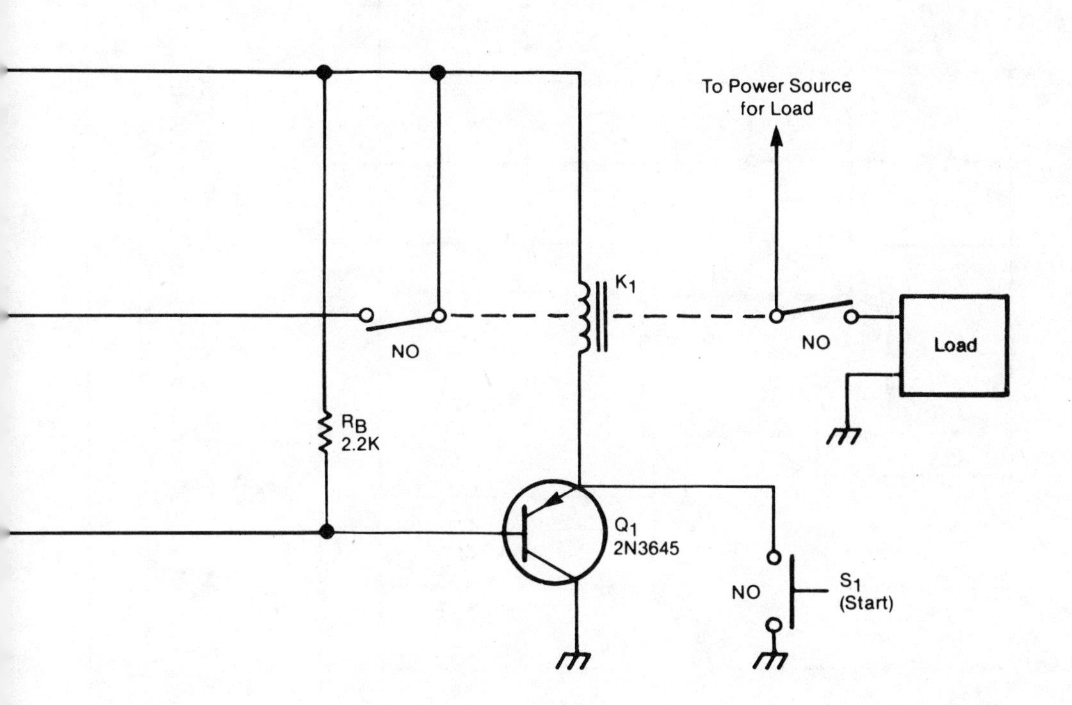

and causing the closure of the switches that supply power to the timer and to the load. The turn-on of V_{CC} at the timer is coupled through C1 to the counter-reset terminal, thus resetting the counter. The initial reset voltage transient is then drained away through R1 to permit normal operation. During the first half cycle of the counter operation, the counter output voltage (at pin 3 in this case) is low. This turns on transistor Q1 so that relay K1 latches on, enabling the timer to continue running even though switch S1 has opened. The oscillator runs while the relay is on. When the number of oscillator cycles reaches the limit, the counter output voltage at pin 3 goes high. This turns off Q1, thereby turning off the relay and returning the system to the original ''power-off'' state to await the next starting command. The timing cycle can also be interrupted and the system turned off by opening normally-closed switch S2.

TIMER WITH ALARM

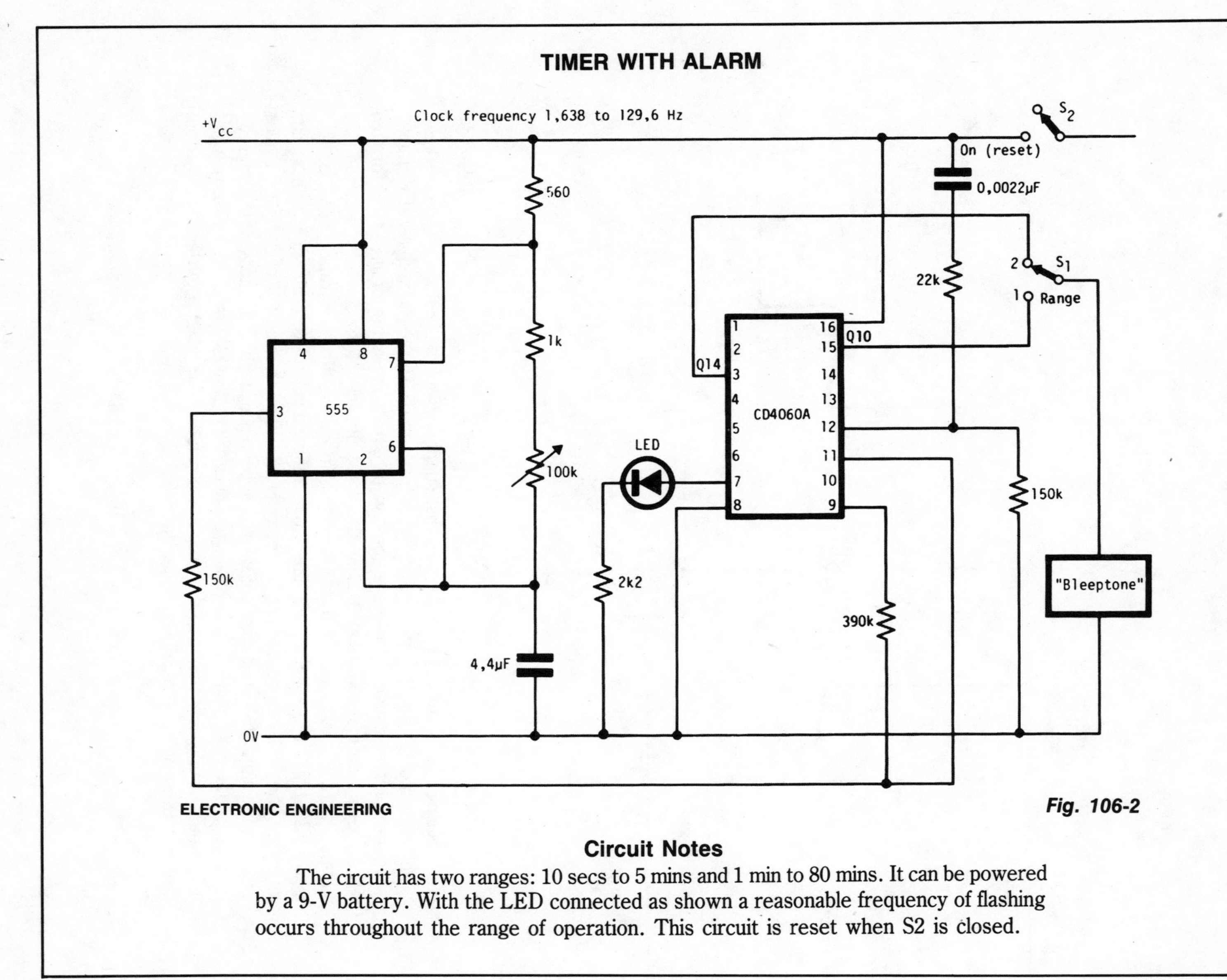

ELECTRONIC ENGINEERING

Fig. 106-2

Circuit Notes

The circuit has two ranges: 10 secs to 5 mins and 1 min to 80 mins. It can be powered by a 9-V battery. With the LED connected as shown a reasonable frequency of flashing occurs throughout the range of operation. This circuit is reset when S2 is closed.

TIMER CIRCUIT

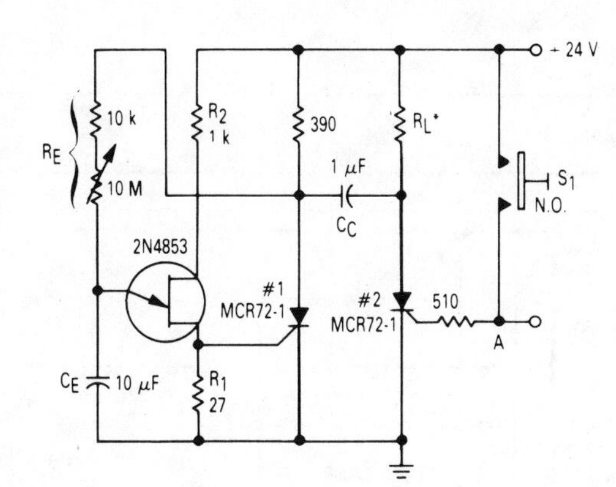

MOTOROLA *Fig. 106-3*

Circuit Notes

After one cycle of operation, SCR 1 will be on, and a low value of voltage is applied to the UJT emitter circuit, interrupting the timing function. When pushbutton S1 is pushed, or a positive going pulse is applied at point A, SCR 2 will turn on, and SCR 1 will be turned off by commutating capacitor CC. With SCR 1 off, the supply voltage will be applied to RE and the circuit will begin timing again. After a period of time determined by the setting of RE, the UJT will fire and turn SCR 1 on and commutate SCR 2 off. The time delay is determined by the charge time of the capacitor.

PUT LONG DURATION TIMER

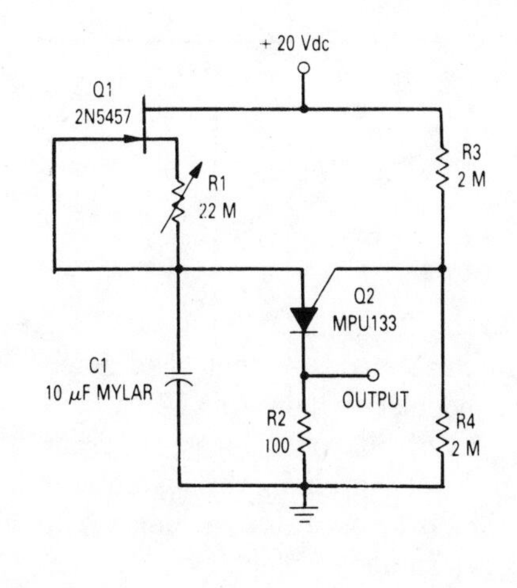

MOTOROLA *Fig. 106-4*

Circuit Notes

The time circuit can provide a time delay of up to 20 minutes. The circuit is a standard relaxation oscillator with a FET current source in which resistor R1 is used to provide reverse bias on the gate-to-source of the JFET. This turns the JFET off and increases the charging time of C1. C1 should be a low leakage capacitor such as a mylar type.

PROGRAMMABLE VOLTAGE CONTROLLED TIMER

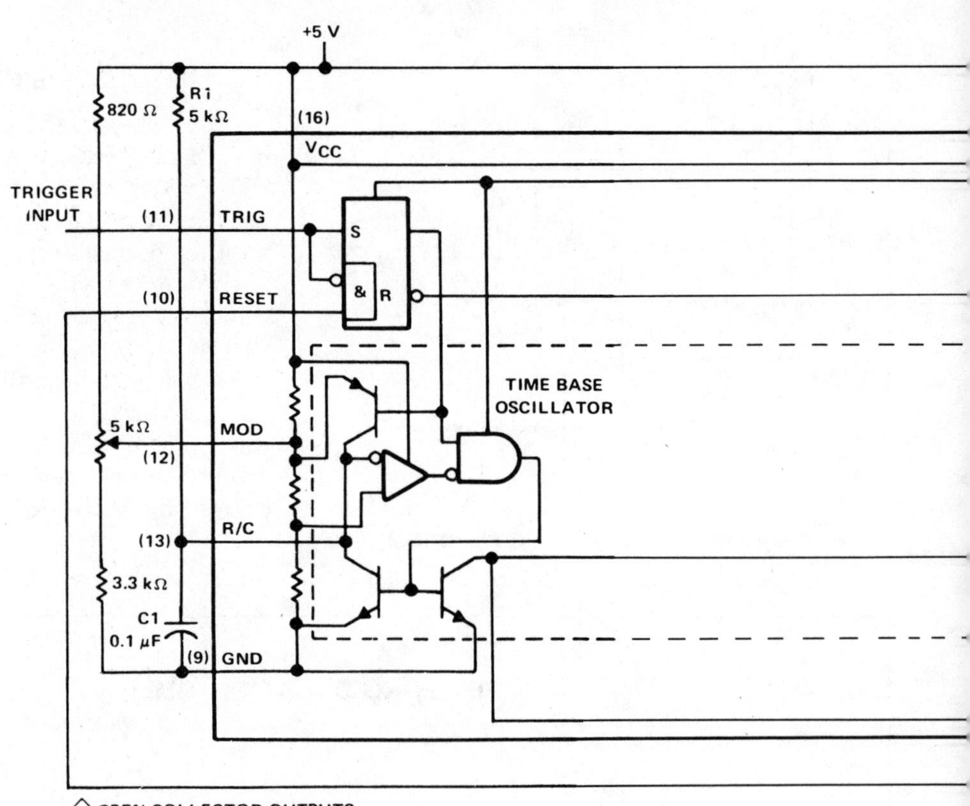

TEXAS INSTRUMENTS

Fig. 106-5

Circuit Notes

The μA2240 may easily be configured as a programmable voltage controlled timer with a minimum number of external components. The modulation input (pin 12), which allows external adjustment of the input threshold level. A variable voltage is applied from the arm of a 10 k ohm potentiometer connected from V_{CC} to ground. A change in the modulation input voltage will result in a change in the time base oscillator frequency and the period of the time base output (TBO). The TBO has an open-collector output that

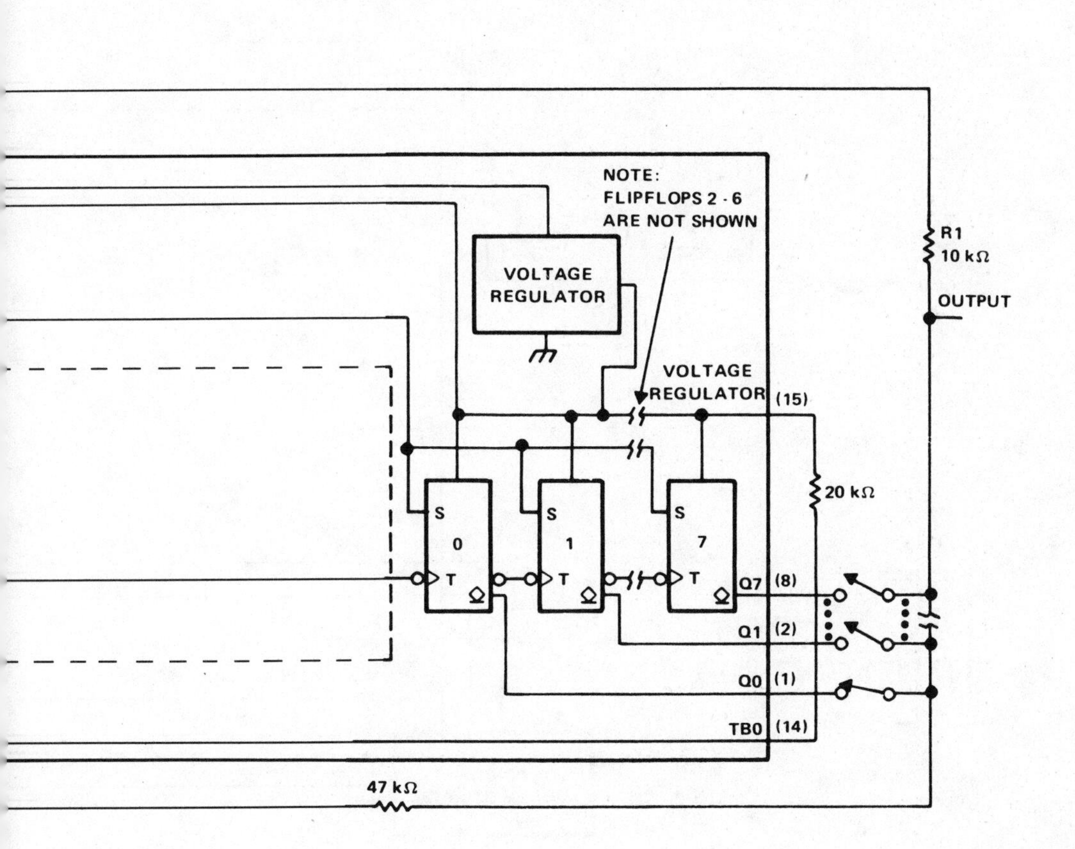

is connected to the regulator output via a 10 k ohm pull-up resistor. The output of the TBO drives the input to the 8-stage counter section.

At start-up, a positive trigger pulse starts the TBO and sets all counter outputs to a low state. The binary outputs are open-collector stages that may be connected together to the 10 k ohm pull-up resistor to provide a "wired-OR" output function. This circuit may be used to generate 255 discrete time delays that are integer multiples of the time-base period. The total delay is the sum of the number of time-base periods, which is the binary sum of the Q outputs connected. Delays from 200 μs to 0.223 s are possible with this configuration.

LOW POWER MICROPROCESSOR PROGRAMMABLE INTERVAL TIMER

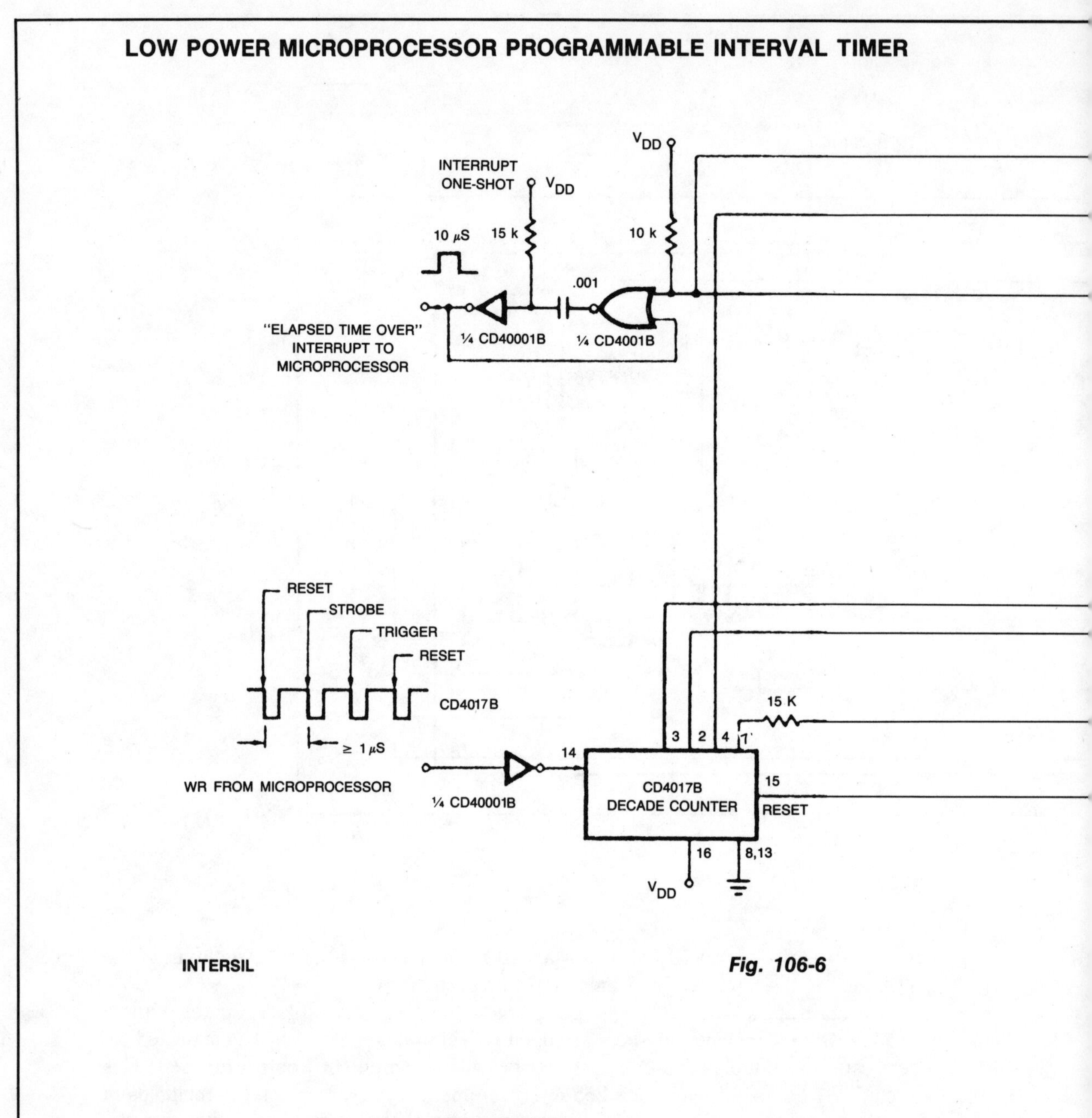

Fig. 106-6

Circuit Notes

The microprocessor sends out an 8-bit binary code on its 8-bit I/O bus (the binary value needed to program the ICM7240), followed by four WRITE pulses into the CD4017B decade counter. The first pulse resets the 8-bit latch, the second strobes the binary value into the 8-bit latch, the third triggers the ICM7240 to begin its timing cycle and the fourth resets the decade counter. The ICM7240 then counts the interval of time

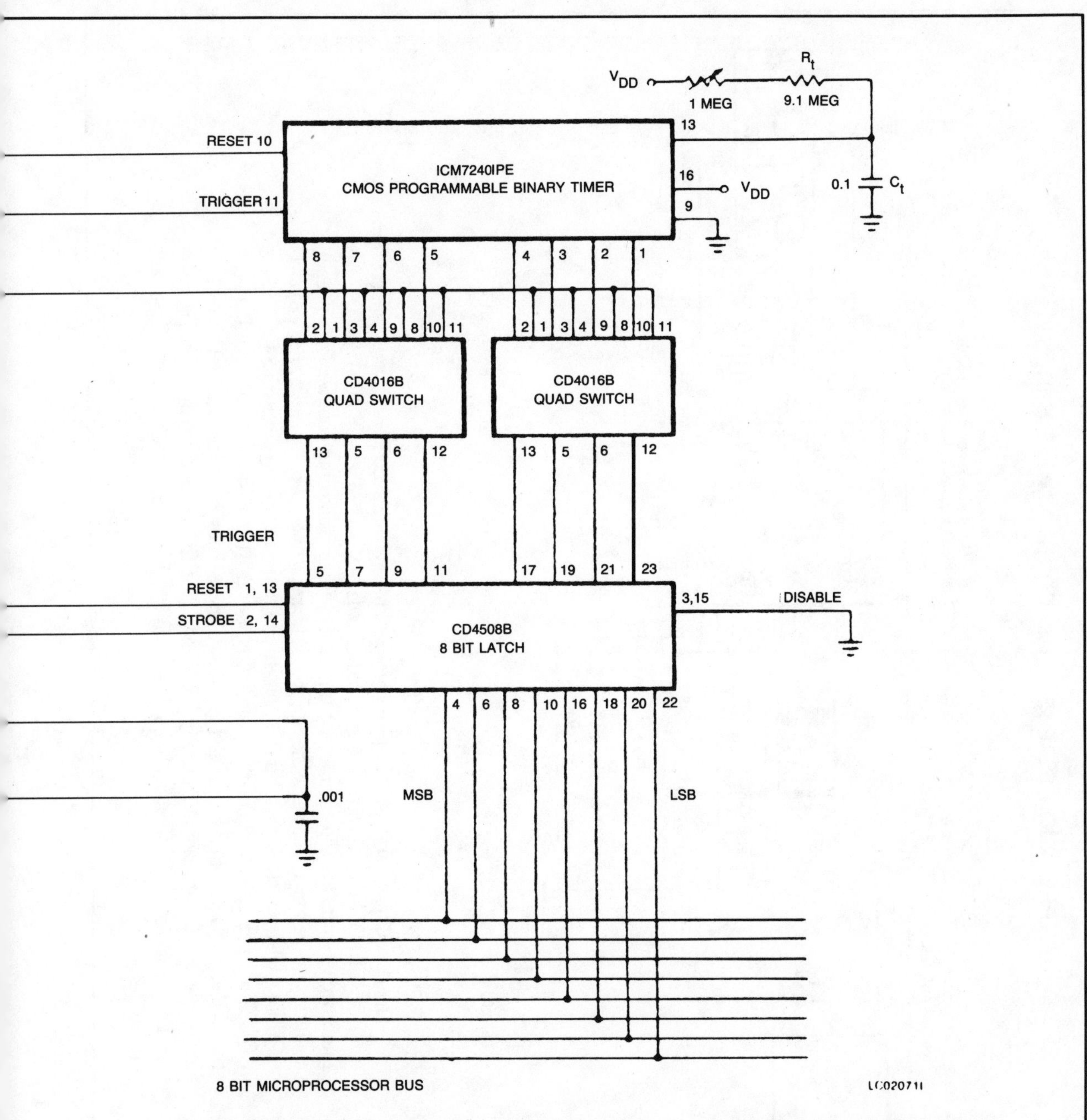

determined by the R-C value on pin 13, and the programmed binary count on pins 1 through 8. At the end of the programmed time interval, the interrupt one-shot is triggered, informing the microprocessor that the programmed time interval is over. With a resistor of approximately 10 M ohm and a can capacitor of 0.1 μF, the time base of the ICM7240 is one second. Thus, a time of 1-255 seconds can be programmed by the microprocessor, and by varying R or C, longer or shorter time bases can be selected.

PRECISION ELAPSED TIME/COUNTDOWN TIMER

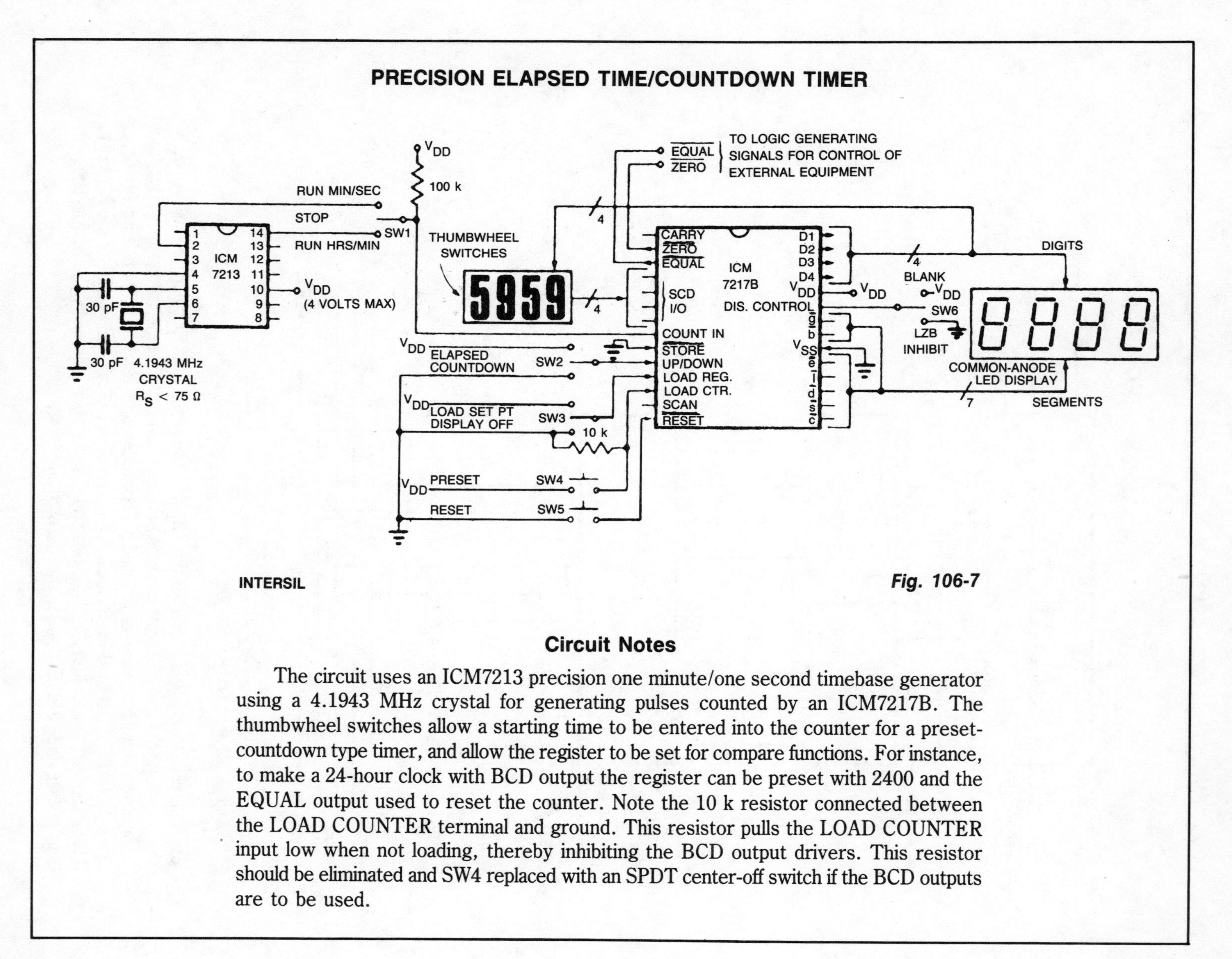

INTERSIL

Fig. 106-7

Circuit Notes

The circuit uses an ICM7213 precision one minute/one second timebase generator using a 4.1943 MHz crystal for generating pulses counted by an ICM7217B. The thumbwheel switches allow a starting time to be entered into the counter for a preset-countdown type timer, and allow the register to be set for compare functions. For instance, to make a 24-hour clock with BCD output the register can be preset with 2400 and the EQUAL output used to reset the counter. Note the 10 k resistor connected between the LOAD COUNTER terminal and ground. This resistor pulls the LOAD COUNTER input low when not loading, thereby inhibiting the BCD output drivers. This resistor should be eliminated and SW4 replaced with an SPDT center-off switch if the BCD outputs are to be used.

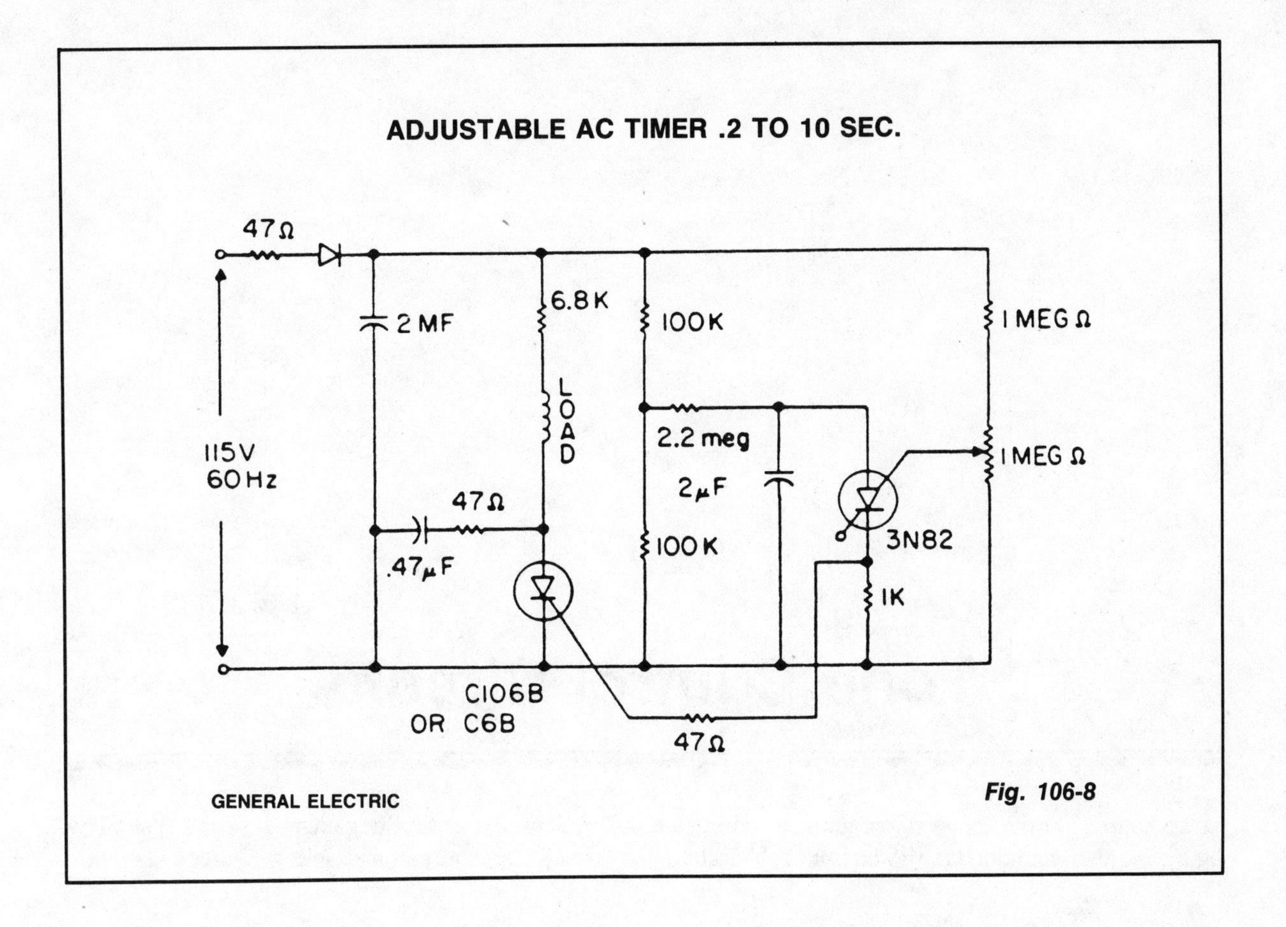
ADJUSTABLE AC TIMER .2 TO 10 SEC.
47Ω
2 MF
6.8K
100K
1MEG Ω
LOAD
115V
60Hz
2.2 meg
1MEG Ω
2μF
47Ω
.47μF
100K
3N82
1K
C106B
OR C6B
47Ω
GENERAL ELECTRIC
Fig. 106-8

107

Tone Control Circuits

The sources of the following circuits are contained in the Sources section beginning on page 694. The figure number contained in the box of each circuit correlates to the source entry in the Sources section.

GUITAR TREBLE BOOST

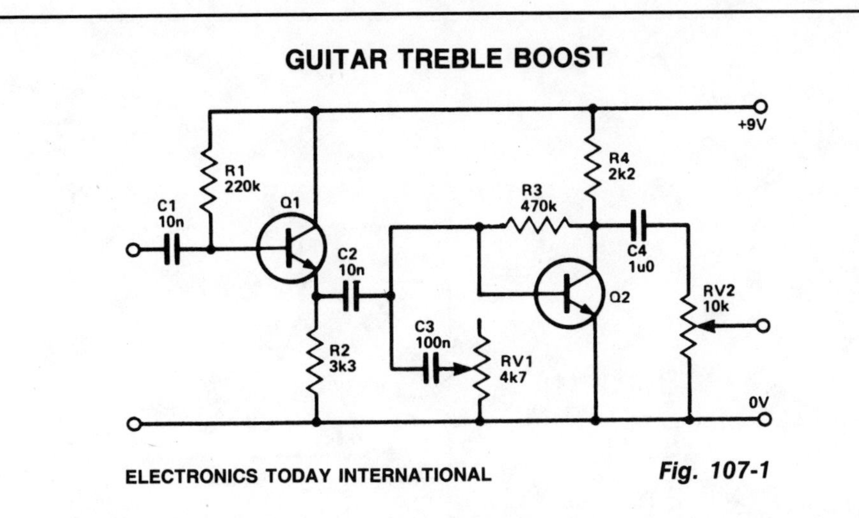

ELECTRONICS TODAY INTERNATIONAL

Fig. 107-1

Circuit Notes

Q1 is connected as an emitter follower in order to present a high input impedance to the guitar. C2, being a relatively low capacitance, cuts out most of the bass, and C3 with RV1 acts as a simple tone control to cut the treble, and hence the amount of treble boost can be altered. Q2 is a simple preamp to recover signal losses in C2, C3, and RV1.

TONE CONTROL

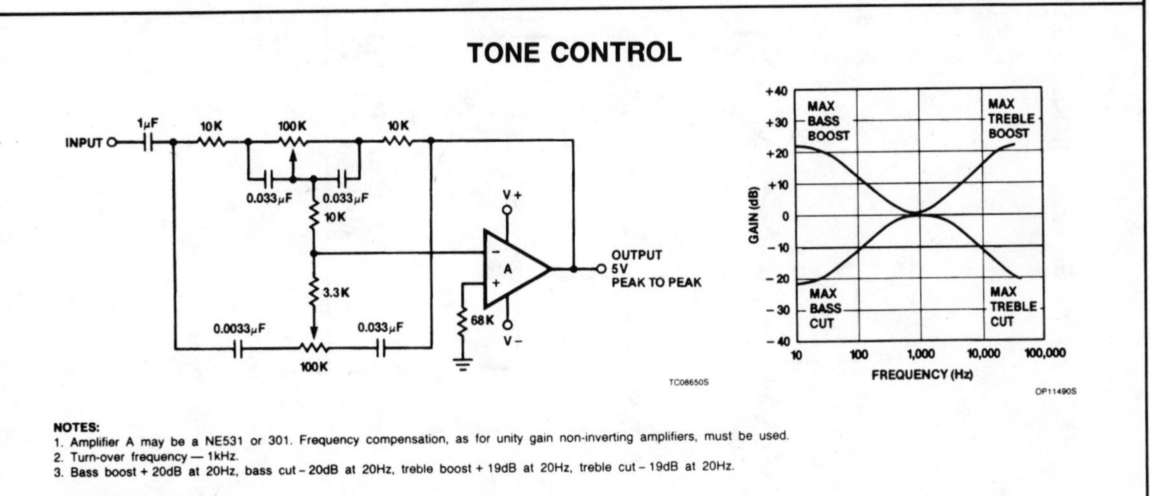

NOTES:
1. Amplifier A may be a NE531 or 301. Frequency compensation, as for unity gain non-inverting amplifiers, must be used.
2. Turn-over frequency — 1kHz.
3. Bass boost + 20dB at 20Hz, bass cut − 20dB at 20Hz, treble boost + 19dB at 20Hz, treble cut − 19dB at 20Hz.

All resistor values are in ohms.

SIGNETICS

Fig. 107-2

Circuit Notes

Tone control of audio systems involves altering the flat response in order to attain more low frequencies or more high ones, dependent upon listener preference. The circuit provides 20 dB of bass or treble boost or cut as set by the variable resistance. The actual response of the circuit is shown also.

TEN BAND GRAPHIC EQUALIZER, USING ACTIVE FILTERS

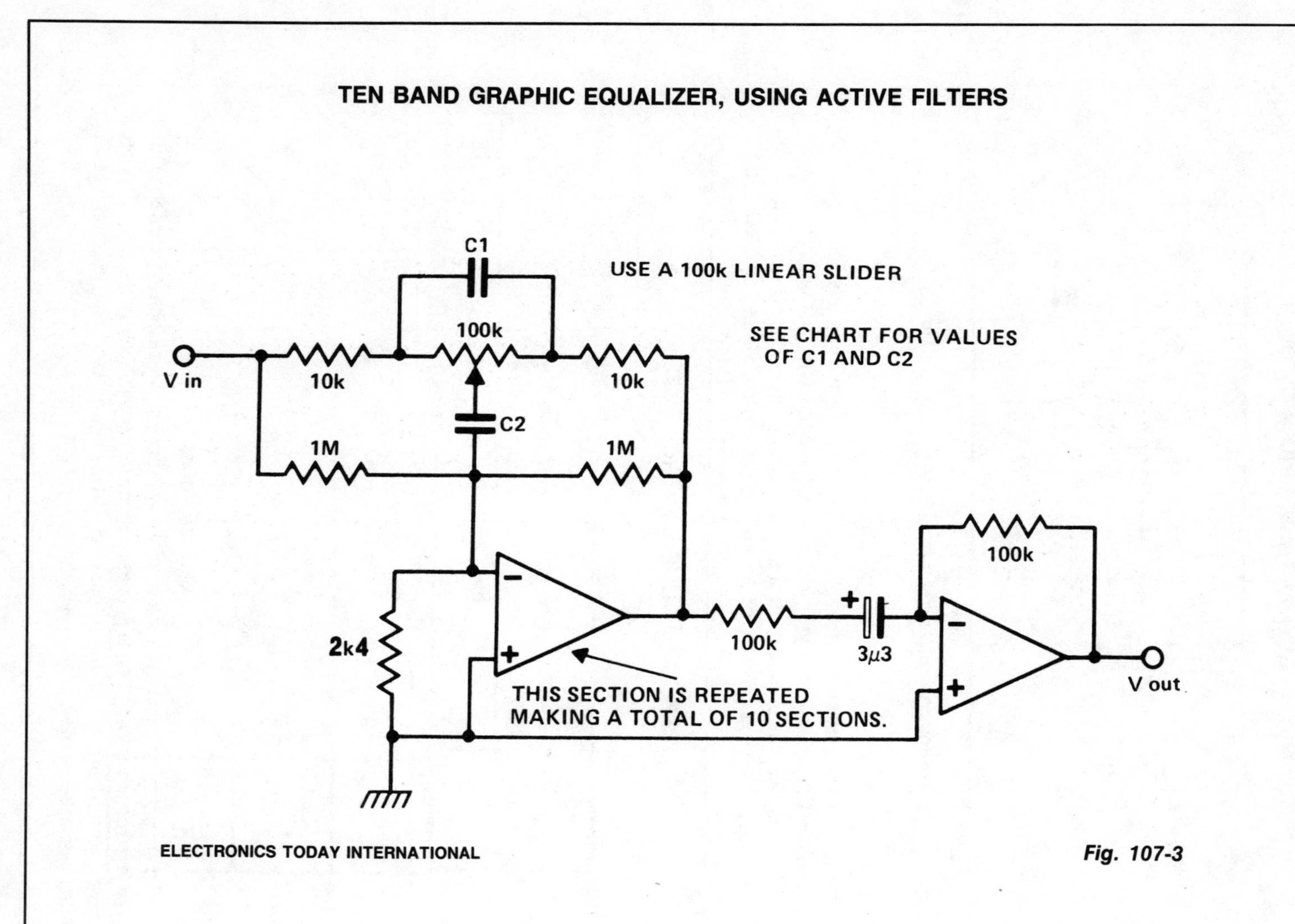

ELECTRONICS TODAY INTERNATIONAL

Fig. 107-3

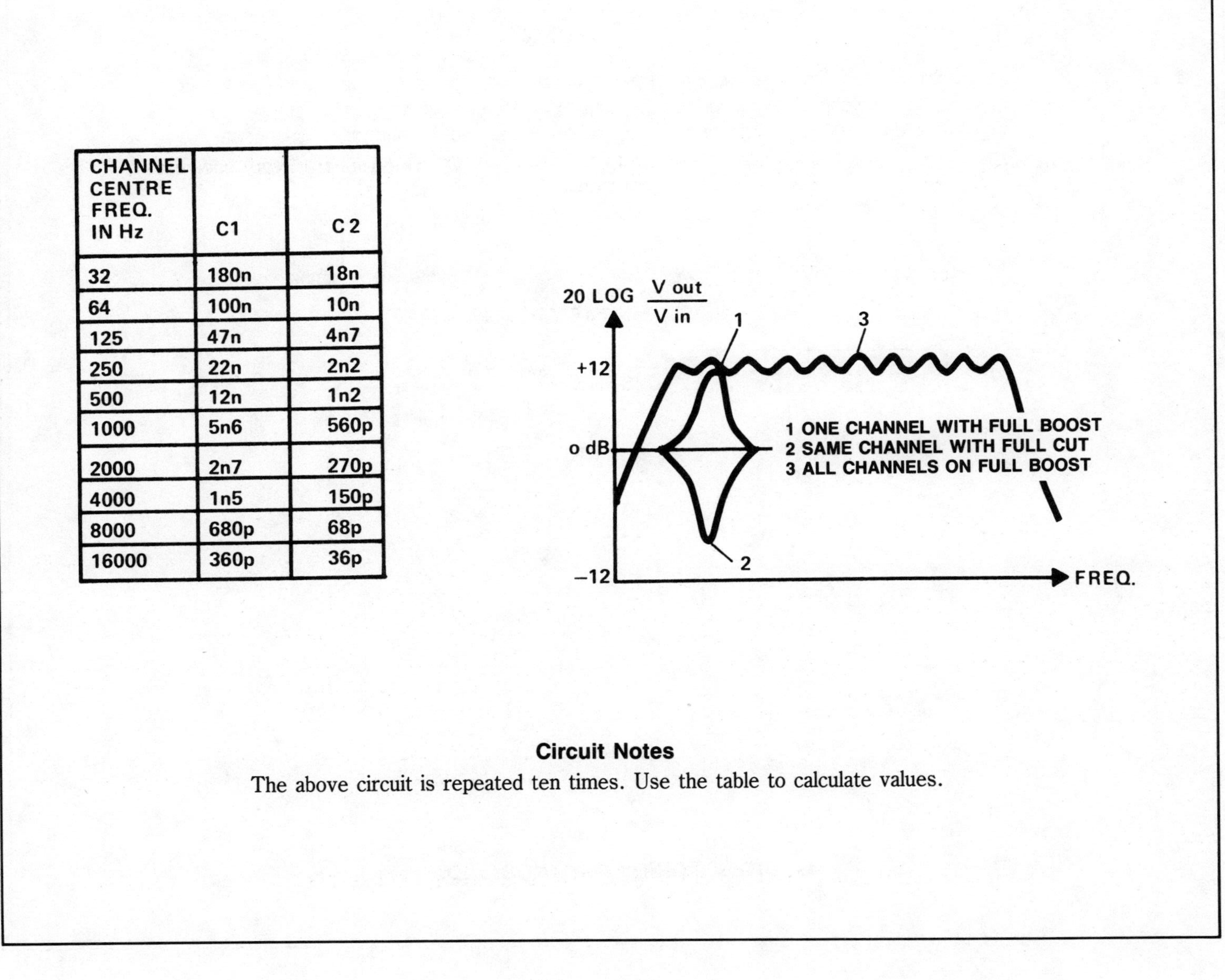

CHANNEL CENTRE FREQ. IN Hz	C1	C2
32	180n	18n
64	100n	10n
125	47n	4n7
250	22n	2n2
500	12n	1n2
1000	5n6	560p
2000	2n7	270p
4000	1n5	150p
8000	680p	68p
16000	360p	36p

Circuit Notes

The above circuit is repeated ten times. Use the table to calculate values.

TONE-CONTROL AUDIO AMPLIFIER

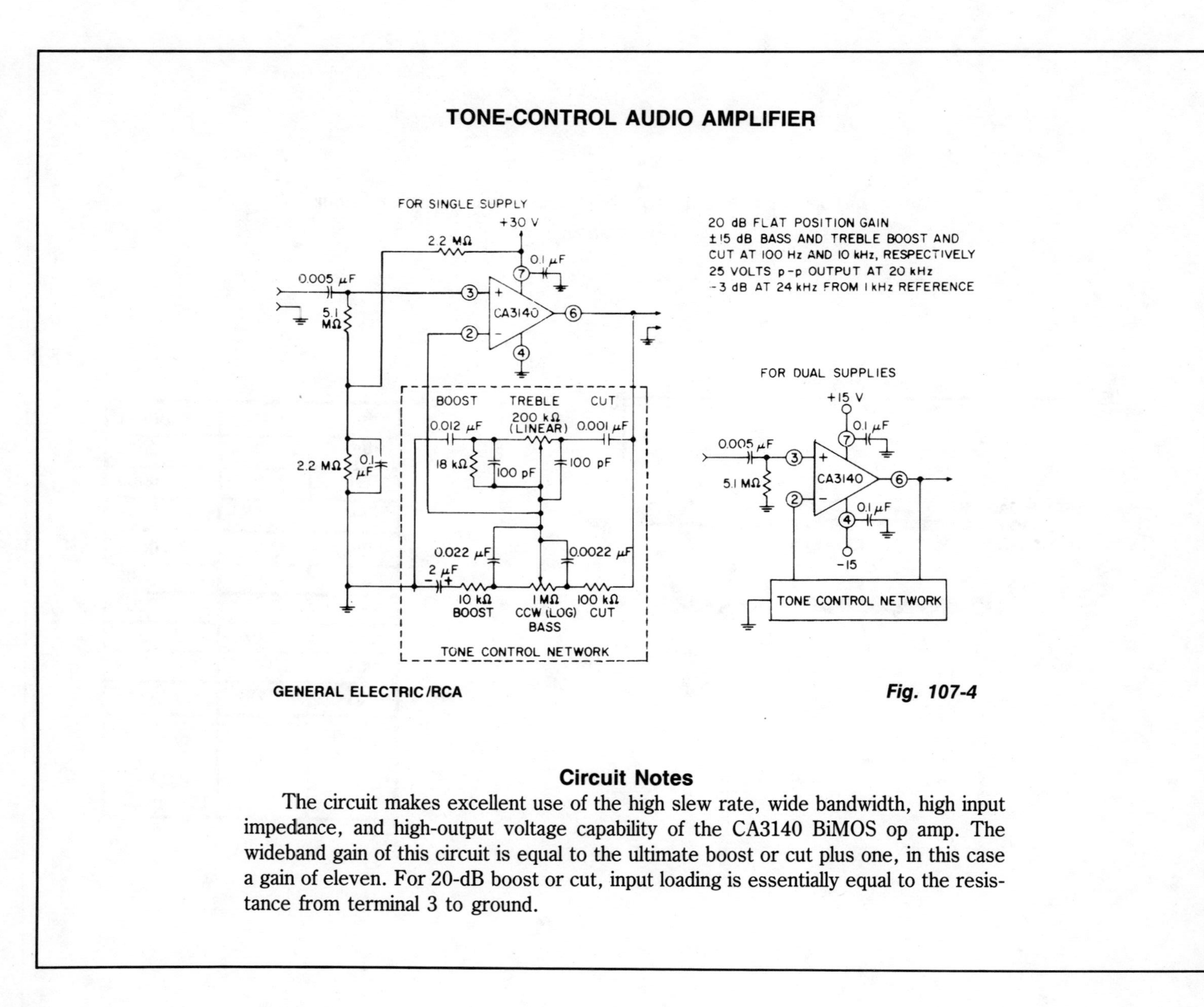

GENERAL ELECTRIC/RCA

Fig. 107-4

Circuit Notes

The circuit makes excellent use of the high slew rate, wide bandwidth, high input impedance, and high-output voltage capability of the CA3140 BiMOS op amp. The wideband gain of this circuit is equal to the ultimate boost or cut plus one, in this case a gain of eleven. For 20-dB boost or cut, input loading is essentially equal to the resistance from terminal 3 to ground.

MIKE PREAMP WITH TONE CONTROL

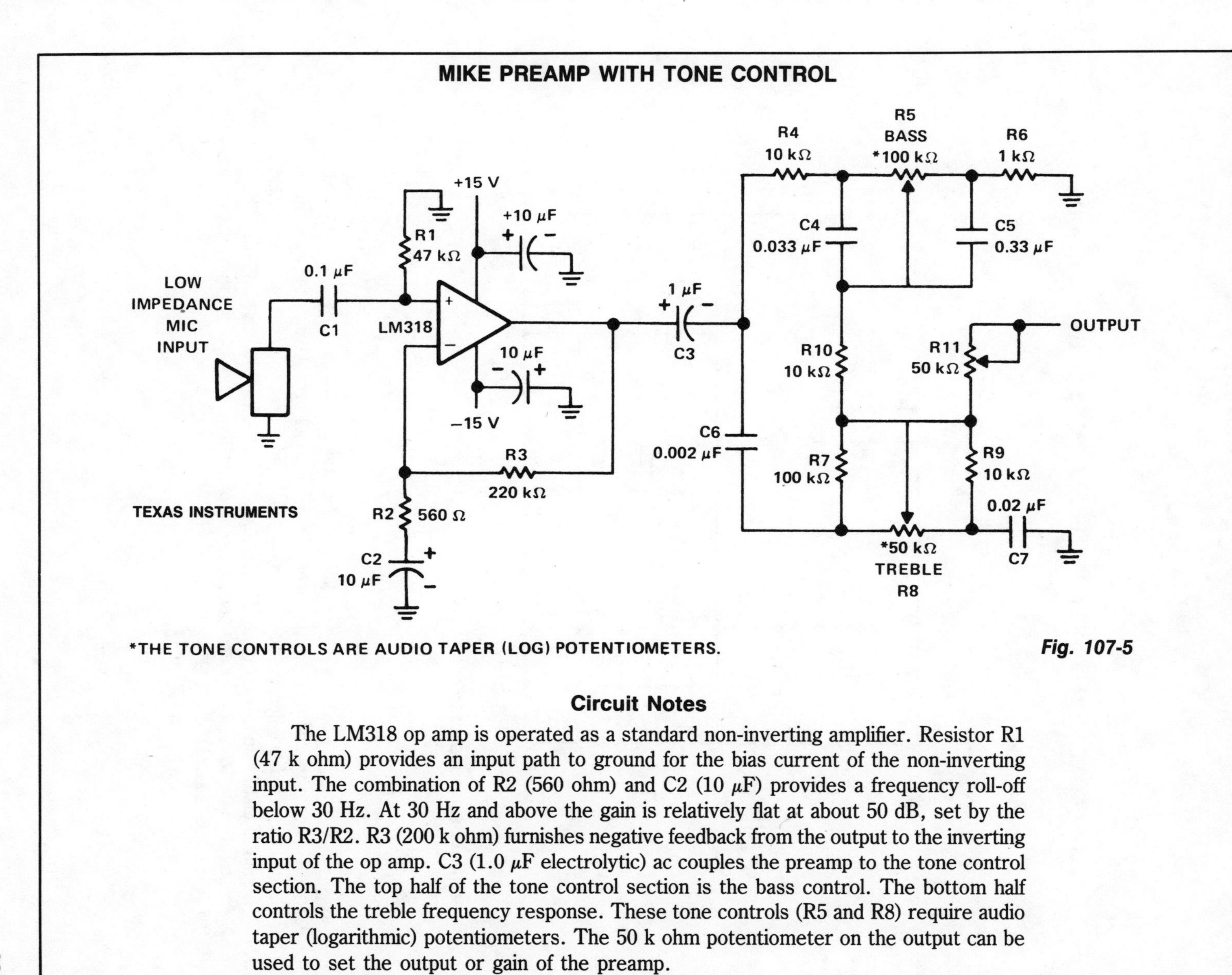

*THE TONE CONTROLS ARE AUDIO TAPER (LOG) POTENTIOMETERS.

Fig. 107-5

Circuit Notes

The LM318 op amp is operated as a standard non-inverting amplifier. Resistor R1 (47 k ohm) provides an input path to ground for the bias current of the non-inverting input. The combination of R2 (560 ohm) and C2 (10 μF) provides a frequency roll-off below 30 Hz. At 30 Hz and above the gain is relatively flat at about 50 dB, set by the ratio R3/R2. R3 (200 k ohm) furnishes negative feedback from the output to the inverting input of the op amp. C3 (1.0 μF electrolytic) ac couples the preamp to the tone control section. The top half of the tone control section is the bass control. The bottom half controls the treble frequency response. These tone controls (R5 and R8) require audio taper (logarithmic) potentiometers. The 50 k ohm potentiometer on the output can be used to set the output or gain of the preamp.

LOW COST HIGH-LEVEL PREAMP AND TONE CONTROL CIRCUIT

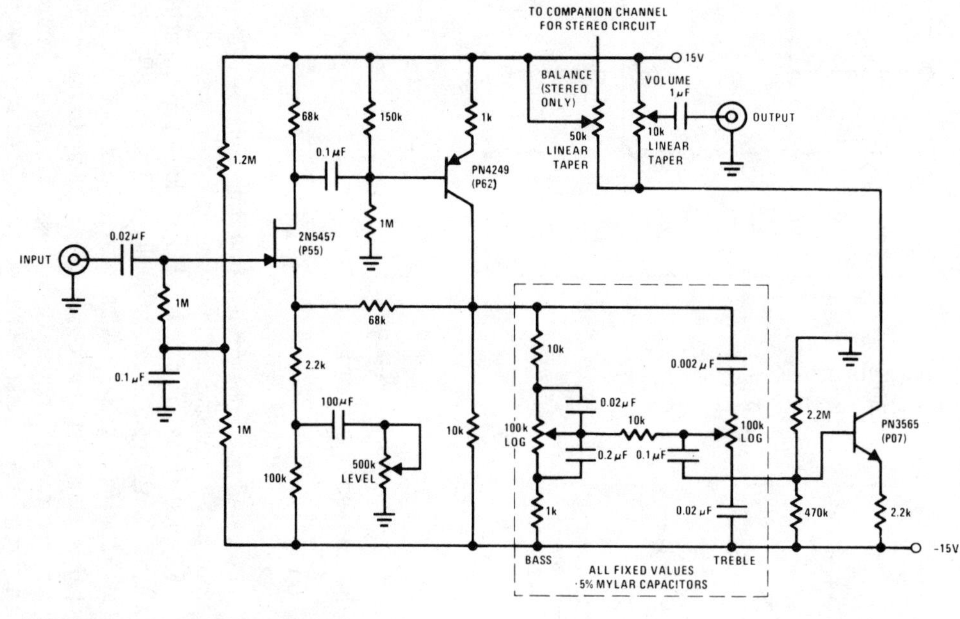

NATIONAL SEMICONDUCTOR CORP. ***Fig. 107-6***

Circuit Notes

This preamp and tone control uses the JFET to its best advantage; as a low noise high input impedance device. All device parameters are noncritical, yet the circuit achieves harmonic distortion levels of less than 0.05% with a S/N ratio of over 85 dB. The tone controls allow 18 dB of cut and boost; the amplifier has a 1-V output for 100-mV input at maximum level.

PASSIVE TONE-CONTROL CIRCUIT

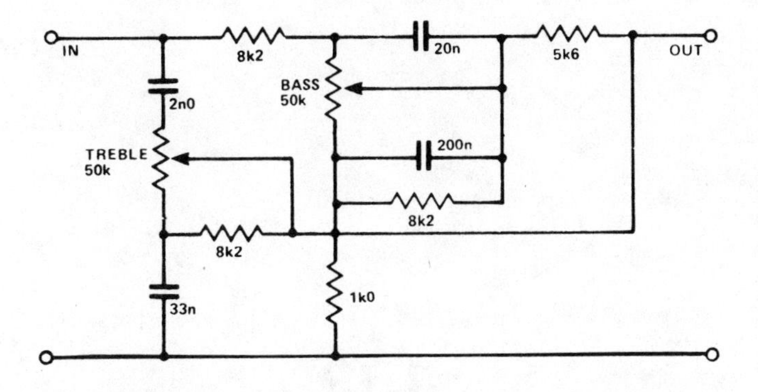

ELECTRONICS TODAY INTERNATIONAL *Fig. 107-7*

Circuit Notes

A simple circuit using two potentiometers and easily available standard value components provides tone control. The impedance level is suitable for low-level transistor or op amp circuitry.

108

Touch-Switch Circuits

The sources of the following circuits are contained in the Sources section beginning on page 694. The figure number contained in the box of each circuit correlates to the source entry in the Sources section.

TOUCH ON/OFF SWITCH

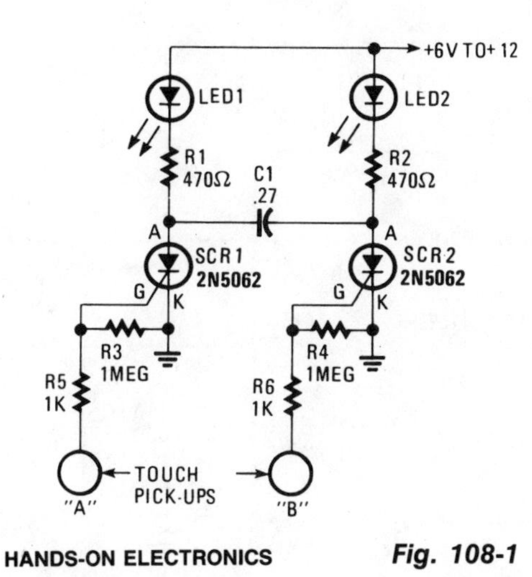

HANDS-ON ELECTRONICS *Fig. 108-1*

Circuit Notes

If a Touch On/Off Switch is desired, this circuit fills the bill. Two sensitive gate SCRs are interconnected, so that when one of the devices is turned on, the other (if on) is forced off. That toggling effect gives an on/off circuit condition for each of the LEDs in the SCR-anode circuits. To turn LED1 on and LED2 off, simply touch the "A" terminal, and to turn LED1 off and LED2 on, the "B" pick-up must be touched. It is possible to simultaneously touch both terminals, causing both SCRs to turn on together. To reset the circuit to the normal one-on/one-off condition, momentarily interrupt the circuit's dc power source. Additional circuitry can be connected to the anode circuit of either or both SCRs to be controlled by the on/off function of the touch switch.

TOUCH SWITCH

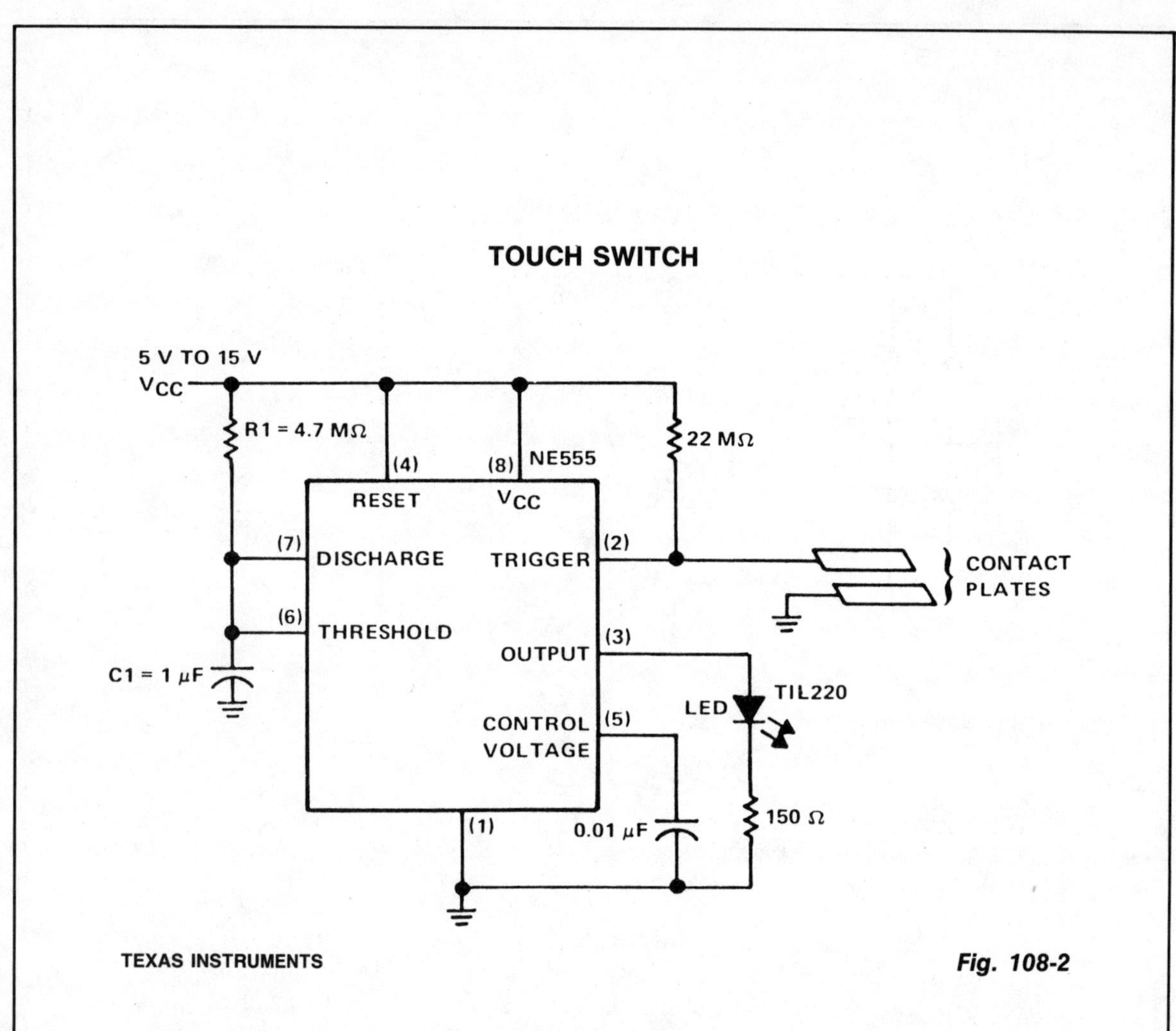

TEXAS INSTRUMENTS

Fig. 108-2

Circuit Notes

The circuit is basically a NE555 monostable, the only major difference being its method of triggering. The trigger input is biased to a high value by the 22 M ohm resistor. When the contact plates are touched, the skin resistance of the operator will lower the overall impedance from pin 2 to ground. This action will reduce the voltage at the trigger input to below the ⅓ V_{CC} trigger threshold and the timer will start. The output pulse width will be T = 1.1 R1C1, in this circuit about 5 seconds. A relay connected from pin 3 to ground instead of the LED and resistor could be used to perform a switching function.

TOUCHOMATIC

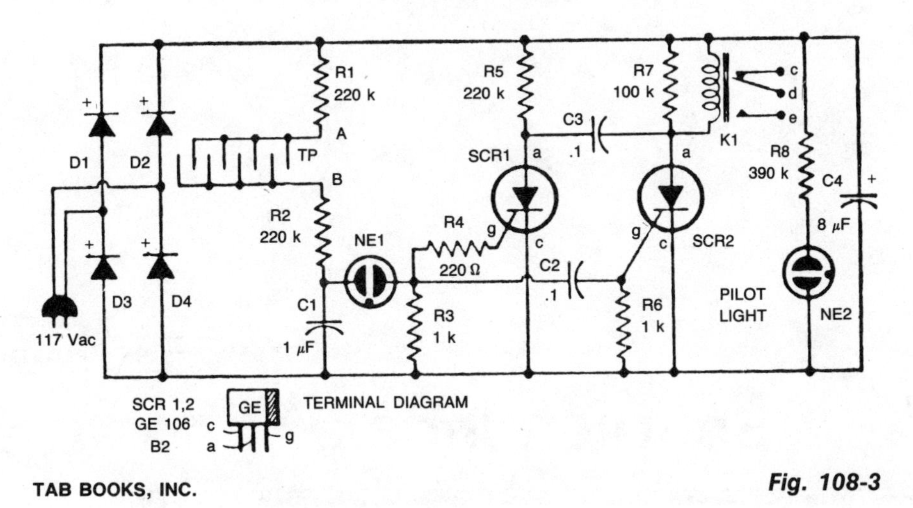

TAB BOOKS, INC.

Fig. 108-3

Circuit Notes

When someone touches the touchplate (TP), the resistance of his finger across points A and B is added in series to the combination of R1 and R2, the capacitor C2 begins to charge. When the voltage across C1 is finally sufficient to fire NE1, C1 will begin to discharge. When NE1 fires, it produces a short between its terminals. Since R3 is connected across C1, they are effectively in series after NE1 fires. A voltage spike will then be passed by C2 and this will act as a positive triggering pulse. The pulse is fed to both SCR gates: SCR2 conducts, thereby closing relay K1. With a finger no longer on the touchplate, no more pulses are forthcoming because the C1 charge path is open. The next contact with the touchplate will produce a pulse which triggers SCR1. SCR2 is now off by capacitor C3 which was charged by current passing through R6 and SCR2. The firing of SCR1 in this way places a negative voltage across SCR2 which momentarily drops the relay current to a point below the holding current value of SCR2. (Holding current is the minimum current an SCR requires to remain in a conducting state once its gate voltage is removed.) With SCR2 turned off, the relay will open and SCR1 will turn off due to the large resistance in series with its anode. Starved in this way SCR1 turns off because of a forced lack of holding current.

Sources Index

Chapter 1

Fig. 1-1. Signetics 1987 Linear Data Manual Vol. 2, 2/87, p.7-65.

Fig. 1-2. General Electric Application Note 90.16, p. 25.

Fig. 1-3. Courtesy, William Sheets.

Fig. 1-4. General Electric Application Note 90.16, p. 25.

Fig. 1-5. General Electric/RCA, BiMOS Operational Amplifiers Circuit Ideas, 1987, p. 27.

Fig. 1-6. R-E Experimenters Handbook, p. 157.

Chapter 2

Fig. 2-1. Linear Technology Corporation, Linear Applications Handbook, 1987, p. AN21-2

Fig. 2-2. Linear Technology Corporation, Linear Databook Supplement, 1988, p. S2-34.

Fig. 2-3. Siliconix, Integrated Circuits Data Book, 3/85, p. 10-154.

Fig. 2-4. Linear Technology Corp., Linear Databook, 1986, p. 2-83.

Fig. 2-5. National Semiconductor Corp., Transistor Databook, 1982, p. 11-23.

Fig. 2-6. Linear Technology Corp., Linear Applications Handbook, 1987, p. AN21-1.

Fig. 2-7. Signetics, 1987 Linear Data Manual, Vol. 2: Industrial, 10/86, p. 4-260.

Fig. 2-8. Electronic Engineering, 11/86, p. 40.

Fig. 2-9. National Semiconductor Corp., Transistor Databook, 1982, p. 11-25.

Fig. 2-10. Siliconix, Integrated Circuits Data Book, 3/85, p. 2-112.

Fig. 2-11. Electronics Engineering, 9/78, p. 17.

Fig. 2-12. Electronics Engineering, 9/84, p. 33.

Fig. 2-13. Signetics, 1987 Linear Data Manual, Vol. 1: Communications, 8/87, p. 4-346.

Fig. 2-14. NASA Tech Briefs, Spring 1983, p. 244.

Fig. 2-15. Radio Electronics, 7/83, p. 74.

Fig. 2-16. Motorola Inc., Linear Integrated Circuits, 1979, p. 6-58.

Fig. 2-17. MR-E Experimenters Handbook, p. 158.

Fig. 2-19. Popular Electronics, 8/68.
Fig. 2-20. Ham Radio, 9/84, p. 24.
Fig. 2-21. General Electric/RCA, BiMOS Operational Amplifiers Circuit Ideas, 1987, p. 20.
Fig. 2-22. Texas Instruments, Linear and Interface Circuits Applications, Vol. 1, 1985, p. 3-2, 3-4.
Fig. 2-23. Signetics, 1987 Linear Data Manual Vol. 1: Communications, 3/87, p. 4-345.
Fig. 2-24. Electronic Engineering, 11/85, p. 32.
Fig. 2-25. NASA Tech Briefs, Sept/Oct 1986, p. 43.

Chapter 3

Fig. 3-1. Intersil, Component Data Catalog, 1987, p. 4-43.
Fig. 3-2. Linear Technology, Application Note 9, p. 16.
Fig. 3-3. General Electric/RCA, BiMOS Operational Amplifiers Circuit Ideas, 1987, p. 26.
Fig. 3-4. Linear Technology Corp., Linear Applications Handbook, 1987, p. AN15-2.
Fig. 3-5. Linear Technology Corp., Linear Databook, 1986, p. 5-17.
Fig. 3-6. Signetics, 1987 Linear Data Manual, Vol. 2: Industrial, 11/86, p. 5-215.
Fig. 3-7. National Semiconductor Corp., 1984 Linear Supplement Databook, p. S5-126.
Fig. 3-8. Signetics, 1987 Linear Data Manual Vol. 2: Industrial, 12/86, p. 4-67.

Chapter 4

Fig. 4-1. Electronic Design, 5/82, p. 214.
Fig. 4-2. R-E Experimenters Handbook, p. 160.
Fig. 4-3. Popular Electronics, 11/73, p. 50.

Chapter 5

Fig. 5-1. Hands-On Electronics, Summer 1984, p. 77.
Fig. 5-2. Courtesy, William Sheets.
Fig. 5-3. Courtesy, William Sheets.
Fig. 5-4. Electronic Engineering, 5/84, p. 44
Fig. 5-5. Texas Instruments, Linear and Interface Circuits Applications, vol. 1, 1985, p. 3-13.
Fig. 5-6. Electronics Today International, 4/85, p. 82.

Chapter 6

Fig. 6-1. Tab Books Inc., 101 Sound, Light, and Power IC Projects.
Fig. 6-2. Radio Electronics, 7/70, p. 38.
Fig. 6-3. Hands-On Electronics, Jul/Aug 1986, p. 16.
Fig. 6-4. 73 Magazine, 12/76, p. 170.
Fig. 6-5. General Electric/RCA, BiMOS Operational Amplifiers Circuit Ideas, 1987, p. 21.
Fig. 6-6. Ibid.
Fig. 6-7. Texas Instruments, Linear and Interface Circuits Applications, Vol. 1, 1985, p. 3-17.
Fig. 6-8. Signetics, Analog Data Manual, 1982, p. 3-90.

Chapter 7

Fig. 7-1. Motorola, TMOS Power FET Design Ideas, 1985.
Fig. 7-2. Hands-On Electronics, 4/87, p. 95.
Fig. 7-3. Tab Books, Inc., The Giant book of Easy-To-Build Electronics Projects, 1982, p. 196.
Fig. 7-4. General Electric, Optoelectronics, Third Edition, p. 151.
Fig. 7-5. Tab Books, Inc., The Build-It Book of Electronic Projects, No. 1498, p. 28.
Fig. 7-6. Radio-Electronics, 6/85, p. 60.
Fig. 7-7. Popular Electronics, 12/74, p. 6.
Fig. 7-8. Popular Electronics, 4/75, p. 68.
Fig. 7-9. Texas Instruments, Linear and Interface Circuits Applications, 1987, p. 10-21.
Fig. 7-10. Texas Instruments, Linear and Interface Circuits Applications, 1985, vol. 1, p. 3-5.
Fig. 7-11. Radio-Electronics, 1979.
Fig. 7-12. Hands-On Electronics, 1/87, p. 30.
Fig. 7-13. Electronic Engineering, 12/75, p. 9.
Fig. 7-14. Electronic Design 18, 9/76, p. 114.
Fig. 7-15. Radio-Electronics, 5/87, p. 10.
Fig. 7-16. Hands-On Electronics, Fall 1984, p. 45.
Fig. 7-17. Hands-On Electronics, 4/87, p. 92.
Fig. 7-18. General Electric, Optoelectronics, Third Edition, p. 105.

Chapter 8

Fig. 8-1. CQ, 7/82, p. 18.
Fig. 8-2. Electronics Design, 7/76, p. 120.
Fig. 8-3. Courtesy, William Sheets.
Fig. 8-4. Moli Energy Limited.
Fig. 8-5. Linear Technology Corp., Linear Applications Handbook, 1987, p. AN6-3.
Fig. 8-6. Courtesy, William Sheets.
Fig. 8-7. Texas Instruments, Linear and Interface Circuits Applications, Vol. 1, p. 6-24.
Fig. 8-8. Linear Technology Corp., Linear Databook Supplement, 1988, p. S5-11.
Fig. 8-9. Motorola, TMOS Power FET Design Ideas, 1985, p. 8.
Fig. 8-10. Radio Electronics, 9/85, p. 44.
Fig. 8-11. Siliconix, MOSpower Applications Handbook, p. 6-176.

Chapter 9

Fig. 9-1. Motorola, TMOS Power FET Ideas, 1985, p. 7.
Fig. 9-2. Electronic Engineering, 2/85, p. 45.
Fig. 9-3. Electronic Engineering, 10/70, p. 17.
Fig. 9-4. Electronic Engineering, Mid5/78, p. 11.
Fig. 9-5. Moli Energy Limited, Publication MEL-126.
Fig. 9-6. Linear Technology Corp., Linear Databook, 1986, p. 2-104.
Fig. 9-7. Courtesy, William Sheets.
Fig. 9-8. Tab Books, Inc., 101 Sound, Light, and Power IC Projects.

Chapter 10

Fig. 10-1. Signetics, 1987 Linear Data Manual, Vol. 2: Industrial, 2/87, p. 5-367.
Fig. 10-2. Texas Instruments, Linear and Interface Circuits Applications, Vol. 1, 1985, p. 3-3, 3-4.
Fig. 10-3. Courtesy, William Sheets.
Fig. 10-4. General Electric/RCA, BiMOS Operational Amplifiers Circuit Ideas, 1987, p. 17.
Fig. 10-5. Siliconix, Small-Signal FET Data Book, 1/86, p. 7-29.
Fig. 10-6. Linear Technology Corp., Linear Databook, 1986, p. 2-83.
Fig. 10-7. Linear Technology Corp., Linear Databook, 1986, p. 2-101.

Chapter 11

Fig. 11-1. Signetics, 1987 Linear Data Manual, Vol. 1: Communications, 2/87, p. 4-312.
Fig. 11-2. Electronic Engineering, 5/86, p. 50.
Fig. 11-3. Electronic Design, 9/73, p. 148.
Fig. 11-4. Intersil, Component Data Catalog, 1987, p. 6-28.
Fig. 11-5. Signetics, 555 Timers, 1973, p. 19.

Chapter 12

Fig. 12-1. Texas Instruments, Linear and Interface Circuits Applications, Vol. 1, 1985, p. 7-21.
Fig. 12-2. Electronic Engineering, 2/85, p. 34.

Chapter 13

Fig. 13-1. Hands-On Electronics, 3/87, p. 25.
Fig. 13-2. Motorola, TMOS Power FET Design Ideas, 1985, p. 17.
Fig. 13-3. Linear Technology Corp., Linear Applications Handbook, 1987, p. AN13-23.
Fig. 13-4. Intersil, Component Data Catalog, 1987, p. 5-113.
Fig. 13-5. Courtesy, William Sheets.
Fig. 13-6. Linear Technology Corp., Linear Databook, 1986, p. 3-23.
Fig. 13-7. Motorola, Thyristor Device Data, Series A, 1985, p. 1-6-57.

Chapter 14

Fig. 14-1. Electronic Engineering, 3/78, p. 38.
Fig. 14-2. Hands-On Electronics, Nov/Dec 1985, p. 4.

Chapter 15

Fig. 15-1. Linear Technology Corp., Linear Databook, 1986, p. 2-82.
Fig. 15-2. Popular Electronics, 9/77, p. 92.
Fig. 15-3. Signetics, 1987 Linear Data Manual Vol. 2: Industrial, 11/86, p. 5-269.
Fig. 15-4. Siliconix, Integrated Circuits Data Book, 3/85, p. 5-8.
Fig. 15-5. Electronic Engineering, 2/85, p. 45.
Fig. 15-6. Electronic Engineering, 2/84, p. 36.
Fig. 15-7. Electronic Design 15, 7/79, p. 120.
Fig. 15-8. Electronic Engineering, 12/78, p. 17.

Fig. 15-9. General Electric/RCA, BiMOS Operational Amplifiers Circuit Ideas, 1987, p. 23.
Fig. 15-10. Electronic Engineering, 11/86, p. 39.
Fig. 15-11. Popular Electronics, 3/79, p. 77.
Fig. 15-12. Electronic Engineering, 7/86, p. 27.
Fig. 15-13. Electronic Engineering, 1/86, p. 37.

Chapter 16

Fig. 16-1. Signetics, 1987 Linear Data Manual Vol. 2: Industrial, 11/86, p. 5-215.
Fig. 16-2. Linear Technology Corp., Linear Databook, 1986, p. 4-15.
Fig. 16-3. Siliconix, Integrated Circuits Data Book, 3/85, p. 2-207.
Fig. 16-4. Siliconix, Integrated Circuits Data Book, 3/85, p. 2-231.
Fig. 16-5. Intersil, Component Data Catalog, 1987, p. 13-51.
Fig. 16-6. Siliconix, Integrated Circuits Data Book, 3/85, p. 3-62.
Fig. 16-7. Datel, Data Conversion Components, p. 4-37.

Chapter 17

Fig. 17-1. Signetics, 1987 Linear Data Manual Vol. 2: Industrial, 2/87, p. 7-62.
Fig. 17-2. Signetics, 1987 Linear Data Manual Vol. 2: Industrial, 11/86, p. 4-136.
Fig. 17-3. Signetics, 1987 Linear Data Manual Vol. 2: Industrial, 11/86, p. 5-269.
Fig. 17-4. Electronic Engineering, 12/77, p. 19.
Fig. 17-5. Siliconix, Integrated Circuits Data Book, 3/85, p. 5-17.
Fig. 17-6. Linear Technology Corp., Linear Databook, 1986, p. 5-17.
Fig. 17-7. Analog Devices, Data Acquisition Databook, 1982, p. 4-56.
Fig. 17-8. Signetics, 1987 Linear Data Manual Vol. 1: Communications, 2/87, p. 4-311.
Fig. 17-9. General Electric/RCA, BiMOS Operational Amplifiers Circuit Ideas, 1987, p. 20.
Fig. 17-10. General Electric/RCA, BiMOS Operational Amplifiers Circuit Ideas, 1987, p. 11.
Fig. 17-11. Siliconix, MOSpower Applications Handbook, p. 6-178.
Fig. 17-12. Signetics, 1987 Linear Data Manual Vol. 2: Industrial, 2/87, p. 5-368.
Fig. 17-13. National Semiconductor Corp., Transistor Databook, 1982, p. 7-27.
Fig. 17-14. National Semiconductor Crop., 1984 Linear Supplemental Databook, p. S5-142.
Fig. 17-15. Electronic Engineering, 8/83, p. 141.

Chapter 18

Fig. 18-1. Intersil, Component Data catalog, 1987, p. 14-70.
Fig. 18-2. General Electric, Application Note 90.16, p. 29.
Fig. 18-3. General Electric, Application Note 90.16, p. 29.
Fig. 18-4. General Electric, Application Note 90.16, p. 28.
Fig. 18-5. Intersil, Component Data Catalog, 1987, p. 14-91.
Fig. 18-6. Intersil, Component Data Catalog, 1987, p. 7-96.
Fig. 18-7. Intersil, Component Data Catalog, 1987, p. 14-121.
Fig. 18-8. Intersil, Databook 1987, p. 7-47.

Chapter 19

Fig. 19-1. Ham Radio, 6/85, p. 23.
Fig. 19-2. Siliconix, MOSpower Design Catalog, 1/83, p. 5-27.
Fig. 19-3. Linear Technology Corp., Linear Databook, 1986, p. 2-104.
Fig. 19-4. QST, 12/85, p. 38.
Fig. 19-5. Linear Technology Corp., Linear Applications Handbook 1987, p. AN20-12.
Fig. 19-6. QST, 2/28, p. 43.
Fig. 19-7. Ham Radio, 2/79, p. 40.
Fig. 19-8. Courtesy, William Sheets.
Fig. 19-9. Courtesy, William Sheets.
Fig. 19-10. Electronic Design 21, 10/75, p. 98.
Fig. 19-11. Ham Radio, 2/79, p. 40.
Fig. 19-12. Ham Radio, 2/79, p. 42.
Fig. 19-13. Tab Books, Inc., The Complete Handbook of Amplifiers, Oscillators, and Multivibrators, No. 1230, p. 328.
Fig. 19-14. Signetics, 1987 Linear Data Manual Vol. 2: Industrial, 11/86, p. 5-269.

Fig. 19-15. Motorola, MECL System Design Handbook, 1983, p. 227.
Fig. 19-16. Electronic Design, 11/69, p. 109.
Fig. 19-17. QST, 1/86, p. 40.
Fig. 19-18. R-E Experimenters Handbook, p. 157.
Fig. 19-19. Electronic Design 23, 11/74, p. 148.

Chapter 20

Fig. 20-1. Linear Technology Corp., 1986 Linear Databook, p. 2-57.
Fig. 20-2. Linear Technology Corp., Linear Applications Handbook, 1987, p. AN3-13.
Fig. 20-3. Intersil, Component Data Catalog, 1987, p. 7-4.
Fig. 20-4. NASA Tech Briefs, Jul/Aug 1986, p. 37.
Fig. 20-5. General Electric/RCA, BiMOS Operational Amplifiers Circuit Ideas, 1987, p. 17.
Fig. 20-6. Linear Technology Corp., Linear Databook, 1986, p. 2-85.
Fig. 20-7. General Electric/RCA, BiMOS Operational Amplifiers Circuit Ideas, 1987, p. 14.

Chapter 21

Fig. 21-1. EXAR, Telecommunications Databook, 1986, p. 9-23.
Fig. 21-2. National Semiconductor Corp., Audio/Radio Handbook, 1980, p. 3-17.
Fig. 21-3. Signetics, 1987 Linear Data Manual, Vol. 1: Communications, 2/87, p. 4-66.
Fig. 21-4. Signetics, 1987 Linear Data Manual Vol. 1: Communications, 11/86, p. 4-263.

Chapter 22

Fig. 22-1. Radio-Electronics, 12/86, p. 57.
Fig. 22-2. Radio-Electronics, 8/87, p. 63.
Fig. 22-3. Radio-Electronics, 8/87, p. 53.
Fig. 22-4. Radio-Electronics, 6/87, p. 12.
Fig. 22-5. EXAR, Telecommunications Databook, 1986, p. 9-23.
Fig. 22-6. Radio-Electronics, 3/86, p. 51.
Fig. 22-7. Signetics, 1987 Linear Data Manual Vol. 1: Communications, 11/86, p. 7-123.
Fig. 22-8. Signetics, 1987 Linear Data Manual Vol. 1: Communications, 11/86, p. 7-123.
Fig. 22-9. Signetics, 1987 Linear Data Manual Vol. 1: Communications, 11/86, p. 4-295.

Chapter 23

Fig. 23-1. General Electric, Application Note 90.16, p. 26.
Fig. 23-2. Siliconix, Integrated Circuits Data Book, 3/85, p. 5-16.
Fig. 23-3. Texas Instruments, Linear and Interface Circuits Applications, Vol. 1, 1985, p. 3-23.
Fig. 23-4. General Electric/RCA, BiMOS Operational Amplifiers Circuit Ideas, 1987, p. 18.
Fig. 23-5. Intersil, Component Data Catalog, 1987, p. 7-44.
Fig. 23-6. Electronic Engineering, 11/86, p. 39.
Fig. 23-7. General Electric, Application Note 90.16, p. 27.
Fig. 23-8. Signetics, 1987 Linear Data Manual Vol. 2: Industrial, 2/87, p. 5-367.
Fig. 23-9. Intersil, Component Data Catalog, 1987, p. 5-112.
Fig. 23-10. GENERAL Electric, Application Note 90.16, p. 26.

Chapter 24

Fig. 24-1. Electronic Engineering, 8/85, p. 30.
Fig. 24-2. Electronic Engineering, 11/86, p. 40.

Chapter 25

Fig. 25-1. CQ, 1/87, p. 36.

Chapter 26

Fig. 26-1. General Electric/RCA, BiMOS Operational Amplifiers Circuit Ideas, 1987, p. 26.
Fig. 26-2. National Semiconductor Corp., Linear Databook, 1982, p. 171.
Fig. 26-3. Electronic Engineering, 9/84, p. 30.
Fig. 26-4. GENERAL Electric/RCA, BiMOS Operational Amplifiers Circuit Ideas, 1987, p. 12.

Chapter 27

Fig. 27-1. Motorola, *TMOS Power FET Design Ideas*, 1985, p. 18.

Fig. 27-2. Electronic Design, 12/87, p. 67.

Fig. 27-3. General Electric/RCA, *BiMOS Operational Amplifiers Circuit Ideas*, 1987, p. 22.

Fig. 27-4. Electronic Engineering, 6/78, p. 32.

Fig. 27-5. Electronic Engineering, 2/83, p. 37.

Chapter 28

Fig. 28-1. Tab Books, Inc., *The Giant Book of Easy-To-Build Electronic Projects*, 1982, p. 53.

Fig. 28-2. Hands-On Electronics, 2/87, p. 38.

Chapter 29

Fig. 29-1. Signetics, 1987 Linear Data Manual Vol. 2: Industrial, 11/86, p. 4-135.

Fig. 29-2. Linear Technology Corp., Linear Databook, 1986, p. 2-82.

Fig. 29-3. Signetics, 1987 Linear Data Manual Vol. 2: Industrial, 11/86, p. 4-135.

Fig. 29-4. Transistor Databook, 1982, p. 11-25.

Chapter 30

Fig. 30-1. Hands-On Electronics, May/Jun 1986, p. 52.

Fig. 30-2. Popular Electronics, 3/67.

Fig. 30-3. Courtesy, William Sheets.

Chapter 31

Fig. 31-1. Linear Technology Corp., Linear Applications Handbook, 1987, p. AN13-22.

Fig. 31-2. General Electric, Optoelectronics, Third Edition, p. 149.

Fig. 31-3. Motorola, Thyristor Device Data, Series A, 1985, p. 1-6-39.

Chapter 32

Fig. 32-1. Courtesy, William Sheets.

Fig. 32-2. Courtesy, William Sheets.

Fig. 32-3. Hands-On Electronics, 8/87, p. 65.

Fig. 32-4. Courtesy, William Sheets.

Fig. 32-5. Hands-On Electronics, 3/87, p. 27.

Fig. 32-6. Ham Radio, 9/86, p. 67.

Chapter 33

Fig. 33-1. Electronic Engineering, 10/48, p. 45.

Fig. 33-2. Electronics Today International, 10/78, p. 26.

Fig. 33-3. Hybrid Products Databook, 1982, p. 17-131.

Fig. 33-4. Siliconix, Integrated Circuits Data Book, 3/85, p. 10-62.

Fig. 33-5. Intersil, Component Data Catalog, 1987, p. 8-102.

Fig. 33-6. 73 for Radio Amateurs, 2/86, p. 10.

Fig. 33-7. Intersil, Component Data Catalog, 1987, p. 7-45.

Fig. 33-8. Linear Technology Corp., 1986 Linear Databook, p. 2-56.

Fig. 33-9. Electronic Engineering, 2/47, p. 47.

Fig. 33-10. Courtesy, William Sheets.

Fig. 33-11. Raytheon, Linear and Integrated Circuits, 1984, p. 6-205.

Fig. 33-12. Texas Instruments, Linear and Interface Circuits Applications, Vol. 1, 1985, p. 3-7.

Fig. 33-13. Motorola, Linear Integrated Circuits, 1979, p. 3-147.

Fig. 33-14. Texas Instruments, Linear and Interface Circuits Applications, Vol. 1, 1985, p. 3-9.

Chapter 34

Fig. 34-1. Motorola, Thyristor Device Data, Series A, 1985, p. 1-6-52.

Fig. 34-2. Popular Electronics, 3/81, p. 100.

Fig. 34-3. Hands-On Electronics, Spring 1986, p. 4.

Fig. 34-4. Hands-On Electronics, Fall 1984, p. 61.

Fig. 34-5. General Electric, SCR Manual, Fourth Edition, p. 85.

Fig. 34-6. Electronic Design 65, 3/73, p. 84.

Fig. 34-7. Electronic Design, 3/69, p. 96.

Fig. 34-8. Electronic Engineering, 676, p. 32.

Fig. 34-9. Popular Electronics, 3/75, p. 78.

Fig. 34-10. General Electric, Optoelectronics, Third Edition.

Fig. 34-11. Radio-Electronics, 2/87, p. 36.

Fig. 34-12. General Electric, Application Note 200.35, p. 16.

Fig. 34-13. General Electric, Application Note 90. 16, p. 27.

Fig. 34-14. National Semiconductor Corp., CMOS Databook, 1981, p. 8-45.
Fig. 34-15. General Electric, Application Note 90.25.
Fig. 34-16. Motorola, Circuit Applications for the Trian (AN-466), p. 11.
Fig. 34-17. Siliconix, MOSpower Applications Handbook, p. 6-181.
Fig. 34-18. Popular Electronics, 3/75, p. 78.

Chapter 35

Fig. 35-1. Linear Technology Corp., Linear Applications Handbook, 1987, p. AN5-6.
Fig. 35-2. Linear Technology Corp., Linear Databook, 1986, p. 2-82.

Chapter 36

Fig. 36-1. Courtesy, William Sheets.
Fig. 36-2. Electronics Today International, 6/76, p. 43.
Fig. 36-3. Radio-Electronics, 4/87, p. 48.
Fig. 36-4. Radio-Electronics, 2/84, p. 97.
Fig. 36-5. Hands-On Electronics, Sep/Oct 1986, p. 24.
Fig. 36-6. Electronic Engineering, 9/86, p. 37.
Fig. 36-7. R-E Experimenters Handbook, p. 162.
Fig. 36-8. Linear Technology Corp., Linear Databook, 1986, p. 2-96.

Chapter 37

Fig. 37-1. Electronic Design, 3/75, p. 68.
Fig. 37-2. EXAR, Telecommunications Databook, 1986, p. 11-38.

Chapter 38

Fig. 38-1. Electronic Design, 8/73, p. 86.
Fig. 38-2. Electronic Design, 12/78, p. 98.
Fig. 38-3. Motorola, Thyristor Device Data, SEries A, 1985, p. 1-6-53.
Fig. 38-4. Signetics, 1987 Linear Data Manual Vol. 2: Industrial, 2/87, p. 7-59.

Chapter 39

Fig. 39-1. National Semiconductor Corp., 1984 Linear Supplement Databook, p. S5-143.

Chapter 40

Fig. 40-1. Electronic Design, 6/79, p. 122.
Fig. 40-2. Electronic Engineering, 9/84, p. 37.
Fig. 40-3. NASA Tech Briefs, 6/87, p. 26.
Fig. 40-4. Electronics Today International, 6/80, p. 68.
Fig. 40-5. Electronic Engineering, 9/87, p. 27.
Fig. 40-6. Texas Instruments, Linear and Interface Circuits Applications, Vol. 1, 1985, p. 7-25.
Fig. 40-7. Radio-Electronics, 5/70, p. 33.
Fig. 40-8. Linear Technology Corp., Linear Databook, 1986, p. 5-78.
Fig. 40-9. Linear Technology Corp., Linear Databook, 1986, p. 8-40.
Fig. 40-10. Electronic Engineering, 2/79, p. 23.
Fig. 40-11. Electronic Engineering, 7/86, p. 30.
Fig. 40-12. Motorola, Application Note AN-294, p. 6.
Fig. 40-13. Motorola, Linear Integrated Circuits, p. 3-139.
Fig. 40-14. Hands-On Electronics, Winter 1985, p. 60.
Fig. 40-15. Texas Instruments, Linear and Interface Circuits Applications, Vol. 1, 1985, p. 7-16.
Fig. 40-16. Signetics, Analog Data Manual, 1982, p. 3-39.
Fig. 40-17. General Electric/RCA, BiMOS Operational Amplifier Circuit Ideas, 1987, p. 10.
Fig. 40-18. Texas Instruments, Linear and Interface Circuits Applications, Vol. 1, 1985, p. 3-20.
Fig. 40-19. National Semiconductor, Linear Brief 23.

Chapter 41

Fig. 41-1. Tab Books, Inc., 101 Sound, Light, and Power IC Projects.
Fig. 41-2. Courtesy, William Sheets.

Chapter 42

Fig. 42-1. Courtesy, William Sheets.
Fig. 42-2. Hands-On Electronics, Sep/Oct 1986, p. 85.
Fig. 42-3. Courtesy, William Sheets.

Chapter 43

Fig. 43-1. Texas Instruments, Linear and Interface Circuits Applications, 1987, p. 12-8.

Fig. 43-2. Texas Instruments, Linear and Interface Circuits Applications, 1987, p. 12-10.

Chapter 44

Fig. 44-1. Linear Technology Corp., Linear Applications Handbook, 1987, p. AN3-7.

Chapter 45

Fig. 45-1. Linear Technology Corp., Linear Databook Supplement, 1988, p. S2-34.

Fig. 45-2. Hands-On Electronics, Jul/Aug 1986, p. 86.

Fig. 45-3. Signetics, Linear Data Manual Vol. 3: Video, p. 5-15.

Fig. 45-4. Tab Books, Inc. Build Your Own Laser, Phaser, Ion Ray Gun, 1983, p. 29.

Fig. 45-5. Hands-On Electronics, Jul/Aug 1986, p. 86.

Fig. 45-6. Electronic Engineering, 8/78, p. 24.

Chapter 46

Fig. 46-1. Hands-On Electronics, 12/86, p. 42.

Fig. 46-2. Linear Technology Corp., Linear Databook, 1986, p. 2-82.

Fig. 46-3. Electronic Engineering, 9/84, p. 33.

Fig. 46-4. Texas Instruments, Linear and Interface Circuits Applications Vol. 1, 1985, p. 3-18.

Fig. 46-5. Linear Technology Corp., Linear Databook, 1986, p. 2-83.

Chapter 47

Fig. 47-1. Electronic Engineering, 7/86, p. 30.

Fig. 47-2. Signetics, Analog Data Manual, 1982, p. 3-73.

Fig. 47-3.National Semiconductor Corp., Data Conversion/Acquisition Databook, 1980, p. 3-30.

Chapter 48

Fig. 48-1. Hands-On Electronics, 5/87, p. 95.

Fig. 48-2. Electronic Design 16, 8/76, p. 76.

Chapter 49

Fig. 49-1. Motorola, Thyristor Device Data, Series A, 1985, p. 1-6-50.

Fig. 49-2. Motorola, TMOS Power FET Design Ideas, 1985, p. 20.

Fig. 49-3. Motorola, TMOS Power FET Design Ideas, 1985, p. 21.

Fig. 49-4. R-E Experimenters Handbook, p. 156.

Fig. 49-5. Motorola, Thyristor Device Data, Series A, 1985, p. 1-6-48.

Fig. 49-6. Motorola, Thyristor Device Data, Series A, 1985, 1-6-55.

Fig. 49-7. Electronic Engineering, 9/84, p. 38.

Fig. 49-8. Tab Books, Inc., 101 Sound, Light, and Power IC Projects.

Fig. 49-9. Motorola, Thyristor Device Data, Series A, 1985, p. 1-6-60.

Fig. 49-10. General Electric, Application Note 200.35, p. 17.

Chapter 50

Fig. 50-1. Tab Books, Inc., Build Your Own Laser, Phasor, Ion Ray Gun, 1983, p. 104.

Fig. 50-2. Electric Engineering, 12/84, p. 34.

Chapter 51

Fig. 51-1. Hands-On Electronics, Sep/Oct 1986, p. 26.

Fig. 51-2. General Electric, Optoelectronics, Third Edition, p. 107.

Fig. 51-3. Courtesy, William Sheets.

Fig. 51-4. Courtesy, William Sheets.

Fig. 51-5. Electronic Engineering, 12/75, p. 15.

Fig. 51-6. Hands-On Electronics, 4/87, p. 94.

Fig. 51-7. Hands-On Electronics, 2/87, p. 87.

Fig. 51-8. Hands-On Electronics, 10/87, p. 92.

Fig. 51-9. Linear Technology Corp., Linear Databook, 1986, p. 2-83.

Fig. 51-10. Radio-Electronics, 11/86, p. 38.

Fig. 51-11. General Electric, Application Note 200.35, p. 15.

Fig. 51-12. Intersil, Component Data Catalog, 1987, p. 7-44.

Fig. 51-13. Radio Electronics, 3/86, p. 32.

Fig. 51-14. Electronic Design, 11/82, p. 172.

Fig. 51-15. Electronic Design, 6/76, p. 120.

Fig. 51-16. Linear Technology Corp., Linear Application Handbook, 1987, p. AN5-3.

Chapter 52

Fig. 52-1. Siliconix, Integrated Circuit Data Book, 3/85, p. 10-85.

Fig. 52-2. Siliconix, Integrated Circuit Data Book, 3/85, p. 10-79.
Fig. 52-3. Siliconix, Integrated Circuit Data Book, 3/85, p. 2-144.
Fig. 52-4. Signetics, 1987 Linear Data Manual, Vol. 2: Industrial, 10/86, p. 4-261.
Fig. 52-5. Siliconix, Integrated Circuit Data Book, 3/85, p. 2-103.

Chapter 53

Fig. 53-1. Signetics, 1987 Linear Data Manual, Vol. 2: Industrial, 2/87, p. 5-350.
Fig. 53-2. Linear Technology Corp., Linear Application Handbook, 1987, p. AN3-9.

Chapter 54

Fig. 54-1. Hands-On Electronics, May/Jun 1986, p. 63.
Fig. 54-2. Electronic Design, 10/73, p. 114.
Fig. 54-3. Electronic Engineering, 7/85, p. 44.
Fig. 54-4. Electronics Today International, 3/80, p. 25.
Fig. 54-5. Popular Electronics, 1/82, p. 76.
Fig. 54-6. Electronic Engineering, 6/87, p. 28.
Fig. 54-7. Popular Electronics, 8/69, p. 71.
Fig. 54-8. Electronics Today International, 1/76, p. 52.
Fig. 54-9. Electronic Engineering, 9/78, p. 20.

Chapter 55

Fig. 55-1. Electronic Engineering, 1/85, p. 39.
Fig. 55-2. Intersil, Component Data Catalog, 1987, p. 7-44.

Chapter 56

Fig. 56-1. Courtesy, William Sheets.
Fig. 56-2. Courtesy, William Sheets.

Chapter 57

Fig. 57-1. Popular Electronics, 6/73.
Fig. 57-2. Courtesy, William Sheets.
Fig. 57-3. Courtesy, William Sheets.
Fig. 57-4. Courtesy, William Sheets.

Chapter 58

Fig. 58-1. Unitrode Corp., 10/86, p. 332.
Fig. 58-2. Unitrode Corp., 10/86, p. 332.
Fig. 58-3. Radio Electronics, 8/82, p. 36.
Fig. 58-4. Hands-On Electronics, Winter 1985, p. 93.
Fig. 58-5. Hands-On Electronics, 9/87, p. 71.
Fig. 58-6. Electronic Engineering, 10/77, p. 23.
Fig. 58-7. Electronic Design, 11/8/69, p. 109.
Fig. 58-8. Radio-Electronics, 11/82, p. 79.
Fig. 58-9. National Semiconductor Corp., Transistor Databook, 1982, p. 11-34.
Fig. 58-10. National Semiconductor Corp., Data Conversion/Acquisition Databook, 1980, p. 2-5.
Fig. 58-11. Linear Technology Corp., Linear Databook, 1986, p. 8-42.
Fig. 58-12. Signetics, Linear Data Manual Vol. 3: Video, p. 11-120.
Fig. 58-13. General Electric, Application Note 90.16, p. 28.
Fig. 58-14. RCA, Digital Integrated Circuits Application Note ICAN-6346, p. 5.
Fig. 58-15. Linear Technology Corp., Linear Databook, 1986, p. 5-15.
Fig. 58-16. General Electric, SCR Manual, Sixth Edition, 1979, p. 204.

Chapter 59

Fig. 59-1. Motorola, TMOS Power FET Design Ideas, 1985, p. 45.
Fig. 59-2. Courtesy, William Sheets.
Fig. 59-3. Signetics, Linear Data Manual, Vol. 3: Video, p. 11-3.
Fig. 59-4. Radio-Electronics, 8/77, p. 33.
Fig. 59-5. Electronic Design, 3/77, p. 76.

Chapter 60

Fig. 60-1.Electronic Engineering, 5/84, p. 43.
Fig. 60-2. General Electric, Optoelectronics, Third Edition, p. 114.
Fig. 60-3. Motorola, TMOS Power FET Design Ideas, 1985, p. 32.
Fig. 60-4. Siliconix, MOSpower Applications Handbook, p. 6-186.
Fig. 60-5. Electronic Engineering, 7/86, p. 34.
Fig. 60-6. Electronic Engineering, 4/85, p. 47.
Fig. 60-7. Motorola, Thyristor Device Data, Series A, 1985, p. 1-6-8.
Fig. 60-8. Motorola, TMOS Power FET Design Ideas, 1985, p. 31.
Fig. 60-9.Sprague Electric Co., Integrated Circuits Databook WR504, p. 4-159.
Fig. 60-10. General Electric, Application Note 200.35, p. 18.
Fig. 60-11. Sprague Electric Co., Integrated Circuits Databook WR504, p. 4-160.

Fig. 60-12. Motorola, TMOS Power FET Design Ideas, 1985, p. 55.
Fig. 60-13. Motorola, TMOS Power FET Design Ideas, 1985, p. 54.
Fig. 60-14. Motorola, TMOS Power FET Design Ideas, 1985, p. 51.
Fig. 60-15. Electronic Engineering, 2/84, p. 23.
Fig. 60-16. National Semiconductor Corp., Linear Application Databook, p. 1066.
Fig. 60-17. General Electric, Optoelectronics, Third Edition, p. 113.

Chapter 61

Fig. 61-1. Fairchild Corp., Linear Databook, 1982, p. 4-72.
Fig. 61-2. Linear Technology Corp., Linear Applications Handbook, 1987, p. AN3-14.

Chapter 62

Fig. 62-1. 73 Magazine, 12/76, p. 170.
Fig. 62-2. Electronics International Today, 1/76, p. 44.
Fig. 62-3. Signetics Analog Data Manual, 1983, p. 10-93.
Fig. 62-4. Electronics Today International, 9/75, p. 66.
Fig. 62-5. CQ, 5/76, p. 26.

Chapter 63

Fig. 63-1. National Semiconductor Corp., Linear Applications Databook, p. 1096.
Fig. 63-2. EXAR, Telecommunications Databook, 1986, p. 7-24.
Fig. 63-3. EXAR, Telecommunications Databook, 1986, p. 7-24.
Fig. 63-4. Electronic Engineering, 12/84, p. 33.
Fig. 63-5. Electronic Engineering, 11/85, p. 31.
Fig. 63-6. Courtesy, William Sheets.
Fig. 63-7. Texas Instruments, Linear and Interface Circuits Applications, Vol. 1, 1985, p. 2-11.
Fig. 63-8. Courtesy, William Sheets.
Fig. 63-9. Courtesy, William Sheets.
Fig. 63-10. Courtesy, William Sheets.

Chapter 64

Fig. 64-1. Electronic Engineering, 6/83, p. 31.
Fig. 64-2. Electronic Design 15, 7/75, p. 68.

Chapter 65

Fig. 65-1. General Electric, Optoelectronics, Third Edition, p. 135.
Fig. 65-2. NASA, Tech Briefs, Summer 1984, p. 446.
Fig. 65-3. General Electric, Optoelectronics, Third Edition, p. 140.
Fig. 65-4. General Electric, Optoelectronics, Third Edition, p. 121.
Fig. 65-5. General Electric, Optoelectronics, Third Edition, p. 120.
Fig. 65-6. General Electric, Optoelectronics, Third Edition, p. 139.
Fig. 65-7. General Electric, Optoelectronics, Third Edition, p. 120.
Fig. 65-8. General Electric, Optoelectronics, Third Edition, p. 134.
Fig. 65-9. General Electric, Optoelectronics, Third Edition, p. 112.
Fig. 65-10. National Semiconductor Corp., Data Conversion/Acquisition Databook, 1980, p. 13-46.
Fig. 65-11. General Electric, Optoelectronics, Third Edition, p. 133.
Fig. 65-12. Courtesy, William Sheets.
Fig. 65-13. General Electric, Optoelectronics, Third Edition, p. 117.
Fig. 65-14. Electronic Engineering, 8/86, p. 36.
Fig. 65-15. General Electric, Optoelectronics, Third Edition, p. 118.

Chapter 66

Fig. 66-1. 73 For Radio Amateurs, 11/85, p. 32.
Fig. 66-2. Radio-Electronics, 5/70, p. 35.
Fig. 66-3. Electronics Today International, 7/78, p. 16.
Fig. 66-4. Electronics Today International, 12/78, p. 15.
Fig. 66-5. General Electric, Semiconductor Data Handbook, Third Edition, p. 513.
Fig. 66-6. Electronic Design, 11/29/84, p. 281.
Fig. 66-7. Unitrode Corp., Databook 1986, p. 51.
Fig. 66-8. Electronic Engineering, 5/77, p. 27.
Fig. 66-9. National Semiconductor Corp., Transistor Databook, 1982, p. 7-19.
Fig. 66-10. Courtesy, William Sheets.
Fig. 66-11. Electronic Design, 10/65.

Fig. 66-12. Hands-On Electronics, Summer 1984, p. 43.
Fig. 66-13. Signetics, Analog Data Manual, 1982, p. 8-10.

Chapter 67

Fig. 67-1. Electronic Design, 5/79, p. 102.
Fig. 67-2. Radio-Electronics, 7/70, p. 36.
Fig. 67-3. Courtesy, William Sheets.

Chapter 68

Fig. 68-1. Texas Instruments, Linear and Interface Circuits Applications, Vol. 1, 1985, p. 3-18.
Fig. 68-2. Popular Electronics, 3/79, p. 78.

Chapter 69

Fig. 69-1. Electronic Engineering, 2/86, p. 38.
Fig. 69-2. Courtesy, William Sheets.
Fig. 69-3. Electronic Engineering, 4/77, p. 13.
Fig. 69-4. Electronic Engineering, 7/85, p. 34.
Fig. 69-5. Electronic Design, 3/77, p. 106.
Fig. 69-6. Electric Engineering, 1/87, p. 25.

Chapter 70

Fig. 70-1. Hands-On Electronics, 10/87, p. 96.
Fig. 70-2. Hands-On Electronics, Spring 1985, p. 82.
Fig. 70-3. General Electric Project G4, p. 131.
Fig. 70-4. Radio Electronics, 12/84, p. 77.
Fig. 70-5. Electronics Today International, 6/75, p. 42.
Fig. 70-6. Electronics Today International, 9/82, p. 42.

Chapter 71

Fig. 71-1. Signetics, 1987 Linear Data Manual Vol. 1: Communications, 11/86, p. 7-251.
Fig. 71-2. National Semiconductor Corp., Linear Applications Databook, p. 1065.
Fig. 71-3. Siliconix, MOSpower Applications Handbook, p. 6-101.
Fig. 71-4. Hands-On Electronics, 5/87, p. 96.
Fig. 71-5. Hands-On Electronics, Spring 1985, p. 36.
Fig. 71-6. Hands-On Electronics, Summer 1984, p. 74.
Fig. 71-7. Radio-Electronics, 3/86, p. 59.
Fig. 71-8. National Semiconductor Corp., Audio/Radio Handbook, 1980, p. 4-20.
Fig. 71-9. National Semiconductor Corp., Linear Databook, 1982, p. 3-187.
Fig. 71-10. Signetics, 1987 Linear Data Manual, Vol. 2: Industrial, 11/86, p. 4-135.

Chapter 72

Fig. 72-1. Texas Instruments, Linear and Interface Circuits Applications, Vol. 1, 1985, , p. 6-35.
Fig. 72-2. Motorola, TMOS Power FET Design Ideas, 1985, p. 43.
Fig. 72-3. Electronic Engineering, 12/84, p. 41.
Fig. 72-4. NASA, Tech Briefs, 9/87, p. 21.
Fig. 72-5. Motorola, TMOS Power FET Design Ideas, 1985, p. 37.
Fig. 72-6. Siliconix, MOSpower Applications Handbook, p. 6-51.
Fig. 72-7. General Electric/RCA, BiMOS Operational Amplifiers Circuit Ideas, 1987, p. 24.
Fig. 72-8. Courtesy, William Sheets.
Fig. 72-9. General Electric/RCA, BiMOS Operational Amplifiers Circuit Ideas, 1987, p. 24.
Fig. 72-10. Radio-Electronics, 6/86, p. 52.
Fig. 72-11. Siliconix, MOSpower Applications Handbook, p. 6-177.
Fig. 72-12. Siliconix, MOSpower Applications Handbook, p. 6-59.
Fig. 72-13. Linear Technology Corp., Linear Databook, 1986, p. 3-23.
Fig. 72-14. Electronic Engineering, 10/84, p. 38.
Fig. 72-15. NASA Tech Briefs, Summer 1985, p. 32.
Fig. 72-16. Electronic Engineering, 1/87, p. 22.
Fig. 72-17. Motorola, TMOS Power FET Design Ideas, 1985, p. 42.
Fig. 72-18. Motorola, Thyristor Device Data, Series A, 1985, p. 1-6-55.
Fig. 72-19. 73 Magazine, 12/70, p. 170.
Fig. 72-20. Signetics, Analog Data Manual, 1983, p. 12-27.
Fig. 72-21. Signetics, 1987 Linear Data Manual, Vol. 2: Industrial, 2/87, p. 8-223.
Fig. 72-22. Electronic Engineering, 1/85, p. 45.
Fig. 72-23. Motorola, Linear Integrated Circuits, p. 3-138.

Fig. 72-24. Electronic Design, 11/29/84, p. 282.

Fig. 72-25. Electronics Today International, 1/70, p. 45.

Fig. 72-26. Linear Technology, 1986 Linear Databook, p. 3-22.

Chapter 73

Fig. 73-1. Electronic Engineering, 10/76, p.17.

Fig. 73-2. Electronic Engineering, 7/77, p. 26.

Fig. 73-3. Popular Electronics, 5/74, p. 24.

Fig. 73-4. Courtesy, William Sheets.

Chapter 74

Fig. 74-1. NASA, Tech Briefs, Winter 1985, p. 52.

Fig. 74-2. Courtesy, William Sheets.

Fig. 74-3. Electronic Engineering, 3/86, p. 34.

Fig. 74-4. Courtesy, William Sheets.

Fig. 74-5. Courtesy, William Sheets.

Chapter 75

Fig. 75-1. Electronic Design 25, 1275, p. 90.

Fig. 75-2. Courtesy, William Sheets.

Fig. 75-3. Electronics Today International, 9/75, p. 66.

Fig. 75-4. Electronic Engineering, 1/85, p. 41.

Fig. 75-5. Hands-On Electronics, Fall 1984, p. 66.

Fig. 75-6. Radio-Electronics, 3/77, p. 76.

Chapter 76

Fig. 76-1. Hands-On Electronics, 11/86, p. 92.

Fig. 76-2. Popular Electronics, 11/77, p. 62.

Chapter 77

Fig. 77-1. Electronic Engineering, 9/86, p. 38.

Fig. 77-2. RCA, Design Guide for Fire Detection Systems, Publication 2M1189, p. 27.

Fig. 77-3. Electronic Engineering, 9/86, p. 34.

Chapter 78

Fig. 78-1. Electronic Engineering, 5/76, p. 17.

Fig. 78-2. Electronic Design, 4/74, p. 114.

Fig. 78-3. Electronic Engineering, 10/86, p. 41.

Chapter 79

Fig. 79-1. Electronic Engineering, 12/75, p. 15.

Fig. 79-2. Radio-Electronics, Experimenters Handbook, p. 122.

Chapter 80

Fig. 80-1. Signetics, 1987 Linear Data Manual, Vol. 1: Communications, 2/87, p. 4-310.

Fig. 80-2.Motorola, Thyristor Device Data, Series A, 1985, p. 1-6-52.

Chapter 81

Fig. 81-1.Signetics, 1987 Linear Data Manual, Vol. 1: Communications, 11/86, p. 7-14.

Fig. 81-2. Hands-On Electronics, 3/87, p. 28.

Fig. 81-3. Hands-On Electronics, 12/86, p. 22.

Chapter 82

Fig. 82-1. General Electric/RCA, BiMOS Operational Amplifiers Circuit Ideas, 1987, p. 11.

Fig. 82-2. Signetics, 1987 Linear Data Manual, Vol. 2: Industrial, 11/86, p. 4-135.

Chapter 83

Fig. 83-1. Electronic Design, 9/69, p. 106.

Fig. 83-2. Motorola, Thyristor Device Data, Series A, 1985, p. 1-6-61.

Fig. 83-3. QST, 7/87, p. 32.

Chapter 84

Fig. 84-1. Electronic Engineering, 4/86, p. 34.

Fig. 84-2. Electronic Engineering, 7/85, p. 44.

Fig. 84-3. Electronic Engineering, 11/86, p. 34.

Fig. 84-4. NASA, Tech Briefs, 1/88, p. 18.

Chapter 85

Fig. 85-1. Motorola, RF Data Manual, 1986, p. 6-141.

Fig. 85-2. Motorola, RF Data Manual, 1986, p. 6-240.

Fig. 85-3. QST, 7/87, p. 31.

Fig. 85-4. QST, 5-86, p. 23.

Fig. 85-5. Motorola, RF Data Manual, 1986, p. 6-181.

Fig. 85-6. Ham Radio, 7/86, p. 50.

Fig. 85-7. Radio Electronics, 3/87, p. 42.

Fig. 85-8. NASA, Tech Briefs, Spring 1984, p. 322.

Fig. 85-9. Motorola, RF Data Manual, 1986, p. 6-232.

Chapter 86

Fig. 86-1. QST, 12/85, p. 39.

Chapter 87

Fig. 87-1. General Electric/RCA, BiMOS Operational Amplifiers circuit Ideas, 1987, p. 14.

Fig. 87-2. Intersil, Component Data Catalog, 1987, p. 7-5.

Fig. 87-3. Linear Technology Corp., Linear Databook, 1986, p. 2-113.

Fig. 87-4. Electronics Today International, 3/78, p. 51.

Fig. 87-5. National Semiconductor Corp., Hybrid Products Databook, 1982, p. 17-149.

Fig. 87-6. Siliconix, Integrated Circuits Data Book, 3/85, p. 10-58.

Chapter 88

Fig. 88-1. General Electric/RCA, BiMOS Operational Amplifiers Circuit Ideas, 1987, p. 8.

Fig. 88-2. Linear Technology Corp., Linear Databook, 1986, p. 2-113.

Fig. 88-3. Texas Instruments, Linear and Interface Circuits Applications, Vol. 1, 1985, p. 3-15, 3-16.

Fig. 88-4. QST, 4/87, p. 48.

Fig. 88-5. Electronic Engineering, 5/85, p. 38.

Fig. 88-6. Electronic Design, 6/81, p. 250.

Fig. 88-7. Ham Radio, 1/87, p. 97.

Fig. 88-8. Electronic Engineering, 2/76, p. 17.

Fig. 88-9. Electronic Design, 2/73, p. 82.

Fig. 88-10. Radio-Electronics, 2/71, p. 37.

Fig. 88-11. Ham Radio, 6/82, p. 33.

Chapter 89

Fig. 89-1. Popular Electronics, 12/74, p. 68.

Fig. 89-2. Electronics Today International, 1/77, p. 49.

Fig. 89-3. Electronics Today International, 1/77, p. 49.

Fig. 89-4. Popular Electronics, 8/74, p. 98.

Fig. 89-5. General Electric/RCA, BiMOS Operational Amplifiers Circuit Ideas, 1987, p. 28.

Fig. 89-6. Electronics Today International, 11/76, p. 45.

Fig. 89-7. Electronics Today International, 2/75, p. 66.

Fig. 89-8. Electronics Today International, 1/77, p. 85.

Fig. 89-9. Electronics Today International, 6/75, p. 63.

Fig. 89-10. Electronics Today International, 1/77, p. 49.

Fig. 89-11. Courtesy, William Sheets.

Fig. 89-12. Electronics Today International, 11/80, p. 43.

Fig. 89-13. Radio-Electronics, 2/75, p. 42.

Fig. 89-14. Courtesy, William Sheets.

Chapter 90

Fig. 90-1. Courtesy, William Sheets.

Fig. 90-2. Radio-Electronics, 12/83, p. 38.

Fig. 90-3. Hands-On Electronics, 8/87, p. 77.

Fig. 90-4. Electronics today International, 6/79, p. 27.

Fig. 90-5. Courtesy, William Sheets.

Chapter 91

Fig. 91-1. Hands-On Electronics, 6/87, p. 40.

Fig. 91-2. Electronics Today International, 8/77, p. 25.

Fig. 91-3. Electronics Today International, 11/80.

Fig. 91-4. Texas Instruments, Complex Sound Generator, Bulletin No. DL-12612, p. 13.

Fig. 91-5. Electronics Today International, 4/82, p. 34.

Fig. 91-6. Electronics Today International, 2/75, p. 66.

Fig. 91-7. Hands-On Electronics, 12/86, p. 42.

Fig. 91-8. Courtesy, William Sheets.

Fig. 91-9. Courtesy, William Sheets.

Fig. 91-10. Texas Instruments, complex Sound Generator, Bulletin No. DL-S 12612, p. 11.

Fig. 91-11. Courtesy, William Sheets.

Chapter 92

Fig. 92-1. Electronic Engineering, 3/82, p. 29.

Fig. 92-2. Electronic Engineering, 1078, p. 17.

Fig. 92-3. EXAR, Telecommunications Databook, 1986, p. 9-24.

Fig. 92-4. Electronic Design, 5/79, p. 100.

Fig. 92-5. Siliconix, Integrated Circuits Data Book, 3/85, p. 5-17.

Fig. 92-6. General Electric/RCA, BiMOS Operational Amplifiers Circuit Ideas, 1987, p. 7.

Fig. 92-7. Electronic Design, 6/69, p. 126.
Fig. 92-8. Electronic Engineering, 8/84, p. 27.
Fig. 92-9. Electronic Engineering, 12/85, p. 35.
Fig. 92-10. Electronic Engineering, 8/84, p. 29.

Chapter 93

Fig. 93-1. General Electric/RCA, Operational Amplifiers Circuit Ideas, 1987, p. 10.
Fig. 93-2. Linear Technology Corp., Linear Databook, 1986, p. 8-42.

Chapter 94

Fig. 94-1. Tab Books, Inc. 303 Dynamic Electronic Circuits, p. 169.
Fig. 94-2. Electronics Today International, 12/77, p. 86.

Chapter 95

Fig. 95-1. Electronics Today International, 10/78, p. 46.
Fig. 95-2. Hands-On Electronics, Fall 1984, p. 65.
Fig. 95-3. Courtesy, William Sheets.

Chapter 96

Fig. 96-1. Electronic Engineering, 6/86, p. 35.
Fig. 96-2. General Electric, SCR Manual, Sixth Edition, 1979, p. 200.

Chapter 97

Fig. 97-1. Electronic Engineering, 9/87, p. 32.
Fig. 97-2. Intersil, Component Data Catalog, 1987, p. 14-67.

Chapter 98

Fig. 98-1. Electronic Engineering, 2/87, p. 40.
Fig. 98-2. Tab Books, Inc. The Giant Book of Easy-To-Build Electronic Projects, 1982, p. 1.
Fig. 98-3. Radio-Electronics, 2/85, p. 90.
Fig. 98-4. EXAR, Telecommunications Databook, 1986, p. 11-38.
Fig. 98-5. Tab Books, Inc. build Your Own Laser, Phaser, Ion gun, 1983, p. 305.
Fig. 98-6. Electronic Design, 12/78, p. 95.
Fig. 98-7. Radio-Electronics, 11/79, p. 53.
Fig. 98-8. Electronic Design, 10/76, p. 194.
Fig. 98-9. Hands-On Electronics, 12/86, p. 22.
Fig. 98-10. Electronics Engineering, 1/79, p. 17.
Fig. 98-11. Radio-Electronics, 12/78, p. 67.
Fig. 98-12. Radio-Electronics, 11/77, p. 45.
Fig. 98-13. Hands-On Electronics, Summer 1985, p. 74.
Fig. 98-14. Hands-On Electronics, Sep/Oct 1986, p. 88.
Fig. 98-15. Hands-On Electronics, Sep/Oct 1986, p. 105.
Fig. 98-16. Hands-On Electronics, Summer 1984, p. 39.
Fig. 98-17. EXAR, Telecommunications Databook, 1986, p. 4-19.
Fig. 98-18. EXAR, Telecommunications Databook, 1986, p. 5-14.
Fig. 98-19. EXAR, Telecommunications Databook, 1986, p. 4-15.
Fig. 98-20. Signetics, 1987 Linear Data Manual, Vol. 1: Communications, 12/2/86, p. 6-50.

Chapter 99

Fig. 99-1. NASA, Tech Briefs, 12/87, p. 28.
Fig. 99-2. Signetics, 1987 Linear Data Manual, Vol. 2: Industrial, 2/87, p. 7-67.
Fig. 99-3. Electronic Design, 8/75, p. 82.
Fig. 99-4. Linear Technology Corp., Linear Application Handbook, 1987, p. AN3-6.
Fig. 99-5. General Electric, Optoelectronics, Third Edition, p. 153.
Fig. 99-6. General Electric, Application Note 200.85, p. 18.
Fig. 99-7. Electronic Design, 8/82, p. 217.
Fig. 99-8. Electronic Design, 8/83, p. 230.
Fig. 99-9. Courtesy, William Sheets.

Chapter 100

Fig. 100-1. General Electric, SCR Manual, Sixth Edition, 1979, p. 222.
Fig. 100-2. Electronic Engineering, 9/85, p. 30.
Fig. 100-3. National Semiconductor Corp., 1984 Linear Supplement Databook, p. S1-41.
Fig. 100-4. National Semiconductor Corp., 1984 Linear Supplement Databook, p. S1-42.
Fig. 100-5. Linear Technology Corp., Linear Databook, 1986, p. 2-101.
Fig. 100-6. Courtesy, William Sheets.

Chapter 101

Fig. 101-1. Linear Technology Corp., Linear Applications Handbook, 1987, p. AN7-2.
Fig. 101-2. Electric Engineering, 7/84, p. 31.

Chapter 102

Fig. 102-1. Hands-On Electronics, 11/86, p. 93.
Fig. 102-2. Hands-On Electronics, 9/87, p. 32.

Chapter 103

Fig. 103-1. Teledyne Semiconductor, Data & Design Manual, 1981, p. 7-17.
Fig. 103-2. Linear Technology Corp., Application Note 9, p. 18.
Fig. 103-3. Radio-Electronics, 9/82, p. 42.
Fig. 103-4. Intersil, Component Data Catalog, 1987, p. 6-8.
Fig. 103-5. Intersil, Component Data Catalog, 1987, p. 6-11.
Fig. 103-6. Intersil, Component Data Catalog, 1987, p. 6-8.

Chapter 104

Fig. 104-1. NASA, Tech Briefs, Spring 1985, p. 40.
Fig. 104-2. NASA, Tech Briefs, Fall/Winter 1981, p. 319.
Fig. 104-3. Popular Electronics, 12/82, p. 82.

Chapter 105

Fig. 105-1. General Electric, Semiconductor Data Handbook, Third Edition, p. 513.
Fig. 105-2. Motorola, Application Note AN294.
Fig. 105-3. Electronic Design, 4/77, p. 120.
Fig. 105-4. Motorola, Application Note AN294.

Chapter 106

Fig. 106-1. NASA, Tech Briefs, Sep/Oct 1986, p. 36.
Fig. 106-2. Electronic Engineering, 9/77, p. 37.
Fig. 106-3. Motorola, Thyristor Device Data, Series A, 1985, p. 1-6-51.
Fig. 106-4. Motorola, Thyristor Device Data, Series A, 1985, p. 1-6-54.
Fig. 106-5. Texas Instruments, Linear and Interface Circuits Applications, Vol. 1, 1985, p. 7-23.
Fig. 106-6. Intersil, Databook, 1987, p. 7-102.
Fig. 106-7. Intersil, Component Data Catalog, 1987, p. 14-67.
Fig. 106-8. General Electric, Application Note 90.16, p. 30.

Chapter 107

Fig. 107-1. Electronics Today International, 11/80.
Fig. 107-2. Signetics, 1987 Linear Data Manual, Vol. 2: Industrial, 2/87, p. 4-107.
Fig. 107-3. Electronics Today International, 9/77, p. 55.
Fig. 107-4. General Electric/RCA, BiMOS Operational Amplifiers Circuit Ideas, 1987, p. 21.
Fig. 107-5. Texas Instruments, Linear and Interface Circuits Applications, Vol. 1, 1985, p. 3-11.
Fig. 107-6. National Semiconductor Corp., Transistor Databook, 1982, p. 11-35.
Fig. 107-7. Electronics Today International, 6/82, p. 61.

Chapter 108

Fig. 108-1. Hands-On Electronics, 9/87, p. 88.
Fig. 108-2. Texas Instruments, Linear and Interface Circuits Applications, Vol. 1, 1985, p. 7-15.
Fig. 108-3. Tab Books, Inc., The Giant book of Easy-To-Build Electronics Projects, 1982, p. 31.

Index

A

B

C

D

E

F

G

H

I

M

N

O

P

Q

R

S

U

V

W

Y

Z

Other Bestsellers From TAB